Features and Benefits

Merrill Integrated Mathematics
Course 2

1. **NCTM *Standards*** As the correlation in the *Teacher's Handbook* shows, this program adheres strictly to the NCTM *Standards* (page T3).

2. **Integration** of content addresses the essence of the NCTM *Standards*.
   - Chapters 1-3 deal with **logic** and **properties** of the real number system.
   - This content is applied to **geometry** in Chapter 4-7.
   - The content on **coordinate geometry** in Chapter 8 and 9 shows the true integration of the algebraic and geometric ideas presented in Chapter 1-7.
   - The algebra and geometry content presented in Chapters 1-9 are applied to **quadratic equations** (Chapter 10), **probability** and **combinatorics** (Chapter 11), **transformation geometry** (Chapter 12), and **trigonometry** (Chapter 13).
   - After completing Chapters 1-9, Chapters 10-13 can be presented in any order.

3. **Proof** Students are introduced to **two-column proofs** in Chapter 1 (page 10) as part of the study of logic. This prepares students for formal proof in Euclidean geometry (Chapters 4-7) and **coordinate proofs** (Chapter 9).

4. **Constructions** Geometric constructions are integrated with the related content (page 226). This type of hands-on activity helps to improve the students' understanding of the related geometric concepts.

5. **Problem Solving and Applications** A complete repertoire of strategies is presented in the problem-solving lessons found in each chapter (page 115). The **Mathematical Excursions** explore interesting mathematical topics and applications (page 100). Each chapter opening page focuses on a real-life application (page 1).

6. **Technology** **Calculator Hints** are provided where the use of the calculator is appropriate (page 171). They also refer students to the **Technology Activities** (pages A2-A9) where students learn to use graphing calculators for various topics in the text. The **Technology Masters** booklet includes graphing calculators and additional scientific calculator applications for each chapter.

7. **Projects** The **Extended Projects** in the appendix offer ways for students to work together on intriguing long-term projects (page A10-A16). Each chapter opening page also presents a class project to complete as a group or as individuals (page 1).

8. **Journal** The *Journal* entry gives students the opportunity to keep a log of their thoughts and ideas about mathematics (page 33).

9. **Portfolio Suggestion** Students are asked to select items from their work that represent different aspects of their mathematical knowledge (page 38).

10. **Performance Assessment** An alternative assessment task is included at the end of each chapter as an additional tool for measuring each student's performance (page 38).

INTEGRATED MATHEMATICS

$P = 2l + 2w$

$F = \frac{9}{5}C + 32$

$-3x > 15$
$x < -5$

$\triangle ABC \cong \triangle FGH$

$x = \dfrac{-b \pm \sqrt{b^2 - 4ac}}{2a}$

The Logical Approach to Geometry, Algebra, and Everything in Between

GLENCOE

Our Integrated Program Contains Just the Right Content Mix of Algebra and Geometry

Integrated Mathematics is a three-course series that provides an effective alternative to teaching algebra and geometry as separate courses. The concepts of algebra and geometry are presented in a logical sequence that clearly shows the relationship between the two in easy-to-follow Student Editions.

For example, motivating and exciting content in Course 1 is presented in the following integrated sequence:

- Logic and algebraic skills and concepts in Chapters 1-4.

- This content is applied to geometry in Chapters 5 and 6 reinforcing the algebra from previous chapters.

Meets NCTM Standards!

- Algebra content is extended to real numbers in Chapter 7. This allows for the presentation of measurement and additional geometry content in Chapter 8.

- Polynomials and factoring are introduced and applied to geometry in Chapters 9 and 10.

- Remaining chapters apply algebra and geometry content to probability (Chapter 11), statistics (Chapter 12), coordinate geometry (Chapter 13), and transformational geometry (Chapter 14).

- The chapters on probability and statistics represent the concern for discrete mathematics, as does the work on graph theory in Chapter 14.

- The integrated sequencing is continued and expanded in Courses 2 and 3.

Courses 2 and 3 Emphasize the Ongoing Relationship Between Algebra and Geometry

Where Course 1 focused on algebra and its connection to geometry, Course 2 focuses on geometry and its connection to algebra. The interweaving of algebra and geometry in Courses 1 and 2 helps students see the importance of studying mathematics. The logical way algebra and geometry are presented builds upon previous knowledge.

Because concepts are presented repeatedly in different contexts, students more readily remember and apply what they have learned. Math becomes more meaningful as students are led logically through the problem-solving applications in each of the three books.

Because of the integration of algebra with geometry in Course 2, less time is required in Course 3 to review previously-learned algebraic concepts. Instead, students are logically and effectively introduced to quadratics, parabolas, and other more advanced mathematical topics.

Course 2
Chapter 1
Working with Logic

Chapter 2
Mathematical Systems

Chapter 3
A System with Two Operations

Chapter 4
Introductions to Euclidean Geometry

Chapter 5
Congruent Triangles

Chapter 6
Applications of Congruent Triangles

Chapter 7
Similarity

Chapter 8
Geometry with Coordinates

Chapter 9
Using Coordinates

Chapter 10
Solving Quadratic Equations

Chapter 11
Probability and Combinatorics

Chapter 12
Transformation Geometry

Chapter 13
Introduction to Trigonometry

Course 3
Chapter 1
Working with Polynomial and Rational Expressions

Chapter 2
Working with Complex Numbers

Chapter 3
Relations and Functions

Chapter 4
An Introduction to Transformation Geometry

Chapter 5
Exponential and Logarithmic Functions

Chapter 6
Circles

Chapter 7
An Introduction to Circular Functions

Chapter 8
Applications of Circular Functions

Chapter 9
Trigonometric Identities and Equations

Chapter 10
More Work with Transformations

Chapter 11
Probability and the Binomial Theorem

Chapter 12
Statistics

Chapter 13
Sequences and Series

Chapter 14
An Introduction to Matrices

Chapter 15
Polynomials, Complex Numbers, and DeMoivre's Theorem

Chapter 16
Mathematical Induction

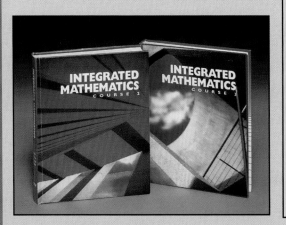

These Vital Activities and Features Are Essential Components of Each Chapter

Communication

Journals provide prompts for students to record their thoughts and ideas about mathematics.

You can learn how to use counters and cups to solve equations involving addition in Lab 2 on pages A4–A5.

Hands-On Activities

Lab Activity references in Course 1 direct students to appropriate activities in the Manipulative Lab Activities on pages A1–A16.

Complete this sentence: "The thing I like best about math class is...." Tell why.

Cooperative Learning

Assessment

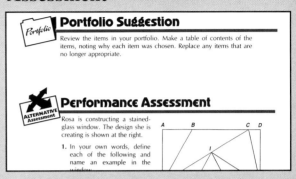

Portfolio Suggestions give students an opportunity to assess their own work and select appropriate pieces from their class work.

The **Performance Assessment** task in each chapter provides an alternative assessment for evaluating students' mathematical understanding.

Technology

Calculator Hints suggests places where this technology is appropriate. In Courses 2 and 3, they also direct students to the appropriate Tecnology Activities

Many calculators have an $\boxed{x!}$ key that calculates the factorial of the last number you entered.

Group and Individual Projects in each chapter opener and **Extended Projects** offer students extensive experience in working together to solve problems.

Review

Mixed Reviews present a quick review of previously-learned concepts in an integrated setting.

Mixed Review

Find each of the following and simplify.
1. $2.9 + (-4.6)$ 2. $17.2 - (-8.8)$ 3. $(9.75)(-0.16)$ 4. $8.12 \div (-0.7)$

Solve and check each equation.
5. $23 - 8x = 17$ 6. $3(2m - 7) = 11m - 6$

Let p represent "$-6y + 2 \leq 14$" and q represent "$(y < -1) \leftrightarrow (y - 6 \geq -5)$." If $y = -2$, determine the truth value of each statement.
7. $p \rightarrow q$ 8. $p \wedge q$ 9. $p \wedge \sim q$ 10. $\sim p \vee q$

Comprehensive Teacher Support Makes It Easy to Organize, Plan, and Teach Each Lesson

COURSE 1

CHAPTER 2 Operations and Numbers

Chapter Overview

Operations and numbers and their properties are the foundation of algebraic techniques. In this chapter, students who need to do so, can review and practice addition, subtraction, multiplication, and division of rational numbers and their properties. The basic ideas and terminology of algebra can then be introduced.

Teaching Suggestions

2.1 The concepts of logic should be applied to the new concepts of using variables in algebraic expressions and working with exponents.

2.2 Grouping symbols and the order of operations are used to simplify expressions. You may wish to use calculators to add interest and emphasis.

2.3 This lesson may be a review for some classes. It introduces graphing integers on a number line.

2.4 An informal approach is used for number line addition. This will be a review for many students. Lab 1 on page A2–A3 shows how you can use counters to model addition of integers.

2.5 Have students find their own short cuts for adding same sign and different sign integers. Absolute value is introduced.

2.6 Discuss opposites and how they can be used to subtract one integer from another. Then introduce the rule for subtraction of integers. Lab 1 on pages A2–A3 shows how you can use counters to model subtraction of integers.

2.7 Lead students to discover the pattern for multiplying and dividing integers with same and different signs. The rules for multiplying and dividing integers should then seem reasonable to them. Discuss division by zero and the term *undefined*.

2.8 Rational numbers are introduced along with practice and review in basic computation of fractions using the least common denominator. Have students demonstrate that whole numbers, fractions, and decimals are rational numbers.

2.9 To emphasize the properties for addition and multiplication of rational numbers, have students show that there are no commutative or associative properties for subtraction or division.

2.10 Emphasize the distinction between like and unlike terms. Have students demonstrate that the distributive property applies to subtraction as well as addition.

Problem Solving Application Students use the guess-and-check strategy to solve problems.

Using Communication

Journals can act as an excellent vehicle for student expression of what they have learned or items they still are uncertain about. Have students keep a log of the new concepts they learn. Also have them include examples. At the end of the chapter, have them review their lists and mark any concepts they still do not understand.

Lesson Planning Guide / Blackline Masters Booklets (pages)

Lesson	Pages	Practice with Integrated Review	Multicultural Activities	Technology	Cooperative Learning	Problem Solving	Enrichment	Manipulative Activities	Evaluation
2.1	38–40	10							
2.2	41–42	11		3					
2.3	43–45	12		4					7 Quiz A
2.4	46–47	13			2				
2.5	48–49	14							
2.6	50–51	15	2						
2.7	52–55	16				16		25–28	8 Quiz B
2.8	56–60	17							
2.9	61–64	18					30		
2.10	65–66	19							
Problem Solving Application	67–68	20							
Review	79–71	Chapter 2 Test Form A							5
Test	72	Chapter 2 Test Form B							6

Time Required
Integrated Mathematics Course: 14 days
Integrated Mathematics with Transformational Geometry Course: 13 days

Other Chapter Resources

Student Edition
Chapter Opener, p. 37
Mixed Review, pp. 45, 60
Journal Entry, p. 41
Portfolio Suggestion, p. 69
Performance Assessment, p. 69

Chapter Summary, pp. 69–70
Manipulative Activity, Lab 1, pp. A2–A3
Extended Project 1, pp. B2–B5

Teacher's Classroom Resources
Transparencies 3, 4, and 5
New York Regents Examination Practice Masters
Tutorial Software (Apple)
Glencoe Mathematics Professional Series

The new **Teacher's Handbooks** in the *Teacher's Annotated Editions* provide information on critical current issues in mathematics education. It also includes the following features:

Chapter Materials
- overview
- teaching suggestions for each lesson
- alternative teaching techniques
- organizational chart coordinating the ancillary materials for the chapter

Additional Course Materials
- Course Planning Guide
- Teacher's Answer Key
- List of Learning Objectives

Outstanding Ancillaries Help You to Systematically Enhance and Present Content

Real-World Applications: Transparencies and Blackline Masters show real-life photographs and current statistical information that students use to explore situations presented on the Student Activity Masters. Complete teaching suggestions and answers are also provided.

Practice Masters with Integrated Review reinforce learning for every lesson including the problem-solving application lessons. An answer key in red makes it easy for you or your students to check their work.

Evaluation Masters for each chapter include two forms of each Chapter Test, two Quizzes, five Cumulative Tests / Reviews, and one Final Test. A line-up answer key makes grading easy!

Multicultural Activity Masters for each chapter relate lessons to culturally diverse situations.

Problem Solving and Enrichment Masters provide multiple opportunities to expand students skills in cooperative learning and logical reasoning techniques.

Technology Masters include scientific calculator, graphing calculator, and computer activities. There are two activities for each chapter.

Manipulative Activity Resources contains easy-to-make manipulative masters and 20 activities correlated to Courses 1, 2, and 3.

Review Problem Software helps students review and practice major concepts. A variety of drill, tutorial, and interactive quiz formats are used.

Choose *Integrated Mathematics* — An Unequalled Program for Teaching Algebra and Geometry

Integrated Mathematics	Course 1	Course 2	Course 3
Student Edition	0-02-824566-0	0-02-824906-2	0-02-824968-2
Teacher's Annotated Edition	0-02-824567-9	0-02-824907-0	0-02-824969-0
Real World Applications Transparencies and Blackline Masters	0-02-824568-7	0-02-824914-3	0-02-824976-3
Practice Masters with Integrated Review	0-02-824569-5	0-02-824908-9	0-02-824970-4
Evaluation Masters	0-02-824570-9	0-02-824909-7	0-02-824971-2
Multicultural Activity Masters	0-02-824571-7	0-02-824910-0	0-02-824972-0
Technology Masters	0-02-824572-5	0-02-824911-9	0-02-824973-9
Problem Solving and Enrichment Masters	0-02-824573-3	0-02-824912-7	0-02-824974-7
Review Problems Software (Apple)	0-02-824565-2	0-02-824913-5	0-02-824975-5
*Manipulative Activity Resources	0-02-824563-6	0-02-824563-6	0-02-824563-6
*New York Regents Exam Practice Masters	0-02-824564-4	0-02-824564-4	0-02-824564-4

*For use with all three courses.

For more information contact your nearest regional office or call 1-800-334-7344.

Northeast Region
Glencoe/McGraw-Hill
15 Trafalgar Square #201
Nashua, NH 03063-1968
603-880-4701
800-424-3451
Fax: 603-595-0204
(CT, MA, ME, NH, NY, RI, VT)

Mid-Atlantic Region
Glencoe/McGraw-Hill
P.O. Box 458
Hightstown, NJ 08520-0458
609-426-5560
800-553-7515
Fax: 609-426-7063
(DC, DE, MD, NJ, PA)

Atlantic-Southeast Region
Glencoe/McGraw-Hill
Brookside Park
One Harbison Way,
Suite 101
Columbia, SC 29212
803-732-2365
800-731-2365
Fax: 803-732-4582
(KY, NC, SC, VA, WV)

Southeast Region
Glencoe/McGraw-Hill
6510 Jimmy Carter Boulevard
Norcross, GA 30071
404-446-7493
800-982-3992
Fax: 404-446-2356
(AL, FL, GA, TN)

Mid-America Region
Glencoe/McGraw-Hill
936 Eastwind Drive
Westerville, OH 43081
614-890-1111
800-848-1567
Fax: 614-899-4905
(IN, MI, OH)

Great Lakes Region
Glencoe/McGraw-Hill
846 East Algonquin Road
Schaumburg, IL 60173
708-397-8448
800-762-4876
Fax: 708-397-9472
(IL, MN, WI)

Mid-Continent Region
Glencoe/McGraw-Hill
846 East Algonquin Road
Schaumburg, IL 60173
708-397-8448
800-762-4876
Fax: 708-397-9472
(IA, KS, MO, ND, NE, SD)

Southwest Region
Glencoe/McGraw-Hill
320 Westway Place,
Suite 550
Arlington, TX 76018
817-784-2113
800-828-5096
Fax: 817-784-2116
(AR, LA, MS, NM, OK)

Texas Region
Glencoe/McGraw-Hill
320 Westway Place,
Suite 550
Arlington, TX 76018
817-784-2100
800-828-5096
Fax: 817-784-2116
(TX)

Western Region
Glencoe/McGraw-Hill
709 E. Riverpark Lane,
Suite 150
Boise, ID 83706
208-368-0300
800-452-6126
Fax: 208-368-0303
(AK, AZ, CO, ID, MT, NV, OR, UT, WA, WY)

California Region
Glencoe/McGraw-Hill
15319 Chatsworth Street
P. O. Box 9609
Mission Hills, CA 91346
818-898-1391
800-423-9534
Fax: 818-365-5489
(CA, HI)

Glencoe Catholic School Region
Glencoe/McGraw-Hill
25 Crescent Street, 1st Floor
Stamford, CT 06906
203-964-9109
800-551-8766
Fax: 203-967-3108

Canada
McGraw-Hill Ryerson Ltd.
300 Water Street
Whitby, Ontario
Canada L1N 9B6
905-430-5000
800-565-5758
Fax: 905-430-5020

International
McGraw-Hill, Inc. International Group
Princeton-Hightstown Road
Hightstown, NJ 08520
609-426-5934
Fax: 609-426-7917

DoDDS and Pacific Territories
McGraw-Hill School Publishing Company
1221 Avenue of the Americas
13th Floor
New York, NY 10020
212-512-6128
Fax: 212-512-6050

GLENCOE
McGraw-Hill

P.O. Box 508
Columbus, Ohio 43216

TEACHER'S ANNOTATED EDITION

INTEGRATED MATHEMATICS
COURSE 2

BUMBY

KLUTCH

COLLINS

EGBERS

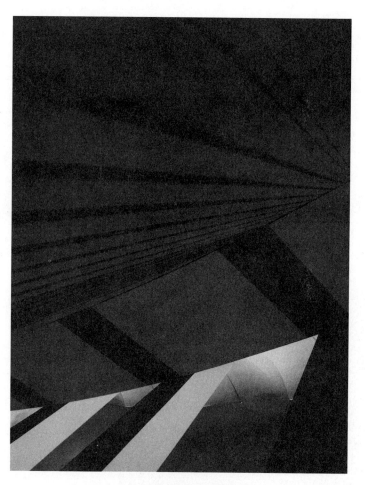

GLENCOE
McGraw-Hill

New York, New York   Columbus, Ohio   Mission Hills, California   Peoria, Illinois

Copyright © 1996 by Glencoe/McGraw-Hill. All rights reserved.
Originally copyrighted by Merrill Publishing Company. Except as permitted under the United States Copyrights Act, no part of this publication may be reproduced or distributed in any form or by any means, or stored in a database or retrieval system, without prior written permission of the publisher.

Send all inquiries to:
Glencoe/McGraw-Hill
936 Eastwind Drive
Westerville, Ohio 43081

ISBN: 0-02-824907-0

Printed in the United States of America.

2 3 4 5 6 7 8 9 10 RRC/LP 03 02 01 00 99 98 97 96 95

Teacher's Handbook

The purpose of this 14-page *Teacher's Handbook* for the **Integrated Mathematics, Course 2, Teacher Annotated Edition.** is to relate some background information on several contemporary issues, such as the use of technology and alternative assessment strategies, facing mathematics educators in the 1990s.

Table of Contents

NCTM Standards	iv–T1
Cooperative Learning	T2–T3
Alternative Assessment	T4–T5
Problem Solving	T6
Using Manipulatives	T7
Using Technology	T8–T10
Meeting Individual Needs	T11
Involving Parents and the Community	T12
Planning Your Course of Study	T13–T40
Teacher's Answer Key	T42–T121
Learning Objectives	T122–T124

NCTM Standards

In 1986, the Board of Directors of the National Council of Teachers of Mathematics (NCTM) established the Commission on Standards for School Mathematics. Three years later, in 1989, this commission published a document entitled *Curriculum and Evaluation Standards for School Mathematics,* based on the principle that all students need to learn more — and sometimes different — mathematics. This document was written to help improve the quality of school mathematics by establishing a broad framework to guide the reform of the mathematics curriculum during the 1990s. The *Standards* offers a vision of what the mathematics curriculum should include in terms of priority and emphasis in content.

In addition to delineating specific mathematics content standards for the K–12 curriculum, the commission presented an entirely new and different set of four standards involving the process of doing mathematics. Unlike the content standards, which vary somewhat for the primary, middle school, and high school levels, these standards apply to all grade levels. They are, in fact, the first four standards for each grade level and are sometimes referred to as process goals.

Standard 1: Mathematics as Problem Solving
Standard 2: Mathematics as Communication
Standard 3: Mathematics as Reasoning
Standard 4: Mathematical Connections

The remaining standards are the content standards, which delineate specific content for certain grade spans. The content standards for grades nine through twelve are as follows.

Standard 5: Algebra
Standard 6: Functions
Standard 7: Geometry from a Synthetic Perspective
Standard 8: Geometry from an Algebraic Perspective
Standard 9: Trigonometry
Standard 10: Statistics
Standard 11: Probability
Standard 12: Discrete Mathematics
Standard 13: Conceptual Underpinnings of Calculus
Standard 14: Mathematical Structure

In ***Integrated Mathematics, Course 2,*** the study of algebra and geometry are integrated along with the topics of logic, statistics, and probability. In this integrated approach, students learn mathematics as a whole and not as compartmentalized areas of instruction. The following is a chapter-by-chapter correlation of Course 2 content to specific NCTM Standards.

Standard 1: Mathematics as Problem Solving
Chapters 1, 2, 3, 4, 5, 6, 7, 8, 9, 10, 11, 12, 13

Standard 2: Mathematics as Communication
Chapters 1, 2, 3, 4, 5, 6, 7, 8, 9, 10, 11, 12, 13

Standard 3: Mathematics as Reasoning
Chapters 1, 2, 3, 4, 5, 6, 7, 8, 9, 10, 11, 12, 13

Standard 4: Mathematical Connections
Chapters 1, 2, 3, 4, 5, 6, 7, 8, 9, 10, 11, 12, 13

Standard 5: Algebra
Chapters 1, 2, 3, 4, 5, 6, 7, 8, 9, 10, 11, 12, 13

Standard 6: Functions
Chapters 12, 13

Standard 7: Geometry from a Synthetic Perspective
Chapters 4, 5, 6, 7, 8, 9, 12, 13

Standard 8: Geometry from an Algebraic Perspective
Chapters 8, 9, 12, 13

Standard 9: Trigonometry
Chapter 13

Standard 11: Probability
Chapter 11

Standard 12: Discrete Mathematics
Chapter 11

Standard 13: Conceptual Underpinnings of Calculus
Chapter 13

Standard 14: Mathematical Structure
Chapters 1, 2, 3, 4

Cooperative Learning

Cooperative learning groups *learn* things together, not just *do* things together. Studies show that cooperative learning promotes more learning than competitive or individual learning experiences regardless of student age, subject matter, or learning activity. More difficult learning tasks, such as problem solving, critical thinking, and conceptual learning, come out far ahead when cooperative strategies are used.

> "*S*tudies show that cooperative learning promotes more learning than competitive or individual learning experiences."

Studies also show that in classroom settings, adolescents learn more from each other about subject matter than from the teacher.

The basic elements of a cooperative learning group are as follows:

- Students must perceive that they "sink or swim together."
- Students are responsible for everyone else in the group, as well as themselves, in learning the assigned material.
- Students must see that they all have the same goal, that they need to divide up the tasks and share the responsibility equally, and that they will be given one evaluation or reward that will apply to all members of the group.

- **Arrange your room.**
  Students in groups should face each other as they work together.

- **Decide on the size of the group.**
  Groups work best as teams of two to five students, as materials or activities dictate.

- **Assign students to groups.**
  Each group should be mixed socially, ethnically, by sex, and by learning abilities.

- **Prepare students for cooperation.**
  Tell students about the rationale, procedures, and expected outcomes of this method of instruction.

- **Plan teaching materials.**
  Provide only one set of materials to the group to encourage working together. Or vary this by providing each member with a worksheet but collect only one of them.

- **Explain group tasks.**
  For most activities, each group will need someone to take notes, someone to summarize progress, and someone to make sure all are involved. Students should rotate tasks with each new activity.

- **Explain the day's lesson.**
  At this time, go over instructions and relate the work to previous learning. Indicate how you will assess their performance as a group. List the specific group skills you will be checking: quiet voices, encouraging each other, eye contact, group decisions, and so on, adding higher level skills as students become more comfortable with cooperative groups.

  What should the teacher be doing while students are working in cooperative groups? In cooperative learning, the teacher's role is more that of a facilitator than of a dispenser of knowledge. You need to monitor group behavior, using a formal observation sheet to count the number of times you observe the behavior expected on an assignment for each group. You may need to intervene to ask a group why it is not being effective and have them work toward a solution.

  You may also need to provide assistance with the task, not in a direct way, but by clarifying instructions and asking open-ended questions to direct the thinking of groups.

After groups have completed their task, students need to summarize what they have learned and relate it to previous experiences. They should also evaluate how they worked together as a group, decide if they achieved the criterion you set up, and suggest ways their group could improve.

Keep in mind that cooperative learning does not just "happen." The first few days may seem like bedlam, with some students upset, others mistrustful, and others off task. Both you and your students will make mistakes. Be patient and keep at it.

For more information on cooperative learning, see *Cooperative Learning in the Mathematics Classroom*, a part of **Glencoe's Mathematics Professional Series.** A cooperative learning activity has been provided for each chapter in ***Integrated Mathematics, Course 2*** in the *Problem Solving and Enrichment Masters* booklet in your **Teacher's Classroom Resources.**

Alternative Assessment Strategies

Most people would agree that the test is often seen as the bottom line of educational enterprises. Many would also agree that the present tests used to determine a student's accomplishment often fall short of the goal of measuring the student's true understanding of the concepts being taught.

Most evaluation materials today include multiple choice or true/false questions or questions that require objective one-word answers. These types of questions act as a mechanical means of evaluation that focus on a student's ability to reiterate memorized and practiced definitions and procedures for solving each type of problem. This often does not accurately reflect the student's true capabilities and understanding.

According to the National Council of Teachers of Mathematics in their *Curriculum and Evaluation Standards for School Mathematics* (1989), "Assessment must be more than testing; it must be a continuous, dynamic, and often informal process." The purpose of assessment is to identify areas of weakness and strengths for individual students, to assign grades, to gather data for teachers to plan their course of instruction, and to evaluate the current course of instruction. The NCTM *Standards* suggests that different types of evaluation materials be used in making assessments. These evaluations are an ongoing process and are used to augment formal assessment, such as the traditional chapter test.

There are numerous strategies whereby the teacher of mathematics may expand his or her repertoire of alternative forms of assessment. *Alternative Assessment in the Mathematics Classroom*, a part of the **Glencoe Mathematics Professional Series,** offers an in-depth study of assessment strategies.

Open-ended questions are one of the easiest ways to employ a new evaluation strategy. In an assessment of this nature, students are forced to think for themselves and to express mathematical ideas in language that corresponds with their mathematical development. When using open-ended questions, there may not be *one* correct answer, but rather an array of answers and explanations that are acceptable.

Classroom Observations are another way of documenting student achievement. While observing your students at work, you often form quite accurate opinions concerning your students' capabilities. By simply implementing some structure into such observations, teachers can capitalize on this information for assessment purposes.

This strategy is especially valuable during cooperative learning activities.

Performance Assessment is used to determine what students know and what they can do when presented with a realistic problem-solving situation. Each chapter in *Integrated Mathematics, Course 2* contains a Performance Assessment task for the students to complete.

Questioning is an excellent tool for assessing the comprehension of students. Dialogue between you and your students is most beneficial in identifying comprehension errors and the use of process skills. The Exploratory exercises at the beginning of each exercise set allow you to evaluate students' understanding before assigning homework.

Self-Evaluation permits students to monitor their own progress. Material at the end of each chapter provides students with an opportunity to test their understanding. Using this review material in a cooperative group setting not only provides students the opportunity to reinforce skills, but allows you to observe the strengths and weaknesses of students in an informal situation. Using tests to which students have contributed questions makes for an excellent review strategy and also provides students with a sense of ownership in the assessment process.

Portfolios are student collections of their work that are representative of their growth as learners. Your students should periodically submit samples of their work to document their progress over the course of the year.

Each chapter in *Integrated Mathematics, Course 2* has portfolio suggestions which will help guide students in what to include in their portfolios.

Journal Entries allow students to express what they are learning. These may involve many different forms of expression such as graphs, drawings, models, and other explanations. Suggestions for journal entries are provided in each chapter of *Integrated Mathematics, Course 2.* Regular perusal of the journals can guide teaching procedures and serve as the basis for individual student-teacher conferences.

Although there are undoubtedly situations when the written test is effective, we must guard against using it to the exclusion of other forms of assessment. The examiner must select, from among the many strategic techniques available, those which will best indicate what his or her students can do.

Problem Solving

According to the NCTM *Standards*, "Problem solving is the process by which students experience the power and usefulness of mathematics in the world around them. It is also a method of inquiry and application...to provide a consistent context for learning and applying mathematics. Problem situations can establish a 'need to know' and foster the motivation for the development of concepts."

One of the challenges of teaching mathematics is helping students learn to solve problems, which involves more than solving routine verbal problems — it also involves the ability to analyze a new problem and discover a way to solve it. **Problem-Solving Application** lessons and examples using the four-step plan can be found throughout the text to help students develop their problem-solving abilities. In addition, you can involve your students with problem solving using the *Problem Solving and Enrichment Masters* and the *Real-World Applications Transparencies and Blackline Masters* included in your **Teacher's Classroom Resources.**

FOUR-STEP PLAN

1. EXPLORE the problem. Read the problem and identify what is given and what is asked. Jot down important facts and draw a chart or diagram if it will help. Think about how the facts are related. If an equation will be used to solve the problem, choose a variable to represent one of the unspecified numbers in the problem. Read the problem again and use the variable in writing expressions for other unspecified numbers in the problem.

2. PLAN the solution. Many different strategies may be used. If an equation will be used to solve the problem, read the problem again. Decide how the unspecified numbers relate to other given information. Write an equation to represent the relationship.

3. SOLVE the problem. This involves doing the mathematics and interpreting the answer. If an equation was written, solve the equation and interpret the solution. State the answer to the verbal problem.

4. EXAMINE the solution. Does the answer makes sense, given the conditions of the problem? If not, check your mathematics again. If the mathematics was correct, a mistake was made in "setting up" the problem. In that case, explore the problem again and try a different approach.

Using Manipulatives in Your Classroom

Most teachers agree that the use of manipulative materials helps students build a solid understanding of mathematical concepts and enhances students' achievement in mathematics. The universal anxiety today about achievement in mathematics should prompt educators to heed results from recent studies which indicate that by using manipulative materials at every level, test scores are dramatically improved.

Many teachers find that demonstrations on an overhead projector ease students with little experience in hands-on learning into working with manipulatives. Simply observing such a demonstration, however, is not enough. Follow-up activities that involve the students in the act of physically manipulating such aids are essential if students are to experience the patterns and relations inherent in mathematics.

Using manipulatives assists students in bridging the gap between their own concrete environment and the abstract concepts of mathematics. **Integrated Mathematics** contains many opportunities for teachers to introduce the use of manipulatives to encourage students in their discovery and exploration of concepts. Activities, as well as masters for easy-to-make manipulatives, can be found in the *Manipulative Activity Resources* booklet, available in your **Teacher's Classroom Resources.**

To make sure that students understand what is expected of them in any manipulative activity, the teacher must discuss its goal and model how the manipulative is to be used to achieve it. Encouraging students to suggest ways in which manipulatives can be used helps them to relate concepts and develop mathematical insights.

The important thing to keep in mind is that most students do not automatically abstract the concepts they explore with materials. They will usually need thoughtful guidance. Summarizing and recording activity results help to focus on relationships and concepts. Used in this way, manipulative materials are a justifiable means to a desirable end.

Using Technology in Your Classroom

"*Students today can't prepare bark to calculate their problems. They depend upon their slates which are more expensive. What will they do when their slate is dropped and it breaks? They will be unable to write!*"

The above quote from a Teachers' Conference in 1703 seems ridiculous today, but it was not back then. Many people may ask, "Couldn't they see enough into the future to know that slates would be required?" But, as the old adage goes, hindsight is twenty-twenty: the difficulty lies in foreseeing the future.

The argument today is in regard to computers and calculators; and while most people agree that technology should be used in mathematics classrooms, ideas about its incorporation, testing, appropriate use, and so forth are often hard to find. Some use this as an argument against using technology; instead, we must be able to recognize a good idea when we see one and adopt it for our classrooms.

There are many different ideas and definitions of "appropriate use of technology." Every class and every teacher is different. What is appropriate and workable for one may not be so for another. Keep in mind that many of your students are just learning the technology. You should refrain from getting "high-tech" with them. Other students may be proficient with technology to the point where they can help others in the classroom. Give both groups opportunities for success early so they can feel comfortable with the technology you use. Starting with easy computations for the first classroom encounter will give the unexperienced a sense of accomplishment and the more advanced a sense of confidence and pride that they already know what is being done.

Technology opens up many new ideas and opportunities. Lab partners and cooperative groups can be used if you have a limited supply of technology. Projection panels for the overhead projector can be used for giving computer demonstrations. Overhead projector calculators are available for most models.

Simple templates of computer or calculator keyboards can be used on the overhead or put on the bulletin board for showing keystrokes. This can help especially in conveying the location of second-function keys on graphing calculators.

Using the Graphing Calculator

Incorporating the graphing calculator into your classroom may seem like an arduous task, but it may be easier than you think. And the results will be very rewarding. Once your students are familiar with using the technology, they may begin to respond to mathematics in a way that you have never seen before. Because the graphing calculator is a visual teaching tool, many of today's students work well with it. You can definitely use this to your advantage. The graphing calculator is also a scientific calculator. Students do not need a separate calculator for routine numerical computations.

It may be necessary for you to familiarize yourself with the graphing calculator before using it in your classroom. The manuals for the calculators can help you get started and can be used as a guide or reference to clear up questions that you have along the way. In addition, a comparison of various calculators, along with keystroke instruction, applications, programs, and much more can be found in *Graphing Calculators in the Mathematics Classroom,* a part of the **Glencoe Mathematics Professional Series.** The *Technology Masters* booklet in your **Teacher's Classroom Resources** also contains a scientific and a graphing calculator activity for each chapter in **Integrated Mathematics, Course 2.** You can also talk to other teachers who are using the calculator if you need help getting off the ground with the technology.

In the beginning, many of your students will be looking at a graphing calculator for the first time in their lives, and they may be apprehensive about what they see. A graphing calculator may seem like something they will not be able to understand or something that will be too difficult for them to use. As a start, have them graph an equation like $y = x$, which uses just a few keystrokes. This way they can see their accomplishments and may realize that the calculator is not as difficult as they thought. Then go step by step with them as they progress into doing more difficult activities with the calculator.

One of the most helpful tools to make teaching the use of the graphing calculator or scientific calculator easier is a transparency or poster of the calculator keyboard. These enlargements will help you show your students where keys are if they are having trouble finding them or if you are explaining a lesson and need to demonstrate your keystrokes. The *Technology Masters* booklet in your **Teacher's Classroom Resources** contains masters for making such a transparency for the TI-34 scientific calculator and the TI-81 graphing calculator.

Flexibility

The graphing calculator offers you much more flexibility than before. Some people may not understand that it is okay to solve problems with a calculator and that most of the time the calculator does not "find the answer" for you but merely helps you to find appropriate solutions to problems. If you do not understand how to interpret the information the calculator provides, it will be of no use to you. For example, if you look at the graph of a line and ask what the slope of the graph is, some students may tell you they do not know how to find it. It is still up to the teacher to make connections with the technology and to teach with it. The calculator simply saves time and effort and allows you to go deeper into mathematical concepts with your students than you could in the past.

Other Advantages

Besides saving time, there are numerous other advantages to using the graphing calculator. One is that students will have a deeper understanding of the concepts being taught. Technology teaches them in a way that piques their interest and leads to questions that show a desire to learn. Another advantage of technology is the cooperative spirit that develops among students which will benefit them in college and in their careers. Since students will probably need to work with others when they become members of the work force, it is best for them to learn to work with others while they are in school.

Role of Technology

Educators sometimes still express the fear that technology will take the place of the teacher. *This will never happen.* There is no way that a computer can do what a teacher does, and in particular, sense what a teacher can. For example, no computer can sense uncertainty of an answer in a student's tone of voice. Few students will be able to turn on a calculator or computer and teach themselves with it. They need teachers to explain the technology and to help make connections.

Technology as a Tool

Computers and calculators are excellent tools for teaching mathematics concepts. Though there are still some who argue their use in the classroom, none can argue their educational advantage. Technology can teach students in ways that were never before possible. In the past, teachers could only dream of being able to do some of the things that they can now do with ease, thanks to technology. While incorporating technology into your mathematics classroom can be difficult at first, in the long run, it will definitely pay off for both you and your students.

Meeting Individual Needs

Multicultural Perspective

There is no doubt that the United States is a multicultural society. Changing demographics, and changing economic and social orders are having a tremendous impact on the schools in this country. The term *multicultural* represents more than just the many cultures we find in our society today. It also reflects the multicultural basis to all knowledge, including mathematics.

From the ancient Egyptians, who used the "Pythagorean" theorem fifteen centuries *before* Pythagoras, to the ancient Chinese, who knew the value of π to ten places twelve centuries *before* the Europeans, mathematics as we know it today has been shaped by many cultures.

What is the role of the mathematics educator in all this? Students should have the opportunity to learn about persons from all cultures who have contributed to the development of mathematics. To this end, the *Multicultural Activity Masters* included in your **Teacher's Classroom Resources** provide the students with a multicultural experience for each chapter in ***Integrated Mathematics, Course 2***.

In addition, mathematics educators also must prepare all students for the new jobs of the future that will require mathematical literacy. The mathematics teacher must be in the vanguard of those who demand that *all* students be given the opportunity to study the more advanced forms of mathematics. It is our responsibility as educators to make every effort to prepare today's students for success in tomorrow's complex world.

Limited English Proficiency Needs

One of the greatest factors contributing to under-achievement in mathematics education for the language-minority student is failure to understand the language of instruction. There are, however, strategies which can be employed to help overcome the obstacles that beset language-minority students.

Ideally, these students would have new concepts and skills reinforced by discussing them in their native tongue, through either a bilingual teacher or tutor. Materials written in the native language can be provided to supplement classroom instruction, as can an oral approach using pictorial materials and/or manipulative devices. Observing the student's verbal interaction with other students can help determine the student's level of understanding.

Students coming from diverse backgrounds lack common educational experiences. Many times, the only way to establish a basis for communication is find content that is familiar to one and all. By integrating the development of language into such familiar content, we may be able to open the door to challenging and exciting world of mathematics for the language-minority student.

Involving Parents and the Community

Research shows that parents exert a strong influence on their children's attitudes toward mathematics and on their performance in the classroom. By keeping parents informed, perhaps through parent conferences and newsletters, you are keeping parents involved.

When parents have the opportunity to meet with teachers, they often ask what they can do to motivate their children. The following ideas, developed by Reginald Clark of the Academy for Educational Development in Washington, D.C., are designed to foster positive attitudes and boost learning.

- Share the fact that the more students watch TV, the less likely they are to do well in school.
- Suggest family discussions about school, education, careers, life skills, rules, and expectations.
- Explain that it's important for parents to use their experiences and share problem-solving strategies with their children.
- Stress the importance of seeing to it that their children complete all homework.
- Encourage parents to monitor after-school activities.
- Urge parents to provide time, space, and materials needed for homework, reading, and hobbies.
- Suggest that the family play board games together.
- Urge parents to introduce their children to other responsible mentors.
- Emphasize the importance of having books and magazines in the home.

Just as parents need to be partners in convincing your students about the importance of mathematics, members of your community also can be influential. Partnerships with local businesses can provide field trip possibilities, project assistance, speakers, and classroom demonstrations. Community members with strong mathematical backgrounds can volunteer to become tutors to assist individuals or small groups. Community service projects can involve your students with the community and benefit both.

Getting parents and the community involved with your students in a positive way can greatly enhance their mathematics education. Additional methods and strategies can be found in *Involving Parents and the Community in the Mathematics Classroom*, a part of the **Glencoe Mathematics Professional Series**.

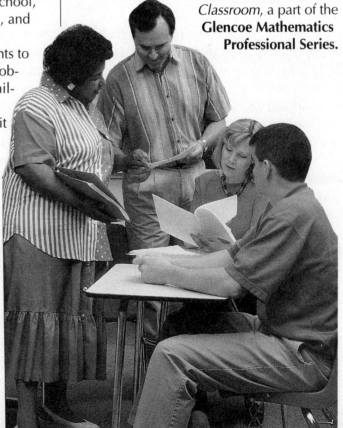

Planning Your Course of Study

At the beginning of each school year, you find new classes of students, each with varying needs and abilities. *Integrated Mathematics, Course 2* offers you various options and teaching aids to design a course of study suitable for each class of students.

The following pages contain information on each chapter as well as the Technology Lab Activities and the Extended Projects, located at the back of the student text. The Overview and Teaching Suggestions provide a glimpse into the content of each chapter. Suggestions for implementing various teaching strategies such as cooperative learning, using manipulatives and models, and using technology are also presented. A complete list of the Learning Objectives are presented on pages T122–124.

Integrated Mathematics, Course 2 offers a wide variety of teaching materials to aid in instruction and enhancement of each chapter. An organizational chart is provided to show what blackline masters are available for each lesson. Additional components that accompany the chapter are also listed.

The Assignment Guide shown below is provided to help in planning your year's work. Two courses are outlined in order to offer flexibility in response to your curriculum requirements and the abilities of your students. This guide offers a 170-day time schedule. It can be adjusted to fit your own school calendar.

CHAPTER	1	2	3	4	5	6	7
Integrated Math Course	13	10	10	19	17	16	18
Integrated Math Course with Transformation Geometry and Trigonometry	11	9	8	17	15	14	16

CHAPTER	8	9	10	11	12	13
Integrated Math Course	16	15	21	15	0	0
Integrated Math Course with Transformation Geometry and Trigonometry	14	13	19	14	12	8

Each lesson in the text includes more exercises than an average class normally could complete. To help you meet the needs of your student, three daily assignments are annotated on the exercise pages for maximum, average, and minimum assignments.

CHAPTER 1
Working with Logic

Chapter Overview
The study of logic can be applied to problem solving in other areas of mathematics as well as in everyday life. In this chapter, students learn to identify the truth value of logic statements and use logic symbols and truth tables to summarize these values. They are introduced to various types of statements including negations, conjunctions, disjunctions, conditionals, biconditionals, converses, inverses, and contrapositives. Students apply laws of logic and use the statement-reason form of proof to show that the indicated conclusion follows from the given premises.

Teaching Suggestions

1.1 Be sure students know how to construct a truth table and are familiar with the truth tables for conjunctions and disjunctions.

1.2 Conditional statements are the basis of many mathematical concepts. Point out that a conditional is true in every case except when the antecedent is true and the consequent is false.

1.3 Formal two-column, statement-reason proofs are introduced in this lesson. Point out to students that this method of proof is also used in geometry and algebra.

1.4 The Law of Detachment can be illustrated by using everyday situations as models.

1.5 Stress that while the converse or inverse of a true statement may or may not be true, the contrapositive is always true.

1.6 This lesson introduces the Law of Syllogism. Emphasize that the conclusion of this law is a conditional statement and cannot be used to prove the truth value of any of the individual components.

1.7 The De Morgan Laws define the negation of a conjunction and the negation of a disjunction. Students should understand these laws and be able to apply them.

1.8 Emphasize the difference between universal and existential quantifiers. Point out that in the study of logic, the word "some" means "at least one."

1.9 Show that negations of universal and existential quantifiers are "opposites" of each other. That is, negating a universally quantified statement produces an existentially quantified statement, and vice versa.

Problem Solving Application Students learn to solve real-life problems using matrix logic to eliminate possibilities.

Using Applications
Logical reasoning is frequently used in many fields such as politics and advertising. Explain to the students how politicians, for example, may use logical reasoning to persuade people to vote for them, but that their reasoning may be incorrect; or how advertisers may use inverses and converses incorrectly to incite people to buy their products. Have students look for examples of logical reasoning in magazines, newspapers, or on television. Have them present their findings to the class and explain whether the use of reasoning is correct or incorrect.

Lesson Planning Guide

Lesson	Pages	Practice with Integrated Review	Multicultural Activities	Technology	Cooperative Learning	Problem Solving	Enrichment	Manipulative Activities	Evaluation
1.1	2–5	1		1			27		
1.2	6–9	2		2					
1.3	10–12	3	1						3 Quiz A
1.4	13–16	4			1				
1.5	17–19	5						21–24	
1.6	20–23	6							4 Quiz B
1.7	24–27	7						85–88	
1.8	30–32	8							
1.9	33–35	9					14		
Problem Solving Application	36–37	10							
Review	38–41			Chapter 1 Test Form A					1
Test	42			Chapter 1 Test Form B					2

Time Required
Integrated Mathematics Course: 13 days
Integrated Mathematics with Transformational Geometry and Trigonometry Course: 11 days

Other Chapter Resources

Student Edition
Chapter Opener, p. 1
Mixed Review, p. 23
Mathematical Excursions, pp. 28–29
Journal Entries, p. 33

Portfolio Suggestion, p. 38
Performance Assessment, p. 38
Chapter Summary, pp. 38–39

Teacher's Classroom Resources
Transparencies 1 and 2
New York Regents Examination Practice Masters
Tutorial Software (Apple)
Glencoe Mathematics Professional Series

CHAPTER 2
Mathematical Systems

Chapter Overview

Properties of the real number system are applied to other systems. In this chapter, students become more familiar with these properties by applying them to the clock system. Students also identify operations on a set and identify groups and their properties.

Teaching Suggestions

2.1 This lesson introduces clock arithmetic. Students who have difficulty adding numbers in a clock system may prefer to use tables for obtaining solutions.

2.2 Be sure students understand the difference between an "ordinary" product and a clock product.

2.3 Students should be familiar with the Z_n notation for "clock n" since it will be used frequently in this chapter and future chapters.

2.4 Students identify an operation on a set. Stress that variables such as a, b, and c do not necessarily represent different numbers, but are elements upon which an operation acts.

2.5 Students should understand the symbolism used for identity and inverse, and be able to express the definition of each using those symbols.

2.6 This lesson introduces another kind of mathematical system called a *group*. Give examples of sets that are groups and sets that are not, to help clarify the definition of group.

2.7 Unlike the previous lesson, this lesson deals with identifying groups from an infinite number of elements. Students should know the properties of the sets N, W, Z, Q, and R under the operations of addition, subtraction, multiplication, and division.

2.8 Proofs are used to verify some group theorems. The exercises will help students become more familiar with the names of the four new theorems and how to use them.

2.9 This lesson may be used for an enrichment activity or for more advanced students. Students learn to compute using non-commutative groups.

Problem Solving Application Students use the guess-and-check strategy to solve problems.

Using Connections

The commutative and associative properties are used in many different areas of mathematics and science where numerical computations are necessary. For example, both of the properties are used when determining probabilities, interpreting statistics or analyzing mixtures in chemistry. Ask students to make a list of different mathematical or scientific applications for the commutative and associative properties.

Lesson Planning Guide

Lesson	Pages	Practice with Integrated Review	Multicultural Activities	Technology	Cooperative Learning	Problem Solving	Enrichment	Manipulative Activities	Evaluation
2.1	44–47	11							
2.2	48–51	12					28		
2.3	52–55	13		3					7 Quiz A
2.4	56–60	14		4		15			
2.5	61–64	15							
2.6	65–67	16	2						8 Quiz B
2.7	68–70	17			2				
2.8	71–75	18							
2.9	76–78	19							
Problem Solving Application	79–80	20							
Review	81–83		Chapter 2 Test Form A						5
Test	84		Chapter 2 Test Form B						6

Time Required
Integrated Mathematics Course: 10 days
Integrated Mathematics with Transformational Geometry and Trigonometry Course: 9 days

Other Chapter Resources

Student Edition
Chapter Opener, p. 43
Mixed Review, pp. 51, 64
Mathematical Excursions, p. 55
Journal Entry, p. 76

Portfolio Suggestion, p. 81
Performance Assessment, p. 81
Chapter Summary, pp. 81–82

Teacher's Classroom Resources
Transparencies 3 and 4
New York Regents Examination Practice Masters
Tutorial Software (Apple)
Glencoe Mathematics Professional Series

CHAPTER 3: A System with Two Operations

Chapter Overview
Properties of a field can be used to solve equations. This chapter includes the study of a two-fold operational system called a *field* and the proofs of field theorems. Students also use the definitions of subtraction and division to simplify expressions and solve equations.

Teaching Suggestions

3.1 This lesson reviews groups and their properties, and how to solve simple linear equations. Students less familiar with the algebraic concepts needed to solve linear equations may need additional review.

3.2 The Distributive Law can be thought of as the "glue" that binds two operations together in a system. Emphasize the fact that multiplication distribution over addition is an important and special mathematical property.

3.3 You may wish to review square roots before assigning exercises in this lesson. Students are to determine whether certain systems are fields, and then solve problems in those fields.

3.4 A negative exponent is used to redefine the multiplicative inverse. Be sure students understand this notation.

3.5 You may wish to omit or place less emphasis on some of the material in this lesson and the following lessons in this chapter. It is more important that students recognize the familiar relationships they learned in algebra rather than try to master all the symbolism and notation.

3.6 Point out that the theorem presented in this lesson has already been used to solve quadratic equations by factoring. It may be helpful to review factoring and multiplying of polynomials, before assigning exercises.

3.7 Discuss the relationship between subtraction and additive inverse; between division and multiplicative inverse.

3.8 Students are to order elements from least to greatest. Be sure they understand how to find the decimal equivalent of any real number.

Problem Solving Application Students learn to solve multi-step problems by identifying the separate subgoals, or steps, of each problem.

Using Logical Reasoning

Have students determine whether the following solutions are correct and have them explain the reasoning to support their answers.

$$4 \times (1 + 6) = (4 \times 1) + (4 \times 6)$$
$$= 4 + 24$$
$$= 28$$
Yes. Multiplication distributes over addition.

$$4 + (1 + 6) = (4 + 1) + (4 + 6)$$
$$= 5 + 10$$
$$= 15$$
No. Addition does not distribute over itself.

Lesson Planning Guide

Lesson	Pages	Practice with Integrated Review	Multicultural Activities	Technology	Cooperative Learning	Problem Solving	Enrichment	Manipulative Activities	Evaluation
3.1	86–87	21	3						
3.2	88–91	22					29		
3.3	92–95	23							11 Quiz A
3.4	96–99	24		5					
3.5	101–103	25			3				
3.6	104–106	26							12 Quiz B
3.7	107–110	27							
3.8	111–114	28		6		16			
Problem Solving Application	115–116	29							
Review	117–119		Chapter 3 Test Form A						9
			Chapter 3 Test Form B						10
Test	120		Cumulative Test/Review (Chapters 1–3)						13–14

Time Required

Integrated Mathematics Course: 10 days
Integrated Mathematics with Transformational Geometry and Trigonometry Course: 8 days

Other Chapter Resources

Student Edition
Chapter Opener, p. 85
Mixed Reviews, pp. 95, 110
Mathematical Excursions, p. 100
Journal Entry, p. 111

Portfolio Suggestion, p. 117
Performance Assessment, p. 117
Chapter Summary, pp. 117–118

Teacher's Classroom Resources
Transparencies 5, 6, and 7
New York Regents Examination Practice Masters
Tutorial Software (Apple)
Glencoe Mathematics Professional Series

CHAPTER 4
Introduction to Euclidean Geometry

Chapter Overview
Concepts of geometry can be found all around us in the real world; the process used to logically organize these concepts, in the form of proofs, can be applied to situations in everyday life. In this chapter, students use familiar geometric ideas to help prove basic theorems and to solve problems. Properties of lines, planes, angles, real numbers, and triangles are introduced and applied to problem solving.

Teaching Suggestions

4.1 This lesson reviews some basic geometric terms. It is important that students know these terms and definitions since they will be used throughout the text.

4.2 Familiar geometric ideas are organized into formal statements called *axioms*. Point out that axioms are statements assumed true, whereas theorems are statements that must be proved true.

4.3 Encourage students to find their own way to remember the difference between complementary and supplementary angles. An example is to remember that c comes before s in the alphabet and 90 comes before 180 in counting.

4.4 A proof is presented to demonstrate the use of real-number properties in geometry. Explain to students that proofs start with a given statement assumed to be true and logically proceed from one statement to the next to reach the desired conclusion.

4.5 The exercises in this lesson help familiarize students with writing proofs by having them prove theorems about angles. Stress that no assumptions should be made on the basis of appearance of a figure other than collinearity and betweenness of points, the location of points relative to an angle, and adjacent angles.

4.6 Remind students that lines in a figure are not perpendicular unless that information is given or the right angle symbol is used.

4.7 Students should be able to identify the different pairs of angles that are formed when a transversal intersects two parallel lines.

4.8 Theorems about parallel lines and interior angles are proved and applied. These theorems are useful in finding the measurements of certain angles in a figure.

4.9 Stress that in an indirect proof, the first step is to assume that the statement to be proved is false. Be sure students understand theorems 4–16 and 4–17 because they are essential to the proof of other theorems.

4.10 Students should be familiar with the theorems about angles of a triangle. These theorems are applied to find measures of particular angles in figures.

4.11 You may consider this lesson optional in this chapter. However, the definition of the term *regular polygon* is important and is used frequently in future chapters. Have students give examples of a regular polygon.

Problem Solving Apppliations Students use equations to solve word problems.

Using Critical Thinking
Have students name several ways to conclude that two angles are congruent. Some possibilities include proving that they are both right angles, showing that they are corresponding angles, and showing that they are each supplementary to the same angle.

Lesson Planning Guide

Lesson	Pages	Practice with Integrated Review	Multicultural Activities	Technology	Cooperative Learning	Problem Solving	Enrichment	Manipulative Activities	Evaluation
4.1	122–124	30		7					
4.2	125–127	31					30		
4.3	128–131	32		8					
4.4	132–134	33	4						17 Quiz A
4.5	135–138	34							
4.6	139–142	35							
4.7	143–146	36							
4.8	147–150	37				17			18 Quiz B
4.9	151–153	38							
4.10	154–156	39							
4.11	157–159	40			4			37–40	
Problem Solving Application	160–161	41							
Review	162–165		Chapter 4 Test Form A						16
Test	166		Chapter 4 Test Form B						17

Time Required
Integrated Mathematics Course: 19 days
Integrated Mathematics with Transformational Geometry and Trigonometry Course: 17 days

Other Chapter Resources

Student Edition
Chapter Opener, p. 121
Mixed Review, p. 131, 150
Journal Entry, p. 143

Portfolio Suggestion, p. 162
Performance Assessment, p. 162
Chapter Summary, pp. 162–163

Teacher's Classroom Resources
Transparencies 8 and 9
New York Regents Examination Practice Masters
Tutorial Software (Apple)
Glencoe Mathematics Professional Series

CHAPTER 5: Congruent Triangles

Chapter Overview
Properties of congruent triangles are useful in finding missing measures of angles and sides of triangles. This chapter shows students how to identify congruent triangles using *SSS, SAS,* and *ASA* axioms. Students also identify special elements of a triangle and apply theorems about isosceles triangles.

Teaching Suggestions

5.1 Students should be able to identify and label corresponding parts of triangles and recognize the "included" sides and angles.

5.2 It is important that students are able to analyze the information given to determine which of the axioms *(SSS, ASA, SAS)* applies.

5.3 The term CPCTC (corresponding parts of congruent triangles are congruent) should become a familiar one to students.

5.4 Show that triangles cannot be proven congruent using AAA or SSA principles by giving some examples.

5.5 Point out that it may be easier to identify the corresponding parts of overlapping triangles by drawing separate diagrams of each triangle. You may also use colored chalk or overlapping transparencies to help students visualize the triangles being discussed.

5.6 The terms *median* and *altitude* are defined as an introduction to the study of isosceles triangles. Students should be familiar with these terms since they are used frequently in future sections.

5.7 Since the Hypotenuse-Leg Theorem is difficult to prove, only a plan for proof is given. Emphasize that this theorem applies to only right triangles.

5.8 It may be helpful to review the properties of inequalities using real numbers before applying the properties to geometry.

5.9 Experiment with different lengths of paper, pipe cleaners, or other materials to demonstrate to students the possible lengths of a side of a triangle, given the measures of two sides.

Problem Solving Application Students learn to recognize information which can and cannot be assumed from a figure, and use figures in solving problems.

Using Models
Have students use materials such as strips of cardboard, straws, pipe cleaners, twist ties, pins, tape, or paper fasteners to demonstrate *SAS, ASA, AAS,* and *SSS.* Have them also show that two triangles with three pairs of congruent angles may or may not be congruent.

Lesson Planning Guide

Lesson	Pages	Practice with Integrated Review	Multicultural Activities	Technology	Cooperative Learning	Problem Solving	Enrichment	Manipulative Activities	Evaluation
5.1	168–170	42							
5.2	171–174	43	5						
5.3	175–179	44							
5.4	180–183	45							21 Quiz A
5.5	184–188	46					31		
5.6	189–191	47		9				73–76	
5.7	192–194	48				18			22 Quiz B
5.8	195–199	49		10	5				
5.9	199–200	50							
Problem Solving Application	201–202	51							
Review	203–205		Chapter 5 Test Form A						19
Test	206		Chapter 5 Test Form B						20

Blackline Masters Booklets (pages) — Problem Solving and Enrichment

Time Required
Integrated Mathematics Course: 17 days
Integrated Mathematics with Transformational Geometry and Trigonometry Course: 15 days

Other Chapter Resources

Student Edition
Chapter Opener, p. 167
Mixed Reviews, p. 179
Journal Entry, p. 195

Portfolio Suggestion, p. 203
Performance Assessment, p. 203
Chapter Summary, pp. 203–204
Technology Activity 1, p. A2

Teacher's Classroom Resources
Transparencies 10 and 11
New York Regents Examination Practice Masters
Tutorial Software (Apple)
Glencoe Mathematics Professional Series

CHAPTER 6: Applications of Congruent Triangles

Chapter Overview
Knowing properties of congruent triangles can be very useful in proving facts about other types of figures. In this chapter, axioms and theorems about congruent triangles are used to develop important properties of parallelograms and trapezoids. Students use a compass to construct basic geometric figures including congruent segments, congruent angles, angle bisectors, perpendicular and parallel lines, and medians and altitudes of a triangle.

Teaching Suggestions

6.1 Point out that congruent triangles are used to prove theorems about parallelograms.

6.2 This lesson presents theorems that prove a quadrilateral is a parallelogram. Where possible, encourage students to develop rationales and/or proofs of these theorems on their own.

6.3 Point out to students that a square is both a rectangle and a rhombus and therefore assumes their properties. Emphasize that the rectangle, rhombus, and square have all the characteristics of a parallelogram presented in the previous two lessons.

6.4 Stress that a trapezoid has one and only one pair of parallel sides, not two.

6.5 Emphasize the difference between a drawing and a construction. Be sure each student has a compass and straightedge and knows how to properly use them.

6.6 This lesson presents four more constructions. Encourage students to think of other constructions they might attempt and point out that some, such as trisecting an angle, are impossible.

6.7 This lesson may be considered optional. However, the theorem presented about a segment that joins the midpoint of two sides of a triangle is useful in the study of geometry.

Problem Solving Application Students learn to solve problems using indirect reasoning.

Using Cooperative Learning
Divide the class into small groups. Have each group choose five theorems from the chapter and write one exercise for each. Have groups exchange exercises and solve each other's exercises.

Lesson	Pages	Practice with Integrated Review	Multicultural Activities	Technology	Cooperative Learning	Problem Solving	Enrichment	Manipulative Activities	Evaluation
6.1	208–211	52		11					
6.2	212–217	53							25 Quiz A
6.3	218–221	54					32		
6.4	222–225	55			6				26 Quiz B
6.5	226–230	56		12				77–80	
6.6	231–234	57				19			
6.7	235–237	58	6						
Problem Solving Application	238–239	59							
Review	240–241	Chapter 6 Test Form A							23
Test	242	Chapter 6 Test Form B							24

Time Required
Integrated Mathematics Course: 16 days
Integrated Mathematics with Transformational Geometry and Trigonometry Course: 14 days

Other Chapter Resources

Student Edition
Chapter Opener, p. 207
Mixed Reviews, p. 225
Journal Entry, p. 235

Portfolio Suggestion, p. 240
Performance Assessment, p. 240
Chapter Summary, p. 240

Teacher's Classroom Resources
Transparencies 12 and 13
New York Regents Examination Practice Masters
Tutorial Software (Apple)
Glencoe Mathematics Professional Series

CHAPTER 7 | Similarity

Chapter Overview
Many geometric figures have the same shape but not the same size. Relationships such as ratios and proportions can be used to show that figures are similar. This chapter presents these relationships, along with properties of similar triangles and a proof of the Pythagorean Theorem.

Teaching Suggestions

7.1 Since proportions are essential for developing the concept of similarity, be sure students are comfortable writing them and obtaining their cross products.

7.2 The mean proportion of two positive numbers is introduced. Students should know how to write radical expressions in simplest form.

7.3 The Challenge exercises provide examples of proportions used in real-life situations.

7.4 Point out the proportional sides that are formed when three parallel lines intersect two transversals.

7.5 When an altitude is drawn to the hypotenuse of a right triangle, two similar triangles are formed that are also similar to the original triangle. Have students draw and label each of the triangles separately to help them visualize the similarity.

7.6 This lesson introduces the Pythagorean Theorem. Students must be able to multiply radical expressions.

7.7 Emphasize that the ratios found in 30°-60°-90° triangles and 45°-45°-90° triangles can be applied to other geometric figures to solve area problems. Students should be familiar with these ratios.

7.8 Construction 12 may be difficult for some students and may require extra practice.

7.9 This lesson and the next can be considered optional. The concepts they introduce unify the various branches of mathematics.

7.10 Illustrate the concepts of reflexive, symmetric, and transitive relations using real-life examples.

Problem Solving Application Students learn to use equations to find the perimeter and area of a polygon.

Using Communication
Draw two similar triangles on the chalkboard or overhead and have students set up ratios of corresponding sides. Bisect two of the corresponding angles in each triangle. Ask which triangles are now similar. Ask students if the angle bisectors are proportional and have them justify their answers.

Lesson Planning Guide

Lesson	Pages	Practice with Integrated Review	Multicultural Activities	Technology	Cooperative Learning	Problem Solving	Enrichment	Manipulative Activities	Evaluation
7.1	244–247	60		13					
7.2	248–251	61					33		
7.3	252–255	62		14				41–44	
7.4	256–260	63				20			29 Quiz A
7.5	261–264	64							
7.6	265–268	65	7						
7.7	269–271	66							30 Quiz B
7.8	272–274	67						81–84	
7.9	275–277	68							
7.10	278–280	69			7				
Problem Solving Application	281–282	70							
Review	283–285	Chapter 7 Test Form A Chapter 7 Test Form B							27 28
Test	286	Cumulative Test/Review (Chapters 4–7) Cumulative Test/Review (Chapters 1–7)							31–32 33–34

Blackline Masters Booklets (pages)

Problem Solving and Enrichment

Time Required
Integrated Mathematics Course: 18 days
Integrated Mathematics with Transformational Geometry and Trigonometry Course: 16 days

Other Chapter Resources

Student Edition
Chapter Opener, p. 243
Mixed Reviews, p. 251
Mathematical Excursions, p. 277
Journal Entry, p. 273

Portfolio Suggestion, p. 283
Performance Assessment, p. 283
Chapter Summary, pp. 283–284
Extended Project 1, pp. A10–A13

Teacher's Classroom Resources
Transparencies 14 and 15
New York Regents Examination Practice Masters
Tutorial Software (Apple)
Glencoe Mathematics Professional Series

CHAPTER 8: Geometry with Coordinates

Chapter Overview
In this chapter, students graph points and lines on the coordinate plane and solve systems of equations graphically and algebraically. Students are introduced to the distance formula, midpoint formula, the slope of a line, and the slope-intercept form of an equation.

Teaching Suggestions

8.1 Emphasize that there is a one-to-one correspondence between the set of ordered pairs of real numbers and the set of all points in the plane.

8.2 Students should know how to use a table of values for x and y to graph different kinds of equations and inequalities. Be sure they are familiar with the form of a linear equation.

8.3 Point out that the substitution and elimination methods give exact solutions to a system of equations, whereas graphing the system of equations gives only an approximate solution.

8.4 Relate the distance formula to the Pythagorean Theorem. Emphasize the importance of keeping the correct order of the coordinates when calculating the distance.

8.5 Most students will have little difficulty understanding this lesson. Students may note that finding the midpoint of a line is like finding the average of two real numbers.

8.6 Emphasize that the slope of a line is independent of the choice of points on the line used to calculate the slope. Give examples of lines that have slopes that are positive, negative, and zero, and lines that have no slope.

8.7 This lesson shows how to determine if lines are parallel, perpendicular, or neither by comparing their slopes. Be sure students understand the concept of negative reciprocal when discussing slopes of perpendicular lines.

8.8 The exercises in this lesson provide students with practice in writing equations of lines in point-slope form and slope-intercept form.

8.9 Students should be familiar with area formulas of various polygons before using coordinates to find areas.

Problem Solving Application Students solve problems by drawing graphs and estimating values.

Using Manipulatives
Have students verify the Pythagorean Theorem by doing the following exercise. Have each student draw a right triangle on grid paper. Next, have them draw a square on each side so that one side of the square is a side of the triangle. Finally, have them cut out the three squares, cut the two smaller squares into pieces, and arrange them so that they cover the larger square.

Lesson Planning Guide

Lesson	Pages	Practice with Integrated Review	Multicultural Activities	Technology	Cooperative Learning	Problem Solving	Enrichment	Manipulative Activities	Evaluation
					Blackline Masters Booklets (pages)		Problem Solving and Enrichment		
8.1	288–290	71					34		
8.2	291–293	72		15	8				
8.3	294–297	73							37 Quiz A
8.4	298–301	74							
8.5	302–305	75							
8.6	306–309	76							38 Quiz B
8.7	310–313	77		16					
8.8	314–318	78				21		65–68	
8.9	319–323	79	8						
Problem Solving Application	324–325	80							
Review	326–327		Chapter 8 Test Form A						35
Test	328		Chapter 8 Test Form B						36

Time Required

Integrated Mathematics Course: 16 days
Integrated Mathematics with Transformational Geometry and Trigonometry Course: 14 days

Other Chapter Resources

Student Edition
Chapter Opener, p. 287
Mixed Review, pp. 297, 309
Mathematical Excursions, p. 313
Journal Entry, p. 323

Portfolio Suggestion, p. 326
Performance Assessment, p. 326
Chapter Summary, p. 326
Technology Activity 2, pp. A3–A4

Teacher's Classroom Resources
Transparencies 16, 17, and 18
New York Regents Examination Practice Masters
Tutorial Software (Apple)
Glencoe Mathematics Professional Series

CHAPTER 9 | Using Coordinates

Chapter Overview

This chapter combines algebraic and geometric ideas to prove theorems about quadrilaterals and triangles. The concept of locus is introduced. Students find that a locus of points, such as a parabola, may satisfy two different equations simultaneously.

Teaching Suggestions

9.1 Emphasize that the placement of a geometric figure in a coordinate plane and the resulting vertex labels are important in proving theorems.

9.2 Students may note that it is possible to prove a theorem about triangles more than one way by positioning the given triangle differently on the coordinate plane.

9.3 Familiarize students with the concept of locus using the examples and exercises provided. Emphasize that a locus is a set of points that may be, for example, a line, a circle, or a sphere.

9.4 Discuss how the Pythagorean Theorem and locus are used to develop the equation of a circle.

9.5 Encourage students to use colors in their graphs to help them visualize the locus being sought.

9.6 Be sure students understand how to find an equation of a parabola, given the locus and directrix.

9.7 This lesson is optional and is intended for students with above average interest or ability. The concept of coordinates in three-dimensional space is introduced.

Problem Solving Application Students learn to use inverse variation to solve real-life problems.

Using Cooperative Groups

Divide the class into small groups and give each group a few problems that involve finding the locus of a figure. They may use any method to help them find each locus. When each group is finished, have students explain what the locus of each problem is and how they found it. Every student in each group should understand all of the solutions.

Lesson Planning Guide

Lesson	Pages	Practice with Integrated Review	Blackline Masters Booklets (pages)						Evaluation
			Multicultural Activities	Technology	Cooperative Learning	Problem Solving	Enrichment	Manipulative Activities	
9.1	330–334	81							
9.2	335–338	82							41 Quiz A
9.3	339–341	83							
9.4	342–346	84		17			35		
9.5	347–350	85			9				42 Quiz B
9.6	351–353	86	9	18					
9.7	354–357	87				22			
Problem Solving Application	358–359	88							
Review	360–361	Chapter 9 Test Form A							39
Test	362	Chapter 9 Test Form B							40

Time Required
Integrated Mathematics Course: 15 days
Integrated Mathematics with Transformational Geometry and Trigonometry Course: 13 days

Other Chapter Resources

Student Edition
Chapter Opener, p. 329
Mixed Reviews, p. 350
Mathematical Excursions, p. 346
Journal Entry, p. 354

Portfolio Suggestion, p. 360
Performance Assessment, p. 360
Chapter Summary, p. 360
Technology Activity 3, p. A5

Teacher's Classroom Resources
Transparencies 19 and 20
New York Regents Examination Practice Masters
Tutorial Software (Apple)
Glencoe Mathematics Professional Series

CHAPTER 10: Solving Quadratic Equations

Chapter Overview
A quadratic equation can be used to represent many kinds of real-life problems. In this chapter, students learn to solve quadratic equations by factoring, graphing, completing the square, and using the quadratic formula. Techniques for simplifying and performing operations on rational expressions are also introduced.

Teaching Suggestions

10.1 Provide students with enough practice in finding the greatest common factor for two or more monomials and finding the difference of two squares.

10.2 Emphasize writing quadratic equations in the standard form, $ax^2 + bx + c = 0$. Students should be able to factor quadratic polynomials whose coefficient is an integer other than one.

10.3 Discuss how the coefficient of x^2 in a quadratic equation determines the width and direction of a parabola and how the axis of symmetry can be used to find the placement of a parabola on the coordinate plane.

10.4 Have students note that a quadratic equation that cannot be factored may still have real roots.

10.5 Be sure students have had plenty of practice with perfect square trinomials before instructing them how to complete the square.

10.6 Emphasize that the quadratic formula can be used to solve any quadratic equation and can also determine the number of real roots without calculating the solution.

10.7 This lesson provides real-life problems for students to solve using quadratic equations. Be sure students check both solutions since one may be an unreasonable answer.

10.8 Point out that each intersection point in a quadratic-linear system is a solution. This solution can also be found algebraically.

10.9 Remind students that a rational expression is not defined when the value of the denominator is zero.

10.10 Students are to simplify the product or quotient of each rational expression using the techniques developed in the previous lesson.

10.11 It may be helpful to review the concepts of prime numbers, prime factors, and least common denominator, before assigning exercises.

Using Technology
A graphing calculator is an excellent tool for studying quadratic equations. Have students graph quadratic equations that have zero, one, and two real roots and sketch their results. Have them graph quadratic-linear systems of equations that have zero, one, and two solutions and sketch their results.

Lesson Planning Guide

Blackline Masters Booklets (pages)

Lesson	Pages	Practice with Integrated Review	Multicultural Activities	Technology	Cooperative Learning	Problem Solving	Enrichment	Manipulative Activities	Evaluation
10.1	364–368	89						57–60	
10.2	369–372	90					36		
10.3	373–376	91		10					45 Quiz A
10.4	377–380	92	10						
10.5	381–384	93							
10.6	385–388	94				23			46 Quiz B
10.7	389–391	95		19					
10.8	394–396	96		20					
10.9	398–399	97							
10.10	400–402	98							
10.11	403–406	99							
Review	407–409			Chapter 10 Test Form A					43
Test	410			Chapter 10 Test Form B					44

Time Required

Integrated Mathematics Course: 21 days
Integrated Mathematics with Transformational Geometry and Trigonometry Course: 19 days

Other Chapter Resources

Student Edition
Chapter Opener, p. 363
Mixed Review, p. 393, 406
Mathematical Excursions, p. 397
Journal Entry, p. 403

Portfolio Suggestion, p. 407
Performance Assessment, p. 407
Chapter Summary, pp. 407–408
Technology Activity 4, pp. A6–A7

Teacher's Classroom Resources
Transparencies 21 and 22
New York Regents Examination Practice Masters
Tutorial Software (Apple)
Glencoe Mathematics Professional Series

CHAPTER 11: Probability and Combinatorics

Chapter Overview
The study of probability and combinatorics has a wide range of real-life applications. In this chapter, students learn to find the probability of an event occurring, apply the counting principle and draw tree diagrams, and use permutations and combinations to solve problems. The chapter concludes with an application lesson using the principles of probability and combinatorics.

Teaching Suggestions

11.1 Discuss the circumstances in which an event has a probability of 0, or a probability of 1. Remind students that no event can ever have a probability that is negative or greater than 1.

11.2 Be sure students are able to recognize inclusive and exclusive events. Real-life examples will help visualize how the probability of two events occurring simultaneously is derived.

11.3 The Counting Principle is used to find the total number of ways two or more events will occur simultaneously. Students should use tree diagrams in solving problems to help visualize the situation.

11.4 Point out that *permutation* is another word for *arrangement*. Students should understand the meaning of factorial and how it is written.

11.5 Students should be familiar with the notation representing "the number of permutations of n things taken r at a time," and understand how to apply the general formula.

11.6 Stress that when some objects in a set are identical, the permutation formula has to be adjusted.

11.7 Emphasize that when finding combinations, the order of the objects does not count, as it does with permutations.

11.8 Discuss the difference between permutations and combinations and how the formulas differ.

11.9 This is an optional lesson which incorporates many of the skills learned throughout the chapter.

Problem Solving Application Students learn to solve problems by listing possibilities and then determining the number of items that satisfy the conditions of the problem.

Using Manipulatives

Give students a box containing 6 blue and 4 white marbles and tell them you will ask them to choose one. In reference to this event, have students write what probabilities are equal to 0, $\frac{3}{5}$, $\frac{2}{5}$, and 1. Expand the lesson to include drawing more than one marble at a time and writing the accompanying probabilities.

Lesson Planning Guide

Lesson	Pages	Practice with Integrated Review	Multicultural Activities	Technology	Problem Solving and Enrichment — Cooperative Learning	Problem Solving and Enrichment — Problem Solving	Problem Solving and Enrichment — Enrichment	Manipulative Activities	Evaluation
11.1	412–415	100		21				61–64	
11.2	416–418	101							
11.3	419–423	102							49 Quiz A
11.4	424–426	103				24			
11.5	427–430	104			11				
11.6	431–433	105	11						50 Quiz B
11.7	434–436	106							
11.8	437–440	107		22					
11.9	441–445	108					37		
Problem Solving Application	447–448	109							
Review	449–451		Chapter 11 Test Form A						47
Test	452		Chapter 11 Test Form B						48
			Cumulative Test/Review (Chapters 8–11)						51–52

Time Required

Integrated Mathematics Course: 15 days
Integrated Mathematics with Transformational Geometry and Trigonometry Course: 14 days

Other Chapter Resources

Student Edition
Chapter Opener, p. 411
Mixed Reviews, pp. 423, 440
Mathematical Excursions, pp. 433, 446
Journal Entry, p. 431

Portfolio Suggestion, p. 449
Performance Assessment, p. 449
Chapter Summary, pp. 449–450
Technology Activity 5, p. A8

Teacher's Classroom Resources
Transparencies 23 and 24
New York Regents Examination Practice Masters
Tutorial Software (Apple)
Glencoe Mathematics Professional Series

CHAPTER 12 | Transformation Geometry

Chapter Overview

Transformations are often seen in nature and architecture. In this chapter, students are introduced to special one-to-one mappings called *transformations.* These transformations include reflections, translations, rotations, half-turns, and dilations.

Teaching Suggestions

12.1 Stress the difference between a one-to-one mapping and a mapping that is not one-to-one. Be sure students understand the symbolism.

12.2 Encourage students to become familiar with the line reflection transformation, as it will help them better understand future lessons.

12.3 Have students experiment with paper models of different figures to help determine lines of symmetry.

12.4 Use transparencies to demonstrate line reflections and other transformations introduced in later lessons.

12.5 Encourage students to learn each transformation's symbolism to avoid confusion in later lessons.

12.6 Point out that a positive angle of rotation indicates a counterclockwise move, while a negative angle indicates a clockwise move.

12.7 Students should know the difference between line symmetry and point symmetry.

12.8 Point out that line reflections, translations, and rotations are examples of isometrics.

12.9 Encourage students to use examples to discover the patterns that occur in a transformation system.

12.10 Stress that the image of a dilation is not congruent to the preimage, but is similar.

Problem Solving Application Students use line reflections to solve problems.

Using Models

Have each student draw and cut out a two-dimensional figure. Have them use that figure to draw examples of the transformations discussed in this chapter: a line reflection, a translation, a rotation, a point reflection, and a dilation.

Lesson	Pages	Practice with Integrated Review	Multicultural Activities	Technology	Cooperative Learning	Problem Solving	Enrichment	Manipulative Activities	Evaluation
12.1	454–457	110		23					
12.2	458–461	111					38		
12.3	462–464	112		24					
12.4	465–468	113				25			55 Quiz A
12.5	469–472	114							
12.6	473–475	115						69–72	
12.7	476–479	116							56 Quiz B
12.8	480–483	117	12						
12.9	484–487	118							
12.10	488–490	119				12			
Problem Solving Application	491–492	120							
Review	493–495		Chapter 12 Test Form A						53
Test	496		Chapter 12 Test Form B						54

Time Required

Integrated Mathematics Course: 0 days
Integrated Mathematics with Transformational Geometry and Trigonometry Course: 12 days

Other Chapter Resources

Student Edition
Chapter Opener, p. 453
Mixed Review, pp. 464, 479
Journal Entry, p. 460

Portfolio Suggestion, p. 493
Performance Assessment, p. 493
Chapter Summary, pp. 493–494
Technology Activity 6, p. A9
Extended Project 2, pp. A14–A16

Teacher's Classroom Resources
Transparencies 25 and 26
New York Regents Examination Practice Masters
Tutorial Software (Apple)
Glencoe Mathematics Professional Series

CHAPTER 13 | Introduction to Trigonometry

Chapter Overview
Knowledge of trigonometry is very useful in the construction of buildings, bridges, canals, and highways. This chapter demonstrates how properties of right triangles and ratios are used to solve problems. The tangent, sine, and cosine ratios are used to find missing measurements of sides and angles of right triangles. The final lesson uses a circle to find values.

Teaching Suggestions

13.1 Students should be familiar with right triangles, the Pythagorean Theorem, similarity, and computation with irrational numbers.

13.2 Point out that the ratio for a given angle measurement remains constant despite the size of the right triangle.

13.3 Be sure students understand that the tangent of either acute angle may be used to find the measure of a leg of a right triangle.

13.4 This lesson uses a right triangle to define the sine and cosine ratios. Students should know the three trigonometric ratios, sine, cosine, and tangent.

13.5 Point out that angle references in this lesson are similar to those of rotations in Chapter 12.

Problem Solving Application Students learn to use trigonometric ratios and the Pythagorean Theorem to solve problems.

Using Communication
Have students draw a right triangle and label the sides and vertices. Have them record in a chart the correct ratios, sine, cosine, and tangent, for both acute angles.

Lesson	Pages	Practice with Integrated Review	Multicultural Activities	Technology	Cooperative Learning	Problem Solving	Enrichment	Manipulative Activities	Evaluation
13.1	498–500	121							
13.2	501–504	122		25					59 Quiz A
13.3	505–510	123			13			89–92	
13.4	511–515	124	13	26		26			60 Quiz B
13.5	517–520	125					39		
Problem Solving Application	521–522	126							
Review	523–525		Chapter 13 Test Form A						57
			Chapter 13 Test Form B						58
Test	526		Cumulative Test/Review (Chapters 1–13)						61–62
			Final Test						63–67

Time Required
Integrated Mathematics Course: 0 days
Integrated Mathematics with Transformational Geometry and Trigonometry Course: 8 days

Other Chapter Resources

Student Edition
Chapter Opener, p. 497
Mixed Review, p. 510
Mathematical Excursions, p. 516
Journal Entry, p. 501

Portfolio Suggestion, p. 523
Performance Assessment, p. 523
Chapter Summary, pp. 523–524

Teacher's Classroom Resources
Transparencies 27 and 28
New York Regents Examination Practice Masters
Tutorial Software (Apple)
Glencoe Mathematics Professional Series

Technology Activities

Overview
Some students can benefit by using technology to perform many of the tedious tasks requiring pencil-and-paper manipulations. The six graphing calculator activities on pages A2–A9 show students how to use a TI–82 graphing calculator to perform many of the graphing and computational tasks presented in this text.

Activity 1 In *Graphing Congruent Triangles*, students use a graphing calculator to graph two triangles on a coordinate grid. They then use their knowledge of congruent triangles to determine whether the triangles they have graphed are congruent.

Activity 2 *Graphing Linear Equations* teaches students how to adjust the viewing window on the graphing calculator so that they see a complete graph. They then learn how to enter an equation that is to be graphed.

Activity 3 The TI–82 graphing calculator has a special feature that allows students to graph a circle on a coordinate plane by listing the center and the radius of the circle to be drawn. In *Graphing Circles*, students use what they have learned about the equations of circles to determine the center and radius and then graph the circle.

Activity 4 The *Quadratic Equations* activity permits students to quickly graph quadratic equations and determine how many and what type of roots they may have. Students use the calculator's special functions to determine the decimal approximations of the real roots.

Activity 5 In *Probability*, students use a graphing calculator program that generates random numbers to simulate throwing a die.

Activity 6 Students use the skills they learned in Activity 1 to extend their graphing capabilities to in *Translations*. They graph a geometric figure using the coordinates of its vertices. Then they repeat the process replacing each vertex coordinate with a mathematical expression to perform the translation.

More Technology
More calculator and computer technology is available in the *Integrated Mathematics Course 2* Technology Masters.

Extended Projects

Overview
There are two extended projects located on pages A10–A16. Each of these projects is a long-term activity that works well in a cooperative group situation. Each of the activities requires students to plan, execute, extend, and draw conclusions about the topics they have been investigating. These activities are excellent opportunities to have students experience the relevance of mathematics to real-world situations and express that relevance using communications skills.

Project 1 *The Geometry of Miniature Golf*

Students discover how geometry is related to real-life as they study the effects of a ball bouncing off of a surface at an angle. This phenomenon is relative to basketball, billiards, racquetball, and table tennis, as well as miniature golf. Students can experiment with the concepts in this project by applying what they learn in an actual game of miniature golf or billiards.

Project 2 *Creating Order from Chaos*

Mathematics is based upon many ordered systems. In Course 2 students learn about types of mathematical systems and how each is organized. In this project, students decide upon a topic to collect data and devise a system in which to organize their data. They apply their organizational skills to other areas of science.

Assessment of Projects
Since these projects have no specific "right answer," you may wonder how they are assessed. Projects are the culminating work of a team of students. As a result, the team should share in the outcome of their project results. Often the results are not given a letter or point grade, but rather a holistic evaluation. The "scores" may be *well-done, acceptable, needs work,* and *unacceptable.*

Every group should present their results to the class. Students should be encouraged to use multiple representations of their findings, including graphs, diagrams, collages, and other pictorial representations, in addition to a written report. They should also be prepared to answer any questions about points that are unclear or questions asked by other students in class. You may ask the group what they might have done differently if they had the opportunity. Ask other groups to evaluate the presentation. Conclude the presentation with a class discussion of the information presented to them.

Teacher's Answer Key

CHAPTER 1 WORKING WITH LOGIC

Page 5 Written **1.** 0.5 is rational and −8 is an integer. true **2.** 0.5 is rational and 15 is not a multiple of 3. false **3.** 0.5 is rational or 15 is not a multiple of 3. true **4.** −8 is an integer or 15 is not a multiple of 3. true **5.** 0.5 is not rational and 15 is not a multiple of 3. false **6.** −8 is an integer and 15 is a multiple of 3. true **7.** 0.5 is not rational or −8 is an integer. true **8.** 15 is a multiple of 3 or −8 is an integer. true

17.

p	q	$\sim q$	$p \wedge \sim q$
T	T	F	F
T	F	T	T
F	T	F	F
F	F	T	F

18.

p	q	$\sim p$	$\sim p \vee q$
T	T	F	T
T	F	F	F
F	T	T	T
F	F	T	T

19.

p	$\sim p$	$p \vee \sim p$
T	F	T
F	T	T

20.

p	$\sim p$	$p \wedge \sim p$
T	F	F
F	T	F

21.

p	q	$\sim p$	$\sim q$	$\sim p \wedge \sim q$
T	T	F	F	F
T	F	F	T	F
F	T	T	F	F
F	F	T	T	T

22.

q	r	$\sim q$	$q \wedge r$	$\sim q \vee (q \wedge r)$
T	T	F	T	T
T	F	F	F	F
F	T	T	F	T
F	F	T	F	T

23.

p	q	$(q \wedge p)$	$p \vee (q \wedge p)$
T	T	T	T
T	F	F	T
F	T	F	F
F	F	F	F

24.

p	q	$\sim p$	$\sim q$	$(\sim q \vee p)$	$\sim p \wedge (\sim q \vee p)$
T	T	F	F	T	F
T	F	F	T	T	F
F	T	T	F	F	F
F	F	T	T	T	T

Page 9 Written **1.** If Labor Day is in November, then 2 is a prime number. true **2.** If 2 is a prime number, then Labor Day is in November. false **3.** If Labor Day is not in November, then $3y + 1$ is a binomial. true **4.** If Labor Day is in November, then 2 is a prime number or $3y + 1$ is a binomial. true **5.** Labor Day is in November if and only if 2 is a prime number. false **6.** 2 is a prime number if and only if $3y + 1$ is not a binomial. false **7.** 2 is not a prime number if and only if Labor Day is in November. true **8.** If Labor Day is in November or 2 is not a prime number, then $3y + 1$ is not a binomial. true **20.** If Brullman is elected senator and Crokus Dam is built. **21.** If there is a heavy turnout on Election Day and Brullman does not win

22.

p	q	$p \rightarrow q$	$q \rightarrow p$
T	T	T	T
T	F	F	T
F	T	T	F
F	F	T	T

23.

p	q	$\sim q$	$p \rightarrow \sim q$	$\sim(p \wedge q)$
T	T	F	F	F
T	F	T	T	T
F	T	F	T	T
F	F	T	T	T

24.

p	q	$\sim p$	$\sim q$	$\sim(p \vee q)$	$\sim p \vee \sim q$
T	T	F	F	F	F
T	F	F	T	F	T
F	T	T	F	F	T
F	F	T	T	T	T

25.

p	q	$\sim p$	$\sim q$	$\sim(p \wedge q)$	$\sim p \wedge \sim q$
T	T	F	F	F	F
T	F	F	T	T	F
F	T	T	F	T	F
F	F	T	T	T	T

Page 12 Written

3.

STATEMENTS	REASONS
1. $\sim S \lor \sim T$	1. Given
2. S	2. Given
3. $\sim T$	3. Law of disjunctive inference

4.

STATEMENTS	REASONS
1. $R \land M$	1. Given
2. R	2. Definition of conjunction
3. $\sim R \lor T$	3. Given
4. T	4. Law of disjunctive inference

5. No conclusion. A disjunction may have two true disjuncts. If one is true, the other may be true or false. Therefore, although $\sim B$ is true, C may be true or false.

6.

STATEMENTS	REASONS
1. $T \lor F$	1. Given
2. $\sim F$	2. Given
3. T	3. Law of disjunctive inference

7. No conclusion. A disjunction may have two true disjuncts. If one is true, the other may be true or false. Therefore, even though it is true that Rose R is red, Violet V may or may not be blue.

8.

STATEMENTS	REASONS
1. $\sim R \lor \sim V$	1. Given
2. R	2. Given
3. $\sim V$	3. Law of disjunctive inference

9.

STATEMENTS	REASONS
1. $P \land A$	1. Given
2. P	2. Definition of conjunction
3. $V \lor \sim P$	3. Given
4. V	4. Law of disjunctive inference

10.

STATEMENTS	REASONS
1. $G \land B$	1. Given
2. G	2. Definition of conjunction
3. $\sim G \lor \sim F$	3. Given
4. $\sim F$	4. Law of disjunctive inference

Page 16 Written

5.

STATEMENTS	REASONS
1. $A \to D$	1. Given
2. A	2. Given
3. D	3. Law of detachment

6. No conclusion follows. If the consequent of a true conditional is true, the antecedent may or may not be true.

7.

STATEMENTS	REASONS
1. $r \to (s \land \sim t)$	1. Given
2. r	2. Given
3. $(s \land \sim t)$	3. Law of detachment

8.

STATEMENTS	REASONS
1. $B \land G$	1. Given
2. $\sim B$	2. Given
3. G	3. Law of disjunctive inference
4. $G \to H$	4. Given
5. H	5. Law of detachment

9.

STATEMENTS	REASONS
1. $Q \lor R$	1. Given
2. $\sim R$	2. Given
3. Q	3. Law of disjunctive inference
4. $Q \to T$	4. Given
5. T	5. Law of detachment

10. No conclusion. B may or may not be true. To use the Law of detachment, B must be true.

11.

STATEMENTS	REASONS
1. $L \to \sim D$	1. Given
2. L	2. Given
3. $\sim D$	3. Law of detachment

12.

STATEMENTS	REASONS
1. $C \to P$	1. Given
2. C	2. Given
3. P	3. Law of detachment

13.

STATEMENTS	REASONS
1. $L \lor F$	1. Given
2. $\sim F$	2. Given
3. L	3. Law of disjunctive inference
4. $L \to P$	4. Given
5. P	5. Law of detachment

14.

STATEMENTS	REASONS
1. $E \to \sim D$	1. Given
2. E	2. Given
3. $\sim D$	3. Law of detachment
4. $D \land F$	4. Given
5. F	5. Law of disjunctive inference

15. No conclusion. Although Crunchy-Wunchies are made from oats, they may or may not be healthful.

16.

STATEMENTS	REASONS
1. H	1. Given
2. $H \to (Y \lor P)$	2. Given
3. $Y \lor P$	3. Law of detachment
4. $\sim Y$	4. Given
5. P	5. Law of disjunctive inference

17.

STATEMENTS	REASONS
1. T	1. Given
2. $T \lor J$	2. Definition of disjunction
3. $(T \lor J) \to P$	3. Given
4. P	4. Law of detachment

18. No conclusion. The fact that Ted is not taking Sally to the dance does not imply that Jack will take her.

Page 16 Challenge 1. No, even though a conditional and its consequent are true, the antecedent of the conditional may be false. **2.** If a conditional is true and its consequent is false, then the antecedent is false.

3.

p	q	$p \to q$	$\sim p$	$\sim q$	$(p \to q) \land \sim q$	$[(p \to q) \land \sim q] \to \sim p$
T	T	T	F	F	F	T
T	F	F	F	T	F	T
F	T	T	T	F	F	T
F	F	T	T	T	T	T

4. The compound statement is true.

Page 18 Exploratory 1. $w \to r$, $\sim r \to \sim w$, $\sim w \to \sim r$ **2.** $\sim w \to t$, $\sim t \to w$, $w \to \sim t$ **3.** $\sim t \to \sim s$, $s \to t$, $t \to s$ **4.** $\sim n \to \sim m$, $m \to n$, $n \to m$ **5.** If $\angle A \cong \angle B$, then $\angle A$ and $\angle B$ are both right angles. If $\angle A$ and $\angle B$ are not right angles, then $\angle A \not\cong \angle B$. If $\angle A \cong \angle B$, then $\angle A$ and $\angle B$ are not right angles. **6.** If John is disappointed, Mei did not show up at the picnic. If Mei shows up at the picnic, John will not be disappointed. If John is not disappointed, Mei showed up at the picnic.

Page 19 Written

1.

p	q	$\sim p$	$\sim q$	$p \to q$	$\sim q \to \sim p$
T	T	F	F	T	T
T	F	F	T	F	F
F	T	T	F	T	T
F	F	T	T	T	T

4.

STATEMENTS	REASONS
1. $A \lor D$	1. Given
2. $\sim A$	2. Given
3. D	3. Law of disjunctive inference
4. $T \to \sim D$	4. Given
5. $D \to \sim T$	5. Law of the contrapositive
6. $\sim T$	6. Law of detachment

5.

STATEMENTS	REASONS
1. $\sim C \vee \sim F$	1. Given
2. C	2. Given
3. $\sim F$	3. Law of disjunctive inference
4. $\sim B \to F$	4. Given
5. $\sim F \to B$	5. Law of the contrapositive
6. B	6. Law of detachment

6.

STATEMENTS	REASONS
1. $\sim C \vee \sim B$	1. Given
2. C	2. Given
3. $\sim B$	3. Law of disjunctive inference
4. $T \to B$	4. Given
5. $\sim B \to \sim T$	5. Law of the contrapositive
6. $\sim T$	6. Law of detachment

7.

STATEMENTS	REASONS
1. $L \wedge M$	1. Given
2. M	2. Definition of conjunction
3. $M \to N$	3. Given
4. N	4. Law of detachment
5. $\sim N \vee \sim D$	5. Given
6. $\sim D$	6. Law of disjunctive inference
7. $R \to D$	7. Given
8. $\sim D \to \sim R$	8. Law of the contrapositive
9. $\sim R$	9. Law of detachment

8.

STATEMENTS	REASONS
1. $\sim B$	1. Given
2. $S \to B$	2. Given
3. $\sim B \to \sim S$	3. Law of the contrapositve
4. $\sim S$	4. Law of detachment
5. $L \to S$	5. Given
6. $\sim S \to \sim L$	6. Law of the contrapositive
7. $\sim L$	7. Law of detachment

9.

STATEMENTS	REASONS
1. R	1. Given
2. $Q \to \sim R$	2. Given
3. $R \to \sim Q$	3. Law of the contrapositve
4. $\sim Q$	4. Law of detachment
5. $\sim Q \to T$	5. Given
6. Suggested conclusion does not follow.	6. Information leads to opposite conclusion.

10. Given: $G \vee T$ Prove: $\sim P$
$P \to \sim T$
$\sim G$

STATEMENTS	REASONS
1. $\sim G$	1. Given
2. $G \vee T$	2. Given
3. T	3. Law of disjunctive inference
4. $P \to \sim T$	4. Given
5. $T \to \sim P$	5. Law of the contrapositive
6. $\sim P$	6. Law of detachment

11. Given: $L \to C$ Prove: $\sim L$
$\sim H \lor \sim C$
H

STATEMENTS	REASONS
1. H	1. Given
2. $\sim H \lor \sim C$	2. Given
3. $\sim C$	3. Law of disjunctive inference
4. $L \to C$	4. Given
5. $\sim C \to \sim L$	5. Law of the contrapositive
6. $\sim L$	6. Law of detachment

12. Given: $W \to I$ Prove: $\sim B$
$B \to W$
$\sim I$

STATEMENTS	REASONS
1. $\sim I$	1. Given
2. $W \to I$	2. Given
3. $\sim I \to \sim W$	3. Law of the contrapositive
4. $\sim W$	4. Law of detachment
5. $B \to W$	5. Given
6. $\sim W \to \sim B$	6. Law of the contrapositive
7. $\sim B$	7. Law of detachment

Pages 21–22 Exploratory 6. no conclusion **7.** His feet are tired. Law of detachment **8.** Alvin is enjoying himself. Law of syllogism **9.** It is not 3 o'clock on Sunday. Law of the contrapositive **10.** no conclusion

Pages 22–23 Written

1.

STATEMENTS	REASONS
1. $A \to B$	1. Given
2. $B \to T$	2. Given
3. $A \to T$	3. Law of syllogism

2.

STATEMENTS	REASONS
1. $A \to T$	1. Given
2. $T \to W$	2. Given
3. $A \to W$	3. Law of syllogism
4. $W \to B$	4. Given
5. $A \to B$	5. Law of syllogism
6. Suggested conclusion does not follow.	6. Insufficient information

3.

STATEMENTS	REASONS
1. $A \to T$	1. Given
2. $T \to Q$	2. Given
3. $A \to Q$	3. Law of syllogism
4. A	4. Given
5. Q	5. Law of detachment

4.

STATEMENTS	REASONS
1. $\sim B$	1. Given
2. $W \to B$	2. Given
3. $\sim B \to \sim W$	3. Law of the contrapositive
4. $\sim W$	4. Law of detachment
5. $A \to W$	5. Given
6. $\sim W \to \sim A$	6. Law of the contrapositive
7. $\sim A$	7. Law of detachment

5.

STATEMENTS	REASONS
1. $A \lor T$	1. Given
2. $\sim A$	2. Given
3. T	3. Law of detachment
4. $L \to \sim T$	4. Given
5. $T \to \sim L$	5. Law of the contrapositive
6. $\sim L$	6. Law of detachment

6.

STATEMENTS	REASONS
1. $B \land T$	1. Given
2. B	2. Definition of conjunction
3. $B \to Q$	3. Given
4. Q	4. Law of detachment

7.

STATEMENTS	REASONS
1. A	1. Given
2. $A \to S$	2. Given
3. $S \to N$	3. Given
4. $A \to N$	4. Law of syllogism
5. N	5. Law of detachment
6. $\sim W \vee \sim N$	6. Given
7. $\sim W$	7. Law of disjunctive inference

8.

STATEMENTS	REASONS
1. $B \to C$	1. Given
2. $\sim C \to \sim B$	2. Law of the contrapositive
3. $\sim C$	3. Given
4. $\sim B$	4. Law of detachment

9.

STATEMENTS	REASONS
1. $\sim T \to \sim S$	1. Given
2. $S \to T$	2. Law of the contrapositive
3. S	3. Given
4. T	4. Law of detachment
5. $T \to A$	5. Given
6. A	6. Law of detachment
7. $A \vee B$	7. Definition of disjunction
8. $(A \vee B) \to Q$	8. Given
9. Q	9. Law of detachment

10.

STATEMENTS	REASONS
1. p	1. Given
2. $n \to \sim p$	2. Given
3. $p \to \sim n$	3. Law of the contrapositive
4. $\sim n$	4. Law of detachment
5. $m \vee n$	5. Given
6. m	6. Law of disjunctive inference
7. $m \to q$	7. Given
8. q	8. Law of detachment

11.

STATEMENTS	REASONS
1. $r \to \sim s$	1. Given
2. $s \to \sim r$	2. Law of the contrapositive
3. s	3. Given
4. $\sim r$	4. Law of detachment
5. $r \vee \sim t$	5. Given
6. $\sim t$	6. Law of disjunctive inference
7. $m \to t$	7. Given
8. $\sim t \to \sim m$	8. Law of the contrapositive
9. $\sim m$	9. Law of detachment

12.

STATEMENTS	REASONS
1. $\sim r \wedge s$	1. Given
2. s	2. Definition of conjunction
3. $\sim s \vee t$	3. Given
4. t	4. Law of disjunctive inference
5. $p \to \sim t$	5. Given
6. $t \to \sim p$	6. Law of the contrapositive
7. $\sim p \to q$	7. Given
8. $t \to q$	8. Law of syllogism
9. q	9. Law of detachment

13. Given: $G \to H$ Prove: $G \to J$
$H \to J$

STATEMENTS	REASONS
1. $G \to H$	1. Given
2. $H \to J$	2. Given
3. $G \to J$	3. Law of syllogism

14. Given: $G \to L$ Prove: C
$L \to C$
G

STATEMENTS	REASONS
1. $G \to L$	1. Given
2. $L \to C$	2. Given
3. $G \to C$	3. Law of syllogism
4. G	4. Given
5. C	5. Law of detachment

15. Given: $P \rightarrow Q$ Prove: $\sim P$
$Q \rightarrow R$
$R \rightarrow S$
$\sim R$

STATEMENTS	REASONS
1. $P \rightarrow Q$	1. Given
2. $Q \rightarrow R$	2. Given
3. $\sim R$	3. Given
4. $P \rightarrow R$	4. Law of syllogism
5. $\sim R \rightarrow \sim P$	5. Law of the contrapositive
6. $\sim P$	6. Law of detachment

16. Given: $L \rightarrow B$ Prove: J
$B \rightarrow C$
$C \vee J$
L

STATEMENTS	REASONS
1. $L \rightarrow B$	1. Given
2. $B \rightarrow C$	2. Given
3. $L \rightarrow C$	3. Law of syllogism
4. L	4. Given
5. C	5. Law of detachment
6. Conclusion does not follow	6. Insufficient information

17. Given: $C \rightarrow L$ Prove: $\sim F$
$L \rightarrow S$
$\sim F \vee \sim S$
C

STATEMENTS	REASONS
1. C	1. Given
2. $C \rightarrow L$	2. Given
3. $L \rightarrow S$	3. Given
4. $C \rightarrow S$	4. Law of syllogism
5. S	5. Law of detachment
6. $\sim F \vee \sim S$	6. Given
7. $\sim F$	7. Law of disjunctive inference

18. Given: $(S \vee O) \rightarrow A$ Prove: A
$G \rightarrow S$
$\sim G \rightarrow \sim L$
L

STATEMENTS	REASONS
1. L	1. Given
2. $\sim G \rightarrow \sim L$	2. Given
3. $L \rightarrow G$	3. Law of the contrapositive
4. $G \rightarrow S$	4. Given
5. $L \rightarrow S$	5. Law of syllogism
6. S	6. Law of detachment
7. $S \vee O$	7. Definition of disjunction
8. $(S \vee O) \rightarrow A$	8. Given
9. A	9. Law of detachment

Page 23 Challenge

2.

p	q	r	$(p \wedge q)$	$(p \wedge q) \vee r$
T	T	T	T	T
T	T	F	T	T
T	F	T	F	T
T	F	F	F	F
F	T	T	F	T
F	T	F	F	F
F	F	T	F	T
F	F	F	F	F

3.

p	q	r	$\sim p$	$(\sim p \vee q)$	$(q \vee r)$	$(\sim p \vee q) \rightarrow (q \vee r)$
T	T	T	F	T	T	T
T	T	F	F	T	T	T
T	F	T	F	F	T	T
T	F	F	F	F	F	T
F	T	T	T	T	T	T
F	T	F	T	T	T	T
F	F	T	T	T	T	T
F	F	F	T	T	F	F

4.

p	q	~q	r	(p ∨ q)	(~q ∨ r)	(p ∨ q) → (~q ∨ r)	(~q ∨ r) → (p ∨ q)	(p ∨ q) ↔ (~q ∨ r)
T	T	F	T	T	T	T	T	T
T	T	F	F	T	F	F	T	F
T	F	T	T	T	T	T	T	T
T	F	T	F	T	T	T	T	T
F	T	F	T	T	T	T	T	T
F	T	F	F	T	F	F	T	F
F	F	T	T	F	T	T	F	F
F	F	T	F	F	T	T	F	F

5.

p	q	r	(p ∧ q)	(p ∧ q) ∧ r	(q ∧ r)	p ∧ (q ∧ r)
T	T	T	T	T	T	T
T	T	F	T	F	F	F
T	F	T	F	F	F	F
T	F	F	F	F	F	F
F	T	T	F	F	T	F
F	T	F	F	F	F	F
F	F	T	F	F	F	F
F	F	F	F	F	F	F

"∧" is associative

6.

p	q	r	(p ∨ q)	(p ∨ q) ∨ r	(q ∨ r)	p ∨ (q ∨ r)
T	T	T	T	T	T	T
T	T	F	T	T	T	T
T	F	T	T	T	T	T
T	F	F	T	T	F	T
F	T	T	T	T	T	T
F	T	F	T	T	T	T
F	F	T	F	T	T	T
F	F	F	F	F	F	F

"∨" is associative

7.

p	q	r	q ∨ r	p ∧ (q ∨ r)	p ∧ q	p ∧ r	(p ∧ q) ∨ (p ∧ r)
T	T	T	T	T	T	T	T
T	T	F	T	T	T	F	T
T	F	T	T	T	F	T	T
T	F	F	F	F	F	F	F
F	T	T	T	F	F	F	F
F	T	F	T	F	F	F	F
F	F	T	T	F	F	F	F
F	F	F	F	F	F	F	F

"∧" distributes over "∨"

8.

p	q	r	q ∨ r	p → (q ∨ r)	p → q	p → r	(p → q) ∨ (p → r)
T	T	T	T	T	T	T	T
T	T	F	T	T	T	F	T
T	F	T	T	T	F	T	T
T	F	F	F	F	F	F	F
F	T	T	T	T	T	T	T
F	T	F	T	T	T	T	T
F	F	T	T	T	T	T	T
F	F	F	F	T	T	T	T

"→" distributes over "∨"

Page 26 Written.

10.
p	q	$(p \wedge q)$	$\sim(p \wedge q)$	$\sim p \wedge \sim q$
T	T	T	F	F
T	F	F	T	F
F	T	F	T	F
F	F	F	T	T

11.
p	q	$(p \vee q)$	$\sim(p \vee q)$	$\sim p \vee \sim q$
T	T	T	F	F
T	F	T	F	T
F	T	T	F	T
F	F	F	T	T

12.
p	q	$p \wedge q$	$\sim(p \wedge q)$	$\sim p \vee \sim q$
T	T	T	F	F
T	F	F	T	T
F	T	F	T	T
F	F	F	T	T

13.
p	q	$p \vee q$	$\sim(p \vee q)$	$\sim p \wedge \sim q$
T	T	T	F	F
T	F	T	F	F
F	T	T	F	F
F	F	F	T	T

Page 27 Written

16.

STATEMENTS	REASONS
1. $\sim(p \vee q)$	1. Given
2. $\sim p \wedge \sim q$	2. DeMorgan Laws
3. $\sim q$	3. Definition of conjunction
4. $\sim q \rightarrow t$	4. Given
5. t	5. Law of detachment

17.

STATEMENTS	REASONS
1. $\sim(\alpha \wedge \beta)$	1. Given
2. $\sim\alpha \vee \sim\beta$	2. DeMorgan Laws
3. α	3. Given
4. β	4. Law of disjunctive inference

18.

STATEMENTS	REASONS
1. $r \rightarrow \sim v$	1. Given
2. $v \rightarrow \sim r$	2. Law of the contrapositive
3. $s \rightarrow v$	3. Given
4. $s \rightarrow \sim r$	4. Law of syllogism
5. s	5. Given
6. $\sim r$	6. Law of detachment
7. $\sim(q \wedge \sim r)$	7. Given
8. $\sim q \vee r$	8. DeMorgan Laws
9. $\sim q$	9. Law of disjunctive inference

19. Given: $\sim(W \vee D)$ Prove: R
$\sim W \rightarrow R$

STATEMENTS	REASONS
1. $\sim(W \vee D)$	1. Given
2. $\sim W \wedge \sim D$	2. DeMorgan Laws
3. $\sim W$	3. Definition of conjuction
4. $\sim W \rightarrow R$	4. Given
5. R	5. Law of detachment

20. Given: $\sim(W \vee D)$ Prove: $\sim A$
$\sim A \rightarrow W$

STATEMENTS	REASONS
1. $\sim(W \vee D)$	1. Given
2. $\sim W \wedge \sim D$	2. DeMorgan Laws
3. $\sim W$	3. Definition of conjunction
4. $\sim A \rightarrow W$	4. Given
5. $\sim W \rightarrow A$	5. Law of the contrapositive
6. A	6. Law of detachment
7. Conclusion does not follow.	7. Information leads to opposite conclusion.

21. Given: $\sim(G \wedge J)$ Prove: D
$\sim G \rightarrow P$
$P \rightarrow D$
J

STATEMENTS	REASONS
1. $\sim(G \wedge J)$	1. Given
2. $\sim G \vee \sim J$	2. DeMorgan Laws
3. J	3. Given
4. $\sim G$	4. Law of disjunctive inference
5. $\sim G \rightarrow P$	5. Given
6. $P \rightarrow D$	6. Given
7. $\sim G \rightarrow D$	7. Law of syllogism
8. D	8. Law of detachment

22. If Tom does not eat salad or does not eat spaghetti, then he does not jog.   **23.** If $r \neq 0$ and $r \leq 4$, then $r \geq n$.   **24.** If $\angle A \not\equiv \angle C$, then $\angle A$ is not supplementary to $\angle B$ or $\angle C$ is not supplementary to $\angle B$.   **25.** If Ted is not wearing his tweeds, then Marsha is not going to the Big Dance or Sally is.   **26.** $(\sim p \vee \sim q) \rightarrow (\sim r \wedge t)$ **27.** $(c \vee \sim d) \rightarrow (\sim t \wedge w)$ **28.** If $m\angle C \neq 90$ or $m\angle D \neq 90$, then $\angle C \not\equiv \angle D$.

Page 27 Challenge   3. $x = 3$ and $x^2 \neq 9$ **4.** $|x| = 5$ and $x \neq 5$ and $x \neq -5$ **5.** It is raining today and Linda does not have her umbrella.   **6.** Mr. Oglethorpe rides the bus if and only if there is not a taxi strike. Mr. Oglethorpe does not ride the bus if and only if there is a taxi strike.   **7.** Miss Weinstein has French toast for breakfast if and only if she is in a hurry or she is having Crunchy-Wunchies. Miss Weinstein does not have French toast if and only if she is not in a hurry and she is not having Crunchy-Wunchies.

Page 29 Mathematical Excursions   1.   1. Assumption   2. Law of double negation   3. Given   4. $W \rightarrow \sim A$ 5. Law of detachment   6. Given   7. N 8. Given   9. Definition of conjunction   10. $\sim(\sim W)$ 11. Law of double negation   **2.**   1. Assumption   2. Given   3. Law of the contrapositive   4. Given   5. Law of syllogism   6. Law of detachment   7. Given   8. Definition of conjunction   9. Given   10. Law of disjunctive inference   11. Definition of conjunction   12. $\sim q$ must be false since it led to a contradiction, $t \wedge \sim t$.   13. Law of double negation

3. Given: $A \rightarrow B$ Prove: A
$\sim B$

STATEMENTS	REASONS
1. $\sim(\sim A)$	1. Assumption
2. A	2. Law of double negation
3. $A \rightarrow B$	3. Given
4. B	4. Law of detachment
5. $\sim B$	5. Given
6. $B \wedge \sim B$	6. Definition of conjunction
7. $\sim[\sim(\sim A)]$	7. $\sim(\sim A)$ must be false since it led to the contradiction $B \wedge \sim B$.
8. $\sim A$	8. Law of double negation

4. Given: $A \rightarrow \sim C$ Prove: D
$C \vee D$
A

STATEMENTS	REASONS
1. $\sim D$	1. Assumption
2. $C \vee D$	2. Given
3. C	3. Law of disjunctive inference
4. $A \rightarrow \sim C$	4. Given
5. $C \rightarrow \sim A$	5. Law of the contrapositive
6. $\sim A$	6. Law of detachment
7. A	7. Given
8. $A \wedge \sim A$	8. Definition of conjunction
9. $\sim(\sim D)$	9. $\sim D$ must be false since it led to the contradiction $A \wedge \sim A$.
10. D	10. Law of double negation

Page 32 Exploratory   1. not quantified   **2.** existentially quantified, some   **3.** universally quantified, all   **4.** existentially quantified, some   **5.** not quantified   **6.** universally quantified, every   **7.** existentially quantified, at least one   **8.** universally quantified, none

Page 32 Written   1. Some whole numbers are less than zero. false   **2.** All integers are rational. true   **3.** All rational numbers are integers. false   **4.** Some integers are prime. true   **5.** Some integers are composite. true   **6.** For every t in the set of rational numbers, $t^2 > t$. false   **11.** All multiples of 3 are rational. true   **12.** Some integers are even. true   **13.** All multiples of 6 are multiples of 3. true   **14.** All multiples of 6 are multiples of 4. false   **15.** Some multiples of 6 are multiples of 3 and multiples of 4. true   **16.** There exists a rational number x such that $x^2 = 2$. false

Page 32 Challenge 1. If a number is a whole number, then the number is an integer. 2. Every rose is a flower. 3. If $x = a$, then $x = b$. 4. If a person is a suspect here, he is not guilty. 5. If a person is a student in this class, he is not over six feet tall.

Page 35 Exploratory 3. No students take Latin. 4. No secretaries live in New Jersey. 5. Some birds do not fly. 6. Some mathematics teachers do not have red hair. 7. No history teacher plays the flute. 8. All dogs bark.

Page 35 Written 1. Some daffodils are not yellow. all, some 2. None of these eggs are "jumbo." some, none 3. No Canadians were born in Kansas. at least one, no 4. Some cheddar in this shop has not been aged over 60 days. every, some 5. All American citizens were born in the United States. some, all 6. All masked men have something to hide. some, all 7. Some rhombuses are not parallelograms. every, some 8. Some of these flowers are roses. none, some 9. All integers are even. some, all 10. At least one even integer is rational. no, at least one 11. There does not exist an x in Q such that $x^2 < 1$. there exists, there does not exist 12. No real number, when squared, equals zero. at least one, no 13. Some roses are not red or violets are not blue. 14. Some elephants are red or some giraffes are green. 15. All elves were not jolly or some trolls were not mean. 16. No integer is even and there exists at least one rose which is red. 17. Some persons in this class are not tall or smart. 18. All children prefer dogs and some dogs are not brown. Some children do not prefer dogs and all dogs are brown.

Page 40 Chapter Review 1. 2 is a prime number and Van Gogh painted the Mona Lisa. false 2. A square is a parallelogram or Van Gogh painted the Mona Lisa. true 3. A square is a parallelogram or 2 is not a prime number. true 4. A square is a parallelogram and Van Gogh did not paint the Mona Lisa. true

5.
p	q	$\sim p$	$\sim p \wedge q$
T	T	F	F
T	F	F	F
F	T	T	T
F	F	T	F

6.
p	q	$\sim p$	$\sim q$	$\sim p \vee \sim q$
T	T	F	F	F
T	F	F	T	T
F	T	T	F	T
F	F	T	T	T

7.
p	q	$\sim q$	$p \vee \sim q$
T	T	F	T
T	F	T	T
F	T	F	F
F	F	T	T

8.
p	q	$\sim p$	$\sim q$	$\sim p \vee q$	$\sim q \vee (\sim p \vee q)$
T	T	F	F	T	T
T	F	F	T	F	T
F	T	T	F	T	T
F	F	T	T	T	T

15. If a square is not a parallelogram, then 2 is a prime number. true 16. If Van Gogh did not paint the Mona Lisa, then 2 is not a prime number. false 17. 2 is a prime number if and only if Van Gogh did not paint the Mona Lisa. true 18. If a square is a parallelogram and 2 is a prime number, then Van Gogh painted the Mona Lisa. false

19.
p	q	$\sim p$	$p \to q$	$\sim p \vee q$
T	T	F	T	T
T	F	F	F	F
F	T	T	T	T
F	F	T	T	T

20.
p	q	$\sim p$	$\sim q$	$p \to \sim q$	$\sim p \vee q$
T	T	F	F	F	T
T	F	F	T	T	F
F	T	T	F	T	T
F	F	T	T	T	T

21. Given: $\sim R \lor \sim V$ Prove: $\sim V$
         R

STATEMENTS	REASONS
1. $\sim R \lor \sim V$	1. Given
2. R	2. Given
3. $\sim V$	3. Law of disjunctive inference

22. Given: $G \land \sim B$ Prove: D
         $B \lor D$

STATEMENTS	REASONS
1. $G \land \sim B$	1. Given
2. $\sim B$	2. Definition of conjunction
3. $B \lor D$	3. Given
4. D	4. Law of disjunctive inference

23. Given: $G \lor D$ Prove: $\sim G$
         D

No conclusion. A disjunction may have two true disjuncts. If one is true, the other may be true or false.

24. Given: $\sim P \to G$ Prove: G
         $\sim P$

STATEMENTS	REASONS
1. $\sim P \to G$	1. Given
2. $\sim P$	2. Given
3. G	3. Law of detachment

25. Given: $T \to S$ Prove: T
         S

No conclusion. Even though the consequent of the conditional and the conditional are true, the antecedent may be false.

26. Given: $F \lor T$ Prove: A
         $F \to A$
         $\sim T$

STATEMENTS	REASONS
1. $F \lor T$	1. Given
2. $\sim T$	2. Given
3. F	3. Law of disjunctive inference
4. $F \to A$	4. Given
5. A	5. Law of detachment

27. $t \to s, \sim s \to \sim t, \sim t \to \sim s$

28. $q \to \sim p, p \to \sim q, \sim q \to p$

29. $\sim B \to \sim A, A \to B, B \to A$

30. $\sim n \to m, \sim m \to n, n \to \sim m$

Page 41   Chapter Review

31.

STATEMENTS	REASONS
1. $S \to \sim T$	1. Given
2. $T \to \sim S$	2. Law of the contrapositive
3. T	3. Given
4. $\sim S$	4. Law of detachment

32.

STATEMENTS	REASONS
1. $A \land B$	1. Given
2. A	2. Definition of conjunction
3. $\sim A \lor C$	3. Given
4. C	4. Law of disjunctive inference
5. $D \to \sim C$	5. Given
6. $C \to \sim D$	6. Law of the contrapositive
7. $\sim D$	7. Law of detachment
8. Suggested conclusion does not follow.	8. Information leads to opposite conclusion

33.

STATEMENTS	REASONS
1. $T \land \sim M$	1. Given
2. T	2. Definition of conjunction
3. Suggested conclusion does not follow.	3. Information leads to opposite conclusion

34.

STATEMENTS	REASONS
1. $T \to {\sim}S$	1. Given
2. ${\sim}S \to U$	2. Given
3. $T \to U$	3. Law of syllogism
4. T	4. Given
5. U	5. Law of detachment

35.

STATEMENTS	REASONS
1. $B \to D$	1. Given
2. $D \to C$	2. Given
3. $B \to C$	3. Law of syllogism
4. B	4. Given
5. C	5. Law of detachment
6. ${\sim}C \lor A$	6. Given
7. A	7. Law of disjunctive inference

36.

STATEMENTS	REASONS
1. ${\sim}r \lor p$	1. Given
2. r	2. Given
3. p	3. Law of disjunctive inference
4. $p \to q$	4. Given
5. $t \to {\sim}q$	5. Given
6. $q \to {\sim}t$	6. Law of the contrapositive
7. $p \to {\sim}t$	7. Law of syllogism
8. ${\sim}t$	8. Law of detachment

37. Given: $C \to E$ Prove: ${\sim}G$
$E \to B$
$B \to {\sim}G$
C

STATEMENTS	REASONS
1. $C \to E$	1. Given
2. $E \to B$	2. Given
3. $C \to B$	3. Law of syllogism
4. C	4. Given
5. B	5. Law of detachment
6. $B \to {\sim}G$	6. Given
7. ${\sim}G$	7. Law of detachment

38. Given: $R \to W$ Prove: W
$R \to S$
S
No conclusion follows. Insufficient information

42.

STATEMENTS	REASONS
1. ${\sim}(v \lor w)$	1. Given
2. ${\sim}v \land {\sim}w$	2. DeMorgan Laws
3. ${\sim}v$	3. Definition of conjunction
4. ${\sim}v \to t$	4. Given
5. t	5. Law of detachment

43.

STATEMENTS	REASONS
1. ${\sim}(a \land {\sim}b)$	1. Given
2. ${\sim}a \lor b$	2. DeMorgan Laws
3. a	3. Given
4. b	4. Law of disjunctive inference
5. $b \to c$	5. Given
6. c	6. Law of detachment

44.

STATEMENTS	REASONS
1. ${\sim}(p \land {\sim}q)$	1. Given
2. ${\sim}p \lor q$	2. DeMorgan Laws
3. ${\sim}(q \lor {\sim}r)$	3. Given
4. ${\sim}q \land r$	4. DeMorgan Laws
5. ${\sim}q$	5. Definition of conjunction
6. ${\sim}p$	6. Law of disjunctive inference
7. $p \lor t$	7. Given
8. t	8. Law of disjunctive inference

45. For every integer y, $y \neq -5$. false **46.** There exists a rational number x such that $x^2 = 2$. false **47.** For every whole number a, a is an integer. true **52.** Some football players weigh less than 200 pounds. all, some **53.** Some choir members play the bassoon. none, some **54.** No flowers are roses. some, no **55.** No teachers have red hair. at least one, no

Page 42 Chapter Test 4. 6 is divisible by 3 and Chicago is the capital of the United States. false 5. 6 is divisible by 3 or February has 30 days. true 6. If 6 is divisible by 3, then February has 30 days. false 7. If Chicago is the capital of the United States, then February does not have 30 days. true

8.
p	q	$\sim q$	$p \rightarrow q$	$\sim q \vee p$
T	T	F	T	T
T	F	T	F	T
F	T	F	T	F
F	F	T	T	T

9.
p	q	$\sim p$	$\sim q$	$\sim q \vee \sim p$	$p \rightarrow \sim q$
T	T	F	F	F	F
T	F	F	T	T	T
F	T	T	F	T	T
F	F	T	T	T	T

20.
STATEMENTS	REASONS
1. $G \vee \sim B$	1. Given
2. B	2. Given
3. G	3. Law of disjunctive inference
4. $G \rightarrow H$	4. Given
5. H	5. Law of detachment

21.
STATEMENTS	REASONS
1. $T \vee R$	1. Given
2. $\sim R$	2. Given
3. T	3. Law of disjunctive inference
4. $\sim B \rightarrow \sim T$	4. Given
5. $T \rightarrow B$	5. Law of the contrapositive
6. B	6. Law of detachment

22.
STATEMENTS	REASONS
1. $A \wedge C$	1. Given
2. A	2. Definition of conjunction
3. $A \rightarrow B$	3. Given
4. $B \rightarrow D$	4. Given
5. $A \rightarrow D$	5. Law of syllogism
6. D	6. Law of detachment

23. Given: $\sim B \vee \sim C$ Prove: A
    $\sim B \rightarrow A$
    C

STATEMENTS	REASONS
1. $\sim B \vee \sim C$	1. Given
2. C	2. Given
3. $\sim B$	3. Law of disjunctive inference
4. $\sim B \rightarrow A$	4. Given
5. A	5. Law of detachment

24. Given: $\sim(T \vee S)$ Prove: V
    $L \rightarrow T$
    $\sim S \rightarrow (L \vee V)$

STATEMENTS	REASONS
1. $\sim(T \vee S)$	1. Given
2. $\sim T \wedge \sim S$	2. DeMorgan Laws
3. $\sim S$	3. Definition of conjunction
4. $\sim S \rightarrow (L \vee V)$	4. Given
5. $L \vee V$	5. Law of detachment
6. $\sim T$	6. Definition of conjunction
7. $L \rightarrow T$	7. Given
8. $\sim T \rightarrow \sim L$	8. Law of the contrapositive
9. $\sim L$	9. Law of detachment
10. V	10. Law of disjunctive inference

CHAPTER 2 MATHEMATICAL SYSTEMS

Page 50 Exploratory 1. Begin at zero and count around the clock by fives three times. The answer is 3.
2. Find the "ordinary" product of the numbers. Then divide this number by 6. The remainder is the answer.
3. For clock system n, find the "ordinary" product of the numbers. Then divide this number by n. The remainder is the answer in clock n.   4. Find two numbers with an "ordinary" product that is a multiple of 100.
5. Find two numbers with an "ordinary" product that is one greater than a multiple of 50.   6. The top row across will always be zeros because these products have no remainder when divided by n.

Page 50 Written

6.

×	0	1	2	3	4	5
0	0	0	0	0	0	0
1	0	1	2	3	4	5
2	0	2	4	0	2	4
3	0	3	0	3	0	3
4	0	4	2	0	4	2
5	0	5	4	3	2	1

Page 51 Written

22.

×	0	1	2	3	4	5	6
0	0	0	0	0	0	0	0
1	0	1	2	3	4	5	6
2	0	2	4	6	1	3	5
3	0	3	6	2	5	1	4
4	0	4	1	5	2	6	3
5	0	5	3	1	6	4	2
6	0	6	5	4	3	2	1

Page 54 Written

1.

−	0	1	2	3	4
0	0	4	3	2	1
1	1	0	4	3	2
2	2	1	0	4	3
3	3	2	1	0	4
4	4	3	2	1	0

Page 64 Mixed Review

1.

STATEMENTS	REASONS
1. $\sim A \rightarrow C$	1. Given
2. $C \rightarrow B$	2. Given
3. $\sim A \rightarrow B$	3. Law of syllogism
4. $\sim B \rightarrow A$	4. Law of the contrapositive
5. $\sim B$	5. Given
6. A	6. Law of detachment

2.

STATEMENTS	REASONS
1. $\sim(T \wedge \sim Q)$	1. Given
2. $\sim T \vee Q$	2. DeMorgan Laws
3. $T \wedge N$	3. Given
4. T	4. Definition of conjunction
5. Q	5. Law of disjunctive inference

Page 70 Written 20. For any two even integers, $2m$ and $2n$, the sum is $2m + 2n = 2(m + n)$, which is also an even integer. 21. $(E, +)$ is an operational system, is associative, and has an identity. To show that an inverse for $2m$ exists, let i be the inverse, and by definition, $2m + i = 0$. Therefore, $i = -2m$. Because the inverse is an even number, $(E, +)$ is a group. 22. A multiple of 3 is any number that has a factor of 3. yes. 23. Let $3m$ and $3n$ be any two multiples of 3. The sum, $3m + 3n$, is also a multiple of 3 because $3m + 3n = 3(m + n)$. 24. $(T, +)$ is an operational system, is associative, and has an identity. To show that an inverse for $3m$ exists, let i be the inverse, and by definition, $3m + i = 0$. Therefore, $i = -3m$. Because the inverse is also a multiple of 3, $(T, +)$ is a group. 25. See exercises 23 and 24. 26. See exercises 23 and 24.

Page 75 Written

32. Given: $(G, *)$ a group; $a, b \in G$; equation $x * a = b$

Prove: $x = b * a_{inv}$ is the solution.

STATEMENTS	REASONS
1. $x * a = b$	1. Given
2. $(x * a) * a_{inv} = b * a_{inv}$	2. Right Operation
3. $x * (a * a_{inv}) = b * a_{inv}$	3. associativity
4. $x * e = b * a_{inv}$	4. definition of inverse
5. $x = b * a_{inv}$	5. definition of identity

Prove: $x = b * a_{inv}$ is the only solution.

STATEMENTS	REASONS
1. Suppose $x = y$ is another solution.	1. Assumption
2. Then $y * a = b$	2. Substitution
3. $(y * a) * a_{inv} = b * a_{inv}$	3. Right Operation
4. $y * (a * a_{inv}) = b * a_{inv}$	4. associativity
5. $y * e = b * a_{inv}$	5. definition of inverse
6. $y = b * a_{inv}$	6. definition of identity

This solution is the same solution as $b * a_{inv}$.

Page 78 Written

1. $\begin{pmatrix} 1 & 2 & 3 \\ 3 & 1 & 2 \end{pmatrix}$ 2. $\begin{pmatrix} 1 & 2 & 3 \\ 1 & 3 & 2 \end{pmatrix}$ 3. $\begin{pmatrix} 1 & 2 & 3 \\ 1 & 2 & 3 \end{pmatrix}$ 4. $\begin{pmatrix} 1 & 2 & 3 \\ 2 & 1 & 3 \end{pmatrix}$ 5. $\begin{pmatrix} 1 & 2 & 3 \\ 1 & 2 & 3 \end{pmatrix}$ 6. $\begin{pmatrix} 1 & 2 & 3 \\ 1 & 2 & 3 \end{pmatrix}$

7. $\begin{pmatrix} 1 & 2 & 3 \\ 1 & 2 & 3 \end{pmatrix}$ 8. $\begin{pmatrix} 1 & 2 & 3 \\ 1 & 2 & 3 \end{pmatrix}$ 9. $\begin{pmatrix} 1 & 2 & 3 \\ 3 & 1 & 2 \end{pmatrix}$ 10. $\begin{pmatrix} 1 & 2 & 3 \\ 2 & 1 & 3 \end{pmatrix}$ 11. $\begin{pmatrix} 1 & 2 & 3 \\ 1 & 3 & 2 \end{pmatrix}$ 12. $\begin{pmatrix} 1 & 2 & 3 \\ 3 & 2 & 1 \end{pmatrix}$

Page 83 Chapter Review 35. yes 36. No, there is no identity element. 37. no, not an operational system 38. No, division is not associative. 39. No, there is no identity element. 40. No, zero has no inverse. 41. no, not an operational system 42. No, there is no identity element.

CHAPTER 3 A SYSTEM WITH TWO OPERATIONS
Page 90 Exploratory

1. $1(2 + 5) = 1 \cdot 2 + 1 \cdot 5$
   $1 \cdot 7 = 2 + 5$
   $7 = 7$

2. $1(5 + 4) = 1 \cdot 5 + 1 \cdot 4$
   $1 \cdot 9 = 5 + 4$
   $9 = 9$

3. $0(13 + 7) = 0 \cdot 13 + 0 \cdot 7$
   $0 \cdot 20 = 0 + 0$
   $0 = 0$

4. $\frac{1}{4}(7 + 5) = \frac{1}{4} \cdot 7 + \frac{1}{4} \cdot 5$
   $\frac{1}{4} \cdot 12 = \frac{7}{4} + \frac{5}{4}$
   $3 = 3$

Page 103 Written

14. Given: $c + a = c + b$ Prove: $a = b$

STATEMENTS	REASONS
1. $c + a = c + b$	1. Given
2. $a + c = b + c$	2. Commutative law in $(F, +)$
3. $a = b$	3. Field Addition Theorem 3

15. Given: $ca = cb$ Prove: $a = b$

STATEMENTS	REASONS
1. $ca = cb$	1. Given
2. $c^{-1}(ca) = c^{-1}(cb)$	2. Left Operation law in $(F, \cdot)$
3. $(c^{-1} \cdot c)a = (c^{-1} \cdot c)b$	3. Associative law in $(F, \cdot)$
4. $1 \cdot a = 1 \cdot b$	4. Definition of multiplicative inverse
5. $a = b$	5. Definition of multiplicative identity

16. Given: $ac = bc$ Prove: $a = b$

STATEMENTS	REASONS
1. $ac = bc$	1. Given
2. $ca = cb$	2. Commutative law in $(F, \cdot)$
3. $a = b$	3. Field Multiplication Theorem 4

17. Given: $b \in F$ Prove: $0 \cdot b = 0$

STATEMENTS	REASONS
1. $(0 + 0)b = 0 \cdot b + 0 \cdot b$	1. Distributive law in $(F, +, \cdot)$
2. $0 + 0 = 0$	2. Definition of additive inverse
3. $0 \cdot b = 0 \cdot b + 0 \cdot b$	3. Substitution principle of equality
4. $0 \cdot b + 0 = 0 \cdot b + 0 \cdot b$	4. Definition of additive identity
5. $0 = 0 \cdot b$	5. Left cancellation law in $(F, +)$

Page 114 Written

16. $z < 3$

17. $z < -8$

18. $z > -\frac{1}{14}$

19. $k < -4$

20. $k > \frac{1}{4}$

21. $k < -1$

22. $k < -\frac{3}{2}$

23. $x > \frac{4}{13}$

24. $x > -12$

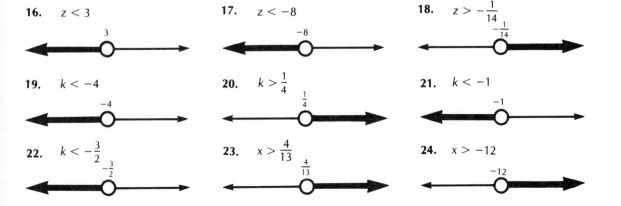

25. Given: $a < b$ Prove: $0 < b - a$

STATEMENTS	REASONS
1. $a < b$	1. Given
2. $a + (-a) < b + (-a)$	2. Additive Property
3. $0 < b + (-a)$	3. Definition of inverse
4. $0 < b - a$	4. Definition of subtraction

26. Given: $c < 0$ Prove: $0 < -c$

STATEMENTS	REASONS
1. $c < 0$	1. Given
2. $c + (-c) < 0 + (-c)$	2. Additive Property
3. $0 < 0 + (-c)$	3. Definition of inverse
4. $0 < -c$	4. Definition of identity

27. Given: $a < b, c < 0$ Prove: $ac > bc$

STATEMENTS	REASONS
1. $a < b$	1. Given
2. $0 < b - a$	2. Exercise 25
3. $c < 0$	3. Given
4. $0 < -c$	4. Exercise 26
5. $0(-c) < (b - a)(-c)$	5. Multiplicative Property
6. $0 < (b - a)(-c)$	6. Field Combination Theorem 2
7. $0 < b(-c) - a(-c)$	7. Field Combination Theorem 6
8. $0 < -bc + ac$	8. Field Combination Theorem 4
9. $bc + 0 < bc + (-bc + ac)$	9. Additive Property
10. $bc < ac$	10. Associative law, inverse, identity
11. $ac > bc$	11. Definition of greater than

28. Given: Field $(F, +, \cdot, <), a \in F$ Prove: $a^2 \geq 0$

STATEMENTS	REASONS
part i	
1. $a < 0$	1. Assumption
2. $a \cdot a > 0 \cdot a$	2. Ordered Field Theorem 1
3. $a^2 > 0 \cdot a$	3. Definition of a^2
4. $a^2 > 0$	4. Field Combination Theorem 2
part ii	
1. $a = 0$	1. Assumption
2. $a \cdot a = 0 \cdot a$	2. Right Operation
3. $a^2 = 0$	3. Definition of a^2, Field Combination Theorem 2
part iii	
1. $a > 0$	1. Assumption
2. $0 < a$	2. Definition of greater than
3. $0 \cdot a < a \cdot a$	3. Multiplicative Property
4. $0 < a^2$	4. Field Combination Theorem 2; Definition of a^2
5. $a^2 > 0$	5. Definition of greater than

CHAPTER 4 INTRODUCTION TO EUCLIDEAN GEOMETRY

Page 124 Exploratory 1. It is almost impossible to define the word "point" without using the word "point." **2.** A line can be defined as a set of points. However, part of the definition should include the fact that the points must be collinear, and the word "collinear" is derived from the word "line." **3.** A plane can be thought of as a flat surface that extends forever in all directions. However, this is not a formal definition. **4.** Point B is between points A and C if the three points are collinear and if A and B lie on opposite sides of C. However, this is not a formal definition.

Page 127 Exploratory 1. 1, Axiom 4-3 **2.** 2, Axiom 4-1 **3.** 1, Axiom 4-2 **4.** yes, definition of congruent segments **5.** no, Axiom 4-5 **6.** 3, Axiom 4-1 **7.** T is the midpoint of $\overline{RS}$, definition of bisector **8.** only if S is between R and T, segment addition axiom

Page 127 Written 1. always; Axiom 4-2 2. sometimes; This is only true if the points are noncollinear. (Axiom 4-3) 3. always; Axiom 4-2 4. always; Axiom 4-6 5. always; Axiom 4-4 6. always; Axiom 4-3 7. never; If the points are collinear, they are in more than one plane. If the points are noncollinear, they are in one and only one plane. 8. always; definition of midpoint 13. No, because if A is between B and C, AB can still be equal to $\frac{1}{2} AC$. 14. $\overline{AB}$ and $\overline{CD}$ intersect at M. 15. No, because $\overline{AB}$ can intersect $\overline{CD}$ at any point.

Page 134 Written 10. Given, Addition axiom, Multiplication axiom 11. Given, Distributive law, Subtraction axiom, Division axiom 12. Given, Subtraction axiom, Addition axiom, Division axiom

13.

STATEMENTS	REASONS
1. $m\angle 1 + m\angle 2 = 90$, $m\angle 1 = m\angle 3$	1. Given
2. $m\angle 3 + m\angle 2 = 90$	2. Substitution

14.

STATEMENTS	REASONS
1. $RT = SV$	1. Given
2. $RS + ST = RT$, $ST + TV = SV$	2. Segment addition axiom
3. $RS + ST = ST + TV$	3. Substitution
4. $RS = TV$	4. Subtraction axiom

15.

STATEMENTS	REASONS
1. $AB = AD$, $BC = DE$	1. Given
2. $AB + BC = AD + DE$	2. Addition axiom
3. $AB + BC = AC$, $AD + DE = AE$	3. Segment addition axiom
4. $AC = AE$	4. Substitution

16.

STATEMENTS	REASONS
1. $m\angle CBE = m\angle ABD$	1. Given
2. $m\angle CBE + m\angle EBD = m\angle ABD + m\angle EBD$	2. Addition axiom
3. $m\angle CBE + m\angle EBD = m\angle CBD$, $m\angle ABD + m\angle EBD = m\angle EBA$	3. Angle addition axiom
4. $m\angle CBD = m\angle EBA$	4. Substitution

17.

STATEMENTS	REASONS
1. $BE = CD$, $FD = FE$	1. Given
2. $BF + FE = BE$, $CF + FD = CD$	2. Segment addition axiom
3. $BF + FE = CF + FD$	3. Substitution
4. $BF = CF$	4. Subtraction axiom

18.

STATEMENTS	REASONS
1. $\overrightarrow{PR}$ bisects $\angle SPT$.	1. Given
2. $m\angle SPR = m\angle RPT$	2. Definition of angle bisector
3. $m\angle SPR + m\angle RPT = m\angle SPT$	3. Angle addition axiom
4. $2(m\angle SPR) = m\angle SPT$	4. Substitution

19.

STATEMENTS	REASONS
1. $PQ = QR$, $PQ = RT$	1. Given
2. $RT = QR$	2. Transitive axiom

20.

STATEMENTS	REASONS
1. $\angle AFC \cong \angle CFE$, $\angle AFD \cong \angle BFE$	1. Given
2. $m\angle AFC + m\angle CFD = m\angle AFD$, $m\angle CFE + m\angle BFC = m\angle BFE$	2. Angle addition axiom
3. $m\angle AFC + m\angle CFD = m\angle CFE + m\angle BFC$	3. Substitution
4. $m\angle CFD = m\angle BFC$	4. Subtraction axiom
5. $\angle BFC \cong \angle CFD$	5. Definition of congruent angles

Page 137 Exploratory 10. If angles are supplements of the same angles, then the angles are congruent. **11.** If two angles are vertical, then they are congruent. **12.** If two planes intersect, then the intersection is a line. **13.** If two angles are adjacent, then they have a common side.

Page 138 Written 12. With a conditional statement, if the antecedent is assumed to be true and the consequent can be proved true, then the statement is true. **13.** If the antecedent is assumed to be true and the consequent is false, then the conditional statement is false.

15. Given: $\angle 1$ is complementary to $\angle 3$; $\angle 2$ is complementary to $\angle 4$; $\angle 3 \cong \angle 4$
Prove: $\angle 1 \cong \angle 2$

16. Given: $\overleftrightarrow{AB}$ and $\overleftrightarrow{CD}$ intersect at X.
Prove: $\angle AXC \cong \angle DXB$; $\angle CXB \cong \angle AXD$

STATEMENTS	REASONS
1. $\angle 1$ is complementary to $\angle 3$. $\angle 2$ is complementary to $\angle 4$.	1. Given
2. $m\angle 1 + m\angle 3 = 90$ $m\angle 2 + m\angle 4 = 90$	2. Definition of complementary angles
3. $m\angle 1 + m\angle 3 = m\angle 2 + m\angle 4$	3. Transitive axiom
4. $\angle 3 \cong \angle 4$ ($m\angle 3 = m\angle 4$)	4. Given
5. $m\angle 1 = m\angle 2$	5. Subtraction axiom
6. $\angle 1 \cong \angle 2$	6. Definition of congruent angles

STATEMENTS	REASONS
1. $\overleftrightarrow{AB}$ and $\overleftrightarrow{CD}$ intersect at X.	1. Given
2. $\angle AXC$ is supplementary to $\angle CXB$. $\angle ACX$ is supplementary to $\angle AXD$. $\angle DXB$ is supplementary to $\angle CXB$.	2. Supplement axiom
3. $\angle AXC \cong \angle DXB$ $\angle CXB \cong \angle AXD$	3. Theorem 4-1

17. a.

STATEMENTS	REASONS
1. $AB = AB$	1. Reflexive axiom
2. $\overline{AB} \cong \overline{AB}$	2. Definition of congruent segments

b.

STATEMENTS	REASONS
1. $\overline{AB} \cong \overline{CD}$	1. Given
2. $AB = CD$	2. Definition of congruent segments
3. $CD = AB$	3. Symmetric axiom
4. $\overline{CD} \cong \overline{AB}$	4. Definition of congruent segments

c.

STATEMENTS	REASONS
1. $\overline{AB} \cong \overline{CD}, \overline{CD} \cong \overline{EF}$	1. Given
2. $AB = CD, CD = EF$	2. Definition of congruent segments
3. $AB = EF$	3. Transitive axiom
4. $\overline{AB} \cong \overline{EF}$	4. Definition of congruent segments

18. a.

STATEMENTS	REASONS
1. $m\angle ABC = m\angle ABC$	1. Reflexive axiom
2. $\angle ABC \cong \angle ABC$	2. Definition of congruent angles

b.

STATEMENTS	REASONS
1. $\angle ABC \cong \angle DEF$	1. Given
2. $m\angle ABC = m\angle DEF$	2. Definition of congruent angles
3. $m\angle DEF = m\angle ABC$	3. Symmetric axiom
4. $\angle DEF \cong \angle ABC$	4. Definition of congruent segments

c.

STATEMENTS	REASONS
1. $\angle ABC \cong \angle DEF$, $\angle DEF \cong \angle GHI$	1. Given
2. $m\angle ABC = m\angle DEF$, $m\angle DEF = m\angle GHI$	2. Definition of congruent angles
3. $m\angle ABC = m\angle GHI$	3. Transitive axiom
4. $\angle ABC \cong \angle GHI$	4. Definition of congruent angles

19.

STATEMENTS	REASONS
1. ∠2 ≅ ∠6	1. Given
2. ∠2 ≅ ∠3, ∠6 ≅ ∠7	2. Theorem 4-4
3. ∠3 ≅ ∠7	3. Substitution

Page 142 Written

10.

STATEMENTS	REASONS
1. ∠1 and ∠2 are supplementary.	1. Given
2. ∠1 ≅ ∠2	2. Theorem 4-4
3. ∠1 and ∠2 are right angles.	3. Theorem 4-8
4. ℓ ⊥ m	4. Definition of perpendicular lines

11.

STATEMENTS	REASONS
1. ∠1 ≅ ∠3	1. Given
2. ∠1 ≅ ∠2, ∠3 ≅ ∠4	2. Theorem 4-4
3. ∠2 ≅ ∠4	3. Substitution

12.

STATEMENTS	REASONS
1. $\overrightarrow{OB} \perp \overrightarrow{OD}$, $\overrightarrow{OC} \perp \overrightarrow{OA}$	1. Given
2. ∠DOB is a right angle. ∠COA is a right angle.	2. Definition of perpendicular lines
3. ∠DOB ≅ ∠COA	3. Theorem 4-7
4. m∠3 + m∠2 = m∠DOB; m∠2 + m∠1 = m∠COA	4. Angle addition axiom
5. m∠3 + m∠2 = m∠2 + m∠4	5. Substitution
6. m∠3 = m∠1	6. Subtraction axiom
7. ∠3 ≅ ∠1	7. Definition of congruent angles

13.

STATEMENTS	REASONS
1. $\overrightarrow{SW} \perp \overrightarrow{RV}, \overrightarrow{TY} \perp \overrightarrow{RV}$	1. Given
2. ∠WST is a right angle. ∠YTS is a right angle.	2. Definition of perpendicular lines
3. ∠WST ≅ ∠YTS	3. Theorem 4-7
4. m∠1 + m∠2 = m∠WST; m∠3 + m∠4 = m∠YTS	4. Angle addition axiom
5. m∠1 + m∠2 = m∠3 + m∠4	5. Substitution
6. m∠2 = m∠3	6. Given
7. m∠1 = m∠4	7. Subtraction axiom

14.

STATEMENTS	REASONS
1. ∠4 ≅ ∠3, ∠2 ≅ ∠1	1. Given
2. ∠2 ≅ ∠3	2. Theorem 4-4
3. ∠4 ≅ ∠1	3. Substitution

15. Given: $\overrightarrow{BA} \perp \overrightarrow{BC}$
Prove: ∠1 and ∠2 are complementary.

STATEMENTS	REASONS
1. $\overrightarrow{BA} \perp \overrightarrow{BC}$	1. Given
2. ∠ABC is a right angle.	2. Definition of perpendicular lines
3. m∠ABC = 90	3. Definition of right angle
4. m∠1 + m∠2 = m∠ABC	4. Angle addition axiom
5. m∠1 + m∠2 = 90	5. Substitution
6. ∠1 and ∠2 are complementary.	6. Definition of complementary

Page 146 Written

14.

STATEMENTS	REASONS
1. ∠1 ≅ ∠3	1. Given
2. $\overline{DE} \parallel \overline{BC}$	2. Axiom 4-15
3. ∠2 ≅ ∠4	3. Axiom 4-14

15.

STATEMENTS	REASONS
1. $\overline{DE} \parallel \overline{BC}$, ∠1 ≅ ∠2	1. Given
2. ∠1 ≅ ∠3, ∠2 ≅ ∠4	2. Axiom 4-14
3. ∠3 ≅ ∠4	3. Substitution

16.

STATEMENTS	REASONS
1. ∠2 ≅ ∠4, ∠4 ≅ ∠3	1. Given
2. $\overline{DE} \parallel \overline{BC}$	2. Axiom 4-15
3. ∠1 ≅ ∠3	3. Axiom 4-14
4. ∠2 ≅ ∠1	4. Substitution

17.

STATEMENTS	REASONS
1. ∠1 is supplementary to ∠2.	1. Given
2. ∠1 is supplementary to ∠3.	2. Supplement axiom
3. ∠2 ≅ ∠3	3. Theorem 4-1
4. $\overline{BC} \parallel \overline{ED}$	4. Axiom 4-15

18. No, the two lines can be skew.

Page 148 Exploratory

1. m∠2 = 130, m∠3 = 130, m∠4 = 50, m∠5 = 130, m∠6 = 50, m∠7 = 50, m∠8 = 130   **2.** m∠1 = 160, m∠2 = 20, m∠3 = 20, m∠4 = 160, m∠6 = 160, m∠7 = 160, m∠8 = 20   **3.** m∠1 = 60, m∠2 = 120, m∠4 = 60, m∠5 = 120, m∠6 = 60, m∠7 = 60, m∠8 = 120   **4.** m∠1 = 112, m∠2 = 68, m∠3 = 68, m∠4 = 112, m∠5 = 68, m∠6 = 112, m∠7 = 112, m∠8 = 68   **5.** m∠1 = 35, m∠2 = 145, m∠3 = 145, m∠5 = 145, m∠6 = 35, m∠7 = 35, m∠8 = 145   **6.** m∠1 = 107, m∠2 = 73, m∠3 = 73, m∠4 = 107, m∠5 = 73, m∠6 = 107, m∠7 = 107   **7.** m∠1 = 156, m∠2 = 24, m∠3 = 24, m∠4 = 156, m∠5 = 24, m∠6 = 156, m∠7 = 156, m∠8 = 24   **8.** m∠1 = 40, m∠2 = 140, m∠3 = 140, m∠4 = 40, m∠5 = 140, m∠6 = 40, m∠7 = 40, m∠8 = 140

Page 149 Written

21. Given: ∠1 ≅ ∠2
Prove: ℓ ∥ m

STATEMENTS	REASONS
1. ∠1 ≅ ∠2	1. Given
2. ∠3 ≅ ∠1	2. Theorem 4-4
3. ∠3 ≅ ∠2	3. Theorem 4-6
4. ℓ ∥ m	4. Axiom 4-15

22. Given: ∠1 is supplementary to ∠2
Prove: ℓ ∥ m

STATEMENTS	REASONS
1. ∠1 is supplementary to ∠2.	1. Given
2. ∠1 is supplementary to ∠3.	2. Supplement axiom
3. ∠2 ≅ ∠3	3. Theorem 4-1
4. ℓ ∥ m	4. Theorem 4-13

23. Given: $\ell \parallel m$
Prove: $\angle 3$ is supplementary to $\angle 5$.
$\angle 4$ is supplementary to $\angle 6$.

STATEMENTS	REASONS
1. $\ell \parallel m$	1. Given
2. $\angle 1 \cong \angle 5, \angle 2 \cong \angle 6$	2. Axiom 4-14
3. $\angle 3$ is supplementary to $\angle 1$. $\angle 4$ is supplementary to $\angle 2$.	3. Supplement axiom
4. $\angle 3$ is supplementary to $\angle 5$. $\angle 4$ is supplementary to $\angle 6$.	4. Substitution

24.
STATEMENTS	REASONS
1. $\ell \parallel m, n \parallel k$	1. Given
2. $\angle 2 \cong \angle 3, \angle 3 \cong \angle 4$	2. Theorem 4-12
3. $\angle 2 \cong \angle 4$	3. Theorem 4-6

25.
STATEMENTS	REASONS
1. $\angle 3 \cong \angle 4, n \parallel k$	1. Given
2. $\angle 2 \cong \angle 3$	2. Theorem 4-12
3. $\angle 2 \cong \angle 4$	3. Theorem 4-6

26.
STATEMENTS	REASONS
1. $\ell \parallel m, n \parallel k$	1. Given
2. $\angle 5 \cong \angle 3, \angle 1 \cong \angle 3$	2. Axiom 4-14
3. $\angle 1 \cong \angle 5$	3. Theorem 4-6

Page 150 Challenge
Given: $\overleftrightarrow{AC} \parallel \overleftrightarrow{DF}$
$\overrightarrow{BD}$ bisects $\angle CBE$.
$\overrightarrow{EA}$ bisects $\angle BEF$.
Prove: $\overrightarrow{EA} \parallel \overrightarrow{BD}$

STATEMENTS	REASONS
1. $\overleftrightarrow{AC} \parallel \overleftrightarrow{DF}$, $\overrightarrow{BD}$ bisects $\angle CBE$, $\overrightarrow{EA}$ bisects $\angle BEF$.	1. Given
2. $\angle CBE \cong \angle BEF$	2. Theorem 4-12
3. $\angle 1 \cong \angle 2, \angle 3 \cong \angle 4$	3. Definition of angle bisector
4. $m\angle 1 + m\angle 2 = m\angle CBE, m\angle 3 + m\angle 4 = m\angle BEF$	4. Angle addition axiom
5. $2(m\angle 1) = 2(m\angle 3)$	5. Substitution
6. $m\angle 1 = m\angle 3$	6. Division axiom
7. $\overrightarrow{EA} \parallel \overrightarrow{BD}$	7. Theorem 4-13

Page 153 Exploratory 7. Jim is not taller than Joe. 8. A, B, and C are noncollinear. 9. Angle X is not an acute angle. 10. If it rains, then it does not thunder.

Page 153 Written 1. Assume $\angle 1 \cong \angle 2$. Therefore, by Theorem 4-13, $\ell \parallel m$. However, this contradicts the given statement that $\ell \nparallel m$. Therefore, if $\angle 1 \ncong \angle 2$, then $\ell \nparallel m$. **2.** Assume $\ell \parallel m$. Therefore, by Theorem 4-12, $\angle 1 \cong \angle 2$. However, this contradicts the given statement that $\angle 1 \ncong \angle 2$. Therefore, if $\ell \nparallel m$, then $\angle 1 \ncong \angle 2$. **3.** Assume $n \perp m$. We are given that $n \perp \ell$. Therefore, by Theorem 4-11, $\ell \parallel m$. But, it is given that $\ell \parallel m$. Therefore, if $\ell \parallel m$ and $n \perp \ell$, then n is not perpendicular to m. **4.** one **5.** one **6.** yes; The line through P and line ℓ can be skew.

7. Given: Line ℓ and point E not on ℓ.
Prove: I. There is one line through E that is perpendicular to ℓ.
II. There is only one line through E that is perpendicular to ℓ.

I.

STATEMENTS	REASONS
1. Choose points A and C on line ℓ.	1. Axiom 4-1
2. Draw $\overleftrightarrow{AE}$ such that $m\angle EAC = 90$.	2. Axiom 4-9
3. $\angle EAC$ is a right angle.	3. Definition of right angle
4. $\overleftrightarrow{AE} \perp \ell$	4. Definition of perpendicular lines

II. Assume that there is more than one line through E perpendicular to ℓ. Suppose one line is $\overleftrightarrow{AE}$ and another is $\overleftrightarrow{CE}$. Then, by Theorem 4-11, $\overleftrightarrow{AE} \parallel \overleftrightarrow{CE}$. However, $\overleftrightarrow{AE}$ and $\overleftrightarrow{CE}$ intersect at E and this contradicts the definition of parallel lines. Thus, $\overleftrightarrow{AE}$ and $\overleftrightarrow{CE}$ must be the same line.

8. Given: $n \parallel m$, $m \parallel \ell$
Prove: $n \parallel \ell$

Assume $n \not\parallel \ell$. Since $n \parallel m$, and $m \parallel \ell$, by Axiom 4-14, $\angle 1 \cong \angle 2$ and $\angle 2 \cong \angle 3$. By substitution, $\angle 1 \cong \angle 3$. Then by Axiom 4-15, $n \parallel \ell$. This contradicts the assumption. Therefore, if $n \parallel m$ and $m \parallel \ell$, then $n \parallel \ell$.

9. Theorem 4–16   **10.** Theorem 4–16
11. Theorem 4–16 with V on the same side of $\overline{PQ}$ as R

Page 155 Exploratory   1. never; An acute triangle always has 3 acute angles.   **2.** never; A right triangle always has 2 acute angles and 1 right angle.   **3.** never; Theorem 4-19   **4.** sometimes; The two acute angles can both have measures of 45.   **5.** sometimes; The two remote interior angles can have a sum less than 90.   **6.** sometimes; The triangle can be a right triangle.

Page 156 Written

12. Given: $\triangle ABC$ with exterior angle ACD.
Prove: $m\angle A + m\angle B = m\angle ACD$

STATEMENTS	REASONS
1. $\triangle ABC$ with exterior angle ACD.	1. Given
2. $m\angle A + m\angle B + m\angle ACB = 180$	2. Theorem 4-19
3. $\angle ACB$ is supplementary to $\angle ACD$.	3. Supplement axiom
4. $m\angle ACB + m\angle ACD = 180$	4. Definition of supplementary
5. $m\angle A + m\angle B + m\angle ACB = m\angle ACB + m\angle ACD$	5. Transitive axiom
6. $m\angle A + m\angle B = m\angle ACD$	6. Subtraction axiom

13. Given: $\angle A \cong \angle X$, $\angle B \cong \angle Y$
Prove: $\angle C \cong \angle Z$

STATEMENTS	REASONS
1. $\angle A \cong \angle X$, $\angle B \cong \angle Y$	1. Given
2. $m\angle A + m\angle B + m\angle C = 180$, $m\angle X + m\angle Y + m\angle Z = 180$	2. Theorem 4-19
3. $m\angle A + m\angle B + m\angle C = m\angle X + m\angle Y + m\angle Z$	3. Transitive axiom
4. $m\angle C = m\angle Z$	4. Subtraction axiom
5. $\angle C \cong \angle Z$	5. Definition of congruent angles

14. Given: △ABC
Prove: I. There can be no more than one right angle in △ABC.
II. There can be no more than one obtuse angle in △ABC.

I. Assume ∠A and ∠B are both right angles. Therefore, by the definition of right angles, m∠A = 90 and m∠B = 90. By Theorem 4-19, m∠A + m∠B + m∠C = 180. By substitution and then the subtraction axiom, m∠C = 0. However, this contradicts the angle measure axiom. So ∠A and ∠B cannot both be right angles. Thus, there can be no more than one right angle in △ABC.

II. Assume ∠A and ∠B are both obtuse. Therefore, by the definition of obtuse angles, m∠A > 90 and m∠B > 90. By Theorem 4-19, m∠A + m∠B + m∠C = 180. However, if both ∠A and ∠B have measures greater than 90, m∠A + m∠B + m∠C ≠ 180. This contradicts Theorem 4-19. So ∠A and ∠B cannot both be obtuse. Thus, there can be no more than one obtuse angle in △ABC.

15. Given: △ABC with right angle B.

Prove: ∠A is complementary to ∠C.

STATEMENTS	REASONS
1. △ABC with right angle B.	1. Given
2. m∠B = 90	2. Definition of right angle
3. m∠A + m∠B + m∠C = 180	3. Theorem 4-19
4. m∠A + 90 + m∠C = 180	4. Substitution
5. m∠A + m∠C = 90	5. Subtraction axiom
6. ∠A is complementary to ∠C.	6. Definition of complementary

16. Given: △ABC with ∠A ≅ ∠B ≅ ∠C

Prove: m∠A = 60, m∠B = 60, m∠C = 60

STATEMENTS	REASONS
1. △ABC with ∠A ≅ ∠B ≅ ∠C	1. Given
2. m∠A + m∠B + m∠C = 180	2. Theorem 4-19
3. 3(m∠A) = 180 3(m∠B) = 180 3(m∠C) = 180	3. Substitution
4. m∠A = 60, m∠B = 60, m∠C = 60	4. Division axiom

Page 156 Challenge

1.

STATEMENTS	REASONS
1. $\overline{BA}$ bisects ∠CBD. $\overline{CA}$ bisects ∠BCE.	1. Given
2. ∠DBA ≅ ∠ABC, ∠BCA ≅ ∠ACE	2. Definition of angle bisector
3. m∠DBA + m∠ABC = m∠DBC, m∠BCA + m∠ACE = m∠BCE	3. Angle addition axiom
4. 2(m∠ABC) = m∠DBC, 2(m∠BCA) = m∠BCE	4. Substitution
5. ℓ ∥ m	5. Given
6. ∠DBC is supplementary to ∠BCE.	6. Theorem 4-15

STATEMENTS	REASONS
7. m∠DBC + m∠BCE = 180	7. Definition of supplementary
8. 2(m∠ABC) + 2(m∠BCA) = 180	8. Substitution
9. m∠ABC + m∠BCA = 90	9. Division axiom
10. m∠ABC + m∠BCA + m∠BAC = 180	10. Theorem 4-19
11. 90 + m∠BAC = 180	11. Substitution
12. m∠BAC = 90	12. Subtraction axiom
13. $\overline{BA}$ ⊥ $\overline{CA}$	13. Definition of perpendicular lines

T66

2. Given: Quadrilateral ABCD

Prove: $m\angle DAB + m\angle ABC + m\angle BCD + m\angle CDA = 360$

STATEMENTS	REASONS
1. Quadrilateral ABCD	1. Given
2. Draw $\overline{DB}$.	2. Axiom 4-2
3. $m\angle ADB + m\angle DBA + m\angle DAB = 180$, $m\angle DBC + m\angle BDC + m\angle BCD = 180$	3. Theorem 4-19
4. $m\angle ADB + m\angle DBA + m\angle DAB + m\angle DBC + m\angle BDC + m\angle BCD = 360$	4. Addition axiom
5. $m\angle ADB + m\angle BDC = m\angle CDA$, $m\angle DBA + m\angle DBC = m\angle ABC$	5. Angle addition axiom
6. $m\angle DAB + m\angle ABC + m\angle BCD + m\angle CDA = 360$	6. Substitution

3. Given: Parallelogram ABCD

Prove: $m\angle A + m\angle D = 180$
$m\angle A + m\angle B = 180$
$m\angle B + m\angle C = 180$
$m\angle C + m\angle D = 180$

STATEMENTS	REASONS
1. Parallelogram ABCD	1. Given
2. $\overline{AB} \parallel \overline{DC}$, $\overline{AD} \parallel \overline{BC}$	2. Definition of parallelogram
3. $\angle A$ is supplementary to $\angle D$. $\angle A$ is supplementary to $\angle B$. $\angle B$ is supplementary to $\angle C$. $\angle C$ is supplementary to $\angle D$.	3. Theorem 4-15
4. $m\angle A + m\angle D = 180$, $m\angle A + m\angle B = 180$, $m\angle B + m\angle C = 180$, $m\angle C + m\angle D = 180$	4. Definition of supplementary

Page 159 Written 21. no; Solve $\frac{180(n-2)}{n} = 50$. Since n is not a whole number, 50 cannot be the measure of an interior angle. **22.** no; Solve $\frac{180(n-2)}{n} = 100$. n is not a whole number. **23.** yes; 6 sides **24.** no; $\frac{180(n-2)}{n} = 180$ cannot be solved. **25.** no; Solve $\frac{180(n-2)}{n} = 181$. n is a negative number.

Page 165 Chapter Review 1. always; meaning of between **2.** never; Vertical angles do not have a common side. **3.** never; the rays overlap **4.** always; definition of vertex of an angle **5.** never; Axioms are accepted as true. **6.** always; Axiom 4-1 **7.** sometimes; if B is between A and C **8.** never; Axiom 4-5 **9.** always; definition of complementary **10.** always; Supplement axiom **11.** never; the bisector of $\angle QRS$ contains R, not Q **12.** always; definition of angle bisector **13.** sometimes; if $a \neq 0$ **14.** always; Transitive axiom **15.** sometimes; if the angles are formed by intersecting lines **16.** sometimes; if $\angle A$ and $\angle B$ are both right angles **17.** always; definition of perpendicular lines **18.** sometimes; A line can pass through the midpoint without being perpendicular to the segment. **19.** sometimes; They can be skew. **20.** sometimes; if the lines are parallel

21.

STATEMENTS	REASONS
1. $\ell \parallel m$, $n \parallel k$	1. Given
2. $\angle 2 \cong \angle 3$, $\angle 3 \cong \angle 5$	2. Axiom 4-15
3. $\angle 2 \cong \angle 5$	3. Theorem 4-6

22.

STATEMENTS	REASONS
1. $\angle 2 \cong \angle 4$, $\ell \parallel m$	1. Given
2. $\angle 3 \cong \angle 4$	2. Theorem 4-12
3. $\angle 2 \cong \angle 3$	3. Theorem 4-6
4. $k \parallel n$	4. Axiom 4-15

23. Assume $n \parallel k$. Therefore, by Axiom 4-14, $\angle 2 \cong \angle 3$. By Theorem 4-4, $\angle 1 \cong \angle 2$. Then, by Theorem 4-6, $\angle 1 \cong \angle 3$. However, this contradicts the Given statement that $\angle 1 \not\cong \angle 3$. Thus, $n \not\parallel k$.

Page 166 Chapter Test

22.

STATEMENTS	REASONS
1. $\angle 2 \cong \angle 3$	1. Given
2. $\angle 1$ is supplementary to $\angle 2$. $\angle 3$ is supplementary to $\angle 4$.	2. Supplement axiom
3. $\angle 1 \cong \angle 4$	3. Theorem 4-2

23.

STATEMENTS	REASONS
1. $\ell \parallel m, n \parallel p$	1. Given
2. $\angle 1 \cong \angle 3$	2. Axiom 4-14
3. $\angle 3 \cong \angle 4$	3. Theorem 4-12
4. $\angle 1 \cong \angle 4$	4. Theorem 4-6

24. Assume $\ell \parallel m$. Therefore, by Theorem 4-12, $\angle 2 \cong \angle 3$. By Theorem 4-4, $\angle 3 \cong \angle 4$ and $\angle 1 \cong \angle 2$. By substitution, $\angle 1 \cong \angle 4$. But, it is given that $\angle 1 \not\cong \angle 4$. Therefore, if $\angle 1 \not\cong \angle 4$, $\ell \not\parallel m$.

CHAPTER 5 CONGRUENT TRIANGLES

Page 170 Written **1.** $\angle BAC \cong \angle DAC$, $\angle B \cong \angle D$, $\angle BCA \cong \angle DCA$, $\overline{AB} \cong \overline{AD}$, $\overline{BC} \cong \overline{DC}$, $\overline{AC} \cong \overline{AC}$
2. $\angle A \cong \angle E$, $\angle ACB \cong \angle ECD$, $\angle B \cong \angle D$, $\overline{AC} \cong \overline{EC}$, $\overline{CB} \cong \overline{CD}$, $\overline{AB} \cong \overline{ED}$ **3.** $\angle BAC \cong \angle DCA$, $\angle B \cong \angle D$, $\angle BCA \cong \angle DAC$, $\overline{AB} \cong \overline{CD}$, $\overline{BC} \cong \overline{DA}$, $\overline{AC} \cong \overline{CA}$ **4.** $\angle CAD \cong \angle ACB$, $\angle ACD \cong \angle CAB$, $\angle CDA \cong \angle ABC$, $\overline{AC} \cong \overline{CA}$, $\overline{CD} \cong \overline{AB}$, $\overline{AD} \cong \overline{CB}$ **5.** $\angle EDA \cong \angle FBC$, $\angle DEA \cong \angle BFC$, $\angle EAD \cong \angle FCB$, $\overline{DE} \cong \overline{BF}$, $\overline{EA} \cong \overline{FC}$, $\overline{DA} \cong \overline{BC}$ **6.** $\angle ECD \cong \angle FAB$, $\angle CED \cong \angle AFB$, $\angle EDC \cong \angle FBA$, $\overline{CE} \cong \overline{AF}$, $\overline{ED} \cong \overline{FB}$, $\overline{CD} \cong \overline{AB}$
7. $\angle CAB \cong \angle DAE$, $\angle ACB \cong \angle ADE$, $\angle CBA \cong \angle DEA$, $\overline{AC} \cong \overline{AD}$, $\overline{CB} \cong \overline{DE}$, $\overline{AB} \cong \overline{AE}$ **8.** $\angle BAD \cong \angle EAC$, $\angle ABD \cong \angle AEC$, $\angle BDA \cong \angle ECA$, $\overline{AB} \cong \overline{AE}$, $\overline{BD} \cong \overline{EC}$, $\overline{AD} \cong \overline{AC}$

11. Given: $\triangle ABC$ Prove: $\triangle ABC \cong \triangle ABC$

STATEMENTS	REASONS
1. $\triangle ABC$	1. Given
2. $\angle A \cong \angle A$, $\angle B \cong \angle B$, $\angle C \cong \angle C$	2. Theorem 4-6
3. $\overline{AB} \cong \overline{AB}$, $\overline{BC} \cong \overline{BC}$, $\overline{AC} \cong \overline{AC}$	3. Theorem 4-5
4. $\triangle ABC \cong \triangle ABC$	4. Definition of congruent triangles

12. Given: $\triangle ABC \cong \triangle DEF$ Prove: $\triangle DEF \cong \triangle ABC$

STATEMENTS	REASONS
1. $\triangle ABC \cong \triangle DEF$	1. Given
2. $\angle A \cong \angle D$, $\angle B \cong \angle E$, $\angle C \cong \angle F$, $\overline{AB} \cong \overline{DE}$, $\overline{BC} \cong \overline{EF}$, $\overline{AC} \cong \overline{DF}$	2. Definition of congruent triangles
3. $\angle D \cong \angle A$, $\angle E \cong \angle B$, $\angle F \cong \angle C$	3. Theorem 4-6
4. $\overline{DE} \cong \overline{AB}$, $\overline{EF} \cong \overline{BC}$, $\overline{DF} \cong \overline{AC}$	4. Theorem 4-5
5. $\triangle DEF \cong \triangle ABC$	5. Definition of congruent triangles

13. Given: $\triangle ABC \cong \triangle DEF$, $\triangle DEF \cong \triangle XYZ$ Prove: $\triangle ABC \cong \triangle XYZ$

STATEMENTS	REASONS
1. $\triangle ABC \cong \triangle DEF$, $\triangle DEF \cong \triangle XYZ$	1. Given
2. $\angle A \cong \angle D$, $\angle B \cong \angle E$, $\angle C \cong \angle F$, $\overline{AB} \cong \overline{DE}$, $\overline{BC} \cong \overline{EF}$, $\overline{AC} \cong \overline{DF}$, $\angle D \cong \angle X$, $\angle E \cong \angle Y$, $\angle F \cong \angle Z$, $\overline{DE} \cong \overline{XY}$, $\overline{EF} \cong \overline{YZ}$, $\overline{DF} \cong \overline{XZ}$	2. Definition of congruent triangles
3. $\angle A \cong \angle X$, $\angle B \cong \angle Y$, $\angle C \cong \angle Z$	3. Theorem 4-6
4. $\overline{AB} \cong \overline{XY}$, $\overline{BC} \cong \overline{YZ}$, $\overline{AC} \cong \overline{XZ}$	4. Theorem 4-5
5. $\triangle ABC \cong \triangle XYZ$	5. Definition of congruent triangles

Page 174 Written

9.

STATEMENTS	REASONS
1. $\overline{AD} \cong \overline{CB}$, $\overline{AB} \cong \overline{CD}$	1. Given
2. $\overline{AC} \cong \overline{CA}$	2. Theorem 4-5
3. $\triangle ABC \cong \triangle CDA$	3. SSS

10.

STATEMENTS	REASONS
1. $\angle BAC \cong \angle DCA$, $\overline{AB} \cong \overline{CD}$	1. Given
2. $\overline{AC} \cong \overline{CA}$	2. Theorem 4-5
3. $\triangle ABC \cong \triangle CDA$	3. SAS

11.

STATEMENTS	REASONS
1. $\overline{AB} \parallel \overline{CD}$, $\overline{AB} \cong \overline{CD}$	1. Given
2. $\angle DCA \cong \angle BAC$	2. Theorem 4-12
3. $\overline{AC} \cong \overline{CA}$	3. Theorem 4-5
4. $\triangle ABC \cong \triangle CDA$	4. SAS

12.

STATEMENTS	REASONS
1. $\overline{BC} \parallel \overline{AD}$, $\overline{AB} \parallel \overline{DC}$	1. Given
2. $\angle BCA \cong \angle DAC$, $\angle BAC \cong \angle DCA$	2. Theorem 4-12
3. $\overline{AC} \cong \overline{CA}$	3. Theorem 4-5
4. $\triangle ABC \cong \triangle CDA$	4. ASA

13.

STATEMENTS	REASONS
1. $\overline{AB} \cong \overline{AD}$, $\overline{AC}$ bisects $\angle BAD$.	1. Given
2. $\angle BAC \cong \angle DAC$	2. Definition of angle bisector
3. $\overline{AC} \cong \overline{AC}$	3. Theorem 4-5
4. $\triangle ABC \cong \triangle ADC$	4. SAS

14.

STATEMENTS	REASONS
1. C is the midpoint of $\overline{BD}$. $\overline{AB} \cong \overline{AD}$	1. Given
2. $\overline{BC} \cong \overline{DC}$	2. Definition of midpoint
3. $\overline{AC} \cong \overline{AC}$	3. Theorem 4-5
4. $\triangle ABC \cong \triangle ADC$	4. SSS

15.

STATEMENTS	REASONS
1. $\overline{AC} \perp \overline{BD}$	1. Given
2. $\angle ACB$ and $\angle ACD$ are right angles.	2. Definition of perpendicular lines
3. $\angle ACB \cong \angle ACD$	3. Theorem 4-7
4. $\overline{AC} \cong \overline{AC}$	4. Theorem 4-5
5. C is the midpoint of $\overline{BD}$.	5. Given
6. $\overline{BC} \cong \overline{DC}$	6. Definition of midpoint
7. $\triangle ABC \cong \triangle ADC$	7. SAS

16.

STATEMENTS	REASONS
1. $\overline{AC}$ bisects $\angle BAD$, $\overline{AC} \perp \overline{BD}$	1. Given
2. $\angle ACB$ and $\angle ACD$ are right angles.	2. Definition of perpendicular lines
3. $\angle ACB \cong \angle ACD$	3. Theorem 4-7
4. $\angle BAC \cong \angle DAC$	4. Definition of angle bisector
5. $\overline{AC} \cong \overline{AC}$	5. Theorem 4-5
6. $\triangle ABC \cong \triangle ADC$	6. ASA

Pages 176–177 Exploratory 1. $\angle M \cong \angle C$, $\angle L \cong \angle B$, $\angle K \cong \angle A$, $\overline{ML} \cong \overline{CB}$, $\overline{LK} \cong \overline{BA}$, $\overline{KM} \cong \overline{AC}$
2. $\angle RST \cong \angle WST$, $\angle SRT \cong \angle SWT$, $\angle RTS \cong \angle WTS$, $\overline{SR} \cong \overline{SW}$, $\overline{RT} \cong \overline{WT}$, $\overline{ST} \cong \overline{ST}$ **3.** $\angle BAC \cong \angle FEC$, $\angle ABC \cong \angle EFC$, $\angle BCA \cong \angle FCE$, $\overline{AB} \cong \overline{EF}$, $\overline{BC} \cong \overline{FC}$, $\overline{AC} \cong \overline{EC}$ **4.** $\angle BAC \cong \angle MLK$, $\angle ABC \cong \angle LMK$, $\angle BCA \cong \angle MKL$, $\overline{AB} \cong \overline{LM}$, $\overline{BC} \cong \overline{MK}$, $\overline{AC} \cong \overline{LK}$ **5.** $\angle BAC \cong \angle LMK$, $\angle ABC \cong \angle MLK$, $\angle BCA \cong \angle LKM$, $\overline{AB} \cong \overline{ML}$, $\overline{BC} \cong \overline{LK}$, $\overline{AC} \cong \overline{MK}$ **6.** $\angle SRT \cong \angle WTR$, $\angle RST \cong \angle TWR$, $\angle STR \cong \angle WRT$, $\overline{RS} \cong \overline{TW}$, $\overline{ST} \cong \overline{WR}$, $\overline{RT} \cong \overline{TR}$ **7.** $x = 4$, $DE = 12$, $EF = SM = 14$, $FD = MR = 16$ **8.** $x = 40$, $MR = 10$, $DE = MS = 200$, $EF = SR = 199$ **9.** $z = 6$, $EF = SM = 12$, $FD = MR = 6$, $DE = RS = 11$ **10.** We are given that $\overline{TQ} \cong \overline{TR}$ and $\angle QTR \cong \angle RTP$. By Theorem 4-5, $\overline{PT} \cong \overline{PT}$. Therefore, by SAS, $\triangle QTP \cong \triangle RTP$. Then by CPCTC, $\overline{PQ} \cong \overline{PR}$. **11.** We are given that $\angle QMP \cong \angle RMP$ and $\angle MPQ \cong \angle MPR$. By Theorem 4-5, $\overline{MP} \cong \overline{MP}$. Therefore, by ASA, $\triangle QMP \cong \triangle RMP$. Then by CPCTC, $\overline{PQ} \cong \overline{PR}$. **12.** We are given that $\overline{PC} \cong \overline{PB}$ and $\angle QCP \cong \angle RBP$. By Theorem 4-4, $\angle QPC \cong \angle RPB$. Therefore, by ASA, $\triangle QPC \cong \triangle RPB$. Then by CPCTC, $\overline{PQ} \cong \overline{PR}$.

Page 178 Written

7.

STATEMENTS	REASONS
1. $\overline{ST} \cong \overline{UT}$, $\angle 3 \cong \angle 4$	1. Given
2. $\overline{RT} \cong \overline{RT}$	2. Theorem 4-5
3. $\triangle TSR \cong \triangle TUR$	3. SAS
4. $\angle 1 \cong \angle 2$	4. CPCTC

8.

STATEMENTS	REASONS
1. $\angle 1 \cong \angle 2$, $\angle 3 \cong \angle 4$	1. Given
2. $\overline{RT} \cong \overline{RT}$	2. Theorem 4-5
3. $\triangle TSR \cong \triangle TUR$	3. ASA
4. $\overline{RS} \cong \overline{RU}$	4. CPCTC

9.

STATEMENTS	REASONS
1. $\overline{RS} \cong \overline{RU}$, $\overline{ST} \cong \overline{UT}$	1. Given
2. $\overline{RT} \cong \overline{RT}$	2. Theorem 4-5
3. $\triangle TSR \cong \triangle TUR$	3. SSS
4. $\angle 1 \cong \angle 2$	4. CPCTC

10.

STATEMENTS	REASONS
1. $\angle 1 \cong \angle 2$, $\overline{RS} \cong \overline{RU}$	1. Given
2. $\overline{RT} \cong \overline{RT}$	2. Theorem 4-5
3. $\triangle TSR \cong \triangle TUR$	3. SAS
4. $\angle S \cong \angle U$	4. CPCTC

11.

STATEMENTS	REASONS
1. L is the midpoint of $\overline{KM}$ and $\overline{JN}$.	1. Given
2. $\overline{KL} \cong \overline{ML}$, $\overline{JL} \cong \overline{NL}$	2. Definition of midpoint
3. $\angle 3 \cong \angle 4$	3. Theorem 4-4
4. $\triangle JKL \cong \triangle NML$	4. SAS
5. $\overline{JK} \cong \overline{NM}$	5. CPCTC

12.

STATEMENTS	REASONS
1. L is the midpoint of $\overline{JN}$. $\angle 2 \cong \angle 6$	1. Given
2. $\overline{JL} \cong \overline{NL}$	2. Definition of midpoint
3. $\angle 3 \cong \angle 4$	3. Theorem 4-4
4. $\triangle JKL \cong \triangle NML$	4. ASA
5. $\overline{KL} \cong \overline{ML}$	5. CPCTC

13.

STATEMENTS	REASONS
1. $\overline{JK} \cong \overline{NM}$, $\overline{JK} \cong \overline{NM}$	1. Given
2. $\angle 2 \cong \angle 6$, $\angle 1 \cong \angle 5$	2. Theorem 4-12
3. $\triangle JKL \cong \triangle NML$	3. ASA
4. $\overline{JL} \cong \overline{NL}$	4. CPCTC

14.

STATEMENTS	REASONS
1. L is the midpoint of $\overline{KM}$ and $\overline{JN}$.	1. Given
2. $\overline{KL} \cong \overline{ML}$, $\overline{JL} \cong \overline{NL}$	2. Definition of midpoint
3. $\angle 3 \cong \angle 4$	3. Theorem 4-4
4. $\triangle JKL \cong \triangle NML$	4. SAS
5. $\angle 1 \cong \angle 5$	5. CPCTC
6. $\overline{JK} \cong \overline{MN}$	6. Theorem 4-13

15.

STATEMENTS	REASONS
1. $\overline{BC} \cong \overline{CD}$, $\angle 2 \cong \angle 4$	1. Given
2. $\overline{AC} \cong \overline{AC}$	2. Theorem 4-5
3. $\triangle CDA \cong \triangle CBA$	3. SAS
4. $\angle D \cong \angle B$	4. CPCTC

16.

STATEMENTS	REASONS
1. $\overline{AB} \cong \overline{CD}$, $\overline{AD} \cong \overline{CB}$	1. Given
2. $\overline{AC} \cong \overline{CA}$	2. Theorem 4-5
3. $\triangle ADC \cong \triangle CBA$	3. SSS
4. $\angle D \cong \angle B$	4. CPCTC

17.

STATEMENTS	REASONS
1. $\angle 1 \cong \angle 4$, $\angle 3 \cong \angle 2$	1. Given
2. $\overline{AC} \cong \overline{CA}$	2. Theorem 4-5
3. $\triangle ADC \cong \triangle CBA$	3. ASA
4. $\overline{AD} \cong \overline{CB}$	4. CPCTC

18.

STATEMENTS	REASONS
1. $\overline{AB} \cong \overline{CD}$, $\overline{AD} \cong \overline{CB}$	1. Given
2. $\overline{AC} \cong \overline{CA}$	2. Theorem 4-5
3. $\triangle ADC \cong \triangle CBA$	3. SSS
4. $\angle 1 \cong \angle 4$	4. CPCTC
5. $\overline{AB} \parallel \overline{CD}$	5. Theorem 4-13

Page 179 Challenge

1.

STATEMENTS	REASONS
1. $\overline{AB} \cong \overline{CD}$, $\overline{BC} \cong \overline{DA}$	1. Given
2. $\overline{AC} \cong \overline{CA}$	2. Theorem 4-5
3. $\triangle ABC \cong \triangle CDA$	3. SSS
4. $\angle BAC \cong \angle DCA$	4. CPCTC
5. $\angle AGE \cong \angle CGF$	5. Theorem 4-4
6. G is the midpoint of $\overline{AC}$.	6. Given
7. $\overline{AG} \cong \overline{CG}$	7. Definition of midpoint
8. $\triangle AGE \cong \triangle CGF$	8. ASA
9. $\overline{EG} \cong \overline{FG}$	9. CPCTC

2.

STATEMENTS	REASONS
1. G is the midpoint of $\overline{DB}$ and $\overline{EF}$.	1. Given
2. $\overline{GD} \cong \overline{GB}$, $\overline{GF} \cong \overline{GE}$	2. Definition of midpoint
3. $\angle DGF \cong \angle BGE$	3. Theorem 4-4
4. $\triangle DGF \cong \triangle BGE$	4. SAS
5. $\angle GDF \cong \angle GBE$ (or $\angle GDC \cong \angle GBA$)	5. CPCTC
6. $\angle DGC \cong \angle BGA$	6. Theorem 4-4
7. $\triangle DGC \cong \triangle BGA$	7. ASA
8. $\overline{AB} \cong \overline{CD}$	8. CPCTC

3.

STATEMENTS	REASONS
1. G is the midpoint of $\overline{AC}$ and $\overline{BD}$.	1. Given
2. $\overline{AG} \cong \overline{CG}$, $\overline{BG} \cong \overline{DG}$	2. Definition of midpoint
3. $\angle AGB \cong \angle CGD$	3. Theorem 4-4
4. $\triangle AGB \cong \triangle CGD$	4. SAS
5. $\angle GAB \cong \angle GCD$ (or $\angle GAE \cong \angle GCF$)	5. CPCTC
6. $\angle EGA \cong \angle FGC$	6. Theorem 4-4
7. $\triangle EGA \cong \triangle FGC$	7. ASA
8. $\overline{EG} \cong \overline{FG}$	8. CPCTC

Page 179 Mixed Review

15.

STATEMENTS	REASONS
1. $\triangle ABC$ with $\overline{AB} \cong \overline{AC}$	1. Given
2. $\overline{AC} \cong \overline{AB}$, $\overline{BC} \cong \overline{CB}$	2. Theorem 4-5
3. $\triangle ABC \cong \triangle ACB$	3. SSS
4. $\angle ABC \cong \angle ACB$	4. CPCTC

Page 182 Exploratory

1. ASA involves two angles and the included side; AAS involves two angles and a nonincluded side. **2.** No, two sides and a nonincluded angle can be congruent without the triangles being congruent.

Page 183 Written

12.

STATEMENTS	REASONS
1. $\angle A$ and $\angle E$ are right angles.	1. Given
2. $\angle A \cong \angle E$	2. Theorem 4-7
3. $\angle ENS \cong \angle ANK$	3. Theorem 4-4
4. N is the midpoint of $\overline{SK}$.	4. Given
5. $\overline{SN} \cong \overline{KN}$	5. Definition of midpoint
6. $\triangle SNE \cong \triangle KNA$	6. AAS

13.

STATEMENTS	REASONS
1. $\overline{AK} \parallel \overline{SE}$, $\overline{SN} \cong \overline{KN}$	1. Given
2. $\angle S \cong \angle K$, $\angle E \cong \angle A$	2. Theorem 4-12
3. $\triangle SNE \cong \triangle KNA$	3. AAS

14.

STATEMENTS	REASONS
1. $\angle E \cong \angle A$, N is the midpoint of $\overline{SK}$.	1. Given
2. $\overline{SN} \cong \overline{KN}$	2. Definition of midpoint
3. $\angle SNE \cong \angle KNA$	3. Theorem 4-4
4. $\triangle SNE \cong \triangle KNA$	4. AAS

16.

STATEMENTS	REASONS
1. $\overline{AB} \parallel \overline{DE}$, $\overline{AC} \cong \overline{EF}$, $\angle ACB \cong \angle EFD$	1. Given
2. $\angle A \cong \angle E$	2. Theorem 4-12
3. $\triangle ABC \cong \triangle EDF$	3. ASA

17.

STATEMENTS	REASONS
1. $\overline{AB} \parallel \overline{DE}$, $\angle B \cong \angle D$, $\overline{BC} \cong \overline{DF}$	1. Given
2. $\angle A \cong \angle E$	2. Theorem 4-12
3. $\triangle ABC \cong \triangle EDF$	3. AAS

20.

STATEMENTS	REASONS
1. $\overline{DB}$ bisects $\angle CDA$, $\angle A \cong \angle C$	1. Given
2. $\angle 3 \cong \angle 2$	2. Definition of angle bisector
3. $\overline{BD} \cong \overline{BD}$	3. Theorem 4-5
4. $\triangle ABD \cong \triangle CBD$	4. AAS

21.

STATEMENTS	REASONS
1. $\overline{BD}$ bisects $\angle ABC$, $\overline{AD} \perp \overline{AB}$, $\overline{BC} \perp \overline{DC}$	1. Given
2. $\angle 1 \cong \angle 4$	2. Definition of angle bisector
3. $\angle A$ and $\angle C$ are right angles.	3. Definition of perpendicular lines
4. $\angle A \cong \angle C$	4. Theorem 4-7
5. $\overline{BD} \cong \overline{BD}$	5. Theorem 4-5
6. $\triangle ABD \cong \triangle CBD$	6. AAS
7. $\overline{AD} \cong \overline{CD}$	7. CPCTC

19.

STATEMENTS	REASONS
1. $\angle 4 \cong \angle 3$, $\angle A \cong \angle C$	1. Given
2. $\overline{BD} \cong \overline{DB}$	2. Theorem 4-5
3. $\triangle ABD \cong \triangle CDB$	3. AAS

Page 183 Challenge

STATEMENTS	REASONS
1. $\overline{TP} \perp \overline{PS}$, $\overline{TP} \parallel \overline{MS}$	1. Given
2. $\overline{MS} \perp \overline{PS}$	2. Theorem 4-10
3. $\angle P$ and $\angle S$ are right angles.	3. Definition of perpendicular lines
4. $\angle P \cong \angle S$	4. Theorem 4-7
5. $\overline{PQ} \cong \overline{SR}$ ($PQ = SR$)	5. Given
6. $PQ + QR = PR$, $SR + RQ = SQ$	6. Segment addition axiom
7. $PQ = PR - QR$, $SR = SQ - RQ$	7. Subtraction axiom
8. $PR - QR = SQ - RQ$	8. Substitution
9. $PR = SQ$ ($\overline{PR} \cong \overline{SQ}$)	9. Addition axiom
10. $\angle T \cong \angle M$	10. Given
11. $\triangle TPR \cong \triangle MSQ$	11. AAS
12. $\overline{TP} \cong \overline{MS}$	12. CPCTC

Page 187   Exploratory   9. $\triangle BDC \cong \triangle CEB$, $\triangle ABE \cong \triangle ACD$, $\triangle DFB \cong \triangle EFC$ **10.** $\overline{AB} \cong \overline{AC}$, $\overline{BE} \cong \overline{CD}$, $\overline{AE} \cong \overline{AD}$, $\angle BAE \cong \angle CAD$, $\angle ABE \cong \angle ACD$, $\angle BEA \cong \angle CDA$ **11.** $\overline{DC} \cong \overline{EB}$, $\overline{CB} \cong \overline{BC}$, $\overline{DB} \cong \overline{EC}$, $\angle CDB \cong \angle BEC$, $\angle DCB \cong \angle EBC$, $\angle CBD \cong \angle BCE$ **12.** $\overline{DF} \cong \overline{EF}$, $\overline{FB} \cong \overline{FC}$, $\overline{DB} \cong \overline{EC}$, $\angle FDB \cong \angle FEC$, $\angle DFB \cong \angle EFC$, $\angle FBD \cong \angle FCE$ **13.** We are given that $\triangle ABE \cong \triangle ACD$. Therefore, by CPCTC, $\angle ABE \cong \angle ACD$. We are also given that $\triangle DCB \cong \triangle EBC$. Therefore, by CPCTC, $\angle CDB \cong \angle BEC$ and $\overline{DB} \cong \overline{EC}$. Thus, by ASA, $\triangle DFB \cong \triangle EFC$. ✔

Pages 187–188   Written

9.

STATEMENTS	REASONS
1. $\overline{BC} \cong \overline{ED}$ ($BC = ED$)	1. Given
2. $BC + CD = CD + ED$	2. Addition axiom
3. $BC + CD = BD$, $CD + ED = EC$	3. Segment addition axiom
4. $BD = EC$ ($\overline{BD} \cong \overline{EC}$)	4. Substitution
5. $\angle A \cong \angle F$, $\angle B \cong \angle E$	5. Given
6. $\triangle ADB \cong \triangle FCE$	6. AAS

10.

STATEMENTS	REASONS
1. $\overline{BC} \cong \overline{DE}$ ($BC = DE$)	1. Given
2. $BC + CD = CD + DE$	2. Addition axiom
3. $BC + CD = BD$, $CD + DE = EC$	3. Segment addition axiom
4. $BD = EC$ ($\overline{BD} \cong \overline{EC}$)	4. Substitution
5. $\overline{AD} \perp \overline{BE}$, $\overline{FC} \perp \overline{BE}$	5. Given
6. $\angle ADB$ and $\angle FCE$ are right angles.	6. Definition of perpendicular lines
7. $\angle ADB \cong \angle FCE$	7. Theorem 4-7
8. $\overline{AD} \cong \overline{FC}$	8. Given
9. $\triangle ADB \cong \triangle FCE$	9. SAS

11.

STATEMENTS	REASONS
1. $\overline{BC} \cong \overline{DE}$ (BC = DE)	1. Given
2. BC + CD = CD + DE	2. Addition axiom
3. BC + CD = BD CD + DE = EC	3. Segment addition axiom
4. BD = EC ($\overline{BD} \cong \overline{EC}$)	4. Substitution
5. $\overline{AD} \perp \overline{BE}, \overline{FC} \perp \overline{BE}$	5. Given
6. $\angle ADB$ and $\angle FCE$ are right angles.	6. Definition of perpendicular lines
7. $\angle ADB \cong \angle FCE$	7. Theorem 4-7
8. $\angle B \cong \angle E$	8. Given
9. $\triangle ABD \cong \triangle FEC$	9. ASA
10. $\overline{AB} \cong \overline{FE}$	10. CPCTC

12.

STATEMENTS	REASONS
1. $\overline{SN} \cong \overline{KA}, \angle 2 \cong \angle 3, \overline{EN} \cong \overline{EA}$	1. Given
2. SN + NA = NA + KA	2. Addition axiom
3. SN + NA = SA NA + KA = KN	3. Segment addition axiom
4. SA = KN ($\overline{SA} \cong \overline{KN}$)	4. Substitution
5. $\triangle SEA \cong \triangle KEN$	5. SAS

13.

STATEMENTS	REASONS
1. $\triangle ENS \cong \triangle EAK$	1. Given
2. $\angle SEN \cong \angle KEA$ ($m\angle SEN = m\angle KEA$)	2. CPCTC
3. $m\angle SEN + m\angle NEA = m\angle NEA + m\angle AEK$	3. Addition axiom
4. $m\angle SEN + m\angle NEA = m\angle SEA$ $m\angle NEA + m\angle AEK = m\angle KEN$	4. Angle addition axiom
5. $m\angle SEA = m\angle KEN$ ($\angle SEA \cong \angle KEN$)	5. Substitution

14.

STATEMENTS	REASONS
1. $\angle 1 \cong \angle 4, \overline{SN} \cong \overline{KA}, EN = EA$ ($\overline{EN} \cong \overline{EA}$)	1. Given
2. $\angle 1$ is supplementary to $\angle 2$. $\angle 3$ is supplementary to $\angle 4$.	2. Supplement axiom
3. $\angle 2 \cong \angle 3$	3. Theorem 4-2
4. SN + NA = NA + KA	4. Addition axiom
5. SN + NA = SA NA + KA = KN	5. Segment addition axiom
6. SA = KN ($\overline{SA} \cong \overline{KN}$)	6. Substitution
7. $\triangle SEA \cong \triangle KEN$	7. SAS

15.

STATEMENTS	REASONS
1. $\overline{SA} \cong \overline{NK}, \angle S \cong \angle K, \angle 2 \cong \angle 3$	1. Given
2. $\triangle SEA \cong \triangle KEN$	2. ASA
3. $\angle SEA \cong \angle KEN$ ($m\angle SEA = m\angle KEN$)	3. CPCTC
4. $m\angle SEN + m\angle NEA = m\angle SEA$; $m\angle NEA + m\angle KEA = m\angle KEN$	4. Angle addition axiom
5. $m\angle SEN + m\angle NEA = m\angle NEA + m\angle KEA$	5. Substitution
6. $m\angle SEN \cong m\angle KEA$ ($\angle SEN \cong \angle KEA$)	6. Subtraction axiom

16.

STATEMENTS	REASONS
1. $\overline{AD} \cong \overline{AE}, \angle CDA \cong \angle BEA$	1. Given
2. $\angle A \cong \angle A$	2. Theorem 4-6
3. $\triangle ADC \cong \triangle AEB$	3. ASA

17.

STATEMENTS	REASONS
1. $\overline{CD} \perp \overline{AB}, \overline{BE} \perp \overline{AC}, \overline{AD} \cong \overline{AE}$	1. Given
2. $\angle ADC$ and $\angle AEB$ are right angles.	2. Definition of perpendicular lines
3. $\angle ADC \cong \angle AEB$	3. Theorem 4-7
4. $\angle A \cong \angle A$	4. Theorem 4-6
5. $\triangle ABE \cong \triangle ACD$	5. ASA
6. $\angle ABE \cong \angle ACD$	6. CPCTC

18.

STATEMENTS	REASONS
1. $\overline{DB} \cong \overline{EC}$, $\angle DBC \cong \angle ECB$	1. Given
2. $\overline{BC} \cong \overline{CB}$	2. Theorem 4-5
3. $\triangle DBC \cong \triangle ECB$	3. SAS
4. $\overline{DC} \cong \overline{EB}$	4. CPCTC

19.

STATEMENTS	REASONS
1. $\overline{AB} \cong \overline{AC}$, $\overline{AD} \cong \overline{AE}$	1. Given
2. $\angle A \cong \angle A$	2. Theorem 4-6
3. $\triangle ABE \cong \triangle ACD$	3. SAS
4. $\angle ABE \cong \angle ACD$	4. CPCTC

20.

STATEMENTS	REASONS
1. $\overline{BD} \cong \overline{CE}$, $\overline{AD} \cong \overline{AE}$	1. Given
2. $BD + AD = CE + AE$	2. Addition axiom
3. $BD + AD = BA$ $CE + AE = CA$	3. Segment addition axiom
4. $BA = CA$	4. Substitution
5. $\angle A \cong \angle A$	5. Theorem 4-6
6. $\triangle BAE \cong \triangle CAD$	6. SAS
7. $\overline{BE} \cong \overline{CD}$	7. CPCTC

21.

STATEMENTS	REASONS
1. $\angle DAE \cong \angle DCE$, $\overline{BD} \perp \overline{AC}$	1. Given
2. $\angle AED$, $\angle CED$, $\angle BEA$, and $\angle BEC$ are right angles.	2. Definition of perpendicular lines
3. $\angle AED \cong \angle CED$ $\angle BEA \cong \angle BEC$	3. Theorem 4-7
4. $\overline{DE} \cong \overline{DE}$, $\overline{BE} \cong \overline{BE}$	4. Theorem 4-5
5. $\triangle AED \cong \triangle CED$	5. AAS
6. $\overline{AE} \cong \overline{CE}$	6. CPCTC
7. $\triangle AEB \cong \triangle CEB$	7. SAS
8. $\overline{AD} \cong \overline{CD}$, $\angle ADE \cong \angle CDE$ (or $\angle ADB \cong \angle CDB$), $\angle ABE \cong \angle CBE$ (or $\angle ABD \cong \angle CBD$)	8. CPCTC
9. $\triangle ABD \cong \triangle CBD$	9. AAS

22.

STATEMENTS	REASONS
1. $\angle BEA \cong \angle BEC$, $\angle ABE \cong \angle CBE$	1. Given
2. $\overline{BE} \cong \overline{BE}$	2. Theorem 4-5
3. $\triangle AEB \cong \triangle CEB$	3. ASA
4. $\overline{AB} \cong \overline{CB}$	4. CPCTC
5. $\overline{BD} \cong \overline{BD}$	5. Theorem 4-5
6. $\triangle ABD \cong \triangle CBD$	6. SAS
7. $\overline{AD} \cong \overline{CD}$ (AC = CD)	7. CPCTC

23.

STATEMENTS	REASONS
1. $\overline{AB} \cong \overline{CB}$, $\angle ABD \cong \angle CBD$	1. Given
2. $\overline{BD} \cong \overline{BD}$, $\overline{BE} \cong \overline{BE}$	2. Theorem 4-5
3. $\triangle ABD \cong \triangle CBD$, $\triangle ABE \cong \triangle CBE$	3. SAS
4. $\overline{AD} \cong \overline{CD}$, $\overline{AE} \cong \overline{CE}$	4. CPCTC
5. $\overline{ED} \cong \overline{ED}$	5. Theorem 4-5
6. $\triangle AED \cong \triangle CED$	6. SSS

24.

STATEMENTS	REASONS
1. $\overline{BD} \perp \overline{AC}$, $\angle BAC \cong \angle BCA$	1. Given
2. $\angle AEB$, $\angle CEB$, $\angle AED$, and $\angle CED$ are right angles.	2. Definition of perpendicular lines
3. $\angle AEB \cong \angle CEB$, $\angle AED \cong \angle CED$	3. Theorem 4-7
4. $\overline{EB} \cong \overline{EB}$, $\overline{ED} \cong \overline{ED}$	4. Theorem 4-5
5. $\triangle AEB \cong \triangle CEB$	5. AAS
6. $\overline{AE} \cong \overline{CE}$	6. CPCTC
7. $\triangle AED \cong \triangle CED$	7. SAS
8. $\angle DAE \cong \angle DCE$ (or $\angle DAC \cong \angle DCA$)	8. CPCTC

Page 188 Challenge

STATEMENTS	REASONS
1. $\angle 1 \cong \angle 2$, $\overline{HB} \cong \overline{TD}$, $\overline{TR} \perp \overline{AD}$, $\overline{HK} \perp \overline{AB}$	1. Given
2. $\angle TRD$ and $\angle HKB$ are right angles.	2. Definition of perpendicular lines
3. $\angle TRD \cong \angle HKB$	3. Theorem 4-7
4. $\triangle HKB \cong \triangle TRD$	4. AAS
5. $\angle H \cong \angle T$	5. CPCTC
6. $HD + DB = HB$ $DB + TB = TD$	6. Segment addition axiom
7. $HD + DB = DB + TB$	7. Substitution
8. $HD = TB$ ($\overline{HD} \cong \overline{TB}$)	8. Subtraction axiom

STATEMENTS	REASONS
9. $\angle HDV$ is supplementary to $\angle 1$. $\angle TBC$ is supplementary to $\angle 2$.	9. Supplement axiom
10. $\angle HDV \cong \angle TBC$	10. Theorem 4-2
11. $\triangle HDV \cong \triangle TBC$	11. ASA
12. $\overline{VD} \cong \overline{CB}$, $\overline{RD} \cong \overline{KB}$	12. CPCTC
13. $VD + RV = RD$ $CB + KC = KB$	13. Segment addition axiom
14. $VD + RV = CB + KC$	14. Substitution
15. $RV = KC$	15. Subtraction axiom

Page 191 Written

8.

STATEMENTS	REASONS
1. $\overline{AB} \cong \overline{AC}$	1. Given
2. $\angle 3 \cong \angle 4$	2. Theorem 5-2
3. $\angle 1$ is supplementary to $\angle 3$. $\angle 2$ is supplementary to $\angle 4$.	3. Supplement axiom
4. $\angle 1 \cong \angle 2$	4. Theorem 4-2

9.

STATEMENTS	REASONS
1. $\overline{AB} \cong \overline{AC}$, $\angle EDB \cong \angle FDC$, D is the midpoint of $\overline{BC}$.	1. Given
2. $\angle 3 \cong \angle 4$	2. Theorem 5-2
3. $\overline{BD} \cong \overline{CD}$	3. Definition of midpoint
4. $\triangle EBD \cong \triangle FCD$	4. ASA
5. $\overline{EB} \cong \overline{FC}$	5. CPCTC

10.

STATEMENTS	REASONS
1. $\angle 3 \cong \angle 4$, $\overline{EB} \cong \overline{FC}$	1. Given
2. $\overline{AB} \cong \overline{AC}$ ($AB = AC$)	2. Theorem 5-3
3. $AE + EB = AB$ $AF + FC = AC$	3. Segment addition axiom
4. $AE + EB = AF + FC$	4. Substitution
5. $AE = AF$ ($\overline{AE} \cong \overline{AF}$)	5. Subtraction axiom

11.

STATEMENTS	REASONS
1. $\angle BDE \cong \angle CDF$, $\overline{DE} \cong \overline{DF}$, $\angle AED \cong \angle AFD$	1. Given
2. $\angle AED$ and $\angle DEB$ are supplementary. $\angle AFD$ and $\angle DFC$ are supplementary.	2. Supplement axiom
3. $\angle DEB \cong \angle DFC$	3. Theorem 4-2
4. $\triangle BDE \cong \triangle CDF$	4. ASA
5. $\angle 3 \cong \angle 4$	5. CPCTC
6. $\overline{AC} \cong \overline{AB}$	6. Theorem 5-3
7. $\triangle ABC$ is isosceles.	7. Definition of isosceles triangle.

12.

STATEMENTS	REASONS
1. $\overline{AB} \cong \overline{AC}$, $\overline{BE}$ bisects $\angle CBA$, $\overline{CD}$ bisects $\angle BCA$.	1. Given
2. $\angle CBA \cong \angle BCA$	2. Theorem 5-2
3. $\angle 1 \cong \angle 3$, $\angle 2 \cong \angle 4$	3. Definition of angle bisector
4. $m\angle 1 + m\angle 3 = m\angle CBA$, $m\angle 2 + m\angle 4 = m\angle BCA$	4. Angle addition axiom
5. $m\angle 1 + m\angle 3 = m\angle 2 + m\angle 4$	5. Substitution
6. $2(m\angle 1) = 2(m\angle 2)$	6. Substitution
7. $m\angle 1 = m\angle 2$ ($\angle 1 \cong \angle 2$)	7. Division axiom

13.

STATEMENTS	REASONS
1. $\overline{AB} \cong \overline{AC}$, $\overline{BD} \cong \overline{CE}$	1. Given
2. $\angle ACB \cong \angle ABC$ (or $\angle ECB \cong \angle DBC$)	2. Theorem 5-2
3. $\overline{CB} \cong \overline{BC}$	3. Theorem 4-5
4. $\triangle DBC \cong \triangle ECB$	4. SAS
5. $\angle 3 \cong \angle 4$	5. CPCTC

14.

STATEMENTS	REASONS
1. $\overline{AD} \cong \overline{AE}$, $\overline{BD} \cong \overline{CE}$	1. Given
2. $\angle AED \cong \angle ADE$	2. Theorem 5-2
3. $\angle AED$ and $\angle DEC$ are supplementary. $\angle ADE$ and $\angle EDB$ are supplementary.	3. Supplement axiom
4. $\angle DEC \cong \angle EDB$	4. Theorem 4-2
5. $\overline{DE} \cong \overline{ED}$	5. Theorem 4-5
6. $\triangle DEC \cong \triangle EDB$	6. SAS
7. $\angle CDE \cong \angle BED$	7. CPCTC

15.

STATEMENTS	REASONS
1. $\overline{BD} \cong \overline{CE}$, $\overline{BE} \cong \overline{CD}$	1. Given
2. $\overline{BC} \cong \overline{CB}$	2. Theorem 4-5
3. $\triangle BDC \cong \triangle CEB$	3. SSS
4. $\angle 3 \cong \angle 4$	4. CPCTC
5. $\overline{FC} \cong \overline{FB}$	5. Theorem 5-3
6. $\triangle BCF$ is isosceles.	6. Definition of isosceles triangle

16. Given: $\triangle ABC$ is equilateral.
Prove: $\triangle ABC$ is equiangular.

STATEMENTS	REASONS
1. $\triangle ABC$ is equilateral.	1. Given
2. $\overline{AB} \cong \overline{BC} \cong \overline{AC}$	2. Definition of equilateral
3. $\angle C \cong \angle B$, $\angle B \cong \angle A$, $\angle C \cong \angle A$	3. Theorem 5-2
4. $\triangle ABC$ is equiangular.	4. Definition of equiangular

17. Given: $\triangle ABC$ is equiangular.
Prove: $\triangle ABC$ is equilateral.

STATEMENTS	REASONS
1. $\triangle ABC$ is equiangular.	1. Given
2. $\angle A \cong \angle B \cong \angle C$	2. Definition of equiangular
3. $\overline{BC} \cong \overline{AC}$, $\overline{AC} \cong \overline{AB}$, $\overline{BC} \cong \overline{AB}$	3. Theorem 5-3
4. $\triangle ABC$ is equilateral.	4. Definition of equilateral

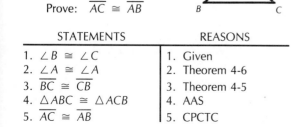

18. Given: $\angle B \cong \angle C$
Prove: $\overline{AC} \cong \overline{AB}$

STATEMENTS	REASONS
1. $\angle B \cong \angle C$	1. Given
2. $\angle A \cong \angle A$	2. Theorem 4-6
3. $\overline{BC} \cong \overline{CB}$	3. Theorem 4-5
4. $\triangle ABC \cong \triangle ACB$	4. AAS
5. $\overline{AC} \cong \overline{AB}$	5. CPCTC

19. Given: $\overline{AB} \cong \overline{AC}$

Prove: $\angle B \cong \angle C$

STATEMENTS	REASONS
1. $\overline{AB} \cong \overline{AC}$	1. Given
2. Draw median $\overline{AD}$.	2. Axiom 4-2
3. D is the midpoint of $\overline{BC}$.	3. Definition of median
4. $\overline{BD} \cong \overline{CD}$	4. Definition of midpoint
5. $\overline{AD} \cong \overline{AD}$	5. Theorem 4-5
6. $\triangle ABD \cong \triangle ACD$	6. SSS
7. $\overline{AB} \cong \overline{AC}$	7. CPCTC

20. Given: $\triangle ABC$ with median $\overline{AD}$ and altitude $\overline{AD}$

Prove: $\triangle ABC$ is isosceles.

STATEMENTS	REASONS
1. $\triangle ABC$ with median $\overline{AD}$ and altitude $\overline{AD}$	1. Given
2. D is the midpoint of $\overline{BC}$.	2. Definition of median
3. $\overline{BD} \cong \overline{CD}$	3. Definition of midpoint
4. $\overline{AD} \perp \overline{BC}$	4. Definition of altitude
5. $\angle ADB$ and $\angle ADC$ are right angles.	5. Definition of perpendicular lines
6. $\angle ADB \cong \angle ADC$	6. Theorem 4-7
7. $\overline{AD} \cong \overline{AD}$	7. Theorem 4-5
8. $\triangle ADB \cong \triangle ADC$	8. SAS
9. $\overline{AB} \cong \overline{AC}$	9. CPCTC
10. $\triangle ABC$ is isosceles.	10. Definition of isosceles triangle

Page 194 Written

2. Given: Isosceles $\triangle ABC$ with altitude $\overline{BD}$

Prove: $\overline{BD}$ bisects $\angle ABC$.

STATEMENTS	REASONS
1. Isosceles $\triangle ABC$ with altitude $\overline{BD}$	1. Given
2. $\overline{AB} \cong \overline{CB}$	2. Definition of isosceles triangle
3. $\overline{BD} \perp \overline{AC}$	3. Definition of altitude
4. $\angle BDA$ and $\angle BDC$ are right angles.	4. Definition of perpendicular lines
5. $\overline{BD} \cong \overline{BD}$	5. Theorem 4-5
6. $\triangle ABD \cong \triangle CBD$	6. HL
7. $\angle ABD \cong \angle CBD$	7. CPCTC
8. $\overline{BD}$ bisects $\angle ABC$.	8. Definition of angle bisector

3. Given: Isosceles $\triangle ABC$ with altitude $\overline{BD}$

Prove: $\overline{BD}$ is median of $\triangle ABC$.

STATEMENTS	REASONS
1. Isosceles $\triangle ABC$ with altitude $\overline{BD}$	1. Given
2. $\overline{AB} \cong \overline{CB}$	2. Definition of isosceles triangle
3. $\overline{BD} \perp \overline{AC}$	3. Definition of altitude
4. $\angle BDA$ and $\angle BDC$ are right angles.	4. Definition of perpendicular lines
5. $\overline{BD} \cong \overline{BD}$	5. Theorem 4-5
6. $\triangle ABD \cong \triangle CBD$	6. HL
7. $\overline{AD} \cong \overline{CD}$	7. CPCTC
8. D is the midpoint of $\overline{AC}$.	8. Definition of midpoint
9. $\overline{BD}$ is a median of $\triangle ABC$.	9. Definition of median

4.

STATEMENTS	REASONS
1. $\overline{MD} \perp \overline{AB}$, $\overline{ME} \perp \overline{AC}$, M is the midpoint of $\overline{BC}$, $\overline{DB} \cong \overline{EC}$	1. Given
2. $\angle BDM$ and $\angle CEM$ are right angles.	2. Definition of perpendicular lines
3. $\overline{BM} \cong \overline{CM}$	3. Definition of midpoint
4. $\triangle DMB \cong \triangle EMC$	4. HL

5.

STATEMENTS	REASONS
1. $\overline{AB} \cong \overline{AC}$, $\overline{MD} \perp \overline{AB}$, $\overline{ME} \perp \overline{AC}$, M is the midpoint of $\overline{BC}$.	1. Given
2. $\angle BDM$ and $\angle CEM$ are right angles.	2. Definition of perpendicular lines
3. $\angle BDM \cong \angle CEM$	3. Theorem 4-7
4. $\overline{BM} \cong \overline{CM}$	4. Definition of midpoint
5. $\angle B \cong \angle C$	5. Theorem 5-2
6. $\triangle BDM \cong \triangle CEM$	6. AAS
7. $\overline{DM} \cong \overline{EM}$	7. CPCTC

6. Given: Isosceles $\triangle ABC$ with $\overline{AB} \cong \overline{CB}$, and altitudes $\overline{CD}$ and $\overline{AE}$
Prove: $\overline{CD} \cong \overline{AE}$

STATEMENTS	REASONS
1. Isosceles $\triangle ABC$ with $\overline{AB} \cong \overline{CB}$ and altitudes $\overline{CD}$ and $\overline{AE}$	1. Given
2. $\overline{CD} \perp \overline{AB}$, $\overline{AE} \perp \overline{CB}$	2. Definition of altitude
3. $\angle ADC$ and $\angle CEA$ are right angles.	3. Definition of perpendicular lines
4. $\angle ADC \cong \angle CEA$	4. Theorem 4-7
5. $\angle BAC \cong \angle BCA$ (or $\angle DAC \cong \angle ECA$)	5. Theorem 5-2
6. $\overline{AC} \cong \overline{CA}$	6. Theorem 4-5
7. $\triangle DAC \cong \triangle ECA$	7. AAS
8. $\overline{CD} \cong \overline{AE}$	8. CPCTC

7.

STATEMENTS	REASONS
1. $\overline{PT} \perp \overline{QS}$, $\overline{RW} \perp \overline{QS}$, $\overline{QT} \cong \overline{SW}$, $\overline{PS} \cong \overline{QR}$	1. Given
2. $\angle PTS$ and $\angle RWQ$ are right angles.	2. Definition of perpendicular lines
3. $QT + TW = TW + SW$	3. Addition axiom
4. $QT + TW = QW$ $TW + SW = ST$	4. Segment addition axiom
5. $QW = ST$ ($\overline{QW} \cong \overline{ST}$)	5. Substitution
6. $\triangle PTS \cong \triangle RWQ$	6. HL

8.

STATEMENTS	REASONS
1. $\overline{PT} \perp \overline{QS}$, $\overline{RW} \perp \overline{QS}$ $\overline{PT} \cong \overline{RW}$, $\overline{PS} \cong \overline{RQ}$	1. Given
2. $\angle PTS$ and $\angle RWQ$ are right angles.	2. Definition of perpendicular lines
3. $\triangle PTS \cong \triangle RWQ$	3. HL
4. $\angle S \cong \angle Q$	4. CPCTC
5. $\overline{PS} \parallel \overline{QR}$	5. Theorem 4-13

9.

STATEMENTS	REASONS
1. $\overline{PS} \parallel \overline{QR}$, $\overline{PT} \parallel \overline{RW}$, $\overline{PT} \cong \overline{RW}$	1. Given
2. $\angle S \cong \angle Q$, $\angle PTS \cong \angle RWQ$	2. Theorem 4-12
3. $\triangle PTS \cong \triangle RWQ$	3. AAS
4. $\overline{ST} \cong \overline{QW}$ ($ST = QW$)	4. CPCTC
5. $SW + TW = ST$ $TW + QT = QW$	5. Segment addition axiom
6. $SW + TW = TW + QT$	6. Substitution
7. $QT = SW$	7. Subtraction axiom

10.

STATEMENTS	REASONS
1. $\overline{PT} \perp \overline{QS}$, $\overline{PT} \parallel \overline{RW}$, $QT = SW$, $PS = RQ$	1. Given
2. $\angle PTS$ is a right angle.	2. Definition of perpendicular lines
3. $\angle PTS \cong \angle RWQ$	3. Theorem 4-12
4. $\angle RWQ$ is a right angle.	4. Substitution
5. $QT + TW = TW + SW$	5. Addition axiom
6. $QT + TW = QW$ $TW + SW = ST$	6. Segment addition axiom
7. $QW = ST$ ($\overline{QW} \cong \overline{ST}$)	7. Substitution
8. $\triangle PST \cong \triangle RQW$	8. HL
9. $\overline{PT} \cong \overline{RW}$ ($PT = RW$)	9. CPCTC

11.

STATEMENTS	REASONS
1. $\overline{RE} \perp \overline{AV}$, $\overline{VL} \perp \overline{AR}$, $\overline{VL} \cong \overline{RE}$	1. Given
2. $\angle VER$ and $\angle RLV$ are right angles.	2. Definition of perpendicular lines
3. $\overline{VR} \cong \overline{RV}$	3. Theorem 4-5
4. $\triangle VER \cong \triangle RLV$	4. HL

12.

STATEMENTS	REASONS
1. $\angle REV \cong \angle REA$, $\angle RLV \cong \angle VLA$, $\overline{EV} \cong \overline{LR}$	1. Given
2. $\angle REV$ and $\angle RLV$ are right angles.	2. Theorem 4-8
3. $\overline{VR} \cong \overline{RV}$	3. Theorem 4-5
4. $\triangle VER \cong \triangle RLV$	4. HL

13.

STATEMENTS	REASONS
1. $\angle REV \cong \angle REA$, $\angle RLV \cong \angle VLA$ $\overline{EV} \cong \overline{LR}$	1. Given
2. $\angle REV$ and $\angle RLV$ are right angles.	2. Theorem 4-8
3. $\overline{VR} \cong \overline{RV}$	3. Theorem 4-5
4. $\triangle VER \cong \triangle RLV$	4. HL
5. $\angle EVR \cong \angle LRV$ (or $\angle AVR \cong \angle ARV$)	5. CPCTC
6. $\overline{AR} \cong \overline{AV}$	6. Theorem 5-3
7. $\triangle VAR$ is isosceles.	7. Definition of isosceles triangle

14.

STATEMENTS	REASONS
1. $\angle REV \cong \angle REA$, $\angle RLV \cong \angle VLA$, $\overline{EV} \cong \overline{LR}$	1. Given
2. $\angle REV$ and $\angle RLV$ are right angles.	2. Theorem 4-8
3. $\overline{VR} \cong \overline{RV}$	3. Theorem 4-5
4. $\triangle VER \cong \triangle RLV$	4. HL
5. $\angle ERV \cong \angle LVR$ (or $\angle KRV \cong \angle KVR$)	5. CPCTC
6. $\overline{KV} \cong \overline{KR}$	6. Theorem 5-3
7. $\triangle VKR$ is isosceles.	7. Definition of isosceles triangle

15.

STATEMENTS	REASONS
1. $\angle REV \cong \angle REA$, $\angle RLV \cong \angle VLA$, $\overline{EV} \cong \overline{LR}$	1. Given
2. $\angle REV$ and $\angle RLV$ are right angles.	2. Theorem 4-8
3. $\angle REV \cong \angle RLV$	3. Theorem 4-7
4. $\angle EKV \cong \angle LKR$	4. Theorem 4-4
5. $\triangle VKE \cong \triangle RKL$	5. AAS

16.

STATEMENTS	REASONS
1. $\angle REV \cong \angle REA$, $\angle RLV \cong \angle VLA$ $\overline{EV} \cong \overline{LR}$	1. Given
2. $\angle REV$, $\angle REA$, $\angle RLV$, and $\angle VLA$ are right angles.	2. Theorem 4-8
3. $\overline{VR} \cong \overline{RV}$	3. Theorem 4-5
4. $\triangle VER \cong \triangle RLV$	4. HL
5. $\overline{ER} \cong \overline{LV}$	5. CPCTC
6. $\angle REA \cong \angle VLA$	6. Theorem 4-7
7. $\angle A \cong \angle A$	7. Reflexive axiom
8. $\triangle VAL \cong \triangle RAE$	8. AAS

17.

STATEMENTS	REASONS
1. $\overline{AV} \cong \overline{AR}$, $\overline{EV} \cong \overline{LR}$	1. Given
2. $\angle AVR \cong \angle ARV$ (or $\angle EVR \cong \angle LRV$)	2. Theorem 5-2
3. $\overline{VR} \cong \overline{RV}$	3. Theorem 4-5
4. $\triangle EVR \cong \triangle LRV$	4. SAS
5. $\overline{ER} \cong \overline{LV}$ ($ER = LV$)	5. CPCTC

Page 197 Exploratory **1.** always; addition property **2.** never; If $a < b$, then $b > a$. **3.** always; addition property **4.** sometimes; if $c < 0$ **5.** always; transitive property **6.** always; multiplication property **7.** always; addition property **8.** always; addition property **9.** always; multiplication property **10.** sometimes; x can be equal to y

Page 198 Written

22.

STATEMENTS	REASONS
1. $BC = DE$	1. Given
2. $AC = AB + BC$	2. Segment addition axiom
3. $AC = AB + DE$	3. Substitution
4. $AC > DE$	4. Definition of greater than

23.

STATEMENTS	REASONS
1. $\triangle PQR$ with exterior $\angle QRS$	1. Given
2. $m\angle PQR = m\angle Q + m\angle P$	2. Exterior Angle Theorem
3. $m\angle PQR > m\angle Q$ $m\angle PQR > m\angle P$	3. Definition of greater than

24.

STATEMENTS	REASONS
1. $QR > QP$	1. Given
2. $m\angle P > m\angle QRP$	2. Theorem 5-5
3. $m\angle SRQ > m\angle P$	3. Exercise 23
4. $m\angle SRQ > m\angle QRP$	4. Transitive property

25.

STATEMENTS	REASONS
1. $\triangle ABC$ with right $\angle C$	1. Given
2. $m\angle C = 90$	2. Definition of right angle
3. $\angle A$ and $\angle B$ are complementary.	3. Theorem 4-23
4. $90 = m\angle A + m\angle B$	4. Definition of complementary
5. $90 > m\angle A$	5. Definition of greater than
6. $m\angle C > m\angle A$	6. Substitution
7. $AB > BC$	7. Theorem 5-6

Page 200 Written

13.

STATEMENTS	REASONS
1. Equilateral $\triangle ABC$	1. Given
2. $\angle CAB \cong \angle B \cong \angle BCA$	2. An equilateral triangle is also equiangular.
3. $m\angle DAB = m\angle DAC + m\angle CAB$	3. Angle addition axiom
4. $m\angle DAB = m\angle DAC + m\angle B$	4. Substitution
5. $m\angle DAB > m\angle B$	5. Definition of greater than
6. $BD > AD$	6. Theorem 5-6

14.

STATEMENTS	REASONS
1. Equilateral $\triangle ABC$	1. Given
2. $\angle CAB \cong \angle B \cong \angle BCA$	2. An equilateral triangle is also equiangular.
3. $m\angle BCA = m\angle D + m\angle DAC$	3. Exterior Angle Theorem
4. $m\angle B = m\angle D + m\angle DAC$	4. Substitution
5. $m\angle B > m\angle D$	5. Definition of greater than
6. $AD > AB$	6. Theorem 5-6

Pages 204–205 Chapter Review

2.

STATEMENTS	REASONS
1. $\overline{ST} \cong \overline{UT}, \angle 1 \cong \angle 2$	1. Given
2. $\overline{RT} \cong \overline{RT}$	2. Theorem 4-5
3. $\triangle RST \cong \triangle RUT$	3. SAS

4.

STATEMENTS	REASONS
1. $\angle 2 \cong \angle 4$, $\overline{JK} \cong \overline{NM}$	1. Given
2. $\overline{JK} \parallel \overline{NM}$	2. Theorem 4-13
3. $\angle 1 \cong \angle 3$	3. Theorem 4-12
4. $\triangle KLJ \cong \triangle MLN$	4. ASA

3.

STATEMENTS	REASONS
1. L is the midpoint of $\overline{KM}, \overline{JK} \perp \overline{KM}, \overline{NM} \perp \overline{KM}$	1. Given
2. $\overline{KL} \cong \overline{ML}$	2. Definition of midpoint
3. $\angle 2$ and $\angle 4$ are right angles.	3. Definition of perpendicular lines
4. $\angle 2 \cong \angle 4$	4. Theorem 4-7
5. $\angle KLJ \cong \angle MLN$	5. Theorem 4-4
6. $\triangle KLJ \cong \triangle MLN$	6. ASA

5.

STATEMENTS	REASONS
1. $\overline{AB} \perp \overline{BD}, \overline{AF} \perp \overline{FD}$, $\overline{AB} \cong \overline{AF}, \angle 1 \cong \angle 4$	1. Given
2. $\angle ABC$ and $\angle AFE$ are right angles.	2. Definition of perpendicular lines
3. $\angle ABC \cong \angle AFE$	3. Theorem 4-7
4. $\triangle ABC \cong \triangle AFE$	4. ASA
5. $\overline{AC} \cong \overline{AE}$	5. CPCTC

6.

STATEMENTS	REASONS
1. $\overline{AB} \cong \overline{AF}, \angle 1 \cong \angle 4, \angle 2 \cong \angle 3$	1. Given
2. $m\angle 1 = m\angle 4, m\angle 2 = m\angle 3$	2. Definition of congruent
3. $m\angle 1 + m\angle 2 = m\angle 4 + m\angle 3$	3. Addition axiom
4. $m\angle 1 + m\angle 2 = m\angle BAD$ $m\angle 4 + m\angle 3 = m\angle FAD$	4. Angle addition axiom
5. $m\angle BAD = m\angle FAD$ ($\angle BAD \cong \angle FAD$)	5. Substitution
6. $\overline{AD} \cong \overline{AD}$	6. Theorem 4-5
7. $\triangle BAD \cong \triangle FAD$	7. SAS
8. $\overline{BD} \cong \overline{FD}$	8. CPCTC

7. yes; by Theorem 4-5, $\overline{AD} \cong \overline{AD}$. Therefore, by SAS, $\triangle ACD \cong \triangle ABD$. Then, by CPCTC, $\overline{AC} \cong \overline{AB}$ (or $AC = AB$).

9.

STATEMENTS	REASONS
1. Median and altitude $\overline{AD}$	1. Given
2. D is the midpoint of $\overline{BC}$.	2. Definition of median
3. $\overline{BD} \cong \overline{CD}$	3. Definition of midpoint
4. $\overline{AD} \perp \overline{CB}$	4. Definition of altitude
5. $\angle ADC$ and $\angle ADB$ are right angles.	5. Definition of perpendicular lines
6. $\angle ADC \cong \angle ADB$	6. Theorem 4-7
7. $\overline{AD} \cong \overline{AD}$	7. Theorem 4-5
8. $\triangle ADC \cong \triangle ADB$	8. SAS
9. $\overline{AC} \cong \overline{AB}$	9. CPCTC
10. $\triangle ABC$ is isosceles.	10. Definition of isosceles triangle

17.

STATEMENTS	REASONS
1. Equilateral $\triangle ABC$	1. Given
2. $\overline{AB} \cong \overline{BC} \cong \overline{CA}$	2. Definition of equilateral
3. $DB = DC + BC$	3. Segment addition axiom
4. $DB > BC$	4. Definition of greater than
5. $DB > CA$	5. Substitution

Page 206 Chapter Test

1.

STATEMENTS	REASONS
1. $\angle Y \cong \angle V$, $\angle YZX \cong \angle VZW$, $\overline{YZ} \cong \overline{VZ}$	1. Given
2. $\triangle XYZ \cong \triangle WYZ$	2. ASA

2.

STATEMENTS	REASONS
1. $\overline{AB} \cong \overline{CD}, \overline{AD} \cong \overline{CB}$	1. Given
2. $\overline{DB} \cong \overline{BD}$	2. Theorem 4-5
3. $\triangle ABD \cong \triangle CDB$	3. SSS
4. $\angle BDA \cong \angle DBC$	4. CPCTC
5. $\overline{AD} \parallel \overline{BC}$	5. Theorem 4-13

3.

STATEMENTS	REASONS
1. $\overline{DA} \perp \overline{AB}$, $\overline{BC} \perp \overline{CD}$, $\overline{AD} \cong \overline{CB}$	1. Given
2. $\angle DAB$ and $\angle BCD$ are right angles.	2. Definition of perpendicular lines
3. $\overline{DB} \cong \overline{BD}$	3. Theorem 4-5
4. $\triangle DAB \cong \triangle BCA$	4. HL
5. $\angle 1 \cong \angle 2$	5. CPCTC

4.

STATEMENTS	REASONS
1. $\overline{AB} \parallel \overline{DC}$, $\overline{AD} \parallel \overline{BC}$	1. Given
2. $\angle 1 \cong \angle 2$, $\angle ADB \cong \angle CBD$	2. Theorem 4-12
3. $\overline{DB} \cong \overline{BD}$	3. Theorem 4-5
4. $\triangle ADB \cong \triangle CBD$	4. ASA
5. $\angle DAB \cong \angle BCD$	5. CPCTC

5.

STATEMENTS	REASONS
1. $\overline{AB} \cong \overline{DC}$, $\overline{AC} \cong \overline{DB}$	1. Given
2. $\overline{BC} \cong \overline{CB}$	2. Theorem 4-5
3. $\triangle ABC \cong \triangle DCB$	3. SSS
4. $\angle A \cong \angle D$	4. CPCTC

6.

STATEMENTS	REASONS
1. $\overline{AB} \cong \overline{DC}$, $\angle ABC \cong \angle DCB$	1. Given
2. $\overline{BC} \cong \overline{CB}$	2. Theorem 4-5
3. $\triangle ABC \cong \triangle DCB$	3. SAS
4. $\overline{AC} \cong \overline{DB}$	4. CPCTC

7.

STATEMENTS	REASONS
1. $\angle DBC \cong \angle ACB$, $\angle ABC \cong \angle DCB$	1. Given
2. $\overline{BC} \cong \overline{CB}$	2. Theorem 4-5
3. $\triangle ABC \cong \triangle DCB$	3. ASA
4. $\overline{AB} \cong \overline{DC}$	4. CPCTC

CHAPTER 6 APPLICATIONS OF CONGRUENT TRIANGLES

Page 211 Written **26.** If quadrilateral ABCD is a parallelogram, then it is a rectangle. false **27.** If quadrilateral ABCD is a square, then it is a rectangle. true **28.** If quadrilateral ABCD is a square, then it is a rhombus. true **29.** If quadrilateral ABCD is a parallelogram, then it is a trapezoid. false **30.** If quadrilateral ABCD is a parallelogram, then it is a square. false **31.** If quadrilateral ABCD is a rectangle, then it is a rhombus. false

32.

STATEMENTS	REASONS
1. $\square ABCD$, $\square EFCG$	1. Given
2. $\angle A \cong \angle C$, $\angle FEG \cong \angle C$	2. Theorem 6-3
3. $\angle A \cong \angle FEG$	3. Theorem 4-6

33.

STATEMENTS	REASONS
1. $\square ABCD$, E is the midpoint of $\overline{DC}$, F is the midpoint of $\overline{AB}$.	1. Given
2. $\overline{DE} \cong \overline{EC}$, $\overline{BF} \cong \overline{FA}$	2. Definition of midpoint
3. $DE + EC = DC$, $BF + FA = AB$	3. Segment addition axiom
4. $\overline{DC} \cong \overline{AB}$	4. Theorem 6-2
5. $DE + EC = BF + FA$	5. Substitution
6. $2(DE) = 2(BF)$	6. Substitution
7. $DE = BF$	7. Division axiom
8. $\overline{AD} \cong \overline{BC}$	8. Theorem 6-2
9. $\angle D \cong \angle B$	9. Theorem 6-3
10. $\triangle ADE \cong \triangle CBF$	10. SAS
11. $\overline{AE} \cong \overline{CF}$	11. CPCTC

34.

STATEMENTS	REASONS
1. $\square ABEF$, $\square BCDE$	1. Given
2. $\overline{AF} \cong \overline{BE}$, $\overline{BE} \cong \overline{CD}$	2. Theorem 6-2
3. $\overline{AF} \cong \overline{CD}$	3. Theorem 4-5

35.

STATEMENTS	REASONS
1. $\square ABCD$, $\overline{AE} \perp \overline{DC}$, $\overline{CF} \perp \overline{AB}$	1. Given
2. $\overline{AB} \parallel \overline{DC}$ (or $\overline{AF} \parallel \overline{EC}$)	2. Definition of a parallelogram
3. $\overline{CF} \perp \overline{DC}$	3. Theorem 4-10
4. $\overline{AE} \parallel \overline{CF}$	4. Theorem 4-11
5. AFCE is a parallelogram	5. Definition of a parallelogram
6. $\overline{AE} \cong \overline{CF}$	6. Theorem 6-2

36.

STATEMENTS	REASONS
1. $\square ABCD$, $\overline{DE} \perp \overline{AE}$, $\overline{BF} \perp \overline{FC}$	1. Given
2. $\angle AED$ and $\angle CFB$ are right angles.	2. Definition of perpendicular lines
3. $\angle AED \cong \angle CFB$	3. Theorem 4-7
4. $\overline{AD} \cong \overline{CB}$	4. Theorem 6-2
5. $\angle D \cong \angle B$	5. Theorem 6-3
6. $\triangle AED \cong \triangle CFB$	6. AAS
7. $\overline{AE} \cong \overline{CF}$	7. CPCTC

Page 216 Exploratory 1. $\overline{AB} \parallel \overline{DC}$, $\overline{AD} \parallel \overline{BC}$, $\overline{AB} \cong \overline{DC}$, $\overline{AD} \cong \overline{BC}$, $\angle ADC \cong \angle ABC$, $\angle DAB \cong \angle DCB$, $\overline{AC}$ bisects $\overline{DB}$, $\overline{DB}$ bisects $\overline{AC}$ 2. both pairs of opposite sides are parallel, two sides that are both parallel and congruent, both pairs of opposite sides are congruent, the diagonals bisect each other, both pairs of opposite angles are congruent 3. true; definition of a parallelogram 4. false; The sides must also be parallel or both pairs of opposite sides must be congruent 5. false; The sides must also be congruent or both pairs of opposite sides must be parallel 6. true; definition of a parallelogram and Theorem 6-5 7. true; definition of a parallelogram 8. true; Theorem 6-4 9. true; CPCTC and Theorem 6-6 10. true; Theorem 6-7 11. false; Both pairs of opposite angles must be congruent 12. Definition of a parallelogram 13. Theorem 6-5 14. Since $\angle 1 \cong \angle 2$, $\overline{AB} \parallel \overline{CD}$ by Theorem 4-13. Thus, by Theorem 6-5, $ABCD$ is a parallelogram. 15. Since $\angle 1 \cong \angle 2$, $\overline{AB} \parallel \overline{CD}$ by Theorem 4-13. Also, since $\angle 3 \cong \angle 4$, $\overline{AD} \parallel \overline{BC}$ by Theorem 4-13. Thus, by the definition of a parallelogram, $ABCD$ is a parallelogram. 16. By the definition of bisector, $\overline{AC}$ and $\overline{BD}$ bisect each other. Thus, by Theorem 6-7, $ABCD$ is a parallelogram. 17. By CPCTC, $\overline{AD} \cong \overline{BC}$ and $\angle 3 \cong \angle 2$. Since $\angle 3 \cong \angle 2$, $\overline{AD} \parallel \overline{BC}$ by Theorem 4-13. Thus, by Theorem 6-5, $ABCD$ is a parallelogram.

Page 217 Written 9. If a quadrilateral is a parallelogram, then the diagonals of the parallelogram separate it into two congruent triangles. If a diagonal of a quadrilateral separates it into two congruent triangles, then the quadrilateral is a parallelogram. true; CPCTC and Theorem 6-5 10. If a quadrilateral is a parallelogram, then both pairs of opposite sides are congruent. If both pairs of opposite sides of a quadrilateral are congruent, then the quadrilateral is a parallelogram. true; Theorem 6-6 11. If a quadrilateral is a parallelogram, then the diagonals bisect each other. If the diagonals of a quadrilateral bisect each other, then the quadrilateral is a parallelogram. true; Theorem 6-7 12. If a quadrilateral has two sides that are both parallel and congruent, then the quadrilateral is a parallelogram. If a quadrilateral is a parallelogram, then it has two sides that are both parallel and congruent. true; Definition of a parallelogram and Theorem 6-2

13. Given: Quadrilateral $ABCD$, $\overline{AC}$ and $\overline{BD}$ bisect each other at M.
Prove: $ABCD$ is a parallelogram.

STATEMENTS	REASONS
1. Quadrilateral $ABCD$, $\overline{AC}$ and $\overline{BD}$ bisect each other at M.	1. Given
2. $\overline{AM} \cong \overline{CM}$ $\overline{BM} \cong \overline{DM}$	2. Definition of segment bisector
3. $\angle AMB \cong \angle CMD$	3. Theorem 4-4
4. $\triangle AMB \cong \triangle CMD$	4. SAS
5. $\overline{AB} \cong \overline{CD}$, $\angle MBA \cong \angle MDC$	5. CPCTC
6. $\overline{AB} \parallel \overline{CD}$	6. Theorem 4-13
7. $ABCD$ is a parallelogram.	7. Theorem 6-5

14. Given: Quadrilateral $ABCD$, $\angle A \cong \angle C$, $\angle B \cong \angle D$.
Prove: $ABCD$ is a parallelogram.

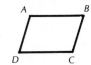

STATEMENTS	REASONS
1. Quadrilateral $ABCD$, $\angle A \cong \angle C$, $\angle B \cong \angle D$	1. Given
2. $m\angle A + m\angle B = m\angle C + m\angle D$	2. Addition axiom
3. $m\angle A + m\angle B + m\angle C + m\angle D = 360$	3. Theorem 4-25
4. $2(m\angle A + m\angle D) = 360$ $2(m\angle A + m\angle B) = 360$	4. Substitution
5. $m\angle A + m\angle D = 180$ $m\angle A + m\angle B = 180$	5. Division axiom
6. $\angle A$ and $\angle D$ are supplementary. $\angle A$ and $\angle B$ are supplementary.	6. Definition of supplementary
7. $\overline{AB} \parallel \overline{DC}$, $\overline{AD} \parallel \overline{BC}$	7. Theorem 4-14
8. $ABCD$ is a parallelogram.	8. Definition of a parallelogram.

15. No; an isosceles trapezoid meets these requirements and is not a parallelogram.

16.

STATEMENTS	REASONS
1. ∠1 is supplementary to ∠4, $\overline{DC} \cong \overline{AB}$	1. Given
2. $\overline{DC} \parallel \overline{AB}$	2. Theorem 4-14
3. ABCD is a parallelogram.	3. Theorem 6-5

18.

STATEMENTS	REASONS
1. ∠1 is supplementary to ∠2.	1. Given
2. ∠1 is supplementary to ∠DAB.	2. Supplement axiom
3. ∠2 ≅ ∠DAB	3. Theorem 4-2
4. $\overline{AD} \parallel \overline{BC}$	4. Axiom 4-15
5. ∠DCB ≅ ∠2	5. Given
6. $\overline{DC} \parallel \overline{AB}$	6. Theorem 4-13
7. ABCD is a parallelogram.	7. Definition of parallelogram

20.

STATEMENTS	REASONS
1. ▱AFCE, $\overline{DE} \cong \overline{BF}$	1. Given
2. $\overline{EC} \cong \overline{AF}$	2. Theorem 6-2
3. EC + DE = AF + BF	3. Addition axiom
4. AF + BF = AB, EC + DE = DC	4. Segment addition axiom
5. AB = DC	5. Substitution
6. $\overline{AF} \parallel \overline{EC}$ (or $\overline{AB} \parallel \overline{DC}$)	6. Definition of a parallelogram
7. ABCD is a parallelogram.	7. Theorem 6-5

21.

STATEMENTS	REASONS
1. ▱ABCD	1. Given
2. $\overline{AB} \parallel \overline{DC}$ (or $\overline{AF} \parallel \overline{EC}$)	2. Definition of a parallelogram
3. $\overline{AB} \cong \overline{DC}$	3. Theorem 6-2
4. E is the midpoint of $\overline{DC}$. F is the midpoint of $\overline{AB}$.	4. Given
5. $\overline{AF} \cong \overline{FB}$, $\overline{DE} \cong \overline{EC}$	5. Definition of midpoint
6. AF + FB = AB, DE + EC = DC	6. Segment addition axiom
7. AF + FB = DE + EC	7. Substitution
8. 2(AF) = 2(EC)	8. Substitution
9. AF = EC	9. Division axiom
10. AFCE is a parallelogram.	10. Theorem 6-5

17.

STATEMENTS	REASONS
1. $\overline{AD} \cong \overline{CE}$, $\overline{CB} \cong \overline{CE}$	1. Given
2. $\overline{AD} \cong \overline{CB}$	2. Theorem 4-5
3. ∠1 is supplementary to ∠2.	3. Given
4. ∠1 is supplementary to ∠DAB.	4. Supplement axiom
5. ∠2 ≅ ∠DAB	5. Theorem 4-2
6. $\overline{AD} \parallel \overline{CB}$	6. Axiom 4-15
7. ABCD is a parallelogram.	7. Theorem 6-5

19.

STATEMENTS	REASONS
1. ∠2 ≅ ∠3, $\overline{AD} \cong \overline{CD}$, $\overline{DC} \cong \overline{AB}$	1. Given
2. $\overline{BC} \cong \overline{EC}$	2. Theorem 5-3
3. $\overline{BC} \cong \overline{AD}$	3. Theorem 4-5
4. ABCD is a parallelogram.	4. Theorem 6-6

22.

STATEMENTS	REASONS
1. ▱ABCD, $\overline{AE}$ bisects ∠DAB, $\overline{CF}$ bisects ∠DCB.	1. Given
2. $\overline{AD} \cong \overline{BC}$	2. Theorem 6-2
3. ∠DAB ≅ ∠DCB	3. Theorem 6-3
4. ∠DAE ≅ ∠EAB, ∠BCF ≅ ∠FCD	4. Definition of angle bisector
5. m∠DAE + m∠EAB = m∠DAB, m∠BCF + m∠FCD = m∠DCB	5. Angle addition axiom
6. 2(m∠EAB) = 2(m∠FCD)	6. Substitution
7. m∠EAB = m∠FCD	7. Division axiom
8. $\overline{AB} \parallel \overline{DC}$	8. Definition of a parallelogram
9. ∠EAB ≅ ∠DEA, ∠FCD ≅ ∠BFC	9. Theorem 4-12
10. ∠DEA ≅ ∠BFC	10. Substitution
11. ∠DEA is supplementary to ∠AEC. ∠BFC is supplementary to ∠AFC.	11. Supplement axiom
12. ∠AEC ≅ ∠AFC	12. Theorem 4-2
13. AFCE is a parallelogram.	13. Theorem 6-8

23.

STATEMENTS	REASONS
1. $\square ABCD$, $\angle DAE \cong \angle DCF$	1. Given
2. $\angle ADE \cong \angle CBF$ $\overline{AD} \cong \overline{CB}$	2. Theorem 6-3
3. $\overline{AD} \cong \overline{CB}$	3. Theorem 6-2
4. $\triangle ADE \cong \triangle CBF$	4. ASA
5. $\overline{AE} \cong \overline{CF}$, $\overline{DE} \cong \overline{BF}$	5. CPCTC
6. $DE + EC = DC$ $AF + BF = AB$	6. Segment addition axiom
7. $\overline{DC} \cong \overline{AB}$	7. Definition of a parallelogram
8. $DE + EC = AF + BF$	8. Substitution
9. $EC = AF$	9. Subtraction axiom
10. $AFCE$ is a parallelogram.	10. Theorem 6-6

24.

STATEMENTS	REASONS
1. E is the midpoint of $\overline{AC}$.	1. Given
2. $\overline{AE} \cong \overline{CE}$	2. Definition of midpoint
3. $\angle AEC \cong \angle CEF$	3. Theorem 4-4
4. $\overline{DE} \cong \overline{FE}$	4. Given
5. $\triangle AEC \cong \triangle CEF$	5. SAS
6. $\overline{AD} \cong \overline{CF}$	6. CPCTC
7. D is the midpoint of $\overline{AB}$.	7. Given
8. $\overline{AD} \cong \overline{BD}$	8. Definition of midpoint
9. $\overline{BD} \cong \overline{CF}$	9. Substitution
10. $\angle DAE \cong \angle FCE$	10. CPCTC
11. $\overline{BA} \parallel \overline{CF}$ (or $\overline{BD} \parallel \overline{CF}$)	11. Theorem 4-13
12. $BCFD$ is a parallelogram.	12. Theorem 6-5

Page 220 Exploratory 5. true; All 4 of the large triangles formed are congruent to each other. Thus, all 4 angles of the parallelogram are congruent. Therefore, the angles must all be right angles, and the parallelogram is a rectangle. **6.** true; All 4 of the small triangles formed are congruent by SAS. Therefore, all of the sides are congruent and the parallelogram is a rhombus. **7.** true; All 4 of the small triangles formed are congruent by AAS. Therefore, all of the sides are congruent and the parallelogram is a rhombus. **8.** true; definition of a rectangle **9.** true; definition of a square **10.** true; definition of a square **11.** false; All rhombuses do not have 4 right angles. **12.** false; All rectangles do not have 4 congruent sides. **13.** false; All rhombuses do not have 4 right angles. **14.** false; The diagonals of a rectangle are not perpendicular unless it is a square. **15.** true; All of the angles are right angles, so any two are supplementary. **16.** true; Theorem 6-4 **17.** false; This is only true for rectangles. **18.** true; Theorem 6-10 **19.** true; Theorem 6-9 **20.** true; The angle opposite the right angle is also a right angle. The other two angles are congruent and have a sum of 180, so they are both right angles. Thus, the parallelogram is a rectangle. **21.** false; This is only true for squares. **22.** true; Theorem 6-2 **23.** true; Theorem 6-1 **24.** false; This is only true for rhombuses. **25.** true; Theorem 6-11 **26.** true; definition of a rectangle and a rhombus **27.** opposite sides parallel and congruent, opposite angles congruent, diagonals bisect each other, each diagonal separates it into two congruent triangles **28.** all properties of a parallelogram, four right angles, congruent diagonals **29.** all properties of a parallelogram, four congruent sides, perpendicular diagonals, each diagonal bisects two opposite angles **30.** all properties of a parallelogram, rectangle, and rhombus **31.** four congruent sides, each diagonal bisects two opposite angles, perpendicular diagonals, See students' work. **32.** four right angles, diagonals congruent

Page 221 Written

7. Given: $ABCD$ is a rectangle.
Prove: $\overline{AC} \cong \overline{BD}$

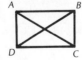

8. Given: $\square ABCD$ is a rhombus.
Prove: $\overline{AC} \perp \overline{BD}$

STATEMENTS	REASONS
1. $ABCD$ is a rectangle.	1. Given
2. $\angle ADC$ and $\angle BCD$ are right angles.	2. Definition of a rectangle
3. $\angle ADC \cong \angle BCD$	3. Theorem 4-7
4. $\overline{AD} \cong \overline{BC}$	4. Theorem 6-2
5. $\overline{DC} \cong \overline{CD}$	5. Theorem 4-5
6. $\triangle ACD \cong \triangle BCD$	6. SAS
7. $\overline{AC} \cong \overline{BD}$	7. CPCTC

STATEMENTS	REASONS
1. $\square ABCD$ is a rhombus.	1. Given
2. $\overline{AD} \cong \overline{AB}$	2. Definition of a rhombus
3. $\overline{DE} \cong \overline{BE}$	3. Theorem 6-4
4. $\overline{AE} \cong \overline{AE}$	4. Theorem 4-5
5. $\triangle ADE \cong \triangle ABE$	5. SSS
6. $\angle 1 \cong \angle 2$	6. CPCTC
7. $\overline{AC} \perp \overline{BD}$	7. Theorem 4-9

9. Given: ▱ABCD is a rhombus.
Prove: $\overline{BD}$ bisects ∠ABC.
$\overline{BD}$ bisects ∠ADC.

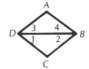

STATEMENTS	REASONS
1. ▱ABCD is a rhombus.	1. Given
2. $\overline{AD} \cong \overline{CD}$, $\overline{BA} \cong \overline{BC}$	2. Definition of a rhombus
3. $\overline{BD} \cong \overline{BD}$	3. Theorem 4-5
4. △ADB ≅ △CDB	4. SSS
5. ∠3 ≅ ∠1, ∠4 ≅ ∠2	5. CPCTC
6. $\overline{BD}$ bisects ∠ABC. $\overline{BD}$ bisects ∠ADC.	6. Definition of angle bisector

10.

STATEMENTS	REASONS
1. Rectangle ABCD	1. Given
2. $\overline{AB} \cong \overline{DC}$	2. Theorem 6-2
3. ∠ABE and ∠DCE are right angles.	3. Definition of a rectangle
4. ∠ABE ≅ ∠DCE	4. Theorem 4-7
5. E is the midpoint of $\overline{BC}$.	5. Given
6. $\overline{BE} \cong \overline{CE}$	6. Definition of midpoint
7. △ABE ≅ △DCE	7. SAS
8. $\overline{AE} \cong \overline{DE}$	8. CPCTC

11.

STATEMENTS	REASONS
1. Rectangle ABCD	1. Given
2. $\overline{DC} \cong \overline{AB}$	2. Theorem 6-2
3. ∠DCE and ∠ABE are right angles.	3. Definition of a rectangle
4. $\overline{DE} \cong \overline{AE}$	4. Given
5. △DCE ≅ △ABE	5. HL
6. ∠EDC ≅ ∠EAB	6. CPCTC

12.

STATEMENTS	REASONS
1. Rectangle ABCD	1. Given
2. $\overline{BC} \cong \overline{AD}$	2. Theorem 6-2
3. $\overline{CD} \cong \overline{DC}$	3. Theorem 4-5
4. $\overline{BD} \cong \overline{AC}$	4. Theorem 6-9
5. △BCD ≅ △ADC	5. SSS
6. ∠BDC ≅ ∠ACD	6. CPCTC

13.

STATEMENTS	REASONS
1. Rectangle ABCD	1. Given
2. $\overline{CD} \cong \overline{BA}$	2. Theorem 6-2
3. $\overline{CA} \cong \overline{BD}$	3. Theorem 6-9
4. $\overline{DA} \cong \overline{AD}$	4. Theorem 4-5
5. △CDA ≅ △BAD	5. SSS
6. ∠CAD ≅ ∠BDA	6. CPCTC

14. If the diagonals of a parallelogram are congruent, then the parallelogram is a rectangle. true

Given: ▱ABCD, $\overline{AC} \cong \overline{BD}$
Prove: ABCD is a rectangle.

STATEMENTS	REASONS
1. ▱ABCD, $\overline{AC} \cong \overline{BD}$	1. Given
2. $\overline{AD} \cong \overline{BC}$	2. Theorem 6-2
3. $\overline{DC} \cong \overline{CD}$	3. Theorem 4-5
4. △ADC ≅ △BCD	4. SSS
5. ∠ADC ≅ ∠BCD	5. CPCTC
6. $\overline{AD} \parallel \overline{BC}$	6. Definition of a parallelogram
7. ∠ADC is supplementary to ∠BCD.	7. Theorem 4-15
8. ∠ADC and ∠BCD are right angles.	8. Theorem 4-8
9. ABCD is a rectangle.	9. Definition of a rectangle

15. If the diagonals of a parallelogram are perpendicular, then the parallelogram is a rhombus. true

Given: ▱ABCD, $\overline{AC} \perp \overline{DB}$
Prove: ABCD is a rhombus.

STATEMENTS	REASONS
1. ▱ABCD, $\overline{AC} \perp \overline{DB}$	1. Given
2. ∠AED, ∠AEB, ∠BEC, and ∠DEC are right angles.	2. Definition of perpendicular lines
3. ∠AED ≅ ∠AEB ≅ ∠BEC ≅ ∠DEC	3. Theorem 4-7
4. $\overline{DE} \cong \overline{BE}$, $\overline{AE} \cong \overline{CE}$	4. Theorem 4-6
5. △DEA ≅ △BEA ≅ △DEC ≅ △BEC	5. SAS
6. $\overline{DA} \cong \overline{BA} \cong \overline{DC} \cong \overline{BC}$	6. CPCTC
7. ABCD is a rhombus.	7. Definition of a rhombus

16. If each diagonal of a parallelogram bisects two opposite angles of the parallelogram, then the parallelogram is a rhombus. true

Given: ▱ABCD, $\overline{DB}$ bisects ∠ADC and ∠ABC.
Prove: ABCD is a rhombus.

STATEMENTS	REASONS
1. ▱ABCD, $\overline{DB}$ bisects ∠ADC and ∠ABC.	1. Given
2. ∠3 ≅ ∠1, ∠4 ≅ ∠2	2. Definition of angle bisector
3. ∠ADC ≅ ∠ABC	3. Theorem 6-3
4. m∠3 + m∠1 = m∠ADC; m∠4 + m∠2 = m∠ABC	4. Angle addition axiom
5. 2(m∠3) = 2(m∠4) 2(m∠1) = 2(m∠2)	5. Substitution
6. m∠3 = m∠4, m∠1 = m∠2	6. Division axiom
7. $\overline{AB} \cong \overline{AD}, \overline{CD} \cong \overline{CB}$	7. Theorem 5-3
8. $\overline{AD} \cong \overline{CB}, \overline{AB} \cong \overline{CD}$	8. Theorem 6-2
9. $\overline{AB} \cong \overline{CB} \cong \overline{CD} \cong \overline{AD}$	9. Substitution
10. ABCD is a rhombus.	10. Definition of rhombus

17. Given: Quadrilateral ABCD with ∠A ≅ ∠B, ∠B ≅ ∠C, ∠C ≅ ∠D, and ∠D ≅ ∠A.
Prove: ABCD is a rectangle.

STATEMENTS	REASONS
1. Quadrilateral ABCD with ∠A ≅ ∠B, ∠B ≅ ∠C, ∠C ≅ ∠D, and ∠D ≅ ∠A.	1. Given
2. m∠A + m∠B + m∠C + m∠D = 360	2. Theorem 4-25
3. 4(m∠A) = 360	3. Substitution
4. m∠A = 90	4. Division axiom
5. m∠B = 90, m∠C = 90, m∠D = 90	5. Substitution
6. ∠A, ∠B, ∠C, and ∠D are right angles.	6. Definition of right angle
7. $\overline{AD} \perp \overline{AB}, \overline{BC} \perp \overline{AB}, \overline{AD} \perp \overline{DC}, \overline{BC} \perp \overline{DC}$	7. Definition of perpendicular lines
8. $\overline{AD} \parallel \overline{BC}, \overline{AB} \parallel \overline{DC}$	8. Theorem 4-11
9. ABCD is a parallelogram.	9. Definition of a parallelogram
10. ABCD is a rectangle.	10. Definition of a rectangle

18.

STATEMENTS	REASONS
1. $\overline{AB}$ bisects ∠EAC, $\overline{CB}$ bisects ∠HCA, $\overline{AD}$ bisects ∠FAC, and $\overline{CD}$ bisects ∠GCA.	1. Given
2. ∠EAB ≅ ∠BAC, ∠HCB ≅ ∠BCA, ∠FAD ≅ ∠DAC, ∠GCD ≅ ∠DCA	2. Definition of angle bisector
3. $\overleftrightarrow{EF} \parallel \overleftrightarrow{HG}$	3. Given
4. ∠EAC ≅ ∠GCA, ∠FAC ≅ ∠HCA	4. Theorem 4-12
5. m∠EAB + m∠BAC = m∠EAC m∠HCB + m∠BCA = m∠HCA m∠FAD + m∠DAC = m∠FAC m∠GCD + m∠DCA = m∠GCA	5. Angle addition axiom
6. 2(m∠BAC) = 2(m∠DCA) 2(m∠BCA) = 2(m∠DAC)	6. Substitution

STATEMENTS	REASONS
7. m∠BAC = m∠DCA m∠BCA = m∠DAC	7. Division axiom
8. $\overline{AB} \parallel \overline{DC}, \overline{DA} \parallel \overline{CB}$	8. Theorem 4-13
9. ABCD is a parallelogram.	9. Definition of a parallelogram
10. ∠EAC is supplementary to ∠HCA.	10. Theorem 4-15
11. m∠EAC + m∠HCA = 180	11. Definition of supplementary
12. 2(m∠BAC) + 2(m∠BCA) = 180	12. Substitution
13. m∠BAC + m∠BCA = 90	13. Division axiom
14. m∠BAC + m∠BCA + m∠ABC = 180	14. Theorem 4-19
15. 90 + m∠ABC = 180	15. Substitution
16. m∠ABC = 90	16. Subtraction axiom
17. ∠ABC is a right angle.	17. Definition of right angle
18. ABCD is a rectangle.	18. Definition of a rectangle

19.

STATEMENTS	REASONS
1. $\overline{AD} \cong \overline{AB}$, $\overline{AC}$ bisects $\angle DAB$, $\overline{AE} \cong \overline{EC}$	1. Given
2. $\angle DAE \cong \angle BAE$	2. Definition of angle bisector
3. $\overline{AE} \cong \overline{AE}$	3. Theorem 4-5
4. $\triangle DAE \cong \triangle BAE$	4. SAS
5. $\overline{DE} \cong \overline{BE}$	5. CPCTC
6. ABCD is a parallelogram.	6. Theorem 6-7
7. $\overline{AB} \cong \overline{DC}$, $\overline{DA} \cong \overline{CB}$	7. Theorem 6-2
8. $\overline{AB} \cong \overline{CB} \cong \overline{DC} \cong \overline{AD}$	8. Substitution
9. ABCD is a rhombus.	9. Definition of a rhombus

20.

STATEMENTS	REASONS
1. $\overline{RT} \parallel \overline{PW}$, $\overline{RT} \parallel \overline{QV}$, $\overline{TX} \parallel \overline{RP}$ (or $\overline{TV} \parallel \overline{RQ}$, $\overline{VX} \parallel \overline{QP}$)	1. Given
2. $\overline{PW} \parallel \overline{QV}$	2. Theorem 4-18
3. RQVT and QPXV are parallelograms.	3. Definition of a parallelogram
4. $\overline{RQ} \cong \overline{TV}$, $\overline{QP} \cong \overline{VX}$	4. Theorem 6-2
5. Q is the midpoint of $\overline{PR}$.	5. Given
6. $\overline{RQ} \cong \overline{QP}$	6. Definition of midpoint
7. $\overline{TV} \cong \overline{VX}$	7. Substitution
8. V is the midpoint of $\overline{TX}$.	8. Definition of midpoint

21.

STATEMENTS	REASONS
1. $\square RTXP$	1. Given
2. $\overline{RT} \parallel \overline{PX}$	2. Definition of a parallelogram
3. $\angle VST \cong \angle YWX$, $\angle STV \cong \angle WXV$	3. Theorem 4-12
4. V is the midpoint of $\overline{SW}$.	4. Given
5. $\overline{SV} \cong \overline{WY}$	5. Definition of midpoint
6. $\triangle SVT \cong \triangle WYX$	6. AAS

22.

STATEMENTS	REASONS
1. $\overline{RT} \parallel \overline{PW}$, $\overline{TX} \parallel \overline{RP}$ (or $\overline{TV} \parallel \overline{RQ}$)	1. Given
2. RTXP is a parallelogram.	2. Definition of a parallelogram
3. $\overline{XT} \cong \overline{PR}$	3. Theorem 6-2
4. Q is the midpoint of $\overline{PR}$. V is the midpoint of $\overline{XT}$.	4. Given
5. $\overline{PQ} \cong \overline{RQ}$, $\overline{XY} \cong \overline{TV}$	5. Definition of midpoint
6. $PQ + RQ = PR$ $XV + TV = XT$	6. Segment addition axiom
7. $2(TV) = 2(RQ)$	7. Substitution
8. $TV = RQ$	8. Division axiom
9. Quadrilateral QRTV is a parallelogram.	9. Theorem 6-5

Pages 224–225 Written

1.

STATEMENTS	REASONS
1. ABCD, an isosceles trapezoid with $\overline{AB} \parallel \overline{DC}$	1. Given
2. $\overline{AD} \cong \overline{BC}$	2. Definition of an isosceles trapezoid
3. $\angle ADC \cong \angle BCD$	3. Theorem 6-12
4. $\overline{DC} \cong \overline{CD}$	4. Theorem 4-5
5. $\triangle ACD \cong \triangle BDC$	5. SAS

2.

STATEMENTS	REASONS
1. ABCD, an isoscles trapezoid with $\overline{AB} \parallel \overline{DC}$	1. Given
2. $\overline{AD} \cong \overline{BC}$	2. Definition of an isosceles trapezoid
3. $\angle ADC \cong \angle BCD$	3. Theorem 6-12
4. $\overline{DC} \cong \overline{CD}$	4. Theorem 4-5
5. $\triangle ACD \cong \triangle BDC$	5. SAS
6. $\angle DAC \cong \angle CBD$ (or $\angle DAE \cong \angle CBE$)	6. CPCTC
7. $\angle AED \cong \angle BEC$	7. Theorem 4-4
8. $\triangle AED \cong \triangle BEC$	8. AAS

3.

STATEMENTS	REASONS
1. ABCD, an isosceles trapezoid with $\overline{AB} \parallel \overline{DC}$	1. Given
2. $\overline{AD} \cong \overline{BC}$	2. Definition of an isosceles trapezoid
3. $\angle ADC \cong \angle BCD$	3. Theorem 6-12
4. $\overline{DC} \cong \overline{CD}$	4. Theorem 4-5
5. $\triangle ACD \cong \triangle BDC$	5. SAS
6. $\angle DAC \cong \angle CBD$ (or $\angle DAE \cong \angle CBE$)	6. CPCTC
7. $\angle AED \cong \angle BEC$	7. Theorem 4-4
8. $\triangle AED \cong \triangle BEC$	8. AAS
9. $\overline{AE} \cong \overline{BE}$	9. CPCTC
10. $\triangle AEB$ is isosceles.	10. Definition of isosceles triangle

4.

STATEMENTS	REASONS
1. ABCD, an isosceles trapezoid with $\overline{AB} \parallel \overline{DC}$	1. Given
2. $\angle ADC \cong \angle BCD$	2. Theorem 6-12
3. $\angle ADC$ is supplementary to $\angle 1$. $\angle BCD$ is supplementary to $\angle 2$.	3. Supplement axiom
4. $\angle 1 \cong \angle 2$	4. Theorem 4-2

5. Because the parallel sides are the bases. Answers may vary. A typical answer is if the trapezoid is isosceles and the parallel sides are not known, the base angles and congruent sides cannot be determined.

6. Given: Isosceles trapezoid ABCD with $\overline{AB} \parallel \overline{DC}$.
Prove: $\overline{AC} \cong \overline{BD}$

STATEMENTS	REASONS
1. Isosceles trapezoid ABCD with $\overline{AB} \parallel \overline{DC}$.	1. Given
2. $\overline{AD} \cong \overline{BC}$	2. Definition of an isosceles trapezoid
3. $\angle ACD \cong \angle BCD$	3. Theorem 6-12
4. $\overline{DC} \cong \overline{CD}$	4. Theorem 4-5
5. $\triangle ADC \cong \triangle BCD$	5. SAS
6. $\overline{AC} \cong \overline{BD}$	6. CPCTC

7.

STATEMENTS	REASONS
1. ABCD is an isosceles trapezoid, $\overline{AB} \parallel \overline{DC}$	1. Given
2. $\overline{AD} \cong \overline{BC}$	2. Definition of an isosceles trapezoid
3. $\angle ADC \cong \angle BCD$ (or $\angle ADE \cong \angle BCF$)	3. Theorem 6-12
4. $\overline{AE} \perp \overline{DC}, \overline{BF} \perp \overline{DC}$	4. Given
5. $\angle AED$ and $\angle BFC$ are right angles.	5. Definition of perpendicular lines
6. $\angle AED \cong \angle BFC$	6. Theorem 4-7
7. $\triangle ADE \cong \triangle BCF$	7. AAS

8.

STATEMENTS	REASONS
1. $\overline{AB} \parallel \overline{DC}, \overline{AE} \perp \overline{DC}, \overline{BF} \perp \overline{DC}$	1. Given
2. ABCD is a trapezoid.	2. Definition of a trapezoid
3. $\overline{AE} \parallel \overline{BF}$	3. Theorem 4-11
4. ABFE is a parallelogram.	4. Definition of a parallelogram
5. $\angle AED$ and $\angle BFC$ are right angles.	5. Definition of perpendicular lines
6. $\angle AED \cong \angle BFC$	6. Theorem 4-7
7. $\angle D \cong \angle C$	7. Given
8. $\triangle AED \cong \triangle BFC$	8. AAS
9. $\overline{AD} \cong \overline{BC}$	9. CPCTC
10. ABCD is an isosceles trapezoid.	10. Definition of an isosceles trapezoid

9.

STATEMENTS	REASONS
1. Isosceles trapezoid ABCD, $\overline{DE}$ bisects $\angle ADC, \overline{AB} \parallel \overline{CD}$	1. Given
2. $\angle ADE \cong \angle EDC$	2. Definition of angle bisector
3. $\angle EDC \cong \angle DEA$	3. Theorem 4-12
4. $\angle ADE \cong \angle DEA$	4. Theorem 4-6
5. $\overline{AD} \cong \overline{AE}$	5. Theorem 5-3

10.

STATEMENTS	REASONS
1. Trapezoid ABED with $\overline{EB} \parallel \overline{DA}$, (or $\overline{CB} \parallel \overline{DA}$), $\overline{BA} \parallel \overline{CD}$, $\angle E \cong \angle 1$	1. Given
2. $\angle 1 \cong \angle 2$	2. Theorem 4-12
3. $\angle 2 \cong \angle E$	3. Theorem 4-6
4. $\overline{ED} \cong \overline{CD}$	4. Theorem 5-3
5. CBAD is a parallelogram.	5. Definition of a parallelogram
6. $\overline{CD} \cong \overline{BA}$	6. Theorem 6-2
7. $\overline{ED} \cong \overline{BA}$	7. Theorem 4-5
8. ABED is isosceles.	8. Definition of an isosceles trapezoid

11.

STATEMENTS	REASONS
1. $\square ABCD$, $\angle E \cong \angle 2$	1. Given
2. $\overline{ED} \cong \overline{CD}$	2. Theorem 5-3
3. $\overline{CD} \cong \overline{BA}$	3. Theorem 6-2
4. $\overline{ED} \cong \overline{BA}$	4. Theorem 4-5
5. $\overline{CB} \parallel \overline{DA}$ (or $\overline{EB} \parallel \overline{DA}$)	5. Definition of a parallelogram
6. ABED is an isosceles trapezoid.	6. Definition of an isosceles trapezoid

12.

STATEMENTS	REASONS
1. $\square ABCD$, $\angle E$ is supplementary to $\angle A$.	1. Given
2. $\angle BCD \cong \angle A$	2. Theorem 6-3
3. $\angle BCD$ is supplementary to $\angle 2$.	3. Supplement axiom
4. $\angle E \cong \angle 2$	4. Theorem 4-2
5. $\overline{ED} \cong \overline{CD}$	5. Theorem 5-3
6. $\overline{CD} \cong \overline{BA}$	6. Theorem 6-2
7. $\overline{ED} \cong \overline{BA}$	7. Theorem 4-5
8. $\overline{CB} \parallel \overline{DA}$ (or $\overline{EB} \parallel \overline{DA}$)	8. Definition of a parallelogram
9. ABED is an isosceles trapezoid.	9. Definition of an isosceles trapezoid

13.

STATEMENTS	REASONS
1. Isosceles trapezoid ABCD with $\overline{AD} \parallel \overline{BC}$	1. Given
2. $\angle A \cong \angle D$	2. Theorem 6-12
3. $\overline{AE} \cong \overline{DE}$	3. Theorem 5-3

14.

STATEMENTS	REASONS
1. Isosceles trapezoid ABCD with $\overline{AD} \parallel \overline{BC}$	1. Given
2. $\angle A \cong \angle D$	2. Theorem 6-12
3. $\angle A \cong \angle EBC$, $\angle D \cong \angle ECB$	3. Axiom 4-14
4. $\angle EBC \cong \angle ECB$	4. Substitution
5. $\overline{EB} \cong \overline{EC}$	5. Theorem 5-3
6. $\triangle BEC$ is isosceles.	6. Definition of an isosceles triangle

15.

STATEMENTS	REASONS
1. $\overline{BE} \cong \overline{CE}$, $\angle EBC \cong \angle EAD$	1. Given
2. $\overline{BC} \parallel \overline{AD}$	2. Axiom 4-15
3. $\angle EBC \cong \angle ECB$	3. Theorem 5-2
4. $\angle ECB \cong \angle EDA$	4. Axiom 4-14
5. $\angle EAD \cong \angle EDA$	5. Substitution
6. $\overline{AE} \cong \overline{DE}$	6. Theorem 5-3
7. $AB + BE = AE$ $DC + CE = DE$	7. Segment addition axiom
8. $AB + BE = DC + CE$	8. Substitution
9. $AB = DC$	9. Subtraction axiom
10. ABCD is an isosceles trapezoid.	10. Definition of an isosceles trapezoid

16.

STATEMENTS	REASONS
1. $\overline{AE} \cong \overline{DE}$, $\overline{BC} \parallel \overline{AD}$	1. Given
2. $\angle A \cong \angle D$	2. Theorem 5-2
3. $\angle EBC \cong \angle A$, $\angle ECB \cong \angle D$	3. Axiom 4-14
4. $\angle EBC \cong \angle ECB$	4. Substitution
5. $\overline{BE} \cong \overline{CE}$	5. Theorem 5-3
6. $AB + BE = AE$ $DC + CE = DE$	6. Segment addition axiom
7. $AB + BE = DC + CE$	7. Substitution
8. $AB = DC$	8. Subtraction axiom
9. ABCD is an isosceles trapezoid.	9. Definition of an isosceles trapezoid

17. If a trapezoid is isosceles, then the base angles are congruent. If the base angles of a trapezoid are congruent, then the trapezoid is isosceles. true

Given: Trapezoid ABCD with $\overline{AB} \parallel \overline{DC}$, $\angle D \cong \angle C$
Prove: Trapezoid ABCD is isosceles.

STATEMENTS	REASONS
1. Trapezoid ABCD with $\overline{AB} \parallel \overline{DC}$	1. Given
2. Draw $\overline{AX} \perp \overline{DC}$ and $\overline{BY} \perp \overline{DC}$.	2. Theorem 4-17
3. $\overline{AX} \parallel \overline{BY}$	3. Theorem 4-11
4. ABYX is a parallelogram.	4. Definition of a parallelogram
5. $\overline{AX} \cong \overline{BY}$	5. Theorem 6-2
6. $\angle AXD$ and $\angle BYC$ are right angles.	6. Definition of perpendicular lines
7. $\angle AXD \cong \angle BYC$	7. Theorem 4-7
8. $\angle D \cong \angle C$	8. Given
9. $\triangle AXD \cong \triangle BYC$	9. AAS
10. $\overline{AD} \cong \overline{BC}$	10. CPCTC
11. Trapezoid ABCD is isosceles.	11. Definition of an isosceles trapezoid

18. If a trapezoid is isosceles, then the diagonals are congruent. If the diagonals of a trapezoid are congruent, then the trapezoid is isosceles. true

Given: Trapezoid ABCD with $\overline{AB} \parallel \overline{DC}$, $\overline{AC} \cong \overline{BD}$
Prove: Trapezoid ABCD is isosceles.

STATEMENTS	REASONS
1. Trapezoid ABCD with $\overline{AB} \parallel \overline{DC}$, $\overline{AC} \cong \overline{BD}$	1. Given
2. Draw $\overline{AX} \perp \overline{DC}$ and $\overline{BY} \perp \overline{DC}$.	2. Theorem 4-7
3. $\overline{AX} \parallel \overline{BY}$	3. Theorem 4-11
4. ABYX is a parallelogram.	4. Definition of a parallelogram
5. $\overline{AX} \cong \overline{BY}$	5. Theorem 6-2
6. $\angle AXC$ and $\angle BYD$ are right angles.	6. Definition of perpendicular lines
7. $\triangle AXC \cong \triangle BYD$	7. HL
8. $\angle ACX \cong \angle BDY$ (or $\angle ACD \cong \angle BDC$)	8. CPCTC
9. $\overline{DC} \cong \overline{CD}$	9. Theorem 4-5
10. $\triangle ACD \cong \triangle BDC$	10. SAS
11. $\overline{AD} \cong \overline{BC}$	11. CPCTC
12. Trapezoid ABCD is isosceles.	12. Definition of an isosceles trapezoid

Page 234 Written 24. Draw $\overline{AC}$, $\overline{AD}$, $\overline{BC}$, and $\overline{BD}$. Let X be the point of intersection of $\overline{AB}$ and $\overline{CD}$. By construction, $\overline{AC} \cong \overline{BC}$ and $\overline{AD} \cong \overline{BD}$. Since $\overline{CD} \cong \overline{CD}$, $\triangle CAD \cong \triangle CBD$ by SSS. $\angle ACD \cong \angle BCD$ by CPCTC. Since $\overline{XC} \cong \overline{XC}$, $\triangle AXC \cong \triangle BXQ$ by SAS. $\overline{AX} \cong \overline{BX}$ by CPCTC. **25.** By construction, $\overline{AF} \perp \overline{BC}$. **26.** By construction, D is the midpoint of $\overline{AB}$. By definition of a median, $\overline{CD}$ is a median of $\triangle ABC$. **27.** By construction, $\angle RPS \cong \angle 1$. Therefore, $\overrightarrow{PR} \parallel \ell$ by Theorem 4-13.

Page 237 Written

20.

STATEMENTS	REASONS
1. $\triangle ABC$, D is the midpoint of $\overline{AB}$, E is the midpoint of $\overline{BC}$, and F is the midpoint of $\overline{AC}$.	1. Given
2. $FE = \frac{1}{2}AB$, $DF = \frac{1}{2}BC$, $DE = \frac{1}{2}AC$	2. Theorem 6-14
3. $FE + DF + DE = \frac{1}{2}AB + \frac{1}{2}BC + \frac{1}{2}AC$	3. Addition axiom
4. $FE + DF + DE = \frac{1}{2}(AB + BC + AC)$	4. Distributive Law
5. The perimeter of $\triangle DEF$ is equal to one-half the perimeter of $\triangle ABC$.	5. Definition of perimeter

21.

STATEMENTS	REASONS
1. $\triangle RQS$, $\overline{QR} \cong \overline{QS}$, T and U are midpoints.	1. Given
2. $\overline{TU} \parallel \overline{RS}$	2. Theorem 6-14
3. $\angle R \cong \angle S$	3. Theorem 5-2
4. $\angle QTU \cong \angle R$, $\angle QUT \cong \angle S$	4. Axiom 4-14
5. $\angle QTU \cong \angle QUT$	5. Substitution
6. $\overline{QT} \cong \overline{QU}$	6. Theorem 5-3
7. $\triangle QTU$ is isosceles.	7. Definition of an isosceles triangle

22. Assume that T is not the midpoint of $\overline{QR}$. We are given that U is the midpoint of $\overline{QS}$. By Theorem 6-14 and Theorem 4-16, we know that the only line that is parallel to $\overline{RS}$ through U also passes through the midpoint of $\overline{QR}$. Thus, $\overline{TU} \not\parallel \overline{RS}$. However, this contradicts the given statement that $\overline{TU} \parallel \overline{RS}$. Therefore, T is the midpoint of $\overline{QR}$.

Page 237 Challenge

1. Given: Quadrilateral $ABCD$, X is the midpoint of $\overline{AB}$, Y is the midpoint of $\overline{BC}$, Z is the midpoint of $\overline{CD}$, and W is the midpoint of $\overline{DA}$.
Prove: $XYZW$ is a parallelogram

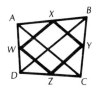

STATEMENTS	REASONS
1. Quadrilateral $ABCD$, X is the midpoint of $\overline{AB}$, Y is the midpoint of $\overline{BC}$, Z is the midpoint of $\overline{CD}$, and W is the midpoint of $\overline{DA}$.	1. Given
2. $\overline{XW} \parallel \overline{DB}$, $\overline{XY} \parallel \overline{AC}$, $\overline{ZY} \parallel \overline{DB}$, $\overline{WZ} \parallel \overline{AC}$	2. Theorem 6-14
3. $\overline{XW} \parallel \overline{ZY}$, $\overline{XY} \parallel \overline{WZ}$	3. Theorem 4-18
4. $XYZW$ is a parallelogram.	4. Definition of a parallelogram

Page 241 Chapter Review

10.

STATEMENTS	REASONS
1. $AFCE$ is a rhombus, $\triangle AED$ and $\triangle BFC$ are equilateral.	1. Given
2. $\overline{AF} \cong \overline{FC} \cong \overline{CE} \cong \overline{AE}$	2. Definition of a rhombus
3. $\overline{AE} \cong \overline{ED} \cong \overline{DA}$, $\overline{BF} \cong \overline{FC} \cong \overline{CB}$	3. Definition of an equilateral triangle
4. $\overline{DA} \cong \overline{CB}$, $\overline{BF} \cong \overline{ED}$	4. Substitution
5. $AF + BF = CE + ED$	5. Addition axiom
6. $AF + BF = AB$, $CE + ED = DC$	6. Segment addition axiom
7. $AB = DC$	7. Substitution
8. $ABCD$ is a parallelogram.	8. Theorem 6-6

11.

STATEMENTS	REASONS
1. $ABCD$ is a parallelogram $\overline{DE} \cong \overline{BF}$	1. Given
2. $\overline{AB} \parallel \overline{DC}$ (or $\overline{AF} \parallel \overline{EC}$)	2. Definition of a parallelogram
3. $\overline{AB} \cong \overline{DC}$	3. Theorem 6-2
4. $AF + BF = AB$, $DE + EC = DC$	4. Segment addition axiom
5. $AF + BF = DE + EC$	5. Substitution
6. $AF = EC$	6. Subtraction axiom
7. $AFCE$ is a parallelogram.	7. Theorem 6-5

12.

STATEMENTS	REASONS
1. $\angle BAE$ and $\angle AFC$ are supplementary, $\triangle ADE \cong \triangle CBF$	1. Given
2. $\overline{AE} \cong \overline{FC}$	2. Theorem 4-14
3. $\overline{AE} \cong \overline{CF}$	3. CPCTC
4. $AFCE$ is a parallelogram.	4. Theorem 6-5
5. $\overline{EC} \cong \overline{AF}$	5. Theorem 6-2
6. $\overline{AD} \cong \overline{CB}$, $\overline{DE} \cong \overline{BF}$	6. CPCTC
7. $EC + DE = AF + BF$	7. Addition axiom
8. $EC + DE = CD$, $AF + BF = AB$	8. Segment addition axiom
9. $CD = AB$	9. Substitution
10. $ABCD$ is a parallelogram.	10. Theorem 6-8

Page 242 Chapter Test

9.

STATEMENTS	REASONS
1. $ABCD$ is an isosceles trapezoid with $\overline{AB} \parallel \overline{DC}$, $\overline{DB}$ bisects $\angle ADC$	1. Given
2. $\angle ABD \cong \angle BDC$	2. Theorem 4-12
3. $\angle ADB \cong \angle BDC$	3. Definition of angle bisector
4. $\angle ABD \cong \angle ADB$	4. Theorem 4-6
5. $\overline{AD} \cong \overline{AB}$	5. Theorem 5-3

10.

STATEMENTS	REASONS
1. $ABCD$ is an isosceles trapezoid with $\overline{AB} \parallel \overline{DC}$.	1. Given
2. $\angle ADC \cong \angle BCD$ (or $\angle EDC \cong \angle ECD$)	2. Theorem 6-12
3. $\angle EAB \cong \angle EDC$, $\angle EBA \cong \angle ECD$	3. Axiom 4-14
4. $\angle EAB \cong \angle EBA$	4. Substitution
5. $\overline{EB} \cong \overline{EA}$	5. Theorem 5-3
6. $\triangle EAB$ is isosceles.	6. Definition of an isosceles triangle

11.

STATEMENTS	REASONS
1. $\triangle EDC$, A is midpoint of $\overline{ED}$, B is midpoint of $\overline{EC}$.	1. Given
2. $\overline{AB} \parallel \overline{DC}$	2. Theorem 6-14
3. $\angle ABD \cong \angle BDC$	3. Theorem 4-12

CHAPTER 7 SIMILARITY

Page 251 Written 27. If $\frac{a}{b} = \frac{c}{d}$, $ad = bc$ by the means-extremes property. By the division axiom, $\frac{ad}{ac} = \frac{bc}{ac}$ and $\frac{d}{c} = \frac{b}{a}$ or $\frac{b}{a} = \frac{d}{c}$. **28.** If $\frac{a}{b} = \frac{c}{d}$, $ad = bc$ by the means-extremes property. By the division axiom, $\frac{ad}{dc} = \frac{bc}{dc}$ and $\frac{a}{c} = \frac{b}{d}$. **29.** If $\frac{a}{b} = \frac{c}{d}$, by the addition axiom, $\frac{a}{b} + \frac{b}{b} = \frac{c}{d} + \frac{d}{d}$. Therefore, $\frac{a+b}{b} = \frac{c+d}{d}$. **30.** If $\frac{a}{b} = \frac{c}{d}$, by the subtraction axiom, $\frac{a}{b} - \frac{b}{b} = \frac{c}{d} - \frac{d}{d}$. Therefore, $\frac{a-b}{b} = \frac{c-d}{d}$.

31. If b is the mean proportional between a and c, then $\frac{a}{b} = \frac{b}{c}$. By the mean-extremes property, $b^2 = ac$ and $b = \sqrt{ac}$.

Pages 254–255 Written

3.

STATEMENTS	REASONS
1. $\overline{JK} \parallel \overline{MN}$	1. Given
2. $\angle K \cong \angle M$, $\angle J \cong \angle N$	2. Theorem 4-12
3. $\triangle JKL \sim \triangle NML$	3. AA

4.

STATEMENTS	REASONS
1. $\overline{JK} \parallel \overline{MN}$	1. Given
2. $\angle K \cong \angle M$, $\angle J \cong \angle N$	2. Theorem 4-12
3. $\triangle JKL \sim \triangle NML$	3. AA
4. $\frac{JL}{LN} = \frac{KL}{LM}$	4. Definition of similar polygons

5.

STATEMENTS	REASONS
1. $\triangle ABC$ with altitudes $\overline{CD}$ and $\overline{AE}$.	1. Given
2. $\overline{CD} \perp \overline{AB}$, $\overline{AE} \perp \overline{BC}$	2. Definition of altitude
3. $\angle FDA$ and $\angle FEC$ are right angles.	3. Definition of perpendicular lines
4. $\angle FDA \cong \angle FEC$	4. Theorem 4-7
5. $\angle AFD \cong \angle CFE$	5. Theorem 4-4
6. $\triangle AFD \sim \triangle CFE$	6. AA

6.

STATEMENTS	REASONS
1. $\triangle ABC$ with altitudes $\overline{CD}$ and $\overline{AE}$.	1. Given
2. $\overline{CD} \perp \overline{AB}$, $\overline{AE} \perp \overline{BC}$	2. Definition of altitude
3. $\angle FDA$ and $\angle FEC$ are right angles.	3. Definition of perpendicular lines
4. $\angle FDA \cong \angle FEC$	4. Theorem 4-7
5. $\angle AFD \cong \angle CFE$	5. Theorem 4-4
6. $\triangle AFD \sim \triangle CFE$	6. AA
7. $\frac{AF}{FC} = \frac{DE}{FE}$	7. Definition of similar polygons

7.

STATEMENTS	REASONS
1. △ABC with altitudes $\overline{CD}$ and $\overline{AE}$.	1. Given
2. $\overline{CD} \perp \overline{AB}$, $\overline{AE} \perp \overline{BC}$	2. Definition of altitude
3. ∠BDC and ∠BEA are right angles.	3. Definition of perpendicular lines
4. ∠BDC ≅ ∠BEA	4. Theorem 4-7
5. ∠ABC ≅ ∠ABC (or ∠CBD ≅ ∠ABE)	5. Theorem 4-6
6. △CBD ~ △ABE	6. AA

8.

STATEMENTS	REASONS
1. △ABC with altitudes $\overline{CD}$ and $\overline{AE}$.	1. Given
2. $\overline{CD} \perp \overline{AB}$, $\overline{AE} \perp \overline{BC}$	2. Definition of altitude
3. ∠BDC and ∠BEA are right angles.	3. Definition of perpendicular lines
4. ∠BDC ≅ ∠BEA	4. Theorem 4-7
5. ∠ABC ≅ ∠ABC (or ∠CBD ≅ ∠ABE)	5. Theorem 4-6
6. △CBD ~ △ABE	6. AA
7. $\dfrac{BC}{AB} = \dfrac{DC}{AE}$	7. Definition of similar polygons

9.

STATEMENTS	REASONS
1. ▱ABCD	1. Given
2. $\overline{AD} \parallel \overline{BC}$	2. Definition of a parallelogram
3. ∠EDA ≅ ∠EBF	3. Theorem 4-12
4. ∠AED ≅ ∠FEB	4. Theorem 4-4
5. △AED ~ △FEB	5. AA

10.

STATEMENTS	REASONS
1. ▱ABCD	1. Given
2. $\overline{AD} \parallel \overline{BC}$	2. Definition of a parallelogram
3. ∠ABF ≅ ∠GCF	3. Axiom 4-14
4. ∠BFA ≅ ∠CFG	4. Theorem 4-4
5. △ABF ~ △GCF	5. AA

11.

STATEMENTS	REASONS
1. ▱ABCD	1. Given
2. $\overline{AB} \parallel \overline{DC}$	2. Definition of a parallelogram
3. ∠EBA ≅ ∠EDG	3. Theorem 4-12
4. ∠AEB ≅ ∠GED	4. Theorem 4-4
5. △AEB ~ △GED	5. AA

12.

STATEMENTS	REASONS
1. ▱ABCD	1. Given
2. $\overline{AD} \parallel \overline{BC}$	2. Definition of a parallelogram
3. ∠ABF ≅ ∠GCF	3. Axiom 4-14
4. ∠BFA ≅ ∠CFG	4. Theorem 4-4
5. △ABF ~ △GCF	5. AA
6. $\dfrac{AF}{GF} = \dfrac{BF}{CF}$	6. Definition of similar polygons

13.

STATEMENTS	REASONS
1. ▱ABCD	1. Given
2. $\overline{AD} \parallel \overline{BC}$	2. Definition of a parallelogram
3. ∠EDA ≅ ∠EBF	3. Theorem 4-12
4. ∠AED ≅ ∠FEB	4. Theorem 4-4
5. △AED ~ △FEB	5. AA
6. $\dfrac{AD}{BF} = \dfrac{AE}{EF}$	6. Definition of similar polygons

14.

STATEMENTS	REASONS
1. ▱ABCD	1. Given
2. $\overline{AB} \parallel \overline{DC}$	2. Definition of a parallelogram
3. ∠EBA ≅ ∠EDG	3. Theorem 4-12
4. ∠AEB ≅ ∠GED	4. Theorem 4-4
5. △AEB ~ △GED	5. AA
6. $\dfrac{EB}{ED} = \dfrac{AB}{GD}$	6. Definition of similar polygons
7. $EB \cdot GD = ED \cdot AB$	7. Means-extremes property

15.

STATEMENTS	REASONS
1. $\triangle ABC$ with $\overline{AB} \cong \overline{AC}$, altitudes $\overline{AE}$ and $\overline{CD}$	1. Given
2. $\angle ACB \cong \angle ABC$ (or $\angle ACE \cong \angle DBC$)	2. Theorem 5-2
3. $\overline{AE} \perp \overline{BC}$, $\overline{CD} \perp \overline{AB}$	3. Definition of altitude
4. $\angle AEC$ and $\angle CDB$ are right angles.	4. Definition of perpendicular lines
5. $\angle AEC \cong \angle CDB$	5. Theorem 4-7
6. $\triangle AEC \sim \triangle CDB$	6. AA
7. $\dfrac{AC}{EC} = \dfrac{CB}{DB}$	7. Definition of similar polygons
8. $AC \cdot DB = CB \cdot EC$	8. Means-extremes property

16.

STATEMENTS	REASONS
1. $\triangle ABC$ with $\overline{AB} \cong \overline{AC}$, altitudes $\overline{AE}$ and $\overline{CD}$	1. Given
2. $\overline{CD} \perp \overline{AB}$, $\overline{AE} \perp \overline{BC}$	2. Definition of altitude
3. $\angle FDA$ and $\angle FEC$ are right angles.	3. Definition of perpendicular lines
4. $\angle FDA \cong \angle FEC$	4. Theorem 4-7
5. $\angle AFD \cong \angle CFE$	5. Theorem 4-4
6. $\triangle AFD \sim \triangle CFE$	6. AA
7. $\dfrac{AF}{AD} = \dfrac{CF}{CE}$	7. Definition of similar polygons
8. $AF \cdot CE = CF \cdot AD$	8. Means-extremes property
9. $\overline{AE}$ is a median.	9. An altitude of an isosceles triangle is also a median
10. $BE = CE$	10. Definition of median
11. $AF \cdot BE = CF \cdot AD$	11. Substitution

17.

STATEMENTS	REASONS
1. $\triangle ABC$ with $\overline{AB} \cong \overline{AC}$, altitudes $\overline{AE}$ and $\overline{CD}$	1. Given
2. $\angle ACB \cong \angle ABC$ (or $\angle ACE \cong \angle DBC$)	2. Theorem 5-2
3. $\overline{AE} \perp \overline{BC}$, $\overline{CD} \perp \overline{AB}$	3. Definition of altitude
4. $\angle AEC$ and $\angle CDB$ are right angles.	4. Definition of perpendicular lines
5. $\angle AEC \cong \angle CDB$	5. Theorem 4-7
6. $\triangle AEC \sim \triangle CDB$	6. AA
7. $\dfrac{AC}{EC} = \dfrac{CB}{DB}$	7. Definition of similar polygons
8. $AC \cdot DB = CB \cdot EC$	8. Means-extremes property
9. $AB \cdot DB = CB \cdot EC$	9. Substitution

18.

STATEMENTS	REASONS
1. $\triangle ABC$ with $\overline{AB} \cong \overline{AC}$, altitudes $\overline{AE}$ and $\overline{CD}$	1. Given
2. $\angle ACB \cong \angle ABC$ (or $\angle ACE \cong \angle DBC$)	2. Theorem 5-2
3. $\overline{AE} \perp \overline{BC}$, $\overline{CD} \perp \overline{AB}$	3. Definition of altitude
4. $\angle AEC$ and $\angle CDB$ are right angles.	4. Definition of perpendicular lines
5. $\angle AEC \cong \angle CDB$	5. Theorem 4-7
6. $\triangle AEC \sim \triangle CDB$	6. AA
7. $\dfrac{AC}{EC} = \dfrac{CB}{DB}$	7. Definition of similar polygons
8. $AC \cdot DB = EC \cdot CB$	8. Means-extremes property
9. $\overline{AE}$ is a median.	9. An altitude of an isosceles triangle is also a median.
10. $BE = EC$	10. Definition of median
11. $AC \cdot DB = BE \cdot CB$	11. Substitution

19. Given: Isosceles triangles ABC and XYZ, ∠A ≅ ∠X
    Prove: △ABC ~ △XYZ

STATEMENTS	REASONS
1. Isosceles triangles ABC and XYZ, ∠A ≅ ∠X	1. Given
2. ∠B ≅ ∠C, ∠Y ≅ ∠Z	2. Theorem 5-2
3. m∠A + m∠B + m∠C = 180 m∠X + m∠Y + m∠Z = 180	3. Theorem 4-19
4. m∠A + 2(m∠B) = m∠X + 2(m∠Y)	4. Substitution
5. 2(m∠B) = 2(m∠Y)	5. Subtraction axiom
6. m∠B = m∠Y (∠B ≅ ∠Y)	6. Division axiom
7. △ABC ~ △XYZ	7. AA

20. Given: Equilateral triangles ABC and XYZ
    Prove: △ABC ~ △XYZ

STATEMENTS	REASONS
1. Equilateral triangles ABC and XYZ	1. Given
2. △ABC and △XYZ are equiangular.	2. If a triangle is equilateral, then it is also equiangular.
3. m∠A = 60, m∠B = 60, m∠X = 60, m∠Y = 60	3. Theorem 4-24
4. m∠A = m∠X, m∠B = m∠Y	4. Transitive axiom
5. △ABC ~ △XYZ	5. AA

21. Given: △ABC with right ∠B.
    △XYZ with right ∠Y.
    ∠A ≅ ∠X

Prove: △ABC ~ △XYZ

STATEMENTS	REASONS
1. △ABC with right ∠B. △XYZ with right ∠Y, ∠A ≅ ∠X	1. Given
2. ∠B ≅ ∠Y	2. Theorem 4-7
3. △ABC ~ △XYZ	3. AA

22. Given: △ABC ~ △XYZ, △XYZ ~ △MNP
    I. Prove: △ABC ~ △ABC

STATEMENTS	REASONS
1. ∠A ≅ ∠A, ∠B ≅ ∠B	1. Theorem 4-6
2. △ABC ~ △ABC	2. AA

II. Prove: △XYZ ~ △ABC

STATEMENTS	REASONS
1. △ABC ~ △XYZ	1. Given
2. ∠A ≅ ∠X, ∠B ≅ ∠Y, ∠C ≅ ∠Z	2. Definition of similar polygons
3. ∠X ≅ ∠A, ∠Y ≅ ∠B, ∠Z ≅ ∠C	3. Theorem 4-6
4. △XYZ ~ △ABC	4. AA

III. Prove: △ABC ~ △MNP

STATEMENTS	REASONS
1. △ABC ~ △XYZ, △XYZ ~ △MNP	1. Given
2. ∠A ≅ ∠X, ∠B ≅ ∠Y, ∠X ≅ ∠M, ∠Y ≅ ∠N	2. Definition of similar polygons
3. ∠A ≅ ∠M, ∠B ≅ ∠N	3. Theorem 4-6
4. △ABC ~ △MNP	4. AA

Pages 259–260 Written

18.

STATEMENTS	REASONS
1. ▱ABCD, $\overline{EG} \parallel \overline{BC}$, $\overline{GF} \parallel \overline{AB}$	1. Given
2. $\overline{BC} \parallel \overline{AD}$, $\overline{AB} \parallel \overline{DC}$	2. Definition of a parallelogram
3. $\overline{EG} \parallel \overline{AD}$, $\overline{GF} \parallel \overline{DC}$	3. Theorem 4-18
4. $\dfrac{BE}{AE} = \dfrac{BG}{DG}$, $\dfrac{BG}{DG} = \dfrac{BF}{FC}$	4. Theorem 7-3
5. $\dfrac{BE}{AE} = \dfrac{BF}{FC}$	5. Transitive axiom

19.

STATEMENTS	REASONS
1. In $\triangle ABC$, $\overline{DE} \parallel \overline{BC}$	1. Given
2. $\angle ADQ \cong \angle ABM$, $\angle AQD \cong \angle AMB$, $\angle AQE \cong \angle AMC$, $\angle AEQ \cong \angle ACM$	2. Axiom 4-14
3. $\triangle ADQ \sim \triangle ABM$, $\triangle AQE \sim \triangle AMC$	3. AA
4. $\dfrac{DQ}{BM} = \dfrac{AQ}{AM}$, $\dfrac{AQ}{AM} = \dfrac{QE}{MC}$	4. Definition of similar polygons
5. $\dfrac{DQ}{BM} = \dfrac{QE}{MC}$	5. Transitive axiom
6. $\dfrac{DQ}{QE} = \dfrac{BM}{MC}$	6. Interchanging property
7. $\overline{AM}$ is a median.	7. Given
8. $BM = MC$	8. Definition of median
9. $\dfrac{BM}{MC} = 1$	9. Division axiom
10. $\dfrac{DQ}{QE} = 1$	10. Substitution
11. $\overline{DQ} \cong \overline{QE}$	11. Multiplication axiom

20.

STATEMENTS	REASONS
1. $\angle 1 \cong \angle 2$	1. Given
2. $\overleftrightarrow{UV} \parallel \overleftrightarrow{RT}$	2. Axiom 4-15
3. $\dfrac{SU}{UR} = \dfrac{SV}{VT}$	3. Theorem 7-2

21.

STATEMENTS	REASONS
1. $\angle 2 \cong \angle 3$	1. Given
2. $\overleftrightarrow{UW} \parallel \overleftrightarrow{ST}$	2. Axiom 4-15
3. $\dfrac{RU}{US} = \dfrac{RW}{WT}$	3. Theorem 7-2
4. $RU \cdot WT = US \cdot RW$	4. Means–extremes property

22.

STATEMENTS	REASONS
1. $\angle 1 \cong \angle 2$, $\angle 2 \cong \angle 3$	1. Given
2. $\overleftrightarrow{UV} \parallel \overleftrightarrow{RT}$, $\overleftrightarrow{UW} \parallel \overleftrightarrow{ST}$	2. Axiom 4-15
3. $\dfrac{SV}{VT} = \dfrac{US}{RU}$, $\dfrac{US}{RU} = \dfrac{WT}{RW}$	3. Theorem 7-2
4. $\dfrac{WT}{RW} = \dfrac{SV}{VT}$	4. Transitive axiom
5. $WT \cdot VT = RW \cdot SV$	5. Means–extremes property

23. Given: $\overleftrightarrow{AD} \parallel \overleftrightarrow{BE} \parallel \overleftrightarrow{CF}$

Prove: $\dfrac{AB}{BC} = \dfrac{DE}{EF}$

STATEMENTS	REASONS
1. $\overleftrightarrow{AD} \parallel \overleftrightarrow{BE} \parallel \overleftrightarrow{CF}$	1. Given
2. Draw $\overline{DC}$.	2. Axiom 4-2
3. $\dfrac{AB}{BC} = \dfrac{DG}{GC}$, $\dfrac{DG}{GC} = \dfrac{DE}{EF}$	3. Theorem 7-2
4. $\dfrac{AB}{BC} = \dfrac{DE}{EF}$	4. Transitive axiom

24. Given: $\overleftrightarrow{AD} \parallel \overleftrightarrow{BE} \parallel \overleftrightarrow{CF}$

Prove: $\dfrac{AB}{BC} = \dfrac{DE}{EF}$

STATEMENTS	REASONS
1. $\overleftrightarrow{AD} \parallel \overleftrightarrow{BE} \parallel \overleftrightarrow{CF}$	1. Given
2. Draw $\overline{AF}$.	2. Axiom 4-2
3. $\dfrac{AB}{BC} = \dfrac{AG}{GF}$, $\dfrac{AG}{GF} = \dfrac{DE}{EF}$	3. Theorem 7-2
4. $\dfrac{AB}{BC} = \dfrac{DE}{EF}$	4. Transitive axiom

25. Case 1: The transversals intersect on either of the outside lines.
Given: $\overleftrightarrow{FA} \parallel \overleftrightarrow{BD} \parallel \overleftrightarrow{CE}$
Prove: $\dfrac{AB}{BC} = \dfrac{AD}{DE}$

STATEMENTS	REASONS
1. $\overleftrightarrow{FA} \parallel \overleftrightarrow{BD} \parallel \overleftrightarrow{CE}$	1. Given
2. $\dfrac{AB}{BC} = \dfrac{AD}{DE}$	2. Theorem 7-2

Case 2. The transversals intersect on the middle line.
Given: $\overleftrightarrow{AD} \parallel \overleftrightarrow{BX} \parallel \overleftrightarrow{CF}$
Prove: $\dfrac{AB}{BF} = \dfrac{DB}{BC}$

STATEMENTS	REASONS
1. $\overleftrightarrow{AD} \parallel \overleftrightarrow{BX} \parallel \overleftrightarrow{CF}$	1. Given
2. $\angle BAD \cong \angle BFC$, $\angle BDA \cong \angle BCF$	2. Theorem 4-12
3. $\triangle ABD \sim \triangle FBC$	3. AA
4. $\dfrac{AB}{BF} = \dfrac{DB}{BC}$	4. Definition of similar polygons

27. Given: $\triangle ABC \sim \triangle XYZ$, altitudes $\overline{BD}$ and $\overline{YW}$
Prove: $\dfrac{BD}{YW} = \dfrac{AB}{XY}$

STATEMENTS	REASONS
1. $\triangle ABC \sim \triangle XYZ$, altitudes $\overline{BD}$ and $\overline{YW}$	1. Given
2. $\overline{BD} \perp \overline{AC}$, $\overline{YW} \perp \overline{XZ}$	2. Definition of altitude
3. $\angle BDA$ and $\angle YWZ$ are right angles.	3. Definition of perpendicular lines
4. $\angle BDA \cong \angle YWZ$	4. Theorem 4-7
5. $\angle A \cong \angle X$	5. Definition of similar polygons
6. $\triangle ABD \sim \triangle XYW$	6. AA
7. $\dfrac{BD}{YW} = \dfrac{AB}{XY}$	7. Definition of similar polygons

26. Given: $\overleftrightarrow{AD} \parallel \overleftrightarrow{BE} \parallel \overleftrightarrow{CF}$
Prove: $\dfrac{AE}{EF} = \dfrac{DB}{BC}$

STATEMENTS	REASONS
1. $\overleftrightarrow{AD} \parallel \overleftrightarrow{BE} \parallel \overleftrightarrow{CF}$	1. Given
2. $\angle XAD \cong \angle XEB$, $\angle XDA \cong \angle XBE$	2. Theorem 4-12
3. $\triangle AXD \sim \triangle EXB$	3. AA
4. $\dfrac{XB}{XE} = \dfrac{XD}{XA}$	4. Definition of similar polygons
5. $XB + XD = DB$, $XE + XA = AE$	5. Segment addition axiom
6. $\dfrac{XB + XD}{XE + XA} = \dfrac{XB}{XE}$	6. Property of proportions (challenge, page 237)
7. $\dfrac{DB}{AE} = \dfrac{XB}{XE}$	7. Substitution
8. $\dfrac{XB}{BC} = \dfrac{XE}{EF}$	8. Theorem 7-2
9. $\dfrac{XB}{XE} = \dfrac{BC}{EF}$	9. Interchanging property
10. $\dfrac{DB}{AE} = \dfrac{BC}{EF}$	10. Transitive property
11. $\dfrac{AE}{EF} = \dfrac{DB}{BC}$	11. Interchanging property, Symmetric axiom

34. Given: $\overline{AD}$ bisects $\angle BAC$.
Prove: $\dfrac{DB}{DC} = \dfrac{AB}{AC}$

STATEMENTS	REASONS
1. $\overline{AD}$ bisects $\angle BAC$	1. Given
2. Draw $\overline{XC}$ parallel to $\overline{AD}$.	2. Theorem 4-16
3. $\angle DAC \cong \angle XCA$	3. Theorem 4-12
4. $\angle BAD \cong \angle AXC$	4. Axiom 4-14
5. $\angle BAD \cong \angle DAC$	5. Definition of angle bisector
6. $\angle XCA \cong \angle AXC$	6. Substitution
7. $\overline{AX} \cong \overline{AC}$	7. Theorem 5-3
8. $\dfrac{DB}{DC} = \dfrac{AB}{AX}$	8. Theorem 7-2
9. $\dfrac{DB}{DC} = \dfrac{AB}{AC}$	9. Substitution

Page 250 Challenge

Given: $\triangle ABC \sim \triangle XYZ$

Prove: $\dfrac{AB + BC + CA}{XY + YZ + ZX} = \dfrac{AB}{XY}$

STATEMENTS	REASONS
1. $\triangle ABC \sim \triangle XYZ$	1. Given
2. $\dfrac{AB}{XY} = \dfrac{BC}{YZ} = \dfrac{CA}{ZX}$	2. Definition of similar polygons
3. $\dfrac{AB + BC + CA}{XY + YZ + ZX} = \dfrac{AB}{XY}$	3. Property of proportions (challenge, page 237)

Pages 263–264 Exploratory

1. $\triangle DGE \sim \triangle FGD$, $\triangle DGE \sim \triangle FDE$, $\triangle DGF \sim \triangle EDF$ **2.** $\triangle DGE \sim \triangle FGD$: $\angle GDE$, $\angle GFD$; $\angle DGE$, $\angle FGD$; $\angle GED$, $\angle GDF$; $\triangle DGE \sim \triangle FDE$: $\angle GDE$, $\angle DFE$; $\angle DGE$, $\angle FDE$; $\angle GED$, $\angle DEF$; $\triangle DGF \sim \triangle EDF$: $\angle GDF$, $\angle DEF$; $\angle DGF$, $\angle EDF$; $\angle GFD$, $\angle DFE$ **3.** $\triangle DGE \sim \triangle FGD$: $\overline{DG}$, $\overline{FG}$; $\overline{GE}$, $\overline{GD}$; $\overline{DE}$, $\overline{FD}$; $\triangle DGE \sim \triangle FDE$: $\overline{DG}$, $\overline{FD}$; $\overline{GE}$, $\overline{DE}$; $\overline{DE}$, $\overline{FE}$; $\triangle DGF \sim \triangle EDF$: $\overline{DG}$, $\overline{ED}$; $\overline{GF}$, $\overline{DF}$; $\overline{DF}$, $\overline{EF}$ **4.** Answers may vary.

26. Theorem 7-6: The measure of the altitude to the hypotenuse of a right triangle is the geometric mean of the measures of the segments of the hypotenuse. Theorem 7-7: If the altitude is drawn to the hypotenuse in a right triangle, the measure of each leg is the geometric mean of the measure of the hypotenuse and the measure of the segment on the hypotenuse that is adjacent to the leg.

Page 264 Written

18. Given: Right $\triangle ACB$; right $\angle C$, $\overline{CD} \perp \overline{AB}$

Prove: $\triangle ABC \sim \triangle ACD$
$\triangle ABC \sim \triangle CBD$
$\triangle ACD \sim \triangle CBD$

STATEMENTS	REASONS
1. Right $\triangle ABC$, right $\angle C$; $\overline{CD} \perp \overline{AB}$	1. Given
2. $\angle CDA$ and $\angle BDC$ are right angles.	2. Definition of perpendicular lines
3. $\angle CDA \cong \angle BDC$, $\angle CDA \cong \angle BCA$, $\angle BDC \cong \angle BCA$	3. Theorem 4-7
4. $\angle BAC \cong \angle BAC$ (or $\angle BAC \cong \angle CAD$) $\angle ABC \cong \angle ABC$ (or $\angle ABC \cong \angle CBD$)	4. Theorem 4-6
5. $\triangle ABC \sim \triangle ACD$, $\triangle ABC \sim \triangle CBD$	5. AA
6. $\triangle ACD \sim \triangle CBD$	6. Theorem 7-1

19. Given: Right $\triangle ABC$, right $\angle C$, $\overline{CD} \perp \overline{AB}$

Prove: $\dfrac{AD}{CD} = \dfrac{CD}{BD}$

STATEMENTS	REASONS
1. Right $\triangle ABC$, right $\angle C$, $\overline{CD} \perp \overline{AB}$	1. Given
2. $\triangle ACD \sim \triangle CBD$	2. Theorem 7-5
3. $\dfrac{AD}{CD} = \dfrac{CD}{BD}$	3. Definition of similar polygons

20. Given: Right $\triangle ABC$ with right $\angle C$, $\overline{CD} \perp \overline{AB}$

Prove: $\dfrac{AB}{AC} = \dfrac{AC}{AD}$, $\dfrac{AB}{BC} = \dfrac{BC}{DB}$

STATEMENTS	REASONS
1. Right $\triangle ABC$ with right $\angle C$, $\overline{CD} \perp \overline{AB}$	1. Given
2. $\triangle ABC \sim \triangle ACD$, $\triangle ABC \sim \triangle CBD$	2. Theorem 7-5
3. $\dfrac{AB}{AC} = \dfrac{AC}{AD}$, $\dfrac{AB}{BC} = \dfrac{BC}{DB}$	3. Definition of similar polygons

Page 268 Challenge

3.

STATEMENTS	REASONS
1. $\triangle ABC$, altitudes $\overline{AF}$, $\overline{BE}$ and $\overline{CD}$ intersect at G.	1. Given
2. $\overline{AF} \perp \overline{BC}$, $\overline{BE} \perp \overline{AC}$, $\overline{CD} \perp \overline{AB}$	2. Definition of altitude
3. $\angle ADG$, $\angle CFG$, $\angle AEG$, $\angle BFG$, $\angle CEG$, and $\angle BDG$ are right angles.	3. Definition of perpendicular lines
4. $\angle ADG \cong \angle CFG$, $\angle AEG \cong \angle BFG$, $\angle CEG \cong \angle BDG$	4. Theorem 4-7
5. $\angle DGA \cong \angle FGC$, $\angle EGA \cong \angle FGB$, $\angle EGC \cong \angle DGB$	5. Theorem 4-4
6. $\triangle ADG \sim \triangle CFG$, $\triangle AEG \sim \triangle BFG$, $\triangle CEG \sim \triangle BDG$	6. AA

STATEMENTS	REASONS
7. $\dfrac{AE}{FB} = \dfrac{AG}{GB}$, $\dfrac{BD}{EC} = \dfrac{GB}{GC}$, $\dfrac{CF}{DA} = \dfrac{CG}{AG}$	7. Definition of similar polygons
8. $\dfrac{AE}{FB} \cdot \dfrac{BD}{EC} \cdot \dfrac{CF}{DA} = \dfrac{AG}{GB} \cdot \dfrac{GB}{GC} \cdot \dfrac{CG}{AG}$	8. Multiplication axiom
9. $\dfrac{AE}{EC} \cdot \dfrac{CF}{FB} \cdot \dfrac{BD}{DA} = \dfrac{AG}{AG} \cdot \dfrac{GB}{GB} \cdot \dfrac{CG}{CG}$	9. Commutative and associative properties
10. $\dfrac{AE}{EC} \cdot \dfrac{CF}{FB} \cdot \dfrac{BD}{DA} = 1$	10. Definition of multiplicative inverse

Page 274 Written 17–20. By labeling each triangle in 11–14 like the one in Construction 12, the justification is the same as the one given for Construction 12.

Page 276 Exploratory 9. {(2, 3), (2, 4), (2, 9), (2, 16), (3, 4), (3, 9), (3, 16), (4, 9), (4, 16), (9, 16)}
10. {(3, 2), (4, 2), (4, 3), (9, 2), (9, 3), (9, 4), (16, 2), (16, 3), (16, 4), (16, 9)} **11.** {(2, 2), (3, 3), (4, 4), (9, 9), (16, 16)} **12.** {(4, 2), (9, 3), (16, 4)} **13.** {(4, 2)} **14.** {(2, 4), (2, 16), (3, 9), (4, 16)}
15. {(2, 4), (3, 9), (4, 16)} **16.** {(2, 3), (3, 4)}

Pages 276–277 Written 1. $T = \{\triangle ADC, \triangle CBA, \triangle AEB, \triangle BEC, \triangle CED, \triangle DEA, \triangle DAB, \triangle BCD\}$
2. $\triangle ADC, \triangle CBA$; $\triangle AEB, \triangle CED$; $\triangle BEC, \triangle DEA$; $\triangle DAB, \triangle BCD$ **3.** $C = \{(\triangle ADC, \triangle ADC), (\triangle ADC, \triangle CBA), (\triangle CBA, \triangle ADC), (\triangle CBA, \triangle CBA), (\triangle AEB, \triangle AEB), (\triangle AEB, \triangle CED), (\triangle CED, \triangle AEB), (\triangle CED, \triangle CED), (\triangle BEC, \triangle BEC), (\triangle BEC, \triangle DEA), (\triangle DEA, \triangle BEC), (\triangle DEA, \triangle DEA), (\triangle DAB, \triangle DAB), (\triangle DAB, \triangle BCD), (\triangle BCD, \triangle DAB), (\triangle BCD, \triangle BCD)\}$
6. {(S, E), (S, F), (S, C), (S, B), (S, L)} **7.** {(J, E), (Z, F), (Z, C)} **8.** {(S, J), (S, Z), (S, M), (S, N), (N, B), (N, L)} **9.** {(M, J), (M, Z), (M, N), (N, J), (N, Z), (F, C), (C, F), (L, B)} **10.** {(J, Z), (J, M), (J, N), (Z, J), (Z, M), (Z, N), (B, L)}
11. {(M, E), (M, F), (M, C), (M, B), (M, L), (N, E), (N, F), (N, C)} **12.** {(J, F), (J, C), (J, B), (J, L), (Z, E), (Z, B), (Z, L)} **13.** {(E, F), (E, C), (E, B), (E, L), (F, E), (F, B), (F, L), (C, E), (C, B), (C, L), (B, E), (B, F), (B, C), (L, E), (L, F), (L, C)} **14.** $R = \{(0, 0), (0, 2), (0, 4), (0, 6), (1, 1), (1, 3), (1, 5), (1, 7), (2, 2), (2, 0), (2, 4), (2, 6), (3, 3), (3, 1), (3, 5), (3, 7), (4, 4), (4, 0), (4, 2), (4, 6), (5, 5), (5, 1), (5, 3), (5, 7), (6, 6), (6, 0), (6, 2), (6, 4), (7, 7), (7, 1), (7, 3), (7, 5)\}$

Page 277 Mathematical Excursions

		X	Y	Z
s	t	$(s^2 - t^2)$	$(2st)$	$(s^2 + t^2)$
2	1	3	4	5
3	2	5	12	13
4	1	15	8	17
4	3	7	24	25
5	2	21	20	29
5	4	9	40	41
6	1	35	12	37
6	5	11	60	61
7	2	45	28	53
7	4	33	56	65
7	6	13	84	85
8	1	63	16	65
8	3	55	48	73
8	5	39	80	89
9	2	77	36	85
9	4	65	72	97

Page 280 Exploratory 1. true; definition of a relation 2. true; definition of a relation 3. false; (4, 3) is not in R. 4. true; R fits the definition of reflexive. 5. false; (3, 4) is in R, but (4, 3) is not in R. 6. false; (2, 3) and (3, 4) are in R, but (2, 4) is not in R. 7. false; R is not symmetric or transitive.

Page 280 Written 7. yes; The relation is reflexive because a lives in the same state as a. The relation is symmetric because if a lives where b lives, b lives where a does. The relation is transitive because if a lives where b lives and b lives where c lives, then a lives where c does 8. yes; Like equality, "has the same weight as" is reflexive, symmetric, and transitive.

Page 286 Chapter Test

15.	STATEMENTS	REASONS
1.	$\overline{UW} \parallel \overline{ST}, \overline{UV} \parallel \overline{RT}$	1. Given
2.	$\dfrac{BU}{US} = \dfrac{RW}{WT}, \dfrac{RU}{US} = \dfrac{VT}{SV}$	2. Theorem 7-2
3.	$\dfrac{RW}{WT} = \dfrac{VT}{SV}$	3. Transitive property
4.	$RW \cdot SV = VT \cdot WT$	4. Means-extremes property

CHAPTER 8 GEOMETRY WITH COORDINATES

Page 290 Written 1–12

Page 293 Written

1.
2.
3.
4.
5.
6.
7.
8.
9.
10.

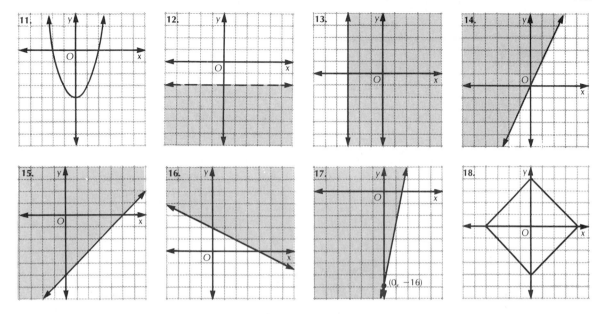

Page 300 Exploratory 1. Form a right triangle with (2, 1) and (5, 9) the endpoints of the hypotenuse and the vertex of the right angle at (5, 1). The legs are 3 units and 8 units in length. Using the Pythagorean Theorem, you find the measure of the segment is $\sqrt{3^2 + 8^2}$ or $\sqrt{73}$. **2.** Use the same procedure as in exercise 1. The vertex of the right angle is at (6, 1). The legs measure 3 and 4 and the segment measures 5. **3.** The vertex of the right angle is (5, 1). The legs measure 4 and 7 and the segment measures $\sqrt{65}$.

Page 301 Written 13. The argument is given for two general points (x_1, y_1) and (x_2, y_2). Their signs have no effect on the result in this consideration. It is obvious that $|x_2 - x_1| = |x_1 - x_2|$ and $|y_2 - y_1| = |y_1 - y_2|$. When the expressions $(x_2 - x_1)$ and $(y_2 - y_1)$ are squared, the results are positive in all cases. The distance is found by taking the square root of two positive numbers.

Page 305 Written 33. (1, 3); (3, 1); $DE = 2\sqrt{2}$; $BC = 4\sqrt{2}$; $DE = \frac{1}{2}BC$ **34.** (0, −1); (3, −2); $DE = \sqrt{10}$; $BC = 2\sqrt{10}$; $DE = \frac{1}{2}BC$ **35.** (−1, 3); (−1, −2); $DE = 5$, $BC = 10$; $DE = \frac{1}{2}BC$ **36.** (−5, 4); (−2, 5$\frac{1}{2}$); $DE = \frac{3}{2}\sqrt{5}$, $BC = 3\sqrt{5}$, $DE = \frac{1}{2}BC$ **38.** (2, 1$\frac{1}{2}$), 2$\frac{1}{2}$; 5; The median is half of the hypotenuse. **39.** (3, 1); $\sqrt{10}$; $2\sqrt{10}$; The median is half the hypotenuse. **40.** (0, −3); $2\sqrt{2}$; $4\sqrt{2}$; The median is half the hypotenuse. **41.** (3, 0); 3; 6; The median is half the hypotenuse.

Page 305 Challenge 5. By the distance formula, $(AP)^2 = \left(\dfrac{bx_1 + ax_2}{a + b} - x_1\right)^2 + \left(\dfrac{by_1 + ay_2}{a + b} - y_1\right)^2$;

$(PB)^2 = \left(x_2 - \dfrac{bx_1 + ax_2}{a + b}\right)^2 + \left(y_2 - \dfrac{by_1 + ay_2}{a + b}\right)^2$ Simplify the expressions on the right and divide the first equation by the second.

$\left(\dfrac{AP}{PB}\right)^2 = \dfrac{(ax_2 - ax_1)^2 + (ay_2 - ay_1)^2}{(bx_2 - bx_1)^2 + (by_2 - by_1)^2}$ or $\dfrac{a^2[(x_2 - x_1)^2 + (y_2 - y_1)^2]}{b^2[(x_2 - x_1)^2 + (y_2 - y_1)^2]}$; $\dfrac{AP}{PB} = \dfrac{a}{b}$

Page 313 Challenge 1. Given: nonvertical lines $\ell \parallel n$. Prove: slope of ℓ = slope of n. If the lines are parallel to the x-axis (horizontal), they both have slope 0. Therefore, their slope is the same. If the lines are not parallel to the x-axis, they intersect the x-axis forming congruent corresponding angles at A and D. Draw lines BC and EF ⊥ x-axis through any points B on ℓ and E on n. Right $\angle BCA \cong$ right $\angle EFD$ since all right angles are congruent. $\triangle BCA \sim \triangle EFD$ by AA. $\dfrac{BC}{AC} = \dfrac{EF}{DF}$ because ratios of measures of corresponding sides of corresponding polygons are equal. But BC is the difference of the y-coordinates of A and B; AC is the difference of the x-coordinates; EF is the difference of the y-coordinates of D and E, and DF is the difference of the x-coordinates. Therefore, $\frac{BC}{AC}$ is the slope of $\overleftrightarrow{AB}$ or ℓ and $\frac{EF}{DF}$ is the slope of $\overleftrightarrow{DE}$ or n. Since $\frac{BC}{AC} = \frac{EF}{DF}$, the slope of ℓ = the slope of n. **2.** Given: nonvertical lines ℓ and n, slope of ℓ = slope of n. Prove: $\ell \parallel n$. Suppose D is a point on line n. Consider the line through $D \parallel \ell$. (Through a point not on a given line, there exists one and only one line parallel to the given line.) This line has the same slope as ℓ by Theorem 8-3. If the line has the same slope as ℓ, it must have the same slope as n (Given). But only one line can be drawn through D with a certain slope, so that line must be n. Therefore, $n \parallel \ell$. **3. Part I** Given: nonvertical lines $\ell_1 \perp \ell_2$. Prove: $m_1 = -\frac{1}{m_2}$ or $m_1 m_2 = -1$. Suppose ℓ_1 and ℓ_2 intersect at $C(x_2, y_2)$; ℓ_1 intersects x-axis at $A(x_1, 0)$; ℓ_2 intersects x-axis at $B(x_3, 0)$. Draw $\overline{CD} \perp \overline{AB}$. (Theorem 4-17: Through a point not on a given line, there exists one and only one line perpendicular to a given line). D has coordinates $(x_2, 0)$. $m_1 = \frac{y_2}{x_2 - x_1}$; $m_2 = \frac{-y_2}{x_3 - x_2}$. Since $\triangle ABC$ is a right triangle and $\overline{CD} \perp \overline{AB}$, $\frac{AB}{CD} = \frac{CD}{DB}$ or $\frac{x_2 - x_1}{y_2} = \frac{y_2}{x_3 - x_2}$. By substitution, $\frac{1}{m_1} = -m_2$. Therefore, $m_1 m_2 = -1$ or $m_1 = -\frac{1}{m_2}$. **Part II** Given: Intersecting nonvertical lines ℓ_1 and ℓ_2 with slopes such that $m_1 m_2 = -1$ or $m_1 = -\frac{1}{m_2}$. Prove: $\ell_1 \perp \ell_2$. Suppose ℓ_1 and ℓ_2 intersect at $C(x_2, y_2)$, ℓ_1 intersects x-axis at $A(x_1, 0)$, and ℓ_2 intersects x-axis at $B(x_3, 0)$. Prove $\triangle ABC$ is a right triangle by using the converse of Pythagorean Theorem. $m_1 = \frac{y_2}{x_2 - x_1}$; $m_2 = \frac{-y_2}{x_3 - x_1}$. Since $m_1 = -\frac{1}{m_2}$ (Given), by substitution, $\frac{y_2}{x_2 - x_1} = -\left(\frac{x_3 - x_2}{-y_2}\right)$ or $\frac{x_3 - x_2}{y_2}$. Cross multiplication produces the following: $y_2^2 = x_2 x_3 - x_2^2 - x_1 x_3 + x_1 x_2$. Multiply by 2: $2y_2^2 = 2x_2 x_3 - 2x_2^2 - 2x_1 x_3 + 2x_1 x_2$. Rearrange terms: $x_2^2 - 2x_1 x_2 + y_2^2 + y_2^2 - 2x_2 x_3 + x_2^2 = -2x_1 x_3$. Add $x_1^2 + x_3^2$ to both sides: $x_2^2 - 2x_1 x_2 + x_1^2 + y_2^2 + y_2^2 + x_3^2 - 2x_2 x_3 + x_2^2 = x_3^2 - 2x_1 x_3 + x_1^2$. Simplify: $(x_2 - x_1)^2 + y_2^2 + y_2^2 + (x_3 - x_2)^2 = (x_3 - x_1)^2$. But, $(AC)^2 = (x_2 - x_1)^2 + y_2^2$ and $(CB)^2 = (x_3 - x_2)^2 + y_2^2$ by Pythagorean Theorem, and $(AB)^2 = (x_3 - x_1)^2$. Hence, $(AC)^2 + (CB)^2 = (AB)^2$. Therefore, $\triangle ABC$ is a right triangle by the converse of the Pythagorean Theorem. $\overline{BC} \perp \overline{AC}$ or $\ell_1 \perp \ell_2$

4. Call slope of $\overline{AB}$, m_1, and slope of $\overline{CD}$, m_2. $m_1 = \frac{y_2 - y_1}{x_2 - x_1}$; $m_2 = \frac{x_1 - x_2}{y_2 - y_1}$ or $-\left(\frac{x_2 - x_1}{y_2 - y_1}\right)$ (Definition of slope). $m_1 = -\frac{1}{m_2}$ by substitution. $\overleftrightarrow{AB} \perp \overleftrightarrow{CD}$ by Theorem 8-5: Two nonvertical lines are perpendicular if their slopes are negative reciprocals of each other.

Page 317 Written

10.

11.

12.

13.

14. **15.** **16.** **17.**

18.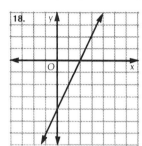

Page 318   Written   63. $(3, 3)$; $y = -x + 6$; $m = -1$; $y = -x + 6$; $y = x$; $(3, 3)$ **64.** $(1, 2)$; $2y + x = 5$; $m = -\frac{1}{2}$; $2y + x = 5$; $y = 2x$; $(1, 2)$

65. $\left(\frac{13}{2}, \frac{5}{2}\right)$; $7x - y = 43$; $m = -3$; $y = -3x + 27$; $3y = x + 1$; $(8, 3)$

66. $\left(-\frac{7}{2}, 3\right)$; $11y - 2x = 40$; $m = \frac{3}{4}$; $4y = 3x + 10$; $3y + 4x = -5$; $(-2, 1)$

67. $\left(1, \frac{3}{2}\right)$; $x = 1$; $m = 10$; $y = 10x - 7$; $10y + x = 16$; $\left(\frac{86}{101}, \frac{153}{101}\right)$

68. $\left(\frac{3}{2}, 4\right)$; $(y = 6x - 5)$; $m = \frac{11}{2}$; $2y = 11x - 9$; $11y + 2x = 47$; $\left(\frac{193}{125}, \frac{499}{125}\right)$

Page 327   Chapter Review

1.–4. **5.** **6.** **7.**

8. **9.** **10.**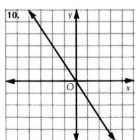

Page 328   Chapter Test

1. **2.** **3.** **4.**

5.   6.

11. $\overline{AB}$: $(-6\frac{1}{2}, 2\frac{1}{2})$; $\overline{AC}$: $(-4, 1)$; $\overline{BC}$: $(-6\frac{1}{2}, -\frac{1}{2})$
12. $\overline{AB}$: $(3, 4\frac{1}{2})$; $\overline{AC}$: $(5\frac{1}{2}, 3)$; $\overline{BC}$: $(6\frac{1}{2}, 4\frac{1}{2})$
13. $\overline{AB}$: $(-3\frac{1}{2}, -3\frac{1}{2})$; $\overline{AC}$: $(-1\frac{1}{2}, -5\frac{1}{2})$; $\overline{BC}$: $(-3, -7)$
14. $\overline{AB}$: $(4, 3)$; $\overline{AC}$: $(5, 0)$; $\overline{BC}$: $(7, 2)$

CHAPTER 9 USING COORDINATES
Page 334 Written

14. By the definition of a rhombus, $AD = DC$.

$$AD = DC$$
$$\sqrt{(a-0)^2 + (0-0)^2} = \sqrt{[(a+b)-a]^2 + (c-0)^2}$$
$$\sqrt{a^2} = \sqrt{b^2 + c^2}$$
$$a^2 = b^2 + c^2 \text{ or } c^2 = a^2 - b^2$$

17. Since $\dfrac{c}{a+b} \cdot \dfrac{c}{b-a} = \dfrac{c^2}{-(a^2-b^2)}$ and $c^2 = a^2 - b^2$, you can substitute c^2 for $a^2 - b^2$. So, $\dfrac{c^2}{-(a^2-b^2)}$ equals $\dfrac{c^2}{-c^2}$ or -1. Since the product of the slopes of $\overline{AC}$ and $\overline{BD}$ is -1, $\overline{AC} \perp \overline{BD}$.   19. Given: Rectangle $ABCD$ with vertices $A(0, 0)$, $B(0, a)$, $C(b, a)$, $D(b, 0)$ Prove: $\overline{AC} \cong \overline{BD}$ Proof: By the distance formula, $AC = \sqrt{(b-0)^2 + (a-0)^2}$ or $\sqrt{a^2 + b^2}$ and $BD = \sqrt{(b-0)^2 + (0-a)^2}$ or $\sqrt{a^2 + b^2}$. By substitution $AC = BD$ or $\overline{AC} \cong \overline{BD}$.   20. Given: Parallelogram $ABCD$ with vertices $A(0, 0)$, $B(b, c)$, $C(a + b, c)$, $D(a, 0)$ Prove: $\overline{AD} \cong \overline{BC}$ and $\overline{AB} \cong \overline{CD}$ Proof: By the distance formula, $AD = \sqrt{(a-0)^2 + (0-0)^2}$ or a and $BC = \sqrt{[(a+b)-b]^2 + (c-c)^2}$ or a. By substitution, $AD = BC$ or $\overline{AD} \cong \overline{BC}$. Also, by the distance formula, $AB = \sqrt{(b-0)^2 + (c-0)^2}$ or $\sqrt{b^2 + c^2}$ and $CD = \sqrt{[a-(a+b)]^2 + (0-c)^2}$ or $\sqrt{b^2 + c^2}$. By substitution, $AB = CD$ or $\overline{AB} \cong \overline{CD}$.   21. Given: Square $ABCD$ with vertices $A(0, 0)$, $B(a, 0)$, $C(a, a)$, $D(0, a)$ Prove: $\overline{AC} \perp \overline{BD}$ Proof: The slope of $\overline{AC}$ is $\dfrac{a-0}{a-0}$ or 1. The slope of $\overline{BD}$ is $\dfrac{a-0}{0-a}$ or $\dfrac{a}{-a}$ or -1. Since the product of the slope is -1, $\overline{AC} \perp \overline{BD}$.   22. Given: Rhombus $ABCD$ with vertices $A(0, 0)$, $B(b, c)$, $C(a + b, c)$, $D(a, 0)$ Prove: $\overline{AC} \perp \overline{BD}$ Proof: The slope of $\overline{AC}$ is $\dfrac{c-0}{a+b-0}$ or $\dfrac{c}{a+b}$. The slope of $\overline{BD}$ is $\dfrac{0-c}{a-b}$ or $\dfrac{-c}{a-b}$. Since the product of the slopes is -1, $\overline{AC} \perp \overline{BD}$.

Page 334 Challenge

1. Since both lines are perpendicular to $y = ax$, $a \neq 0$, they each have slope $-\dfrac{1}{a}$. This is true because two nonvertical lines are perpendicular if and only if the product of their slope is -1. Since both lines have the same slope, they must be parallel.   2. Given: Quadrilateral $ABCD$ with vertices $A(0, 0)$, $B(2a, 2b)$, $C(2d, 2e)$, $D(2c, 0)$; $\overline{AC}$ and $\overline{BD}$ intersect at M and $AM = MC$ and $BM = MD$. Prove: Quadrilateral $ABCD$ is a parallelogram.   Proof: Since $AM = MC$ and $BM = MD$, M is the midpoint of $\overline{AC}$ and $\overline{BD}$. By the midpoint formula, M can have the following coordinates: $\left(\dfrac{2d+0}{2}, \dfrac{2e+0}{2}\right)$ or (d, e) and $\left(\dfrac{2a+2c}{2}, \dfrac{2b+0}{2}\right)$ or $(a + c, b)$. Therefore, $a + c = d$ and $b = e$ so c has coordinates $(2(a + c), 2b)$. $BC = \sqrt{[2(a+c)-2a]^2 + (2b-2b)^2}$ or $\sqrt{(2c)^2}$ or $2c$ $AD = \sqrt{(2c-0)^2 + (0-0)^2}$ or $\sqrt{(2c)^2}$ or $2c$. Therefore, $AD = BC$ or $\overline{AD} \cong \overline{BC}$. The slope of $BC = \dfrac{2b-2b}{2(a+c)-2a}$ or 0, and the slope of $AD = \dfrac{0-0}{2c-0}$ or 0. Since two nonvertical lines are parallel if and only if they have the same slope, $\overline{AD} \parallel \overline{BC}$. Since two sides of quadrilateral $ABCD$ are congruent and parallel, quadrilateral $ABCD$ is a parallelogram.

Pages 337–338   Written   7. The midpoint of $\overline{AB}$ is $E(-1, 6)$. The midpoint of $\overline{BC}$ is $F\left(\frac{5}{2}, 5\right)$. The slope of $\overline{EF}$ is $\dfrac{6-5}{-1-\frac{5}{2}}$ or $\dfrac{-2}{7}$. The slope of $\overline{AC}$ is $\dfrac{3-5}{4-(-3)}$ or $\dfrac{-2}{7}$. Since the slopes of $\overline{EF}$ and $\overline{AC}$ are equal, $\overline{EF} \parallel \overline{AC}$.

8. By the distance formula, $EF = \sqrt{\left[\frac{5}{2}-(-1)\right]^2 + (5-6)^2} = \sqrt{\frac{49}{4}+1} = \sqrt{\frac{53}{4}}$ or $\dfrac{\sqrt{53}}{2}$ and $AC = \sqrt{[4-(-3)]^2 + (3-5)^2} = \sqrt{49+4}$ or $\sqrt{53}$. Since $\dfrac{\sqrt{53}}{2} = \dfrac{1}{2} \cdot \sqrt{53}$, it follows that $EF = \dfrac{1}{2}AC$.

16. The axes could be chosen so that any one of the vertices was at the origin and the side adjacent to it on the x-axis. The result would be the same.   **17.** $A(0, 0)$, $B(2b, 2c)$, $C(2a, 0)$. The midpoint M_1 of $\overline{AB}$ is (b, c), the midpoint M_2 of BC is $(a + b, c)$. The measure of $\overline{AC}$ is $2a$. The measure of $\overline{M_1 M_2}$ is $\sqrt{(a+b-b)^2 + (c-c)^2}$ or $\sqrt{(a+0)^2 + (0)^2}$ or $\sqrt{a^2}$ or a. The measure of $\overline{M_1 M_2}$ is $\dfrac{1}{2}(2a)$ or a.

18. Place x-axis on the base and y-axis on the vertex. The coordinates of the vertex are $(0, c)$ and the endpoints of the base of $(-a, 0)$ and $(a, 0)$. Justification: altitude from the vertex of an isosceles triangle divides it into two congruent triangles by HL. Or as an alternative, place origin at a base angle vertex and the x-axis on the base. The coordinates are $(0, 0)$, (b, c), and $(2b, 0)$.   **19.** Given: Right triangle ABC with vertices $A(0, 2b)$, $B(2a, 0)$, $C(0, 0)$ Prove: $AM = BM = CM$ where M is the midpoint of $\overline{AB}$ Proof: By the midpoint formula, M has coordinates (a, b). $AM = \sqrt{(a-0)^2 + (b-2b)^2}$ or $\sqrt{a^2+b^2}$. $BM = \sqrt{(a-2a)^2 + (b-0)^2}$ or $\sqrt{a^2+b^2}$. $CM = \sqrt{(a-0)^2 + (b-0)^2}$ or $\sqrt{a^2+b^2}$. Therefore, $AM = BM = CM$.   **20.** Given: Isosceles triangle ABC with vertices $A(0, 0)$ $B(2b, 2c)$, $C(4b, 0)$. Note that $AB = 2\sqrt{b^2+c^2}$ and $BC = 2\sqrt{b^2+c^2}$ Prove: $\triangle M_1 M_2 M_3$ is isosceles where M_1 is the midpoint of $\overline{AB}$, M_2 is the midpoint of $\overline{BC}$, and M_3 is the midpoint of $\overline{AC}$.   Proof: M_1 has coordinates (b, c), M_2 has coordinates $(3b, c)$, and M_3 has coordinates $(2b, 0)$. $M_1 M_3 = \sqrt{(2b-b)^2 + (0-c)^2}$ or $\sqrt{b^2+c^2}$ and $M_2 M_3 = \sqrt{(2b-3b)^2 + (0-c)^2}$ or $\sqrt{b^2+c^2}$. Since $M_1 M_3 = M_2 M_3$, $\triangle M_1 M_2 M_3$ is isosceles.   **21.** Given: Isosceles triangle ABC with vertices $A(0, 0)$, $B(2a, 0)$, $C(a, b)$ Prove: $\overline{CM} \perp \overline{AB}$ where M is the midpoint of $\overline{AB}$.   Proof: M has coordinates $(a, 0)$. The slope of $\overline{CM}$ is $\dfrac{0-b}{a-a}$ or none. $\overline{AB}$ has slope $\dfrac{0-0}{2a-0}$ or 0. Since $\overline{CM}$ lies on a vertical line and $\overline{AB}$ lies on a horizontal line, $\overline{CM} \perp \overline{AB}$.   **22.** Given: Isosceles triangle ABC with vertices $A(2a, 2b)$, $B(0, 0)$, $C(4a, 0)$ Prove: $\overline{BM_1} \cong \overline{CM_2}$ where M_1 is the midpoint of $\overline{AC}$ and M_2 is the midpoint of $\overline{AB}$.   Proof: M_1 has coordinates $(3a, b)$ and M_2 has coordinates (a, b). $BM_1 = \sqrt{(3a-0)^2 + (b-0)^2}$ or $\sqrt{9a^2+b^2}$ and $CM_2 = \sqrt{(a-4a)^2 + (b-0)^2}$ or $\sqrt{9a^2+b^2}$. Since $BM_1 = CM_2$, it follows that $\overline{BM_1} \cong \overline{CM_2}$.

Page 338   Challenge   1. Given: Trapezoid ABCD with vertices $A(0, 0)$, $B(2b, 2c)$, $C(2d, 2c)$, $D(2e, 0)$, $\overline{BC} \parallel \overline{AD}$ Prove: $MN = \frac{1}{2}(AD + BC)$ where M is the midpoint of $\overline{AB}$ and N is the midpoint of $\overline{CD}$.   Proof: M has coordinates (b, c) and N has coordinates $(d + e, c)$. $MN = \sqrt{[(d+e)-b]^2 + (c-c)^2}$ or $d + e - b$. $AD = 2e$. $BC = \sqrt{(2d-2b)^2 + (2c-2c)^2}$ or $2d - 2b$. Then $AD + BC = 2e + 2d - 2b$ and $\frac{1}{2}(AD + BC) = e + d - b$ or MN. Therefore, $MN = \frac{1}{2}(AD + BC)$.   **2.** Given: Quadrilateral ABCD with vertices $A(0, 0)$, $B(2a, 2b)$, $C(2c, 2d)$, $D(2e, 0)$; M is the midpoint of $\overline{AB}$, N is the midpoint of $\overline{BC}$, P is the midpoint of $\overline{CD}$, and Q is the midpoint of $\overline{AD}$.   Prove: MNPQ is a parallelogram. The coordinates of the midpoints are $M(a, b)$, $N(a + c, b + d)$, $P(c + e, d)$, and $Q(e, 0)$.   Proof: $MN = \sqrt{[(a+c)-a]^2 + [(b+d)-b]^2}$ or $\sqrt{c^2+d^2}$. $PQ = \sqrt{[(c+e)-e]^2 + (d-0)^2}$ or $\sqrt{c^2+d^2}$. Since $MN = PQ$, $\overline{MN} \cong \overline{PQ}$. The slope of $\overline{MN} = \dfrac{b+d-b}{a+c-a}$

or $\frac{d}{c}$. The slope of $\overline{PQ} = \frac{0-d}{e-(c+e)} = \frac{d}{c}$. Since the slopes of $\overline{MN}$ and $\overline{PQ}$ are equal, $\overline{MN} \parallel \overline{PQ}$. MNPQ is a parallelogram since one pair of sides is congruent and parallel. **3.** *Given:* Isosceles trapezoid ABCD with vertices $A(0, 0)$, $B(2b, 2c)$, $C(2d, 2c)$, $D(2b + 2d, 0)$. M is the midpoint of $\overline{AB}$, N is the midpoint of $\overline{BC}$, P is the midpoint of $\overline{CD}$, and Q is the midpoint of $\overline{AD}$. *Prove:* MNPQ is a rhombus. The coordinates of the midpoints are $M(b, c)$, $N(b + d, 2c)$, $P(b + 2d, c)$, and $Q(b + d, 0)$.
Proof: $MN = \sqrt{[(b + a) - b]^2 + (2c - c)^2}$ or $\sqrt{c^2 + d^2}$. $NP = \sqrt{[(b + 2d) - (b + d)] + (c - 2c)^2}$ or $\sqrt{c^2 + d^2}$. $PQ = \sqrt{[(b + d) - (b + 2d)]^2 + (0 - c)^2}$ or $\sqrt{c^2 + d^2}$. $MQ = \sqrt{[(b + d) - b]^2 + (0 - c)^2}$ or $\sqrt{c^2 + d^2}$. From Challenge exercise 2, quadrilateral MNPQ must be a parallelogram. Since $MN = NP = PQ = MQ = \sqrt{c^2 + d^2}$, MNPQ is a rhombus. **4.** The coordinates are $A(0, 0)$, $B(a, b)$, $C(a + c, b)$, $D(c, 0)$ By the Distance Formula: $(AB)^2 = (a - 0)^2 + (b - 0)^2$ or $a^2 + b^2$, $(BC)^2 = (a + c - a)^2 + (b - b)^2$ or c^2, $(CD)^2 = [c - (a - c)]^2 + (0 - b)^2$ or $a^2 + b^2$, $(AD)^2 = (c - 0)^2 + (0 - 0)^2$ or c^2, $(AC)^2 = (a + c - 0)^2 + (b - 0)^2 + (b - 0)^2$ or $a^2 + 2ac + c^2 + b^2$, and $(BD)^2 = (c - a)^2 + (0 - b)^2$ or $c^2 - 2ac + a^2 + b^2$. By the Addition Axiom: $(AB)^2 + (BC)^2 + (CD)^2 + (AD)^2 = 2(a^2 + b^2 + c^2)$ and $(AC)^2 + (BD)^2 = 2(a^2 + b^2 + c^2)$. By substitution: $(AB)^2 + (BC)^2 + (CD)^2 + (AD)^2 = (AC)^2 + (BD)^2$.

Page 341 Exploratory For exercises 1–5, see students' work. **1.** a circle with center at point P and a radius of 3 cm **2.** two parallel lines on either side of line k and each is 5 cm from it **3.** the perpendicular bisector of $\overline{PQ}$ **4.** a line parallel to both line m and line n and halfway between lines m and n **5.** a pair of lines each of which bisects pairs of vertical angles formed by lines r and s

Page 341 Written For exercises 1–8 and 10–18, see students' work. **1.** two parallel lines on either side of the given line and each is 4 cm from it **2.** the angle bisector **3.** the intersection point of the diagonals of the square **4.** a line parallel to both given lines and 5 cm from each line **5.** the perpendicular bisector of the line segment having the two given points as endpoints **6.** two circles with the same center as the given circle and radii of 2 cm and 10 cm **7.** the perpendicular bisector of the diagonal connecting the two vertices **8.** diagonal $\overline{BD}$, but not points B and D **9.** See the diagram. The circular portions are quarter circles with a vertex as center and a 2 in. radius. The line segments in the figure are 5 in. long and parallel to the sides of the square.

10. a sphere with radius of 8 cm and center at the given point **11.** a sphere with radius of 5.5 cm and center at the given point **12.** a plane parallel to the two given planes and halfway between them **13.** two planes each containing the line of intersection of the two given planes where each plane bisects the "space angles" formed by the given planes **14.** a "cylindrical shape" around the line and 4 cm from the line **15.** a zigzag path of points equidistant from points on the random paths of the floats **16.** a line parallel to a line on the road at a distance equal to the radius of the wheel **17.** a spiral not including the center of a circle and points within the circle whose radius is the same as the radius of the label

Page 341 Challenge For exercises 1–5, see students' work. **1.** a sphere with A as center and radius of 4 cm **2.** a "cylindrical shape" around C and 4 cm from C **3.** a plane parallel to the lines at a distance of 2 cm from each line **4.** two planes both parallel to plane $\mathcal{A}$ one in either side and 4 cm from the plane $\mathcal{A}$ **5.** A circle with center Q and radius equal to the sum of the measures of the radii of the two original circles. **6.** A circle with center Q and radius equal to the difference of the measures of the radii of the two circles. **7.** An epicycloid, a curve that curves around the large circle like mouseketeer ears. A hypocycloid, a curve that curves inside the large circle.

Page 345 Written **1.** a circle with a radius of 1 unit and center at $(0, 0)$ **2.** a circle with a radius of 3 units and center at $(0, 0)$ **3.** a circle with a radius of 2 units and center at $(0, 0)$ **4.** a circle with a radius of $\sqrt{10}$ units and center at $(0, 0)$ **5.** a circle with a radius of $2\sqrt{3}$ units and center at $(0, 0)$ **6.** a circle with a radius of 5 units and center at $(0, 0)$ **22.** $(2, 4)$, 5 units **23.** $(6, 3)$, 8 units **24.** $(-5, 3)$, 11 units **25.** $(3, -8)$, $\sqrt{5}$ units **26.** $(-6, -10)$, $3\sqrt{2}$ units **27.** $(0, -7)$, $2\sqrt{2}$ units **28.** $(x - 5)^2 + (y - 5)^2 = 1$ **29.** $(x - 4)^2 +$

$(y - 6)^2 = 81$ **30.** $(x - 4)^2 + y^2 = 16$ **31.** $(x - 1)^2 + (y + 2)^2 = 169$ **32.** $(x - 6)^2 + y^2 = 6$
33. $(x + 4)^2 + (y + 6)^2 = 11$ **34.** 25.12 units, 50.24 units² **35.** 81.64 units, 530.66 units² **36.** 15.38 units, 18.84 units² **37.** 20.83 units, 34.54 units²

Page 349 Exploratory For exercises 1–8, see students' work. **1.** the 4 intersection points of a circle with a radius of 6 cm and center at the given point on the line and the two lines on either side of the given line 4 cm from the given line **2.** the 2 intersection points of the perpendicular bisector of the segment whose endpoints are the two given points and the two lines on either side of the given line 5 cm from the given line **3.** the two intersection points of a circle with a radius of 3 cm and center at the given point and the two lines on either side of the given line 3 cm from the given line (The locus are points of tangency.) **4.** the 2 intersection points of the perpendicular bisector of $\overline{RS}$ and the two lines on either side of $\overline{RS}$ 5 cm from $\overline{RS}$ **5.** the 2 intersection points of a circle with a radius of 5 cm and center at P and the line 4 cm from ℓ on the same side of ℓ as P **6.** the intersection point of a circle with a radius of 5 cm and center at P and the line 3 cm from ℓ on the same side of ℓ as P. **7.** the 3 intersection points of the circle with a radius of 11 cm and center at P and the two lines 3 cm from ℓ on each side of ℓ (One of the points of the locus is a point of tangency.) **8.** the 4 intersection points of a circle with a radius of 15 cm and center at P and the two lines on either side of ℓ 6 cm from ℓ

Page 349 Written For exercises 1–10, see students' work. **1.** the 2 intersection points of a circle with a radius of 5 cm and center at point A and the line parallel to and halfway between lines ℓ and m **2.** the 2 intersection points of a circle with a radius of 7 cm and center at point A and the perpendicular bisector of $\overline{AB}$ **3.** the two intersection points of a circle with a radius of 13 cm and center at point Q and a circle with a radius of 5 cm and center at point R **4.** the 2 intersection points of the circle with a radius of 5 cm and center at point B and the extended bisector of ∠ABC **5.** the 2 intersection points of the circle with a radius of 2 cm and center at point A and a circle with radius 2 cm and center at point B **6.** the two intersection points of the perpendicular bisector of $\overline{AB}$ and of the two lines parallel to and 4 cm from line ℓ **7.** the intersection point of the perpendicular bisector of $\overline{AC}$ and the bisector of ∠BAC **8.** the one intersection point of the perpendicular bisectors of $\overline{RS}$ and $\overline{RT}$ **9.** The perpendicular bisectors are parallel. **10.** the one intersection point of the diagonals of rhombus ABCD

Page 353 Written
3.
$$y + 1 = \sqrt{(x - 3)^2 + (y - 2)^2}$$
$$(y + 1)^2 = (x - 3)^2 + (y - 2)^2$$
$$y^2 + 2y + 1 = x^2 - 6x + 9 + y^2 - 4y + 4$$
$$2y + 1 = x^2 - 6x - 4y + 13$$
$$6y = x^2 - 6x + 12$$
$$y = \frac{1}{6}x^2 - x + 2$$

4. $14 \stackrel{?}{=} \frac{1}{6} \cdot 12^2 - 12 + 2$
$14 \stackrel{?}{=} 24 - 12 + 2$
$14 = 14$

5. $y = \frac{1}{4}x^2$ **6.** $y = \frac{1}{6}x^2 - \frac{1}{3}x - \frac{1}{3}$ **7.** $y = \frac{1}{4}x^2 - 2x + 4$ **8.** $y = \frac{1}{4}x^2 - \frac{3}{2}x + \frac{17}{4}$
9. $y = \frac{1}{10}x^2 - \frac{1}{5}x - \frac{2}{5}$ **10.** $y = -\frac{1}{6}x^2 - \frac{2}{3}x - \frac{13}{6}$

13.
$$x + 2 = \sqrt{(x - 2)^2 + y^2}$$
$$(x + 2)^2 = (x - 2)^2 + y^2$$
$$x^2 + 4x + 4 = x^2 - 4x + 4 + y^2$$
$$8x = y^2$$
$$x = \frac{1}{8}y^2$$

14. $x = \frac{1}{4}y^2$ **15.** $x = \frac{1}{8}y^2 - \frac{1}{4}y + \frac{1}{8}$
16. $x = \frac{1}{4}y^2 - \frac{3}{2}y + \frac{9}{4}$ **17.** $x = \frac{1}{2}y^2 - 4y + \frac{23}{2}$
18. $x = \frac{1}{2}y^2 - y + 1$ **19.** $x = -\frac{1}{10}y^2 + \frac{2}{5}y + \frac{11}{10}$

Page 353 Challenge
1.
$$y + p = \sqrt{x^2 + (y - p)^2}$$
$$(y + p)^2 = x^2 + (y - p)^2$$
$$y^2 + 2py + p^2 = x^2 + y^2 - 2py + p^2$$
$$4py = x^2$$
$$y = \frac{1}{4p}x^2$$

2.
$$x + p = \sqrt{(x - p)^2 + y^2}$$
$$(x + p)^2 = (x - p)^2 + y^2$$
$$x^2 + 2px + p^2 = x^2 - 2px + p^2 + y^2$$
$$y^2 = 4px \text{ or } x = \frac{1}{4p}y^2$$

Page 357 Exploratory For exercises 1–8, see students' work. **1.** Always true; this is illustrated by a folded piece of paper. The two parts represent two planes and the fold, a line. **2.** Always true; think of a pile of sheets of paper. The top sheet is parallel to one in the middle and the bottom sheet. The middle and bottom sheets are parallel. **3.** Always true; there is only one plane through a line and a point not on the line. This plane also contains the one line through the given point parallel to the first line. **4.** Sometimes true; it is possible for two lines to be skew, neither parallel nor intersecting (consider the edges of a cube). **5.** Always true; the plane containing a line and a point outside the line also contains a line through the point intersecting the line. **6.** Always true; in many geometries this statement is considered self-evident. Any two points will determine $\overleftrightarrow{DE}$, $\overleftrightarrow{EF}$, and $\overleftrightarrow{FD}$. Since any two of these lines intersect, they determine a unique plane. This plane contains all three points. **7.** Sometimes true; an illustration of this is a wobbly four-legged chair. Because three points determine a plane, no plane containing a fourth point outside the first plane will contain all four points. **8.** Always true; a line is parallel to a plane if and only if it does not intersect the plane. Therefore, if it is not parallel, it must intersect in at least one point. If it intersects the plane but in two points, it lies wholly in the plane. But it is not in the plane. Hence it must intersect in exactly one point.

Page 357 Written 7. Answers will vary. Typical answers are $(5, 0, -7)$, $(-1, 0, 2)$, and $(a, 0, b)$. **8.** Horizontal planes are usually pictured as parallelograms. The rectangle that is the top of the box has two pairs of parallel sides. Since the rectangle is horizontal, a parallelogram is used to indicate the visual appearance of right angles in the horizontal plane. Accurate perspective drawing is not used, however, in order to picture the edges as parallel segments. **9.** The diagonals intersect at $(4, 6, 3)$ and are congruent because they are diagonals of a rectangle and diagonals of a rectangle are congruent. **11.** False; an illustration is the pages of a book. **12.** True; the y-axis has x- and z-coordinates 0. **13.** True; the coordinate planes have the equations $z = 0$ (xy plane), $y = 0$ (xz plane), and $x = 0$ (yz plane) **14.** True; the plane $x = 1$ is the plane midway between the planes $x = 0$ and $x = 2$ and parallel to them. Any points with x-coordinates 0 and 2 and the same z-coordinate are in these planes.

Page 361 Chapter Review
3. *Given:* Isosceles trapezoid $A(0, 0)$, $B(a, 0)$, $C(a - b, c)$, $D(b, c)$ *Prove:* $\overline{AC} \cong \overline{BD}$
Proof: $AC = \sqrt{(a - b)^2 + c^2} = \sqrt{a^2 + b^2 + c^2 - 2ab}$ and $BD = \sqrt{(b - a)^2 + c^2} = \sqrt{a^2 + b^2 + c^2 - 2ab}$. Since $AC = BD$, $\overline{AC} \cong \overline{BD}$. **9.** The slope of $\overline{AB}$ is 0 and $\overline{AC}$ has no slope. Since AB and AC lie on a horizontal and a vertical line, they are perpendicular and form a right angle. Therefore, ABC is a right triangle. For exercises 10–13, see students' work. **10.** a circle with center at the origin and a radius of 8 cm **11.** the perpendicular bisector of the line segment whose endpoints are the two given points **12.** the center of the given circle and a circle having the same center and a radius of 20 cm **13.** a sphere with a radius of 5 cm and center at the given point. For exercises 21 and 22, see students' work. **21.** the 2 intersection points of a circle with center at point A and a radius of 6 cm and the line parallel to and halfway between lines r and s **22.** the 2 intersection points of a circle with center at point C and a radius of 8 cm and a circle with center at point D and a radius of 6 cm

Page 362 Chapter Test 4. Distance from $(3, 2)$ to $(11, 17)$: $\sqrt{(11 - 3)^2 + (17 - 2)^2} = \sqrt{64 + 225} = \sqrt{289}$ or 17 units. Distance from $(20, 2)$ to $(3, 2)$: 17 units. Triangle is isosceles. **10.** *Given:* Square ABCD with vertices $A(0, 0)$, $B(a, 0)$, $C(a, a)$, $D(a, 0)$ *Prove:* $\overline{AC} \perp \overline{BD}$ *Proof:* The slope of AC is 1. The slope of BD is -1. Since the product of these slopes is -1, $AC \perp BD$. For exercises 11–16, see students' work. **11.** a circle with center at point B and a radius of 5 cm **12.** the perpendicular bisector of $\overline{AB}$ **13.** a circle with a radius of 4 units and center at $(2, 5)$ **14.** a line halfway between and parallel to the two given lines **15.** two parallel lines on either side of line ℓ and each is 3 cm from line ℓ **16.** a pair of lines each of which bisects pairs of vertical angles formed by the two intersecting lines

CHAPTER 10 SOLVING QUADRATIC EQUATIONS

Page 367 Exploratory 1. a polynomial with two unlike terms **2.** a polynomial with three unlike terms **3.** $(a + b)(a - b) = a^2 - b^2$ **4.** results in all prime factors **5.** $a(b + c) = ab + ac$

Pages 367–368 Written 1. $8c - 16$ 2. $5x^2 - 5x$ 3. $2y^2 - 8y^3$ 4. $x^5 - 4x^3y$ 5. $4x^2y^2 + 4x^3y$ 6. $15x^3y^3 + 18x^4yz^2$ 7. $x^2 + 7x + 10$ 8. $x^2 + 2x - 8$ 9. $x^2 - 81$ 10. $c^2 + 8c - 9$ 11. $x^2 - 25$ 12. $x^2 - 18x + 65$ 13. $2w^2 + 19w + 35$ 14. $3n^2 - 14n - 24$ 15. $8x^2 - 2x - 1$ 16. $36x^2 - 1$ 17. $6y^2 - 7y - 20$ 18. $8 - 2y - 3y^2$ 19. $8(c - 2)$ 20. $4(c - d)$ 21. $5x(x - 1)$ 22. $4(x^2 + 2y^2 - 4z^2)$ 23. $2xy(1 + y)$ 24. $3y(y - 3 - y^2)$ 25. $13xy(z - 2)$ 26. $5x^2(-y + 2)$ 27. $xy(y^2z + 3x)$ 28. $abc(c + b - 1)$ 29. $100x^2(1 - x)$ 30. $xy(-1 + 6xy)$ 31. $(x - 2)(x + 2)$ 32. $(x + 5)(x - 1)$ 33. $(m + 1)^2$ 34. $(x - 3)(x - 2)$ 35. $(x - 9)(x + 9)$ 36. $(b + 0.1)(b - 0.1)$ 37. $(x - 5)(x + 3)$ 38. $(x + 21)(x - 1)$ 39. $(2c - 1)(2c + 1)$ 40. $(x + 6)(x - 5)$ 41. $(x - 8)(x + 3)$ 42. $(y + 5)(y + 12)$ 43. $(2c + 1)(c + 5)$ 44. $(3n + 4)(n + 2)$ 45. $(5x - 1)(x + 4)$ 46. $(2y - 3)(y + 1)$ 47. $(2x - 3)(x - 4)$ 48. $(2x - 3)(4x + 3)$ 49. $9(y - 1)(y + 1)$ 50. $3(c - 21)(c + 1)$ 51. $x(x - 2)$ 52. $2(x - 2)(x - 3)$ 53. $3(b - 4)(b + 4)$ 54. $14x(x - 2)$ 55. $-2(x^2 - x - 15)$ 56. $15x(2x - 1)$ 57. $h^2(h - 1)$ 58. $x^3(1 - x)(1 + x)$ 59. $2m^4(m + 1)(m - 1)$ 60. $3(2w + 3)(w - 1)$ 61. $4x(d - 3)(d + 2)$ 62. $(x - 1)(x + 1)(x^2 + 1)$ 63. $y^4(y - 1)(y + 1)$ 64. $a^2 - 9 = -1(9 - a^2)$ 65. The terms are the same, but they are "reversed." 66. The Distributive Law is used to multiply many polynomials and factoring is the "opposite" of multiplying. 67. $(x + r)(x + s) = x(x + s) + r(x + s) = x^2 + xs + rx + rs = x^2 + (r + s)x + rs$ 68. It cannot be factored in the set of real numbers.

Pages 371–372 Exploratory 1. $x = 0$ or $x - 6 = 0$; $\{0, 6\}$ 2. $m = 0$ or $m + 3 = 0$; $\{-3, 0\}$ 3. $2y = 0$ or $y + 4 = 0$; $\{-4, 0\}$ 4. $a = 0$ or $4a + 1 = 0$; $\left\{-\frac{1}{4}, 0\right\}$ 5. $x + 1 = 0$ or $x + 5 = 0$; $\{-5, -1\}$ 6. $b + 4 = 0$ or $b - 2 = 0$; $\{-4, 2\}$ 7. $c - 5 = 0$ or $c - 3 = 0$; $\{3, 5\}$ 8. $r - 6 = 0$ or $r + 4 = 0$; $\{-4, 6\}$ 9. $2y + 1 = 0$ or $y - 2 = 0$; $\left\{-\frac{1}{2}, 2\right\}$ 10. $4z - 1 = 0$ or $z + 2 = 0$; $\left\{-2, \frac{1}{4}\right\}$ 11. $3x - 4 = 0$ or $4x - 3 = 0$; $\left\{\frac{3}{4}, \frac{4}{3}\right\}$ 12. $4y + 3 = 0$ or $2y - 2 = 0$; $\left\{-\frac{3}{4}, 1\right\}$ 13. $3x^2 - 3x - 1 = 0$ 14. $x^2 - 12x - 13 = 0$ 15. $x^2 + 5x + 1 = 0$ 16. $x^2 + 4x - 4 = 0$ 17. $x^2 + 5x - 13 = 0$ 18. $3x^2 - 10x + 24 = 0$

Page 372 Written 46. $x^2 + 6x - 7 = 0$ 47. $x^2 + 4x + 3 = 0$ 48. $x^2 - 100x + 900 = 0$ 49. $x^2 - 3x + 2 = 0$ 50. $x^2 - \frac{5}{2}x + 1 = 0$ 51. $x^2 - \frac{9}{2}x - \frac{5}{2} = 0$ 52. $x^2 - \frac{5}{6}x + \frac{1}{6} = 0$ 53. $x^2 - \frac{5}{12}x + \frac{1}{24} = 0$ 54. $x^2 - \frac{2}{15}x - \frac{8}{15} = 0$ 55. $x^2 + \frac{2}{3}x + \frac{1}{12} = 0$ 56. $x^2 - 12x + 36 = 0$ 57. $x^2 + 2x + 1 = 0$

Page 376 Exploratory 1. the line about which a parabola is symmetric 2. an equation that will result in a graph called a parabola 3. the equation of the axis of symmetry of a parabola 4. the coefficient of x^2 is negative 5. when a is positive; when a is negative 6. the graph of $y = x^2 + 5$ is located 5 units above the graph of $y = x^2$ 7. $x = 0$

Page 376 Written

Page 376 Challenge

1.
2.
3.

Page 360 Challenge

2.

Pages 379–380 Exploratory 1. Graph the associated parabola. Then the x-coordinates of the points where the parabola crosses the x-axis are the solutions. 2. Yes, the real roots can be approximated. 3. if $ax^2 + bx + c$ is a perfect square trinomial 4. c

Pages 383–384 Exploratory 1. it does not always give exact solutions; to get exact solutions 2. positive, the principal square root 3. $x = \pm\sqrt{121}$ 4. $\pm\sqrt{7}$ means $\sqrt{7}$ or $-\sqrt{7}$ 5. $c = \left(\dfrac{m}{2}\right)^2$ 6. $m = 2(\pm\sqrt{c})$ 7. (a) Write the equation. (b) Subtract 3 from both sides. (c) $\left(\dfrac{8}{2}\right)^2$ or 16 will make the left side a perfect square trinomial. Add 16 to both sides. (d) Factor and add. (e) Take the square root of both sides. (f) Subtract 4 from both sides. (g) Write the solution set.

Page 384 Challenge For the graphs of exercises 1–4, see students' work. 1. a circle with center $(3, -5)$ and radius 7 2. a circle with center $(-3, -2)$ and radius 3 3. a circle with center $(-2, -3)$ and radius 4 4. a circle with center $(-5, 7)$ and radius $\sqrt{124}$ (or $2\sqrt{31}$)

Pages 387–388 Exploratory 1. If $ax^2 + bx^2 + c = 0$, then $x = \dfrac{-b \pm \sqrt{b^2 - 4ac}}{2a}$. the coefficient x^2, the coefficient of x, and the constant term, respectively 2. $b^2 - 4ac$ 3. The discriminant tells you if the equation has real roots. For example, the equation $x^2 + 3x + 4 = 0$ has no real roots because the discriminant is negative. 4. $4x^2 - 4x + 1 = 0$, $a = 4$, $b = -4$, $c = 1$ 5. $2x^2 - 5x - 1 = 0$, $a = 2$, $b = -5$, $c = -1$ 6. $18x^2 - 1 = 0$, $a = 18$, $b = 0$, $c = -1$ 7. $x^2 + 12x - 2 = 0$, $a = 1$, $b = 12$, $c = -2$ 8. $x^2 + 8x - 3 = 0$, $a = 1$, $b = 8$, $c = -3$ 9. $x^2 + 2x = 0$, $a = 1$, $b = 2$, $c = 0$ 10. -3, no real roots 11. -7, no real roots 12. 9, has real roots 13. 0, has real roots 14. 25, has real roots 15. 72, has real roots 16. There is only one real root. 17. The roots are rational. 18. sum = 5, product = -7 19. sum = $\dfrac{13}{2}$, product = 1 20. sum = $-\dfrac{1}{2}$, product = $-\dfrac{1}{2}$ 21. sum = $\dfrac{4}{3}$, product = $\dfrac{5}{3}$ 22. sum = 0, product = $-\dfrac{1}{5}$ 23. sum = -4, product = -20

Page 388 Written 1. $\{-5, 6\}$ 2. $\{-8, -2\}$ 3. $\{-5, 3\}$ 4. $\{-7, -6\}$ 5. $\{4, 6\}$ 6. $\{-8, 3\}$ 7. $\{1, 4\}$ 8. $\left\{-\dfrac{4}{5}, 1\right\}$ 9. $\left\{-\dfrac{5}{3}, 4\right\}$ 10. $\left\{-\dfrac{1}{4}, 3\right\}$ 11. $\left\{-\dfrac{3}{2}, \dfrac{5}{3}\right\}$ 12. $\left\{-\dfrac{1}{4}, \dfrac{5}{6}\right\}$ 13. $\left\{-\dfrac{5}{2}, \dfrac{1}{7}\right\}$ 14. $\left\{-\dfrac{3}{2}, -\dfrac{5}{3}\right\}$ 15. $\left\{-\dfrac{2}{5}, \dfrac{1}{4}\right\}$ 16. $\left\{-\dfrac{5}{3}, -\dfrac{3}{5}\right\}$ 17. $\left\{-\dfrac{3}{4}, \dfrac{5}{6}\right\}$ 18. $\left\{-4, \dfrac{7}{3}\right\}$ 19. $\left\{\dfrac{1 \pm \sqrt{113}}{4}\right\}$ 20. $\left\{\dfrac{9 \pm \sqrt{17}}{8}\right\}$ 21. $\left\{\dfrac{8 \pm 2\sqrt{29}}{13}\right\}$ 22. $\left\{\dfrac{1}{2}, 2\right\}$ 23. $\{0, 13\}$ 24. $\{0, 4\}$ 25. $\left\{\dfrac{7 \pm 3\sqrt{21}}{14}\right\}$ 26. $\left\{\dfrac{-5 \pm \sqrt{30}}{10}\right\}$ 27. $\left\{-4, \dfrac{8}{7}\right\}$ 28. $\left\{-\dfrac{2}{3}, \dfrac{3}{2}\right\}$ 29. $\left\{\pm\dfrac{\sqrt{30}}{2}\right\}$ 30. $\left\{\dfrac{11 \pm \sqrt{265}}{24}\right\}$ 31. has real roots 32. no real roots

Page 388 Challenge 1. (a) general form of a quadratic equation (b) Subtract c from both sides. (c) Multiply both sides by $4a$. (d) Add b^2 to both sides. (e) Factor. (f) Take the square root of both sides. (g) Subtract b from both sides. (h) Divide both sides by $2a$. **2.** $\dfrac{-b+\sqrt{b^2-4ac}}{2a} + \dfrac{-b-\sqrt{b^2-4ac}}{2a} = \dfrac{-b+\sqrt{b^2-4ac}-b-\sqrt{b^2-4ac}}{2a} = \dfrac{-2b}{2a} = -\dfrac{b}{a}$ **3.** $\left(\dfrac{-b+\sqrt{b^2-4ac}}{2a}\right)\left(\dfrac{-b-\sqrt{b^2-4ac}}{2a}\right) = \dfrac{b^2+b\sqrt{b^2-4ac}-b\sqrt{b^2-4ac}-(b^2-4ac)}{4a^2} = \dfrac{b^2-b^2+4ac}{4a^2} = \dfrac{4ac}{4a^2} = \dfrac{c}{a}$

Page 392 Exploratory 1. Let w represent the width of the rectangle. $w(w+5)=24$ **2.** Let w represent the width of the rectangle. $w(w+1)=90$ **3.** Let w represent the width of the rectangle. $w(4w+1)=18$ **4.** Let w represent the width of the rectangle. $w\left(\dfrac{16}{3}w\right)=12$ **5.** Let x represent the smaller number. $x(x+2)=35$ **6.** Let x represent the smaller number. $x^2+(x+3)^2=45$ **7.** Let x represent the smaller number. $x(x+5)=24$ **8.** Let x represent the smaller integer. $x(x+1)=72$

Page 396 Exploratory 1. to find all values of x and y that make both equations true at the same time **2.** 2, Answers may vary. **3.** The solution set is $\emptyset$. Answers may vary. **4.** a system of equations with one linear equation and one quadratic equation **5.** You will get a negative discriminant.

Page 396 Written 16. $\{(0,0),(2,4)\}$ **17.** $\{(-3,9),(0,0)\}$ **18.** $\{(0,0),(2,2)\}$ **19.** $\{(-1,5),(0,4)\}$ **20.** $\{(-2,-5),(1,-8)\}$ **21.** $\{(0,2),(5,-3)\}$ **22.** $\{(1,-3),(6,2)\}$ **23.** $\emptyset$ **24.** $\left\{\left(-1\tfrac{1}{2},-1\tfrac{3}{4}\right),(0,-1)\right\}$ **25.** $\{(2,5),(4,9)\}$ **26.** $\emptyset$ **27.** $\emptyset$

Page 396 Challenge 1. $\left\{\left(\dfrac{5-\sqrt{7}}{3},\dfrac{8+2\sqrt{7}}{9}\right),\left(\dfrac{5+\sqrt{7}}{3},\dfrac{8-2\sqrt{7}}{9}\right)\right\}$ **2.** $\left\{\left(-2\tfrac{1}{2},9\right),(1,2)\right\}$ **3.** $\left\{\left(\dfrac{17-\sqrt{161}}{8},\dfrac{49-\sqrt{161}}{32}\right),\left(\dfrac{17+\sqrt{161}}{8},\dfrac{49+\sqrt{161}}{32}\right)\right\}$ **6.** No, the graph does not cross either axis.

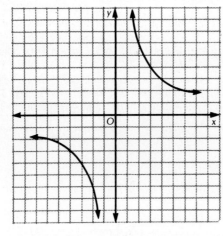

x	y
1	12
2	6
3	4
4	3
6	2
-1	-12
-2	-6
-3	-4
-4	-3
-6	-2

Page 397 Mathematical Excursions 1. $\dfrac{y-x}{x}=\dfrac{x}{y}$

2.
$\dfrac{y-x}{x}=\dfrac{x}{y}$
$x^2=y(y-x)$
$x^2=y^2-yx$
$x^2+yx-y^2=0$
Let $a=1$, $b=y$, and $c=-y^2$.
$x=\dfrac{-y\pm\sqrt{y^2-4(1)(-y^2)}}{2}$
$=\dfrac{-y\pm\sqrt{y^2+4y^2}}{2}$
$=\dfrac{-y\pm\sqrt{5y^2}}{2}$
$=\dfrac{-y\pm y\sqrt{5}}{2}$
$\dfrac{x}{y}=\dfrac{-1+\sqrt{5}}{2}=\dfrac{1}{\dfrac{2}{-1+\sqrt{5}}}\approx\dfrac{1}{1.618}$

Page 399 Written 1. $\dfrac{1}{2a}$ 2. $\dfrac{2}{p}$ 3. $-5n$ 4. $x+4y$ 5. $1-2q$ 6. $2(4w-z)$ 7. $x+1$ 8. $\dfrac{2(y-1)}{y+1}$
9. 4 10. $\dfrac{1}{k}$ 11. $\dfrac{-3}{1+2b}$ 12. $\dfrac{4}{x^2}$ 13. $\dfrac{-1}{c+2}$ 14. $\dfrac{n+4}{n-4}$ 15. $-\dfrac{2(x-1)}{x+1}$ 16. $7y+x$ 17. $\dfrac{p-q}{p+q}$
18. $\dfrac{x+3}{x+5}$ 19. $\dfrac{-5x^4}{2x-5}$ 20. $\dfrac{p+3}{p+7}$ 21. $\dfrac{4}{x+3}$ 22. $\dfrac{t+1}{t-3}$ 23. $\dfrac{j+k}{j-k}$ 24. $\dfrac{2y(x-y)}{3x(3x+2y)}$ 25. $-\dfrac{a+3}{a+2}$
26. $-\dfrac{x+5}{x+6}$ 27. -1 28. -1 29. $\dfrac{4y-1}{8y-1}$ 30. $\dfrac{2x+1}{x-2}$

Page 402 Written 1. $\dfrac{135x^2}{8}$ 2. $\dfrac{4m}{15}$ 3. $\dfrac{1}{4}$ 4. $-y$ 5. $\dfrac{45x^3}{128y^2}$ 6. $-\dfrac{a+b}{a}$ 7. $\dfrac{(x-1)^2(x^2-1)}{xy}$
8. $-a(a+b)$ 9. $\dfrac{x+1}{x-1}$ 10. $\dfrac{2(3b+c)}{b+3c}$ 11. $-\dfrac{2(x-2)}{x}$ 12. $2(x+y)$ 13. $\dfrac{4b(x+15)}{x(x+10)}$
14. $\dfrac{(x-2)(x-3y)}{2x(x-3)(x+1)}$ 15. $8(y-1)$ 16. $\dfrac{2a^2(a+b)}{3b(a-6b)}$ 17. $\dfrac{2(x+3)}{(x-3)(2x-1)}$ 18. $\dfrac{2x(x+3)}{3(x-3)}$
19. $\dfrac{(x-6)(x+2)(x+3)}{x+12}$ 20. $\dfrac{y-2}{y+2}$ 21. $(a-b)^2$ 22. $\dfrac{(y-4)^2(y+3)}{(y+12)(y-6)(y+2)}$ 23. $\dfrac{w-4}{w-10}$ 24. $\dfrac{x-1}{x+3}$

Page 405 Written 1. $\dfrac{19}{24}$ 2. $\dfrac{5}{8}$ 3. $\dfrac{3b-4a}{ab}$ 4. $\dfrac{14+5x}{7x}$ 5. $\dfrac{29x}{28}$ 6. $\dfrac{2}{y}$ 7. $\dfrac{14n-5}{6n^2}$ 8. $\dfrac{x+15}{2x}$
9. $\dfrac{k(5x+1)}{7x^2}$ 10. $\dfrac{a-b}{a+b}$ 11. $\dfrac{x^2+3}{x-5}$ 12. $\dfrac{7m+2n}{12}$ 13. $\dfrac{13x+3y}{18}$ 14. $-\dfrac{7x+1}{8x(x-1)}$ 15. $\dfrac{2(x^2+4)}{x^2-4}$
16. $\dfrac{y^2-6y+7}{y^2-36}$ 17. $\dfrac{27-5x}{x^2-7x+12}$ 18. $-\dfrac{3b+19}{b^2-16}$ 19. $\dfrac{y^2+3y+6}{y^3-4y^2}$ 20. $-\dfrac{2(b+8)}{b^2-16}$ 21. $\dfrac{2(5-3x)}{4x^2-1}$
22. $\dfrac{8(2a-1)}{a^2-4}$ 23. $-\dfrac{x^2-5x+1}{(x-1)^2}$ 24. $\dfrac{(x-y)^2-(x+y)^3}{(x+y)^2(x-y)}$ 25. $\dfrac{a^3+7a^2+8a+12}{(a+2)^2(a-2)}$
26. $\dfrac{3(3m+2)}{(m+2)^2(m+1)}$ 27. $\dfrac{10a^2+11a+1}{(3a-2)(2a+1)}$ 28. $\dfrac{x^2+12x+2}{(x-1)^2(x+4)}$

Page 409 Chapter Review 35. $\dfrac{4ab^2c^4-12a^3bc+13}{2c}$ 36. $\dfrac{1}{4k-5}$ 37. $\dfrac{x+3}{x(x-6)}$ 38. $\dfrac{x-8}{x-6}$ 39. $\dfrac{3axy}{10}$
40. $6b(a-b)$ 41. $\dfrac{(x+4)^2}{(x+2)^2}$ 42. $x-1$ 43. $\dfrac{7ab(x-9)}{3(x-5)}$ 44. $\dfrac{(m+7)^2}{m+3}$ 45. 2 46. $\dfrac{1}{x-1}$
47. $\dfrac{10axy-3}{6x^2y}$ 48. $\dfrac{x^2-7x-15}{x^2+x-6}$ 49. $\dfrac{8x-9}{x^2-4}$ 50. $\dfrac{7y+11}{(y-5)(y+1)(y+3)}$

Page 410 Chapter Test

7.
8.
9.
10.

32. $\dfrac{2(x-6)}{x+12}$ **33.** $\dfrac{3x^2 + 4x + 28}{(x+4)(x+1)}$ **34.** $\dfrac{x+8}{x+5}$ **35.** $\dfrac{4x+3}{x+4}$

CHAPTER 11   PROBABILITY OF COMBINATORICS

Page 415 Challenge 2. {(1, 1, 1), (1, 1, 2), (1, 1, 3), (1, 1, 4), (1, 2, 1), (1, 2, 2), (1, 2, 3), (1, 2, 4), (1, 3, 1), (1, 3, 2), (1, 3, 3), (1, 3, 4), (1, 4, 1), (1, 4, 2), (1, 4, 3), (1, 4, 4), (2, 1, 1), (2, 1, 2), (2, 1, 3), (2, 1, 4), (2, 2, 1), (2, 2, 2), (2, 2, 3), (2, 2, 4), (2, 3, 1), (2, 3, 2), (2, 3, 3), (2, 3, 4), (2, 4, 1), (2, 4, 2), (2, 4, 3), (2, 4, 4), (3, 1, 1), (3, 1, 2), (3, 1, 3), (3, 1, 4), (3, 2, 1), (3, 2, 2), (3, 2, 3), (3, 2, 4), (3, 3, 1), (3, 3, 2), (3, 3, 3), (3, 3, 4), (3, 4, 1), (3, 4, 2), (3, 4, 3), (3, 4, 4), (4, 1, 1), (4, 1, 2), (4, 1, 3), (4, 1, 4), (4, 2, 1), (4, 2, 2), (4, 2, 3), (4, 2, 4), (4, 3, 1), (4, 3, 2), (4, 3, 3), (4, 3, 4), (4, 4, 1), (4, 4, 2), (4, 4, 3), (4, 4, 4)}

Page 422 Written

6.

12.

16.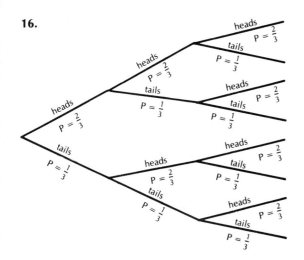

Page 436 Written 12. 9; The symbol $_9C_1$ names the number of combinations of 9 things taken 1 at a time. **13.** The number of combinations of n things taken n at a time **14.** The number of combinations of 4 things taken 4 at a time; 1; There is obviously only one way to make the selection. **15.** There is only one way to choose nothing from n things.

Page 439 Written 4. apples, oranges; apples, dates; apples, grapes; apples, strawberries; oranges, dates; oranges, grapes; oranges, strawberries; dates, grapes; dates, strawberries; grapes, strawberries **5.** A, B, C, D; A, B, C, E; A, B, C, F; A, B, D, E; A, B, D, F; A, B, E, F; A, C, D, E; A, C, D, F; A, C, E, F; A, D, E, F; B, C, D, E; B, C, D, F; B, C, E, F; B, D, E, F; C, D, E, F

Page 440 Challenge 4. $_nC_{n-r} = \dfrac{n!}{(n-r)![n-(n-r)]!} = \dfrac{n!}{(n-r)!\,r!} = {}_nC_r$ **6.** $_nC_{n-1} = \dfrac{n!}{(n-1)![n-(n-1)]!} = \dfrac{n!}{(n-1)!} = n$ Finding the number of combinations of n things taken $(n-1)$ at a time means that a different item is *not* chosen each time for a total of n combinations.

CHAPTER 12 TRANSFORMATIONAL GEOMETRY

Pages 456–457 Exploratory 1. Yes; domain = {100, 35, 20, 0}, range = {212, 95, 68, 32} **2.** No **3.** No **4.** Yes; domain = {Kenton, McGuffey, Alger}, range = {22,358, 4,225} **5.** Yes; domain = {5, 9, 17}, range = {8, 6} **6.** No

Page 457 Written

17.
g
$-3 \longrightarrow -6$
$-2 \longrightarrow -4$
$-1 \longrightarrow -2$
$0 \longrightarrow 0$
$1 \longrightarrow 2$
$2 \longrightarrow 4$
$3 \longrightarrow 6$
range = {−6, −4, −2, 0, 2, 4, 6}

18.
f
$-3 \longrightarrow -8$
$-2 \longrightarrow -7$
$-1 \longrightarrow -6$
$0 \longrightarrow -5$
$1 \longrightarrow -4$
$2 \longrightarrow -3$
$3 \longrightarrow -2$
range = {−8, −7, −6, −5, −4, −3, −2}

19.
m
$-3 \longrightarrow 9$
$-2 \longrightarrow 4$
$-1 \longrightarrow 1$
$0 \longrightarrow 0$
1
2
3
range = {0, 1, 4, 9}

20.
t
$-3 \longrightarrow 13$
$-2 \longrightarrow 7$
$-1 \longrightarrow 3$
$0 \longrightarrow 1$
1
2
3
range = {1, 3, 7, 13}

21.
s
$-3 \longrightarrow -5$
$-2 \longrightarrow -4$
$-1 \longrightarrow -3$
$0 \longrightarrow -2$
$1 \longrightarrow -1$
$2 \longrightarrow 0$
$3 \longrightarrow 1$
range = {−5, −4, −3, −2, −1, 0, 1}

22.
f
$-3 \longrightarrow 1$
$-2 \longrightarrow 0$
$-1 \longrightarrow -1$
$0 \longrightarrow -2$
1
2
3
range = {1, 0, −1, −2}

27.
$$7 \xrightarrow{g} 3 \xrightarrow{f} 7$$
$$9 \xrightarrow{g} 5 \xrightarrow{f} 11$$
$$11 \xrightarrow{g} 4 \xrightarrow{f} 9$$
$$13 \xrightarrow{g} 6 \xrightarrow{f} 13$$

28.
$$3 \xrightarrow{f} 7 \xrightarrow{g} 3$$
$$4 \xrightarrow{f} 9 \xrightarrow{g} 5$$
$$5 \xrightarrow{f} 11 \xrightarrow{g} 4$$
$$6 \xrightarrow{f} 13 \xrightarrow{g} 6$$

Page 460 Written 8. Using a protractor, find point M on m such that $\overleftrightarrow{PM} \perp m$. Using a ruler, find point P' on $\overleftrightarrow{PM}$ such that $PM = MP'$. P' is the image of point P. **18.** A, B, C, and D are collinear as are their images A', B', C', and D'. **19.** B is between A and C as B' is between A' and C'. C is between B and D as C' is between B' and D'. **20.** Yes, because the image of an angle is congruent to the angle. This can be seen because given $\angle ABC$ and $\angle A'B'C'$, $\triangle ABC \cong \triangle A'B'C'$ by SSS. Therefore, $\angle ABC \cong \angle A'B'C'$.

Page 463 Written 3. No. Lines of symmetry must divide figures into two congruent figures. If there are two lines of symmetry of a triangle there must be two sets of congruent triangles. Hence, two angles must be congruent to the third angle. In which case the triangle is equilateral and has three lines of symmetry.

Page 468 Written 13. $A'(-4, 2)$, $B'(-1, 5)$, $AB = 3\sqrt{2}$, $A'B' = 3\sqrt{2}$ **14.** $A'(0, 3)$, $B'(-3, -1)$, $AB = 5$, $A'B' = 5$ **15.** $A'(3, 4)$, $B'(-5, 6)$, $AB = 2\sqrt{17}$, $A'B' = 2\sqrt{17}$ **16.** $A'(2, 3)$, $B'(-5, 1)$, $AB = \sqrt{53}$, $A'B' = \sqrt{53}$ **17.** $A'(4, -2)$, $B'(1, -5)$, $AB = 3\sqrt{2}$, $A'B' = 3\sqrt{2}$ **18.** $A'(0, -3)$, $B'(3, 1)$, $AB = 5$, $A'B' = 5$ **19.** $A'(-3, -4)$, $B'(5, -6)$, $AB = 2\sqrt{17}$, $A'B' = 2\sqrt{17}$ **20.** $A'(-2, -3)$, $B'(5, -1)$, $AB = \sqrt{53}$, $A'B' = \sqrt{53}$ **21.** $A'(2, 4)$, $B'(5, 1)$, $AB = 3\sqrt{2}$, $A'B' = 3\sqrt{2}$ **22.** $A'(3, 0)$, $B'(-1, 3)$, $AB = 5$, $A'B' = 5$ **23.** $A'(4, -3)$, $B'(6, 5)$, $AB = 2\sqrt{17}$, $A'B' = 2\sqrt{17}$ **24.** $A'(3, -2)$, $B'(1, 5)$, $AB = \sqrt{53}$, $A'B' = \sqrt{53}$

Page 468 Written

38.–39.

Page 471 Exploratory

Page 472 Written

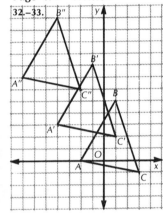

35. collinearity, betweenness of points, congruence, angle measure, distance measure, orientation, parallelism **36.** orientation; See exercises 37 and 38.

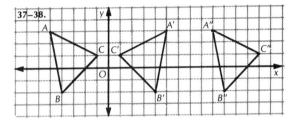

39. yes; $T_{14,0}$; In $R_{y\text{-axis}}$ $(x, y) \longrightarrow (-x, y)$. In $R_{x=7}$ $(x, y) \longrightarrow (14 - x, y)$, So $(-x, y)$ becomes $(14 + x, y)$. This is also the transformation, $T_{14,0}$, that takes (x, y) into $(14 + x, y)$.

Page 482 Exploratory

1. line reflection R_ℓ followed by line reflection R_m **2.** line reflection R_m followed by 90° counterclockwise rotation **3.** a half-turn around the origin followed by translation $T_{2,3}$ **8.** Yes. In $T_{c,d}$ $(x, y) \rightarrow (x + c, y + d)$. In $T_{a,b}$ $(x, y) \rightarrow (x + a, y + b)$. Therefore, in $T_{a,b} \circ T_{c,d}$ $(x, y) \rightarrow (x + c + a, y + d + b)$. In $T_{c,d} \circ T_{a,b}$ $(x, y) \rightarrow (x + a + c, y + b + d)$. Hence, the transformations are equivalent. **9.** No. If the lines are not perpendicular the images of composite reflections are different depending upon the order of the reflections. **10.** Yes. The rotation of a figure through an angle whose measure is the sum of two angle measures will result in the same figure as the rotation through the angle whose measure is the sum of the same two angle measures in different order. This is because the addition of angle measure is commutative.

Page 482 Written

1. $A'(6, -4)$, $B'(3, -2)$, $C'(1, 2)$ **2.** $A'(6, -4)$, $B'(3, -2)$, $C'(1, 2)$ **3.** $A'(-6, -4)$, $B'(-3, -2)$, $C'(-1, 2)$ **4.** $A'(-4, 9)$, $B'(-1, 7)$, $C'(1, 3)$ **5.** $A'(-4, 9)$, $B'(-1, 7)$, $C'(1, 3)$ **6.** $A'(-8, -8)$, $B'(-5, -6)$, $C'(-3, -2)$ **7.** $A'(-8, 0)$, $B'(-5, 2)$, $C'(-3, 6)$ **8.** $A'(8, -9)$, $B'(5, -7)$, $C'(3, -3)$ **9.** $A'(4, 1)$, $B'(1, 3)$, $C'(-1, 7)$ **10.** $A'(6, 11)$, $B'(3, 9)$, $C'(1, 5)$ **11.** $A'(6, 11)$, $B'(3, 9)$, $C'(1, 5)$ **12.** $A'(-6, 4)$, $B'(-3, 2)$, $C'(-1, -2)$

Page 483 Written

38. A glide reflection is an isometry because it is a composite of a line reflection and a translation both of which are isometries and hence preserve distance. **39.** No. The line reflection is not a direct isometry though a translation is. The composition will not be a direct isometry. **40.** Yes. The composition of two reflections in parallel lines will result in a figure whose order is returned to the original order since each reflection separately reverses the order. **41.** Yes. Each translation is a direct isometry. Any number taken consecutively will therefore preserve orientation. **42.** Yes. Each rotation is a direct isometry and preserves order, so any number taken one after the other will preserve order.

Page 483 Challenge Call the reflected image of P in the line ℓ, P''. ℓ is the bisector of $\overline{PP''}$ by definition of line reflection. In $\triangle PP''P'$, $\overline{P''P'} \parallel \ell$ since the direction of the translation is in the direction of line ℓ. Line ℓ is the bisector of $\overline{PP'}$ because if a line is parallel to one side of a triangle and it intersects the other two sides, then it divides them proportionally.

Page 487 Exploratory 1. Since $P \rightarrow P'$ so that ℓ is the perpendicular bisector of $\overline{PP'}$ and $P' \rightarrow P''$ so that ℓ is the perpendicular bisector of $\overline{P'P''}$, P and P'' must be the same point. **2.** A half-turn with center P is a rotation about point P of 180°. One followed by another is a rotation of 360° about P or a full turn. Each point rotated through a full turn maps into itself. **3.** In $T_{a,b}$ $(x, y) \rightarrow (x + a, y + b)$. In $T_{-a,-b}$ $(x, y) \rightarrow (x - a, y - b)$. In $T_{-a,-b} \circ T_{a,b}$ $(x, y) \rightarrow (x + a - a, y + b - b)$ or $(x, y) \rightarrow (x, y)$. Hence every point maps into itself. **4.** The composite of these rotations is a clockwise rotation with center P through an angle of 120° + 240° or 360°, or a full turn. Each point rotated clockwise about P through a full-turn maps into itself. **5.** Yes, $T_{0,0}$. In $T_{0,0}$ $(x, y) \rightarrow (x + 0, y + 0)$ or $(x, y) \rightarrow (x, y)$. **6.** Yes. $Rot_{P,0°}$. In this rotation every point of the plane maps into itself. **7.** No. In a line reflection only points on the line map into themselves. Since the whole plane must be reflected it is impossible for a single reflection to map points into themselves.

Page 487 Written

13.

°	$T_{0,0}$	R_{ℓ_1}	R_{ℓ_2}	R_{ℓ_3}	$Rot_{O,-120°}$	$Rot_{O,-240°}$
$T_{0,0}$	$T_{0,0}$	R_{ℓ_1}	R_{ℓ_2}	R_{ℓ_3}	$Rot_{O,-120°}$	$Rot_{O,-240°}$
R_{ℓ_1}	R_{ℓ_1}	$T_{0,0}$	$Rot_{O,-240°}$	$Rot_{O,-120°}$	R_{ℓ_3}	R_{ℓ_2}
R_{ℓ_2}	R_{ℓ_2}	$Rot_{O,-120°}$	$T_{0,0}$	$Rot_{O,-240°}$	R_{ℓ_1}	R_{ℓ_3}
R_{ℓ_3}	R_{ℓ_3}	$Rot_{O,-240°}$	$Rot_{O,-120°}$	$T_{0,0}$	R_{ℓ_2}	R_{ℓ_1}
$Rot_{O,-120°}$	$Rot_{O,-120°}$	R_{ℓ_2}	R_{ℓ_3}	R_{ℓ_1}	$Rot_{O,-240°}$	$T_{0,0}$
$Rot_{O,-240°}$	$Rot_{O,-240°}$	R_{ℓ_3}	R_{ℓ_1}	R_{ℓ_2}	$T_{0,0}$	$Rot_{O,-120°}$

25. Yes, because for every element a of the set there is an element a^{-1} such that $a \circ a^{-1} = I$. **26.** Yes, it satisfies the four requirements of a group: 1. The set with ° is an operational system. 2. The operation ° is associative. 3. I is the identity element. 4. Every element has an inverse with respect to °. It is not commutative because $Rot_{O,-120°} \circ R_{\ell_1} = R_{\ell_2}$ but $R_{\ell_1} \circ Rot_{O,-120°} = R_{\ell_3}$. $Rot_{O,-120°} \circ R_{\ell_1} \neq R_{\ell_1} \circ Rot_{O,-120°}$. **27.** Yes. See page 74, Section 2.9. The set $S = \{e, p, q, r, s, t\}$ with operation ° is equivalent to this set with $e = T_{0,0}$, $p = Rot_{O,-120°}$, $q = Rot_{O,-240°}$, $r = R_{\ell_1}$, $s = R_{\ell_2}$, $t = R_{\ell_3}$.

Page 490 Written 8. $(x, y) \xleftarrow{D_{O,k} \circ D_{O,\ell}} (\ell kx, \ell ky)$ By first performing the dilation with scale factor ℓ followed by the dilation with scale factor k produces the result of a dilation with scale factor ℓk. **9.** In an isometry distance is preserved. Distance can by preserved in a dilation only with scale factor 1 or -1. **10.** Since $(x, y) \xleftarrow{D_{O,k}} (kx, ky)$, if $k = -1$ $(x, y) \xleftarrow{D_{O,-1}} (-x, -y)$ which is exactly the rule for a half-turn. **17.** False. Isometries preserve distance. **18.** False. A dilation with scale factor 1 preserves distance. **19.** False. Transformation t may be any isometry. **20.** False. Transformation t might be a dilation with scale factor not equal to 1 and the transformation t would not preserve distance.

Page 496 Chapter Test

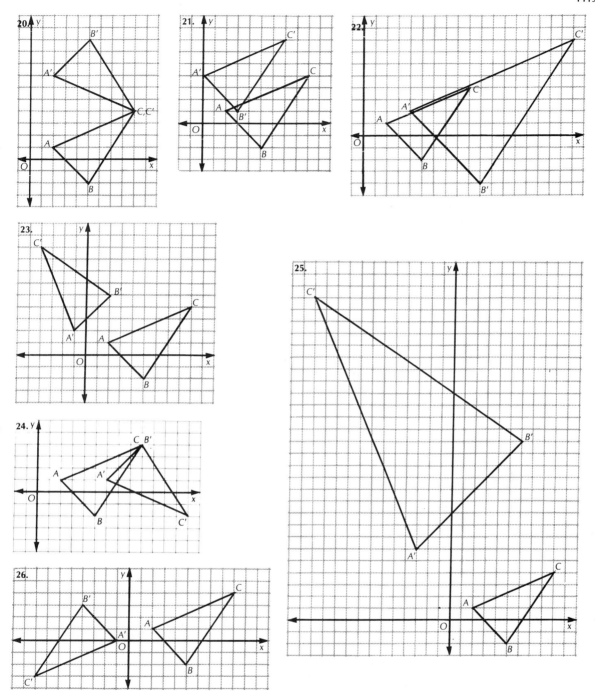

31. dilation with center A and scale factor of 2   **32.** reflection over line through C parallel to $\overline{AB}$ **33.** rotation of 180° about midpoint of $\overline{AC}$ **34.** translation of length AA' **35.** reflection in line parallel to $\overline{AB}$ and $\overline{A'B'}$ midway between them   **36.** translation of length AA' **37.** rotation of 180° about midpoint of $\overline{BD}$.
38. dilation with center at C with scale factor between 0 and -1

CHAPTER 13   INTRODUCTION TO TRIGONOMETRY

Page 504   Written   6. $\overline{EC}, \overline{CG}$ **7.** $\overline{AD}, \overline{DG}$ **8.** $\overline{DK}, \overline{KG}$ **9.** $\overline{BE}, \overline{AB}$ **10.** $\overline{AB}, \overline{BE}$ **11.** $\overline{KG}, \overline{DK}$ **12.** $\overline{AK}, \overline{KD}$ **13.** $\overline{AK}, \overline{KD}$ **14.** $\overline{DG}, \overline{AD}$

Page 515   Challenge   4. Since $\tan A = \dfrac{\text{side opposite}}{\text{side adjacent}}$, and the measure of the side opposite can be infinitely greater than the measure of the side adjacent, the tangent of an angle can get greater and greater without bound. As the measure of the angle gets closer to 90, the measure of the side opposite gets closer to the measure of the hypotenuse, but never surpasses it. Hence, the sine of the angle approaches 1, and is 1 when the angle measures 90, but never surpasses 1. Likewise, the cosine of an angle, as it decreases in measure getting closer to 0, becomes closer to 1 because the measure of the adjacent side gets closer to the measure of the hypotenuse.

Page 520   Challenge

Page 525   Chapter Review   46. For a 45° angle, $\cos a = \sin a$. Since $\cos^2 a + \sin^2 a = 1$, and by substitution $\cos^2 a + \cos^2 a = 1$, $2\cos^2 a = 1$, $\cos^2 a = \dfrac{1}{2}$, $\sqrt{\cos^2 a} = \sqrt{\dfrac{1}{2}}$, and $\cos a = \dfrac{\sqrt{2}}{2}$. So, $\sin 45° = \cos 45° = \dfrac{\sqrt{2}}{2}$.   **47.** $\sin 90° = 1$, $\sin 180° = 0$, $\sin 270° = -1$, $\sin 360° = 0$, $\cos 90° = 0$, $\cos 180° = -1$, $\cos 270° = 0$, $\cos 360° = 1$ **49.** For any angle of measure a, $\tan a = \dfrac{\text{side opposite}}{\text{side adjacent}}$, $\sin a = \dfrac{\text{side opposite}}{\text{hypotenuse}}$, $\cos a = \dfrac{\text{side adjacent}}{\text{hypotenuse}}$, and therefore $\tan a = \dfrac{\frac{\text{side opposite}}{\text{hypotenuse}}}{\frac{\text{side adjacent}}{\text{hypotenuse}}} = \dfrac{\sin a}{\cos a}$.

ANSWERS TO CHAPTER OPENERS

Page 85, Chapter 3

	↑	↕	120°	↗	↙	240°
↑	↑	↕	120°	↗	↙	240°
↕	↕	↑	↙	240°	120°	↗
120°	120°	↗	240°	↙	↕	↑
↗	↗	120°	↕	↑	240°	↙
↙	↙	240°	↗	120°	↑	↕
240°	240°	↙	↑	↕	↗	120°

Page 167, Chapter 5

- Three sides
- any two sides and their included angle
- any two angles and the included side
- any two angles and the side opposite one of those angles
- given it's a right triangle, any leg and the hypotenuse
- given it's a right triangle, any angle and the hypotenuse
- given it's a right triangle, the two legs
- given it's a right triangle, a leg and an angle

LEARNING OBJECTIVES

1 Working with Logic

1.1 Identify statements and open sentences and find solution sets.
1.2 Write a conditional and find its truth value.
1.3 Write a conclusion by using the Law of Disjunctive Inference.
1.4 Write a conclusion by using the Law of Detachment.
1.5 Write a converse, inverse, and contrapositive of the conditional.
1.6 Write a conclusion by using the Law of Syllogism.
1.7 Write the negation of a statement.
1.8 Determine if sentences are quantified and to write sentences using $\forall$ or $\exists$ notation.
1.9 Write a negation of a quantified statement and to identify quantifiers.

2 Mathematical Systems

2.1 Solve addition problems using clock arithmetic.
2.2 Solve multiplication problems using clock arithmetic.
2.3 Solve subtraction and division problems using clock arithmetic.
2.4 Identify an operation on a set and to identify operations as commutative or associative.
2.5 Find the identities and inverses of elements.
2.6 Identify a group from a finite number of elements.
2.7 Identify a group from an infinite number of elements.
2.8 Use theorems to prove facts about groups.
2.9 Compute using non-commutative groups.

3 A System with Two Operations

3.1 Solve equations using real numbers, group properties, and group theorems.
3.2 Solve equations using the Distributive Law.
3.3 Determine whether a system is a field and solve equations in the rational and real number system.
3.4 Determine whether a system is finite field and use the properties of a field to solve equations.
3.5 Prove field theorems by using additive and multiplicative language.
3.6 Solve equations using Field Combination Theorem 3.
3.7 Solve equations using the definitions of subtraction and division.
3.8 Order a collection of rational numbers.

4 Introduction to Euclidean Geometry

4.1 Identify undefined and defined terms and use basic geometric terms.
4.2 Use the properties of lines, planes, and segments to solve problems.
4.3 Use the properties of angles to solve problems.
4.4 Identify the axiom that justifies a given statement.
4.5 Prove theorems about angles.
4.6 Prove and apply theorems about right angles and perpendicular lines.
4.7 Prove and apply theorems about parallel lines.
4.8 Prove and apply theorems about parallel lines and interior angles.
4.9 Prove theorems using an indirect method of proof.
4.10 Prove and apply theorems about angles of triangles.

4.11	Apply theorems about the angles of a polygon.

5 Congruent Triangles

5.1	Identify corresponding parts of congruent triangles.
5.2	Identify congruent triangles using SSS, SAS, or ASA.
5.3	Apply congruence axioms.
5.4	Identify congruent triangles using AAS.
5.5	Identify congruent triangles that overlap.
5.6	Identify special elements of a triangle and apply theorems about isosceles triangles.
5.7	Prove that two right triangles are congruent.
5.8	Use theorems of inequalities to prove that the measures of angles or sides of triangles are unequal.
5.9	Determine the possible measures of the third side of a triangle given the measures of two sides.

6 Applications of Congruent Triangles

6.1	Prove and apply theorems about parallelograms.
6.2	Prove when a quadrilateral is a parallelogram.
6.3	Prove and apply theorems about rectangles and rhombuses.
6.4	Prove and apply theorems about trapezoids.
6.5	construct basic geometric figures.
6.6	Construct basic geometric figures.
6.7	Prove and apply a theorem about the segment that joins the midpoints of two sides of a triangle.

7 Similarity

7.1	Apply and solve ratios and proportions.
7.2	Use properties of proportions to solve problems.
7.3	Determine if two triangles are similar and apply the properties of similar triangles.
7.4	Prove and apply theorems about proportions and triangles.
7.5	Prove and apply theorems about right triangles and similarity.
7.6	Prove and apply the Pythagorean Theorem and its converse.
7.7	Use the Pythagorean Theorem to find missing measures of sides of special right triangles.
7.8	Construct geometric figures based on similarities.
7.9	Identify and write relations.
7.10	Identify a relation as reflexive, symmetric, or transitive, or as an equivalence relation.

8 Geometry with Coordinates

8.1	Name and locate points on a plane.
8.2	Graph equations and inequalities.
8.3	Solve systems of equations using the substitution method and the elimination method.
8.4	Find the distance between two points and to apply the distance formula.
8.5	Find the midpoint of a line segment.
8.6	Find the slope of a line.
8.7	Determine if lines are parallel, perpendicular, or neither by comparing their slopes.
8.8	Write equations for nonvertical lines.
8.9	Find the area of polygons graphed on a coordinate plane.

9 Using Coordinates

9.1	Use coordinates to prove theorems involving quadrilaterals.
9.2	Use coordinates to prove theorems involving triangles.
9.3	Describe and draw a locus of points that satisfies certain conditions.

9.4	Write an equation of a locus of points that satisfies certain conditions.	11.8	Use combinations to count possible groupings of objects.
9.5	Find a locus of points that satisfies two different conditions simultaneously.	11.9	Apply various principles of probability and combinatorics.
9.6	Write an equation of a parabola.		
9.7	Name a point in three-dimensional space using three coordinates.		

10 Solving Quadratic Equations

10.1	Multiply and factor polynomials.
10.2	Solve quadratic equations by factoring.
10.3	Identify and graph parabolas.
10.4	Solve quadratic equations by graphing.
10.5	Solve quadratic equations by making perfect square trinomials.
10.6	Solve quadratic equations by using the quadratic formula.
10.7	Solve problems using quadratic equations.
10.8	Solve systems of equations graphically or algebraically.
10.9	Simplify rational expressions.
10.10	Multiply and divide rational expressions.
10.11	Add and subtract rational expressions.

11 Probability and Combinatorics

11.1	Find the outcome set of a situation and the probability that an event will occur.
11.2	Find the probability of mutually exclusive or inclusive events.
11.3	Use the Counting Principle and draw tree diagrams.
11.4	Use permutations to count possible orders of objects.
11.5	Use permutations to count possible orders of objects that do not include all available objects.
11.6	Use permutations to count possible orders of objects when some objects are identical.
11.7	Determine if the number of groupings of objects is found using a combination or a permutation.

12 Transformation Geometry

12.1	Identify mappings, one-to-one mappings, and to draw arrow diagrams.
12.2	Find the image of a figure over a line of reflection.
12.3	Find the line of symmetry in a figure.
12.4	Find the image of a figure over a line of reflection in the coordinate plane.
12.5	Find the image of a figure for a translation.
12.6	Find the image of a figure for a rotation.
12.7	Find the image of a figure for a half-turn (point reflection).
12.8	Find the image of a figure for a composition of transformations.
12.9	Relate the composition of transformations to a mathematical system.
12.10	Find the image of a figure for a dilation.

13 Introduction to Trigonometry

13.1	Use the Pythagorean Theorem and the properties of similarity to solve problems.
13.2	Use the tangent ratio to find the measures of the angles of a right triangle.
13.3	Use the tangent ratio to find missing measures of sides in a right triangle.
13.4	Find the sine, cosine, or tangent of angles in a right triangle.
13.5	Find the values of the sine or cosine or an angle using a unit circle.

INTEGRATED MATHEMATICS
C O U R S E 2

BUMBY

KLUTCH

COLLINS

EGBERS

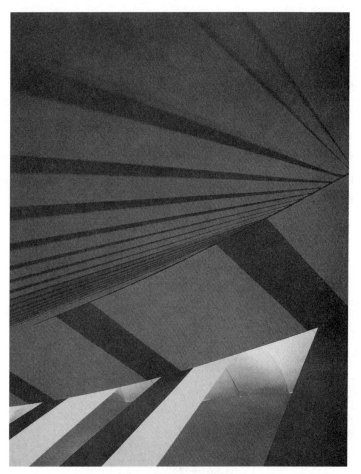

GLENCOE
McGraw-Hill

New York, New York   Columbus, Ohio   Mission Hills, California   Peoria, Illinois

Reviewers

Carole Bickel
Mathematics Department Chair
Gahanna Lincoln High School
Gahanna, Ohio

Sue Boice
Mathematics Teacher
Wilson Magnet School
Rochester, New York

Bill Bonney
Mathematics Teacher
Ballard High School
Seattle, Washington

Bill Collins
Supervisor: Mathematics and
  Computer Training
Syracuse City School District
Syracuse, New York

Thomas Ford
Mathematics Teacher
Alhambra High School
Martinez, California

Barbara Lapetina
Mathematics Department Head
Brentwood High School
Brentwood, New York

Jane S. Prochazka
Mathematics Teacher
Calabasas High School
Calabasas, California

Donald G. Sexton
Assistant Principal
  Supervision — Mathematics
Adlai E. Stevenson High School
Bronx, New York

William K. Somerville
Mathematics Teacher
Massapequa High School
Massapequa, New York

Jeanette Tomasullo
Assistant Principal
  Supervision — Mathematics
Eastern District High School
Brooklyn, New York

Copyright © 1996 by Glencoe/McGraw-Hill. All rights reserved.
Originally copyrighted by Merrill Publishing Company. Except as permitted under the United States Copyrights Act, no part of this publication may be reproduced or distributed in any form or by any means, or stored in a database or retrieval system, without prior written permission of the publisher.

Send all inquiries to:
Glencoe/McGraw-Hill
936 Eastwind Drive
Westerville, Ohio 43081

ISBN: 0-02-824906-2 (Student Edition)
ISBN: 0-02-824907-0 (Teacher's Annotated Edition)

Printed in the United States of America.

2 3 4 5 6 7 8 9 10 RRC/LP 03 02 01 00 99 98 97 96 95

Authors

Douglas R. Bumby is chairman of the mathematics department at Scarsdale High School, Scarsdale, New York. Dr. Bumby taught mathematics at Hunter College High School and was Clinical Associate in Mathematics Education at Teachers College, Columbia University, where he taught graduate courses and advised doctoral candidates. Dr. Bumby's research interests are in the areas of mathematics curriculum development and readability of mathematics textbooks. He was instrumental in developing a model integrated mathematics program at Hunter College High School in New York City.

Richard J. Klutch taught mathematics at Freeport High School in New York and now teaches mathematics at the Hunter College Campus Schools of the City University of New York. He has participated in the development of major mathematics curriculum projects and played a major role in developing such programs at Hunter College High School and elsewhere. He has served as consultant to the Bureau of Mathematics Education of the State of New York. He is one of the authors of the new syllabus for the three-year sequence for high school mathematics for the State of New York.

Donald W. Collins is an assistant professor of mathematics education at Western Kentucky University in Bowling Green, Kentucky. Dr. Collins taught mathematics at Elmhurst Public Schools in Elmhurst, Illinois and Abilene Public School in Abilene, Texas. He has many years of experience in the field of mathematical education, including over 20 years of work in the mathematics textbook publishing industry.

Elden B. Egbers was the Supervisor of Mathematics for the state of Washington. Mr. Egbers taught mathematics and was the department chairman at Queen Anne High School in Seattle, Washington. Most recently, he directed the writing of the Washington State Curriculum Guidelines. Mr. Egbers was active in giving workshops on integrated mathematics to mathematics teachers and advisors.

Table of Contents

Chapter 1 — Working with Logic

- **1.1** Conjunctions and Disjunctions 2
- **1.2** Conditionals and Bi-Conditionals 6
- **1.3** Beginning to Reach Conclusions 10
- **1.4** The Law of Detachment 13
- **1.5** The Law of the Contrapositive 17
- **1.6** The Law of Syllogism 20
- **1.7** The De Morgan Laws 24
- **1.8** Quantifiers 30
- **1.9** Negating Quantified Statements 33
  Problem Solving Application:
    Eliminating Possibilities 36
  Chapter Summary 38
  Chapter Review 40
  Chapter Test 42

Chapter 2 — Mathematical Systems

- **2.1** Clock Arithmetic 44
- **2.2** Multiplication in a Clock System 48
- **2.3** Subtraction and Division 52
- **2.4** Operations 56
- **2.5** Identities and Inverses 61
- **2.6** The Group 65
- **2.7** Infinite Groups 68
- **2.8** Some Theorems About Groups 71
- **2.9** A Non-Commutative Group 76
  Problem Solving Application:
    Guess and Check 79
  Chapter Summary 81
  Chapter Review 83
  Chapter Test 84

High Interest Features

Journal Entries
33, 76

Mixed Reviews
23, 51, 64

Mathematical Excursions
28-29, 55

Portfolio Suggestions
38, 81

Performance Assessment
38, 81

Chapter 3 — A System with Two Operations

- 3.1 Introduction 86
- 3.2 The Distributive Law 88
- 3.3 A Field 92
- 3.4 Finite Fields..96
- 3.5 Some Field Theorems.................................101
- 3.6 An Important Field Theorem 104
- 3.7 Subtraction and Division............................107
- 3.8 Ordering the Rationals 111
- **Problem Solving Application:**
  - Identifying Subgoals 115
- Chapter Summary 117
- Chapter Review..118
- Chapter Test..120

Chapter 4 — Introduction to Euclidean Geometry

- 4.1 Undefined Terms and Defined Terms............................122
- 4.2 Axioms About Lines, Planes, and Segments..................125
- 4.3 Axioms About Angles....................................128
- 4.4 Properties of Real Numbers..........................132
- 4.5 Theorems About Angles 135
- 4.6 Right Angles and Perpendicular Lines 139
- 4.7 Parallel Lines and Angles 143
- 4.8 Parallel Lines and Interior Angles 147
- 4.9 Indirect Proof...151
- 4.10 Angles of a Triangle 154
- 4.11 Angles of a Polygon 157
- **Problem Solving Application:**
  - Using Equations 160
- Chapter Summary 162
- Chapter Review 165
- Chapter Test 166

High Interest Features

Journal Entries
112, 143

Mixed Reviews
95, 110, 131, 150

Mathematical Excursions
100

Portfolio Suggestions
117, 162

Performance Assessment
117, 162

Chapter 5 — Congruent Triangles

5.1	Introduction	168
5.2	Some Congruence Axioms	171
5.3	Proving Corresponding Parts Congruent	175
5.4	The AAS Congruence Theorem	180
5.5	Triangles That Overlap	184
5.6	Isosceles Triangles, Special Segments	189
5.7	Proving Right Triangles Congruent	192
5.8	Inequalities	195
5.9	Another Triangle Inequality	199

Problem Solving Application:
Using Figures 201
Chapter Summary 203
Chapter Review 204
Chapter Test 206

Chapter 6 — Applications of Congruent Triangles

6.1	All About Parallelograms	208
6.2	When is a Quadrilateral a Parallelogram?	212
6.3	Rectangles and Rhombuses	218
6.4	Trapezoids	222
6.5	Some Constructions	226
6.6	More Constructions	231
6.7	Another Triangle Relationship	235

Problem Solving Application:
Indirect Reasoning 238
Chapter Summary 240
Chapter Review 241
Chapter Test 242

High Interest Features

Journal Entries
195, 235

Mixed Reviews
179, 225

Portfolio Suggestions
203, 240

Performance Assessment
203, 240

Chapter 7 — Similarity

7.1	Ratio and Proportion	244
7.2	Some Properties of Proportions	248
7.3	Similar Triangles	252
7.4	Proportions and Triangles	256
7.5	Right Triangles and Similarity	261
7.6	The Pythagorean Theorem	265
7.7	Special Right Triangles	269
7.8	More Constructions	272
7.9	Relations	275
7.10	Equivalence Relations	278
	Problem Solving Application:	
	Perimeter and Area	281
	Chapter Summary	283
	Chapter Review	285
	Chapter Test	286

High Interest Features

Journal Entries
273, 323

Mixed Reviews
251, 297, 309

Mathematical Excursions
277, 313

Portfolio Suggestions
283, 326

Performance Assessment
283, 326

Chapter 8 — Geometry with Coordinates

8.1	Naming Points in a Plane	288
8.2	Graphing Linear Equations	291
8.3	Algebraic Solutions	294
8.4	Distance Between Two Points	298
8.5	The Midpoint Formula	302
8.6	Slope	306
8.7	Parallel and Perpendicular Lines	310
8.8	Equations of Lines	314
8.9	Area Using Coordinates	319
	Problem Solving Application:	
	Graphs and Estimation	324
	Chapter Summary	326
	Chapter Review	327
	Chapter Test	328

Chapter 9 Using Coordinates

9.1	Using Coordinates to Prove Theorems	330
9.2	More Coordinate Proofs	335
9.3	Locus	339
9.4	Locus, Coordinates, and Circles	342
9.5	Some Locus Problems	347
9.6	A Special Locus: Parabola	351
9.7	Coordinates in Space	355
	Problem Solving Application:	
	Using Inverse Variation	358
	Chapter Summary	360
	Chapter Review	361
	Chapter Test	362

High Interest Features

Journal Entries
354, 403

Mixed Reviews
350, 393, 406

Mathematical Excursions
346, 397

Portfolio Suggestions
360, 407

Performance Assessment
360, 407

Chapter 10 Solving Quadratic Equations

10.1	Multiplying and Factoring Polynomials	364
10.2	Solving Quadratics by Factoring	369
10.3	Graphing $y = ax^2 + bx + c$	373
10.4	Solving Quadratic Equations by Graphing	377
10.5	Completing the Square	381
10.6	The Quadratic Formula	385
10.7	Problem Solving with Quadratic Equations	389
10.8	Systems of Equations	394
10.9	Simplifying Rational Expressions	398
10.10	Multiplying and Dividing Rational Expressions	400
10.11	Adding and Subtracting Rational Expressions	403
	Chapter Summary	407
	Chapter Review	408
	Chapter Test	410

Chapter 11 Probability and Combinatorics

11.1	Introduction	412
11.2	"Or" Statements	416
11.3	Applications of the Counting Principle	419
11.4	Permutations	424
11.5	$_nP_r$	427
11.6	Permutations with Repetitions	431
11.7	Picking a Committee	434
11.8	$_nC_r$	437
11.9	Applications	441
	Problem Solving Application: List Possibilities	447
	Chapter Summary	449
	Chapter Review	450
	Chapter Test	452

Chapter 12 Transformation Geometry

12.1	Mappings	454
12.2	Line Reflections	458
12.3	Line Symmetry	462
12.4	Line Reflections in the Coordinate Plane	465
12.5	Translations	469
12.6	Rotations	473
12.7	Half-Turns or Point Reflections	476
12.8	Composing Transformations	480
12.9	A Mathematical System	484
12.10	Dilations	488
	Chapter Summary	491
	Chapter Review	494
	Chapter Review	496

High Interest Features

Journal Entries
433, 460

Mixed Reviews
423, 440, 464, 479

Mathematical Excursions
433, 446

Portfolio Suggestions
449, 493

Performance Assessment
449, 493

Chapter 13 — Introduction to Trigonometry

13.1	The Right Triangle	498
13.2	The Tangent Ratio	501
13.3	Using the Tangent Ratio	505
13.4	The Sine and Cosine Ratios	511
13.5	The Unit Circle	517
	Problem Solving Application: Using Right Triangles	521
	Chapter Summary	523
	Chapter Review	524
	Chapter Test	526

Technology Activities and Extended Projects A1–A16

High Interest Features

Journal Entries
501

Mixed Reviews
510

Mathematical Excursions
516

Portfolio Suggestions
523

Performance Assessment
523

CHAPTER 1
Working with Logic

Application in Law

Logical thinking is critical to lawyers when presenting cases and also to judges who hear those cases. Constance Baker Motley is a former civil rights attorney and a current U.S. District Court judge in New York State. Appointed by President Lyndon Johnson, Ms. Motley is the first African-American woman appointed to serve as a judge in a Federal Court.

Individual Project: *Business*

Advertisers often use arguments to try to convince you to buy their products. Look through some magazines or newspapers for conditional statements and uses of arguments in advertising. Identify the antecedent and the consequent in each conditional statement. Are the laws of logic used in any of the arguments you found? Create a fully-illustrated magazine advertisement for an imaginary product. Be sure to use the laws of logic.

A lawyer presented the following argument. "The law states that if a driver proceeds through a red traffic light that is in proper working order, that driver is subject to a $50 fine. The defendant was seen driving through a red traffic light at the corner of Washington and Elm. The traffic computer shows no indication that the signal was down. Therefore, the defendant is guilty and subject to a $50 fine." What law of logic did the lawyer use?
Law of Detachment

1.1 Conjunctions and Disjunctions

Objective: To identify statements and open sentences and find solution sets.

Miss Scarlett is called to testify in the trial of Professor Plum. She tells the court, "I saw Professor Plum in the library at noon on August 25." Miss Scarlett is making a statement. The statement may be true or false.

In mathematics, a **statement** is any declarative sentence that is either *true* or *false,* but *not* both. The truth or falsity of a statement is called its **truth value.**

If p is a simple statement, then $\sim p$, read "not p," is its **negation.** Study the truth table at the left. The table shows that when p is true, its negation, $\sim p$, is false. When the original statement p is false, its negation, $\sim p$, is true.

Words or phrases that connect two or more simple statements to form a compound statement are called **connectives.** Two basic connectives are "and" and "or."

Truth Table for Negation

p	$\sim p$
T	F
F	T

Definition of Conjunction

> A compound statement formed by joining two statements with the word *and* is called a **conjunction.** Each of the statements is called a **conjunct.**

Truth Table for Conjunction

p	q	$p \wedge q$
T	T	T
T	F	F
F	T	F
F	F	F

The symbol for the conjunction "p and q" is "$p \wedge q$."

The conjunction $p \wedge q$ is true whenever, and only whenever, both of the individual statements, p and q, are true. This information is summarized in the truth table at the left.

Definition of Disjunction

> A compound statement formed by joining two statements with the word *or* is called a **disjunction.** Each of the statements is called a **disjunct.**

Truth Table for Disjunction

p	q	$p \vee q$
T	T	T
T	F	T
F	T	T
F	F	F

The symbol for the disjunction "p or q" is "$p \vee q$."

A disjunction $p \vee q$ is true when p is true, or q is true, or they are both true. This information is summarized in the truth table at the left.

Examples

1 Let p represent "George Washington was the fifth U.S. president" and q represent "Thomas Jefferson was the third U.S. president." Write the statement represented by each of the following. Then determine the truth value.

    a. $p \wedge q$ George Washington was the fifth U.S. president and Thomas Jefferson was the third U.S. president. Because George Washington was *not* the fifth U.S. president, the statement is false.

    b. $q \vee \sim p$ Thomas Jefferson was the third U.S. president or George Washington was *not* the fifth U.S. president. Because at least one of the disjuncts is true, the statement is true.

2 Suppose p is false and q is true. Determine the truth value of each of the following.

    a. $\sim p \wedge q$ Because p is false, $\sim p$ is true. Therefore, $\sim p \wedge q$ is true because both $\sim p$ and q are true.

    b. $\sim q \vee (\sim p \wedge q)$ Here is a disjunction whose first disjunct is $\sim q$ and whose second is $(\sim p \wedge q)$. The first disjunct, $\sim q$, is false, because q is true. The second disjunct is a conjunction. It is true because both of its conjuncts are true. The compound statement, $\sim q \vee (\sim p \wedge q)$ is true because it is a disjunction with at least one true disjunct.

3 Construct a truth table for $\sim p \wedge (p \vee q)$.

First, make columns with headings p, q, $\sim p$, $p \vee q$, and $\sim p \wedge (p \vee q)$. Next, list all the possible combinations of truth values for p and q. Then, write the truth values for $\sim p$ and $p \vee q$. Finally, write the truth values for $\sim p \wedge (p \vee q)$.

p	q	$\sim p$	$p \vee q$	$\sim p \wedge (p \vee q)$
T	T	F	T	F
T	F	F	T	F
F	T	T	T	T
F	F	T	F	F

An **open sentence** is a sentence that *cannot* be described as true or false. It becomes a statement that is either true or false when its variable is replaced with an element from some replacement set. The replacement set is called the **domain.** The set of all replacements from the domain that make an open sentence true is called the **solution set.**

For algebraic problems, the domain is frequently $\mathcal{R}$, the set of real numbers. Other possible domains are $\mathcal{N}$ (the set of natural numbers), $\mathcal{W}$ (the set of whole numbers), $\mathcal{Z}$ (the set of integers), or $\mathcal{Q}$ (the set of rational numbers).

Examples

4 Find the solution set of $2x - 5 = 11$ if the domain is $\mathcal{R}$.

$$2x - 5 = 11$$
$$2x - 5 + 5 = 11 + 5 \qquad \text{Add 5 to both sides.}$$
$$2x = 16$$
$$\frac{2x}{2} = \frac{16}{2} \qquad \text{Divide both sides by 2.}$$
$$x = 8$$

The solution set is {8}.

5 Find the solution set of $\frac{y}{3} + 7 < -2$ if the domain is $\mathcal{Z}$.

$$\frac{y}{3} + 7 < -2$$
$$\frac{y}{3} + 7 - 7 < -2 - 7 \qquad \text{Subtract 7 from both sides.}$$
$$\frac{y}{3} < -9$$
$$\frac{y}{3} \cdot 3 < -9 \cdot 3 \qquad \text{Multiply both sides by 3.}$$
$$y < -27$$

The solution set is {. . ., −30, −29, −28}.

Exercises

Max: 1-6, 1-40; Avg: 1-6, 1-23 odds, 25-28, 29-39 odds; Min: 1-6, 1-37 odds

Exploratory State the truth value of each of the following.
1. $4 < 7$ and $8 > 0$ true
2. $-2 < 0$ and $-6 > -3$ false
3. June has 30 days or Valentine's Day is not in February. true
4. $85\% \geq 0.85$ and water does not freeze at $0°C$. false
5. 2 is prime or 8 is odd. true
6. 5 is prime but not even. true

Written Let p represent "0.5 is rational," q represent "-8 is an integer," and r represent "15 is not a multiple of 3." Write the statement represented by each of the following. Then determine the truth value. For answers to exercises 1-8, see the Teacher's Guide.

1. $p \wedge q$
2. $p \wedge r$
3. $p \vee r$
4. $q \vee r$
5. $\sim p \wedge r$
6. $q \wedge \sim r$
7. $\sim p \vee q$
8. $\sim r \vee q$

Suppose p is false, q is false, and r is true. Determine the truth value of each of the following.

9. $p \vee q$ false
10. $p \vee \sim q$ true
11. $\sim p \wedge r$ true
12. $\sim(p \vee q)$ true
13. $\sim p \wedge \sim q$ true
14. $p \wedge (q \vee r)$ false
15. $\sim p \wedge (q \wedge r)$ false
16. $(p \wedge r) \vee (\sim p \wedge q)$ false

Construct a truth table for each of the following.
For answers to exercises 17-24, see the Teacher's Guide.

17. $p \wedge \sim q$
18. $\sim p \vee q$
19. $p \vee \sim p$
20. $p \wedge \sim p$
21. $\sim p \wedge \sim q$
22. $\sim q \vee (q \wedge r)$
23. $p \vee (q \wedge p)$
24. $\sim p \wedge (\sim q \vee p)$

State whether each of the following is *sometimes, always,* or *never* true.

25. A conjunction has exactly one true conjunct. never
26. The sentence $p \wedge \sim p$ is true. never
27. A disjunction has exactly one false disjunct. sometimes
28. A statement of the form $q \vee (\sim q \vee r)$ is true. always

Find the solution set for each of the following. The domain is Z.

29. $2y - 5 = 9$ {7}
30. $3m + 5 = 38$ {11}
31. $-11 = 3x - 2$ {-3}
32. $4 + 8x = 20$ {2}
33. $1 + 2t = -31$ {-16}
34. $\frac{y}{4} + 6 = 10$ {16}
35. $\frac{2x}{3} - 1 = 3$ {6}
36. $10 - 2x = -2$ {6}
37. $2y + 1 > -3$ {-1, 0, 1, 2, ...}
38. $-2y < -16$ {9, 10, 11, ...}
39. $3(3x + 2) > 7x - 2$ {-3, -2, -1, 0, ...}
40. $1 + 2(x + 4) \geq 1 + 3(x + 2)$ {..., -1, 0, 1, 2}

1.2 Conditionals and Bi-Conditionals

Objective: To write a conditional and find its truth value.

Two statements can be connected with the words "if" and "then" to form a new statement.

Definition of Conditional

A compound statement of the form "if p, then q" is called a **conditional**. The part of the statement logically associated with "if" is called the **antecedent**. The part associated with "then" is called the **consequent**.

In symbols, "if p, then q" can be written

$$p \to q.$$

Other ways of writing the conditional $p \to q$ are

p implies q and q if p.

In all these ways of writing "if p, then q," p is still the antecedent and q the consequent. The antecedent does *not* have to be written first.

The statement "Senator Smiley wins the election" is the antecedent of both of the following conditionals.

If Senator Smiley wins the election, then the new dam will be built.

Your property taxes will go up next year if Senator Smiley wins the election.

Truth Table for the Conditional

p	q	$p \to q$
T	T	T
T	F	F
F	T	T
F	F	T

A conditional is true in every case *except when the antecedent is true and the consequent is false*. The truth table at the left shows this situation.

Sometimes the following "shortcuts" are helpful.

If the antecedent of a conditional is false, then the conditional is true.

If the consequent of a conditional is true, then the conditional is true.

Example

1 Let *p* represent "General Grant fought in World War II" and *q* represent "Tom Sawyer is a fictional character." Write the statement represented by each of the following. Then determine the truth value.

a. $p \rightarrow q$ — If General Grant fought in World War II, then Tom Sawyer is a fictional character. Because the antecedent is false, the conditional is true.

b. $q \rightarrow \sim p$ — If Tom Sawyer is a fictional character, General Grant did *not* fight in World War II. Because the consequent is true, the conditional is true.

Definition of Bi-Conditional

A **bi-conditional** is the conjunction of two conditionals: $p \leftrightarrow q$ means $(p \rightarrow q) \wedge (q \rightarrow p)$.

Other ways of saying "$p \leftrightarrow q$" are

p if and only if q and p iff q.

As the truth table below shows, the bi-conditional is true whenever, and only whenever, *p* and *q* have the same truth value.

Truth Table for the Bi-Conditional

p	q	$p \rightarrow q$	$q \rightarrow p$	$p \leftrightarrow q$ $(p \rightarrow q) \wedge (q \rightarrow p)$
T	T	T	T	T
T	F	F	T	F
F	T	T	F	F
F	F	T	T	T

Example

2 Suppose p is true and q and r are both false. Determine the truth value of each of the following.

a. $p \leftrightarrow q$ Because p is true and q is false, the statement is false.

b. $p \leftrightarrow (q \lor \sim r)$ The truth value of $\sim r$ is true and thus the truth value of $q \lor \sim r$ is true. Because p is true and $q \lor \sim r$ is true, the statement is true.

If $p \leftrightarrow q$ is an equivalence, then p and q are said to be equivalent.

Of special significance in the study of mathematics are those bi-conditionals that are true for any replacement of the p's, q's, or other statements, of which they are made. These bi-conditionals are called **equivalences**.

One example of an equivalence is $p \leftrightarrow \sim(\sim p)$. This statement is called the **Law of Double Negation**.

Example

3 Construct truth tables to determine whether $p \rightarrow q$ and $\sim p \lor q$ are equivalent.

p	q	$p \rightarrow q$	$\sim p$	$\sim p \lor q$	$(p \rightarrow q) \leftrightarrow (\sim p \lor q)$
T	T	T	F	T	T
T	F	F	F	F	T
F	T	T	T	T	T
F	F	T	T	T	T

Since $(p \rightarrow q) \leftrightarrow (\sim p \lor q)$ is an equivalence, the statements are equivalent.

Exercises

Max: 1-8, 1-25; Avg: 1-8, 1-25; Min: 1-8, 1-14, 22-25

Exploratory Identify the antecedent and the consequent for each statement. Then determine the truth value. antecedent circled and consequent underlined
1. (If $2 + 3 = 4$) then $3 + 3 = 6$. true
2. (If $2 + 2 = 4$) then $5 + 6 = 56$. false
3. (If $-2 < -7$) then $0.15 < 0.14$. true
4. (If $(0.06)^2 = 0.036$) then $(0.3)^2 = 0.9$. true
5. (If $-2 - (-5) = -7$) then Jane Austen is the author of *Macbeth*. true
6. (If $6 > -7$) then Scott Joplin is the author of *Macbeth*. false
7. (If $x^2 + 9 = (x + 3)^2$) then $\frac{2}{x}$ is a monomial. true
8. (If Jane Austen is the author of *Pride and Prejudice*, and Shakespeare is the author of *The Tempest*, then) $\frac{\pi}{3} < \frac{21}{20}$. true

Written Let *p* represent "Labor Day is in November," *q* represent "2 is a prime number," and *r* represent "$3y + 1$ is a binomial." Write the statement represented by each of the following. Then determine the truth value. For answers to exercises 1-8, see the Teacher's Guide.
1. $p \to q$
2. $q \to p$
3. $\sim p \to r$
4. $p \to (q \vee r)$
5. $p \leftrightarrow q$
6. $q \leftrightarrow \sim r$
7. $\sim q \leftrightarrow p$
8. $(p \vee \sim q) \to \sim r$

Suppose that *p* and *q* are false and *s* is true. Determine the truth value of each of the following.
9. $\sim p \to \sim s$ false
10. q if $\sim s$ true
11. $(q \vee s) \to \sim p$ true
12. p iff $(q \vee s)$ false
13. $(p \vee q) \leftrightarrow (p \wedge \sim s)$ true
14. $\sim [\sim s \leftrightarrow (p \vee \sim q)]$ true

In each of the following find a correct heading for the third column of the given truth table. Some problems may have more than one answer.

15.
p	q	?
T	T	T
T	F	F
F	T	T
F	F	T

$p \to q, \sim p \vee q$

16.
p	q	?
T	T	F
T	F	T
F	T	T
F	F	T

$p \to \sim q, \sim p \vee \sim q, \sim (p \wedge q)$

17.
p	q	?
T	T	T
T	F	T
F	T	F
F	F	T

$p \vee \sim q, q \to p$

18.
p	q	?
T	T	F
T	F	F
F	T	F
F	F	T

$\sim (p \vee q), \sim p \wedge \sim q$

Answer the question.
19. Suppose that a given exercise from numbers 15 to 18 does have two answers. What does this mean about those two answers? The answers are equivalent

Election time is here. Tell what circumstances make each of the following false.
20. The Crokus Dam will not be built if you elect Brullman senator.
21. If there is a heavy turnout on Election Day, then Brullman will win.

For answers to exercises 20-25, see the Teacher's Guide.
Construct truth tables to determine whether the two statements are equivalent.
22. $p \to q; q \to p$ not equivalent
23. $p \to \sim q; \sim (p \wedge q)$ equivalent
24. $\sim (p \vee q); \sim p \vee \sim q$ not equivalent
25. $\sim (p \wedge q); \sim p \wedge \sim q$ not equivalent

1.3 Beginning To Reach Conclusions

Objective: To write a conclusion by using the Law of Disjunctive Inference.

Suppose a detective knows that either Mr. Green, or Mr. Brown, or both, are guilty of some crime. Later, evidence is obtained that shows that Mr. Green could *not* possibly have been involved in the crime. The detective concludes that Mr. Brown is guilty.

She can represent this situation using logical symbols. Let p represent "Mr. Green is guilty" and q represent "Mr. Brown is guilty." The information the detective had at first can be represented by $p \lor q$.

The new information that Mr. Green is *not* guilty, can be represented as $\sim p$. The conclusion, that Mr. Brown is the guilty one, is the statement q. In this case, the pieces of information

1. $p \lor q$ and **2.** $\sim p$

lead to the conclusion, q.

Now go back to the investigations about Mr. Green and Mr. Brown. If Green or Brown is guilty and evidence proves that Brown is *not* guilty, then Green is the culprit. In symbols, this would mean that the two pieces of information

1. $p \lor q$ and **2.** $\sim q$

lead to the conclusion, p.

The argument presented above is an example of a rule of inference. This rule is called the **Law of Disjunctive Inference.**

Law of Disjunctive Inference

> If a disjunction is true and one of its disjuncts is false, then the other disjunct is true.

In logic, statements that are assumed true are called **premises.** A statement that follows from the premises is called the **conclusion.**

A **formal proof** may be used to demonstrate that a given set of premises leads to a certain conclusion. The next example illustrates the statement-reason or two-column form of proof.

Example

1 The following premises are given.
   Tom or Felix is going to take Gwen to the Halloween Dance.
   Felix is *not* taking her.

Use the statement-reason form of proof to show that Tom is taking Gwen to the dance.

Let T represent "Tom is taking Gwen to the dance."
Let F represent "Felix is taking Gwen to the dance."

Given: $T \lor F$ **Prove:** T
$\sim F$

Proof:

STATEMENTS	REASONS
1. $T \lor F$	1. Given
2. $\sim F$	2. Given
3. T	3. Law of disjunctive inference; 1, 2
The last statement in a proof must be the desired conclusion.	The "1, 2" in step 3 tells what previous steps in the proof you are using with the Law of Disjunctive Inference.

Exercises

Max: 1-12, 1-13; Avg: 1-12, 1-11; Min: 1-10, 1-8

Exploratory For each of the following pairs of statements, state what conclusion, if any, can be inferred using the Law of Disjunctive Inference.

1. $r \lor t$ t
   $\sim r$
2. $B \lor C$ c
   $\sim B$
3. $q \lor r$ q
   $\sim r$
4. $m \lor t$ no
            t conclusion
5. $\sim d \lor e$ e
   d
6. $\sim s \lor \sim t$ $\sim s$
   t
7. $\sim p$ no conclusion
   $\sim p \lor q$
8. m $\sim n$
   $\sim m \lor \sim n$
9. Binky or Sally dated Felix last weekend.
   Binky did *not* date Felix last weekend. Sally dated Felix last weekend.
10. Sally or Gwen came in first in the 200-meter dash. Sally came in first
    Gwen did *not* come in first in the 200-meter dash. in the 200-meter dash.

11. What can be inferred from $\sim p \lor q$ and p? How is the Law of Double Negation used here? Explain. $q, \sim(\sim p) \leftrightarrow p$

12. What can be inferred from the sentences, "Charlene is going to the Big Dance or Tom isn't" and "Tom is going to the Big Dance"?
Charlene is going to the Big Dance.

Written In each of the following proofs, provide the missing statements or reasons.

1. **Given:** $\sim P \land R$ **Prove:** T
   $P \lor T$

   Proof:

STATEMENTS	REASONS
1. $\sim P \land R$	1. Given
2. $\sim P$	2. Definition of conjunction; 1
3. $P \lor T$	3. Given
4. T	4. Law of disjunctive inference; 2, 3

2. **Given:** $L \lor M$ **Prove:** Q
   $\sim L \lor Q$
   $\sim M$

   Proof:

STATEMENTS	REASONS
1. $\sim L \lor Q$	1. Given
2. $L \lor M$	2. Given
3. $\sim M$	3. Given
4. L	4. Law of disjunctive inference; 2, 3
5. Q	5. Law of disjunctive inference; 1, 4

For each of the following, use the statement-reason form of proof to show that the indicated conclusion follows from the given premises. If the conclusion does not follow, explain why it does not. For exercises 6–10, use the letters indicated.
For answers to exercises 3–10, see the Teacher's Guide.

3. **Given:** $\sim S \lor \sim T$
   S
   Conclusion: $\sim T$

4. **Given:** $R \land M$
   $\sim R \lor T$
   Conclusion: T

5. **Given:** $\sim A \land \sim B$
   $\sim B \lor C$
   Conclusion: $\sim C$

6. **Given:** Tom or Felix owns that old Ford. Felix does *not* own it.
   Conclusion: Tom owns the old Ford. (T, F)

7. **Given:** Rose R is red or Violet V is blue. Rose R is red.
   Conclusion: Violet V is *not* blue. (R, V)

8. **Given:** Rose R is *not* red or Violet V is *not* blue. Rose R is red.
   Conclusion: Violet V is *not* blue. (R, V)

9. **Given:** Violet V is blue or it is *not* a prize winner. Violet V is a prize winner, and its owner, Alice Potts, is very proud.
   Conclusion: Violet V is blue. (V, P, A)

10. **Given:** Gwen and Bonny are going to the party. Gwen is *not* going to the party or Gwen is *not* going to the football game.
    Conclusion: Gwen is *not* going to the football game. (G, B, F)

For each pair of statements, state what conclusion can be inferred using the Law of Disjunctive Inference.

11. $0 < a < 6$
    $a < 0$ or $a > 2$ $a > 2$

12. $1 < x < 7$
    $x = 8$ or $x \neq 3$ $x \neq 3$

13. $-1 < m < 3$
    $m < 0$ or $m \geq 3$ $m < 0$

1.4 The Law of Detachment

Objective: To write a conclusion by using the Law of Detachment.

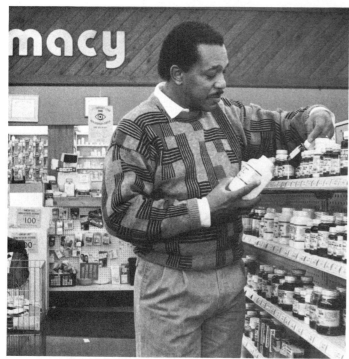

Suppose a detective has the following information: "If Professor Plum was seen at Irv's Pharmacy, then he committed the crime." Remember, this does not mean that Professor Plum was seen at Irv's Pharmacy. It means that if he was seen at the pharmacy, then he is guilty.

But suppose the detective learns that, indeed, Professor Plum was seen at Irv's Pharmacy. Then his conclusion is that Professor Plum is guilty.

To put this information into logical symbols, let p represent, "Professor Plum was seen at Irv's Pharmacy." Let q represent, "Professor Plum is guilty." The original information is the conditional $p \rightarrow q$. The new information is the statement p. The final conclusion is the statement q.

Restating this information more compactly, the two pieces of information

1. $p \rightarrow q$ and **2.** p

lead to the conclusion, q.

Even more efficiently, the information can be written:

If $(p \rightarrow q) \land p$ is true, then q is true.

This rule of inference is called the **Law of Detachment.**

Law of Detachment

If $p \rightarrow q$ is a true conditional and p is true, then q is true.

Examples

1 The following premises are given.

    If Bonny goes canoeing over the weekend, she will *not* be at the dance Friday night.
    Bonny is going canoeing over the weekend.

Use the statement-reason form of proof to show that Bonny will *not* be at the dance Friday night.

Let B represent "Bonny is going canoeing over the weekend."
Let D represent "Bonny will be at the dance Friday night."

Given: $B \rightarrow \sim D$ **Prove:** $\sim D$
           B

Proof:

STATEMENTS	REASONS
1. $B \rightarrow \sim D$	1. Given
2. B	2. Given
3. $\sim D$	3. Law of detachment; 1, 2

2 The following premises are given.

    Either Tom is a jogger or he eats green salad.
    If Tom eats green salad, then he is *not* a fool.
    Furthermore, Tom is *not* a jogger.

Use the statement-reason form of proof to show that Tom is *not* a fool.

Let J represent "Tom is a jogger."
Let G represent "Tom eats green salad."
Let F represent "Tom is a fool."

Given: $J \vee G$ **Prove:** $\sim F$
           $G \rightarrow \sim F$
           $\sim J$

Proof:

STATEMENTS	REASONS
1. $J \vee G$	1. Given
2. $\sim J$	2. Given
3. G	3. Law of disjunctive inference; 1, 2
4. $G \rightarrow \sim F$	4. Given
5. $\sim F$	5. Law of detachment; 3, 4

CHAPTER 1 15

Exercises

Max: 1-10, 1-18, 1-4;   Avg: 1-10, 1-18;   Min: 1-10, 1-17 odds

Exploratory For each of the following pairs of statements, state what conclusion, if any, can be inferred using the Law of Detachment.

1. $p \to q$ q
   p
2. $r \to \sim s$ $\sim s$
   r
3. $\sim q \to m$ m
   $\sim q$
4. $t \to s$ no conclusion
   s
5. $k \to \sim n$
   $\sim n$ no conclusion
6. r s ∨ t
   $r \to (s \lor t)$
7. r iff q r
   q
8. $(g \lor m) \to (k \lor s)$
   $k \lor s$ no conclusion

State the law of reasoning that justifies each conclusion.

9. **Given:** $p \lor q$ Law of disjunctive inference
   $\sim q$
   Conclusion: p

10. **Given:** $\sim t \to s$ Law of detachment
    $\sim t$
    Conclusion: s

Written In each of the following proofs, supply the missing statements or reasons.

1. **Given:** $\sim A \to B$
   $\sim A \land M$

   Prove: B

   Proof:

STATEMENTS	REASONS
1. $\sim A \to B$	1. Given
2. $\sim A \land M$	2. Given
3. $\sim A$	3. Definition of conjunction; 2
4. B	4. Law of detachment; 1, 3

2. **Given:** $T \lor \sim W$
   W
   $T \to R$

   Prove: R

   Proof:

STATEMENTS	REASONS
1. $T \to R$	1. Given
2. $T \lor \sim W$	2. Given
3. W	3. Given
4. T	4. Law of disjunctive inference; 2, 3
5. R	5. Law of detachment; 1, 4

3. **Given:** $A \to (B \land C)$
   $\sim B \lor \sim G$; A

   Prove: $\sim G$

   Proof:

STATEMENTS	REASONS
1. $A \to (B \land C)$	1. Given
2. A	2. Given
3. $B \land C$	3. Law of detachment; 1, 2
4. B	4. Definition of conjunction; 3
5. $\sim B \lor \sim G$	5. Given
6. $\sim G$	6. Law of disjunctive inference; 4, 5

4. **Given:** $(C \lor D) \to (E \lor F)$
   C; $\sim F$

   Prove: E

   Proof:

STATEMENTS	REASONS
1. $(C \lor D) \to (E \lor F)$	1. Given
2. C	2. Given
3. $C \lor D$	3. Definition of disjunction; 2
4. $E \lor F$	4. Law of detachment; 1, 3
5. $\sim F$	5. Given
6. E	6. Law of disjunctive inference; 4, 5

For answers to exercises 5-18, see the Teacher's Guide.

For each of the following, use the statement-reason form of proof to show the indicated conclusion follows from the given premises. If the conclusion does not follow, explain why it does not. In exercises 11–18, use the letters indicated.

5. Given: $A \to D$
   A
   Conclusion: D

6. Given: $B \to C$
   C
   Conclusion: B

7. Given: $r \to (s \wedge \sim t)$
   r
   Conclusion: $(s \wedge \sim t)$

8. Given: $G \to H$
   $G \vee B$
   $\sim B$
   Conclusion: H

9. Given: $Q \vee R$
   $\sim R$
   $Q \to T$
   Conclusion: T

10. Given: A
    $A \vee \sim B$
    $B \to C$
    Conclusion: C

11. If Luisa goes to the Big Dance, then Diego will *not*. Luisa is going to the Big Dance.
    Conclusion: Diego will *not* go to the Big Dance. (B, T)

12. Tom will pass Social Studies if he completes his book report on time. Tom completes his book report on time.
    Conclusion: He will pass Social Studies. (P, C)

13. Tom likes logic or he likes French. Tom does *not* like French. If Tom likes logic, he is popular with all the girls.
    Conclusion: Tom is popular with all the girls. (L, F, P)

14. If Lu Chan exercises an hour a day, then he does *not* drink a lot of milk. Lu Chan drinks a lot of milk or he eats a lot of fruit. Lu Chan exercises an hour a day.
    Conclusion: Lu Chan eats a lot of fruit. (E, M, F)

15. Crunchy-Wunchies are made from oats or they are *not* healthful. If they are *not* healthful, then they will *not* be sold at Gwen's Market. As it happens, Crunchy-Wunchies are made from oats.
    Conclusion: Crunchy-Wunchies are *not* sold at Gwen's Market. (O, H, G)

16. If Carita is playing handball, then she is playing at the YWCA or she is playing in Poughkeepsie. Carita is playing handball, but she is *not* playing at the YWCA.
    Conclusion: Carita is playing handball in Poughkeepsie. (H, Y, P)

17. If Ted or Jack is taking Sally to the dance, then her mother will be pleased. Ted is taking Sally to the dance.
    Conclusion: Her mother will be pleased. (T, J, P)

18. If Ted or Jack is taking Sally to the dance, then her mother will be pleased. Ted is *not* taking Sally to the dance.
    Conclusion: Her mother will be pleased. (T, J, P)

For answers to exercises 1-4, see the Teacher's Guide.

Challenge

1. If $(p \to q) \wedge p$ is true, then q is true. Determine if there is a corresponding law "If $(p \to q) \wedge q$ is true, then p is true." Why or why not?

2. Explain the compound statement $[(p \to q) \wedge \sim q] \to \sim p$ in words.

3. Use a truth table to investigate the compound statement in exercise 2.

4. State the conclusions you can draw from exercise 3.

1.5 The Law of the Contrapositive

Objective: To write a converse, inverse, and contrapositive of the conditional.

The conditional $p \rightarrow q$ has three related statements, each of which is also a conditional.

Consider the conditional $p \rightarrow q$. A new, different conditional can be formed by reversing the order of the antecedent and the consequent. This new conditional is $q \rightarrow p$ and is called the **converse** of the original.

The conditional formed by keeping the p and q in the same places, but negating them both, is called the **inverse** of the original. The inverse is $\sim p \rightarrow \sim q$.

Finally, the antecedent and consequent can be both reversed and negated. This new conditional, $\sim q \rightarrow \sim p$, is called the **contrapositive** of the original.

Example

1 Write the converse, inverse, and contrapositive of the conditional, "If $x = y$, then $x^2 = y^2$."

CONVERSE: If $x^2 = y^2$, then $x = y$.
INVERSE: If $x \neq y$, then $x^2 \neq y^2$.
CONTRAPOSITIVE: If $x^2 \neq y^2$, then $x \neq y$.

In this example, the original conditional is true. Which, if any, of the other conditionals are also true? Look carefully at each conditional.

The converse and the inverse are both false. If $x = 5$ and $y = -5$ are substituted in the converse and the inverse, the statements are false. However, the contrapositive is still true even with these replacements.

This is a good example of the following logical principles.
   1. The converse of a true conditional may be false.
   2. The inverse of a true conditional may be false.
   3. The contrapositive *always* has the same truth value as the original conditional.

Working with Logic

Law of the Contrapositive

A conditional and its contrapositive are equivalent.
$(p \rightarrow q) \leftrightarrow (\sim q \rightarrow \sim p)$

Example

2 Use a statement-reason form of proof to show $\sim R$ follows from the given premises.

Given: $A \rightarrow \sim C$ Prove: $\sim R$
$A \lor W$
$R \rightarrow \sim W$
C

Proof:

STATEMENTS	REASONS
1. $A \rightarrow \sim C$	1. Given
2. $C \rightarrow \sim A$	2. Law of the contrapositive; 1
3. C	3. Given
4. $\sim A$	4. Law of detachment; 2, 3
5. $A \lor W$	5. Given
6. W	6. Law of disjunctive inference; 4, 5
7. $R \rightarrow \sim W$	7. Given
8. $W \rightarrow \sim R$	8. Law of the contrapositive; 7
9. $\sim R$	9. Law of detachment; 6, 8

Exercises

Max: 1-10, 1-12; Avg: 1-10, 1-12; Min: 1-10, 1-9 odds, 10-12

Exploratory State the converse, inverse, and contrapositive of the given conditional.

1. $r \rightarrow w$
2. $t \rightarrow \sim w$
3. $\sim s \rightarrow \sim t$
4. $\sim m \rightarrow \sim n$
5. If $\angle A$ and $\angle B$ are both right angles, then $\angle A \cong \angle B$.
6. John will be disappointed if Mei does not show up at the picnic.

For answers to exercises 1-6, see the Teacher's Guide.

State whether each pair of statements is equivalent.

7. $t \rightarrow w$
   $w \rightarrow t$
   not equivalent

8. $r \rightarrow s$
   $\sim s \rightarrow \sim r$
   equivalent

9. $m \rightarrow \sim n$
   $\sim m \rightarrow n$
   not equivalent

10. $t \rightarrow \sim m$
    $m \rightarrow t$
    not equivalent

Written Complete the following.

1. Construct a truth table to show that $p \to q$ is equivalent to $\sim q \to \sim p$.
For the answer to exercise 1, see the Teacher's Guide.

In each of the following proofs, provide the missing statements or reasons.

2. Given: $B \to \sim Q$ Prove: Q
$T \vee \sim B$
$\sim T$

Proof:

STATEMENTS	REASONS
1. $B \to \sim Q$	1. Given
2. $T \vee \sim B$	2. Given
3. $\sim T$	3. Given
4. $\sim B$	4. Law of disjunctive inference; 2, 3
5. $Q \to \sim B$	5. Law of the contrapositive; 1
6. Suggested conclusion does *not* follow.	6. Insufficient information

3. Given: $A \to N$ Prove: W
$\sim A \to W$
$\sim N$

Proof:

STATEMENTS	REASONS
1. $A \to N$	1. Given
2. $\sim N \to \sim A$	2. Law of the contrapositive; 1
3. $\sim N$	3. Given
4. $\sim A$	4. Law of detachment; 2, 3
5. $\sim A \to W$	5. Given
6. W	6. Law of detachment; 4, 5

For answers to exercises 4–12, see the Teacher's Guide.

For each of the following, use the statement-reason form of proof to show that the indicated conclusion follows from the given premises. If the conclusion does not follow, explain why it does not. For exercises 10–12, use the letters indicated.

4. Given: $A \vee D$
$\sim A$
$T \to \sim D$
Conclusion: $\sim T$

5. Given: $\sim C \vee \sim F$
C
$\sim B \to F$
Conclusion: B

6. Given: $T \to B$
$\sim C \vee \sim B$
C
Conclusion: $\sim T$

7. Given: $L \wedge M$
$M \to N$
$\sim N \vee \sim D$
$R \to D$
Conclusion: $\sim R$

8. Given: $S \to B$
$L \to S$
$\sim B$
Conclusion: $\sim L$

9. Given: $Q \to \sim R$
$\sim Q \to T$
R
Conclusion: $\sim T$

10. The general will win the battle or a truce will be declared. If there are political complications, a truce will *not* be declared. It is known that the general will *not* win the battle.
Conclusion: There will *not* be political complications. (G, T, P)

11. If Inez is good at logic, then she is good at chemistry. Inez is *not* good at history or she is *not* good at chemistry. But Inez is quite good at history.
Conclusion: Inez is *not* good at logic. (L, C, H)

12. If Fred walks home from school, then he buys an ice cream cone at Earl's Ice Cream Shoppe. If Fred stays after school for basketball practice, then he walks home from school. Today, Fred is not buying an ice cream cone at Earl's Ice Cream Shoppe.
Conclusion: Fred is *not* staying after school for basketball practice today. (W, I, B)

1.6 The Law of Syllogism

Objective: To write a conclusion by using the Law of Syllogism.

Suppose a detective knows the following statements are true.
1. If Nurse Wilkins saw Colonel Mustard at the picnic, then he was in the city Saturday afternoon.
2. If Colonel Mustard was in the city Saturday afternoon, then he is *not* guilty of the crime.

Reread these sentences carefully, noting especially the position of the phrase "Colonel Mustard was in the city Saturday afternoon."

Now, the detective has *not* spoken to Nurse Wilkins yet, so he does *not* know whether she did or did not see Colonel Mustard at the picnic. But, he is led to believe that "If Nurse Wilkins saw Colonel Mustard at the picnic, then he is not guilty of the crime."

Let W represent "Nurse Wilkins saw Colonel Mustard at the picnic." Let M represent "Colonel Mustard was in the city Saturday afternoon." Let C represent "Colonel Mustard is guilty of the crime." In this case, the two pieces of information

1. $W \rightarrow M$ and 2. $M \rightarrow \sim C$

lead to the conclusion $W \rightarrow \sim C$. That is,

If $[(W \rightarrow M) \wedge (M \rightarrow \sim C)]$ is true, then $(W \rightarrow \sim C)$ is true.

Being a good detective, he is careful to observe that his final conclusion is itself a conditional. He still does *not* know if W is true or if $\sim C$ is true. But he does know that IF W is true, THEN $\sim C$ is true.

The argument presented above is an example of the **Law of Syllogism**.

Law of Syllogism If $p \rightarrow q$ and $q \rightarrow r$ are true, then $p \rightarrow r$ is true.

Example

1 State a conclusion that can be inferred from the following premises.

If it rains Saturday, then Jill will go to the movies. If Jill goes to the movies, then Jill will eat too much popcorn.

The Law of Syllogism may be used to conclude that "If it rains Saturday, then Jill will eat too much popcorn."

Example

2 Use a statement-reason form of proof to show C follows from the given premises.

Given: $A \to H$ **Prove:** C
$H \to C$
A

Proof:

STATEMENTS	REASONS
1. $A \to H$	1. Given
2. $H \to C$	2. Given
3. $A \to C$	3. Law of syllogism; 1, 2
4. A	4. Given
5. C	5. Law of detachment; 3, 4

The general procedure of using the Law of Syllogism in combination with the Law of Detachment is common in logic. In example 2, can C be inferred from the given information without using the Law of Syllogism?
Yes. Use the Law of detachment twice.

Exercises

Max: 1-10, 1-18, 1-8:   Avg: 1-10, 1-17 odds, 1-4;   Min: 1-8, 1-17 odds

Exploratory State a conclusion that can be inferred from each pair of premises.

1. $p \to q$ $p \to r$
   $q \to r$
2. $s \to \sim t$ $s \to w$
   $\sim t \to w$
3. $n \to m$ $p \to m$
   $p \to n$
4. $\sim q \to w$
   $p \to \sim q$
   $p \to w$
5. If Flora is good at chess, then she is good at logic.
   If Flora is good at logic, then she gets good grades.
   If Flora is good at chess, then she gets good grades.

State a conclusion, if any, that can be inferred from the given premises. State the law of reasoning that justifies your conclusion. For answers to exercises 6-10, see the Teacher's Guide.

6. If it is 3 o'clock on Sunday, then Alvin is at the museum. Alvin is at the museum.
7. If it is 3 o'clock on Sunday, then Alvin is at the museum. If Alvin is at the museum, his feet are tired. Alvin is at the museum.

8. If it is 3 o'clock on Sunday, then Alvin is at the museum. If Alvin is at the museum, he is enjoying himself. It is 3 o'clock on Sunday.
9. If it is 3 o'clock on Sunday, then Alvin is at the museum. If Alvin is at the museum, then Alvin is standing in the Hall of Modern Art. Alvin is *not* in the Hall of Modern Art.
10. If it is 3 o'clock on Sunday, or Tuesday afternoon, then Alvin is at the museum. If Alvin is at the museum, then he is working on his Fine Arts paper. It is *not* Tuesday afternoon.

For answers to exercises 1-18, see the Teacher's Guide.

Written In each of the following, use the statement-reason form of proof to show that the indicated conclusion follows from given premises. If the conclusion does not follow, explain why it does not. For exercises 13–18, use the letters indicated. In some cases, there may be some "extra" information. In real life, a person rarely receives exactly the information he or she needs.

1. Given: $A \to B$
   $B \to T$
   Conclusion: $A \to T$

2. Given: $A \to T$
   $T \to W$
   $W \to B$
   Conclusion: B

3. Given: $A \to T$
   $T \to Q$
   A
   Conclusion: Q

4. Given: $A \to W$
   $W \to B$
   $\sim B$
   Conclusion: $\sim A$

5. Given: $A \lor T$
   $\sim A$
   $L \to \sim T$
   Conclusion: $\sim L$

6. Given: $B \land T$
   $B \to Q$
   $\sim N$
   Conclusion: Q

7. Given: $A \to S$
   $S \to N$
   $\sim W \lor \sim N$
   A
   Conclusion: $\sim W$

8. Given: $A \to B$
   $B \to C$
   $C \to D$
   $\sim C$
   Conclusion: $\sim B$

9. Given: $(A \lor B) \to Q$
   $T \to A$
   $\sim T \to \sim S$
   S
   Conclusion: Q

10. Given: $m \lor n$
    $n \to \sim p$
    p
    $m \to q$
    Conclusion: q

11. Given: $r \to \sim s$
    $r \lor \sim t$
    s
    $m \to t$
    Conclusion: $\sim m$

12. Given: $\sim r \land s$
    $\sim s \lor t$
    $p \to \sim t$
    $\sim p \to q$
    Conclusion: q

13. If Tom eats green salad, then he is healthy. If Tom is healthy, then he jogs an hour a day.
    Conclusion: If Tom eats green salad, then he jogs an hour a day.   (G, H, J)
14. If Tom eats green salad, then he enjoys logical studies. If he enjoys logical studies, then he is a credit to his family. It is known that Tom eats green salad.
    Conclusion: Tom is a credit to his family.   (G, L, C)
15. If $a = b$, then $c = d$. If $c = d$, then $d = n$. If $d = n$, then $a < 1$. But $d \neq n$.
    Conclusion: $a \neq b$.   (P, Q, R, S)
16. If Tom likes logic, then he is good at backgammon. If he is good at backgammon, then he is good at chess. Tom is good at chess or he jogs an hour a day. In fact, Tom likes logic.
    Conclusion: Tom jogs an hour a day.   (L, B, C, J)

17. If Ella is good at chess, then she is good at logic. If she is good at logic, then she is good at soccer. Either she is *not* good at French or she is *not* good at soccer. As it happens, Ella is good at chess. **Conclusion:** Ella is *not* good at French.   (C, L, S, F)
18. If Lynn is good at swimming, or good at soccer, then she is a genuine athlete. If she drinks a pint of grapefruit juice every day, then she is good at swimming. If she does *not* drink a pint of grapefruit juice every day, then she is *not* good at logic. However, Lynn is good at logic. **Conclusion:** Lynn is a genuine athlete.   (S, O, A, G, L)

Challenge Answer the question.
 1. The Law of Syllogism is the first rule to involve three separate statements, p, q, r. If a truth table is used to show the truth of the Law, how many horizontal rows would this table have? (Note that the tables with p and q have four rows across.) 8

Use truth tables to determine when each of the compound statements is true.
 2. $(p \wedge q) \vee r$
 3. $(\sim p \vee q) \rightarrow (q \vee r)$
 4. $(p \vee q) \leftrightarrow (\sim q \vee r)$

Use truth tables to determine if each of the following is true.
 5. $\wedge$ is associative.
 6. $\vee$ is associative.
 7. $\wedge$ distributes over $\vee$.
 8. $\rightarrow$ distributes over $\vee$.

For answers to exercises 2-8, see the Teacher's Guide.

Mixed Review

Let p represent "x is divisible by 3," q represent "x is divisible by 5," and r represent "x is not divisible by 6." For the given value of x, determine the truth value of each statement.
 1. $x = 16$; $p \vee q$ false
 2. $x = 54$; $p \wedge \sim r$ true
 3. $x = 21$; $q \rightarrow r$ true
 4. $x = 35$; $\sim q \leftrightarrow p$ true
 5. $x = 45$; $q \wedge (r \leftrightarrow p)$ true
 6. $x = 30$; $(\sim p \vee q) \rightarrow r$ false

Determine whether the indicated conclusion can be inferred from the given premises.
 7. If y is not composite, then $t = 4$. t is a negative integer. $d = 8$ or $y = 7$.
    Conclusion: $d = 8$ yes
 8. Reggie or Charlena went to the concert on Saturday. If Reggie did not go to the concert on Saturday, then he went to the movies. If Reggie went to the movies, then Charlena did not study on Saturday night. Charlena studied on Saturday night.
    Conclusion: Charlena did not go to the concert on Saturday.   no
 9. If Maria passes calculus, then she will attend Syracuse or Seattle University. If Maria attends Syracuse University, then she will not receive a scholarship. Maria passed calculus and she received a scholarship.
    Conclusion: Maria will attend Seattle University.   yes
10. If b is divisible by 3, then 6 is odd. If a is even, then $b = 9$. $a = 20$ or $c = 1$.
    Conclusion: $c = 1$ yes

Objective: To write the negation of a statement.

1.7 The De Morgan Laws

Suppose a crafty detective finds the following note slipped under her door: "Colonel Mustard committed the crime and Mrs. Peacock planned it."

Under what circumstances is this message true?

Any conjunction is true when both conjuncts are true. If the detective's investigations show that Colonel Mustard did *not* commit the crime *or* that Mrs. Peacock did *not* plan it, then the message she received is false.

To put this into logical symbols, let p represent "Colonel Mustard committed the crime" and let q represent "Mrs. Peacock planned it." Then, the information on the original message is $p \wedge q$.

The statement is false if either p or q or both is false. That is the same as saying that either $\sim p$, or $\sim q$, or both, is true. Thus, the negation of $p \wedge q$ is

$$\sim p \vee \sim q.$$

Put another way, the negation of the conjunction of p and q is the disjunction of the negation of p and the negation of q.

$$\sim (p \wedge q) \leftrightarrow (\sim p \vee \sim q)$$

Example

1 **Write the negation of $t \wedge \sim w$.**

The negation is $\sim t \vee \sim(\sim w)$. This is equivalent to $\sim t \vee w$.

Suppose that the detective's informant gives her this message: "Colonel Mustard committed the crime or Professor Plum did." This time the message is correct if *either* person is guilty, or if they are both guilty. Only in the case that Colonel Mustard is not guilty *and* Professor Plum is not guilty, is the statement false.

In symbols, the negation of $p \vee q$ is $\sim p \wedge \sim q$. The negation of a disjunction $p \vee q$ is the conjunction of the negation of p and the negation of q.

These two rules of inference are called the **De Morgan Laws,** after the nineteenth century British mathematician Augustus De Morgan.

Augustus De Morgan

De Morgan Laws

1. $\sim(p \wedge q) \leftrightarrow (\sim p \vee \sim q)$
2. $\sim(p \vee q) \leftrightarrow (\sim p \wedge \sim q)$

Examples

2 Write the negation of $\sim p \vee \sim t$.

The negation is $\sim(\sim p) \wedge \sim(\sim t)$, or $p \wedge t$.

3 Write the negation of $p \wedge (q \vee r)$.

Complete the procedure in two steps, using a different De Morgan Law in each step.

Step 1 $\quad \sim[p \wedge (q \vee r)] \leftrightarrow \sim p \vee \sim(q \vee r)$

Now, using De Morgan Laws on "$\sim(q \vee r)$,"

Step 2 $\quad \sim p \vee \sim(q \vee r) \leftrightarrow \sim p \vee (\sim q \wedge \sim r)$

The correct negation of $p \wedge (q \vee r)$ is $\sim p \vee (\sim q \wedge \sim r)$.

4 Write the contrapositive of "If r is even and r is prime, then $r = 2$."

Let E represent "r is even," P represent "r is prime," and T represent "$r = 2$." The original statement is

$$(E \wedge P) \rightarrow T$$

The contrapositive is

$$\sim T \rightarrow \sim(E \wedge P) \quad \text{or} \quad \sim T \rightarrow (\sim E \vee \sim P).$$

The contrapositive is "If $r \neq 2$, then r is *not* even or r is *not* prime."

Exercises

Max: 1-10, 1-28, 1-7;   Avg: 1-10, 1-28;   Min: 1-9 odds, 1-27 odds

Exploratory   Write the negation of each of the following.

1. $a = b$ or $c = d$ $a \neq b$ and $c \neq d$
2. $c = d$ and $k = m$ $c \neq d$ or $k \neq m$
3. $c < d$ and $d < e$ $c \geq d$ or $d \geq e$
4. $n = 1$ or $n < 0$ $n \neq 1$ and $n \geq 0$
5. m is positive or $m = -1$ m is not positive and $m \neq -1$
6. $k = 0$ and $m < 0$ $k \neq 0$ or $m \geq 0$
7. Binky is good at Latin and Mary is good at French.   Binky is not good at Latin or Mary is not good at French.
8. Bonny is *not* good at backgammon and she is *not* good at chess.   8. Bonny is good at backgammon or she is good at chess.
9. Otto is going to be class treasurer or Gwen is.   Otto is not going to be class treasurer and Gwen is not.
10. Tom jogs an hour a day or he does *not* have soda with lunch.   Tom does not jog an hour a day and he has soda with lunch.

Written   Write the negation of each of the following.

1. $r \wedge s$ $\sim r \vee \sim s$
2. $\sim m \vee t$ $m \wedge \sim t$
3. $\sim p \wedge \sim q$ $p \vee q$
4. $k \vee g$ $\sim k \wedge \sim g$
5. $r \wedge \sim t$ $\sim r \vee t$
6. $m \vee (t \wedge s)$ $\sim m \wedge (\sim t \vee \sim s)$
7. $(r \vee \sim s) \wedge t$ $(\sim r \wedge s) \vee \sim t$
8. $\sim r \vee (s \wedge \sim t)$ $r \wedge (\sim s \vee t)$
9. $(\sim p \vee q) \wedge (t \vee \sim s)$ $(p \wedge \sim q) \vee (\sim t \wedge s)$

Complete each of the following. For answers to exercises 10-13, see the Teacher's Guide.

10. Use a truth table to show $\sim(p \wedge q)$ and $\sim p \wedge \sim q$ are *not* equivalent.
11. Use a truth table to show $\sim(p \vee q)$ and $\sim p \vee \sim q$ are *not* equivalent.
12. Use a truth table to show $\sim(p \wedge q)$ and $\sim p \vee \sim q$ are equivalent.
13. Use a truth table to show $\sim(p \vee q)$ and $\sim p \wedge \sim q$ are equivalent.

In each of the following proofs, provide the missing reasons.

14. Given: $\sim(A \vee B)$ Prove: $\sim T$
    $T \rightarrow A$

Proof:

STATEMENTS	REASONS
1. $\sim(A \vee B)$	1. Given
2. $\sim A \wedge \sim B$	2. DeMorgan Laws; 1
3. $\sim A$	3. Definition of conjunction; 2
4. $T \rightarrow A$	4. Given
5. $\sim A \rightarrow \sim T$	5. Law of the contrapositive; 4
6. $\sim T$	6. Law of detachment; 3, 5

15. Given: $\sim(A \wedge \sim B)$ Prove: H
    A
    $B \rightarrow C$
    $C \rightarrow H$

Proof:

STATEMENTS	REASONS
1. $\sim(A \wedge \sim B)$	1. Given
2. $\sim A \vee B$	2. DeMorgan Laws; 1
3. A	3. Given
4. B	4. Law of disjunctive inference; 2
5. $B \rightarrow C$	5. Given
6. $C \rightarrow H$	6. Given
7. $B \rightarrow H$	7. Law of syllogism; 5
8. H	8. Law of detachment;

In each of the following, use the statement-reason form of proof to show that the indicated conclusion follows from the given premises. If the conclusion does not follow, explain why it does not. For exercises 19–21, use the letters indicated.

For answers to exercises 16-28, see the Teacher's Guide.

16. **Given:** $\sim(p \lor q)$
    $\sim q \to t$
    Conclusion: t

17. **Given:** $\sim(\alpha \land \beta)$
    α
    Conclusion: $\sim\beta$

18. **Given:** $s \to v$
    $r \to \sim v$
    s
    $\sim(q \land \sim r)$
    Conclusion: $\sim q$

19. It is false that either the general will win the battle or that a truce will be declared. If the general does *not* win the battle, he will be relieved of command.
    Conclusion: The general will be relieved of command. (W, D, R)

20. It is false that either the general will win the battle or that a truce will be declared. If the enemy's air support is *not* decisive, then the general will win the battle.
    Conclusion: The enemy's air support will *not* be decisive. (W, D, A)

21. It is *not* the case that Tom eats green salad and that he jogs an hour a day. If he does *not* eat green salad, then he drinks a pint of unsweetened grapefruit juice a day. If he drinks a pint of unsweetened grapefruit juice a day, then he has a sour disposition. It is known, however, that Tom jogs an hour a day.
    Conclusion: Tom has a sour disposition. (G, J, P, D)

Write the contrapositive of each of the following.

22. If Tom jogs, then he eats salad and spaghetti.

23. If $r < n$, then $r = 0$ or $r > 4$.

24. If $\angle A$ is supplementary to $\angle B$ and $\angle C$ is supplementary to $\angle B$, then $\angle A \cong \angle C$.

25. If Marsha is going to the Big Dance and Sally is *not*, then Ted is wearing his tweeds.

Write the inverse of each of the following.

26. $(p \land q) \to (r \lor \sim t)$

27. $(\sim c \land d) \to (t \lor \sim w)$

28. $\angle C \cong \angle D$ if $m\angle C = 90$ and $m\angle E = 90$.

Challenge In this section, the negations of conjunctions and disjunctions are discussed. But what about the negation of conditionals and bi-conditionals? The following exercises help answer this question.

1. Which, if any, of the following can be the negation of $p \to q$? Explain.
   a. $\sim p \to q$ no
   b. $p \to \sim q$ no
   c. $p \land \sim q$ yes
   d. $\sim p \land q$ no

2. Which, if any, of the following can be the negation of $p \leftrightarrow q$? Explain.
   a. $p \leftrightarrow \sim q$ yes
   b. $\sim p \leftrightarrow q$ yes
   c. $p \land \sim q$ no
   d. $\sim p \land q$ no

Each negation has a truth value that is different from the truth value of the original conditional or biconditional.

Negate each of the following. For answers to exercises 3-7, see the Teacher's Guide.

3. If $x = 3$, then $x^2 = 9$.
4. If $|x| = 5$, then $x = 5$ or $x = -5$.
5. If it rains today, Linda has her umbrella.
6. Mr. Oglethorpe rides the bus if and only if there is a taxi strike.
7. Miss Weinstein has French toast for breakfast if and only if she is *not* in a hurry and she is *not* having Crunchy-Wunchies.

INDIRECT PROOF

Mathematical Excursions

Very often mathematicians find that they *cannot* prove a certain statement directly, and so they try what is called an **indirect proof.** One such argument is to show that a statement is true by showing that its negation is false.

Here is the method of proving that a statement *p* is true.

Step 1 Assume that $\sim p$ is true.
Step 2 Show that this assumption leads to some statement that is clearly a contradiction.
Step 3 Conclude from this that the assumption, $\sim p$, must have been wrong.
Step 4 Conclude that if $\sim p$ is false, then $\sim(\sim p)$, or simply *p*, is true.

Example The following premises are given.

> Suzanne eats tacos or she drinks grape juice.
> If Suzanne drinks grape juice, then she eats cheddar cheese.
> Suzanne does *not* eat tacos.

Use an indirect proof to show that Suzanne eats cheddar cheese.

Let *T* represent "Suzanne eats tacos."
Let *G* represent "Suzanne drinks grape juice."
Let *C* represent "Suzanne eats cheddar cheese."

Given: $T \vee G$ **Prove:** C
$G \rightarrow C$
$\sim T$

Proof:

STATEMENTS	REASONS
1. $\sim C$	1. Assumption
2. $G \rightarrow C$	2. Given
3. $\sim C \rightarrow \sim G$	3. Law of the contrapositive; 2
4. $\sim G$	4. Law of detachment; 1, 3
5. $T \vee G$	5. Given
6. T	6. Law of disjunctive inference; 4, 5
7. $\sim T$	7. Given
8. $T \wedge \sim T$	8. Definition of conjunction; 6, 7
9. $\sim(\sim C)$	9. $\sim C$ must be false since it led to the clear contradiction, $T \wedge \sim T$.
10. C	10. Law of double negation

Thus, it has been shown that Suzanne eats cheddar cheese.

This method of proof is still called by its original Latin name, *Reductio ad Absurdum*. It is often used to prove theorems related to parallel lines, inequality relationships in triangles, and other geometric relationships related to measurements.

For answers to exercises 1-4, see the Teacher's Guide.

Exercises
In each proof, provide the missing statement or reason.

1. Given: $\sim A \to N$
$A \to \sim W$
$\sim N$

Prove: $\sim W$

Proof:

STATEMENTS	REASONS
1. $\sim(\sim W)$	1. _____
2. W	2. _____
3. $A \to \sim W$	3. _____
4. _____	4. Law of the contrapositive; 3
5. $\sim A$	5. _____
6. $\sim A \to N$	6. _____
7. _____	7. Law of detachment; 5, 6
8. $\sim N$	8. _____
9. $N \wedge \sim N$	9. _____
10. _____	10. $\sim(\sim W)$ must be false since it led to a contradiction, $N \wedge \sim N$.
11. $\sim W$	11. _____

2. Given: $\sim r \wedge s$
$\sim s \vee t$
$p \to \sim t$
$\sim p \to q$

Prove: q

Proof:

STATEMENTS	REASONS
1. $\sim q$	1. _____
2. $\sim p \to q$	2. _____
3. $\sim q \to p$	3. _____
4. $p \to \sim t$	4. _____
5. $\sim q \to \sim t$	5. _____
6. $\sim t$	6. _____
7. $\sim r \wedge s$	7. _____
8. s	8. _____
9. $\sim s \vee t$	9. _____
10. t	10. _____
11. $t \wedge \sim t$	11. _____
12. $\sim(\sim q)$	12. _____
13. q	13. _____

Use an indirect proof, written in statement-reason form, to show that the indicated conclusion follows from the given premises. Use the letters indicated.

3. If Janice likes apples, then Sarah likes bananas. Sarah does *not* like bananas.
Conclusion: Janice does *not* like apples. (A, B)

4. Peter will *not* take Jill to the concert if Juan takes Olivia to the amusement park. Peter will take Jill to the concert or Raul will take Tara to the dance. Patricia told Steven that she saw Juan and Olivia together at the amusement park.
Conclusion: Raul will take Tara to the dance. (C, A, D)

1.8 Quantifiers

Objective: To determine if sentences are quantified and to write sentences using ∀ or ∃ notation.

Study the following statements.

1. All elephants are gray.
2. Some roses are red.
3. Not every rational number is an integer.
4. No zebras have a sense of humor.
5. At least one zebra has a sense of humor.
6. There exists a perfect square that is even.

Statements such as 1–6 are said to be quantified. The word, or group of words, that acts to quantify the statement is called a **quantifier.** The quantifier tells to what part of the total "population" the sentence refers. For example, a sentence might refer to "all" elephants, or the sentence might refer to "some" elephants. Or, the sentence might refer to "no" elephants.

The quantifiers in sentences 1–6 above are, for sentence 1, "all"; for sentence 2, "some"; for sentence 3, "not every"; for sentence 4, "no"; for sentence 5, "at least one"; and for sentence 6, "there exists a."

There are two kinds of quantifiers.

The first kind is called a **universal quantifier.** This kind of quantifier tells about an entire population of something. Sentences 1 and 4 above are universally quantified. The first sentence tells something about the color of all elephants. Sentence 4 tells something about the sense of humor of all zebras—in this case that all zebras do not have one.

A symbol used for universal quantification is ∀.

Example

1 Write $\forall x \in E$ (x is gray) in words.

$\forall x \in E$ (x is gray) means "For all x in the set E, x is gray" or "Every x is gray."

Example

2 Write "Every integer is rational" using ∀ notation. State whether the sentence is true or false.

"Every integer is rational" can be written "For every z in the set of integers ($\mathcal{Z}$), z is in the set of rational (Q). Using symbols, this statement is written

$$\forall z \in \mathcal{Z}\, (z \in Q).$$

The statement is true.

The second kind of quantifier is the **existential quantifier.** This kind of quantifier tells something about only *part* of a population. These are words such as "some," "not every," "at least one," or "there exists a." Sentences 2, 3, 5, and 6 in the list on the previous page are existentially quantified.

Note that in the study of logic, the word "some" means "at least one."

The symbol for existential quantification is ∃.

Examples

3 Write $\exists x \in \mathcal{Z}\, (x \text{ is even})$ in words. State whether the sentence is true or false.

$\exists x \in \mathcal{Z}\, (x \text{ is even})$ means "For some x in the set of integers ($\mathcal{Z}$), x is even" or "There exists at least one even integer." This statement is true.

4 Write "There exists at least one element t in the set T such that t has scales" using ∃ notation.

The statement can be written as

$$\exists\, t \in T\, (t \text{ has scales}).$$

32 WORKING WITH LOGIC

In the last example, *T* might be the set of animals, or some other set. The truth of the statement depends upon the sets involved. Suppose *T* is the set of reptiles. Since "some" means "at least one," the sentence "some reptiles have scales" is true even though it is also true that *all* reptiles have scales.

Exercises

Max: 1-8, 1-16, 1-5; Avg: 1-8, 1-16; Min: 1-8, 1-15 odds

Exploratory Tell if each of the following sentences is quantified. If it is, tell whether it is universally or existentially quantified and identify the word or words that act as quantifiers.

1. Baton Rouge is the capital of Louisiana.
2. Some triangles contain three acute angles.
3. All squares are rectangles.
4. Some roses are *not* red.
5. Mr. Oglethorpe's prize rose is red.
6. Every inscribed angle is acute.
7. At least one inscribed angle is a right angle.
8. None of these polygons is a pentagon.

For answers to exercises 1-8, see the Teacher's Guide

Written Write each sentence in words. State whether it is true or false.

1. $\exists x \in \mathcal{W}$ ($x < 0$)
2. $\forall y \in \mathcal{Z}$ (y is rational)
3. $\forall x \in \mathcal{Q}$ (x is an integer)
4. $\exists x \in \mathcal{Z}$ (x is prime)
5. $\exists s \in \mathcal{Z}$ (s is composite)
6. $\forall t \in \mathcal{Q}$ ($t^2 > t$)

For answers to exercises 1-6, see the Teacher's Guide.

Write each of these sentences using $\forall$ or $\exists$ notation. State whether it is true or false.

7. There exists an integer x such that $x^2 > 0$. $\exists x \in \mathcal{Z}$ ($x^2 > 0$), true
8. There exists an integer y such that $y^2 < 1$. $\exists y \in \mathcal{Z}$ ($y^2 < 1$), true
9. Some integers are even. $\exists x \in \mathcal{Z}$ (x is even), true
10. All real numbers are integers. $\forall x \in \mathcal{R}$ ($x \in \mathcal{Z}$), false

Write each of the following in words. Determine the truth value of each sentence. For these exercises, let $T = \{$multiples of 3$\}$, $F = \{$multiples of 4$\}$, and $S = \{$multiples of 6$\}$.

11. $\forall x \in T$ (x is rational)
12. $\exists z \in \mathcal{Z}$ (z is even)
13. $\forall x \in S$ ($x \in T$)
14. $\forall x \in S$ ($x \in F$)
15. $\exists x \in S$ ($x \in T$ and $x \in F$)
16. $\exists x \in \mathcal{Q}$ ($x^2 = 2$)

For answers to exercises 11-16, see the Teacher's Guide.

Challenge The sentence, "Every integer is rational" can also be expressed, "If a number is an integer, then the number is rational" or as "If x is an integer, then x is rational." For each of the following, write the sentence in "the other" form.

1. Every whole number is an integer.
2. If r is a rose, then r is a flower.
3. All a's are b's.
4. None of the suspects here is guilty.
5. No student in this class is over six feet tall.

For answers to exercises 1-5, see the Teacher's Guide.

1.9 Negating Quantified Statements

Objective: To write a negation of a quantified statement and to identify quantifiers.

Consider the following universally quantified statement.

All students in this high school study French.

Some people might be tempted to say that the negation of this statement is "No student in this high school studies French." Is this negation correct? Think about it for a while.

The original statement is almost surely false, because there is undoubtedly someone in the high school who does *not* study French. But what of the suggested negation, "No student in this high school studies French?" If there is even one student in the high school who studies French, then this statement is false also. The negation of a false statement must be a true statement.

The example suggests that the real negation of the original statement is shown below at the right.

Original:
All students in this high school study French.

Negation:
Some students in this high school do *not* study French.

> **Journal**
>
> Complete this sentence: "The thing I like best about math class is. . ." Tell why.

To negate a universally quantified statement, proceed as follows:

Original: For all x, (a statement is true about x)
Negation: For some x, (the statement is false about x)

Here is another universally quantified statement: "No roses are yellow." Can the negation be "All roses are yellow?" No, because again, the original and the suggested negation are both false. It happens that some roses are yellow and some roses are *not* yellow.

The negation of

No roses are yellow

is

Some roses are yellow.

"No roses are yellow" can be written as "All roses are *not* yellow." Thus, this is the same as the example about studying French.

Negation of a Universal Quantifier

The negation of a universally quantified statement is an existentially quantified statement.

Consider the following existentially quantified statement.

Some animals have eight legs.

Is the negation "Some animals do *not* have eight legs?" Careful now—in this case, the original and the suggested negation are both true.

Under what circumstances would it be false that "Some animals have eight legs?" Well, the only case where this statement would be false is if

No animal has eight legs.

Here is another case.

Original: Some animals do not have eight legs.

Negation: All animals have eight legs.

To negate an existentially quantified statement, proceed as follows:

Original: For some x, (a statement about x is true.)
Negation: For all x, (the statement about x is false.)

Negation of an Existential Quantifier

The negation of an existentially quantified statement is a universally quantified statement.

Example

1 Select the negation of "Some integers are even" from among the choices given, and explain your answer.
a. All integers are even. b. No integers are even.
c. Some integers are not even. d. Not all integers are even.

The negation is choice b. Choice a is wrong because even though its truth value is opposite that of the original, it is possible to find sentences of the same structure where this would *not* be so. For example, "Some people are less than 20 feet tall" and "All people are less than 20 feet tall" are both true. Choice c is wrong because it and the original are both true. Choice d means the same as choice c.

Exercises

Max: 1-8, 1-26; Avg: 1-8, 1-12, 13-25 odds; Min: 1-8, 1-25 odds

Exploratory Complete each of the following.

1. From the choices given, select the negation of $\forall x \in T$ (x is green) and explain your choice. c, This choice states that there is at least one x that is not green.
   a. $\exists x \in T$ (x is green)   b. $\forall x \in T$ (x is *not* green)
   c. $\exists x \in T$ (x is *not* green)

2. From the choices given, select the negation of $\exists x \in T$ (x is red) and explain your choice. b, This choice states that no x is red.
   a. $\exists x \in T$ (x is *not* red)   b. $\forall x \in T$ (x is *not* red)
   c. $\forall x \in T$ (x is red)

State the negation of each statement. For answers to exercises 3-8, see the Teacher's Guide.

3. Some students take Latin.
4. Some secretaries live in New Jersey.
5. All birds fly.
6. Every mathematics teacher has red hair.
7. At least one history teacher plays the flute.
8. Some dogs do *not* bark.

For answers to exercises 1-12, see the Teacher's Guide.

Written Write a negation of each of the following quantified statements. In each case, identify the quantifiers in the original statement and in the negation.

1. All daffodils are yellow.
2. Some of these eggs are "jumbo."
3. There exists at least one Canadian who was born in Kansas.
4. Every cheddar in this shop has been aged over 60 days.
5. Some U.S. citizens were *not* born in the United States.
6. Some masked men have something to hide.
7. Every rhombus is a parallelogram.
8. None of these flowers is a rose.
9. Some integers fail to be even.
10. No even integer is rational.
11. There exists an x in Q such that $x^2 < 1$.
12. At least one real number, when squared, equals zero.

Negate each sentence. Read them carefully first. For answers to exercises 13-18, see the Teacher's Guide.

13. All roses are red and all violets are blue.
14. No elephants are red and no giraffes are green.
15. Some elves were jolly and all trolls were mean.
16. At least one integer is even or there exist no roses that are red.
17. Everyone in this class is tall and smart.
18. All children prefer dogs if and only if all dogs are brown.

Negate each sentence, using $\exists$ or $\forall$ notation.

19. $\forall x \in T$ (x is rational) $\exists x \in T$ (x is not rational)
20. $\forall y \in T$ (y is rational) $\exists y \in T$ (y is not rational)
21. $\exists z \in Z$ (z is even) $\lor$ $z \in \sim$ (z is not even)
22. $\exists x \in S$ ($x \in T$) $\forall x \in S$ ($x \notin T$)
23. $\forall x \in S$ ($x \in T$) $\exists x \in S$ ($x \notin T$)
24. $\exists a \in T$ ($a \in F$) $\forall a \in T$ ($a \notin F$)
25. $\forall x \in S$ ($x \in F$) $\exists x \in S$ ($x \notin F$)
26. $\exists x \in S$ ($x \in T$ and $x \in F$) $\forall x \in S$ ($x \notin T$ or $x \notin F$)

Problem Solving Application: Eliminating Possibilities

Objective: To solve problems by using matrix logic and eliminating possibilities.

A matrix is an array of items in rows and columns.

Certain real-world problems can be solved by using a method sometimes referred to as *matrix logic*. To solve one of these problems, a system of possibilities is set up using a matrix or grid. As each condition or clue for the problem is evaluated, the grid is marked to indicate a logical conclusion or a possibility that can be eliminated. Often a ✓ is used to indicate a logical conclusion, and an **X** is used to indicate a possibility that can be eliminated.

Example

1. A shipment of compact disks and cassette tapes is packed in three boxes. The boxes are labeled "compact disks" (CD), "tapes" (T), and "compact disks and tapes" (CDT). Due to a distribution center error, *none* of the boxes is packed correctly. Determine the contents of each box given that a compact disk is found in the "compact disks and tapes" box.

We can list the possibilities using a grid, as shown at the right.

<u>CDs</u> is an abbreviation for compact disks.

	CD	T	CDT
CDs			
tapes			
CDs and tapes			

Now we reason as follows.

a. None of the boxes is packed correctly. Thus, we can place **X**s in the grid to indicate that these possibilities are eliminated.

	CD	T	CDT
CDs	X		
tapes		X	
CDs and tapes			X

b. A compact disk was found in the "CDT" box. From this clue and the results of step a, we can conclude that the "CDT" box contains compact disks only. All other possibilities for the "CDT" box and compact disks can now be eliminated.

	CD	T	CDT
CDs	X	X	✓
tapes		X	X
CDs and tapes			X

c. The grid can now be completed.

We can conclude that the "CD" box contains tapes only and the "T" box contains compact disks and tapes.

	CD	T	CDT
CDs	X	X	✓
tapes	✓	X	X
CDs and tapes	X	✓	X

CHAPTER 1   37

Exercises

Max: 1-5; Avg: 1-5; Min: 1-3, 5

Written   Solve each problem.

1. Fred, Ted, and Ed are taking Mary, Cari, and Teri to the homecoming dance. Use these clues to determine the couples that will be attending the homecoming dance.
   a. Mary is Ed's sister and lives on Fifth Avenue.
   b. Ted drives a car to school each day.
   c. Ed is taller than Teri's date.
   d. Cari and her date ride their bicycles to school every day.
   e. Fred's date lives on State Street.   Ed and Cari, Fred and Teri, Ted and Mary

2. Pedro, Patti, and Paulette each ate a pair of pears. The six different varieties of pear consumed were Anjou, Bosc, Bartlett, Clapp Favorite, Comice, and Seckel. Use these clues to determine who ate which pair of pears.
   a. One of the girls was the only person who ate two pears whose varieties started with the same letter.   Pedro: Anjou, Bartlett; Patti: Bosc, Seckel; Paulette: Comice,
   b. Pedro ate the Bartlett pear, but not the Seckel.   Clapp Favorite
   c. Paulette ate the Comice pear.

3. Angie, Brenda, Carol, and Darlene have these jobs: cook, carpenter, cab driver, and lawyer. Use these clues to determine each person's job.
   a. Angie's job is not in the food industry.   Angie: carpenter; Brenda: cab driver;
   b. Brenda and Carol live next door to the carpenter.   Carol: lawyer; Darlene: cook
   c. Carol and the cook are sisters.
   d. Darlene and the cab driver are *not* sisters.
   e. The carpenter and the cook are *not* neighbors.

4. Rae, Melanie, and Bryan are neighbors. Their hobbies are sculpting, fixing cars, and gardening. Their occupations are doctor, teacher, and computer programmer. Use these clues to determine each person's hobby and occupation.
   a. The gardener and the teacher both graduated from the same college.
   b. Both the computer programmer and Rae have poodles, as does the sculptor.
   c. The doctor bandaged the sculptor's broken thumb.   Bryan: sculpting, teacher;
   d. Bryan and the computer programmer live next door to each other.   Melanie: fixing cars,
   e. Melanie beat both Bryan and the gardener in tennis.   computer programmer; Rae: gardening, doctor

5. Five automobile salesmen are named Marty, Mervin, Mike, Morey, and Murray. Their last names are Hood, Tyre, Chassie, Frame, and Enjin. Each man drives a different color car. The colors are blue, red, black, white, and tan. Use these clues to determine the first and the last name of each salesman and the color of his car.
   a. Morey is *not* Mr. Chassie, and he does *not* own a black car.
   b. Mervin, Murray, and Mr. Frame do *not* own red cars or white cars.
   c. Of Marty and Murray, one's name is Hood and one owns a tan car.
   d. The owner of the red car and Mr. Enjin are *not* named Morey or Mike.
   e. Either Mr. Tyre or Mr. Chassie, who owns a white car, is named Murray.
   Marty Hood, red; Morey Frame, blue; Marvin Enjin, black; Murray Tyre, tan; Mike Chassie, white

Portfolio Suggestion

Your portfolio contains representative samples of your work, collected over a period of time. Begin your portfolio by selecting an item that shows something new you learned in this chapter.

Performance Assessment

In your own words, explain the meaning of *deductive reasoning*. Give an example of the correct use of deductive reasoning to reach a conclusion, and tell why the reasoning is correct. Then give an example of incorrect reasoning using conditional statements. **See students' work.**

Chapter Summary

1. A **statement** is any declarative sentence that is either true or false, but *not* both. The truth or falsity of a statement is called its **truth value.** (2)
2. If p is a simple statement, then $\sim p$, read "not p," is its **negation.** If p is true, then $\sim p$ is false; if p is false, then $\sim p$ is true. (2)
3. A compound statement formed by joining two statements with the word *and* is called a **conjunction.** Each statement is called a **conjunct.** The symbol for the conjunction "p and q" is "$p \wedge q$." (2)
4. A compound statement formed by joining two statements with the word *or* is called a **disjunction.** Each of the statements is a **disjunct.** The symbol for the disjunction "p or q" is "$p \vee q$." (2)
5. An **open sentence** is a sentence that cannot be described as true or false. It becomes a statement that is either true or false when its variable is replaced with an element from some replacement set.

The replacement set is called the **domain.** The set of all replacements from the domain that make an open sentence true is called the **solution set.** (4)
6. A compound statement of the form "if p, then q" is called a **conditional.** The part of the statement logically associated with "if" is the **antecedent.** The part associated with "then" is the **consequent.** In symbols, "if p, then q" can be written $p \to q$. (6)
7. A **bi-conditional** is the conjunction of two conditionals. The statement $p \leftrightarrow q$ means $(p \to q) \land (q \to p)$. (7)
8. Bi-conditionals that are true for any replacement of the p's, q's, or other statements, of which they are made, are called **equivalences.** (8)
9. If a disjunction is true and one of its disjuncts is false, then the other disjunct is true. This rule is the **Law of Disjunctive Inference.** (10)
10. Statements that are assumed true are called **premises.** A statement that follows from the premises is called the **conclusion.** A **formal proof** may be used to demonstrate that a given set of premises leads to a certain conclusion. (10)
11. If $p \to q$ is a true conditional and p is true, then q is true. This rule is called the **Law of Detachment.** (13)
12. Consider the conditional $p \to q$. If the order of the antecedent and consequent are reversed, the new conditional $q \to p$ is called the **converse** of the original. If the antecedent and consequent are negated but the order remains the same as the original, the new conditional $\sim p \to \sim q$ is called the **inverse** of the original. If the antecedent and consequent are both reversed and negated, the new conditional $\sim q \to \sim p$ is called the **contrapositive** of the original. (17)
13. A conditional and its contrapositive are equivalent. This rule is called the **Law of the Contrapositive.** (18)
14. If $p \to q$ and $q \to r$ are true, then $p \to r$ is true. This rule is called the **Law of Syllogism.** (20)
15. The **De Morgan Laws** are $\sim(p \land q) \leftrightarrow (\sim p \lor \sim q)$ and $\sim(p \lor q) \leftrightarrow (\sim p \land \sim q)$. (25)
16. A **quantifier** tells to what part of the total "population" the sentence refers. A **universal quantifier** (whose symbol is $\forall$) tells about an entire population of something. An **existential quantifier** (whose symbol is $\exists$) tells something about only part of a population. (30, 31)
17. The negation of a universally quantified statement is an existentially quantified statement. The negation of an existentially quantified statement is an universally quantified statement. (33, 34)

Chapter Review

For answers to exercises 1-8, see the Teacher's Guide.

1.1 Let *p* represent "A square is a parallelogram," *q* represent "2 is a prime number," and *r* represent "Van Gogh painted the Mona Lisa." Write the statement represented by each of the following. Then determine the truth value.

1. $q \wedge r$
2. $p \vee r$
3. $p \vee \sim q$
4. $p \wedge \sim r$

Construct a truth table for each of the following.

5. $\sim p \wedge q$
6. $\sim p \vee \sim q$
7. $p \vee \sim q$
8. $\sim q \vee (\sim p \vee q)$

Find the solution set for each of the following. The domain is Z.

9. $5x + 7 = 32$ {5}
10. $4n - 3 = 13$ {4}
11. $\dfrac{t}{3} + 1 = -5$ {-18}
12. $3(2y - 4) = 18$ {5}
13. $-7x \geq 21$ {..., -5, -4, -3}
14. $4y - 1 < 7$ {..., -1, 0, 1}

1.2 Let *p*, *q*, and *r* represent the statements used for exercises 1–4. Write the statement represented by each of the following. Then determine the truth value.

15. $\sim p \to q$
16. $\sim r \to \sim q$
17. $q \leftrightarrow \sim r$
18. $(p \wedge q) \to r$

For answers to exercises 15-30, see the Teacher's Guide.

Construct truth tables to determine whether the two statements are equivalent.

19. $p \to q$; $\sim p \vee q$ equivalent
20. $p \to \sim q$; $\sim p \vee q$ not equivalent

1.3 For each of the following, use the statement reason form of proof to show that the indicated conclusion follows from the given premises. If the conclusion does not follow, explain why it does not. Use the letters indicated.

21. Rose *R* is *not* red or violet *V* is *not* blue. Rose *R* is red.
    Conclusion: Violet *V* is not blue. (R, V)
22. Gwen belongs to the chess club and Bonny does *not* belong to the chess club. Bonny belongs to the chess club or Bonny belongs to the drama club.
    Conclusion: Bonny belongs to the drama club. (G, B, D)
23. Otto is going to the grocery store or the drugstore. Otto is going to the drugstore.
    Conclusion: Otto is *not* going to the grocery store. (G, D)

1.4 24. If Mrs. Apple was *not* seen at the party then she is guilty of the crime. Mrs. Apple was *not* seen at the party.
    Conclusion: Mrs. Apple is guilty of the crime. (P, G)
25. If the time is 6:00 PM, Binky is eating supper. Binky is eating supper.
    Conclusion: The time is 6:00 PM. (T, S)
26. Felix or Tom is treasurer of the student council. If Felix is the treasurer of the student council, then Felix is good at arithmetic. Tom is *not* the treasurer of the student council. **Conclusion:** Felix is good at arithmetic. (F, T, A)

1.5 State the converse, inverse, and contrapositive of the given conditional.

27. $s \to t$
28. $\sim p \to q$
29. $\sim A \to \sim B$
30. $\sim n$ if *m*

For answers to exercises 31-38, see the Teacher's Guide.

For each of the following, use the statement-reason form of proof to show that the indicated conclusion follows from the given premises. If the conclusion does not follow, explain why it does not. For exercises 37 and 38, use the letters indicated.

31. Given: $S \rightarrow \sim T$
    T
    Conclusion: $\sim S$

32. Given: $A \wedge B$
    $\sim A \vee C$
    $D \rightarrow \sim C$
    Conclusion: D

33. Given: $M \rightarrow N$
    $T \wedge \sim M$
    Conclusion: $\sim T$

1.6
34. Given: $T \rightarrow \sim S$
    $\sim S \rightarrow U$
    T
    Conclusion: U

35. Given: $B \rightarrow D$
    $D \rightarrow C$
    $\sim C \vee A$
    B
    Conclusion: A

36. Given: $p \rightarrow q$
    $t \rightarrow \sim q$
    $\sim r \vee p$
    r
    Conclusion: $\sim t$

37. If Ella eats carrots, then she has good eyesight. If Ella has good eyesight, then she can read the bottom row on the eye chart. If Ella can read the bottom row on the eye chart, then she does *not* need glasses. Ella eats carrots.
    Conclusion: Ella does *not* need glasses. (C, E, B, G)

38. If Albert is a good runner, then he will win the race. If Albert is a good runner, then he wears good running shoes. Albert wears good running shoes.
    Conclusion: Albert will win the race. (R, W, S)

1.7 Write the negation of each of the following.

39. $\sim a \wedge \sim b$ a v b
40. $\sim m \wedge (n \vee p)$
    m v (~n ∧ ~p)
41. $x \vee (\sim y \wedge \sim z)$
    ~ x ∧ (y v z)

In each of the following, use the statement-reason form of proof to show that the indicated conclusion follows from the given premises. If the conclusion does not follow, explain why it does not. For answers to exercises 42-47, see the Teacher's Guide.

42. Given: $\sim(v \vee w)$
    $\sim v \rightarrow t$
    Conclusion: t

43. Given: $\sim(a \wedge \sim b)$
    $b \rightarrow c$
    a
    Conclusion: c

44. Given: $\sim(p \wedge \sim q)$
    $\sim(q \vee \sim r)$
    $p \vee t$
    Conclusion: t

1.8 Write each sentence in words. State whether it is true or false.

45. $\forall y \in \mathcal{Z} \ (y \neq -5)$
46. $\exists x \in Q \ (x^2 = 2)$
47. $\forall a \in \mathcal{W} \ (a \in \mathcal{Z})$

Write each sentence using $\forall$ or $\exists$ notation. State whether it is true or false.

48. There exists a rational number x such that $2x + 5 = 2$. $\exists x \in Q \ (2x + 5 = 2)$, true
49. For every rational number t, $t + 1 > t$. $\forall t \in Q \ (t + 1 > t)$, true
50. Some whole numbers are prime. $\exists x \in \mathcal{W} \ (x \text{ is prime})$, true
51. All whole numbers are prime. $\forall x \in \mathcal{W} \ (x \text{ is prime})$, false

1.9 Write a negation of each of the following quantified statements. In each case, identify the quantifiers in the original statement and in the negation.
For answers to exercises 52-55, see the Teacher's Guide.

52. All football players weigh at least 200 pounds.
53. None of the choir members plays the bassoon.
54. Some flowers are roses.
55. At least one teacher has red hair.

Chapter Test

Find the solution set of each of the following. The domain is Z.

1. $6x - 5 = 67$ {12}
2. $2 - 5y = -13$ {3}
3. $3t - 4 \leq -10$ $\{\ldots, -4, -3, -2\}$

For answers to exercises 4-9, see the Teacher's Guide.

Let a represent "6 is divisible by 3," b represent "Chicago is the capital of the United States," and c represent "February has 30 days." Write the statement represented by each of the following. Then determine the truth value.

4. $a \wedge b$
5. $a \vee c$
6. $a \rightarrow c$
7. $b \rightarrow \sim c$

Construct truth tables to determine whether the two statements are equivalent.

8. $p \rightarrow q$; $\sim q \vee p$ not equivalent
9. $\sim q \vee \sim p$; $p \rightarrow \sim q$ equivalent

State the converse, inverse, and contrapositive of the given conditional.

10. $m \rightarrow n$ $n \rightarrow m, \sim m \rightarrow \sim n, \sim n \rightarrow \sim m$
11. $\sim s \rightarrow t$ $t \rightarrow \sim s, s \rightarrow \sim t, \sim t \rightarrow s$

Write each sentence in words. State whether it is true or false.

12. $\exists x \in Q$ $(x \in \mathcal{R})$ There exists a rational number x such that x is a real number. true
13. $\forall y \in Z$ $(y > 0)$ For every integer y, y > 0. false

Write each sentence using $\forall$ or $\exists$ notation. State whether it is true or false.

14. There exists an integer x such that $3x - 4 = 7$. $\exists x \in Z$ $(3x - 4 = 7)$, false
15. For every rational number y, $y + 0 = y$. $\forall y \in Q$ $(y + 0 = y)$, true

Write the negation of each of the following.

16. $\sim p \vee q$ $p \wedge \sim q$
17. $\sim r \wedge (s \vee t)$ $r \vee (\sim s \wedge \sim t)$
18. All sailors must be able to swim. Some sailors must not be able to swim.
19. Some cows are brown. No cows are brown.

In each of the following, use the statement-reason form of proof to show that the indicated conclusion follows from the given premises. If the conclusion does not follow, explain why it does not. For exercises 23 and 24, use the letters indicated.

20. Given: $G \rightarrow H$
    $G \vee \sim B$
    B
    Conclusion: H

21. Given: $T \vee R$
    $\sim R$
    $\sim B \rightarrow \sim T$
    Conclusion: B

22. Given: $A \rightarrow B$
    $A \wedge C$
    $B \rightarrow D$
    Conclusion: D

23. Charlene is no good at backgammon or she is no good at chess. If she is no good at backgammon, then she is good at archery. But Charlene is awfully good at chess.
    Conclusion: Charlene is good at archery. (B, C, A)

24. It is false that Tom likes tennis or that Tom likes soccer. If Tom likes lacrosse, then he likes tennis. If Tom does not like soccer, then he likes lacrosse or he likes volleyball. **Conclusion:** Tom likes volleyball. (T, S, L, V)

For answers to exercises 20-24, see the Teacher's Guide.

25. Write the contrapositive of "If $x \leq 5$ and x is not prime, then $x = 4$."
    If $x \neq 4$, then $x > 5$ or x is prime.

CHAPTER 2

Mathematical Systems

Application in Business

The word *code* makes most people think of spy movies and secret messages. But thanks to technology, codes are now being used in everyday business. The Uniform Product Code, or UPC, is the coded set of bars printed on packages to identify products. Used with a scanner, a UPC code can tell a computer the product manufacturer, the specific type of product, and other information about the item like the color and the price.

UPC codes use a special **number system** called *binary numbers.* In binary numbers, the place value positions represent the powers of 2 instead of the powers of 10 like the number system that we usually use. The first digit to the right of a decimal point would represent 2^0 or 1s, the next digit to the left is 2^1 or 2s, the next digit is 2^2 or 4s, and so on. Find the value of the binary number 100010.

Individual Project:
Postal Service

The United States Postal Service has incorporated codes into their sorting system. A postal code can tell the computer the place of delivery for a piece of mail as specific as the correct side of the street. Research the USPS code system. How is the code displayed? Does the code use binary numbers? Present your findings in a report to the class.

44 MATHEMATICAL SYSTEMS

Objective: To solve addition problems using clock arithmetic.

2.1 Clock Arithmetic

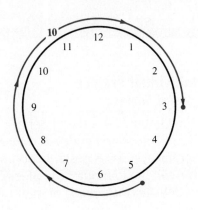

Suppose the time is five o'clock. What will the time be in ten hours?

If the time is five o'clock, then in ten hours, the hour hand will pass twelve o'clock and "start again." It will go to one, then two, then three o'clock. In ten hours, the time will be three o'clock. This means that in the counting system represented by an ordinary clock, the following statement is true.

$$5 + 10 = 3$$

We will call this system the clock 12 system. What is $9 + 4$ in clock 12? To answer this question, begin at 9 and move four units in the direction shown—*clockwise*. Thus, $9 + 4 = 1$ in clock 12.

What is $3 + 7$ in clock 12? As in ordinary arithmetic, $3 + 7 = 10$ in clock 12.

A table showing all the possible sums in a clock system can make addition easier. At the right is a table for addition in clock 12. This table contains all the possible sums that can be obtained when adding any two numbers in clock 12. The only numbers that are in the set are 1, 2, 3, 4, 5, 6, 7, 8, 9, 10, 11, 12. These are listed down the left side and along the top. The operation, +, appears in the upper left corner.

+	12	1	2	3	4	5	6	7	8	9	10	11
12	12	1	2	3	4	5	6	7	8	9	10	11
1	1	2	3	4	5	6	7	8	9	10	11	12
2	2	3	4	5	6	7	8	9	10	11	12	1
3	3	4	5	6	7	8	9	10	11	12	1	2
4	4	5	6	7	8	9	10	11	12	1	2	3
5	5	6	7	8	9	10	11	12	1	2	3	4
6	6	7	8	9	10	11	12	1	2	3	4	5
7	7	8	9	10	11	12	1	2	3	4	5	6
8	8	9	10	11	12	1	2	3	4	5	6	7
9	9	10	11	12	1	2	3	4	5	6	7	8
10	10	11	12	1	2	3	4	5	6	7	8	9
11	11	12	1	2	3	4	5	6	7	8	9	10

Examples

1 **Use the table to find the sum of 11 + 7.**

Find the first number to be added, 11, in the leftmost column, and the second number 7, along the top row. Then use the table as shown to conclude that 11 + 7 = 6. This result can be checked by counting, or by asking, "If it is eleven o'clock what will the time be in seven hours?"

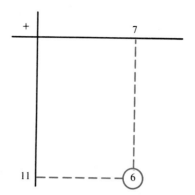

2 **Add 4 + 11 in clock 12.**

Using the table, 4 + 11 = 3.

3 **Add each of the following in clock 12.**
 a. 12 + 9   **b.** 2 + 12

Use the table or count.
 a. 12 + 9 = 9   **b.** 2 + 12 = 2

Look carefully at example 3. In clock 12, does the number 12 appear to act in a special way? Notice that when 12 is added to another number in clock 12, the answer is the other number.

+	0	1	2	3	4	5	6	7	8	9	10	11
0	0	1	2	3	4	5	6	7	8	9	10	11
1	1	2	3	4	5	6	7	8	9	10	11	0
2	2	3	4	5	6	7	8	9	10	11	0	1
3	3	4	5	6	7	8	9	10	11	0	1	2
4	4	5	6	7	8	9	10	11	0	1	2	3
5	5	6	7	8	9	10	11	0	1	2	3	4
6	6	7	8	9	10	11	0	1	2	3	4	5
7	7	8	9	10	11	0	1	2	3	4	5	6
8	8	9	10	11	0	1	2	3	4	5	6	7
9	9	10	11	0	1	2	3	4	5	6	7	8
10	10	11	0	1	2	3	4	5	6	7	8	9
11	11	0	1	2	3	4	5	6	7	8	9	10

Is there another number that acts in the same way in systems like the integers or the rationals? In other systems, *zero* is the number that leaves any number unchanged when the numbers are added. For this reason, you can rename the 12 in clock 12, as 0. The set of numbers for clock 12 becomes {0, 1, 2, 3, 4, 5, 6, 7, 8, 9, 10, 11}. A new table showing this change is at the left.

Example

4 Add each of the following in clock 12.
    a. 5 + 7   **b.** 0 + 3   **c.** 10 + 0   **d.** 6 + 6

Use the new table.
    a. 5 + 7 = 0   **b.** 0 + 3 = 3   **c.** 10 + 0 = 10   **d.** 6 + 6 = 0

Now consider clock 4. A clock for a clock 4 system is shown in Figure 1. Notice that adding 4 to any number produces the original number. Thus, 4 is the "zero" element. A clock showing this change appears in Figure 2. A table with all possible sums in clock 4 is shown below at the right. Check all the entries in this table.

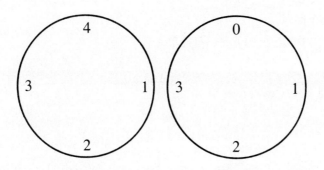

Figure 1   **Figure 2**

Examples

5 Add each of the following in clock 4.
    a. 2 + 3   **b.** 1 + 3   **c.** (3 + 1) + 2

Use the table or the clock.
    a. 2 + 3 = 1   **b.** 1 + 3 = 0   **c.** (3 + 1) + 2 = 0 + 2 = 2

6 Solve the equation $x + 2 = 1$ in clock 4.

The x represents the number that must be added to 2 to obtain 1. As the table shows, x must represent 3. Thus, the solution set is {3}.

Exercises

Max: 1-12, 1-58, 1-2; Avg: 1-12, 1-57 odds; Min: 1-11 odds, 1-45 odds

Exploratory The time is eight o'clock. State the time it will be in each of the following number of hours.

1. 1 hour 9:00
2. 6 hours 2:00
3. 12 hours 8:00
4. 3 hours 11:00
5. 10 hours 6:00
6. 7 hours 3:00
7. 8 hours 4:00
8. 11 hours 7:00
9. 5 hours 1:00
10. 4 hours 12:00
11. 9 hours 5:00
12. 15 hours 11:00

Written Complete the following.

1. Draw a clock that could be used for solving equations in clock 5.

Add each in clock 5.

2. $1 + 2$ 3
3. $4 + 3$ 2
4. $3 + 3$ 1
5. $4 + 1$ 0
6. $0 + 2$ 2
7. $4 + 4$ 3
8. $4 + (3 + 2)$ 4
9. $(3 + 3) + 3$ 4
10. $(4 + 2) + 4$ 0
11. $(4 + 4) + 4$ 2
12. $(1 + 3) + 4$ 3
13. $1 + (3 + 4)$ 3

Solve each in clock 5.

14. $y + 2 = 3$ 1
15. $3 + t = 2$ 4
16. $3 + p = 1$ 3
17. $x + 4 = 1$ 2
18. $0 = 3 + n$ 2
19. $4 = a + 4$ 0
20. $1 = 2 + c$ 4
21. $2 = q + 1$ 1
22. $4 = 2 + k$ 2
23. $1 + m = 0$ 4
24. $2 = d + 4$ 3
25. $g + g = 1$ 3, 3

Find each in clock 8. Draw a clock diagram, if needed.

26. $6 + 5$ 3
27. $7 + 7$ 6
28. $5 + 4$ 1
29. $1 + (7 + 7)$ 7
30. $(2 + 2) + (5 + 1)$ 2
31. $(3 + 1) + (0 + 5)$ 1
32. $(6 + 2) + (3 + 1)$ 4
33. $(5 + 1) + (1 + 1)$ 0
34. $(7 + 7) + (6 + 6)$ 2

Solve each in clock 8.

35. $r + 3 = 5$ 2
36. $6 + n = 0$ 2
37. $y + 6 = 2$ 4
38. $p + 5 = 1$ 4
39. $4 = x + 6$ 6
40. $5 = 7 + z$ 6
41. $3 = q + 5$ 6
42. $4 = d + 7$ 5
43. $5 + t = 2$ 5
44. $3 + w = 0$ 5
45. $0 = y + y$ 4
46. $x = x + x$ 0

Solve each in clock 9.

47. $x + 8 = 8$ 0
48. $y + 8 = 0$ 1
49. $3 + t = 1$ 7
50. $a + 4 = 1$ 6
51. $2 = 6 + m$ 5
52. $6 = 7 + y$ 8
53. $3 = s + 5$ 7
54. $5 = 6 + f$ 8
55. $2 = d + 3$ 8
56. $3 = h + h$ 6
57. $n + n = 1$ 5
58. $k + k = 5$ 7

Challenge Answer the questions.

1. How many entries would an addition table for clock 24 have? 576
2. How many times would the element 0 appear in the table in exercise 1? Where would it appear? Explain. Name five sums that would produce the entry 0 in such a table. 24, in the upper left corner of the table and along a diagonal; Answers may vary.

2.2 Multiplication in a Clock System

Objective: To solve multiplication problems using clock arithmetic.

Suppose today is Thursday. What will the day be in 5,000 days? One way to solve this problem is to count the days one by one. This method is very time-consuming. But a better method exists to solve this problem, which will be discussed later in this section. First, consider multiplication in a clock system. What is the product 3 × 5 in clock 12?

In "ordinary" arithmetic, 3 × 5 means 5 + 5 + 5. If you use the same interpretation of multiplication in clock 12, then

$$3 \times 5 = 5 + 5 + 5 = 10 + 5 = 3.$$

One way to check this answer is to start at zero (12 o'clock) and think of three successive five-hour time periods. The last would end at 3 o'clock.

Definition of Multiplication

> **Multiplication** in clock arithmetic, as in ordinary arithmetic, is repeated addition.

Example

1 Compute each of the following in clock 12.

  a. 4 × 7   **b.** 5 × 8   **c.** 10 × 6

Use the clock and count, or use the table on page 45, to obtain each answer.

  a. 4 × 7 = 4   **b.** 5 × 8 = 4   **c.** 10 × 6 = 0

Is there an easier way to handle clock multiplication? You know that in clock 12 every time you pass 0 you start again. If you pass twice, you start again; if you pass 100 times, you start again. For example, consider the problem 9 × 4 in clock 12. You start at 0. The first three 4s take you back to 0. At the end of the second three 4s, you are back at zero again. Then, after nine 4s, you are back again at 0. Thus, in clock 12, 9 × 4 = 0. Ten times 4 is 4, because this is one more 4 than 9 × 4.

How about 9 × 7 in clock 12? In ordinary arithmetic, 9 × 7 = 63. To multiply 9 × 7 by counting around the clock, start at 0 and go around 63 units. But the first 12 + 12 + 12 + 12 + 12, or 60, would get you back to zero. There is 63 − 60 remaining. So, 9 × 7 = 3 in clock 12.

Multiplication in Clock Arithmetic

> To multiply numbers in clock n,
> 1. Find the "ordinary" product of the numbers.
> 2. Divide this number by n.
> 3. The remainder is the answer in clock n.

Examples

2 Multiply 3×3 in clock 4.

In ordinary arithmetic, $3 \times 3 = 9$. In this case, $n = 4$. Divide 4 into 9. This produces a remainder of 1. Thus, $3 \times 3 = 1$ in clock 4.

3 Multiply 19×26 in clock 30.

In ordinary arithmetic, $19 \times 26 = 494$. Dividing 30 into 494 leaves a remainder of 14. Therefore, $19 \times 26 = 14$ in clock 30.

4 Today is Thursday. What will the day be in 5,000 days?

Think about it first. Set up a clock 7 system. Let Thursday be 0. What will the day be in 7 days? In 14 days? In 77 days? In 78 days? To finish the problem, divide 5,000 by 7. The quotient is 714 with a remainder of 2. Therefore, the day will be Saturday.

Note that Saturday is two days after Thursday.

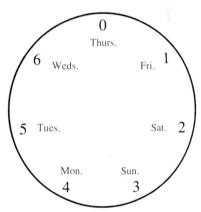

×	0	1	2	3
0	0	0	0	0
1	0	1	2	3
2	0	2	0	2
3	0	3	2	1

Setting up tables for multiplication in clock systems sometimes provides a few surprises. To the left is a table for multiplication in clock 4. Check every entry very carefully. Answer the following questions.

1. Why is every entry along the top and down the left side 0?
2. Why aren't all the entries in every row and column different, as in addition?

You can use this table to solve some equations in clock 4.

Examples

5 Solve $3x = 1$ in the clock 4 system.

Use the table to find which element, when multiplied by 3, produces 1. The solution set is {3}.

6 Solve $2x = 2$ in the clock 4 system.

Check the table. Note that there are *two* elements that make this open sentence true. The solution set is {1, 3}.

7 Solve $2y = 3$ in the clock 4 system.

The table shows that there is *no* element that, when multiplied by 2, yields 3. The solution set is ∅.

The last two examples illustrate important items to watch for when solving linear equations in clock systems. Sometimes the solution set contains *more than one* element. Or, it can contain *no* elements.

Exercises

Max: 1-10, 1-44, 1-2; Avg: 1-10, 1-16, 22-44; Min: 1-10, 1-6, 7-21 odds, 22-28

Exploratory Explain how to do each of the following.

1. Multiply 5 by 3 in clock 6 by counting around the clock.
2. Multiply any two numbers in clock 6 without "counting."
3. Multiply any two numbers in any clock system without "counting."
4. Find two numbers with a product of zero in clock 100.
5. Find two numbers with a product of 1 in clock 50.
6. Know, in advance, what the row across the top of any multiplication table for a clock system will be.

For answers to exercises 1-6, see the Teacher's Guide.

Find each in clock 3.

7. $2 \times 1 2$
8. $0 \times 2 0$
9. $1 \times 1 1$
10. $2 \times 2 1$

Written Multiply each in clock 5.

1. $2 \times 2 4$
2. $1 \times 4 4$
3. $3 \times 3 4$
4. $4 \times 4 1$
5. $3 \times 4 2$

Complete. For the answer to exercise 6, see the Teacher's Guide.

6. Set up and complete a multiplication table for clock 6.

Multiply each in clock 6.

7. 2 × 3 0   **8.** 3 × 3 3   **9.** 0 × 1 0   **10.** 4 × 5 2   **11.** 5 × 3 3

Solve each in clock 6.

12. $1y = 4$ {4}   **13.** $3t = 3$ {1,3,5}   **14.** $2x = 4$ {2,5}   **15.** $2a = 1$ ∅   **16.** $4n = 2$ {2,5}
17. $3c = 0$ {0,2,4}   **18.** $5r = 0$ {0}   **19.** $5m = 4$ {2}   **20.** $x^2 = 1$ {1,5}   **21.** $y^2 = y$ {0,1,3,4}

Complete.

22. Set up and complete a multiplication table for clock 7. For the answer to exercise 22, see the Teacher's Guide.

Multiply each in clock 7.

23. 3 × 6 4   **24.** 6 × 3 4   **25.** 5 × 4 6   **26.** 4 × 5 6
27. 3 × (2 × 4) 3   **28.** (3 × 2) × 4 3   **29.** 6 × (4 × 3) 2   **30.** (6 × 4) × 3 2

Answer the questions.

31. Is multiplication commutative in clock systems? Why or why not?
31. Multiplication is commutative in clock systems because it is based on "ordinary" multiplication which is commutative.

32. Is multiplication associative in clock systems? Why or why not?
32. Multiplication is associative in clock systems because it is based on "ordinary" multiplication which is associative.

Multiply each in clock 57.

33. 19 × 23 38   **34.** 29 × 52 26   **35.** 50 × 50 49   **36.** 49 × 33 21

Today is Monday. What will the day be in each number of days?

37. 9 days Wednesday   **38.** 900 days Friday   **39.** 9,000 days Saturday   **40.** 11,352 days Saturday

This month is October. What will the month be in each number of months?

41. 3 months January   **42.** 30 months April   **43.** 300 months October   **44.** 3,003 months January

Mixed Review

Solve each equation in the indicated clock system.

1. $x + 7 = 5$, clock 8 {6}   **2.** $r + 3 = 0$, clock 5 {2}   **3.** $7 = a + 8$, clock 9 {8}
4. $m + m = m$, clock 6 {0}   **5.** $3y + 5 = 4$, clock 7 {2}   **6.** $2y + 1 = 3$, clock 4 {1, 3}

7. If $5 + 3 = 2$ in clock x, find x. 6   **8.** If $2(6) = 2$ in clock y, find y. 10
9. Write the inverse of "If $n \geq 3$ and $n < 5$, then $n = 3$ or $n = 4$." If $n < 3$ or $n \geq 5$, then $x \neq 3$ and $x \neq 4$.
10. Write the contrapositive of "If all squares are rectangles, then at least one integer is odd." If no integers are even, then some squares are not rectangles.

Determine if each pair of statements is equivalent.

11. $\sim(p \wedge \sim q)$; $\sim p \wedge q$ no   **12.** $\sim[n \vee (\sim t \vee s)]$; $\sim n \wedge (t \wedge \sim s)$ yes

State the conclusion that can be inferred from the given premises.

13. If $a < 2$, then $x = 3$. If $x = 3$, then $y \neq 7$. $z = 5$ or $y = 7$. But $a = 1$. $z = 5$
14. If $a < 2$, then $x = 3$. If $x = 3$, then $y \neq 7$. $z = 5$ or $y = 7$. But $z = 4$. $a \geq 2$

2.3 Subtraction and Division

Objective: To solve subtraction and division problems using clock arithmetic.

It is 5 o'clock. What was the time 7 hours ago? This situation corresponds to the subtraction problem 5 − 7 = ? in clock 12.

One way to answer the problem is to count the hours on an ordinary clock—but this time, count *counterclockwise*. This procedure is shown at the left. If the time is five o'clock now, then seven hours ago it was 10 o'clock. Thus, 5 − 7 = 10 in clock 12. How can you check this result?

In ordinary subtraction, you can check your answer by setting up the related addition problem. For example, to check 20 − 13 = 7, ask "Is 7 the number that must be added to 13 to produce 20?"

20 − 13 = 7 is equivalent to 20 = 7 + 13.

To check 5 − 20 = −15, ask "Is 5 = 20 + −15 true?" This method may be used to check a subtraction problem in clock 12. If 5 − 7 = 10 is true, then 5 = 7 + 10 has to be true. Indeed, 5 is equal to 7 + 10 in clock 12. Thus, 5 − 7 = 10 in clock 12.

Definition of Subtraction

In clock arithmetic, as well as in ordinary arithmetic, $a - b = c$ means $a = b + c$

Examples

1 Subtract 1 − 3 in clock 4.

Use the clock at the right. Notice that 1 − 3 = 2. To check this answer, ask "Is it the case that 3 + 2 = 1 in clock 4?" Because the answer to this is *yes*, the answer to the subtraction problem is correct.

2 The table shown is for subtraction in clock 4. Find the missing entries.

1 − 3 = 2 because 1 = 3 + 2
2 − 3 = 3 because 2 = 3 + 3
0 − 2 = 2 because 0 = 2 + 2
2 − 0 = 2 because 2 = 0 + 2

−	0	1	2	3
0	0		3	1
1	1	0	3	
2			1	0
3	3	2	1	0

Examples

3 **Solve $x - 2 = 5$ in clock 6.**

Because $x - 2 = 5$ is equivalent to $x = 5 + 2$, the solution set in clock 6 is {1}.

4 **Solve $3 - x = 5$ in clock 6.**

The problem $3 - x = 5$ is equivalent to $3 = x + 5$. In clock 6, 4 must be added to 5 to produce 3. Thus, the solution set for $3 - x = 5$ is {4}.

Division has the same relationship to multiplication as subtraction does to addition.

Definition of Division

In clock arithmetic, as well as in ordinary arithmetic,

$$\frac{a}{b} = c \quad \text{means} \quad a = bc$$

Examples

5 **Find $3 \div 1$ in clock 4.**

Suppose $3 \div 1 = n$. This means that $3 = 1 \cdot n$. Using the chart on page 49, $n = 3$. Therefore, $3 \div 1 = 3$ in clock 4.

6 **Find $3 \div 0$ in clock 4.**

If $3 \div 0 = n$, then $3 = 0 \cdot n$. But there is not any n such that $3 = 0 \cdot n$. In fact, in any clock system, just as in an ordinary arithmetic, zero times any number equals zero.

Division by Zero

In clock arithmetic, as in ordinary arithmetic, division by zero is meaningless.

Example

7 Find $\frac{3}{2}$ in clock 4.

If $\frac{3}{2} = n$, then $2n = 3$, but there is no n satisfying this condition. In clock 4, 2 times any number is either 2 or 0. Thus, $\frac{3}{2}$ is meaningless in clock 4.

From example 7, we see that in clock arithmetic, division is not always possible.

The symbol Z_n is often used for clock n. Thus, Z_6 is clock 6 and names the set {0, 1, 2, 3, 4, 5}. This notation will be used in this book.*

Exercises

Max: 1-16, 1-34, 1-5; Avg: 1-16, 1-29; Min: 1-15 odds, 1-33 odds

Exploratory Write an addition statement that is equivalent to each subtraction statement.

1. $10 - 3 = 7$
   $10 = 7 + 3$
2. $52 - 14 = 38$
   $52 = 38 + 14$
3. $3x - x = 2x$
   $3x = 2x + x$
4. $(2 + 1) - 1 = 2$
   $(2 + 1) = 2 + 1$

Use a drawing of an ordinary clock to find each in clock 12.

5. $9 - 2 \quad 7$
6. $9 - 9 \quad 12$
7. $6 - 11 \quad 7$
8. $(3 - 8) - 9 \quad 10$

Write a multiplication statement that is equivalent to each division statement.

9. $12 \div 3 = 4$
   $12 = 4 \times 3$
10. $\frac{51}{3} = 17$
    $51 = 17 \times 3$
11. $8y \div 2y = 4$
    $8y = 4 \cdot 2y$
12. $\frac{3t - 9}{3} = t - 3$
    $(3t - 9) = 3(t - 3)$

Find each in Z_3. If the instruction is meaningless, state that and explain.

13. $2 \div 1 \quad 2$
14. $\frac{1}{2} \quad 2$
15. $2 \div 0$ meaningless
16. $\frac{0}{1} \quad 0$

Written Complete. For the answer to exercise 1, see the Teacher's Guide.

1. Set up and complete a subtraction table for clock 5.

Subtract each in clock 5.

2. $1 - 2 \quad 4$
3. $4 - 4 \quad 0$
4. $0 - 1 \quad 4$
5. $0 - 3 \quad 2$
6. $2 - 4 \quad 3$

*Sometimes the phrase "integers modulo n" is used for Z_n. For example, "integers modulo 4" means Z_4 or clock 4.

Subtract each in Z_{10}.

7. $3 - 4$ 9
8. $9 - 7$ 2
9. $2 - 6$ 6
10. $5 - 7$ 8
11. $1 - 5$ 6

Solve each in clock 9.

12. $8 - x = 3$ {5}
13. $0 - y = 5$ {4}
14. $t - 3 = 4$ {7}
15. $4 - x = x$ {2}

Solve each in Z_{15}.

16. $5 - n = 2$ {3}
17. $x - 4 = 11$ {0}
18. $9 - y = 4$ {5}
19. $1 - a = 6$ {10}

Find each in clock 5.

20. $2 \div 1$ 2
21. $\dfrac{3}{2}$ 4
22. $\dfrac{4}{3}$ 3
23. $1 \div 4$ 4
24. $\dfrac{2}{3}$ 4

Find each in Z_6. If the instruction is meaningless, state that and explain.

25. $3 \div 1$ 3
26. $\dfrac{1}{5}$ 5
27. $5 \div 0$ meaningless
28. $4 \div 2$ 2,5
29. $2 \div 4$ 2,5
30. $\dfrac{3}{4}$ meaningless
31. $0 \div 3$ 0,2,4
32. $\dfrac{4}{3}$ meaningless
33. $\dfrac{2}{5}$ 4
34. $2 \div 3$ meaningless

Challenge Answer the question.

1. In which clock systems is the division by numbers *other than zero* not always possible?
Clock n systems in which n is not a prime number

Use Z_n notation to name at least two sets in which each statement is true. If there are fewer than two, explain.

2. $3 + 4 = 7$
Z_n, where $n > 7$

3. $10 - 0 = 0$
Z_n, where $n > 10$

4. $9 \times 9 = 1$
Z_n, where $9 < n \leq 80$ and n is a factor of 80

5. If $3 \cdot k = 0$, then $k = 6$
Z_{18}; If $3 \cdot k = 0$ and $k = 6$, then $3 \cdot 6 = 0$. This is only true in Z_{18}.

Mathematical Excursions

Number Systems

Number systems are important in mathematics.

Exercises Refresh your memory by defining or identifying each system.

1. integers $\mathbb{Z} = \{\ldots, -3, -2, -1, 0, 1, 2, 3, \ldots\}$
2. rationals $Q = \{\dfrac{a}{b}$ such that $a, b \in \mathbb{Z}, b \neq 0\}$
3. reals $\mathcal{R} = \{\text{all rational numbers and all irrational numbers}\}$

Give three examples of each of the following. Answers may vary.

4. rational numbers that are *not* integers
5. real numbers that are not rational
6. numbers greater than 1 and less than 2 that are not rational

State whether each is *sometimes, always,* or *never* true. Defend your answer.

7. The sum of two irrational numbers is irrational. sometimes
8. The product of two irrational numbers is irrational. sometimes
9. The product of a rational and an irrational number is irrational. always

2.4 Operations

Objective: To identify an operation on a set and to identify operations as commutative or associative.

You are familiar with the operations of arithmetic—addition, subtraction, multiplication, and division. In algebra, operations are more formally defined.

Operation on a Set

> Let S be a set. Then $*$ is **an operation on set** S if, for every a and b in S, there exists a single element c, also in S, such that $a * b = c$.

The element c does not have to be different from a or b. It just has to be a member of set S and it has to be the only result of combining a and b.

Addition (+) is an operation on the set $\mathcal{W}$ of whole numbers. The sum of any two numbers from $\mathcal{W}$ is a unique third element in $\mathcal{W}$. For example, $3 + 7 = 10$ and $100 + 305 = 405$.

Examples

1 Is subtraction (−) an operation on $\mathcal{W}$?

No, because it is possible to choose two numbers a and b from $\mathcal{W}$ such that subtracting $a - b$ does not produce an element in $\mathcal{W}$. For example, $2 - 5 = -3$ and $-3 \notin \mathcal{W}$.

2 Is subtraction an operation on Z_4?

Recall your work with subtraction in Z_4 from the previous section. Any element in Z_4 can be subtracted from any other element in Z_4 and the answer is an element of Z_4. Hence, subtraction is an operation on Z_4.

3 Is division (÷) an operation on Q?

Careful! Is the result of dividing one rational by another rational always a rational? Well, it is *almost* always rational. There is one element in Q—namely, zero— by which division is *not* possible. Because of this exception, division is *not* an operation on Q. However, division is an operation on the special subset {rationals without zero}.

The symbol $Q/\{0\}$ means *the rational numbers without zero*. In a similar manner, the symbol A/B means *all the elements in a set A except those in set B*. For example if $A = \{1, 2, 3, 4, 5\}$ and $B = \{2, 4\}$, then $A/B = \{1, 3, 5\}$.

Examples

4 Let $\mathbb{A}$ be the symbol for averaging two numbers. That is, $a \mathbin{\mathbb{A}} b = \dfrac{a + b}{2}$. When is $\mathbb{A}$ an operation?

Averaging of two numbers, $\mathbb{A}$, is an operation on the set of rationals, but *not* on the set of integers or the set of whole numbers. Can you see why not? The average of any two integers is not always an integer.

5 Let & be the symbol for assigning a third number to *a* and *b* in the following way: $a \mathbin{\&} b$ = any number greater than $a + b$. Is & an operation on Q?

The process & is not an operation on the rational numbers because the result of using & is not unique. For example, 3 & 4 could equal $7\tfrac{1}{2}$, or it could equal $8\tfrac{1}{4}$, or 25.1, or 50,000, or any other number greater than 7.

Ordered pair notation is used to refer to a set together with an operation. For example, $(\mathcal{W}, +)$ names the set $\mathcal{W}$ of whole numbers with the process of addition. Also, $(Z_6, \cdot)$ names the set Z_6 with multiplication. Such pairs are called *operational systems*.

For ∗ to be an operation on a set, the set must be <u>closed</u> under ∗.

Sometimes an operation is said to be *closed* on the set of elements on which it is defined.

Definition of Closure

> Let S be a set. Then S is closed under an operation ∗ if, for every a and b in S, $a ∗ b$ is an element in S.

The whole numbers are closed with respect to addition and multiplication, but not with respect to division or subtraction.

Two other properties of operations that are familiar to us are those of commutativity and associativity.

58 MATHEMATICAL SYSTEMS

Definition of Commutative

Let S be a set. The operation $*$ is **commutative** on S if for every a and b in S, $a * b = b * a$.

The operations of addition and multiplication are commutative on the set Z of integers. Subtraction is not commutative on Z. For instance, $8 - 3 \neq 3 - 8$.

Examples

6 In $(Q/\{0\}, \div)$ is $\div$ commutative?

Division is *not* commutative on $Q/\{0\}$ because, for example, $10 \div 2 \neq 2 \div 10$.

7 Is the operation $\mathbb{A}$, averaging of two numbers, commutative on the reals? Is it commutative on the rationals?

The operation $\mathbb{A}$ is commutative on the reals and on the rationals. Can you explain why? This operation uses addition which is commutative on the reals and on the rationals.

Definition of Associative

Let S be a set. The operation $*$ is **associative** on S if for every a, b, and c in S, $a * (b * c) = (a * b) * c$.

As you know, addition and multiplication are associative on the reals, rationals, integers, and whole numbers.

Example

8 Are addition and multiplication associative in clock systems?

Try a few and see what happens. Although this kind of experimentation does *not* prove anything, it should be enough to convince you that addition and multiplication are both associative in clock systems. In future work, addition and multiplication in clock systems will be considered associative.

Examples

9 Is $\mathbb{A}$, averaging, associative on the rationals?

Investigate this question. Examples are not proofs. However, a counterexample can prove that a statement is false.

Let $a = 2$, $b = 4$, and $c = 6$. Is $2 \mathbb{A} (4 \mathbb{A} 6) = (2 \mathbb{A} 4) \mathbb{A} 6$ true?

$$2 \mathbb{A} (4 \mathbb{A} 6) \stackrel{?}{=} (2 \mathbb{A} 4) \mathbb{A} 6$$
$$2 \mathbb{A} \quad 5 \quad \stackrel{?}{=} \quad 3 \quad \mathbb{A} \ 6$$
$$3\frac{1}{2} \neq 4\frac{1}{2}$$

The computation shows that the answer is *no*. Averaging, although commutative, is *not* associative on the rationals.

10 Let # be an operation defined on the rationals as follows: For every a and b in Q, $a \# b = a + b + 3$. For instance, $2 \# 5 = 2 + 5 + 3 = 10$, and $\frac{1}{2} \# \frac{1}{3} = \frac{1}{2} + \frac{1}{3} + 3 = 3\frac{5}{6}$. Is # associative on Q?

You can analyze this question as shown.

$$a \# (b \# c) \stackrel{?}{=} (a \# b) \# c$$
$$a \# (b + c + 3) \stackrel{?}{=} (a + b + 3) \# c$$
$$a + (b + c + 3) + 3 \stackrel{?}{=} (a + b + 3) + c + 3$$
$$a + b + c + 6 = a + b + c + 6$$

Because $a \# (b \# c)$ and $(a \# b) \# c$ always name the same quantity, # is associative on Q.

Exercises

Max: 1-12, 1-60; Avg: 1-12, 1-59 odds; Min: 1-11 odds, 1-53 odds

Exploratory State the meaning of each of the following.

the rationals without 1, -1

1. $\mathbb{Z}/\{0\}$ the integers without 0
2. $\mathcal{R}/\{1\}$ the reals without 1
3. $Q/\{1, -1\}$
4. $Z_5/\{0\}$ clock 5 without 0
5. $\mathcal{W}/\{n: n \leq 5\}$ the whole numbers greater than 5
6. $Z_7/\{0\}$ clock 7 without 0
7. C/D where $C = \{1, 2, 3, 4, 5\}$ and $D = \{2, 4\}$ {1,3,5}

8. Is division an operation on $\mathcal{R}$? no
9. Is division an operation on $\mathcal{R}/\{0\}$? yes
10. Show that multiplication is an operation on Z_3. Any element in Z_3 can be multiplied by any other element in Z_3 and the answer is an element of Z_3.
11. Show that subtraction is *not* commutative on Z_7. Answers may vary.

12. Give three examples that show that multiplication is associative on Z_3. Answers may vary.

Written For each, decide if the process is an operation. If it is, decide if it is *commutative, associative, both,* or *neither.* Justify each answer.

1. $(Z_5, +)$ yes, both   **2.** $(Z/\{0\}, \div)$ no   **3.** $(Z, -)$ yes, neither   **4.** $(Z, \cdot)$ yes, both   **5.** $(\mathcal{R}, \cdot)$ yes, both
6. $(Z_7, \div)$ no   **7.** $(\mathcal{R}, -)$ yes, neither   **8.** $(Q, +)$ yes, both   **9.** $(Z/W, +)$ yes, both   **10.** $(Z_6/\{0\}, \div)$ no

Solve each using the definition of $\mathbb{A}$, averaging, given in this section. You are in Q.

11. 3 $\mathbb{A}$ 6 $4\frac{1}{2}$ **12.** 78 $\mathbb{A}$ 80   79   **13.** $\frac{1}{2}$ $\mathbb{A}$ $\frac{1}{4}$ $\frac{3}{8}$
14. 3.2 $\mathbb{A}$ 3.3   3.25   **15.** 0 $\mathbb{A}$ 5.6   2.8   **16.** 8.123 $\mathbb{A}$ 3.404   5.7635

Determine if each is defined in the set of integers.

17. 3 $\mathbb{A}$ 5   yes   **18.** 4 $\mathbb{A}$ 100   yes   **19.** -5 $\mathbb{A}$ 5   yes   **20.** 3 $\mathbb{A}$ 6   no   **21.** 1 $\mathbb{A}$ -35 yes

State the simplest name for each of the following. You are in $\mathcal{R}$.

22. $2\sqrt{2}$ $\mathbb{A}$ $4\sqrt{2}$ $3\sqrt{2}$ **23.** $-\sqrt{3}$ $\mathbb{A}$ $\sqrt{3}$ 0   **24.** $\sqrt{5}$ $\mathbb{A}$ $\sqrt{45}$ $2\sqrt{5}$
25. $\frac{\pi}{4}$ $\mathbb{A}$ $\frac{\pi}{3}$ $\frac{7\pi}{24}$ **26.** 1 $\mathbb{A}$ $\sqrt{2}$ $\frac{1+\sqrt{2}}{2}$ **27.** $(\pi + 1)$ $\mathbb{A}$ $(\pi - 1)$ π

Find a rational approximation for each of the following.

28. 2 $\mathbb{A}$ $2\sqrt{2}$ 2.41   **29.** $\sqrt{3}$ $\mathbb{A}$ $7\sqrt{3}$ 6.93   **30.** 0 $\mathbb{A}$ $\sqrt{200}$ 7.07   **31.** 4π $\mathbb{A}$ $\sqrt{24}$ 8.73

Find each using the operation # as given in this section.

32. 5 # 3   11   **33.** 0 # 0   3   **34.** 18 # 32   53   **35.** -1 # 57   59   **36.** -6 # -42 -£

Let @ be defined as follows: For every a and b in the real numbers, $a @ b = (a + 2b)(a + 2b) - 1$. Find each of the following.

37. 3 @ 2   48   **38.** 2 @ 1   15   **39.** 3 @ $-\frac{3}{2}$ -1
40. -5 @ 1   8   **41.** $\frac{1}{4}$ @ $\frac{1}{2}$ $\frac{9}{16}$ **42.** $\sqrt{3}$ @ $2\sqrt{3}$ 74

Let $ be defined as follows: For every a and b in the real numbers, $a \$ b = a^2 + b$. Solve each of the following.

43. 5 $ -1 24   **44.** 3 $ 7   16   **45.** $\sqrt{2}$ $ 0   2
46. -6 $ 4   40   **47.** $3\sqrt{5}$ $ 5   50   **48.** $2\sqrt{3}$ $ 1   13

Use the definitions in this section to answer the questions.

49. Is $\mathbb{A}$ an operation on $\mathcal{R}$? yes
50. Is $\mathbb{A}$ commutative on $\mathcal{R}$? yes
51. Is $\mathbb{A}$ associative on $\mathcal{R}$? no
52. Is # an operation on $\mathcal{R}$? yes
53. Is # commutative on $\mathcal{R}$? yes
54. Is # associative on $\mathcal{R}$? yes
55. Is @ an operation on $\mathcal{R}$? yes
56. Is @ commutative on $\mathcal{R}$? no
57. Is @ associative on $\mathcal{R}$? no
58. Is $ an operation on $\mathcal{R}$? yes
59. If $ commutative on $\mathcal{R}$? no
60. Is $ associative on $\mathcal{R}$? no

2.5 Identities and Inverses

Objective: To find the identities and inverses of elements.

Identity elements and inverses are two important ideas about operations.

Definition of Identity Element

> Let S be a set and $*$ be an operation defined on S. Then the element e, in S, is the **identity element** for $*$ in S if for every a in S,
> $$a * e = e * a = a.$$

That is, e is an identity element if operating with it leaves the other element unchanged.

As you know, the number 0 is the identity element for addition on the sets W, Z, Q, and $\mathcal{R}$. For instance, $3 + 0 = 0 + 3 = 3$. The number 1 is the identity element for ordinary multiplication in the same sets. The product of any number and 1 is that number.

Study the tables below. In clock systems, 0 is the identity for addition and 1 is the identity for multiplication.

+	0	1	2	3	4
0	0	1	2	3	4
1	1	2	3	4	0
2	2	3	4	0	1
3	3	4	0	1	2
4	4	0	1	2	3

$(Z_5, +)$

·	0	1	2	3	4
0	0	0	0	0	0
1	0	1	2	3	4
2	0	2	4	1	3
3	0	3	1	4	2
4	0	4	3	2	1

$(Z_5, \cdot)$

A table may be used to find the identity element in a mathematical system. The table at the left is the table for the system $(B, ☼)$ where $B = \{\triangle, \heartsuit, \clubsuit\}$ and $☼$ is the operation defined in the table. What is the identity? The element $\heartsuit$ reproduces the row and column at the head or side of the table. It is the identity.

Definition of Inverse

> Let S be a set with $*$ an operation defined on S. Let e be the identity element for $*$ in S. Then a_{inv} is the **inverse** for a, with respect to $*$, if
> $$a * a_{inv} = a_{inv} * a = e$$

That is, a_{inv} is the inverse of a if the combination of a and a_{inv} is the identity.

Note that every element may have its *own inverse*, but there is only one identity in a set. Remember that before you go hunting for inverses, you must know what your identity is. In $(Z, +)$—that is, the integers with addition—the identity is 0. The inverse of 3 is -3. The inverse of -17 is 17. The inverse of 0 is 0.

Examples

1 **In the system $(W, +)$, find the inverse for 3.**
The identity in $(W, +)$ is 0. What element in W added to 3 produces zero? There is no such element. In fact, no element except 0 itself has an inverse in $(W, +)$.

2 **Find the inverse of 4 in $(Z_{15}, +)$.**
By definition, the inverse of 4 is the element that must be added to 4 to produce the identity, 0. Because $4 + 11 = 0$ in Z_{15}, $4_{inv} = 11$.

3 **In $(Q, \cdot)$, find the inverse of 3.**
The identity in $(Q, \cdot)$ is 1. Because $3 \cdot \frac{1}{3} = 1$, $3_{inv} = \frac{1}{3}$ in $(Q, \cdot)$. Furthermore, $\left(\frac{1}{3}\right)_{inv} = 3$.

4 **For each of the systems for which a table is shown below, find the inverse for each element.**

a. table for $(Z_4, \cdot)$

$\cdot$	0	1	2	3
0	0	0	0	0
1	0	1	2	3
2	0	2	0	2
3	0	3	2	1

b. table for $(H, \lozenge)$; $H = \{a, b, c, d\}$

$\lozenge$	a	b	c	d
a	b	d	a	c
b	d	c	b	a
c	a	b	c	d
d	c	a	d	b

a. The identity is 1. Find 0_{inv}. Check the table to see what element must be multiplied by 0 to give 1. There is none. Thus, 0 has no inverse. The inverse of 1 is 1. The element 2 has no inverse. The inverse of 3 is 3.

b. The identity is element c. Find a_{inv}. The table shows that $a \lozenge d = c$. So, $a_{inv} = d$. Further analysis shows that:
$$b_{inv} = b$$
$$c_{inv} = c$$
$$d_{inv} = a$$

Generally, if the operation under consideration is addition, a_{inv} is written $-a$. For instance, in $(Z_{15}, +)$, 4_{inv} is written -4 and equals 11. In Z_5, $-3 = 2$.

If the operation is multiplication, a_{inv} is written $\frac{1}{a}$ or a^{-1}.

Look at the table for $(Z_5, \cdot)$ on page 61. In Z_5, $\frac{1}{2} = 2^{-1} = 3$ and $\frac{1}{4} = 4^{-1} = 4$.

Exercises

Max: 1-10, 1-44, 1-2; Avg: 1-10, 1-43 odds; Min: 1-9 odds, 1-37 odds

Exploratory Find the additive inverse in the real numbers for each of the following.

1. 7 -7
2. -28 28
3. $\frac{3}{4}$ $-\frac{3}{4}$
4. $\sqrt{7}$ $-\sqrt{7}$
5. $2 - \sqrt{11}$ $-2 + \sqrt{11}$

Find the multiplicative inverse in the real numbers for each of the following.

6. 5 $\frac{1}{5}$
7. $-\frac{5}{2}$ $-\frac{2}{5}$
8. 0.05 20
9. -101 $-\frac{1}{101}$
10. $\sqrt{2}$ $\frac{\sqrt{2}}{2}$

Written In the system Z_5, name each of the following.

1. the additive identity 0
2. the multiplicative identity 1
3. the additive inverse of 4 1
4. the multiplicative inverse of 2 3
5. -2 3
6. $\frac{1}{3}$ 2
7. -0 0
8. $\frac{1}{4}$ 4
9. $-(-4)$ 4

In Z_7, find the simplest name for each of the following.

10. -3 4
11. $\frac{1}{6}$ 6
12. $\frac{1}{4}$ 2
13. -2 5
14. $-(5 + 6)$ 3

Find each in Z_8 or write *does not exist*.

15. -5 3
16. $\frac{1}{2}$ does not exist
17. $\frac{1}{5}$ 5
18. $\frac{1}{6}$ does not exist
19. -7 1

Find each in Z_9 or write *does not exist*.

20. $\frac{1}{2}$ 5
21. $\frac{1}{3}$ does not exist
22. -3 6
23. $\frac{1}{4}$ 7
24. $-(6 + 1)$ 2

In the chart shown, find each of the following.

25. the ☼-identity ◠
26. ♀ inv ♀
27. ◠ inv ◠
28. ☺ inv ~
29. ~ inv ☺
30. (☺ ☼ ☺) inv ♀
31. ♀ inv ☼ ~ inv ~

For Exercises 32–43, let $⩓$, #, !, $, %, *max*, and ? be operations defined as follows: For every a and b in the real numbers,

$$a ⩓ b = \frac{a + b}{2}$$
$$a \# b = a + b + 3$$
$$a \,!\, b = a + b - ab$$
$$a \% b = (a - b)^2 - (a + b)^2$$
$$a \text{ max } b = a \text{ if } a \geq b, \text{ or } b \text{ if } a < b$$
$$a \,?\, b = a \text{ if } |a| \geq |b|, \text{ or } b \text{ if } |a| < |b|$$

Find the identity element, if it exists, for each operation.

32. $⩓$ does not exist
33. #   -3
34. !   0
35. % $-\frac{1}{4}$
36. *max*   does not exist
37. ?   0

Find 2_{inv}, if it exists, for each operation.

38. $⩓$ does not exist
39. #   -8
40. !   2
41. % $\frac{1}{32}$
42. *max*   does not exist
43. ?   does not exist

44. Let S be a set and $*$ be an operation defined on S. If there is *no* identity element for $*$ in S, can be an element in S have an inverse? no

Challenge

1. Can you devise a system in which the identity for the system is also the inverse for every element? See students' work.

2. If you have found a system described in Exercise 1, is it closed with respect to its operation? associative? commutative?

Answers may vary.

Mixed Review

In each of the following, use the statement-reason form of proof to show that the indicated conclusion follows from the given premises.

1. Given: $\sim A \to C$ Conclusion: A
           $C \to B$
           $\sim B$

2. Given: $\sim(T \land \sim Q)$ Conclusion: Q
           $T \land N$

For answers to exercises 1-2, see the Teacher's Guide.

Find each of the following using the definitions given in Exercises 32–43 above.

3. 1 ! 4   1
4. -3 ! 2   5
5. -1 ! -5 -11
6. 1 % 4   -16
7. -3 % 2   24
8. -1 % -5 -20
9. 1 max 4   4
10. -3 max 2   2
11. -1 max -5 -1
12. 1 ? 4   1
13. -3 ? 2   -3
14. -1 ? -5 -5

15. Is ! commutative on $\mathcal{R}$? yes
16. Is % commutative on $\mathcal{R}$? yes
17. Is max commutative on $\mathcal{R}$? yes
18. Is ? commutative on $\mathcal{R}$? yes
19. Which of the operations given above are operations on W? max, ?

Solve each equation in the indicated clock system.

20. $x - 5 = 4$, clock 7   {2}
21. $5 - x = 4$, clock 7   {1}
22. $5 - x = x$, clock 7   {6}
23. $y^2 = y$, clock 9   {0, 1}
24. $3y - 1 = 4$, clock 5   {0}
25. $0 - y^2 = 2$, clock 8   ∅

2.6 The Group

In general, a *group* is considered a collection of objects that are alike in some way. A very important kind of mathematical system is also called a **group**.

Objective: To identify a group from a finite number of elements.

Definition of Group

A group is a set G together with an operation $*$ that satisfies these four requirements:
1. $(G, *)$ is an operational system.
2. The operation $*$ is associative on G.
3. There is an identity element for $*$ in G.
4. Every element in G has an inverse with respect to $*$ in G.

Example

1 **Show that clock 4 with addition—that is, $(Z_4, +)$—is a group.**

To be a group, $(Z_4, +)$ must satisfy the four requirements.

1. *Operational system.* The table for $(Z_4, +)$ shows this condition is satisfied. For every pair of elements in Z_4, their sum is in Z_4.
2. *Associativity.* To prove that an operation is associative is often time-consuming. In this case, for every a, b, and c in Z_4, $a + (b + c)$ must be shown to equal $(a + b) + c$. However, in Section 2.4 it was agreed that addition is associative in clock systems.
3. *Identity element.* The element 0 is the identity. For all elements a in Z_4, $a + 0 = 0 + a = a$. For instance, if $a = 2$, we have $2 + 0 = 0 + 2 = 2$.
4. *Inverses for every element.* This condition is satisfied, as shown below.
   The inverse of 0 is 0 because $0 + 0 = 0 + 0 = 0$.
   The inverse of 1 is 3 because $1 + 3 = 3 + 1 = 0$.
   The inverse of 2 is 2 because $2 + 2 = 2 + 2 = 0$.
   The inverse of 3 is 1 because $3 + 1 = 1 + 3 = 0$.

Thus, $(Z_4, +)$ is a group.

Examples

2 **Decide if clock 4 with multiplication—that is, $(Z_4, \cdot)$—is a group.**

Check the four requirements.

$\cdot$	0	1	2	3
0	0	0	0	0
1	0	1	2	3
2	0	2	0	2
3	0	3	2	1

1. *Operational system*. The table shows that $(Z_4, \cdot)$ is an operational system.
2. *Associativity*. Multiplication is associative in all the clock systems.
3. *Identity Element*. The identity is 1, as the table shows. Multiplication of any element a in Z_4 by 1 produces the element. For instance, if $a = 2$, then $2 \cdot 1 = 1 \cdot 2 = 2$.
4. *Inverse for every element*. What is the inverse of 0? There is no element in Z_4 that gives 1 when multiplied by 0. So, zero has no inverse. Therefore, $(Z_4, \cdot)$ is not a group because not all four conditions are satisfied. Is there another element that fails to have an inverse in $(Z_4, \cdot)$?   Yes, 2

3 **Decide if the system for which the table is shown is a group.**

1. The table shows that the system is an operational system.
2. Associativity is difficult to check, but analysis will show that the system is associative.
3. The identity in the system is ♡. This can be seen from the table.
4. The *inverse* of △ is ♣. The inverse of ♡ is itself. The inverse of ♣ is △.

The system is a group.

Groups do *not* have to be commutative. The groups studied in this section are, in fact, commutative. That makes them **commutative groups**.

Exercises

Max: 1-4, 1-25; Avg: 1-4, 1-25; Min: 1-4, 1-17 odds, 18-19, 21-25 odds

Exploratory Decide if each system is a group. If the system is not a group, explain why not.

1. $(Z_3, +)$ yes
2. $(Z_3, -)$ No, subtraction is not associative.
3. $(Z_3, \cdot)$ No, zero has no inverse.
4. $(Z_4, \div)$ No, zero has no inverse.

Written Decide if each system is a group. If the system is not a group, explain why not.

1. $(Z_5, \cdot)$ No, zero has no inverse.
2. $(Z_6, +)$ yes
3. $(Z_{18}, +)$ yes
4. $(Z_5, \div)$ No, zero has no inverse.
5. $(Z_{30}, \cdot)$ No, zero has no inverse.
6. $(Z_2, +)$ yes
7. $(Z_4, -)$ No, subtraction is not associative.
8. $(Z_4/\{0\}, +)$ No, there is no identity element.
9. set: $\{1, -1\}$; method of combination: addition — no, not an operational system
10. set: $\{1, -1\}$; method of combination: multiplication — yes
11. set: $\{1, 2, \frac{1}{2}\}$; method of combination: multiplication — no, not an operational system

Using the four group requirements, determine if each set and operation is a group. If it is, show how it satisfies the four requirements. If not, give at least one reason it is not.

12. $Z_5/\{0\}$; multiplication in clock 5 — yes
13. $Z_3/\{0\}$; multiplication in clock 3 — yes
14. $Z_4/\{0\}$; multiplication in clock 4 — No, 2 has no inverse.
15. $Z_6/\{0\}$; multiplication in clock 6 — No, 2, 3, and 4 have no inverse.
16. $Z_7/\{0\}$; multiplication in clock 7 — yes
17. $Z_8/\{0\}$; multiplication in clock 8 — No, 2, 4, and 6 have no inverse.

Answer the questions.

18. Analyze your work in exercises 12–17. Can you draw any conclusions? Be careful. Exercise 19 may change your mind. Answers may vary.

19. Is $Z_9/\{0\}$ with multiplication in clock 9 a group? Explain. No, 3 and 6 have no inverse. Only clock n systems in which n is a prime number are groups.

Decide if each table represents a group. If the system is a group, show how it satisfies the four requirements. Assume associativity after checking three cases successfully. If it is not a group, show why.

20.

▽	d	f	g
d	d	f	g
f	f	g	f
g	g	f	g

no, f and g have no inverse.

21.

✕	k	m	p
k	k	m	p
m	m	p	k
p	p	k	m

yes

22.

□	☺	◇	∩
☺	☺	◇	∩
◇	◇	∩	☺
∩	∩	☺	◇

yes

23.

♂♀	♀	♂	☺
♀	♂	☺	♀
♂	☺	♀	♂
☺	♀	♂	☺

yes

24.

△	α	β	γ	φ
α	α	β	γ	φ
β	β	α	φ	γ
γ	γ	φ	α	β
φ	φ	γ	β	α

yes

25.

Σ	ψ	θ	ν	κ
ψ	θ	κ	ψ	ν
θ	κ	ψ	θ	ψ
ν	ψ	θ	ν	κ
κ	ν	ψ	κ	ψ

No, θ has no inverse

2.7 Infinite Groups

Objective: To identify a group from an infinite number of elements.

All the groups dealt with in the last section have one item in common: the sets all contained a *finite number of elements*. In this section, the sets will involve *an infinite number of elements*.

Examples

1 **Show that the integers with addition, the system (Z, +), is a group.**
 1. The integers with addition is an *operational system*. When two integers are added, the sum is one and only one integer.
 2. Addition is *associative* on the integers.
 3. The additive *identity* is 0. For any integer n, $n + 0 = 0 + n = n$.
 4. Every element n in the integers has an additive *inverse,* named $-n$. For instance, if $n = 5$, then $-n = -5$. If $n = -75$, then $-n = -(-75)$ or 75. The inverse of 0 is 0.

 Thus, the integers with addition form a group.

2 **Decide if the whole numbers with addition, (W, +), is a group.**
 1. The whole numbers with addition is an *operational system*.
 2. Addition is *associative* on W.
 3. Zero is the additive *identity*.
 4. Except for zero, the elements in W *do not have additive inverses.*

 Therefore, (W, +) is *not* a group.

3 **Let E = {even integers}. Determine if (E, +) is a group.**
 1. Check to see if (E, +) is an *operational system*. Is the sum of two even integers an even integer? In the exercises, you will be asked to give an argument that the sum of any two even integers is an even integer. For now, assume the statement is true.
 2. Addition is *associative* on the set of all the integers, so it must be associative on any subset of them.
 3. Zero is the additive *identity*. But is zero an even integer? Yes it is, so set E has an identity element.
 4. Does every element in E have an inverse that is also in E? What is the inverse of 4? Of -50? Of 1,000,000? For now, accept that the additive inverse of an even integer is another even integer. You will be asked to give an argument in the exercises.

 Therefore, the even integers with addition is a group.

Example

4 Which of these sets, along with multiplication, forms a group?
   a. integers   b. rationals   c. rationals without zero

  a. The integers with multiplication is an operational system. Multiplication is associative on the integers. Furthermore, the multiplicative identity, 1, is an integer. However, most integers do not have integer multiplicative inverses. For example, the number by which 5 must be multiplied to obtain 1 is $\frac{1}{5}$, and $\frac{1}{5}$ is not an integer. Thus $(\mathcal{Z}, \cdot)$ is not a group.
  b. The rationals with multiplication almost meet the requirements for a group. There is one element, 0, that does not have a multiplicative inverse. There is no single number that when multiplied by 0, equals 1. Anything multiplied by 0 equals 0. So, the rationals with multiplication is not a group.
  c. In this system, the 0 has been removed. Now every element in the system does have its own inverse. Thus, the rationals except 0, with multiplication, is a group. This group can be symbolized $(Q/\{0\}, \cdot)$.

Exercises

Max: 1-5, 1-28, 1-16; Avg: 1-5, 1-15 odds, 16-28, 1-12; Min: 1-5, 1-15 odds, 16-21

Exploratory Consider the set Q, the rational numbers, together with the operation $+$. Answer each of the following.

1. Give an argument that $(Q, +)$ is an operational system.
   When two rational numbers are added, the sum is one and only one rational number.

2. Check the associativity of $+$ in Q by choosing three examples, and showing that the associative law holds, at least in those three cases. *Answers may vary.*

3. Is there an identity in $(Q, +)$? What is it? *yes, 0*

4. Does every element in $(Q, +)$ have an inverse? Give an argument. What is the additive inverse of 5? Of $\frac{1}{4}$? Of .0021?
   Yes, the inverse of element a is -a; -5, $-\frac{1}{4}$, -.0021

5. Give an argument to show that $(Q, +)$ is a group.
   $(Q, +)$ is a group because it is an operational system, addition is associative, there is an identity element, and every element has an inverse.

Written Decide if each system is a group. Justify your answer.

1. $(Q/\{0\}, +)$ no
2. $(Q, -)$ no
3. $(Q, \cdot)$ no
4. $(Q/\{0\}, \cdot)$ yes
5. $(Q, \div)$ no
6. $(\mathcal{R}, +)$ yes
7. $(\mathcal{R}, -)$ no
8. $(\mathcal{R}, \cdot)$ no
9. $(\mathcal{R}/\{1\}, +)$ no
10. $(\mathcal{R}/\{0\}, +)$ no
11. $(\mathcal{R}/\{0\}, \cdot)$ yes
12. $(\mathcal{R}/\{2, -2\}, +)$ no
13. (Even integers, $\cdot$) no
14. (Odd integers, $+$) no
15. (Odd integers, $\cdot$) no

Answer the following.

16. Define: even number. An even number has a factor of 2.

17. Inez says, "If I have two even numbers, then I can name one of them $2m$ and the other $2n$, where m and n are integers. These names guarantee the evenness of numbers." Is she right? yes

18. Use Inez's method to rename the following even numbers: 6, 10, 50, −50. $2 \cdot 3$, $2 \cdot 5$, $2 \cdot 25$, $2(-25)$

19. Use Inez's method and the Distributive Law to show that the sum of any two even numbers is even. You will have to show that this sum can be written in the form $2 \cdot$ (something that is an integer). $2m + 2n = 2(m + n)$

For answers to exercises 20-26, see the Teacher's Guide.

20. Use the answer to exercise 19 to show that addition is an operation on the set of even integers.

21. See example 3 in this section. Offer a more formal justification that $(E, +)$ is a group.

22. Define: multiple of 3. Is zero a multiple of 3?

23. Show that addition is an operation on the set $T = \{$multiples of 3$\}$.

24. Let $T = \{$multiples of 3$\}$. Show that $(T, +)$ is a group.

25. Show that the multiples of 4 with addition is a group.

26. Show that the multiples of 7 with addition is a group.

27. Let $H = \{$multiples of 4$\}$ and $J = \{$multiples of 7$\}$. What is $H \cap J$? Is $(H \cap J, +)$ a group? multiples of 28, yes

28. Let n be any positive integer. Generalize the results of exercises 24–26 for multiples of n. If $T = \{$multiples of $n\}$, $(T, +)$ is a group.

Challenge Recall that for all $x \neq 0$ and all $n \in \mathcal{N}$, $x^{-n} = \dfrac{1}{x^n}$. **Simplify the following.**

1. 3^{-2} $\frac{1}{9}$
2. 4^{-3} $\frac{1}{64}$
3. $3 \cdot 4^{-1}$ $\frac{3}{4}$
4. 15^0 1
5. $\left(\dfrac{1}{4}\right)^{-2}$ 16
6. $\left(\dfrac{2}{3}\right)^{-2}$ $\frac{9}{4}$
7. $\left(\dfrac{1}{2}\right)^{-1}$ 2
8. $(3 \cdot 4)^{-1}$ $\frac{1}{12}$

Write each product as a power of 2.

9. $2^9 \cdot 2^{-3}$ 2^6
10. $2^{12} \cdot 2^{-11}$ 2^1
11. $2^0 \cdot 2^{-3}$ 2^{-3}
12. $2^k \cdot 2^m$ 2^{k+m}

Answer the following.

13. Let P be the set of integer powers of 2. That is, $P = \{$all numbers of the form 2^n, where $n \in \mathcal{Z}\}$. Name five members of P greater than 1 and five less than 1. Answers may vary.

14. Let P be defined as in exercise 13. Show that P, with multiplication, is a group. Remember the rules for exponents. P is an operational system and multiplication is associative. An identity, 2^0, exists, and each element, 2^n, has an inverse, 2^{-n}. Therefore, P, with multiplication is a group.

15. If $(G, *)$ is a group, what do you think a subgroup of $(G, *)$ is? A group that is contained in $(G, *)$.

16. Look for any subgroup of any of these groups: $(\mathcal{Z}, +)$, (Even integers, $+$), (Multiples of 3, $+$). What other subgroups of other infinite groups can you find? Answers may vary.

2.8 Some Theorems About Groups

Objective: To use theorems to prove facts about groups.

An important reason for studying groups is that all groups have the same structure. Whether groups have an "ordinary" operation or a "made-up" one, or are finite or infinite, they all share the *group structure*. Thus, if something is proven to be true about groups in general, then this fact is true about every system that is shown to be a group.

Four theorems about groups are presented in this section. A **theorem** is a statement that can be proven. Statements that are accepted as true without proof are called **axioms**. Following are two axioms that will be needed in this section.

Axiom 2–1

> Every number, or element, is equal to itself. This is called the **Reflexive Property of Equality**.

Axiom 2–2

> If two elements are equal, then one may be substituted for the other in any expression at any time. This is called the **Substitution Property of Equality**.

The first theorem about groups we will prove is sometimes called **Right Operation**.

Theorem 2–1 (Right Operation)

> Let $(G, *)$ be a group. Then for all a, b, c in G, if $a = b$, then $a * c = b * c$.

Given: $(G, *)$ a group; $a, b, c \in G$; $a = b$
Prove: $a * c = b * c$
Proof:

STATEMENTS	REASONS
1. $a = b$	1. Given
2. $a * c = a * c$	2. Reflexive Property of Equality
3. $a * c = b * c$	3. Substitution Property of Equality (substituting b for a on right)

You have used Right Operation many times when solving equations. For example, suppose $x + 3 = 8$. To solve the equation, add the quantity -3 to both sides of the equation.

$$(x + 3) + (-3) = 8 + (-3)$$

Because groups need *not* be commutative, the statements that are true about $a * c$ and $b * c$ are *not* automatically true about $c * a$ and $c * b$. For this reason, Theorem 2–2, **Left Operation**, is needed. It is similar to Right Operation.

Theorem 2–2 (Left Operation)

> Let $(G, *)$ be a group. For all a, b, c in G, if $a = b$, then $c * a = c * b$.

Theorems 2–1 and 2–2 are true in group systems, but are also true in many systems that are not groups. Theorems 2–3 and 2–4 are a bit different.

Suppose Clara and Jack are asked to think of a number in Z_6. Then, they are told to add 2 to the numbers. Clara and Jack tell their results and they have the same result. If the two numbers are x for Jack and y for Clara, then $x + 2 = y + 2$. Jack and Clara must have chosen the same number. For example, if they both obtained 4 after the addition, they both started with 2.

Now suppose Jack and Clara each take another number from Z_6. This time they multiply their numbers by 2. Again they get the same result. In other words, $2x = 2y$. Did they start with the same number this time? Suppose Jack started with 2 and Clara started with 5. After multiplying their numbers by 2, they would both get 4.

What happened? Well, Z_6 with addition is a group, and Z_6 with multiplication is not. Some important conclusions can be drawn from the group structure that might not be possible elsewhere.

Theorem 2-3 is one of the *Cancellation Laws*. It justifies the use of "canceling" in a group. Remember that canceling is *not* always possible outside the group structure.

Theorem 2–3 (Right Cancellation)

> Let $(G, *)$ be a group. For all a, b, c in G, if $a * c = b * c$, then $a = b$.

Given: $(G, *)$ a group; $a, b, c \in G$; $a * c = b * c$
Prove: $a = b$
Proof:

STATEMENTS	REASONS
1. $a * c = b * c$	1. Given
2. c_{inv} exists.	2. In a group, every element has its own inverse.
3. $(a * c) * c_{inv} = (b * c) * c_{inv}$	3. Right operation
4. $a * (c * c_{inv}) = b * (c * c_{inv})$	4. Associativity of $*$ in the group
5. $a * e = b * e$	5. Definition of inverse
6. $a = b$	6. Definition of identity

Theorem 2–4   Let $(G, *)$ be a group. For all a, b, c in G, if $c * a = c * b$,
(Left Cancellation)   then $a = b$.

The proof is left for you.
    Following are points to remember about the Cancellation Laws.
1. The proofs of these laws make use of a variety of reasons to justify each step. Look over the proof of Theorem 2–3. Note the use of previously proven theorems and the use of the properties of groups.
2. The Cancellation Laws are the converses of Theorems 2–1 and 2–2. But, while Theorems 2–1 and 2–2 often hold in systems that are not groups, the Cancellation Laws often do not.
3. The laws are formal justifications of processes often used in mathematics. When you "cancel" on both sides of some equations, you are using either Theorem 2–3 or Theorem 2–4.

Example

1   In $(Q/\{0\}, \cdot)$, solve $3x = \dfrac{3}{8}$.

Rewrite the equation as $3 \cdot x = 3 \cdot \dfrac{1}{8}$. Then use Theorem 2–4, Left Cancellation, to conclude that $x = \dfrac{1}{8}$.

Exercises

Max: 1-10, 1-32;   Avg: 1-10, 1-6, 7-19 odds, 20-32;   Min: 1-10, 1-23 odds, 24-32

Exploratory   Answer the questions.

1. Why are both Theorem 2–1 (Right Operation) and Theorem 2–2 (Left Operation) needed? The group may not be commutative.
2. Does the Left Cancellation Law work on every choice of a, b, c in Q, with multiplication as the operation? no
3. Is $(Q, \cdot)$ a group? Why or why not? No, zero has no inverse.
4. How does an axiom differ from a theorem? An axiom is accepted as true, a theorem must be proven.
5. If $0 \cdot a = 0 \cdot b$, can you conclude that $a = b$? no
6. Is $(Q/\{0\}, \cdot)$ a group? Can the Cancellation Laws be used in $(Q/\{0\}, \cdot)$? yes, yes

7. Is there any way to tell if the Cancellation Laws hold in a given system by examining the table for the system? Yes, if an element of the system appears more than once in any row or column, the Cancellation Law will not hold.

Solve each equation in the rationals.

8. $x + 2 = 7$ {5}
9. $y - 5 = 1$ {6}
10. $2z = 10$ {5}

Written Solve each equation in the rationals. State when and if Right or Left Operation or Right or Left Cancellation are used.

1. $0.07 + t = -0.021$ {-0.091}
2. $\frac{1}{2} + m = 5\frac{1}{4}$ {$4\frac{3}{4}$}
3. $\frac{3}{5} + a = -\frac{1}{2}$ {$-1\frac{1}{10}$}
4. $10 \cdot c = -\frac{1}{4}$ {$\frac{1}{40}$}
5. $x \cdot 0.7 = 63$ {90}
6. $-1.2 = 8w$ {-0.15}

Suppose that $(G, *)$ is a commutative group with identity e and that a, b, c, d are in G. Justify each of the following statements by citing one or more of the properties of a group, or the theorems of this section.

7. $a * c = c * a$ (G, *) is a commutative group.
8. d_{inv} exists. In a group, every element has an inverse.
9. $e * c_{inv} = c_{inv}$ definition of identity
10. $(a * a_{inv}) * c = c$ definition of inverse, definition of identity
11. $(a * b) * d_{inv} = a * (b * d_{inv})$ associativity of * in a group
12. If $a = c$, then $b * a = b * c$. Left Operation
13. If $a = d_{inv}$, then $a * c_{inv} = d_{inv} * c_{inv}$. Right Operation
14. $(c * b) = (c * d)$ implies $b = d$. Left Cancellation
15. $(a = d) \rightarrow (a * a = d * a)$ Right Operation
16. If $c * d = b * d$, then $c = b$. Right Cancellation
17. $(b * c) * c_{inv} = b$ Associativity, definition of inverse, definition of identity
18. $(a * a_{inv}) * (d * d_{inv}) = e$ definition of inverse, definition of identity
19. If $c = (a * b)_{inv}$, then $c = (b * a)_{inv}$. (G, *) is a commutative group.

The Cancellation Laws do not hold in $(Z_6, \cdot)$ because, for example, $2 \cdot 1 = 2 \cdot 4$ and clearly $1 \neq 4$. Give three examples to show that the Cancellation Laws do not hold in each of these systems. For exercises 20-23, answers may vary

20. $(Z_9, \cdot)$
21. $(Z_{10}, \cdot)$
22. $(Z_{12}, \cdot)$
23. $(Z_{600}, \cdot)$

Supply the reason for each statement in the following proof.

Theorem 2–5 Let $(G, *)$ be a group. For all $a \in G$, $(a_{inv})_{inv} = a$.

24. **Given:** $(G, *)$ a group
 $a \in g$
 Prove: $(a_{inv})_{inv} = a$
 Proof:

STATEMENTS	REASONS
1. $a_{inv} * a = e$	1. definition of inverse
2. $a_{inv} * (a_{inv})_{inv} = e$	2. definition of inverse
3. $a_{inv} * (a_{inv})_{inv} = a_{inv} * a$	3. Substitution Property
4. $(a_{inv})_{inv} = a$	4. Left Cancellation

For exercises 25-30, answers may vary.
Give an example of the use of Theorem 2–5 in each system.

25. $(Z, +)$ **26.** $(Z_5, +)$ **27.** $(Z_{30}, +)$
28. $(Q/\{0\}, \cdot)$ **29.** $(Z_5/\{0\}, \cdot)$ **30.** $(\mathcal{R}, +)$

Supply the reasons for the following proof.

Theorem 2–6 Let a, b be elements of G and $(G, *)$ be a group. Then the equation $a * x = b$ has the unique solution $x = a_{inv} * b$.

31. Notice that this theorem states three points.
    i. The equation HAS a solution.
    ii. The name of the solution is $a_{inv} * b$.
    iii. $a_{inv} * b$ is its *only* solution.
    First prove facts **i** and **ii**, and then prove **iii**.

Proof of parts i and ii

Given: $(G, *)$ a group   **Prove:** $x = a_{inv} * b$ is the solution.
       $a, b \in G$; equation $a * x = b$

Proof:

STATEMENTS	REASONS
1. $a * (a_{inv} * b) = (a * a_{inv}) * b$	1. associativity
2. $a * (a_{inv} * b) = e * b$	2. definition of inverse
3. $a * (a_{inv} * b) = b$	3. definition of identity
4. $a_{inv} * b$ is a solution for $a * x = b$.	4. Substitution Property

Proof of part iii

Prove: $x = a_{inv} * b$ is the only solution.

STATEMENTS	REASONS
1. Suppose that $x = y$ is another solution.	1. Assumption
2. Then $a * y = b$.	2. Substitution Property
3. $a_{inv} * (a * y) = a_{inv} * b$	3. Left Operation
4. $(a_{inv} * a) * y = a_{inv} * b$	4. associativity
5. $e * y = a_{inv} * b$	5. definition of inverse
6. $y = a_{inv} * b$	6. definition of identity

This "other" solution is the same solution as $a_{inv} * b$.

Prove the following theorem. For the answer to exercise 32, see the Teacher's Guide.

32. Theorem 2–7 Let a, b be elements of G and $(G, *)$ be a group. Then the equation $x * a = b$ has the unique solution $x = b * a_{inv}$.

2.9 A Non-Commutative Group

Objective: To compute using non-commutative groups.

So far in this chapter, all the groups have been commutative. You may be wondering if any non-commutative groups exist.

∘	e	p	q	r	s	t
e	e	p	q	r	s	t
p	p	q	e	s	t	r
q	q	e	p	t	r	s
r	r	t	s	e	q	p
s	s	r	t	p	e	q
t	t	s	r	q	p	e

Consider the set $S = \{e, p, q, r, s, t\}$ and the operation "little circle." The table showing how ∘ is defined on set S is shown at the left. Is $(S, \circ)$ a group?

The table shows that ∘ is an operational system. Some checking will show that ∘ is associative on S. (To *prove* associativity would take a long time.) The element e appears to be the identity of the system.

For inverses, check the table to see that $e_{inv} = e$; $p_{inv} = q$; $q_{inv} = p$; $r_{inv} = r$; $s_{inv} = s$; and $t_{inv} = t$.

So, $(S, \circ)$ is a group. But is it commutative? Note, for example, that $s \circ p \neq p \circ s$. The group, therefore, is non-commutative.

This new, non-commutative group can be described in several ways. The following is one way of developing this group.

Journal

How is clock arithmetic different from the type of arithmetic with which you are familiar?

Consider a dance or athletic training procedure or a game involving three persons. They occupy three places, called Place 1, Place 2, and Place 3. At one point in the process, they are required to change to new places. Suppose, for example, that the person in the first place must go to Place 3, the person in second place must go to Place 1, and the person in Place 3 must go to Place 2. This change is shown below in Figure 1. It is shown in a more "mathematical" way in Figure 2.

$$\begin{pmatrix} 1 & 2 & 3 \\ 3 & 1 & 2 \end{pmatrix}$$

Figure 1   **Figure 2**

In another case, the instruction $\begin{pmatrix} 1 & 2 & 3 \\ 1 & 3 & 2 \end{pmatrix}$ means that the person in Place 1 stays there, and the persons in Places 2 and 3 switch places.

Now recall your work with arrangements, or *permutations*. If there are three symbols, or three persons, there are $3 \cdot 2 \cdot 1 = 6$ different ways of arranging them.

Here are the six ways. Each change is given a letter name.

$$e = \begin{pmatrix} 1 & 2 & 3 \\ 1 & 2 & 3 \end{pmatrix} \quad p = \begin{pmatrix} 1 & 2 & 3 \\ 2 & 3 & 1 \end{pmatrix} \quad q = \begin{pmatrix} 1 & 2 & 3 \\ 3 & 1 & 2 \end{pmatrix}$$

$$r = \begin{pmatrix} 1 & 2 & 3 \\ 1 & 3 & 2 \end{pmatrix} \quad s = \begin{pmatrix} 1 & 2 & 3 \\ 3 & 2 & 1 \end{pmatrix} \quad t = \begin{pmatrix} 1 & 2 & 3 \\ 2 & 1 & 3 \end{pmatrix}$$

Now suppose the three persons in the game are given two instructions, one after the other. They will move (or stay put) *twice*. These *two* changes of place will put the three persons in one of the six positions shown above. Remember, the result of any two moves can be obtained with just one move.

For example, $\begin{pmatrix} 1 & 2 & 3 \\ 2 & 3 & 1 \end{pmatrix}$ followed by $\begin{pmatrix} 1 & 2 & 3 \\ 2 & 1 & 3 \end{pmatrix}$ yields $\begin{pmatrix} 1 & 2 & 3 \\ 1 & 3 & 2 \end{pmatrix}$ as shown in the diagram at the left. Using the names of these instructions, this fact is abbreviated $p \circ t = r$. Note that the "little circle" in this case means "followed by."

Examples

1 Compute $\begin{pmatrix} 1 & 2 & 3 \\ 3 & 2 & 1 \end{pmatrix}$ followed by $\begin{pmatrix} 1 & 2 & 3 \\ 3 & 2 & 1 \end{pmatrix}$.

Person 1 moves first to Place 3 and then to Place 1. So, ultimately, 1 goes to 1. Person 2 stays in place, then stays in place again, so 2 goes to 2. Finally, 3 goes first to Place 1 and then to Place 3. So, $\begin{pmatrix} 1 & 2 & 3 \\ 3 & 2 & 1 \end{pmatrix}$ followed by $\begin{pmatrix} 1 & 2 & 3 \\ 3 & 2 & 1 \end{pmatrix} = \begin{pmatrix} 1 & 2 & 3 \\ 1 & 2 & 3 \end{pmatrix}$.

2 Compute $\begin{pmatrix} 1 & 2 & 3 \\ 1 & 3 & 2 \end{pmatrix} \circ \begin{pmatrix} 1 & 2 & 3 \\ 2 & 1 & 3 \end{pmatrix}$.

Use the method of example 1. The answer is $\begin{pmatrix} 1 & 2 & 3 \\ 2 & 3 & 1 \end{pmatrix}$.

Use the names for the six instructions given at the beginning of this page and see if you can produce the group that opened this section.

Exercises

Max: 1-6, 1-12, 1-5; Avg: 1-6, 1-12; Min: 1-6, 1-11 odds

Exploratory Give the symbol used in this section that corresponds to each of the following.

1. $\begin{pmatrix} 1 & 2 & 3 \\ 2 & 1 & 3 \end{pmatrix}$, t

2. $\begin{pmatrix} 1 & 2 & 3 \\ 2 & 3 & 1 \end{pmatrix}$, p

3. $\begin{pmatrix} 1 & 2 & 3 \\ 1 & 2 & 3 \end{pmatrix}$, e

Write each of the following using "little circle" notation.

4. $\begin{pmatrix} 1 & 2 & 3 \\ 3 & 2 & 1 \end{pmatrix} \circ \begin{pmatrix} 1 & 2 & 3 \\ 2 & 1 & 3 \end{pmatrix}$

5. $\begin{pmatrix} 1 & 2 & 3 \\ 1 & 3 & 2 \end{pmatrix} \circ \begin{pmatrix} 1 & 2 & 3 \\ 1 & 3 & 2 \end{pmatrix}$

6. $\begin{pmatrix} 1 & 2 & 3 \\ 1 & 2 & 3 \end{pmatrix} \circ \begin{pmatrix} 1 & 2 & 3 \\ 3 & 2 & 1 \end{pmatrix}$

Written Compute each of the following. Recall that ∘ means "followed by."

For answers to exercises 1-12, see the Teacher's Guide.

1. $\begin{pmatrix} 1 & 2 & 3 \\ 3 & 2 & 1 \end{pmatrix} \circ \begin{pmatrix} 1 & 2 & 3 \\ 2 & 1 & 3 \end{pmatrix}$

2. $\begin{pmatrix} 1 & 2 & 3 \\ 3 & 1 & 2 \end{pmatrix} \circ \begin{pmatrix} 1 & 2 & 3 \\ 3 & 2 & 1 \end{pmatrix}$

3. $\begin{pmatrix} 1 & 2 & 3 \\ 1 & 2 & 3 \end{pmatrix} \circ \begin{pmatrix} 1 & 2 & 3 \\ 1 & 2 & 3 \end{pmatrix}$

4. $\begin{pmatrix} 1 & 2 & 3 \\ 1 & 3 & 2 \end{pmatrix} \circ \begin{pmatrix} 1 & 2 & 3 \\ 2 & 3 & 1 \end{pmatrix}$

5. $\begin{pmatrix} 1 & 2 & 3 \\ 3 & 1 & 2 \end{pmatrix} \circ \begin{pmatrix} 1 & 2 & 3 \\ 2 & 3 & 1 \end{pmatrix}$

6. $\begin{pmatrix} 1 & 2 & 3 \\ 2 & 3 & 1 \end{pmatrix} \circ \begin{pmatrix} 1 & 2 & 3 \\ 3 & 1 & 2 \end{pmatrix}$

7. $\begin{pmatrix} 1 & 2 & 3 \\ 3 & 1 & 2 \end{pmatrix} \circ \begin{pmatrix} 1 & 2 & 3 \\ 2 & 3 & 1 \end{pmatrix}$

8. $\begin{pmatrix} 1 & 2 & 3 \\ 2 & 1 & 3 \end{pmatrix} \circ \begin{pmatrix} 1 & 2 & 3 \\ 2 & 1 & 3 \end{pmatrix}$

9. $\begin{pmatrix} 1 & 2 & 3 \\ 2 & 3 & 1 \end{pmatrix} \circ \begin{pmatrix} 1 & 2 & 3 \\ 2 & 3 & 1 \end{pmatrix}$

10. $\begin{pmatrix} 1 & 2 & 3 \\ 3 & 2 & 1 \end{pmatrix} \circ \begin{pmatrix} 1 & 2 & 3 \\ 3 & 1 & 2 \end{pmatrix}$

11. $\begin{pmatrix} 1 & 2 & 3 \\ 3 & 2 & 1 \end{pmatrix} \circ \begin{pmatrix} 1 & 2 & 3 \\ 2 & 3 & 1 \end{pmatrix}$

12. $\begin{pmatrix} 1 & 2 & 3 \\ 1 & 3 & 2 \end{pmatrix} \circ \begin{pmatrix} 1 & 2 & 3 \\ 3 & 1 & 2 \end{pmatrix}$

Challenge Answer the questions.

1. How would you interpret the symbol $\begin{pmatrix} 1 & 2 & 3 & 4 \\ 2 & 4 & 3 & 1 \end{pmatrix}$?

2. How many permutations are there of four symbols? 24

3. How many entries would be listed along the top of a table for the group of arrangements of four persons? 24
   How many entries would be listed along the side? 24
   How many entries would the table have? 576

Compute each of the following.

4. $\begin{pmatrix} 1 & 2 & 3 & 4 \\ 2 & 1 & 4 & 3 \end{pmatrix} \circ \begin{pmatrix} 1 & 2 & 3 & 4 \\ 1 & 2 & 3 & 4 \end{pmatrix} \begin{pmatrix} 1 & 2 & 3 & 4 \\ 2 & 1 & 4 & 3 \end{pmatrix}$

5. $\begin{pmatrix} 1 & 2 & 3 & 4 \\ 2 & 3 & 4 & 1 \end{pmatrix} \circ \begin{pmatrix} 1 & 2 & 3 & 4 \\ 4 & 3 & 2 & 1 \end{pmatrix} \begin{pmatrix} 1 & 2 & 3 & 4 \\ 3 & 2 & 1 & 4 \end{pmatrix}$

Problem Solving Application: Guess and Check

Objective: To solve problems using the guess and check problem-solving strategy.

Many real-world problems are solved by applying the following four-step problem-solving plan.

Make sure students understand the steps in this problem-solving plan.

Explore Read the problem carefully and explore the situation described. Identify what information is given and what you are asked to find.

Plan Plan how to solve the problem. Your plan may involve solving equations or inequalities, drawing a picture, or making a guess.

Solve Carry out your plan for solving the problem.

Examine Check your results. Be sure your solution satisfies the conditions of the given problem. Ask yourself, "Is the answer reasonable?"

Remind students to reread the original problem when examining their solution.

This problem-solving strategy is often called <u>guess and check</u>.

One method of solving problems is to make a guess about the solution. Then check your guess to see if it satisfies the conditions of the problem. If the first guess is incorrect, continue guessing and checking until you find the correct answer. Use the results of previous guesses to make better guesses.

Example

1 Dana wants to give 20 baseball cards to each of her card-collecting friends, but she has 10 too few cards. If she decides to give each friend only 15 baseball cards, then she will have 20 extra cards. How many baseball cards and card-collecting friends does Dana have?

Explore Dana has 10 baseball cards less than the number she needs if she gives each of her friends 20 cards.
Dana also has 20 baseball cards more than the number she needs if she gives each of her friends 15 cards.

Plan To solve this problem, guess the number of card-collecting friends that Dana has. Use this guess to determine her initial number of cards before trying to give each friend 20 cards *and* before trying to give each friend 15 cards. The guess will be correct when these two numbers are the same. Why?

Solve Try 4 friends.

Dana would have 10 less than 20(4) or 70 cards. **and** Dana would have 20 more than 15(4) or 80 cards.

This guess is incorrect since the initial number of cards is not the same.

Try 3 friends.

Dana would have 10 less than 20(3) or 50 cards. and Dana would have 20 more than 15(3) or 65 cards.

This guess is also incorrect.

Notice that this guess led to a larger difference between the values for the initial number of cards (15 instead of 10). Thus, our next guess must be greater than the first guess, 4.

Try 6 friends.

Dana would have 10 less than 20(6) or 110 cards. and Dana would have 20 more than 15(6) or 110 cards.

Since the numbers are the same, this guess is correct.

Dana has 110 baseball cards and 6 card-collecting friends.

Examine Since 110 baseball cards is 10 less than 6 times 20 cards and is 20 more than 6 times 15 cards, the solution is correct.

Exercises

Max: 1-10; Avg: 1-10; Min: 1-8

Written Solve each problem.

1. Terry wants to give $250 to each of his favorite charities, but is $187 short. If he gives each charity $180, he will have $163 left over. How much money does Terry have? $1063

2. Richard needs 146 bottles of soda for his party. Bottled soda comes in packages of 6 or 8. If Richard bought 20 packages of soda, how many of each size package did he buy? 7 packages of 6 and 13 packages of 8

3. Tricia has $5.90 worth of quarters and dimes. She has exactly 38 coins. How many quarters and dimes does she have? 24 dimes and 14 quarters

4. Using each of the digits 1, 2, 3, 4, 5, and 6 exactly once, find two integers whose product is as large as possible. 631,542

5. The square of a certain number is 13,924. Find the number. 118

6. The cube of a certain number is approximately 4000. Find the number. 16

7. Sherry bought fifteen plain and glazed doughnuts for $4.65. Each glazed doughnut costs 10¢ more than each plain doughnut. What is the cost of each kind of doughnut? How many of each kind did Sherry buy?

8. Write an eight-digit number using the digits 1, 2, 3, and 4 each twice so that the 1's are separated by 1 digit, the 2's are separated by 2 digits, the 3's are separated by 3 digits, and the 4's are separated by 4 digits. 41312432 or 23421314

9. Find the least prime number greater than 840. 853

10. Copy the figure at the right. Then fill in the digits 1, 2, 3, 4, 5, 6, 7, and 8 in such a way that no two consecutive numbers are in boxes that touch at a point or side.

	6	4	
8	1	9	2
	5	7	

7. 12 glazed at 33¢, 3 plain at 23¢;
6 glazed at 27¢, 9 plain at 37¢

Portfolio Suggestion

Select one of the assignments from this chapter that you found especially challenging and place it in your portfolio.

Performance Assessment

Create a mathematical system that uses five different characters. Explain what operations are valid in your system and whether the system represents a group. **See students' work.**

Chapter Summary

1. When adding in a **clock system**, use a diagram of a clock. Start at the first number and go clockwise. (43, 44)
2. Multiplication in clock arithmetic, as in ordinary arithmetic, is repeated addition. (48)
3. To multiply numbers in clock n or Z_n, (49)
    1. Find the "ordinary product" of the numbers.
    2. Divide this number by n.
    3. The remainder is the answer in clock n.
4. In clock arithmetic, as well as in ordinary arithmetic, $a - b = c$ means $a = b + c$. (52)
5. In clock arithmetic, as well as in ordinary arithmetic, $\frac{a}{b} = c$ means $a = bc$. (53)
6. In clock arithmetic, as well as in ordinary arithmetic, division by zero is meaningless. (53)
7. In clock arithmetic, division is not always possible. (54)
8. The symbol Z_n is often used for clock n. (54)
9. Let S be a set. Then $*$ is an **operation on the set** S if, for every a and b in S, there exists a single element c, also in S, such that $a * b = c$. (56)

10. Let S be a set. The operation $*$ is **commutative** on S if for every a and b in S, $a * b = b * a$. (58)
11. Let S be a set. The operation $*$ is **associative** on S if for every a, b, and c in S, $a * (b * c) = (a * b) * c$. (58)
12. Let S be a set and $*$ be an operation defined on S. Then the element e, in S, is the **identity element** for $*$ in S if for every a in S, $a * e = e * a = a$. (61)
13. Let S be a set with $*$ an operation defined on S. Let e be the identity element for $*$ in S. Then a_{inv} is the **inverse** for a with respect to $*$ if $a * a_{inv} = a_{inv} * a = e$. (61)
14. A **group** is a set G together with an operation $*$ that satisfies these four requirements. (65)
    1. $(G, *)$ is an operational system.
    2. The operation $*$ is associative on G.
    3. There is an identity element for $*$ in G.
    4. Every element in G has an inverse with respect to $*$ in G.
15. Axiom 2–1: Every number, or element, is equal to itself. This is called the **Reflexive Property of Equality.** (71)
16. Axiom 2–2: If two elements are equal, then one may be substituted for the other in any expression at any time. This is called the **Substitution Property of Equality.** (71)
17. Theorem 2–1 (**Right Operation**): Let $(G, *)$ be a group. Then for all a, b, c in G, if $a = b$, then $a * c = b * c$. (71)
18. Theorem 2–2 (**Left Operation**): Let $(G, *)$ be a group. For all a, b, c in G, if $a = b$, then $c * a = c * b$. (72)
19. Theorem 2–3 (**Right Cancellation**): Let $(G, *)$ be a group. For all a, b, c in G, if $a * c = b * c$, then $a = b$. (72)
20. Theorem 2–4 (**Left Cancellation**): Let $(G, *)$ be a group. For all a, b, c in G, if $c * a = c * b$, then $a = b$. (73)
21. Theorem 2–5: Let $(G, *)$ be a group. For all $a \in G$, $(a_{inv})_{inv} = a$. (74)
22. Theorem 2–6: Let a, b be elements of G and $(G, *)$ be a group. Then the equation $a * x = b$ has the unique solution $x = a_{inv} * b$. (75)
23. Theorem 2–7: Let a, b be elements of G and $(G, *)$ be a group. Then the equation $x * a = b$ has the unique solution $x = b * a_{inv}$. (75)
24. Some groups are non-commutative. (76)

Chapter Review

2.1 Add each in clock 6.

1. $3 + 0$ ³ 2. $5 + 5$ ⁴ 3. $4 + 2$ ⁰ 4. $5 + 3$ ²

Solve each in clock 6.

5. $t + 3 = 1$ {4} 6. $5 + m = 3$ {4} 7. $1 + p = 0$ {5} 8. $x + 4 = 2$ {4}

2.2 Multiply each in clock 8.

9. 3×2 ⁶ 10. 3×3 ¹ 11. 5×6 ⁶ 12. 4×6 ⁰

Solve each in clock 5.

13. $2t = 4$ {2} 14. $3d = 2$ {4} 15. $4y = 1$ {4} 16. $3w = 0$ {0}

2.3 Subtract each in clock 6.

17. $3 - 4$ ⁵ 18. $0 - 5$ ¹ 19. $2 - 5$ ³ 20. $1 - 4$ ³

Solve each in Z_5.

21. $3 \overline{} x = 1$ {2} 22. $v \overline{} 4 = 3$ {2} 23. $1 \overline{} c = 3$ {3} 24. $2 \overline{} p = 4$ {3}

2.4 For each of the following, decide if the process is an operation. If it is, state *commutative*, *associative*, *both*, or *neither*. Justify each answer.

25. $(Z_4, +)$ yes, both
26. $(Z_4, -)$ yes, neither
27. $(Z_4, \cdot)$ yes, both
28. $(Z_4, \div)$ no
29. $(Z_4/\{0\}, \div)$ no

2.5 Find each in Z_6 or write *does not exist*.

30. -4 2 31. $\frac{1}{4}$ does not exist 32. $\frac{1}{5}$ 5 33. -5 1 34. -2 4

For answers to exercises 35-42, see the Teacher's Guide.

2.6 Decide if each system is a group. If it is not, explain why.

35. $(Z_5, +)$ 36. $(Z_5/\{0\}, +)$ 37. $(Z_5, \div)$ 38. $(Z_5/\{0\}, \div)$

2.7 39. $(\mathcal{N}, +)$ 40. $(Q, \cdot)$ 41. $(\mathcal{Z}/\{0\}, \div)$ 42. (Even integers, $\cdot$)

2.8 Solve each in the rationals.

43. $1.34 + m = -4.4$ {-5.74}
44. $4\frac{1}{3} + g = \frac{1}{2}$ {-3$\frac{5}{6}$}
45. $\frac{3}{4}k = \frac{2}{3}$ {$\frac{8}{9}$}

Suppose that $(G, *)$ is a commutative group with identity e and that a, b, c, d are in G. Justify each statement by citing one or more of the properties of a group or the theorems in this chapter.

46. $a * (b * c) = (a * b) * c$. associativity
47. If $b = c$, then $a * b = a * c$. Left Operation
48. $(a * c) * e = a * c$. definition of identity
49. If $a * b = c * b$, then $a = c$. Right Cancellation

2.9 Compute each of the following. Recall that ∘ means "followed by."

50. $\begin{pmatrix} 1 & 2 & 3 \\ 3 & 2 & 1 \end{pmatrix} \circ \begin{pmatrix} 1 & 2 & 3 \\ 1 & 2 & 3 \end{pmatrix}$ $\begin{pmatrix} 1 & 2 & 3 \\ 3 & 2 & 1 \end{pmatrix}$

51. $\begin{pmatrix} 1 & 2 & 3 \\ 1 & 3 & 2 \end{pmatrix} \circ \begin{pmatrix} 1 & 2 & 3 \\ 1 & 3 & 2 \end{pmatrix}$ $\begin{pmatrix} 1 & 2 & 3 \\ 1 & 2 & 3 \end{pmatrix}$

52. $\begin{pmatrix} 1 & 2 & 3 \\ 3 & 1 & 2 \end{pmatrix} \circ \begin{pmatrix} 1 & 2 & 3 \\ 1 & 3 & 2 \end{pmatrix}$ $\begin{pmatrix} 1 & 2 & 3 \\ 2 & 1 & 3 \end{pmatrix}$

Chapter Test

Find each in the clock system indicated. If the expression is meaningless, explain why.
1. $6 + 6$, clock 8   4
2. $5 + 2$, clock 6   1
3. $1 + 7$, clock 8   0
4. $4 \cdot 3$, clock 6   0
5. $6 \cdot 0$, clock 8   0
6. $5 \cdot 7$, clock 10   5
7. $7 - 1$, clock 9   6
8. $6 - 7$, clock 10   9
9. $0 - 3$, clock 7   4
10. $3 \div 3$, clock 9   1, 4, 7
11. $\frac{2}{5}$, clock 7   6
12. $6 \div 7$, clock 9   6

Solve each in Z_8.
13. $y + 4 = 1$ {5}
14. $6 + c = 3$ {5}
15. $4q = 5$ ∅
16. $3z = 7$ {5}
17. $0 = 6a$ {0, 4}
18. $b - 7 = 2$ {1}
19. $7 - s = 1$ {6}
20. $1 - x = 3$ {6}

Find each in Z_6 or write *does not exist*.
21. -5 1
22. $\frac{1}{3}$ does not exist
23. $\frac{1}{2}$ does not exist
24. -3 3
25. -1 5

Decide if each system is a group. If it is not, explain why.
26. $(Z_7, +)$ yes
27. $(Z_2, \cdot)$ No, zero has no inverse
28. $(W, +)$ No, no elements have an inverse.
29. $(\mathcal{R}, \cdot)$ No, zero has no inverse.
30. set: $\{0, 1, -1\}$; method of combination: multiplication No, zero has no inverse.

Solve each in the rationals.
31. $x + 0.08 = 1.15$ {1.07}
32. $0.01a = 4.2$ {420}
33. $-\frac{1}{2}g = 6$ {-12}

Compute each of the following. Recall that ∘ means "followed by."
34. $\begin{pmatrix} 1 & 2 & 3 \\ 3 & 2 & 1 \end{pmatrix} \circ \begin{pmatrix} 1 & 2 & 3 \\ 1 & 3 & 2 \end{pmatrix}$ $\begin{pmatrix} 1 & 2 & 3 \\ 2 & 3 & 1 \end{pmatrix}$
35. $\begin{pmatrix} 1 & 2 & 3 \\ 1 & 3 & 2 \end{pmatrix} \circ \begin{pmatrix} 1 & 2 & 3 \\ 2 & 1 & 3 \end{pmatrix}$ $\begin{pmatrix} 1 & 2 & 3 \\ 2 & 3 & 1 \end{pmatrix}$

Use the chart at the right for Exercises 36–40.
36. Find the Ω-identity. A
37. Find T_{inv}. N
38. What is the value of $N\,\Omega\,(A\,\Omega\,T)$? A
39. Find x if $N\,\Omega\,x = I$. N
40. Find y if $y\,\Omega\,y = A$. I, A

Ω	N	I	T	A
N	I	T	A	N
I	T	A	N	I
T	A	N	I	T
A	N	I	T	A

CHAPTER 3

A System with Two Operations

Application in Biology

Have you ever watched the sun reflect the bright colors on a butterfly as it moves from flower to flower? If you were to draw a line down the middle of a photograph of a butterfly, the two halves would match. When you can do this, the object has **line symmetry**.

An equilateral triangle also displays symmetry. Study the actions shown.

Given:

I

120°

240°

Create a multiplication table with each of these actions along the top and the side of the table. Perform the top action first, and then the side operation to determine the single operation that is performed on the triangle for a cell.
See Teacher's Answer Key, p. T120.

Individual Project: *History*

The Commutative Law is critical in the study of groups and fields. Sir William Rowan Hamilton was the first to propose that it may not always apply. Emmy Noether, George Boole, and Augustus De Morgan also made contributions to this branch of mathematics. Choose one of these mathematicians and research his or her life and work. Present your findings in a poster display.

3.1 Introduction

Objective: To solve equations using real numbers, group properties, and group theorems.

From your study of a mathematical group you learned that a group is a set together with one operation satisfying four requirements. Many kinds of groups exist. Some of the familiar ones use addition and multiplication.

Groups Using Addition

Rationals $(Q, +)$
Clock 5 $(Z_5, +)$

Groups Using Multiplication

Rationals without zero $(Q/\{0\}, \cdot)$
Clock 5 without zero $(Z_5/\{0\}, \cdot)$

Examples

1 Solve the following equations using rational numbers. Use group properties and group theorems that have been proven.

a.
$$x + 29 = 15$$
$$x + 29 + (-29) = 15 + (-29)$$
$$x = -14$$

Add the additive inverse of 29 to both sides.

b.
$$14x = 3$$
$$\frac{1}{14} \cdot 14x = \frac{1}{14} \cdot 3$$
$$x = \frac{3}{14}$$

Multiply both sides by $\frac{1}{14}$, which is the multiplicative inverse of 14.

2 Solve $x + 2 = 1$ using $(Z_5, +)$. Use group properties and group theorems that have been proven.

$$x + 2 = 1$$
$$(x + 2) + 3 = 1 + 3$$
$$x + (2 + 3) = 1 + 3$$
$$x = 4$$

Add the additive inverse of 2 to both sides.
Use the associative property.
$(2 + 3) = 0$ in Z_5.

Exercises

Max: 1-24, 1-38;   Avg: 1-6, 7-23 odds, 5-16, 21-37 odds;   Min: 1-23 odds, 5-12, 21-31 odds

Exploratory   Determine which of the following systems are groups. Justify your answers.
1. $(\mathcal{W}, +)$ No, there are no inverses.
2. $(\mathcal{Z}, +)$ yes
3. $(Z_6, +)$ yes
4. $(Q/\{0\}, \cdot)$ yes
5. $(Z_7, \cdot)$ No, zero has no inverse.
6. $(\mathcal{R}, \div)$
6. no, not an operational system

Solve each equation in $(Q, +)$. State which properties of system $(Q, +)$ are used.
7. $x + 2 = 5$ 3
8. $5 + x = 7$ 2
9. $x - 100 = 27$ 127
10. $5 = 9 - x$ 4
11. $\frac{1}{2} + y = \frac{1}{4}$ $-\frac{1}{4}$
12. $\frac{2}{3} + w = \frac{1}{5}$ $-\frac{7}{15}$
13. $x + 0.001 = 1$ 0.999
14. $-10 = x + 1$ -11
15. $x - 3 = -30$ -27

Solve each equation in $(Q/\{0\}, \cdot)$. State which properties of $(Q/\{0\}, \cdot)$ are used.
16. $2x = 1$ $\frac{1}{2}$
17. $3x = 2$ $\frac{2}{3}$
18. $0.2x = 15$ 75
19. $\frac{1}{2}x = 80$ 160
20. $\frac{1}{4}x = 180$ 720
21. $\frac{1}{5}y = \frac{1}{3}$ $\frac{5}{3}$
22. $\frac{1}{30}y = 5$ 150
23. $\frac{1}{5}x = \frac{2}{7}$ $\frac{10}{7}$
24. $\frac{1}{2}x = 3$ 6

Written   Solve each equation in the system indicated.

In $(Z_5, +)$:
1. $x + 2 = 4$ 2
5. $4 - x = 1$ 3
9. $x + 2 = 1$ 4
13. $x + 3 = 4$ 1
17. $3 - x = 2$ 1

In $(Z_5/\{0\}, \cdot)$:
2. $2x = 1$ 3
6. $3x = 1$ 2
10. $4x = 3$ 2
14. $3x = 4$ 3
18. $2x = 2$ 1

In $(Z_{12}, +)$:
3. $x + 9 = 10$ 1
7. $7 + x = 4$ 9
11. $x + x = 0$ 0, 6
15. $8 + x = 5$ 9
19. $10 + x = 7$ 9

In $(Z_7/\{0\}, \cdot)$:
4. $3x = 2$ 3
8. $5x = 1$ 3
12. $3 = 6x$ 4
16. $4x = 3$ 6
20. $6x = 1$ 6

Solve each of the following equations in the domain of real numbers. Use ordinary addition and multiplication and the equation-solving techniques already learned.
21. $3x + 1 = 10$ 3
22. $x - 5 = 2x$ -5
23. $\frac{1}{2}x + 1 = 6$ 10
24. $x + 1 = 2x - 1$ 2
25. $2x + 1 = x + 2$ 1
26. $k + 1 = -13k$ $-\frac{1}{14}$
27. $2k - 3 = 5k - 1$ $-\frac{2}{3}$
28. $4y - 7 = y + 8$ 5
29. $4k + 7 = -10k$ $-\frac{1}{2}$
30. $x + 2 = 3x - 4$ 3
31. $3k - 6 = k + 4$ 5
32. $5y - 2 = y + 16$ $\frac{9}{2}$
33. $3y - \frac{1}{4} = y + \frac{3}{8}$ $\frac{5}{16}$
34. $\frac{1}{4}w - 1 = w - \frac{1}{2}$ $-\frac{2}{3}$
35. $\frac{1}{3}x - 1 = \frac{1}{6}x + 1$ 12
36. $6x + \frac{1}{5} = 5x + \frac{1}{2}$ $\frac{3}{10}$
37. $4y + 5 = y + 23$ 6
38. $\frac{1}{2}z + 10 = z + 4$ 12

3.2 The Distributive Law

Objective: To solve equations using the Distributive Law.

It is often necessary to use two different operations to solve an equation. In a mathematical system that contains any two operations there may be a rule that tells us how the operations are related to each other. This rule can be thought of as the "glue" that binds the two operations together in the system.

Consider the way the expression $2(x + 3)$ is simplified.

This expression calls on us first to add x and 3, and then to multiply that sum by 2. You know from past work that $2(x + 3)$ equals $2x + 6$. The **Distributive Law** is used to obtain this result.

The Distributive Law tells us how addition and multiplication are related in the systems with which we are familiar. The Distributive Law is the "glue" that binds these two operations together in a way we can use in equation solving.

Here is the general statement of the Distributive Law.

Distributive Law | For all a, b, c, in Q, $a(b + c) = ab + ac$.

Example

1 Use the Distributive Law to solve the equation $3(x + 5) = 21$ in the rational numbers.

$$3(x + 5) = 21$$
$$3x + 15 = 21$$
$$3x = 6$$
$$x = 2$$

The solution set for $3(x + 5) = 21$ is $\{2\}$.

The following are other everyday examples that use the Distributive Law.

$$3(a + b) = 3a + 3b \qquad \frac{1}{2}(-7 + y) = -\frac{7}{2} + \frac{1}{2}y$$

$$-5(2a + 3b + c^2) = -10a + (-15b) + (-5c^2)$$

Because $a \cdot (b + c) = (a \cdot b) + (a \cdot c)$, we say *multiplication distributes over addition*. If the following is true, we say that multiplication is *back* distributive over addition.

$$(b + c)a = ba + ca$$

Are the rational and real number systems back distributive?

Examples

2 **Does addition distribute over multiplication?**
Here is a test case. If addition did distribute over multiplication, then for all a, b, c in Q, $a + (bc) = (a + b)(a + c)$.
Let $a = 2, b = 3, c = 5$.

$$a + (bc) = (a + b)(a + c)$$
$$2 + (3 \cdot 5) \stackrel{?}{=} (2 + 3)(2 + 5)$$
$$2 + 15 \stackrel{?}{=} (5)(7)$$
$$17 \neq 35$$

No. Addition does not distribute over multiplication.

3 **Does addition distribute over itself?**
If it did, then for all a, b, c in Q, $a + (b + c) = (a + b) + (a + c)$.
Let $a = 2, b = 3, c = 5$.

$$a + (b + c) = (a + b) + (a + c)$$
$$2 + (3 + 5) \stackrel{?}{=} (2 + 3) + (2 + 5)$$
$$2 + 8 \stackrel{?}{=} 5 + 7$$
$$10 \neq 12$$

No. Addition does not distribute over addition.

4 **In the integers, does multiplication distribute over *max*?**
If it did, then for all integers a, b, c, there is $a \cdot (b \text{ max } c) = (a \cdot b) \text{ max } (a \cdot c)$.
Remember that *max* is defined as follows:
   $a \text{ max } b = a$ if $a \geq b$ or $a \text{ max } b = b$ if $a < b$.
Look at these examples.
a. Let $a = 5, b = 3, c = 2$. $5(3 \text{ max } 2) = (5 \cdot 3) \text{ max } (5 \cdot 2)$
$$5 \cdot 3 \stackrel{?}{=} 15 \text{ max } 10$$
Yes, it does work in this case. $15 = 15$

b. Let $a = 0$, $b = 5$, $c = 10$. $0 \cdot (5 \text{ max } 10) = (0 \cdot 5) \text{ max } (0 \cdot 10)$
$$0 \cdot 10 \stackrel{?}{=} 0 \text{ max } 0$$
$$0 = 0$$
Yes, it does work again.

c. Let $a = -5$, $b = 1$, $c = -2$. $-5(1 \text{ max } -2) = (-5 \cdot 1) \text{ max } (-5 \cdot -2)$
$$-5 \cdot 1 \stackrel{?}{=} -5 \text{ max } 10$$
$$-5 \neq 10$$
No. Multiplication does not distribute over *max*.
One case was found that did not work.

These investigations lead to the conclusion that when one operation distributes over another it is a special case. The fact that multiplication distributes over addition is an important and special mathematical property.

Does the Distributive Law hold in Z_5 with the operations of addition and muliplication? Look at these specific cases.

In Z_5: $4(3 + 2) = 4(3) + 4(2)$
$$4(0) \stackrel{?}{=} 2 + 3$$
$$0 = 0$$

In Z_5: $2(4 + 4) = 2(4) + 2(4)$
$$2(3) \stackrel{?}{=} 3 + 3$$
$$1 = 1$$

These examples do not constitute a proof. However, they do suggest that the Distributive Law does hold in Z_5 with the operations of clock 5 addition and clock 5 multiplication. Without checking all cases, it is assumed in this book that the Distributive Law holds in the clock systems we have studied.

Exercises

Max: 1-12, 1-18, 1-6; Avg: 1-12, 1-18; Min: 1-11 odds, 1-9, 11-17 odds

Exploratory Show that the Distributive Law holds true by substituting the values of a, b, c into $a(b + c) = ab + ac$ and simplifying. For answers to exercises 1-4, see the Teacher's Guide.

1. $a = 1$
   $b = 2$
   $c = 5$

2. $a = 1$
   $b = 5$
   $c = 4$

3. $a = 0$
   $b = 13$
   $c = 7$

4. $a = \dfrac{1}{4}$
   $b = 7$
   $c = 5$

Use the Distributive Law to simplify these expressions.

5. $3(z^2 + b) \quad 3z^2 + 3b$
6. $\frac{1}{2}\left(y^3 + \frac{1}{2}\right) \quad \frac{1}{2}y^3 + \frac{1}{4}$
7. $-5(a + b) \quad -5a - 5b$
8. $4(x + 2y) \quad 4x + 8y$
9. $-5(a^2 + 2b) \quad -5a^2 - 10b$
10. $x(x + y) \quad x^2 + xy$
11. $7x(x + 1) \quad 7x^2 + 7x$
12. $6\left(2x^3 + \frac{1}{2}y\right) \quad 12x^3 + 3y$

Written For the given a, b, c in Q, test whether the first operation distributes over the second operation.

1. multiplication over multiplication
   a. $a = 2, b = 3, c = 4$ no
   b. $a = 4, b = 3, c = 0$ yes
   c. $a = -2, b = 6, c = -4$ no

2. multiplication over subtraction
   a. $a = -3, b = 2, c = 1$ yes
   b. $a = 0, b = 2, c = 1$ yes
   c. $a = 2, b = 3, c = 4$ yes

For the given a, b, c in Q, test the Distributive Law when written as $(b + c)a = ba + ca$.

3. $a = 2$ $(3 + 4)2 = 3 \cdot 2 + 4 \cdot 2$
   $b = 3$ $7 \cdot 2 = 6 + 8$
   $c = 4$ $14 = 14$

4. $a = 0$ $(2 + 3)0 = 2 \cdot 0 + 3 \cdot 0$
   $b = 2$ $5 \cdot 0 = 0 + 0$
   $c = 3$ $0 = 0$

5. $a = -2$
   $b = 3$
   $c = 4$
   $(3 + 4)(-2) = 3(-2) + 4(-2) \quad 7(-2) = -6 + -8 \quad -14 = -14$

Use the Distributive Law and other principles as needed to solve each of the equations in the real numbers.

6. $3(x + 1) = 30 \quad x = 9$
7. $4\left(x + \frac{1}{2}\right) = 10 \quad x = 2$
8. $3x - 4(x + 2) = 1 \quad x = -9$
9. $12(x + 2) + 7 = 15 \quad x = -\frac{4}{3}$

Let the following symbols represent the operations indicated. Then perform the given computations.

$\mathbb{A}$ represents averaging; that is, $a \mathbb{A} b = \frac{a + b}{2}$

$a \# b = a + b + 3$
$a @ b = a^2 + b^2$

10. $3 \mathbb{A} 5$ 4
11. $\frac{1}{2} \mathbb{A} \frac{3}{4}$ $\frac{5}{8}$
12. $0 \# 2$ 5
13. $2 @ 5$ 29
14. $-5 \mathbb{A} 5$ 0
15. $-5 @ 5$ 50
16. $-16 \# 1$ -12
17. $-2 \# -17$ -16
18. $13 \# -2$ 14
19. $0 @ -2$ 4
20. $1 @ 3$ 10
21. $7 \mathbb{A} 0$ 3.5

Challenge Using the three operations, $\mathbb{A}$, $\#$, and $@$, as defined above, state whether multiplication distributes over each operation. Justify your answer.

1. $\mathbb{A}$ yes
2. $\#$ no
3. $@$ no

Using the three operations, $\mathbb{A}$, $\#$, and $@$, as defined above, state whether max distributes over each operation. Justify your answer.

4. $\mathbb{A}$ no
5. $\#$ no
6. $@$ no

3.3 A Field

Objective: To determine whether a system is a field and to solve equations in the rational and real number systems.

A mathematical system that uses two operations is called a two-fold operational system. Notation such as shown below is used to represent such systems.

Two-Fold Operational Systems	
Examples of System	Notation
The rationals with addition and multiplication	$(Q, +, \cdot)$
Clock 8 numbers with clock 8 addition and clock 8 multiplication	$(Z_8, +, \cdot)$
Real numbers with addition and max	$(\mathcal{R}, +, max)$

Suppose we build the specific two-fold operational system denoted by

$$(F, +, \cdot)$$

Any two-fold operational system consists of a set and two operations. This one has the following.

Set: F Operations: $+$ and $\cdot$

Suppose *group* properties of the set and each operation are maintained. This means

$$(F, +) \text{ is a group}$$

and $(F, \cdot)$ should be a group. The problem of multiplication systems that would have been groups if it were not for zero now arises. Remember that $(Q, \cdot)$ and $(\mathcal{R}, \cdot)$ failed to be groups because zero had no inverse. We avoid this problem by using $F/\{0\}$. Note

$$(F/\{0\}, \cdot) \text{ is a group.}$$

In this case the two groups, $(F, +)$ and $(F/\{0\}, \cdot)$, are both commutative groups.

Now, a way should exist to connect the two operations. Using the idea of "glue," it is possible to connect the two operations with the Distributive Law. For all a, b, and c in F

$$a \cdot (b + c) = (a \cdot b) + (a \cdot c)$$

The two-fold operational system that has just been described is called a **field**. The letter F was chosen to represent the word "field."

Definition of a Field

A **field** is a two-fold operational system, $(F, +, \cdot)$, that satisfies these requirements.
1. $(F, +)$ is a commutative group.
2. $(F/\{0\}, \cdot)$ is a commutative group.
3. $\cdot$ distributes over $+$.

Examples

1 Determine if the system $(Z, +)$ is a field.

$(Z, +)$ is not a field because it is not a two-fold operational system. Although $(Z, +)$ is a group, it cannot be a field.

2 Determine if the system $(Z, +, \cdot)$ is a field.

$(Z, +)$ is a commutative group. This has been determined previously.

$(Z/\{0\}, \cdot)$ is *not* a commutative group. There are elements in $Z/\{0\}$ that do not have multiplicative inverses.

No need exists to check the third requirement because we have shown that $(Z, +, \cdot)$ is not a field in the second requirement.

3 Determine if the system $(Q, +, \cdot)$ is a field.

It has been established that $(Q, +)$ is a commutative group.
$(Q/\{0\}, \cdot)$ is also a commutative group.
In the rationals, multiplication distributes over addition.
The three requirements are met. $(Q, +, \cdot)$ is a field.

4 Determine if the real numbers with addition and multiplication, $(\mathcal{R}, +, \cdot)$, is a field.

Yes, it is. Think of the previous work done with real numbers.
The real numbers with addition and multiplication form a field.

Because a field requires two groups, it is now possible "officially" to use two operations and all the group theorems that were proven previously to solve equations in the rational and real number systems.

Examples

5 Solve $-3x - 1 = x + 37$ in $(Q, +, \cdot)$.

$$-3x - 1 = x + 37$$
$$-3x = x + 38 \quad \text{Add 1 to both sides.}$$
$$-4x = 38 \quad \text{Add } -x \text{ to both sides.}$$
$$x = -\frac{38}{4} \quad \text{Multiply both sides by } -\frac{1}{4}.$$

This result can be reduced and placed in proper solution set form.

6 Solve $\sqrt{12} + x = 5\sqrt{3} - 2x$ in $(\mathcal{R}, +, \cdot)$.

$$\sqrt{12} + x = 5\sqrt{3} - 2x$$
$$2\sqrt{3} + x = 5\sqrt{3} - 2x \quad \text{Simplify the radical expression } \sqrt{12}.$$
$$x = 3\sqrt{3} - 2x \quad \text{Add } -2\sqrt{3} \text{ to both sides.}$$
$$3x = 3\sqrt{3} \quad \text{Add } 2x \text{ to both sides.}$$
$$x = \sqrt{3} \quad \text{Multiply both sides by } \frac{1}{3}, \text{ or}$$

use Left Cancellation (of the 3s) in $(Q/\{0\}, \cdot)$.

Exercises

Max: 1-8, 1-30, 1-5; Avg: 1-8, 1-30; Min: 1-8, 1-29 odds

Exploratory Determine whether each of the following is a field. Write *yes* or *no*. Then discuss the reason for your answer.

1. $(Q, +)$ no, only one operation
2. $(Z, \cdot)$ no, only one operation
3. $(Z, +, \cdot)$ no, no multiplicative inverse
4. $(\mathcal{R}/\{5\}, +, \cdot)$ no, no additive inverse for -5
5. $(Q, +, \cdot)$ yes
6. $(\mathcal{R}, \cdot)$ no, only one operation
7. $(\mathcal{R}, +, \cdot)$ yes
8. $(Q/\{0\}, \cdot)$ no, only one operation

Written Solve each equation in $(Q, +, \cdot)$.

1. $x + 1 = 3x$ $\{\frac{1}{2}\}$
2. $2x - 1 = \frac{1}{4}$ $\{\frac{5}{8}\}$
3. $x + 3 = 2x + 15$ $\{-12\}$
4. $B + 2 = -3B$ $\{-\frac{1}{2}\}$
5. $\alpha + 2 = 2\alpha - 1$ $\{3\}$
6. $8\left(x + \frac{1}{4}\right) = -4$ $\{-\frac{3}{4}\}$
7. $3\left(y + \frac{5}{6}\right) = 5y - 2$ $\{\frac{9}{4}\}$
8. $\frac{1}{3}(10 + 7N) = \frac{2}{5}(2N + 3)$ $\{-\frac{32}{23}\}$
9. $3.4(6q + 1.1) = 2.8q$ $\{-0.2125\}$
10. $x^2 = 16$ $\{-4, 4\}$
11. $x^2 = \frac{1}{4}$ $\{-\frac{1}{2}, \frac{1}{2}\}$
12. $x^2 + 33 = 58$ $\{-5, 5\}$
13. $x^2 = 13$ $\emptyset$
14. $x^2 = -9$ $\emptyset$
15. $x^2 = 2$ $\emptyset$

Solve each equation in $(\mathcal{R}, +, \cdot)$.

16. $3x - \pi = \dfrac{11}{2}$ $\{\dfrac{11}{6} + \dfrac{\pi}{3}\}$
17. $2x = 5x$ $\{0\}$
18. $\dfrac{1}{2}(y + 3) = \dfrac{7}{3}$ $\{\dfrac{5}{3}\}$
19. $11(4 - 3x) + 1 = 12x$ $\{1\}$
20. $2(x - 1.4) = 1.2(3x + 1)$ $\{-2.5\}$
21. $3B + \sqrt{5} = 64\sqrt{5}$ $\{21\sqrt{5}\}$
22. $11z - \sqrt{18} = \sqrt{3} - z$
23. $\sqrt{2}(\sqrt{8}x + 3) = 7x - 1$ $\{\sqrt{2}\}$
24. $\sqrt{3}(x + 1) = x(\sqrt{3} + 3)$ $\{\dfrac{\sqrt{3}}{3}\}$
25. $x^2 = 13$ $\{-\sqrt{13}, \sqrt{13}\}$
26. $x^2 = \dfrac{\sqrt{3}}{3}$
   $x^2 = -9\emptyset$
27. $x^2 = 2$ $\{-\sqrt{2}, \sqrt{2}\}$
28. $x^2 = 2.89$ $\{-1.7, 1.7\}$
29. $x^2 = \dfrac{4}{9}$ $\{-\dfrac{2}{3}, \dfrac{2}{3}\}$
30. $x^2 = \dfrac{27}{32}$ $\{-\dfrac{3\sqrt{6}}{8}, \dfrac{3\sqrt{6}}{8}\}$

Challenge Determine whether each of the following is a field.

1. $(J, +, \cdot)$ where $J = \{$odd integers$\}$ no
2. $(Q, \div, -)$ no
3. $(T, +, \cdot)$ where $T = \{$all real numbers between -1 and 1, inclusive$\}$ no
4. $(D, +, \cdot)$ where $D = \{$all numbers of the form $a + b\sqrt{2}$ where a and b are rational numbers$\}$ yes
5. $(C, +, \cdot)$ where $C = \{$all numbers of the form $a + bi$ where a and b are real numbers and $i^2 = -1\}$ Assume that $bi + di = (b + d)i$, $c(bi) = (cb)i$, and $(bi)(di) = (bd)i^2$.
   yes

Mixed Review

Solve each equation in $(\mathcal{R}, +, \cdot)$.

1. $2x + 7 = -5x$ $\{-1\}$
2. $5x - \sqrt{12} = 4x + \sqrt{27}$ $\{5\sqrt{3}\}$
3. $3(x + 7) = 11 - 3x$ $\{-\dfrac{5}{3}\}$
4. $10\left(2x + \dfrac{3}{4}\right) = 7x + \dfrac{9}{2}$ $\{-\dfrac{3}{13}\}$
5. $\dfrac{2}{3}(x + 5) = \dfrac{4}{3}(4 + 5x)$ $\{-\dfrac{1}{3}\}$
6. $\dfrac{3}{8}x + \dfrac{11}{12} = \dfrac{5}{6}x - \dfrac{13}{18}$ $\{\dfrac{118}{33}\}$
7. $3(6x + \sqrt{24}) + 2\sqrt{3}x = 2(\sqrt{3}x + \sqrt{6})$ $\{-\dfrac{2\sqrt{6}}{9}\}$
8. $3.6(2x + 21) = 8(-2.7 - 0.3x)$ $\{-10.125\}$
9. Write the negation of "$x = 5$ or $y = 7$." $x \neq 5$ and $y \neq 7$
10. Write the negation of "n is prime, or n is not odd and n is divisible by 5." n is composite, and n is odd or n is not divisible by 5.
11. Write the contrapositive of "If a is not divisible by 3, then a is even or $b \neq 0$."
    If a is odd and $b = 0$, then a is divisible by 3.

In the table shown, find each of the following.

12. the Ω-identity D
13. A_{inv} A
14. C_{inv} C
15. E_{inv} F
16. D_{inv} D
17. $A \Omega B_{inv}$ F
18. $(A \Omega B)_{inv}$ E
19. $A_{inv} \Omega B_{inv}$ F
20. $(A \Omega C) \Omega F$ D
21. $A \Omega (C \Omega F)$ D
22. $(E \Omega E) \Omega B$ A
23. $(A \Omega C) \Omega F$ D

Ω	A	B	C	D	E	F
A	D	F	E	A	C	B
B	E	D	F	B	A	C
C	F	E	D	C	B	A
D	A	B	C	D	E	F
E	B	C	A	E	F	D
F	C	A	B	F	D	E

24. Is Ω commutative? no
25. Does this system represent a group? yes

3.4 Finite Fields

Objective: To determine whether a system is a finite field and to use the properties of a field to solve equations.

A set is said to be finite if, given enough time, you can count every element. For example, the set of classrooms in a school is finite. The set of integers is not.

You have studied two important examples of a field—the field of rationals and the field of reals, both with addition and multiplication. In the following examples, finite fields are investigated. A finite field consists of a countable set of numbers and two operations that meet the field requirements.

Examples

1 Decide if $(Z_4, +, \cdot)$ is a field.

a. *Requirement 1* Is $(Z_4, +)$ a commutative group? Yes, it is. It was shown to be the case when groups were studied.

b. *Requirement 2* Is $(Z_4/\{0\}, \cdot)$ a commutative group? Here is a table for $(Z_4/\{0\}, \cdot)$. In order for this to be a group, every element must have an inverse. Find the identity. In this multiplicative system, the identity is 1. The element 2 does not have an inverse. This means $(Z_4/\{0\}, \cdot)$ is not a group.

$\cdot$	1	2	3
1	1	2	3
2	2	0	2
3	3	2	1

c. *Requirement 3* Distributive Law. There is no need to check this one. $(Z_4, +, \cdot)$ cannot be a field because it already failed in Requirement 2.

2 Decide if $(Z_6, +, \cdot)$ is a field.

A table for $(Z_6/\{0\}, \cdot)$ is at the right. It is obvious that there are elements of $(Z_6/\{0\}, \cdot)$ that do not have multiplicative inverses. Therefore, $(Z_6, +, \cdot)$ does not satisfy Requirement 2.

$\cdot$	1	2	3	4	5
1	1	2	3	4	5
2	2	4	0	2	4
3	3	0	3	0	3
4	4	2	0	4	2
5	5	4	3	2	1

Example

3 Decide if $(Z_5, +, \cdot)$ is a field.

Test the three requirements.

a. *Requirement 1* It previously has been determined that $(Z_5, +)$ is a commutative group.

b. *Requirement 2* In the table for $(Z_5/\{0\}, \cdot)$ at the right, the identity is 1. Every element in $(Z_5/\{0\}, \cdot)$ does have an inverse. Therefore, $(Z_5/\{0\}, \cdot)$ is a commutative group.

$\cdot$	1	2	3	4
1	1	2	3	4
2	2	4	1	3
3	3	1	4	2
4	4	3	2	1

c. *Requirement 3* The Distributive Law, which states that multiplication distributes over addition, has been determined to hold true in clock systems.

Therefore, we conclude that $(Z_5, +, \cdot)$ is a field.

Now that you know Z_5 with addition and multiplication is a field, you can use its properties to solve equations in Z_5.

Before attempting the solutions, some notation must be understood.

Examples

4 Find the equivalent of -3 in Z_5.

The symbol -3 names the additive inverse of 3 in any additive system. In Z_5, the additive inverse of 3 is 2. Thus $-3 = 2$.

5 Find the equivalent of 4^{-1} in Z_5.

The symbol "4^{-1}" names the multiplicative inverse of 4. In Z_5, the multiplicative inverse of 4 is 4 because $4 \cdot 4 = 1$ in Z_5. Thus $4^{-1} = 4$.

6 Evaluate the following in $(Z_5, +, \cdot)$.

a. $-(-3)$ This is "the additive inverse of the additive inverse of 3", which is 3 in any system. Recall the theorem from groups: $(a_{inv})_{inv} = a$. This example is one application of that theorem.

b. $-\dfrac{1}{3}$ This calls for the additive inverse of $\dfrac{1}{3}$. But $\dfrac{1}{3}$ is the multiplicative inverse of 3. So, the multiplicative inverse of 3 in clock 5 must be found first. The multiplicative inverse of 3 in clock 5 is 2. Next, the additive inverse of 2 is 3. Therefore, we conclude that $-\dfrac{1}{3} = 3$.

c. $(-3)^{-1}$ This example calls for the multiplicative inverse of the additive inverse of 3. Start with the additive inverse of 3, which is 2 in Z_5. Then find the multiplicative inverse of 2, which is 3. So, $(-3)^{-1} = 3$.

d. $\dfrac{1}{\frac{1}{4}}$ Again two inverses are involved. This time it calls for the multiplicative inverse of the multiplicative inverse of 4. Begin with the latter. In Z_5, 4 is the multiplicative inverse of 4. Thus:

$$\dfrac{1}{\frac{1}{4}} = \dfrac{1}{4} = 4$$

The following examples use the properties of a field to solve the problems.

Examples

7 Solve $4x = 2$ in Z_5.

$\begin{aligned} 4x &= 2 \\ \tfrac{1}{4} \cdot 4x &= \tfrac{1}{4} \cdot 2 \quad &\text{Using left operation, multiply both sides by } \tfrac{1}{4}, \text{ which is the multiplicative inverse of 4.} \\ x &= 4 \cdot 2 \quad &\text{In } Z_5, \tfrac{1}{4} = 4. \\ x &= 3 \quad &\text{Multiplying in } Z_5, 4 \cdot 2 = 3. \end{aligned}$

8 Solve $3x + 1 = 2$ in Z_5.

$\begin{aligned} 3x + 1 &= 2 \\ (3x + 1) + (-1) &= 2 + (-1) \quad &\text{Add } (-1) \text{ to both sides.} \\ 3x &= 1 \quad &\text{In } Z_5, (-1) = 4. \text{ Add } 2 + 4 \text{ in } Z_5. \\ \tfrac{1}{3} \cdot 3x &= \tfrac{1}{3} \cdot 1 \quad &\text{Multiply both sides by } \tfrac{1}{3}. \\ x &= 2 \quad &\tfrac{1}{3} = 2 \text{ in } Z_5. \text{ Thus, } 2 \cdot 1 = 2. \end{aligned}$

Exercises

Max: 1-8, 1-18, 1;   Avg: 1-8, 1-18;   Min: 1-8, 1-17 odds

Exploratory   Give an example for each of the following.
1. finite field Answers may vary.
2. infinite field Answers may vary.

Determine whether each of the following is a field.
3. $(Z_5, \cdot)$ no
4. $(Z_3, +, \cdot)$ yes
5. $(Z_7, +, \cdot)$ yes
6. $(Z_{11}, +, \cdot)$ yes
7. $(Z_5/\{2\}, +, \cdot)$ no
8. $(Z_5/\{0\}, +, \cdot)$ no

Written   Give reasons for each step in each solution.

1. In Z_5: $3x = 2$ Multiply both sides by $\frac{1}{3}$.
$$3x = 2$$
$$\frac{1}{3}(3x) = \frac{1}{3}(2) \quad \text{In } Z_5, \frac{1}{3} = 2.$$
$$\text{Multiplying in } Z_5, 2 \cdot 2 = 4.$$
$$x = 2(2)$$
$$x = 4$$

2. In Z_5: $4x = 1$ Multiply both sides by $\frac{1}{4}$.
$$4x = 1$$
$$\frac{1}{4} \cdot 4x = \frac{1}{4}(1) \quad \text{In } Z_5, \frac{1}{4} = 4.$$
$$\text{Multiplying in } Z_5, 4 \cdot 1 = 4.$$
$$x = 4(1)$$
$$x = 4$$

3. In Z_{13}: $x + 8 = 3$ Add (-8) to both sides.
$$(x + 8) + (-8) = 3 + (-8) \quad \text{Associative rule}$$
$$x + [8 + (-8)] = 3 + (-8) \quad \text{Inverse}$$
$$x + 0 = 3 + 5$$
$$x = 8 \quad \text{Identity; In } Z_{13}, 3 + 5 = 8.$$

4. In Z_7: $y - 5 = 4$ Add 5 to both sides.
$$(y - 5) + 5 = 4 + 5 \quad \text{Associative rule}$$
$$y + (-5 + 5) = 4 + 5 \quad \text{Inverse}$$
$$y + 0 = 2$$
$$y = 2 \quad \text{Identity; In } Z_7, 4 + 5 = 2.$$

Solve these equations in $(Z_5, +, \cdot)$.
5. $3x = 1 + x$ $x = 3$
6. $2x = 4$ $x = 2$
7. $x + 4 = 3$ $x = 4$
8. $x - 1 = 4$ $x = 0$

Solve these equations in $(Z_7, +, \cdot)$.
9. $3x = 1$ $x = 5$
10. $4x = 2 - x$ $x = 6$
11. $x - 3 = 5$ $x = 1$
12. $4 + x = 1$ $x = 4$

Solve each equation *or* explain why it cannot be solved in the system indicated.
13. Z_7: $3x = 2$ $x = 3$
14. Z_{11}: $x - 8 = 2$ $x = 10$
15. Z_8: $2x = 3$
16. Z_{52}: $13x = 8$ Z_{52} is not a field.
17. Z_6: $3x = 2$ Z_6 is not a field.
18. Z_{11}: $3x = 1$ $x = 4$

15. Z_8 is not a field.

Challenge   Solve.

Let $L = \{☼, 🎩, ？, ∼, ☾\}$. Is $(L, □, △)$ a field? yes

ART AND MATH

Mathematical Excursions

This design was constructed from a mathematical system with which you are familiar. For example, ■ represents one symbol, ▦ another and so forth. Can you figure out what system it is, using the following hints?   (Z_9, ·)

  Hints: Does every symbol appear in every row? Is the "table" more likely an addition or a multiplication table? If multiplication, does one element appear to have been left out?   no, multiplication, yes, zero

This artistic design is one example of many that you can create for your art or mathematics class, or as a special project combining both disciplines. Of course, colors, systems chosen, how they are represented, artistic medium, and such can vary. See what you can do.

3.5 Some Field Theorems

Objective: To prove field theorems by using additive and multiplicative language.

Previously proven group theorems hold true for all systems that are groups. This also applies to fields. One of the major reasons for studying fields is to prove some of the field theorems used in the rational and real number systems.

Some of the properties of the field $(F, +, \cdot)$ exist because $(F, +)$ and $(F/\{0\}, \cdot)$ are groups. The following is a list of proven group theorems that have already been used.

$(G, *)$ a group for all a, b, and $c \in G$	
Group Theorems	**Name**
If $a = b$, then $a * c = b * c$.	Right Operation
If $a = b$, then $c * a = c * b$.	Left Operation
If $a * c = b * c$, then $a = b$.	Right Cancellation
If $c * a = c * b$, then $a = b$.	Left Cancellation

Examples

1 **Rewrite the Right Operation Theorem in the additive language of $(F, +)$.**

Substitute $+$ for $*$ and the result is if $a = b$, then $a + c = b + c$.
A statement of the theorem in a field would be,
"In a field $(F, +, \cdot)$ for all a, b, and c, if $a = b$, then $a + c = b + c$."

2 **Rewrite the Left Cancellation Theorem in multiplicative language.**

Recall that the multiplicative group is $(F/\{0\}, \cdot)$.
When $\cdot$ is substituted for $*$ the zero must be considered. The resulting multiplicative theorem states,
"for a, b, and $c \neq 0$, if $a \cdot c = b \cdot c$, then $a = b$."

The first ten theorems in fields are group theorems interpreted properly for the additive or multiplicative part of the field.

In a field $(F, +, \cdot)$:	
Field Addition Theorems	**Field Multiplication Theorems**
1. If $a = b$, then $a + c = b + c$.	1. If $a = b$, then $ac = bc$.
2. If $a = b$, then $c + a = c + b$.	2. If $a = b$, then $ca = cb$.
3. If $a + c = b + c$, then $a = b$.	3. If $ac = bc$ and $c \neq 0$, then $a = b$.
4. If $c + a = c + b$, then $a = b$.	4. If $ca = cb$ and $c \neq 0$, then $a = b$.
5. $-(-a) = a$	5. $(a^{-1})^{-1} = a$, for $a \neq 0$.

Some theorems combine a knowledge of $+$ and $\cdot$, the two field operations. The first shows how zero, the additive identity, behaves under multiplication. Remember, zero is still in the field; it just doesn't have a multiplicative inverse. The proof uses various properties of fields so the theorem becomes a Field Combination Theorem.

Field Combination Theorem 1

For all $a \in F$, $a \cdot 0 = 0$.

STATEMENTS	REASONS
1. $a(0 + 0) = a \cdot 0 + a \cdot 0$	1. Distributive law in $(F, +, \cdot)$
2. $0 + 0 = 0$	2. Definition of additive inverse
3. $a \cdot 0 = a \cdot 0 + a \cdot 0$	3. Substitution principle of equality: 0 was substituted for $0 + 0$ on the left.
4. $(a \cdot 0) + 0 = a \cdot 0 + a \cdot 0$	4. Definition of additive identity
5. $0 = a \cdot 0$	5. Left cancellation law in $(F, +)$

Because there is commutativity of multiplication in a field, Field Combination Theorem 2 follows immediately from the first one.

Field Combination Theorem 2

For all $b \in F$, $0 \cdot b = 0$.

Exercises

Max: 1-10, 1-17;   Avg: 1-10, 1-17;   Min: 1-9 odds, 1-13

Exploratory Use the two Field Combination Theorems to simplify the following.
1. $17 \cdot 0$ 0
2. $-\dfrac{2}{19} \cdot 0$ 0
3. $0 \cdot 18.3$ 0
4. $0 \cdot \left(-\dfrac{1}{4}\right)$ 0

Find the simplest name for each of the following in the clock system indicated.

In Z_5:
5. -4 1
8. $\dfrac{1}{2}$ 3

In Z_7:
6. -4 3
9. $\dfrac{1}{4}$ 2

In Z_{13}:
7. -4 9
10. $\dfrac{1}{6}$ 11

Written For each group expression for group $(G, *)$, write the expression in additive language.
1. $a * b$ $a + b$
2. $(a * b)_{inv}$ $-(a + b)$
3. $a * (b * c)$ $a + (b + c)$
4. $c_{inv} * (b * c)$ $-c + (b + c)$
5. $(b * b) * b$ $(b + b) + b$
6. $a_{inv} * (c * d_{inv})_{inv}$ $-a - (c - d)$

For each of the following expressions for group $(G, *)$, give the equivalent in multiplicative language.
7. $a * b$ $a \cdot b$
8. $(a * b)_{inv}$ $(a \cdot b)^{-1}$ for $a \neq 0$, $b \neq 0$
9. $a * (b * c)$ $a \cdot (b \cdot c)$
10. $a_{inv} * (b * c)$ $a^{-1} \cdot (b \cdot c)$ for $a \neq 0$
11. $c_{inv} * c_{inv}$ $c^{-1} \cdot c^{-1}$ for $c \neq 0$
12. $(a * b) * c_{inv}$ $(a \cdot b) \cdot c^{-1}$ for $c \neq 0$

Give the reasons for the following proof of the Field Addition Theorem 3, which is right cancellation. Note that additive language is used throughout.

13. **Given:** $a + c = b + c$ **Prove:** $a = b$

    Proof:

STATEMENTS	REASONS
1. $a + c = b + c$	1. Given
2. $(a + c) + (-c) = (b + c) + (-c)$	2. Right Operation
3. $a + [c + (-c)] = b + [c + (-c)]$	3. Associativity
4. $a + 0 = b + 0$	4. Definition of additive inverse
5. $a = b$	5. Definition of additive identity

Prove the following theorems, giving statements and reasons. Be sure to use additive or multiplicative language as appropriate. For answers to exercises 14-17, see the Teacher's Guide.
14. Field Addition Theorem 4
15. Field Multiplication Theorem 4
16. Field Multiplication Theorem 3
17. Field Combination Theorem 2 without using Field Combination Theorem 1

3.6 An Important Field Theorem

Objective: To solve equations using Field Combination Theorem 3.

One theorem you have used before is pinpointed as one of the most important of all field theorems.

Remember, in $(Z_4, \cdot)$ there is a product equal to 0 (it is circled) in which *neither* factor is 0. In the integers, the rationals, or other systems of "ordinary arithmetic," this can never be true. Recall, in the reals, *if the product of two numbers is 0, then at least one of those numbers must be 0.*

$\cdot$	0	1	2	3
0	0	0	0	0
1	0	1	2	3
2	0	2	⓪	2
3	0	3	2	1

Here is the table for $(Z_5, \cdot)$. Note that this situation does not occur in Z_5.

Do you think there is a case in $(Z_6, \cdot)$ where two nonzero elements have the product zero? Yes, $2 \cdot 3 = 0$ in Z_6. Also, $4 \cdot 3 = 0$.

$\cdot$	0	1	2	3	4
0	0	0	0	0	0
1	0	1	2	3	4
2	0	2	4	1	3
3	0	3	1	4	2
4	0	4	3	2	1

Consider Z_8, Z_{12}, Z_7 and Z_{31}. Do you think that in these systems two nonzero elements can have a product equal to 0? Yes, in Z_8 and Z_{12}. No, in Z_7 and Z_{31}.

You may have come to the conclusion that in a field, if the product of two numbers is 0 at least one of the two numbers must be 0. This is the third Field Combination Theorem.

Field Combination Theorem 3

> In a field $(F, +, \cdot)$, for all a, b, if $ab = 0$, then $a = 0$ or $b = 0$.

This theorem is often called the Zero Product Property.

Given: Field $(F, +, \cdot)$, two elements $a, b \in F$ such that $a \cdot b = 0$

Prove: $a = 0$ or $b = 0$

STATEMENTS	REASONS
1. $a \cdot b = 0$	1. Given
2. If $a = 0$, you are finished.	2. The proof was to show $a = 0$ or $b = 0$.
3. If $a \neq 0$, $\dfrac{1}{a}$ exists.	3. In $(F/\{0\}, \cdot)$, every element must have a multiplicative inverse.
4. $\dfrac{1}{a} \cdot (a \cdot b) = \dfrac{1}{a} \cdot 0$	4. Left operation
5. $\dfrac{1}{a} \cdot (a \cdot b) = 0$	5. Field combination theorem 1 ($a \cdot 0 = 0$)

6. $\left(\dfrac{1}{a} \cdot a\right) \cdot b = 0$
7. $1 \cdot b = 0$
8. $b = 0$

6. Associativity of multiplication in $(F/\{0\}, \cdot)$
7. Definition of multiplicative inverse
8. Definition of multiplicative identity

Thus, if a is not 0, then b must be 0.

Here are some examples that use this theorem. You will recall that this theorem has been used in the past.

Examples

1 Solve $(x + 2)(x + 5) = 0$ in $\mathcal{R}$.

If the product of $x + 2$ and $x + 5 = 0$, then either $x + 2$ is 0, or $x + 5$ is 0. If $x + 2 = 0$, then $x = -2$. If $x + 5 = 0$, then $x = -5$. The solution set is $\{-5, -2\}$.

2 Solve $(y + 7)(y + 12) = 6$ in $\mathcal{R}$.

Remember you cannot use the method of example 1 unless a zero is on one side of the equals sign. Use trial and error. The solution set is $\{-6, -13\}$.

3 Solve $x^2 - 2x - 15 = 0$ in $\mathcal{R}$.

Rewrite the left side as a product.

$$(x + 3)(x - 5) = 0$$

If the product of these two binomials is zero, then one or the other must be zero. The solution set is $\{-3, 5\}$.

Remember that Field Combination Theorem 3 has been used previously to solve quadratic equations by factoring.

Example

4 Solve $x^2 - 16 = 0$ in $\mathcal{R}$.

Remember $a^2 - b^2$ factors into $(a + b)(a - b)$, which was called factoring the difference of two squares. So, $x^2 - 16 = (x + 4)(x - 4) = 0$. The solution set is $\{-4, 4\}$.

Examples

5 Solve $x^2 - 13 = 0$ in $\mathcal{R}$.

This expression factors over the reals as $(x + \sqrt{13})(x - \sqrt{13}) = 0$. The solution set is $\{-\sqrt{13}, \sqrt{13}\}$.

6 Solve the equation, $x^2 + 4x - 5 = 0$ in $(Z_7, +, \cdot)$. Assume that in a field, "a^2" means $a \cdot a$.

$x^2 + 4x - 5 = (x + 5)(x - 1) = 0$. Therefore, $x + 5 = 0$ or $x - 1 = 0$. Now, $x = -5$ or $x = 1$. But in Z_7, $-5 = 2$. The solution set is $\{2, 1\}$.

Exercises

Max: 1-9, 1-22; Avg: 1-9, 1-22; Min: 1-9, 1-21 odds

Exploratory Give a specific instance in each of the following systems that shows that Field Combination Theorem 3 does not hold in that system. For exercises 1-6, answers may vary.

1. $(Z_{10}, +, \cdot)$
2. $(Z_{12}, +, \cdot)$
3. $(Z_{16}, +, \cdot)$
4. $(Z_{18}, +, \cdot)$
5. $(Z_{96}, +, \cdot)$
6. $(Z_{51}, +, \cdot)$

Explain how Field Combination Theorem 3 is used to solve these equations in Q.

7. $(x + 1)(x - 9) = 0$
   $x + 1 = 0$ or $x - 9 = 0$
8. $x^2 + 7x + 12 = 0$ $(x + 4)(x + 3) = 0$
   $x + 4 = 0$ or $x + 3 = 0$
9. $x^2 - 3x = 4$ $(x - 4)(x + 1) = 0$
   $x - 4 = 0$ or $x + 1 = 0$

Written Use Field Combination Theorem 3 to solve each of these quadratic equations in the field $(Q, +, \cdot)$.

1. $(x + 2)(x - 9) = 0$ $\{-2, 9\}$
2. $x^2 + 11x + 24 = 0$ $\{-3, -8\}$
3. $x^2 + 11x + 30 = 0$ $\{-5, -6\}$
4. $x^2 - x - 30 = 0$ $\{6, -5\}$
5. $x^2 + 19x + 90 = 0$ $\{-9, -10\}$
6. $x^2 + 21x = -20$ $\{-20, -1\}$
7. $0 = x^2 - 25$ $\{5, -5\}$
8. $x^2 = x + 42$ $\{7, -6\}$
9. $x^2 + 13x - 48 = 0$ $\{3, -16\}$
10. $x^2 - 7x - 60 = 0$ $\{12, -5\}$
11. $x^2 - \dfrac{4}{81} = 0$ $\{\frac{2}{9}, -\frac{2}{9}\}$
12. $h^2 + 3h = 130$ $\{-13, 10\}$

Solve each of the following equations in the system indicated.

13. $(x + 1)(x - 4) = 0$ in Z_5 $\{4\}$
14. $x^2 - x - 6 = 0$ in Z_7 $\{3, 5\}$
15. $(x + 10)(x + 3) = 0$ in Z_{11} $\{1, 8\}$
16. $(x + 6)(x + 4) = 0$ in Z_{11} $\{5, 7\}$

Solve each equation in the system indicated, if possible. If not possible, explain why.

17. $x^2 - 2x - 3 = 0$ in Z_5 $\{3, 4\}$
18. $x^2 + 9x + 20 = 0$ in $\mathcal{R}$ $\{-4, -5\}$
19. $x^2 - 2x - 3 = 0$ in Z_8 Z_8 is not a field.
20. $(x - 2)(x - 8) = 0$ in Z_{17} $\{2, 8\}$
21. $x^2 - 9 = 0$ in Z_{15} Z_{15} is not a field.
22. $x^2 - 10x + 16 = 0$ in Z_{91} Z_{91} is not a field.

3.7 Subtraction and Division

Objective: To solve equations using the definitions of subtraction and division.

Mary is nine years old, which is one year more than twice her sister's age. How old is her sister?

The equation is $2x + 1 = 9$. Which way do you proceed?

a. $\qquad 2x + 1 = 9$
$2x + 1 + (-1) = 9 + (-1)$
Addition is used in **a**.

b. $\qquad 2x + 1 = 9$
$2x + 1 - 1 = 9 - 1$
Subtraction is used in **b**.

Similarly, in the equation $2x = 8$, both sides can by multiplied by $\frac{1}{2}$ or divided by 2.

But subtraction and division are *not* field operations. Thus, it is necessary to extend our concept of the field to include these two familiar operations. From previous work with subtraction and division, definitions are available.

Definition of Subtraction

> In a field $(F, +, \cdot)$, $a - b = a + (-b)$.

That is, to subtract b from a, add the additive inverse of b to a.

Definition of Division

> In a field $(F, +, \cdot)$, if $b \neq 0$, $a \div b = a \cdot \frac{1}{b}$.

Alternative Notation of Division

> In a field $(F, +, \cdot)$, if $b \neq 0$, $a \div b = a \cdot (b^{-1})$.

That is, to divide a by b, multiply a by the multiplicative inverse of b.

Examples

1 Use the definition of subtraction to simplify these expressions.
  a. In the rationals, $15 - 12 = 15 + (-12) = 3$.
  b. In the reals, $3\sqrt{2} - 7\sqrt{2} = 3\sqrt{2} + (-7\sqrt{2}) = -4\sqrt{2}$.
  c. In Z_5, $2 - 4 = 2 + (-4) = 2 + 1 = 3$.
  d. In Z_{17}, $4 - 13 = 4 + (-13) = 4 + 4 = 8$.

2 Use the definition of division to simplify these expressions.
  a. In the reals, $6 \div 2 = 6 \cdot \frac{1}{2} = 6 \cdot 2^{-1} = 3$.

b. In Z_7, $6 \div 2 = 6 \cdot \dfrac{1}{2} = 6 \cdot 4 = 3$ because the multiplicative inverse of 2 is 4. Therefore $6 \cdot \dfrac{1}{2}$ is equal to 3.

3 In the reals, compute $\sqrt{50} \div \sqrt{2}$.

$$\begin{aligned}\sqrt{50} \div \sqrt{2} &= \sqrt{50} \cdot \dfrac{1}{\sqrt{2}} \\ &= \sqrt{25}\sqrt{2} \cdot \dfrac{1}{\sqrt{2}} \\ &= 5\sqrt{2} \cdot \dfrac{1}{\sqrt{2}} \text{ or } 5\end{aligned}$$

Look at the following example. To multiply $10(4 - \pi)$, we proceed as follows:

$$\begin{aligned}10(4 - \pi) &= 10[4 + (-\pi)] \\ &= 40 + (-10\pi) \text{ or } 40 - 10\pi.\end{aligned}$$

The definition of subtraction was used to turn $4 - \pi$ into a *sum*.

The proof of multiplication distributing over subtraction uses the following theorem.

Field Combination Theorem 4

a. $-(a \cdot b) = (-a)(b)$
b. $-(a \cdot b) = (a)(-b)$
c. $(-a)(-b) = a \cdot b$

This theorem officially states how the additive inverse of two elements in a field $(F, +, \cdot)$ behave under multiplication. A step-by-step proof of Field Combination Theorem 4a is in the exercises. It is used to prove the important theorem, Field Combination Theorem 5.

Field Combination Theorem 5

In a field $(F, +, \cdot)$ for all a, b, c in F, $a(b - c) = ab - ac$.

STATEMENTS	REASONS
1. $a(b - c) = a[b + (-c)]$	1. Definition of subtraction
2. $ = ab + a(-c)$	2. Distributive law in $(F, +, \cdot)$
3. $ = ab + (-ac)$	3. Field combination theorem 4b
4. $ = ab - ac$	4. Definition of subtraction

Field Combination Theorem 6

In a field $(F, +, \cdot)$ for all a, b, c in F, $(b - c)a = ba - ca$.

Examples

Solve each equation.

4
$$3(x - 14) = -1$$
$$3x - 3(14) = -1$$
$$3x - 42 = -1$$
$$3x = 41$$
$$x = \frac{41}{3}$$

5
$$4\left(k - \frac{1}{2}\right) - 1 = 3\left(k - \frac{2}{3}\right)$$
$$4k - 4\left(\frac{1}{2}\right) - 1 = 3k - 3\left(\frac{2}{3}\right)$$
$$(4k - 2) - 1 = 3k - 2$$
$$4k - 3 = 3k - 2$$
$$4k = 3k + 1$$
$$k = 1$$

Exercises

Max: 1-11, 1-29; Avg: 1-11, 1-15, 17-29 odds; Min: 1-11, 1-29 odds

Exploratory Rewrite each of the following subtraction expressions as addition expressions. For example, $x - 2a = x + (-2a)$.

1. $7 - 2$ $\quad$ 2. $5 - 1$ $\quad$ 3. $9 - (-3)$ $\quad$ 4. $3x - 9$ $\quad$ 5. $\pi r^2 - 1$
   $7 + (-2)$ $\quad\quad$ $5 + (-1)$ $\quad\quad$ $9 + 3$ $\quad\quad$ $3x + (-9)$ $\quad\quad$ $\pi r^2 + (-1)$

Rewrite each of these division expressions as equivalent multiplication expressions.

6. $6 \div 3$ $\quad 6 \cdot \frac{1}{3}$ $\quad$ 7. $x^2 \div y$ $\quad x^2 \cdot \frac{1}{y}$ $\quad$ $(x^2 - 4) \cdot \frac{1}{(x+2)}$ 8. $(x^2 - 4) \div (x + 2)$
9. $x^3 \div x$ $\quad x^3 \cdot \frac{1}{x}$ $\quad$ 10. $5 \div -1$ $\quad 5 \cdot (-\frac{1}{1})$ $\quad$ 11. $6^{-1} \div 7^{-1}$ $\quad 6^{-1} \cdot 7$

Written Find the simplest name for each of the following in the system given.

1. In Z_5, $\frac{2}{2}$ $\quad$ 1
2. In Z_5, $\frac{3}{2}$ $\quad$ 4
3. In Z_5, $\frac{2}{3} \cdot \frac{4}{3}$ $\quad$ 2
4. In Z_5, $\frac{1}{3}$ $\quad$ 2
5. In Z_5, $\frac{4}{3}$ $\quad$ 3
6. In Z_5, $-\frac{2}{4}$ $\quad$ 2
7. In Z_7, $\frac{1}{6}$ $\quad$ 6
8. In Z_7, $\left(-\frac{2}{3}\right) \cdot 4$ $\quad$ 2
9. In Z_7, $\frac{2}{3} \cdot (-4)$ $\quad$ 2
10. In Z_7, $3 \div \frac{2}{5}$ $\quad$ 4
11. In Z_7, $\frac{5}{6} \cdot \frac{3}{4}$ $\quad$ 5
12. In Z_7, $\frac{5}{6} \div \frac{3}{4}$ $\quad$ 5
13. In $\mathcal{R}$, $\frac{\sqrt{48}}{2}$ $\quad 2\sqrt{3}$
14. In $\mathcal{R}$, $\frac{\sqrt{75}}{10}$ $\quad \frac{\sqrt{3}}{2}$
15. In $\mathcal{R}$, $\frac{\sqrt{243}}{9}$ $\quad \sqrt{3}$
16. In $\mathcal{R}$, $\frac{\sqrt{96}}{6}$ $\quad \frac{2\sqrt{6}}{3}$
17. In $\mathcal{R}$, $\frac{\sqrt{120}}{\sqrt{10}}$ $\quad 2\sqrt{3}$
18. In $\mathcal{R}$, $\frac{\sqrt{816}}{\sqrt{34}}$ $\quad 2\sqrt{6}$
19. In $\mathcal{R}$, $\frac{\sqrt{91}}{\sqrt{28}}$ $\quad \frac{\sqrt{13}}{2}$
20. In $\mathcal{R}$, $\frac{\sqrt{864}}{\sqrt{300}}$ $\quad \frac{6\sqrt{2}}{5}$

Solve each equation in $\mathcal{R}$.

21. $-5(2x - 8) = 22$ $\{\frac{9}{5}\}$
22. $7 - 6(8 - x) = 15x$ $\{-\frac{41}{9}\}$
23. $3\left(3x - \frac{2}{5}\right) + x = 5\left(5x - \frac{2}{3}\right)$ $\{\frac{32}{225}\}$
24. $\frac{3}{8}(4x - 15) = \frac{1}{2}x - 9\left(\frac{1}{4} - \frac{5}{6}x\right)$ $\{-\frac{27}{52}\}$
25. $3.4(3x - 4) = 21.5 - 0.6x$ $\{-3.25\}$
26. $x - 8(1.5x - 2.8) = 2.7(7 - 2x)$
27. $12(2x - \sqrt{6}) = \sqrt{54} - 3(\sqrt{24} - x)$ $\{\frac{3\sqrt{6}}{7}\}$
28. $\sqrt{8}(3x - \sqrt{12}) = \sqrt{2}(5x - \sqrt{3})$

26. $\{0.625\}$
28. $\{3\sqrt{3}\}$

29. Give a reason for each step of the proof of Field Combination Theorem 4a.
    Prove: In a field $(F, +, \cdot)$, for all a and b in F, $-(a \cdot b) = (-a) \cdot b$

STATEMENTS	REASONS
1. $(-a + a) \cdot b = 0 \cdot b$	1. Definition of additive inverse
2. $(-a + a) \cdot b = 0$	2. Field Combination Theorem 2
3. $(-a + a) \cdot b = -(a \cdot b) + (a \cdot b)$	3. Definition of additive inverse
4. $(-a) \cdot b + (a \cdot b) = -(a \cdot b) + (a \cdot b)$	4. Distributive law in $\{F, +, \cdot\}$
5. $(-a) \cdot b = -(a \cdot b)$	5. Right Cancellation

Mixed Review

Choose the best answer.

b **1.** Which statement is the negation of $\sim p \lor (q \land r)$?
  a. $p \lor (\sim q \land \sim r)$ **b.** $p \land (\sim q \lor \sim r)$ **c.** $\sim p \land (\sim q \lor \sim r)$ **d.** $p \land (\sim q \lor r)$

d **2.** What conclusion can be inferred from the premises "If $x = 7$, then y is odd" and "y is odd?"
  a. $x = 7$ **b.** $x \neq 7$ **c.** $y = 7$ **d.** no conclusion

c **3.** What is the solution set for $2x + 6 = 4$ in clock 12?
  a. $\{10\}$ **b.** $\{5\}$ **c.** $\{5, 11\}$ **d.** $\emptyset$

a **4.** What is the value of $3 - 5$ in clock 7?
  a. 5 **b.** 2 **c.** 6 **d.** 1

b **5.** What is the simplest name for $\frac{1}{8}$ in Z_{15}?
  a. 8 **b.** 2 **c.** 1 **d.** does not exist

c **6.** If p represents "$2x = 2$ in clock 8" and q represents "$x^2 = 7$ in clock 9," for which value of x is $\sim(\sim p \to q)$ true?
  a. 5 **b.** 4 **c.** 7 **d.** 1

b **7.** What is the solution set for $4x - 7(x + 4) = 12 - 5x$ in $\mathcal{R}$?
  a. $\{5\}$ **b.** $\{20\}$ **c.** $\{-5\}$ **d.** $\{8\}$

c **8.** What is the solution set for $x^2 - 24 = 10x$ in $\mathcal{R}$?
  a. $\{4, 6\}$ **b.** $\{-4, -6\}$ **c.** $\{-2, 12\}$ **d.** $\{2, -12\}$

3.8 Ordering the Rationals

Objective: To order a collection of rational numbers.

Everyone knows that 5 is less than 10. Everyone knows that $\frac{1}{8}$ is less than $\frac{3}{4}$ and that -5.095 is less than 3.2. In fact, whenever there are two different rational numbers, it can always be said that one is less than the other.

A system in which this is true is said to be *ordered,* or have an *order relation* defined on it. The rationals, reals, and integers are ordered.

Here is a formal discussion of what it means for the numbers in a field to be ordered.

Definition of an Order Relation

Suppose $(F, +, \cdot)$ is a field. Then $<$ is an order relation on $(F, +, \cdot)$ if all four of the following properties hold:
1. For every a and b in F either
   i. $a < b$ or ii. $a = b$ or iii. $b < a$.   Trichotomy Property
2. If $a < b$ and $b < c$, then $a < c$.   Transitive Property
3. If $a < b$, then $a + c < b + c$.   Additive Property
4. If $a < b$, and $0 < c$ then $ac < bc$.   Multiplicative Property

This particular order relation, $<$, is called *less than*. If a field does have such an order relation defined on it, then it can be written $(F, +, \cdot, <)$. In this case F is an ordered field.

Definition of Greater Than

If $a < b$, then $b > a$.

Example

1 The field of rationals is ordered. It is written $(Q, +, \cdot, <)$. Some examples of each of the four properties in the definition above are given using the rationals.

a. Either $2 < 7$ or $2 = 7$ or $7 < 2$.   Obviously, $2 < 7$.
b. If $2 < 7$ and $7 < 56.1$, then $2 < 56.1$.
c. 2 is less than 7. This implies $2 + 3\frac{1}{4}$ is less than $7 + 3\frac{1}{4}$.
d. $2 < 7$ and 5 is positive. This implies $2(5) < 7(5)$ or $10 < 35$.

Example

2 Solve $2x + 1 < 5$ in $(\mathcal{R}, +, \cdot)$.

$2x + 1 < 5$	
$(2x + 1) + (-1) < 5 + (-1)$	Add -1 to both sides (Ordered Field Property, Additive).
$2x + [1 + (-1)] < 4$	Associativity in $(F, +)$, Simplify $5 + (-1)$
$2x + 0 < 4$	Definition of Additive Inverse
$2x < 4$	Definition of Additive Identity
$x < 2$	Multiply both sides by $\frac{1}{2}$ (Ordered Field Property, Multiplicative).

This could be shortened to: $2x + 1 < 5$.
$$2x < 4$$
$$x < 2$$

The solution can be written as {all numbers less than 2}. It can be graphed on the number line below. The open circle means that 2 is not included in the solution.

Journal

Make a list of all the new things you have learned in this chapter. Also make a list of questions you still have about any of the concepts.

Suppose $x < y$. Consider $-2x$ compared to $-2y$. Is $-2x < -2y$? Look at these examples.

1. $3 < 5$. Multiplying both sides of the inequality by -2 gives us $-2 \cdot 3 = -6$ and $-2 \cdot 5 = -10$. However -6 is greater than -10.
2. $-4 < 11$. Again, multiply both sides by -2. The result is 8 and -22. Of course, 8 is greater than -22.
3. $0 < \frac{1}{4}$. But $0 > -\frac{1}{2}$, as a result of $-2 \cdot 0$ and $-2 \cdot \frac{1}{4}$.

These examples lead us to the conclusion of Ordered Field Theorem 1.

Ordered Field Theorem 1

> If $a < b$, and $c < 0$, then $ac > bc$.

This theorem is proven in the exercises. Ordered Field Theorem 1 states that when both sides of an inequality are multiplied by a negative number, the direction of the inequality must be changed.

Example

3 Solve $-3y < 42$.

$$-3y < 42$$
$$-\frac{1}{3}(-3y) > -\frac{1}{3}(42)$$

Use Ordered Field Theorem 1. Because $-\frac{1}{3}$ is negative, the direction of the inequality must be changed.

$$y > -\frac{42}{3}$$
$$y > -14$$

The solution set is {all numbers greater than -14}.

Exercises

Max: 1-15, 1-34; Avg: 1-15, 1-32; Min: 1-9, 1-24

Exploratory Order these collections of rational numbers from least to greatest.

1. $-2, -3, 0$ $-3, -2, 0$
2. $\frac{1}{2}, \frac{2}{3}, 0.4$ $.4, \frac{1}{2}, \frac{2}{3}$
3. $\frac{1}{2}, 0.\overline{5}, -\frac{1}{2}$ $-\frac{1}{2}, \frac{1}{2}, .\overline{5}$
4. $0.4\overline{1}, 0.41, 0.412$ $.41, .4\overline{1}, .412$
5. $-3, -\frac{31}{10}, -(1.6)^2$ $-\frac{31}{10}, -3, -(1.6)^2$
6. $\frac{1}{10}, 0.101, \frac{2}{21}$
7. $\frac{502}{1,000}, 0.002, \frac{5,002}{10,000}$ $.002, \frac{5,002}{10,000}, \frac{502}{1,000}$
8. $(0.\overline{3})^2, 0.0\overline{8}, \frac{1}{12}$ $\frac{1}{12}, .0\overline{8}, (.3)^2$
9. $\frac{13}{41}, \frac{1}{3}, \frac{2}{7}$ $\frac{2}{7}, \frac{13}{41}, \frac{1}{3}$

6. $\frac{2}{21}, \frac{1}{10}, .101$

Explain each of these terms.

10. order relation on a set — Given two different numbers, one is less than the other.
11. ordered field — a field with an order relationship
12. trichotomy property — For every a and b in a field, either a < b, a = b, or b < a.
13. additive property — If a < b, a + c < b + c.
14. transitive property — If a < b and b < c, then a < c.
15. multiplicative property — If a < b and 0 < c, then ac < bc.

Written Give a reason for each step in the following procedure used to solve $2x + 3 < 17$.

1. a. $2x + 3 < 17$
   b. $(2x + 3) + (-3) < 17 + (-3)$ Additive Property
   c. $2x + [3 + (-3)] < 14$ Associative law
   d. $2x + 0 < 14$ Definition of additive inverse
   e. $2x < 14$ Definition of additive identity
   f. $\frac{1}{2}(2x) < \frac{1}{2}(14)$ Left Operation
   g. $\left(\frac{1}{2} \cdot 2\right)x < 7$ Associative law
   h. $1 \cdot x < 7$ Definition of multiplicative inverse
   i. $x < 7$ Definition of multiplicative identity

A SYSTEM WITH TWO OPERATIONS

Order these collections of real numbers from least to greatest.

2. $\sqrt{2}$, 1.4, 1.44
   1.4, $\sqrt{2}$, 1.44

3. $1.\overline{4}$, 1.45, $\sqrt{2}$
   $\sqrt{2}$, $1.\overline{4}$, 1.45

4. $2\sqrt{2}$, 2.8, $\frac{29}{10}$ 2.8, $2\sqrt{2}$, $\frac{29}{10}$

5. π, $\sqrt{2}+1$, 3.1
   $\sqrt{2}+1$, 3.1, π

6. 10π, 313, $\left(\frac{7}{2}\right)^2$
   $\left(\frac{7}{2}\right)^2$, 10π, 313

7. $\sqrt{5}$, 2.25, $2.\overline{5}$
   $\sqrt{5}$, 2.25, $2.\overline{5}$

Solve each of these inequalities using the method in exercise 1.

8. $3x + 1 < 16$ $x < 5$
9. $3 + 4x < 9$ $x < \frac{3}{2}$
10. $50 + 10 < x + 41$ $x > 19$
11. $7 < 5a + 8$ $a > -\frac{1}{5}$

Using symbols of logic, follow the directions in each statement.

12. Rewrite the assertion $r \leq 5$ using the logical connective $\vee$. $(r < 5) \vee (r = 5)$

13. Rewrite the negation of $c < d$ using $\geq$. $c \geq d$

14. Rewrite this rule in logic. "If $a < b$, and $c < 0$, then $ac > bc$."
    $(a < b \wedge c < 0) \rightarrow ac > bc$

15. Write the contrapositive of this rule. "If F is an ordered field, then all the ordered field requirements are satisfied." If all of the ordered field requirements are not satisfied, then F is not an ordered field.

Use the four ordered field principles and Ordered Field Theorem 1 to solve these inequalities. Graph each of the solutions on a number line. For answers to exercises 16-28, see the Teacher's Guide

16. $2z - 1 < 5$
17. $z + 7 < -1$
18. $14z + 3 > 2$
19. $2k + 1 < k - 3$
20. $8k + 1 > 3$
21. $-4k + 1 < -5k$
22. $3\left(k + \frac{2}{3}\right) < k - 1$
23. $\frac{1}{4}x - 1 > -3x$
24. $-2x + 1 < 13 - x$

Prove the following statements in an ordered field.

25. Prove: If $a < b$, then $0 < b - a$.
26. Prove: If $c < 0$, then $0 < -c$.
27. Prove: If $a < b$, and $c < 0$, then $ac > bc$. You may use exercises 25 and 26.
28. Prove: In an ordered field $(F, +, \cdot, <)$, $a^2 \geq 0$ for all $a \in F$.

Suppose that the field $(Z_5, +, \cdot)$ is ordered. Answer the following questions.

29. What must be true about the relationship of the elements 2, 4? either $2 < 4$, $2 = 4$, or $2 > 4$
30. Assume $2 < 4$. Add 2 to both sides. What does it produce? $4 < 1$
31. Use the transitive property on the two statements in exercise 30. $2 < 1$
32. Multiply the result of exercise 31 by 2. What do you conclude? $4 < 2$; The field $(Z_5, +, \cdot)$ is not ordered.

Using the given elements, answer exercises 29 through 32 in each field. Assume each field is ordered. Write a statement about ordering these finite fields.

33. Field $(Z_7, +, \cdot)$; elements 4, 6.
    The field $(Z_7, +, \cdot)$ is not ordered.

34. Field $(Z_{13}, +, \cdot)$; elements 10, 1.
    The field $(Z_{13}, +, \cdot)$ is not ordered.

Problem Solving Application: Identifying Subgoals

Objective: To solve problems that require several steps by first identifying subgoals.

Sometimes finding the solution to a problem requires several steps. An important strategy for solving such problems is to *identify subgoals*. This strategy involves taking steps that will either produce part of the solution or make the problem easier to solve.

Emphasize the importance of using the four-step problem-solving plan.

Example

1 Robots are used to assemble toy cars. Three robots can assemble 10 cars in 21 minutes. How many cars can 14 robots assemble in 72 minutes if all the robots assemble cars at the same rate?

Explore We know that 3 robots can assemble 10 cars in 21 minutes and that all robots assemble cars at the same rate.
We want to know how many cars can be assembled by 14 robots in 72 minutes.
Let x = the number of cars.

Plan Writing an equation to represent this problem will be easier if we develop the equation in steps rather than directly from the given information. The first step (subgoal) is to determine the rates at which the cars are assembled for 3 robots and for 14 robots. Then these rates can be used to find the rate for 1 robot.

$$\text{Assembly Rate} = \frac{\text{number of cars}}{\text{time}}$$

$$\text{Assembly Rate for 3 robots} = \frac{10}{21} \qquad \text{Assembly Rate for 14 robots} = \frac{x}{72}$$

$$\text{Assembly Rate for 1 robot} = \frac{10}{21} \div 3 \quad \text{or} \quad \frac{10}{63} \qquad \text{Assembly Rate for 1 robot} = \frac{x}{72} \div 14 \quad \text{or} \quad \frac{x}{1008}$$

Since all robots assemble cars at the same rate, the assembly rates for 1 robot given above must be equal.

$$\frac{10}{63} = \frac{x}{1008}$$

Solve
$$10080 = 63x$$
$$x = 160$$

Therefore, 14 robots can assemble 160 cars in 72 minutes.

Examine There are about five times as many robots working for about three times as long. Hence, they should assemble 5 · 3 or 5 times as many cars. Thus, the solution appears to be reasonable.

Exercises

Max: 1-4, 1-11; Avg: 1-4, 1-11; Min: 1-4, 1-5, 7-11 odds

Exploratory Solve each problem.

1. $\dfrac{x}{8} = \dfrac{28}{32}$ {7}
2. $\dfrac{11}{x} = \dfrac{33}{42}$ {14}
3. $\dfrac{w}{50} = \dfrac{11}{20}$ $\{\dfrac{55}{2}\}$
4. $\dfrac{12}{5} = \dfrac{15}{2x}$ $\{\dfrac{25}{4}\}$

Written Solve each problem. 6. 7B, 6B and 2A, 5B and 4A, 4B and 6A, 3B and 8A, 2B and 10A, 1B and 12A, or 14A

1. Two cows need to graze on 3 acres of grass per month for proper nutrition. How many cows can graze 90 acres of grass for 6 months? 5 cows

2. Four packing machines can pack 6 boxes every 5 seconds. How long would it take 3 packing machines to pack 180 boxes at that rate? 100 seconds

3. Three painters can paint 4 houses in 5 days. How long would it take 5 painters to paint 18 houses if all the painters worked at the same rate? 13.5 days

4. How many houses could be painted by 3 painters in 2 months if the painters worked 22 days a month and if they worked at the same rate as the painters in exercise 3? about 35 houses (35.2)

5. A baseball player hit 714 homeruns over 23 seasons. He averaged playing 140 games per season. How long would it have taken him to hit that many homeruns if he had averaged playing 162 games per season? about 20 seasons

6. Three bricklayers and one apprentice can lay 12 bundles of bricks in 3 days. How many bricklayers and apprentices would it take to lay 40 bundles of bricks in 5 days if a bricklayer lays brick twice as fast as an apprentice?

7. How many whole numbers less than 100 have digits whose sum is 10? 9

8. How many whole numbers less than 1000 have digits whose sum is 10? 63

9. Find all the whole numbers between 10 and 1000 that stay the same when the digits are written in reverse order. For example, 686 has this property. These numbers are called palindromes. There are 99 such numbers.

10. Suppose the scoring in football is simplified to 7 points for a touchdown and 3 points for a field goal. What scores are impossible to achieve? 1, 2, 4, 5, 8, 11

11. When changing a dollar, you can give two coins (2 half-dollars), three coins (1 half-dollar and 2 quarters), four coins (4 quarters), and so forth. What is the smallest number of coins that is impossible to give as change for a dollar? (For example, it is impossible to make change for a dollar using 81 coins.) 77 coins

Portfolio Suggestion

Select an item from your work in this chapter that shows your problem-solving skills.

Performance Assessment

The formula $F_c = \dfrac{mv^2}{r}$ can be used to find the centripetal force F_c upon an object moving in a circle, where m is the mass of the object, v is the velocity, and r is the radius of the circular path. Solve the equation for m. Explain each step in your equation. Would there be any limitations for the value of each variable? If so, explain the limitations. $m = \dfrac{F_c r}{v^2}, r \neq 0, v \neq 0$

Chapter Summary

1. The **Distributive Law** states that for all a, b, c in Q, $a(b + c) = ab + ac$. (88)
2. A two-fold operational system is a mathematical system that uses two operations. In a two-fold operational system that is a field the group properties of the set and each operation are maintained. (92)
3. A **field** is a two-fold operation system, $(F, +, \bullet)$, which satisfies the following requirements. (92, 93)
   a. $(F, +)$ is a commutative group.
   b. $(F/\{0\}, \bullet)$ is a commutative group.
   c. Multiplication distributes over addition.
4. The multiplicative inverse of a is represented by both a^{-1} and $\dfrac{1}{a}$. (102)
5. Proven group theorems hold true for all systems made up of groups. (102)

6. The first ten theorems in fields include five addition theorems and five multiplication theorems from the group theorems. (102)
7. Field Combination Theorem 1: For all $a \in F$, $a \bullet 0 = 0$. (102)
8. Field Combination Theorem 2: For all $b \in F$, $0 \bullet b = 0$. (102)
9. Field Combination Theorem 3: In a field $(F, +, \bullet)$, for all a, b, if $ab = 0$, then $a = 0$ or $b = 0$. (104)
10. In a field $(F, +, \bullet)$, $a - b = a + (-b)$. (107)
11. In a field $(F, +, \bullet)$, if $b \neq 0$, $a \div b = a \bullet \frac{1}{b} = a \bullet (b^{-1})$. (107)
12. Field Combination Theorem 4: (108)
    a. $-(a \bullet b) = (-a) \bullet (b)$ **b.** $-(a \bullet b) = (a) \bullet (-b)$
    c. $(-a)(-b) = a \bullet b$
13. Field Combination Theorem 5: In a field $(F, +, \bullet)$, for all a, b, c in F, $a(b - c) = ab - ac$. (108)
14. Field Combination Theorem 6: In a field $(F, +, \bullet)$, for all a, b, c in F, $(b - c)a = ba - ca$. (109)
15. Definition of an Order Relation: Suppose $(F, +, \bullet)$ is a field. Then $<$ is an order relation on $(F, +, \bullet)$ if all four of the following properties hold. (111)
    1. Trichotomy Property: For every a and b in F one of the following is true: **i.** $a < b$ **ii.** $a = b$ **ii.** $b < a$.
    2. Transitive Property: If $a < b$ and $b < c$, then $a < c$.
    3. Additive Property: If $a < b$, then $a + c < b + c$.
    4. Multiplicative Property: If $a < b$, and $0 < c$, then $ac < bc$.
16. Ordered Field Theorem 1: If $a < b$, and $c < 0$, then $ac > bc$, which means when both sides of an inequality are multiplied by a negative number the direction of the inequality must be changed. (112)

Chapter Review

3.1 State which of the following systems are groups. Justify your answers.

1. $(Z_5, +)$ yes
2. $(W, \bullet)$ no, no inverse
3. $(Q, \bullet)$ No, zero has no inverse.
4. $(\mathcal{R}, \max)$ no, no inverse or identity

3.2 Use the Distributive Law to simplify the following expressions.

5. $3(x + 4)$
   $3x + 12$
6. $175(2 + x)$
   $350 + 175x$
7. $-3[x + (-4)]$
   $-3x + 12$
8. $2[x + (-9)]$
   $2x - 18$

3.3 State which of the following systems are fields.

9. $(\mathcal{Z}, +, \bullet)$ no
10. $(Q/\{0\}, +, \bullet)$ no
11. $(Z_{10}, +, \bullet)$ n

3.4 In Z_7, find the equivalent of the following.

12. -6 $_1$ 13. -1 $_6$ 14. 4^{-1} $_2$ 15. -4 $_3$

Give the reasons for each step in each of the following solutions.

16. In $(Z_7, +, \cdot)$:
    a. $x + 6 = 3$
    b. $x + 6 + (-6) = 3 + (-6)$
       Right Operation
    c. $x + [6 + (-6)] = 3 + (-6)$
       Associative law
    d. $ x + 0 = 3 + 1$
       Inverse
    e. $ x = 4$
       Identity; In Z_7, $3 + 1 = 4$.

17. In $(Q, +, \cdot)$:
    a. $4x = 1$
    b. $\frac{1}{4}(4x) = \frac{1}{4}(1)$ Left Operation
    c. $\left(\frac{1}{4} \cdot 4\right)x = \frac{1}{4}(1)$ Associative law
    d. $1 \cdot x = \frac{1}{4}$ Inverse
    e. $x = \frac{1}{4}$ Identity

3.5 Write these group expressions of group $(G, *)$ in multiplicative language.

18. $(a * b)_{inv}$ $\frac{1}{a \cdot b}$
19. $a * (b * c)$
20. $C_{inv} * C_{inv}$ $\frac{1}{c} \cdot \frac{1}{c}$
21. $(a * b) * c$ $a \cdot (b \cdot c)$
   $(a \cdot b) \cdot c$

3.6 Solve these quadratic equations in the field $(\Re, +, \cdot)$ using theorems learned in this chapter.

22. $(x + 2)(x - 15) = 0$ $\{-2, 15\}$
23. $x^2 + 16x + 60 = 0$ $\{-10, -6\}$
24. $0 = x^2 - 49$ $\{-7, 7\}$
25. $x^2 = x + 42$ $\{7, -6\}$

3.7 Rewrite each of the following subtraction expressions as addition expressions.

26. $1 - x$ $1 + (-x)$
27. $xy^4 - x^2y^2$ $xy^4 + (-x^2y^2)$
28. $7 - (-2)$ $7 + 2$

Rewrite each of the following division expressions as multiplication expressions.

29. $7 \div 4$ $7 \cdot \frac{1}{4}$
30. $2x \div 4y$ $2x \cdot \frac{1}{4y}$
31. $6 \div \frac{1}{2}$ $6 \cdot 2$

3.8 Solve these inequalities using principles and theorems in this chapter. Graph the solutions on a number line.

32. $z + 7 < -1$ $z < -8$
33. $3x + 13 < 4$ $x < -3$
34. $-1{,}500y < \frac{1}{2} - 1{,}300y$ $y > -\frac{1}{400}$

$\xleftarrow{}\underset{-8}{\circ}\xrightarrow{}$ $\xleftarrow{}\underset{-3}{\circ}\xrightarrow{}$ $\xleftarrow{}\underset{-\frac{1}{400}}{\circ}\xrightarrow{}$

Order these collections of real numbers from least to greatest.

35. $\frac{1}{8}$, 0.1251, $\frac{11}{79}$
    $\frac{1}{8}$, $.1251$, $\frac{11}{79}$

36. $3\sqrt{2}$, 4.21, $2\sqrt{3} + 0.8$
    4.21, $3\sqrt{2}$, $2\sqrt{3} + .8$

Chapter Test

Indicate which of the following systems is *a group*, *a field*, or *neither* a group nor a field.

1. $(Z_7, +)$ group
2. $(Q/\{0\}, +, \cdot)$ neither
3. $(\mathcal{R}, +, \cdot)$ field
4. $(Z_8, +, \cdot)$ neither
5. $(Q, +)$ group
6. $(Z, +, \cdot)$ neither

Indicate which line in the following proof of Field Combination Theorem 1 uses the Distributive Law and which line uses the Property of Additive Inverses.

7. For all $a \in F$, $a \cdot 0 = 0$.
   a. $a(0 + 0) = a \cdot 0 + a \cdot 0$ Distributive Law
   b. $0 + 0 = 0$ Property of Additive Inverses
   c. $a \cdot 0 = a \cdot 0 + a \cdot 0$
   d. $(a \cdot 0) + 0 = a \cdot 0 + a \cdot 0$
             $0 = a \cdot 0$

Solve the following equations in $(\mathcal{R}, +, \cdot)$.

8. $3x - 1 = -x + 3$ $x = 1$
9. $\left(x + \dfrac{1}{3}\right) = 2x + 2$ $x = -\dfrac{5}{3}$
10. $2(x + \sqrt{7}) = \sqrt{7}$ $x = \dfrac{-\sqrt{7}}{2}$
11. $-(x - 2) = -5$ $x = 7$
12. $\dfrac{1}{2}(\sqrt{5} + 4x) = 2\sqrt{20}$ $x = \dfrac{7\sqrt{5}}{4}$
13. $x^2 + \dfrac{3}{5} = \dfrac{17}{20}$ $x = \dfrac{1}{2}, -\dfrac{1}{2}$
14. $9 = x^2 - 8x$ $x = 9, -1$
15. $x^2 + 3x = 4$ $x = -4, 1$

Solve the following equations in $(Z_7, +, \cdot)$.

16. $x + 2 = 4$ $x = 2$
17. $4 = x + 6$ $x = 5$
18. $4x = 2$ $x = 4$
19. $4(x + 2) = 1$ $x = 0$
20. $x^2 - 4 = 2$ $\emptyset$
21. $x^2 + x - 6 = 0$ $x = 2, 4$

Solve these inequalities. Graph the solutions on a number line.

22. $x + 7 < -1$
23. $8x + 1 > 3$
24. $-2x + 1 < 13 - x$
25. $\dfrac{1}{4}x - 1 > -3x$

22. $x < -8$

-8

23. $x > \dfrac{1}{4}$

$\dfrac{1}{4}$

24. $x > -12$

-12

25. $x > \dfrac{4}{13}$

$\dfrac{4}{13}$

CHAPTER 4

Introduction to Euclidean Geometry

Application in Air Traffic Control

One of the responsibilities of an air traffic controller is to assign pilots of airplanes their cruising altitudes as they head for their destinations. Each airplane is to fly at a certain level so that it may be tracked and so that it will not collide with other airplanes in the area. Since the assigned altitude levels do not intersect, they can be described as **parallel**.

Air traffic controller Bill Randell assigns eastbound airplanes that are flying above 30,000 feet cruising altitudes of 33,000 feet, 37,000, 41,000 feet, and so on. Westbound airplanes above 30,000 feet are assigned cruising altitudes of 31,000 feet, 35,000 feet, 39,000 feet, and so on. What is the closest possible vertical distance between two airplanes that pass over each other? What are some of the advantages of this system? **2,000 feet; Sample answer: There is less worry about collisions, and it is easier to keep track of which planes are headed east and west.**

Group Project: *Architecture*

The study of geometry is intertwined with the field of architecture. Even before the rules of geometry were formally proved or written down, the principles were being used to build structures. Work with your group to choose a culture that is of interest, such as Native American, Japanese, Egyptian, Greek, Chinese, or German. Research the traditional and modern architecture of the culture. What are the characteristics that define the style? Does the climate where the people live have an effect on the architecture? Present your findings to the class using a news commentary format.

4.1 Undefined Terms and Defined Terms

Objective: To identify undefined and defined terms and to use basic geometric terms.

The first step in organizing ideas in geometry is to agree on precise definitions for the words and phrases used. However, not every word and phrase can be defined. Some words' meanings must be assumed. The words or phrases whose meanings are assumed to be known are called **undefined terms**.

In geometry, the basic undefined terms are **point, line, plane,** and **between**. Although these terms are undefined, we do know a lot about them.

For example, we agree that a line is a set of points. Also, a line is a "straight line" that extends forever in both directions. Recall that the symbol for line AB is $\overleftrightarrow{AB}$.

A plane can be thought of as a flat surface that extends forever in all directions. Part of plane $\mathcal{P}$ is shown in the figure.

When we say that point B is between point A and point C, we agree that A, B, and C must lie on the same line in the order shown in the diagram.

Once we have agreed on the words that are accepted as undefined, these words can be used to define other terms.

Definitions of Collinear and Noncollinear

Two or more points are said to be **collinear** if and only if they are contained in the same line. Points that are not contained in the same line are **noncollinear**.

Definitions of Coplanar and Noncoplanar

Two or more points are said to be **coplanar** if and only if they are contained in the same plane. Points that are not contained in the same plane are **noncoplanar**.

Points K, M, and G are collinear.
Points H, I, and J are noncollinear.
Points G, H, I, and J are coplanar.
Points H, I, J, and K are noncoplanar.

Definition of Line Segment

A **line segment** is part of a line consisting of two points on the line and all points between them. The two points are called **endpoints**.

The part of $\overleftrightarrow{AB}$ in black is line segment AB, which is symbolized $\overline{AB}$.

CHAPTER 4   123

Definition of Ray   A **ray** is part of a line consisting of one endpoint and all points of the line on one side of the endpoint.

If point B is between points A and C, then ray BA and ray BC are called **opposite rays**. Ray BA is symbolized $\overrightarrow{BA}$.

Definition of Angle   An **angle** is the union of two rays that have a common endpoint called the **vertex**.

The diagram shows the *interior* and *exterior* of angle CBA. Angle CBA is symbolized $\angle CBA$. $\overrightarrow{BC}$ and $\overrightarrow{BA}$ are the *sides* of $\angle CBA$. B is the vertex.

Definition of Adjacent Angles   Two angles in a plane that have a common vertex, a common side, and no interior points in common are called **adjacent angles**.

$\angle 1$ and $\angle 2$ are adjacent angles. The sides that are not common to adjacent angles are called *exterior sides*. $\overrightarrow{OA}$ and $\overrightarrow{OC}$ are exterior sides.

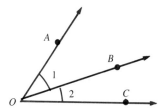

Definition of Vertical Angles   Two nonadjacent angles formed by two intersecting lines are called **vertical angles**.

Example

1 Name two pairs of vertical angles.

$\overleftrightarrow{AB}$ and $\overleftrightarrow{CD}$ are intersecting lines and $\angle AOD$ and $\angle COB$ are not adjacent. Therefore, $\angle AOD$ and $\angle COB$ are vertical angles, as are $\angle AOC$ and $\angle DOB$.

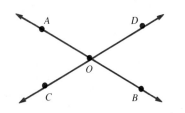

124 INTRODUCTION TO EUCLIDEAN GEOMETRY

Exercises

Max: 1-18, 1-19; Avg: 1-18, 1-19; Min: 1-4, 5-17 odds, 1-19 odds

Exploratory Explain the difficulty in trying to make up definitions for the following geometric terms. For answers to exercises 1-4, see the Teacher's Guide.

1. point
2. line
3. plane
4. between

Refer to the figure below. Tell whether each statement is *true* or *false*. Justify your answer.

5. A, B, and D are collinear. true; definition of collinear
6. A and E are collinear.
7. A, B, and E are coplanar. true; definition of coplanar
8. A, B, and E are collinear.
9. B is between A and D. true; B lies on $\overline{AD}$.
10. E is between A and D. false; E does not lie on $\overline{AD}$.
11. C lies on $\overline{AB}$. false; C is not between A and B.
12. $\overrightarrow{BA} \cup \overrightarrow{BD} = \overleftrightarrow{AD}$. true; $\overrightarrow{BA}$ and $\overrightarrow{BD}$ are opposite rays.
13. $\overrightarrow{EC} \cap \overrightarrow{CF} = \overline{EF}$. false; $\overrightarrow{EC} \cap \overrightarrow{CF}$ = point C
14. $\overrightarrow{DA}$ and $\overrightarrow{DK}$ are opposite rays. true; D is between A and K.
15. $\overrightarrow{CE}$ and $\overrightarrow{EF}$ are opposite rays. false; E is not between C and F.
16. $\overrightarrow{CB}$ and $\overrightarrow{CA}$ name the same ray. true; They have the same endpoint and both contain points B and A.
17. ∠KCF and ∠FCA are adjacent angles. true; definition of adjacent angles
18. ∠ACF and ∠DCF are adjacent angles. true; definition of adjacent angles

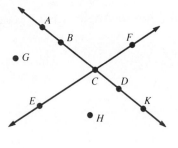

6. true; A line can be drawn through A and E.   8. false; A, B, and E are noncollinear.

Written Refer to the figure below. Name each of the following.

Answers to exercises 2-8 and 10 may vary.

1. one point in the interior of ∠CAD J
2. two points in the exterior of ∠DAE
3. two adjacent angles
4. two vertical angles
5. two line segments
6. two angles that are neither adjacent nor vertical
7. two pairs of opposite rays
8. a point in the interior of ∠BAE and also in the exterior of ∠EAD
9. a point in the interior of both ∠FAB and ∠DAF G
10. the exterior sides of two adjacent angles

A line separates a plane into two regions. Each region is called a *half-plane*. The line is called the *edge* of the half-plane. Name the points from the figure above that are in the half-plane determined by each of the following.

11. $\overleftrightarrow{AB}$ and point N G,F,N
12. $\overleftrightarrow{AB}$ and point C H,C,J,D,K,E,L
13. $\overleftrightarrow{AC}$ and point H N,
14. $\overleftrightarrow{FE}$ and point M N,L,M
15. $\overleftrightarrow{FE}$ and point H G,B,H,C,J,D,K
16. $\overleftrightarrow{BM}$ and point F G,
17. $\overleftrightarrow{AC}$ and point M M,L,E,K,D,J
18. $\overleftrightarrow{BM}$ and point L L,E,K,D,J,C,H
19. $\overleftrightarrow{AD}$ and point E N,

4.2 Axioms About Lines, Planes, and Segments

Objective: To use the properties of lines, planes, and segments to solve problems.

The next step in organizing the study of geometry is to agree on the rules, or statements, accepted as true. Such statements are called **postulates** or **axioms.**

Once we have these axioms, they are used to prove other statements that follow logically from them. A statement that can be *proved* true is called a **theorem.** Thus, the organization of geometry consists of *undefined terms, defined terms, axioms,* and *theorems.*

This section contains some of the basic axioms. They are written as more formal statements of familiar ideas.

Axiom 4–1
> A line contains at least two points. A plane contains at least three noncollinear points.

Axiom 4–2
> For any two points, there exists one and only one line containing both points.

Axiom 4–3
> For any three noncollinear points, there exists one and only one plane containing all three points.

Axiom 4–4
> If two points are in a plane, then the line containing the two points is also in that plane.

Axiom 4–5
> If two planes intersect, then their intersection is a line.

Planes X and Y intersect in line AB.

Axiom 4–6
> For any two points on a line and a given unit of measure, there is a unique positive number called the measure of the distance between the two points.

Notice that $\overline{AB}$ is a line segment and AB is a positive number.

Of course, the measure of the distance between A and B varies depending upon the unit of measure used. The measure of $\overline{AB}$ is symbolized by AB. Note that AB and BA denote the same measure.

126 INTRODUCTION TO EUCLIDEAN GEOMETRY

Definition of Congruent Segments

> Two segments are said to be congruent if and only if they have the same measure. That is, $\overline{AB} \cong \overline{CD}$ if and only if $AB = CD$.

The next axiom is used so frequently that it is given a special name.

Axiom 4–7 Segment Addition Axiom

> If point B is between points A and C, then $AB + BC = AC$.

Sometimes Axiom 4–7 is stated more informally as "The whole is equal to the sum of its parts."

Definition of Midpoint

> If point M on $\overline{CD}$ lies between C and D such that $\overline{CM} \cong \overline{MD}$, then M is called the **midpoint** of $\overline{CD}$.

It is also true that $CM = MD$.

How many midpoints does $\overline{CD}$ have? Your answer should suggest Axiom 4–8.

Axiom 4–8

> A line segment has one and only one midpoint.

Definition of Bisector

> A line, ray, line segment, or plane that intersects $\overline{CD}$ a its midpoint is called a **bisector** of the line segment.

M is the midpoint of $\overline{CD}$.

$\overleftrightarrow{AB}$ bisects $\overline{CD}$.
$\overline{AB}$ also bisects $\overline{CD}$.
$\overrightarrow{AB}$ also bisects $\overline{CD}$.

Example

1 State whether the following statement is *always, sometimes,* or *never* true.

If plane $\mathcal{P}$ intersects $\overline{AB}$ at point M, plane $\mathcal{P}$ is a bisector of $\overline{AB}$.

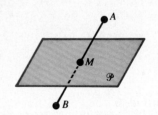

If M is the midpoint of $\overline{AB}$, then plane $\mathcal{P}$ is a bisector of $\overline{AB}$ by definition of bisector. However, if M is *not* the midpoint of $\overline{AB}$, then plane $\mathcal{P}$ is *not* a bisector of $\overline{AB}$. The statement is sometimes true.

Exercises

Max: 1-8, 1-22; Avg: 1-8, 1-22; Min: 1-8, 1-21 odds

Exploratory Answer each of the following. State the axiom or definition that justifies your answer. For the answers to exercises 1-8, see the Teacher's Guide.

1. How many planes contain the three noncollinear points A, B, and C?
2. At least how many points are contained in a line?
3. How many lines contain X and Y?
4. If DE = FG, is $\overline{DE} \cong \overline{FG}$?
5. Can two planes intersect in a single point?
6. At least how many points are contained in a plane?
7. If line ℓ bisects $\overline{RS}$ at T, what can you conclude?
8. If R, S, and T are collinear, must RS + ST = RT?

Written State whether each statement is *always*, *sometimes*, or *never* true. Justify your answer. For answers to exercises 1-8 and 13-15, see the Teacher's Guide.

1. Two points are contained in exactly one line.
2. Three points determine one and only one plane.
3. Two points are collinear.
4. AB is a positive number.
5. If A and B are both contained in plane X, then $\overleftrightarrow{AB}$ is contained in X.
6. There is only one plane containing three noncollinear points.
7. Points A, B, and C are noncoplanar.
8. If $\overline{TM} \cong \overline{MS}$, then M is the midpoint of $\overline{TS}$.

Suppose M is the midpoint of $\overline{AB}$ and AB = 8. Find each of the following.

9. AM 4
10. MB 4
11. $\frac{1}{2}$MB 2
12. AM + MB 8

Answer each of the following.

13. Suppose A, B, and C are collinear. $AB = \frac{1}{2}AC$. Does it follow that B is the midpoint of $\overline{AC}$? Explain.
14. Suppose the midpoint of $\overline{AB}$ is M. Suppose $\overline{CD}$ bisects $\overline{AB}$. What can you say about the intersection of $\overline{AB}$ and $\overline{CD}$?
15. If $\overline{DC}$ bisects $\overline{AB}$, does it follow that $\overline{AB}$ bisects $\overline{CD}$? Explain.

Refer to the figure at the right for exercises 16–22.
Name each of the following.

16. three collinear points A,E,B; C,E,D
17. four coplanar points Answers may vary.
18. the intersection of two planes $\overleftrightarrow{CD}$
19. three noncoplanar points none
20. three noncollinear points Answers may vary.
21. the intersection of $\overleftrightarrow{AB}$ and $\overleftrightarrow{CD}$ E
22. the edge of a half-plane $\overleftrightarrow{AB}$, $\overleftrightarrow{CD}$

4.3 Axioms About Angles

Objective: To use the properties of angles to solve problems.

Angles are often measured in units called *degrees*. The Angle Measure Axiom determines a degree measure for any angle.

**Axiom 4–9
Angle Measure
Axiom**

For every angle, there is one and only one real number between 0 and 180 called the degree measure of the angle.

When using a protractor to draw an angle with degree measure of 70, you assume there is only one ray, $\overrightarrow{RS}$, in the half-plane above $\overleftrightarrow{RQ}$, such that the degree measure of $\angle QRS$ is 70. This idea is stated in Axiom 4–10 as the Protractor Axiom. The degree measure of $\angle QRS$ is written as $m \angle QRS$. For simplicity, "the measure of an angle" is used to mean the degree measure of that angle.

**Axiom 4–10
Protractor
Axiom**

Suppose $\mathcal{H}$ is a half-plane determined by $\overrightarrow{OA}$. Then for any real number r such that $0 < r < 180$, there is one and only one ray, $\overrightarrow{OB}$, such that B is in $\mathcal{H}$ and $m \angle AOB = r$.

**Definition of
Congruent Angles**

Two angles are said to be congruent if and only if they have the same measure. That is, $\angle ABC \cong \angle DEF$ if and only if $m \angle ABC = m \angle DEF$.

In the figure at the right, notice that $m \angle AOB + m \angle BOC = m \angle AOC$. This idea is stated formally in Axiom 4–11, the Angle Addition Axiom.

**Axiom 4–11
Angle Addition
Axiom**

If B is in the interior of $\angle AOC$, then $m \angle AOB + m \angle BOC = m \angle AOC$.

Sometimes the Angle Addition Axiom is stated more informally as "The whole is equal to the sum of its parts."

Definition of Angle Bisector	A ray, $\overrightarrow{OB}$, is a **bisector** of $\angle AOC$ if and only if B is in the interior of the angle and $\angle AOB \cong \angle BOC$.

$\overrightarrow{OB}$ bisects $\angle AOC$, if $\angle AOB \cong \angle BOC$.

Axiom 4–12	An angle has one and only one bisector.

Recall the following definitions.

Definition of Right Angle	A **right angle** is an angle whose degree measure is 90. The symbol ⌐ indicates a right angle.
Definition of Supplementary	Two angles are **supplementary** if and only if the sum of their degree measures is 180. Each angle is called the supplement of the other.
Definition of Complementary	Two angles are **complementary** if and only if the sum of their degree measures is 90. Each angle is called the complement of the other.

Example

1 Find the measure of an angle whose complement is one third of its supplement.

Let x = the measure of the angle.
Then, $180 - x$ = the measure of the supplement,
and $90 - x$ = the measure of the complement.

$$90 - x = \frac{1}{3}(180 - x) \quad \text{complement} = \frac{1}{3} \text{ of the supplement}$$
$$90 - x = 60 - \frac{1}{3}x$$
$$30 = \frac{2}{3}x$$
$$45 = x \quad \text{The measure of the angle is 45.}$$

130 INTRODUCTION TO EUCLIDEAN GEOMETRY

Suppose ∠AOB and ∠COB are adjacent angles and that $\overrightarrow{OA}$ and $\overrightarrow{OC}$ are opposite rays. What is true about the sum of measures of ∠AOB and ∠COB?

Axiom 4-13 Supplement Axiom: If the exterior sides of two adjacent angles are opposite rays, then the angles are supplementary.

Exercises

Max: 1-6, 1-36; Avg: 1-6, 1-18, 19-35 odds; Min: 1-6, 1-35 odds

Exploratory Refer to the figure below. State whether each statement is *true* or *false*. Justify your answer.

1. ∠ABC is a right angle. true; indicated in figure
2. m∠DBA = m∠DBE + m∠EBA true; angle addition axiom
3. $\overline{BE}$ bisects ∠ABD. false; ∠ABE ≠ ∠EBD
4. ∠ABC and ∠ABD are supplementary. true; supplement axiom
5. If m∠DBE = m∠EBA, then ∠DBE ≅ ∠EBA. true; definition of congruent angles
6. m∠CBA + m∠DBE = 180 false; m∠DBE < 90

Written Use a protractor to draw an angle for each degree measure.

1. 30
2. 90
3. 140
4. 170

For exercises 1-8, see students' work.

Illustrate each of the following.

5. two adjacent complementary angles
6. two nonadjacent supplementary angles
7. $\overrightarrow{SQ}$ that bisects ∠PSR
8. $\overrightarrow{AP}$ that bisects ∠MLR and ∠NLQ

Suppose that ∠1 and ∠2 are complementary. Solve for x and find m∠1 and m∠2.

9. m∠1 = x + 6
    m∠2 = x + 4 46, 44
10. m∠1 = 65 − 2x
    m∠2 = 7x 55, 35
11. m∠1 = 3(x − 5)
    m∠2 = 2(x + 5) 42, 48

Suppose that ∠1 and ∠2 are supplementary. Solve for x and find m∠1 and m∠2.

12. m∠1 = 2x
    m∠2 = 3x 72, 108
13. m∠1 = 3x − 4
    m∠2 = 4x − 5 77, 103
14. m∠1 = 2(4x + 5)
    m∠2 = 3(2x − 4) 114, 66

Solve.

15. The measure of the supplement of an angle is 20 more than four times the measure of the angle. Find the measure of the angle. 32
16. The measure of the complement of an angle is 30 more than twice the measure of the angle. Find the measure of the angle. 20
17. The measure of the supplement of an angle is 35 less than four times the measure of the angle. Find the measure of the angle. 43
18. The measure of the complement of an angle is 15 less than twice the measure of the angle. Find the measure of the angle. 35

Suppose $m\angle ABC = 7x + 6$. Find the values of x given the following conditions.
19. $\angle ABC$ is a right angle. $x = 12$
20. $\angle ABC$ is an acute angle. $x < 12$
21. $\angle ABC$ is an obtuse angle. $12 < x < 24\frac{6}{7}$
22. $\overrightarrow{BD}$ bisects $\angle ABC$ and $m\angle ABD = 37$. $x = 9\frac{5}{7}$
23. The measure of the supplement of $\angle ABC$ is 139. $x = 5$
24. The measure of the supplement of $\angle ABC$ is four times the measure of its complement. $x = 7\frac{5}{7}$

Suppose $\overrightarrow{OB}$ bisects $\angle AOC$. Find the values of x given the following conditions.
25. $m\angle AOB = 37$, $m\angle BOC = 2x - 3$ $x = 20$
26. $m\angle AOC = 132$, $m\angle AOB = 5x + 11$ $x = 11$
27. $m\angle BOC = 3x - 16$, $m\angle AOC = 3x - 11$ $x = 7$
28. $m\angle AOB = x^2$, $m\angle BOC = 8x + 33$ $x = -3, 11$

In each figure, find the value of x.

Mixed Review

Solve each equation in $(\mathcal{R}, +, \cdot)$.
1. $4(5x + 3) = x - 7$ $\{-1\}$
2. $3(x - 11) = 4 - 2(5 - 2x)$ $\{-27\}$
3. $\frac{7}{4}(x - 3) = 4x - 3$ $\{-1\}$
4. $\frac{2}{3}(x + 7) - x = \frac{5}{3}(3 - 8x)$ $\{\frac{1}{39}\}$
5. $5(x - 0.7) = 1.3 - 3x$ $\{0.6\}$
6. $1.8(3x + 2) = 0.7 - 5(x - 1.1)$ $\{0.25\}$
7. $x^2 - 5x - 6 = 0$ $\{-1, 6\}$
8. $2x^2 + 9x + 10 = 0$ $\{-\frac{5}{2}, 2\}$

Use the figure at the right for exercises 9–15.

9. Name a point not on $\overleftrightarrow{PQ}$. S or T
10. Name a point not in plane $\mathcal{A}$. T
11. Name all points common to $\overrightarrow{SP}$ and plane $\mathcal{B}$. P
12. Name all points common to planes $\mathcal{A}$ and $\mathcal{B}$. $\overleftrightarrow{PQ}$
13. Name all points common to $\overleftrightarrow{SP}$ and $\overleftrightarrow{QT}$. none
14. Are points S, P, and T coplanar? Explain. yes, Axiom 4-3
15. If $\overline{QT}$ bisects $\overline{PR}$, $PQ = 5x - 18$, and $PR = 3x + 13$, find QR. 17

4.4 Properties of Real Numbers

Objective: To identify the axiom that justifies a given statement.

Much of our work in geometry deals with numbers. For example, the measures of line segments and angles are real numbers. Therefore, we can use the properties of real numbers in geometry. The properties, or axioms, to be used most often are listed below. In each axiom, assume all variables represent real numbers.

Addition Axiom	If $a = b$ and $c = d$, then $a + c = b + d$.
Subtraction Axiom	If $a = b$ and $c = d$, then $a - c = b - d$.
Multiplication Axiom	If $a = b$ and $c = d$, then $ac = bd$.
Division Axiom	If $a = b$ and $c \neq 0$, then $\dfrac{a}{c} = \dfrac{b}{c}$.
Reflexive Axiom	Any number is equal to itself, or $a = a$.
Symmetric Axiom	If $a = b$, then $b = a$.
Transitive Axiom	If $a = b$ and $b = c$, then $a = c$.
Substitution Axiom	If $a = b$, then a can be replaced by b (and b by a).

Example

1. Prove that if $AB = CD$, then $AC = BD$.

Given: $AB = CD$
Prove: $AC = BD$
Proof:

STATEMENTS	REASONS
1. $AB = CD$	1. Given
2. $BC = BC$	2. Reflexive Axiom
3. $AB + BC = BC + CD$	3. Addition Axiom
4. $AB + BC = AC$ $BC + CD = BD$	4. Segment Addition Axiom
5. $AC = BD$	5. Substitution

Exercises

Max: 1-14, 1-20; Avg: 1-11 odds, 13, 14, 1-18; Min: 1-13 odds, 1-19 odds

Exploratory State the axiom that justifies each statement.
1. If $x + 3 = 5$, then $x = 2$. subtraction
2. $y = y$ reflexive
3. If $AB = CD$ and $CD = EF$, then $AB = EF$. transitive
4. If $AB + BC = AC$, then $AB = AC - BC$. subtraction
5. $CD = CD$ reflexive
6. If $m\angle 1 = m\angle 2$ and $m\angle 2 = m\angle 3$, then $m\angle 1 = m\angle 3$. transitive
7. If $x + y = 15$ and $y = 2x$, then $x + 2x = 15$. substitution
8. If $2(m\angle 1) = 2(m\angle 2)$, then $m\angle 1 = m\angle 2$. division
9. If $a = 6$, then $6 = a$. symmetric
10. If $y - 7 = 12$, then $y = 19$. addition
11. If $m\angle A = m\angle B$, then $\frac{1}{2} m\angle A = \frac{1}{2} m\angle B$. multiplication
12. If $QR = QT + TR$ and $TR = AB$, then $QR = QT + AB$. substitution

Answer the following.
13. Explain why $c \neq 0$ in the Division Axiom. Division by 0 is undefined.
14. Give an example to differentiate between the Substitution Axiom and the Transitive Axiom. Answers may vary.

Written State the axiom or definition that justifies each statement.
1. If R is between A and B, then $AR + RB = AB$. segment addition axiom
2. If $m\angle 1 + m\angle 2 = 180$, then $m\angle 1 = 180 - m\angle 2$. subtraction axiom
3. If $m\angle ABC = m\angle DEF$ and $m\angle 3 = m\angle 4$, then $m\angle ABC + m\angle 3 = m\angle DEF + m\angle 4$. addition axiom
4. If $\overline{RS}$ is in the interior of $\angle QRT$, then $m\angle QRS + m\angle SRT = m\angle QRT$. angle addition axiom
5. If P is the midpoint of $\overline{AB}$, then $\overline{AP} \cong \overline{PB}$. definition of midpoint
6. If $AB = CD$, then $\overline{AB} \cong \overline{CD}$. definition of congruent segments
7. If $m\angle 1 + m\angle 2 = 90$, then $\angle 1$ and $\angle 2$ are complementary angles. definition of complementary

Copy and complete each proof.

8. Given: $m\angle 1 = m\angle 2$
   $m\angle 3 = m\angle 4$
   Prove: $\angle CAB \cong \angle ACB$
   Proof:

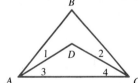

STATEMENTS	REASONS
1. $m\angle 1 = m\angle 2$, $m\angle 3 = m\angle 4$	1. Given
2. $m\angle 1 + m\angle 3 = m\angle 2 + m\angle 4$	2. Addition Axiom
3. $m\angle 1 + m\angle 3 = m\angle CAB$ $m\angle 2 + m\angle 4 = m\angle ACB$	3. Angle Addition Axiom
4. $m\angle CAB = m\angle ACB$	4. Substitution
5. $\angle CAB \cong \angle ACB$	5. Definition of Congruent Angles

9. **Given:** B is the midpoint of $\overline{AC}$.
   Prove: $AB = \frac{1}{2}AC$

Proof:

STATEMENTS	REASONS
1. B is the midpoint of $\overline{AC}$.	1. Given
2. $AB = BC$	2. Definition of Midpoint
3. $AB + BC = AC$	3. Segment Addition Axiom
4. $AB + AB = AC$ or $2AB = AC$	4. Substitution
5. $AB = \frac{1}{2}AC$	5. Multiplication Axiom

Justify each step in the solution of each equation. For answers to exercises 10-20, see the Teacher's Guide.

10. $\frac{1}{2}x - 3 = 11$
    $\frac{1}{2}x = 14$
    $x = 28$

11. $3(x + 2) = 24$
    $3x + 6 = 24$
    $3x = 18$
    $x = 6$

12. $10x - 15 = 6x + 12$
    $4x - 15 = 12$
    $4x = 27$
    $x = \frac{27}{4}$

Write a two-column proof for the following.

13. **Given:** $m\angle 1 + m\angle 2 = 90$
    $m\angle 1 = m\angle 3$
    Prove: $m\angle 3 + m\angle 2 = 90$

14. **Given:** $RT = SV$
    Prove: $RS = TV$

15. **Given:** $AB = AD$
    $BC = DE$
    Prove: $AC = AE$

16. **Given:** $m\angle CBE = m\angle ABD$
    Prove: $m\angle CBD = m\angle EBA$

17. **Given:** $BE = CD$
    $FD = FE$
    Prove: $BF = CF$

18. **Given:** $\overrightarrow{PR}$ bisects $\angle SPT$.
    Prove: $2(m\angle SPR) = m\angle SPT$

19. **Given:**
    $PQ = QR$
    $PQ = RT$
    Prove:
    $RT = QR$

20. **Given:**
    $\angle AFC \cong \angle CFE$
    $\angle AFD \cong \angle BFE$
    Prove:
    $\angle BFC \cong \angle CFD$

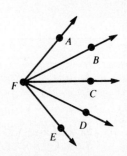

4.5 Theorems About Angles

Objective: To prove theorems about angles.

In the preceding section, we used axioms and principles of logic to write a proof. Some of the following theorems may be familiar or seem obvious. However, the objective here is to show how the theorems can be proved.

Theorem 4–1

> If two angles are supplementary to the same angle, then they are congruent to each other.

Note that the theorem is a conditional statement. Recall that a conditional is false only when its antecedent is true and its consequent is false.

To prove Theorem 4–1, assume that its antecedent is true. Then show that its consequent is also true. The following example shows a proof of Theorem 4–1. Notice that the antecedent of the theorem appears as the given. The consequent appears as the statement to be proved.

Example

1 Prove Theorem 4–1.

Given: ∠1 is supplementary to ∠3.
∠2 is supplementary to ∠3.
Prove: ∠1 ≅ ∠2
Proof:

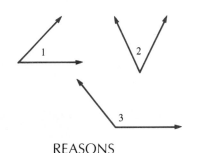

STATEMENTS	REASONS
1. ∠1 is supplementary to ∠3. ∠2 is supplementary to ∠3.	1. Given
2. $m\angle 1 + m\angle 3 = 180$ $m\angle 2 + m\angle 3 = 180$	2. Definition of supplementary angles
3. $m\angle 1 + m\angle 3 = m\angle 2 + m\angle 3$	3. Substitution Axiom
4. $m\angle 1 = m\angle 2$	4. Subtraction Axiom
5. ∠1 ≅ ∠2	5. Definition of congruent angles

This proof shows how the final statement follows from the statements that precede it. Statements are written in the left column. The reason that justifies each statement is written in the right column. The only reasons that may be used in a proof are the given information, definitions, axioms, and theorems that have been proved previously. You will be asked to prove the following theorems in the exercises.

Theorem 4–2 If two angles are supplementary to congruent angles, then they are congruent to each other.

Theorem 4–3 If two angles are complementary to the same or congruent angles, then they are congruent to each other.

Theorem 4–4 If two lines intersect, the vertical angles formed are congruent.

Congruence of segments and angles is related to equality axioms of real numbers. You will be asked to prove the following theorems in exercises 17 and 18.

Theorem 4–5 Congruence of segments is reflexive, symmetric, and transitive.

Theorem 4–6 Congruence of angles is reflexive, symmetric, and transitive.

In this text, the Reflective Property is sometimes used to signify that congruence of segments, or angles, is reflexive.

Exercises

Max: 1-14, 1-19; Avg: 1-14, 1-11 odds, 12-19; Min: 1-13 odds, 1-11 odds, 12-19

Exploratory State the theorem that justifies each statement.
1. $\overline{RS} \cong \overline{RS}$
2. If $\overline{PQ} \cong \overline{RS}$, then $\overline{RS} \cong \overline{PQ}$. Theorem 4-5
3. $\angle R \cong \angle R$
4. If $\angle A \cong \angle B$ and $\angle B \cong \angle C$, then $\angle A \cong \angle C$. Theorem 4-6

<u>1</u>. Theorem 4-5   <u>3</u>. Theorem 4-6

In the figure $m \angle ABD = 27$, $m \angle DBF = 90$, and $m \angle FBC = 63$. Find the measure of the following angles.

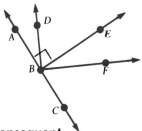

5. $\angle ABE$ 90
6. $\angle DBE$ 63
7. $\angle EBF$ 27
8. $\angle EBC$ 90
9. $\angle ABC$ 180

For answers to exercises 10-14, see the Teacher's Guide.

Restate each sentence in if-then form. Identify the antecedent and consequent.

10. Supplements of the same angle are congruent.
11. Vertical angles are congruent.
12. The intersection of two planes is a line.
13. Two adjacent angles have a common side.
14. Complements of congruent angles are congruent.

Written In the figure, $\angle QVR \cong \angle RVS$, $m \angle PVQ = 72$, and $m \angle TVS = 70$. Find the measure for each angle.

1. $\angle QVS$ 38
2. $\angle QVR$ 19
3. $\angle PVR$ 91
4. $\angle TVR$ 89
5. $\angle PVS$ 110
6. $\angle TVQ$ 108
7. $\angle PVW$ 70
8. $\angle WVZ$ 38
9. $\angle TVW$ 110
10. $\angle TVZ$ 72

Copy and complete the following proof.

11. **Theorem 4-2**   If two angles are supplementary to congruent angles, then they are congruent to each other.

   Given: $\angle 1$ is supplementary to $\angle 3$.
   $\angle 2$ is supplementary to $\angle 4$.
   $\angle 3 \cong \angle 4$

   Prove: $\angle 1 \cong \angle 2$
   Proof:

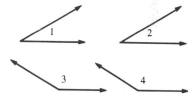

STATEMENTS	REASONS
1. $\angle 1$ is supplementary to $\angle 3$. $\angle 2$ is supplementary to $\angle 4$.	1. Given
2. $m \angle 1 + m \angle 3 = 180$ $m \angle 2 + m \angle 4 = 180$	2. Definition of supplementary
3. $m \angle 1 + m \angle 3 = m \angle 2 + m \angle 4$	3. Substitution
4. $\angle 3 \cong \angle 4$	4. Given
5. $m \angle 3 = m \angle 4$	5. Definition of congruent angles
6. $m \angle 1 = m \angle 2$	6. Subtraction axiom
7. $\angle 1 \cong \angle 2$	7. Definition of congruent angles

For answers to exercises 12, 13, and 15-19, see the Teacher's Guide.

Answer each of the following.

12. Explain why the statements that appear in the *Given* are assumed true.

13. Explain why the *Prove* statement must be shown to be true.

Copy and complete the following proof.

14. **Theorem 4–3** (Part I) If two angles are complementary to the same angle, then they are congruent to each other.

    Given: $\angle 1$ is complementary to $\angle 3$.
    $\angle 2$ is complementary to $\angle 3$.
    Prove: $\angle 1 \cong \angle 2$
    Proof:

STATEMENTS	REASONS
1. $\angle 1$ is complementary to $\angle 3$. $\angle 2$ is complementary to $\angle 3$.	1. Given
2. $m\angle 1 + m\angle 3 = 90$ $m\angle 2 + m\angle 3 = 90$	2. Definition of complementary
3. $m\angle 1 + m\angle 3 = m\angle 2 + m\angle 3$	3. Transitive axiom
4. $m\angle 1 = m\angle 2$	4. Subtraction axiom
5. $\angle 1 \cong \angle 2$	5. Definition of congruent angles

Write a two-column proof of each of the following. Include the *Given* and *Prove* statements, and a diagram.

15. **Theorem 4–3** (Part II) If two angles are complementary to congruent angles, then they are congruent to each other. (Hint: Modify the proof in exercise 11.)

16. **Theorem 4–4** If two lines intersect, the vertical angles formed are congruent.

Prove these theorems by writing a two-column proof for each part.

17. **Theorem 4–5**
    a. $\overline{AB} \cong \overline{AB}$
    b. If $\overline{AB} \cong \overline{CD}$, then $\overline{CD} \cong \overline{AB}$.
    c. If $\overline{AB} \cong \overline{CD}$ and $\overline{CD} \cong \overline{EF}$, then $\overline{AB} \cong \overline{EF}$.

18. **Theorem 4–6**
    a. $\angle ABC \cong \angle ABC$
    b. If $\angle ABC \cong \angle DEF$, then $\angle DEF \cong \angle ABC$.
    c. If $\angle ABC \cong \angle DEF$ and $\angle DEF \cong \angle GHI$, then $\angle ABC \cong \angle GHI$.

Prove the following.

19. Given: $\angle 2 \cong \angle 6$
    Prove: $\angle 3 \cong \angle 7$

4.6 Right Angles and Perpendicular Lines

Objective: To prove and apply theorems about right angles and perpendicular lines.

All right angles are congruent because they have the same measure.

Theorem 4–7 If two angles are right angles, then they are congruent.

Example

1 Prove Theorem 4–7.

Given: $\angle 1$ and $\angle 2$ are right angles.
Prove: $\angle 1 \cong \angle 2$
Proof:

STATEMENTS	REASONS
1. $\angle 1$ is a right angle. $\angle 2$ is a right angle.	1. Given
2. $m \angle 1 = 90, m \angle 2 = 90$	2. Definition of right angle
3. $m \angle 1 = m \angle 2$	3. Substitution Axiom
4. $\angle 1 \cong \angle 2$	4. Definition of congruent angles

Suppose two angles are both congruent and supplementary. What must be true about the angles? Your answer should suggest Theorem 4–8.

Theorem 4–8 If two angles are congruent and supplementary, then each is a right angle.

The proof of Theorem 4–8 will be completed in exercise 8.

Recall that **perpendicular lines** are lines that intersect to form right angles. When two intersecting lines form one right angle, they form four right angles. Therefore, to prove two lines perpendicular, you need only show that they intersect to form one right angle.

140 INTRODUCTION TO EUCLIDEAN GEOMETRY

In the figure at the left, the two lines are perpendicular. The symbol ⊥ means "is perpendicular to." $\overleftrightarrow{NS} \perp \overleftrightarrow{WE}$.

Parts of lines that intersect are perpendicular to each other if the lines containing them are perpendicular. For example, a ray can be perpendicular to a line segment. In the figure at the left, $\overrightarrow{CD} \perp \overline{AB}$.

Suppose two intersecting lines form two congruent adjacent angles. What must be true about the two lines? Theorem 4–9 provides an answer.

Theorem 4–9

> If two intersecting lines form two congruent adjacent angles, then the lines are perpendicular.

This theorem will be proved in exercise 9.

Definition of Perpendicular Bisector

> A **perpendicular bisector** of a line segment is a line, line segment, or ray that is perpendicular to the line segment at the midpoint of the line segment.

In the figure at the left, line ℓ is the perpendicular bisector of $\overline{CD}$ because M is the midpoint of $\overline{CD}$ and $\ell \perp \overline{CD}$.

Exercises

Max: 1-13, 1-15; Avg: 1-4, 5-13 odds, 1-9 odds, 10-15; Min: 1-13 odds, 1-15 odds

Exploratory Find the measure of ∠x in each figure. Justify your answer.

1.

2.

3.

4.

Refer to the figure at the right. Tell whether each statement is *true* or *false*.

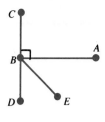

5. ∠ABC is a right angle. true
6. $\overline{AB} \perp \overline{CD}$ true
7. ∠DBE and ∠ABE are complementary. true
8. $\overline{BE} \perp \overline{AB}$ false
9. $\overline{CD}$ is the perpendicular bisector of $\overline{AB}$. false
10. If $\overline{CB} \cong \overline{BD}$, then $\overline{AB}$ is the perpendicular bisector of $\overline{CD}$. true
11. ∠CBA and ∠DBE are congruent adjacent angles. false
12. $\overline{DB} \perp \overline{BA}$ true
13. If m∠ABE = 45, then $\overline{BE}$ bisects ∠DBA. true

Answers to exercises 2-7 may vary.

Written Refer to the figure at the right.
$\overleftrightarrow{EB} \perp \overleftrightarrow{AD}$. Name each of the following.

1. four right angles ∠EOD, ∠DOB, ∠BOA, ∠AOE
2. two angles adjacent to ∠AOF
3. two complementary angles
4. two supplementary angles
5. two supplements of ∠BOC
6. a pair of vertical angles
7. two angles that are both congruent and supplementary

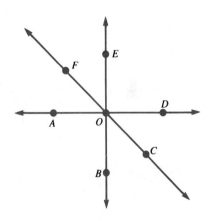

Copy and complete each proof.

8. **Theorem 4–8** If two angles are congruent and supplementary, then each is a right angle.

   Given: ∠1 ≅ ∠2
   ∠1 and ∠2 are supplementary.
   Prove: ∠1 is a right angle.
   ∠2 is a right angle.
   Proof:

STATEMENTS	REASONS
1. ∠1 and ∠2 are supplementary.	1. Given
2. m∠1 + m∠2 = 180	2. Definition of supplementary
3. ∠1 ≅ ∠2	3. Given
4. m∠1 = m∠2	4. Definition of congruent angles
5. m∠1 + m∠1 = 180 or 2(m∠1) = 180	5. Substitution
6. m∠1 = 90	6. Division axiom
7. m∠2 = 90	7. Substitution
8. ∠1 and ∠2 are right angles.	8. Definition of right angle

9. **Theorem 4-9** If two intersecting lines form two congruent adjacent angles, then the lines are perpendicular.

Given: ℓ and m intersect.
$\angle 1 \cong \angle 2$
Prove: $\ell \perp m$
Proof:

STATEMENTS	REASONS
1. $\angle 1$ and $\angle 2$ are supplementary.	1. Supplement axiom
2. $\angle 1 \cong \angle 2$	2. Given
3. $\angle 1$ and $\angle 2$ are right angles.	3. Theorem 4-8
4. $\ell \perp m$	4. Theorem 4-9

For answers to exercises 10-15, see the Teacher's Guide.
Write a two-column proof of each of the following.

10. **Given:** $\angle 1$ and $\angle 2$ are supplementary.
    Prove: $\ell \perp m$

11. **Given:** $\angle 1 \cong \angle 3$
    Prove: $\angle 2 \cong \angle 4$

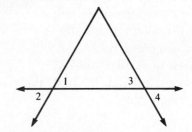

12. **Given:** $\overrightarrow{OB} \perp \overrightarrow{OD}, \overrightarrow{OC} \perp \overrightarrow{OA}$
    Prove: $\angle 1 \cong \angle 3$

13. **Given:** $\overrightarrow{SW} \perp \overrightarrow{RV}, m\angle 2 = m\angle 3$
    $\overrightarrow{TY} \perp \overrightarrow{RV}$
    Prove: $m\angle 1 = m\angle 4$

14. **Given:** $\angle 4 \cong \angle 3, \angle 2 \cong \angle 1$
    Prove: $\angle 4 \cong \angle 1$

15. If the noncommon sides of two adjacent acute angles are perpendicular, then the angles are complementary.

4.7 Parallel Lines and Angles

Objective: To prove and apply theorems about parallel lines.

The rails of a railroad track or the yard markings on a football field suggest the familiar idea of **parallel lines.**

A line is considered parallel to itself.

Definition of Parallel Lines

> Two distinct lines are parallel if and only if they lie in the same plane and they do not intersect.

Journal

Name examples of parallel lines and perpendicular lines that you see in your neighborhood. Sketch a picture to accompany each item.

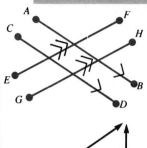

Line segments and rays are parallel if they are contained in parallel lines. The symbol ∥ means "is parallel to." In figures, arrowheads are often used to indicate parallel lines, line segments, or rays. In the figure at the left $\overline{AB} \parallel \overline{CD}$ and $\overline{EF} \parallel \overline{GH}$.

Can two lines have no points in common and yet not be parallel? Consider lines ℓ and m in the cube at the left. They are examples of **skew lines.**

Definition of Skew Lines

> Two lines are skew lines if they are *not* coplanar.

Definition of Transversal

> In a plane, a **transversal** is a line that intersects two or more other lines in different points.

The intersection of a transversal and two lines forms eight angles. In the figure at the left, k is a transversal of h and j. The eight angles have special names.

Interior Angles	∠3, ∠4, ∠5, ∠6
Alternate Interior Angles	∠3 and ∠5, ∠4 and ∠6
Exterior Angles	∠1, ∠2, ∠7, ∠8
Alternate Exterior Angles	∠1 and ∠7, ∠2 and ∠8
Corresponding Angles	∠1 and ∠5, ∠2 and ∠6, ∠4 and ∠8, ∠3 and ∠7

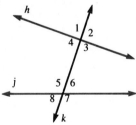

In the figure, ℓ and m are parallel lines cut by transversal n. Compare the measures of any pair of corresponding angles, such as ∠1 and ∠5. The corresponding angles appear to have the same measure. Because this cannot be proved, it will be stated as an axiom.

Axiom 4–14 If two parallel lines are cut by a transversal, then the corresponding angles are congruent.

Consider the converse of Axiom 4–14. Lines ℓ and m are cut by transversal n. Suppose $\angle 1 \cong \angle 2$. It seems that ℓ and m are parallel. However, we cannot prove this true, so we accept it as an axiom.

Axiom 4–15 In a plane, if two lines are cut by a transversal such that the corresponding angles are congruent, then the two lines are parallel.

Axioms 4–14 and 4–15 enable us to prove Theorems 4–10 and 4–11.

Theorem 4–10 If a transversal is perpendicular to one of two parallel lines, it is perpendicular to the other.

Example

1 Prove Theorem 4–10.

Given: $\ell \parallel m$
n is a transversal.
$n \perp \ell$

Prove: $n \perp m$

Proof:

STATEMENTS	REASONS
1. $n \perp \ell$, $\ell \parallel m$	1. Given
2. $\angle 1$ is a right angle.	2. Definition of perpendicular lines
3. $m\angle 1 = 90$	3. Definition of right angle
4. $\angle 1 \cong \angle 2$	4. If two parallel lines are cut by a transversal, corresponding angles are congruent.
5. $m\angle 1 = m\angle 2$	5. Definition of congruent angles
6. $m\angle 2 = 90$	6. Substitution Axiom
7. $\angle 2$ is a right angle.	7. Definition of right angle
8. $n \perp m$	8. Definition of perpendicular lines

CHAPTER 4 145

| Theorem 4–11 | In a plane, if two lines are perpendicular to the same line, then they are parallel. |

Theorem 4–11 will be proved in exercise 13.

Exercises

Max: 1-17, 1-19; Avg: 1-8, 9-17 odds, 1-18; Min: 1-17 odds, 1-17 odds

Exploratory Consider all the lines determined by the vertices of this cube. Name each of the following. Answer for exercises 1–3 may vary.

1. three pairs of parallel lines
2. three pairs of intersecting lines
3. three pairs of skew lines

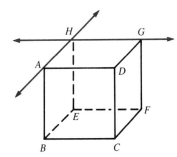

Refer to the figure at the left. Name each of the following.

4. the transversal ℓ
5. two pairs of alternate interior angles ∠2, ∠8; ∠4, ∠5
6. two pairs of alternate exterior angles ∠3, ∠6; ∠1, ∠7
7. four pairs of corresponding angles ∠1, ∠5; ∠2, ∠6; ∠3, ∠8; ∠4, ∠7
8. four pairs of vertical angles ∠2, ∠3; ∠1, ∠4; ∠5, ∠7; ∠6, ∠8

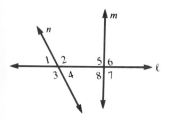

In the following, determine which axioms or theorems are needed to prove $\ell \parallel m$.

9. Axiom 4-15

10. Axiom 4-15

11. Theorem 4-11

12.

13. Theorem 4-4, Axiom 4-15

14. Theorem 4-4, Axiom 4-15

15. Axiom 4-15

16. Theorem 4-4, Axiom 4-15

17. Theorem 4-4, Axiom 4-15

146 INTRODUCTION TO EUCLIDEAN GEOMETRY

Written Refer to the figure at the right. State the axiom or theorem that justifies each conclusion.

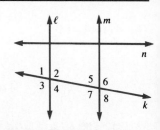

1. Given: $\ell \parallel m$; Conclusion: $\angle 2 \cong \angle 6$ Axiom 4-14
2. Given: $\angle 4 \cong \angle 8$; Conclusion: $\ell \parallel m$ Axiom 4-15
3. Given: $\ell \parallel m, n \perp m$; Conclusion: $n \perp \ell$ Theorem 4-10
4. Given: $n \perp \ell, m \perp n$; Conclusion: $\ell \parallel m$ Theorem 4-11

Refer to the figure at the left. Tell which pairs of lines, if any, are parallel when the following pairs of angles are congruent. Justify your answer.

5. $\angle 1 \cong \angle 13$ 6. $\angle 2 \cong \angle 6$ $\ell \parallel m$; Axiom 4-15
7. $\angle 3 \cong \angle 15$ 8. $\angle 9 \cong \angle 13$
9. $\angle 3 \cong \angle 1$ none   10. $\angle 7 \cong \angle 11$
11. $\angle 1 \cong \angle 11$ none   12. $\angle 8 \cong \angle 16$ none

5. $n \parallel p$; Axiom 4-15   8. $\ell \parallel m$; Axiom 4-15
7. $n \parallel p$; Axiom 4-15   10. $n \parallel p$; Axiom 4-15

Copy and complete the following proof.

13. **Theorem 4–11** In a plane, if two lines are perpendicular to the same line, then they are parallel.

    Given: $\ell \perp n$
            $m \perp n$
    Prove: $\ell \parallel m$
    Proof:

STATEMENTS	REASONS
1. $\ell \perp n, m \perp n$	1. Given
2. $\angle 1$ and $\angle 2$ are right angles.	2. Definition of perpendicular line
3. $\angle 1 \cong \angle 2$	3. Theorem 4-7
4. $\ell \parallel m$	4. Axiom 4-15

Write a two-column proof of each of the following. For answers to exercises 14–18, see the Teacher's Guide.

Exercises 14–16

14. Given: $\angle 1 \cong \angle 3$
    Prove: $\angle 2 \cong \angle 4$
15. Given: $\overline{DE} \parallel \overline{BC}$
            $\angle 1 \cong \angle 2$
    Prove: $\angle 3 \cong \angle 4$
16. Given: $\angle 2 \cong \angle 4$
            $\angle 4 \cong \angle 3$
    Prove: $\angle 2 \cong \angle 1$

Exercise 17

17. Given: $\angle 1$ is supplementary to $\angle 2$.
    Prove: $\overline{BC} \parallel \overline{ED}$

18. If the phrase "in a plane" is removed from the statement of Theorem 4–11, is the resulting statement true? Explain.

4.8 Parallel Lines and Interior Angles

Objective: To prove and apply theorems about parallel lines and interior angles.

When a transversal intersects two parallel lines, pairs of alternate interior angles appear to be congruent.

Theorem 4–12 If two parallel lines are cut by a transversal, then each pair of alternate interior angles is congruent.

Example

1 Prove Theorem 4–12.

Given: $\ell \parallel m$
n is a transversal.
Prove: $\angle 1 \cong \angle 3$
Proof:

STATEMENTS	REASONS
1. $\ell \parallel m$, n is a transversal.	1. Given
2. $\angle 1 \cong \angle 2$	2. Vertical angles are congruent.
3. $\angle 2 \cong \angle 3$	3. If two parallel lines are cut by a transversal, then corresponding angles are congruent.
4. $\angle 1 \cong \angle 3$	4. Congruence of angles is transitive.

Theorem 4–13 In a plane, if two lines are cut by a transversal so that the alternate interior angles are congruent, then the two lines are parallel.

The proof of the following theorem depends on Theorem 4-13.

Theorem 4-14

> In a plane, if two lines are cut by a transversal so that the interior angles on the same side of the transversal are supplementary, then the lines are parallel.

The converse of Theorem 4-14 is also true.

Theorem 4-15

> If two parallel lines are cut by a transversal, then the interior angles on the same side of the transversal are supplementary.

You will be asked to prove these theorems in the exercises.

Exercises

Max: 1-6, 1-35, Challenge; Avg: 1-6, 1-26, 27-35 odds; Min: 1-6, 1-35 odds

Exploratory Refer to the figure at the right. Given the following measures, find the measures of the remaining angles if $\ell \parallel m$. (Hint: In exercises 7-8, solve for x first.) For answers to exercises 1-8, see the Teacher's Guide.

1. $m\angle 1 = 50$
2. $m\angle 5 = 20$
3. $m\angle 3 = 120$
4. $m\angle 6 = 112$
5. $m\angle 4 = 35$
6. $m\angle 8 = 73$
7. $m\angle 2 = x + 16$
   $m\angle 5 = 3x$
8. $m\angle 4 = x + 20$
   $m\angle 7 = 2x$

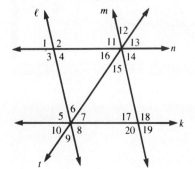

Written Refer to the figure at the left. Given the following information, state which lines are parallel. If none are parallel, write *none*.

1. $\angle 4 \cong \angle 8$ $n \parallel k$
2. $\angle 4 \cong \angle 17$ none
3. $\angle 4 \cong \angle 11$ $\ell \parallel m$
4. $\angle 16 \cong \angle 7$ $n \parallel k$
5. $\angle 11 \cong \angle 1$ $\ell \parallel m$
6. $\angle 5 \cong \angle 18$ none
7. $\angle 11 \cong \angle 19$ $n \parallel k$
8. $\angle 16 \cong \angle 6$ none

In exercises 9–12, $\overline{AB} \parallel \overline{CD}$. State whether or not ∠1 and ∠2 are congruent alternate interior angles. Justify your answer.

9. ∠1 ≅ ∠2; Theorem 4-12

10. ∠1 ≅ ∠2; Theorem 4-12

11. ∠1 ≇ ∠2; They are not alternate interior angles.

12. ∠1 ≅ ∠2; Theorem 4-12

Refer to the figure at the right. For each of the following, state the axiom or theorem that justifies the conclusion.

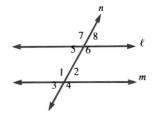

13. Given: ∠6 ≅ ∠1; Conclusion: $\ell \parallel m$ Theorem 4-13
14. Given: ∠6 ≅ ∠4; Conclusion: $\ell \parallel m$ Axiom 4-15
15. Given: $\ell \parallel m$; Conclusion: ∠1 ≅ ∠7 Axiom 4-14

In the figure at the left, $\overleftrightarrow{AB} \parallel \overleftrightarrow{CD}$. Answer each of the following.

16. Find an angle congruent to ∠2. ∠4
17. Find an angle congruent to ∠8. ∠4 + ∠3, ∠6
18. Find two angles congruent to ∠1. ∠5, ∠7
19. If m∠1 = 70 and m∠3 = 50, find m∠2. 60
20. If m∠6 = 120 and m∠2 = 40, find m∠3. 80

For answers to exercises 21–26, see the Teacher's Guide.
Write a two-column proof of each of the following.

21. Theorem 4–13   22. Theorem 4–14   23. Theorem 4–15
24. Given: $\ell \parallel m$
             $n \parallel k$
    Prove: ∠2 ≅ ∠4
25. Given: ∠3 ≅ ∠4
             $n \parallel k$
    Prove: ∠2 ≅ ∠4
26. Given: $\ell \parallel m$
             $n \parallel k$
    Prove: ∠1 ≅ ∠5

Exercises 24–28

27. The intersection of lines ℓ, m, n, and k forms 16 angles. What is the probability of choosing one that is an alternate interior angle with ∠1? $\frac{1}{16}$

28. The intersection of lines ℓ, m, n, and k forms 16 angles. What is the probability of choosing one that is an alternate interior angle with ∠3? $\frac{1}{8}$

For each figure, find the value of x so that ℓ ∥ m.

29. 21

30. 36

31. 14

32. 6

33. 20

34. 10

35. In the figure at the right, find the values of x and y if $\overline{AF} \parallel \overline{BC}$ and $\overline{AE} \parallel \overline{CD}$.

x = 49, y = 76

Challenge Write a two-column proof of the following statement.

If two parallel lines are cut by a transversal, then the bisectors of the alternate interior angles are parallel.

Mixed Review

Solve each inequality in $(\mathcal{R}, +, \cdot)$.

1. $x - 7 < -1$
   x < 6
2. $3a + 14 > 5$
   a > -3
3. $5n + 24 > 9n$
   n < 6
4. $7 - 2t < 4t - 11$
   t > 3

Find $m\angle 1$ and $m\angle 2$ if the following conditions are true.

5. $\angle 1$ and $\angle 2$ are corresponding angles formed when parallel lines ℓ and m are cut by transversal t; $m\angle 1 = 12x + 11$; $m\angle 2 = 7x + 36$ 71, 71

6. $\angle 1$ and $\angle 2$ are alternate interior angles formed when parallel lines ℓ and m are cut by transversal t; $m\angle 1 = 91 - 5x$; $m\angle 2 = 8x + 65$ 81, 81

7. $\angle 1$ and $\angle 2$ are interior angles on the same side of transversal t, which intersects parallel lines ℓ and m; $m\angle 1 = 91 - 5x$; $m\angle 2 = 8x + 65$ 51, 129

8. $\angle 1$ and $\angle 2$ are supplementary angles; $m\angle 1 = 67 - 2x$; $m\angle 2 = 6x + 11$ 16, 164

9. $\angle 1$ and $\angle 2$ are vertical angles; $m\angle 1 = 67 - 2x$; $m\angle 2 = 6x + 11$ 53, 53

10. $\angle 1$ and $\angle 2$ are complementary angles; $m\angle 1 = 67 - 2x$; $m\angle 2 = 6x + 11$ 61, 29

4.9 Indirect Proof

Objective: To prove theorems using an indirect method of proof.

The theorems proved thus far have used a direct method of proof. Some theorems, however, are more easily proved using an indirect method of proof.

Steps for an Indirect Proof

1. Assume that the statement is not true.
2. Show that this assumption leads to a contradiction of given information, a definition, an axiom, or a previously proved theorem.
3. Conclude that the statement must be true.

Example

1 Prove that if lines ℓ and m are *not* parallel, then $\angle 1 \not\cong \angle 2$.

Given: $\ell \not\parallel m$
Prove: $\angle 1 \not\cong \angle 2$
Proof:

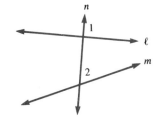

Assume $\angle 1 \cong \angle 2$. Therefore, by Axiom 4–15, if $\angle 1 \cong \angle 2$, then $\ell \parallel m$. However, this contradicts the given statement that $\ell \not\parallel m$. Because the assumption that $\angle 1 \cong \angle 2$ has led to a contradiction, we conclude that $\angle 1 \not\cong \angle 2$. Therefore, if lines ℓ and m are *not* parallel, then $\angle 1 \not\cong \angle 2$.

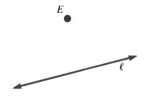

Study the figure at the left. How many lines through point E are parallel to line ℓ? How many lines through point E are perpendicular to line ℓ? Your answers should suggest Theorems 4–16 and 4–17.

Theorem 4–16

Through a point *not* on a given line, there exists one and only one line parallel to the given line.

Example

2 **Prove Theorem 4–16.**

Given: Line ℓ and point E *not* on ℓ
Prove: Two statements must be proved.
  I. There is one line through E that is parallel to ℓ.
  II. There is only one line through E that is parallel to ℓ.

Proof:
Part I

STATEMENTS	REASONS
1. Choose point A on line ℓ.	1. A line contains at least two points.
2. Draw $\overleftrightarrow{AE}$.	2. For any two points, there exists one and only one line containing both of them.
3. Draw $\overleftrightarrow{EC}$ so that $m\angle CEA = m\angle 2$.	3. Protractor Axiom
4. $\angle CEA \cong \angle 2$	4. Definition of congruent angles
5. $\overleftrightarrow{CE} \parallel \ell$	5. In a plane, if two lines are cut by a transversal so that the alternate interior angles are congruent, the two lines are parallel.

Part II

Assume that there is more than one line through E parallel to ℓ. Suppose one line is $\overleftrightarrow{EP}$ and another is $\overleftrightarrow{EC}$. Then $\angle 2 \cong \angle 1$ and $\angle 3 \cong \angle 1$, because they are corresponding angles of parallel lines. By the Transitive Property $\angle 2 \cong \angle 3$. However, $\overrightarrow{EP}$ and $\overrightarrow{EC}$ are different rays. Therefore, $\angle 2$ cannot be congruent to $\angle 3$.

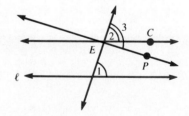

Thus, the assumption that there is more than one line through E parallel to ℓ has led to a contradiction. Therefore, through a point E *not* on ℓ, there exists one and only one line parallel to ℓ.

| Theorem 4–17 | Through a point *not* on a line, there exists one and only one line perpendicular to the given line. |

Theorem 4–18 follows from Theorem 4–16.

| Theorem 4–18 | In a plane, if two lines are parallel to the same line, then they are parallel to each other. |

You will prove these theorems in the exercises.

Exercises

Max: 1-10, 1-11; Avg: 1-10, 1-11; Min: 1-9 odds, 1-11

Exploratory State the first step in an indirect proof of each of the following.
1. $x = y$ $x \neq y$
2. $\ell \parallel m$ $\ell \not\parallel m$
3. $\overline{AB} \perp \overline{CD}$ $\overline{AB} \not\perp \overline{CD}$
4. $\overline{AB} \cong \overline{CD}$ $\overline{AB} \not\cong \overline{CD}$
5. $a \neq b$ $a = b$
6. $a \leq b$ $a > b$
7. Jim is taller than Joe.
8. A, B, and C are collinear.
9. Angle x is an acute angle.
10. If it rains, then it thunders.

For answers to exercises 7-10, see the Teacher's Guide.

Written Write an indirect proof of each of the following.

1. Given: $\ell \not\parallel m$
   Prove: $\angle 1 \not\cong \angle 2$
2. Given: $\angle 1 \not\cong \angle 2$
   Prove: $\ell \not\parallel m$
3. Given: $\ell \not\parallel m$
   $n \perp \ell$
   Prove: n is not perpendicular to m.

For answers to exercises 1-11, see the Teacher's Guide.

Answer the following.

4. How many lines that are parallel to line ℓ can be drawn through P?
5. How many lines that are perpendicular to ℓ can be drawn through P?
6. Sara claims that it is possible for a line to go through P, not parallel to ℓ, yet not intersect ℓ. Is she correct? Explain.

Prove the following theorems using indirect proofs.

7. Theorem 4–17
8. Theorem 4–18

For each statement, name a theorem that justifies the statement.

9. Draw $\overleftrightarrow{QT}$ so that $\overline{QT} \parallel \overline{PR}$.
10. Draw $\overline{SV}$ so that $\overline{PR} \parallel \overline{SV}$.
11. Draw $\overline{SV}$ so that $\angle QSV \cong \angle SPR$.

4.10 Angles of a Triangle

Objective: To prove and apply theorems about angles of a triangle.

Can a triangle contain more than one obtuse angle? More than one right angle? From past experience, you probably know that the answer is no. You can now *prove* that the answer is no.

Theorem 4–19 The sum of the degree measures of the angles of a triangle is 180.

Example

1 Prove Theorem 4–19.

Given: $\triangle ABC$
Prove: $m\angle 1 + m\angle 2 + m\angle 3 = 180$
Proof:

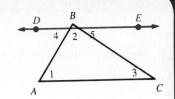

STATEMENTS	REASONS
1. Through B, draw $\overleftrightarrow{DE} \parallel \overline{AC}$.	1. Through a point not on a given line, there exists one and only one line parallel to the given line.
2. $\angle 1 \cong \angle 4$, $\angle 3 \cong \angle 5$	2. If two parallel lines are cut by a transversal, then each pair of alternate interior angles is congruent.
3. $\angle 4$ and $\angle ABE$ are supplementary.	3. Supplement Axiom
4. $m\angle 4 + m\angle ABE = 180$	4. Definition of supplementary angles
5. $m\angle ABE = m\angle 2 + m\angle 5$	5. Angle Addition Axiom
6. $m\angle 4 + m\angle 2 + m\angle 5 = 180$	6. Substitution Axiom
7. $m\angle 1 = m\angle 4$, $m\angle 3 = m\angle 5$	7. Definition of congruent angles
8. $m\angle 1 + m\angle 2 + m\angle 3 = 180$	8. Substitution Axiom

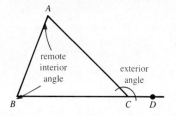

Suppose $\overrightarrow{CD}$ and $\overrightarrow{CB}$ are opposite rays that meet at vertex C of $\triangle ABC$. Angle ACD is called an **exterior angle** of $\triangle ABC$. The two angles, $\angle A$ and $\angle B$, that are not adjacent to $\angle ACD$ are called **remote interior angles**.

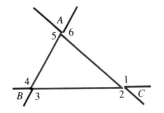

The figure shows that extending each side of a triangle produces *six exterior angles* of a triangle. The angles numbered 1 through 6 are the exterior angles. With respect to ∠6, ∠ABC and ∠ACB are *remote interior angles*.

**Theorem 4–20
Exterior Angle
Theorem**

The measure of an exterior angle of a triangle is equal to the sum of the measures of its two remote interior angles.

**Definition of
Corollary**

A **corollary** is a theorem that follows directly from another theorem.

Theorem 4–20 and the following theorems are corollaries to Theorem 4–19.

Theorem 4–21

If two angles of one triangle are congruent to two angles of another triangle, then the third pair of angles is congruent.

Theorem 4–22

In a triangle, there can be *no* more than one right angle or one obtuse angle.

Theorem 4–23

The acute angles of a right triangle are complementary.

Theorem 4–24

Each angle of an equiangular triangle has a measure of 60.

Exercises

Max: 1-6, 1-16, 1-3; Avg: 1-6, 1-16; Min: 1-6, 1-15 odds

Exploratory State whether each statement is *always, sometimes,* or *never* true. Justify your answers. For answers to exercises 1-6, see the Teacher's Guide.

1. An acute triangle has exactly one acute angle.
2. A right triangle has an obtuse angle.
3. A triangle has two right angles.
4. A right triangle has two congruent acute angles.
5. An exterior angle of a triangle is acute.
6. An exterior angle of a triangle is right.

Written For each of the following, use the information given about △ABC to find each angle of the triangle.

1. $m\angle C = 90, m\angle A = 40, m\angle B = x$ m∠B = 50
2. $m\angle C = 90, m\angle A = x + 20, m\angle B = x$ m∠A = 55, m∠B = 35
3. $m\angle C = 2x - 10, m\angle A = 2x, m\angle B = x$ m∠C = 66, m∠A = 76, m∠B = 38
4. Exterior $\angle CAD$ has a measure of $4(45 - x), m\angle B = x + 6, m\angle C = x$.
   m∠B = 35, m∠C = 29, m∠A = 116

Use the figures below to find each of the following.

5. Find $m\angle C$.
   64

6. Find $m\angle B$.
   80

7. Find $m\angle ABD$.
   Find $m\angle CBD$.
   m∠ABD = 25
   m∠CBD = 52

8. $\overleftrightarrow{CD} \parallel \overline{AB}$
   Find $m\angle A$.
   Find $m\angle B$.
   m∠A = 78
   m∠B = 65

9. $\overline{AB} \parallel \overline{DE}$
   Find $m\angle 1$.
   96
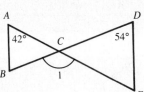

10. $\overline{AB} \parallel \overline{DC}$
    Find $m\angle ADB$.
    55
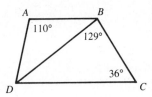

11. $\overline{AB} \parallel \overline{DE}, \overline{BD} \parallel \overline{AE}$
    $m\angle 1 = m\angle 2$
    Find $m\angle ACB$. 50

For answers to exercises 12-16, see the Teacher's Guide.
Write a two-column proof of each theorem. Indirect proofs may be necessary.

12. Theorem 4–20
13. Theorem 4–21
14. Theorem 4–22
15. Theorem 4–23
16. Theorem 4–24

For answers to exercises 1-3, see the Teacher's Guide.
Challenge Write a two-column proof of each of the following.

1. Given: $\ell \parallel m$
   $\overline{BA}$ bisects $\angle CBD$.
   $\overline{CA}$ bisects $\angle BCE$.
   Prove: $\overline{BA} \perp \overline{CA}$

2. The sum of the measures of the angles of a quadrilateral is 360.

3. Any two angles that share a side of a parallelogram are called *consecutive angles*. Prove that the sum of the measures of two consecutive angles of a parallelogram is 180.

Objective: To apply theorems about the angles of a polygon.
4.11 Angles of a Polygon

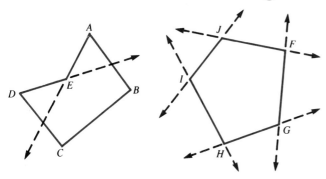

In polygon *ABCDE*, a line containing $\overline{DE}$ or $\overline{AE}$ will pass through the interior of the polygon. This does *not* happen in polygon *FGHIJ*. Polygon *FGHIJ* is an example of a **convex polygon**.

Definition of Convex Polygon

A polygon is convex if and only if any line containing a side of the polygon does *not* contain a point interior to the polygon.

You already know that the sum of the measures of the angles of a triangle is 180. Do you think the sum of the measures of the angles of a quadrilateral is the same for all quadrilaterals? Is the angle sum the same for all pentagons? For octagons? Consider the following polygons.

Quadrilateral

Pentagon

Hexagon

Heptagon

Each polygon has been divided into triangles. The sum of the measures of the angles of each polygon is the same as the sum of the measures of the angles of the triangles. The results for each polygon are shown in the following chart.

Polygon	Number of Triangles	Sum of Measures of Angles
Quadrilateral	2	2(180) = 360
Pentagon	3	3(180) = 540
Hexagon	4	4(180) = 720
Heptagon	5	5(180) = 900
.	.	.
.	.	.
.	.	.
n-gon	*n* − 2	(*n* − 2)180

In advanced mathematics, the following theorem can be proved.

Theorem 4–25

The sum of the measures of the interior angles of any polygon of n sides is $(n - 2)180$.

Definition of Regular Polygon

A **regular polygon** is a polygon that is both equiangular and equilateral.

Example

1 **A dodecagon has 12 sides. Find the measure of an angle of a regular dodecagon.**

sum of the measures $= (n - 2)(180)$
$= (12 - 2)(180)$
$= 10(180)$
$= 1800$

Because a regular polygon is equiangular, the measure of an angle of a regular dodecagon is $1800 \div 12 = 150$.

In the figure at the left, each side of the polygon has been extended to form an exterior angle at each vertex. Use a protractor to find the measure of each exterior angle. Find the sum of the measures of the exterior angles. Repeat this procedure for several different polygons. Are you surprised at your results?

Theorem 4–26

Suppose one exterior angle is drawn at each vertex of any convex polygon. The sum of the measures of these angles is 360.

The following is an argument for the validity of this theorem. At each vertex of any polygon with n sides, the sum of the measures of an interior and an exterior angle is 180. Therefore, because there are n vertices, the sum of the measures of all the interior and exterior angles is $180n$. For convenience, let *sum* refer to "the sum of the degree measures."

$$\text{sum of interior angles} + \text{sum of exterior angles} = 180n$$
$$\text{sum of exterior angles} = 180n - \text{sum interior angles}$$
$$= 180n - (n-2)180$$
$$= 180n - 180n + 360$$
$$= 360$$

Exercises

Max: 1-9, 1-25; Avg: 1-9, 1-25 odds; Min: 1-9, 1-25 odds

Exploratory State whether or not each polygon is convex. Justify your answers.

1.
yes; definition of convex polygon

2.
no; definition of convex polygon

3.
yes; definition of convex polygon

4.
yes; definition of convex polygon

5.
5. no; definition of convex polygon

Use the figure at the right to answer each of the following.

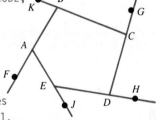

6. Name interior angles of the polygon. ∠ABC, ∠BCD, ∠CDE,
7. Name the exterior angles shown.   ∠DEA, ∠EAB
   7. ∠ABK, ∠GCB, ∠CDH, ∠DEJ, ∠EAF
8. At vertex A, another exterior angle can be formed by extending $\overline{AE}$. Copy this figure and draw in the other five exterior angles.
9. What is true about the measures of the two exterior angles at each vertex? Explain. The measures are equal because the angles are vertical.

8. See students' work.

Written Find the sum of the measures of the interior angles of each *n*-sided polygon.

1. n = 8   1080
2. n = 11   1620
3. n = 20   3240
4. n = 30   5040
5. n = 45   7740

Find the measure of an interior angle of each *n*-sided regular polygon.

6. n = 3   60
7. n = 5   108
8. n = 8   135
9. n = 10   144
10. n = n $\frac{180(n-2)}{n}$

The measures of exterior angles of *n*-sided regular polygons are given. Find *n* for each polygon.

11. 45   8
12. 60   6
13. 36   10
14. 72   5
15. 40   9

Find the measure of an exterior angle of each *n*-sided regular polygon.

16. n = 3   120
17. n = 4   90
18. n = 12   30
19. n = 15   24
20. n = 18   20

Determine whether or not a regular polygon can have an interior angle of the given measure. **Justify your answers.** For answers to exercises 21-25, see the Teacher's Guide.

21. 50
22. 100
23. 120
24. 180
25. 181

Problem Solving Application: Using Equations

Objective: To solve verbal problems by using equations.

More than one equation may be needed to represent the situation described in a problem.

Equations are often used to solve verbal problems. First the problem must be translated into mathematical language, using variables to represent unknown values. Then an equation is written to represent the situation described in the problem. Be sure to check that the solution to this equation satisfies the conditions of the original problem.

Emphasize the importance of using the four-step problem-solving plan.

Examples

1 The sum of the measures of two angles is 127. If the measure of one angle is 43 more than twice the measure of the other, find the measure of each angle.

Explore Let x = the measure of the smaller angle. *Define a variable.*
Then $2x + 43$ = the measure of the larger angle.

Plan We can write an equation to represent the situation described in this problem.

Measure of the smaller angle plus measure of the larger angle is 127.

$$x + 2x + 43 = 127$$

Solve
$$3x + 43 = 127 \quad \text{Solve the equation.}$$
$$3x = 84$$
$$x = 28$$

The measures of the angles are 28 and $2 \cdot 28 + 43$ or 99.

Examine Since 99 is 43 more than 2 times 28 and since $28 + 99 = 127$, the solution is correct.

2 Point B lies on $\overline{AC}$ such that $\overline{AB}$ is three times as long as $\overline{BC}$. If $\overline{AB}$ is 14 cm longer than $\overline{BC}$, find the length of $\overline{AC}$.

Explore Let x = the length of $\overline{BC}$. *Define a variable.*
Then $3x$ = the length of $\overline{AB}$.

Plan Length of $\overline{AB}$ is 14 cm more than length of $\overline{BC}$. *Write an equation.*

$$3x = 14 + x$$

Solve
$$3x = 14 + x \quad \text{Solve the equation.}$$
$$2x = 14$$
$$x = 7$$

The length of $\overline{BC}$ is 7 cm and the length of $\overline{AB}$ is 3(7) or 21 cm. Since B lies on $\overline{AC}$, we know $AB + BC = AC$. Thus, the length of $\overline{AC}$ is 7 + 21 or 28 cm.

Examine Since 3 times 7 cm is 21 cm, and since 21 cm is 14 cm more than 7 cm, the solution is correct.

Exercises

Max: 1–15; Avg: 1–13; Min: 1–15 odds

Written Solve each problem.

1. Point Y lies on $\overline{XZ}$ such that YZ is 9 less than three times XY. If XZ = 25, find XY and YZ. 8.5, 16.5
2. $\overline{RT}$ is in the interior of $\angle QRS$ and $m\angle QRT$ is 4 ($m\angle TRS$). If $m\angle TRS$ is 39 less than $m\angle QRT$, find $m\angle QRS$. 65
3. The measure of the supplement of an angle is 4 less than three times the measure of its complement. Find the measure of the angle. 43
4. The measure of the complement of an angle is 5 more than one-sixth the measure of its supplement. Find the measure of the angle. 66
5. The measure of an angle is 9 more than 20% of its complement. Find the measure of the angle. 22.5
6. The measure of $\angle A$ is 40% of the sum of the measures of its complement and supplement. Find the measure of $\angle A$. 60
7. Find three consecutive integers whose sum is 13 more than the least integer. 5, 6, 7
8. Find three consecutive integers whose sum is 4 times the greatest integer. -5, -4, -3
9. Maria took five English tests. Her scores were 78%, 89%, 98%, 67%, and 90%. What must she score on her sixth test so that her average will be 85%? 88%
10. Felipe bought a television on sale for 33% of its original price. The sale price was $123.75 less than the original price. Find the sale price. $251.25
11. Jenny bought some apples for 79¢ per pound and some bananas for 39¢ per pound. She bought 15 pounds of fruit. If her total bill was $7.85, how many pounds of each fruit did she buy?
12. Roland bought 20 amusement park tickets. Adult tickets cost $25.50 and childrens tickets cost $17.50. If his total bill was $414, how many of each type of ticket did he buy? 8 adult, 12 children
13. Twelve years ago, Theo was twice as old as Tina. In four years, the sum of Theo's age and Tina's age will be 44. How old are Theo and Tina now? Theo: 20, Tina: 16
14. Two hours after a truck leaves Rochester traveling 45 miles per hour, a car, traveling 55 miles per hour, leaves to overtake the truck. In how many hours will the car catch the truck? 9 hours
15. New York City and Seattle are 2961 miles apart by train. An express train leaves New York City at the same time a passenger train leaves Seattle. The express train travels 15 miles per hour faster than the passenger train. The two trains pass each other in 21 hours. How fast is each train traveling? passenger: 63 mph, express: 78 mph

11. 5 pounds of apples, 10 pounds of bananas

Portfolio Suggestion

Select an item from this chapter that you feel shows your best work and place it in your portfolio. Explain why you selected it.

Performance Assessment

Terry is building a fence around his yard. The design of the back fence is shown below. **Sample answers are given.**

b. ∠BDA and ∠ADF are adjacent and supplementary angles.
a. Name two adjacent angles. ∠**BDA**, ∠**ADC**
b. Give two ways that ∠BDA and ∠ADF are related.
c. Name two supplementary angles. ∠**ADB**, ∠**ADF**
d. If $\overline{AG} \parallel \overline{BH}$, name two pairs of alternate interior angles. ∠**CED**, ∠**EDF**
e. Name a right angle, an acute angle, and an obtuse angle.
f. It appears that ∠DEF ≅ ∠HEF. Can you assume that this is true? Explain.
No; you cannot assume that the angles are right angles unless they are marked.
e. acute: ∠**ADB**, right: ∠**ABD**, obtuse: ∠**ADF**

Chapter Summary

1. Axiom 4–1: A **line** contains at least two **points**. A **plane** contains at least three noncollinear points. (125)
2. Axiom 4–2: For any two points, there exists one and only one line containing both of them. (125)
3. Axiom 4–3: For any three **noncollinear** points, there exists one and only one plane containing all three points. (125)
4. Axiom 4–4: If two points are in a plane, then the line containing the two points is also in that plane. (125)
5. Axiom 4–5: If two planes intersect, then their **intersection** is a line. (125)

6. Axiom 4–6: For any two points on a line and a given unit of measure, there is a unique positive number called the measure of the distance between the two points. (125)
7. Axiom 4–7 (**Segment Addition Axiom**): If point B is **between** points A and C, then $AB + BC = AC$. (126)
8. Axiom 4–8: A **line segment** has one and only one **midpoint.** (126)
9. Axiom 4–9 (**Angle Measure Axiom**): For every angle, there is one and only one real number between 0 and 180 called the **degree measure** of the angle. (128)
10. Axiom 4–10 (Protractor Axiom): Suppose $\mathcal{H}$ is a **half-plane** determined by $\overleftrightarrow{OA}$. Then for any real number r such that $0 < r < 180$, there is one and only one ray, $\overrightarrow{OB}$, such that B is in $\mathcal{H}$ and $m\angle AOB = r$. (128)
11. Axiom 4–11 (**Angle Addition Axiom**): If B is in the interior of $\angle AOC$, then $m\angle AOB + m\angle BOC = m\angle AOC$. (128)
12. Axiom 4–12: An angle has one and only one **bisector.** (129)
13. Axiom 4–13 (Supplement Axiom): If the exterior sides of two **adjacent angles** are opposite rays, then the angles are **supplementary.** (130)
14. Addition Axiom: If $a = b$ and $c = d$, then $a + c = b + d$. (132)
15. Subtraction Axiom: If $a = b$ and $c = d$, then $a - c = b - d$. (132)
16. Multiplication Axiom: If $a = b$ and $c = d$, then $ac = bd$. (132)
17. Division Axiom: If $a = b$ and $c \neq 0$, then $\dfrac{a}{c} = \dfrac{b}{c}$. (132)
18. Reflexive Axiom: Any number is equal to itself, or $a = a$. (132)
19. Symmetric Axiom: If $a = b$, then $b = a$. (132)
20. Transitive Axiom: If $a = b$ and $b = c$, then $a = c$. (132)
21. Substitution Axiom: If $a = b$, then a can be replaced by b (and b by a).
22. Theorem 4–1: If two angles are supplementary to the same angle, then they are **congruent** to each other. (135)
23. Theorem 4–2: If two angles are supplementary to congruent angles, then they are congruent to each other. (136)
24. Theorem 4–3: If two angles are **complementary** to the same or congruent angles, then they are congruent to each other. (136)
25. Theorem 4–4: If two lines intersect, the **vertical angles** formed are congruent. (136)
26. Theorem 4–5: Congruence of segments is reflexive, symmetric, and transitive. (136)
27. Theorem 4–6: Congruence of angles is reflexive, symmetric, and transitive. (136)
28. Theorem 4–7: If two angles are **right angles,** then the angles are congruent. (139)
29. Theorem 4–8: If two angles are congruent and supplementary, then each is a right angle. (139)

30. Theorem 4-9: If two intersecting lines form two congruent adjacent angles, then the lines are **perpendicular.** (140)
31. Axiom 4-14: If two **parallel lines** are cut by a **transversal,** then the **corresponding angles** are congruent. (144)
32. Axiom 4-15: In a plane, if two lines are cut by a transversal so that the corresponding angles are congruent, then the two lines are parallel. (144)
33. Theorem 4-10: If a transversal is perpendicular to one of two parallel lines, it is perpendicular to the other. (144)
34. Theorem 4-11: In a plane, if two lines are perpendicular to the same line, then they are parallel. (145)
35. Theorem 4-12: If two parallel lines are cut by a transversal, then each pair of the **alternate interior angles** is congruent. (147)
36. Theorem 4-13: In a plane, if two lines are cut by a transversal so that the alternate interior angles are congruent, then the two lines are parallel. (147)
37. Theorem 4-14: In a plane, if two lines are cut by a transversal so that the **interior angles** on the same side of the transversal are supplementary, then the lines are parallel. (148)
38. Theorem 4-15: If two parallel lines are cut by a transversal, then the interior angles on the same side of the transversal are supplementary. (148)
39. Theorem 4-16: Through a point *not* on a given line, there exists one and only one line parallel to the given line. (151)
40. Theorem 4-17: Through a point *not* on a line, there exists one and only one line perpendicular to the given line. (153)
41. Theorem 4-18: In a plane, if two lines are parallel to the same line, then they are parallel to each other. (153)
42. Theorem 4-19: The sum of the degree measures of the angles of a triangle is 180. (154)
43. Theorem 4-20 (Exterior Angle Theorem): The measure of an **exterior angle** of a triangle is equal to the sum of the measures of its two remote interior angles. (155)
44. Theorem 4-21: If two angles of one triangle are congruent to two angles of another triangle, then the third pair of angles is congruent. (155)
45. Theorem 4-22: In a triangle, there can be no more than one right angle or one **obtuse angle.** (155)
46. Theorem 4-23: The **acute angles** of a right triangle are complementary. (155)
47. Theorem 4-24: Each angle of an **equiangular triangle** has a measure of 60. (153)
48. Theorem 4-25: The sum of the measures of the interior angles of any **polygon** of n sides is $(n - 2)180$. (158)
49. Theorem 4-26: Suppose one exterior angle is drawn at each **vertex** of any **convex polygon.** The sum of the measures of these angles is 360. (158)

Chapter Review

For answers to exercises 1-23, see the Teacher's Guide.

4.1 State whether each of the following is *always, sometimes,* or *never* true. Justify your answer.

1. If A is between B and C, then A, B, and C are collinear.
2. ∠ABC and ∠CBD are vertical angles.
3. $\overrightarrow{AB}$ and $\overrightarrow{BA}$ are opposite rays.
4. S is the vertex of ∠RST.

4.2
5. Axioms must be proved.
6. Three points are coplanar.
7. AB + BC = AC
8. The intersection of two planes is a line segment.

4.3
9. If m∠ABC = 42 and m∠CBF = 48, then ∠ABC and ∠CBF are complementary.
10. If A, B, and C are collinear, then ∠ABD and ∠DBC are supplementary.
11. $\overrightarrow{QT}$ bisects ∠QRS.
12. If $\overrightarrow{MR}$ bisects ∠LMN, then ∠LMR ≅ ∠RMN.

4.4
13. If $t = x$, then $\dfrac{t}{a} = \dfrac{x}{a}$.
14. If m∠A = m∠B and m∠B = m∠C, then ∠A ≅ ∠C.

4.5
15. If m∠ABC = m∠DBF, then ∠ABC and ∠DBF are vertical angles.
16. If ∠A and ∠B are supplementary and ∠B and ∠C are supplementary, then m∠A = m∠B.

4.6
17. If two lines are perpendicular, they form four right angles.
18. A bisector of a line segment is perpendicular to the line segment.

4.7
19. If $\overrightarrow{AB}$ and $\overleftrightarrow{CD}$ do *not* intersect, they are parallel.
20. Corresponding angles are congruent.

4.8 Prove each of the following.
21. Given: ℓ ∥ m, n ∥ k
    Prove: ∠2 ≅ ∠5
22. Given: ∠2 ≅ ∠4, ℓ ∥ m
    Prove: k ∥ n

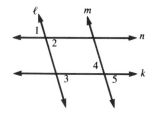

4.9
23. Given: ∠1 ≇ ∠3
    Prove: n ∦ k

4.10 Find a value for *x*, given the following information about △ABC.

24. m∠C = 2x + 5, m∠A = 2x, m∠B = x   35
25. Exterior ∠DCA has a measure of 130, m∠B = 3x, m∠A = x   32.5

4.11 Find the number of sides of a convex polygon given the following sum of the measures of interior angles.

26. 1080   8
27. 900   7
28. 1440   10

Chapter Test

Refer to the figure at the right. Name each of the following.
Answers to exercises 1–3 and 5–10 may vary.
1. three noncollinear points
2. two vertical angles
3. two opposite rays
4. two parallel lines $\overleftrightarrow{EG}, \overleftrightarrow{AB}$
5. a pair of complementary angles
6. two alternate interior angles
7. two corresponding angles
8. an exterior angle of a triangle
9. a right angle
10. two angles that are both congruent and supplementary.

Suppose T is the midpoint of $\overline{KL}$, and $TL = 7$. Find each of the following.
11. KT 7
12. KL 14
13. $KT + TL$ 14

State the axiom that justifies each of the following statements.
14. If $m\angle A = m\angle B$, then $m\angle B = m\angle A$.
    Symmetric axiom
15. If $AC = BD$, then $3AC = 3BD$.
    Multiplication axiom

Answer each of the following.
16. The measure of the supplement of an angle is 60 more than three times the measure of the angle. Find the measure of the angle. 30
17. The measure of the complement of an angle is 30 less than twice the measure of the angle. Find the measure of the angle. 40
18. The measures of the angles of a triangle are $x + 30$, $3x$, and x. Find the measure of each angle of the triangle. 60, 90, 30
19. The measure of an exterior angle of a triangle is $5x + 10$. The measures of the remote interior angles are $2x$ and 40. Find the measure of each angle of the triangle. 20, 40, 120
20. Find the sum of the measures of the interior angles of a polygon with 25 sides. 4140
21. Find the measure of an exterior angle of a regular polygon with five sides. 72
22. Find the value of x if $\overline{AB} \perp \overline{BC}$, $\overline{BQ}$ bisects $\angle ABC$, and $m\angle QBC = 4x - 11$. 19

Prove each of the following. For answers to exercises 23–25, see the Teacher's Guide.

23. **Given:** $\angle 2 \cong \angle 3$
    Prove: $\angle 1 \cong \angle 4$

24. **Given:** $\ell \parallel m$, $n \parallel p$
    Prove: $\angle 1 \cong \angle 4$

25. **Given:** $\angle 1 \not\cong \angle 4$
    Prove: $\ell \not\parallel m$

Congruent Triangles

CHAPTER 5

Application in Sports

The America's Cup race is the most famous international yachting competition. The race was originally called the Hundred-Guinea Cup, but was renamed for the *America*, the first yacht from the United States to win the competition. In 1987, *Stars and Stripes* bought the America's Cup back to the United States after it had been lost to Australia in 1983. The mainsail for *Stars and Stripes*, like the mainsails for most modern sailboats, is shaped like a **right triangle**.

Donna Owens is having a new foresail made for her sailboat to replace the old foresail whose dimensions are shown at the right. What is the minimum amount of information that Ms. Owens could give the sailmaker so that he could make a new sail in the same size and shape as the old one?   **See Teacher's Answer Key, p. T120.**

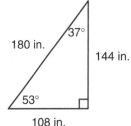

Individual Project: Manufacturing

When products are mass produced, each piece must be interchangeable, so they must have the exact same size and shape. Investigate the impact that mass production had on business at its introduction. Who had the first assembly line? How did the use of the assembly line affect competition? What types of products are best suited to assembly line production? Report your findings in a one-page paper.

5.1 Introduction

Objective: To identify corresponding parts of congruent triangles.

Some ideas about congruence in geometry may be familiar to you. To begin, consider the two triangles shown at the right. Are they congruent? You might think they are **congruent triangles** if you could place one of the triangles over the other such that they fit "exactly." In this case, it is possible to place $\triangle ABC$ over $\triangle DEF$ so that vertex A corresponds to vertex D, vertex B corresponds to vertex E, and vertex C corresponds to vertex F.

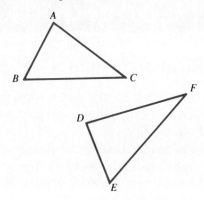

Each angle and each side of $\triangle ABC$ corresponds to an angle and a side of $\triangle DEF$.

Corresponding Angles	Corresponding Sides
$\angle A$ and $\angle D$	$\overline{AB}$ and $\overline{DE}$
$\angle B$ and $\angle E$	$\overline{BC}$ and $\overline{EF}$
$\angle C$ and $\angle F$	$\overline{AC}$ and $\overline{DF}$

Clearly, for the two triangles to be congruent, all pairs of **corresponding parts** must be congruent.

Definition of Congruent Triangles

> Two triangles are congruent if, and only if, there is a correspondence between the vertices such that each side and each angle of one triangle is congruent to the corresponding part of the other triangle.

$\angle R \cong \angle W$
$\angle S \cong \angle X$
$\angle T \cong \angle Y$

$\overline{RS} \cong \overline{WX}$
$\overline{ST} \cong \overline{XY}$
$\overline{RT} \cong \overline{WY}$

To show the pairs of corresponding parts in two triangles, you can mark them with the same number of slashes, as shown on the left.

Also, when writing $\triangle RST \cong \triangle WXY$, it is understood that point R corresponds to point W, S corresponds to X, and T corresponds to Y. Therefore, it is *not* correct to write $\triangle RST \cong \triangle WYX$, because S does not correspond to Y, and T does not correspond to X.

When a statement like $\overline{RS} \cong \overline{WX}$, or $\angle R \cong \angle W$, is written, we know from our definition of congruence of segments and angles that $RS = WX$ and $m\angle R = m\angle W$. That is, $\overline{RS} \cong \overline{WX}$ is logically equivalent to $RS = WX$, and $\angle R \cong \angle W$ is logically equivalent to $m\angle R = m\angle W$. For convenience, often $RS = WX$ is written when it is clear that $\overline{RS} \cong \overline{WX}$, and vice versa.

Similarly, $m\angle R = m\angle W$ and $\angle R \cong \angle W$ are often used interchangeably.

Recall also that an angle of a triangle whose two rays contain the sides of the triangle is called an *included angle* of the two sides. In the figure shown on the preceding page, $\angle S$ is included between $\overline{RS}$ and $\overline{ST}$. A side of a triangle is said to be an included side of two angles if the endpoints are the vertices of the angles. In the figure, $\overline{ST}$ is included between $\angle S$ and $\angle T$.

Previously, it was proved that congruence of segments and congruence of angles are reflexive, symmetric, and transitive. Do you think it is the case that congruence of triangles is also reflexive, symmetric, and transitive? The answer is left for exercises 11–13.

Exercises

Max: 1-23, 1-23; Avg: 1-3, 5-23 odds, 1-19; Min: 1-23 odds, 1-7 odds, 9-13, 15-19 odds

Exploratory   Answer the following.

1. Explain what it means for two figures to be *congruent*. Congruent figures have the same shape and the same size.
2. What is meant by the term *corresponding parts*? logical pairings of angles and sides
3. What relationship exists between corresponding parts of congruent triangles? Corresponding parts of congruent triangles are congruent.

Suppose $\triangle ABC \cong \triangle KML$. For each of the following, name the corresponding part.

4. $\angle A$ $\angle K$ 5. $\overline{ML}$ $\overline{BC}$ 6. $\angle C$ $\angle L$ 7. $\overline{KL}$ $\overline{AC}$ 8. $\angle B$ $\angle M$ 9. $\overline{KM}$ $\overline{AB}$

Refer to the figure. Then name the following.

10. three pairs of corresponding angles $\angle X \cong \angle N$, $\angle W \cong \angle Q$, $\angle Y \cong \angle M$
11. three pairs of corresponding sides $\overline{XW} \cong \overline{NQ}$, $\overline{WY} \cong \overline{QM}$, $\overline{XY} \cong \overline{NM}$
12. the included side of $\angle M$ and $\angle N$ $\overline{NM}$
13. the included angle for $\overline{XW}$ and $\overline{YX}$ $\angle X$

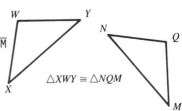
$\triangle XWY \cong \triangle NQM$

The two triangles shown are congruent. On the basis of the figure, tell whether the following statements are *true* or *false*.

14. $\triangle RSQ \cong \triangle UTM$ true
15. $\triangle SQR \cong \triangle TMU$ true
16. $\triangle RQS \cong \triangle TUM$ false
17. $\triangle QRS \cong \triangle MUT$ true
18. $\overline{QS}$ corresponds to $\overline{MT}$. true
19. $\overline{TM} \cong \overline{SR}$ false
20. $\angle S$ corresponds to $\angle T$. true
21. $\overline{QR} \cong \overline{MU}$ true
22. $\overline{QR}$ is the included side of $\angle Q$ and $\angle R$. true
23. $\angle R$ is the included angle of $\overline{SR}$ and $\overline{QR}$. true

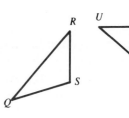

170 CONGRUENT TRIANGLES

For answers to exercises 1–8, see the Teacher's Guide.

Written In exercises 1–8, two congruent triangles are given. Name the corresponding parts of the two triangles.

1. △ABC ≅ △ADC

2. △ACB ≅ △ECD

3. △ABC ≅ △CDA

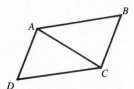

4. △ACD ≅ △CAB
5. △DEA ≅ △BFC
6. △CED ≅ △AFB

7. △ACB ≅ △ADE
8. △ABD ≅ △AEC

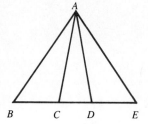

For exercises 9 and 10, see students' work.

9. Draw two triangles with sides measuring 5 cm and 3 cm and an included angle with a measure of 50. What seems to be true about the two triangles? They seem to be congruent.
10. Draw two triangles with two angles with measures of 120 and 15 and an included side of 4 cm. What seems to be true? They seem to be congruent.

For answers to exercises 11–13, see the Teacher's Guide.

Write a two-column proof for the following.

11. Prove that congruence of triangles is reflexive.
12. Prove that congruence of triangles is symmetric.
13. Prove that congruence of triangles is transitive. Explain carefully.

Suppose △DEF ≅ △PQR. Find the value of x given the following conditions.

14. $DE = 10$, $EF = 13$, $DF = 16$, $PQ = 3x - 8$ 6
15. $PQ = 21$, $QR = 15$, $PR = 17$, $FD = 2x + 11$ 3
16. $m\angle DEF = 36$, $m\angle EFD = 76$, $m\angle FDE = 68$, $m\angle QRP = 5x + 6$ 14
17. $m\angle PQR = 113$, $m\angle QRP = 35$, $m\angle QPR = 32$, $m\angle FED = 13x + 22$ 7
18. $m\angle PQR = 42$, $m\angle QRP = 66$, $m\angle RPQ = 72$, $m\angle EDF = 17x - x^2$ 9 or 8
19. $DE = 2x$, $EF = 5x$, $DF = 10x$, $RQ = x^2 - 24$ 8

20. Suppose a triangle can be drawn using three given segments. Can a triangle different from this triangle be drawn using the same three segments? no
21. Repeat Exercise 20 if two segments and the angle included between them are given. no
22. Repeat Exercise 20 if two angles and the segment included between them are given. no
23. Repeat Exercise 20 if two segments and the angle opposite one of them are given. yes

5.2 Some Congruence Axioms

Objective: To identify congruent triangles using SSS, SAS, or ASA.

When two triangles are congruent, there are six pairs of corresponding parts that are congruent. Do you think it is necessary to show that all six pairs of corresponding parts are congruent in order to conclude that two triangles are congruent?

Calculator Hint

You can learn how to graph congruent triangles on a graphing calculator in Activity 1 on page A2.

To investigate this question, consider some examples.

Suppose a triangle is drawn having sides with lengths 3 cm, 4 cm, and 6 cm. Do you think the triangle drawn will necessarily be congruent to △JKL shown in the figure?

If you said yes, then you understand that the size and shape of a triangle are determined by the lengths of its sides. Since this cannot be proved, we will accept it as an axiom.

Axiom 5–1

> If three sides of one triangle are congruent to the corresponding parts of another triangle, then the triangles are congruent.

This axiom is often abbreviated SSS, which stands for *side, side, side*.

If we know that $\overline{AB} \cong \overline{DE}$, $\overline{BC} \cong \overline{EF}$, and $\overline{AC} \cong \overline{DF}$, then, using the SSS axiom, it may be concluded that △ABC ≅ △DEF.

Suppose a triangle is drawn that has one side 3 cm and another side 5 cm. Suppose the angle included between these two sides has a measure of 35. Once the two sides and the angle are drawn, the remaining parts of the triangle appear to be determined. That is, any triangle that satisfies these conditions will be congruent to △WXY, shown above. This idea is stated in the next axiom.

Axiom 5–2 If two sides and the included angle of one triangle are congruent to the corresponding parts of another triangle, then the triangles are congruent.

This axiom is often abbreviated SAS, which stands for *side, angle, side*.

If $\overline{AB} \cong \overline{DE}$, $\overline{BC} \cong \overline{EF}$, and $\angle B \cong \angle E$, then, using the SAS axiom, it may be concluded that $\triangle ABC \cong \triangle DEF$.

This time, begin with a line segment 5 cm long. At one endpoint, draw an angle whose measure is 62. At the other endpoint, draw an angle whose measure is 37.

It appears that only one such triangle with the given dimensions can be drawn.

This is an example of the next congruence axiom.

Axiom 5–3 If two angles and the included side of one triangle are congruent to the corresponding parts of another triangle, then the triangles are congruent.

This axiom is often abbreviated ASA. As you may have guessed by now, this stands for *angle, side, angle*.

 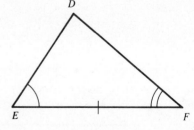

Suppose $\angle B \cong \angle E$, $\overline{BC} \cong \overline{EF}$, and $\angle C \cong \angle F$. Then, using the ASA axiom, it may be concluded that $\triangle ABC \cong \triangle DEF$.

The following example shows the new congruence axioms at work.

Example

1 Write a two-column proof.

Given: $\overline{AC}$ and $\overline{BD}$ bisect each other at M.
Prove: $\triangle ABM \cong \triangle CDM$

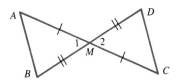

Proof:

STATEMENTS	REASONS
1. $\overline{AC}$ and $\overline{BD}$ bisect each other at M.	1. Given
2. $\overline{AM} \cong \overline{MC}$, $\overline{BM} \cong \overline{MD}$.	2. Definition of bisect
3. $\angle 1 \cong \angle 2$	3. If two lines intersect, the vertical angles formed are congruent.
4. $\triangle ABM \cong \triangle CDM$	4. SAS Axiom

Exercises

Max: 1-11, 1-16; Avg: 1-11, 1-15 odds; Min: 1-11, 1-15 odds

Exploratory Explain what is said in the following congruence axioms.
1. the SSS Axiom See page 171. 2. the SAS Axiom See page 172. 3. the ASA Axiom See page 172.

In exercises 4–8, given conditions are stated for $\triangle ABC$ and $\triangle DEF$. State the axiom that allows you to conclude that $\triangle ABC \cong \triangle DEF$.

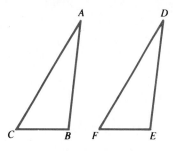

4. $\overline{AC} \cong \overline{DF}$, $\overline{AB} \cong \overline{DE}$, $\overline{CB} \cong \overline{FE}$ SSS
5. $\overline{AB} \cong \overline{DE}$, $\angle B \cong \angle E$, $\overline{CB} \cong \overline{FE}$ SAS
6. $\angle C \cong \angle F$, $\overline{AC} \cong \overline{DF}$, $\overline{CB} \cong \overline{FE}$ SAS
7. $\angle A \cong \angle D$, $\overline{CA} \cong \overline{FD}$, $\overline{AB} \cong \overline{DE}$ SAS
8. $\angle B \cong \angle E$, $\angle A \cong \angle D$, $\overline{BA} \cong \overline{ED}$ ASA

In each figure, congruent parts have been marked. Tell whether we may conclude that the two triangles are congruent. Justify your answer.

9. yes; ASA

10. yes; SAS

11. No, sides can be different lengths.

174 CONGRUENT TRIANGLES

Written In each of exercises 1–6, tell what additional information you must have in order to use the indicated axiom to prove that △ABC ≅ △DEF.

Sample: Given: $\overline{AB} \cong \overline{DE}$. Use SSS.
To use SSS, we must show that $\overline{BC} \cong \overline{EF}$ and $\overline{AC} \cong \overline{DF}$.

1. Given: $\angle A \cong \angle D$. Use SAS. $\overline{AB} \cong \overline{DE}, \overline{AC} \cong \overline{DF}$
2. Given: $\angle A \cong \angle D$. Use ASA
3. Given: $\overline{BC} \cong \overline{EF}$. Use ASA. $\angle B \cong \angle E, \angle C \cong \angle F$
4. Given: $\overline{BC} \cong \overline{EF}$. Use SAS.
5. Given: $\angle A \cong \angle D, \overline{AC} \cong \overline{DF}$. Use ASA. $\angle C \cong \angle F$
6. Given: $\angle A \cong \angle D, \overline{AC} \cong \overline{DF}$. Use SAS. $\overline{AB} \cong \overline{DE}$

2. $\overline{AC} \cong \overline{DF}, \angle C \cong \angle F$ or
   $\overline{AB} \cong \overline{DE}, \angle B \cong \angle E$

4. $\angle C \cong \angle F, \overline{AC} \cong \overline{DF}$ or
   $\angle B \cong \angle E, \overline{AB} \cong \overline{DE}$

Supply reasons for each proof in exercises 7–8.

7. Given: $\angle BAC \cong \angle DAC$
          $\overline{AB} \cong \overline{AD}$
   Prove: △ABC ≅ △ADC

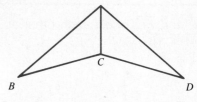

Proof:

STATEMENTS	REASONS
1. $\overline{AB} \cong \overline{AD}, \angle BAC \cong \angle DAC$	1. Given
2. $\overline{AC} \cong \overline{AC}$	2. Theorem 4-5
3. △ABC ≅ △ADC	3. SAS

8. Given: $\overline{AB} \cong \overline{CB}$
          $\overrightarrow{BD}$ bisects $\angle B$.
   Prove: △ABD ≅ △CBD

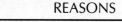

Proof:

STATEMENTS	REASONS
1. $\overline{AB} \cong \overline{CB}, \overrightarrow{BD}$ bisects $\angle B$.	1. Given
2. $\angle CBD \cong \angle ABD$	2. Definition of angle bisector
3. $\overline{BD} \cong \overline{BD}$	3. Theorem 4-5
4. △ABD ≅ △CBD	4. SAS

For answers to exercises 9–16, see the Teacher's Guide.

Write a two-column proof to prove △ABC ≅ △CDA.

9. Given: $\overline{AD} \cong \overline{CB}, \overline{AB} \cong \overline{CD}$
10. Given: $\angle BAC \cong \angle DCA, \overline{AB} \cong \overline{CD}$
11. Given: $\overline{AB} \parallel \overline{CD}, \overline{AB} \cong \overline{CD}$
12. Given: $\overline{BC} \parallel \overline{AD}, \overline{AB} \parallel \overline{DC}$

Write a two-column proof to prove △ABC ≅ △ADC.

13. Given: $\overline{AB} \cong \overline{AD}, \overline{AC}$ bisects $\angle BAD$.
14. Given: C is the midpoint of $\overline{BD}, \overline{AB} \cong \overline{AD}$
15. Given: $\overline{AC} \perp \overline{BD}$, C is the midpoint of $\overline{BD}$.
16. Given: $\overline{AC}$ bisects $\angle BAD, \overline{AC} \perp \overline{BD}$

5.3 Proving Corresponding Parts Congruent

Objective: To apply congruence axioms.

In many situations we wish to prove that a pair of line segments, or angles, are congruent. One way to do this is to show that the line segments or angles are corresponding parts of congruent triangles.

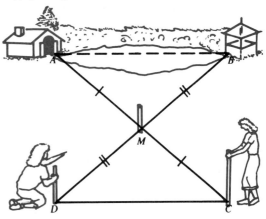

Sara and Jane decide to have a rowing contest across the Pine City Lake, from the boathouse at point A to the gazebo at point B. Both girls are wondering, however, what distance they will row. They decide to use their knowledge of geometry to find out.

First, they choose a convenient place on the shore and mark the spot with a stake. This is point M in the diagram. Then they measure $\overline{AM}$ and $\overline{MB}$. Next, they sight along $\overrightarrow{AM}$ and locate C such that AM = MC. Finally, they sight along $\overrightarrow{BM}$ and locate point D such that BM = MD. They conclude that the length of $\overline{DC}$, which they can measure, is the same as the length of $\overline{AB}$. Are they right? Here is a formal proof that they are correct.

Example

1 Write a two-column proof.

Given: $\overline{AM} \cong \overline{MC}$
$\overline{BM} \cong \overline{MD}$

Prove: AB = DC

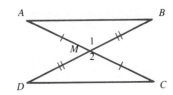

Proof:

STATEMENTS	REASONS
1. $\overline{AM} \cong \overline{MC}$, $\overline{BM} \cong \overline{MD}$	1. Given
2. $\angle 1 \cong \angle 2$	2. If two lines intersect, the vertical angles formed are congruent.
3. $\triangle ABM \cong \triangle CDM$	3. SAS Axiom
4. $\overline{AB} \cong \overline{DC}$	4. CPCTC*
5. AB = DC	5. Definition of congruent segments

*Corresponding Parts of Congruent Triangles are Congruent.

Example

2 Write a two-column proof.

Given: $\overline{AB} \cong \overline{CD}$
$\overline{BC} \cong \overline{DA}$
Prove: $\overline{AB} \parallel \overline{DC}$

Plan: Note that $\overleftrightarrow{BD}$ can be considered as a transversal cutting $\overline{AB}$ and $\overline{CD}$. Angles 1 and 2 are alternate interior angles. But angles 1 and 2 are also corresponding parts of $\triangle ABD$ and $\triangle CDB$, which are congruent by the SSS Axiom. Thus, $\overline{AB} \parallel \overline{CD}$.

Proof:

STATEMENTS	REASONS
1. $\overline{AB} \cong \overline{CD}$, $\overline{BC} \cong \overline{DA}$	1. Given
2. $\overline{BD} \cong \overline{BD}$	2. Congruence of segments is reflexive.
3. $\triangle ABD \cong \triangle CDB$	3. SSS Axiom
4. $\angle 1 \cong \angle 2$	4. CPCTC
5. $\overline{AB} \parallel \overline{DC}$	5. If two lines are cut by a transversal and the alternate interior angles are congruent, then the two lines are parallel.

Exercises

Max: 1-12, 1-21, 1-3; Avg: 1-12, 1-6, 7-21 odds; Min: 1-11 odds, 1-21 odds

Exploratory Name six corresponding parts. In every case, Triangle *I* ≅ Triangle *II*.

1.

2.

3.

4.

5.

6.

For answers to exercises 1-12, see the Teacher's Guide.

CHAPTER 5   177

If △DEF ≅ △RSM, find the value of x or z and the measure of each side of each triangle.

7.   **8.**   **9.**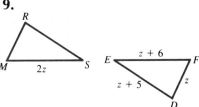

Explain, in detail, how to prove $\overline{PQ} \cong \overline{PR}$ in each of the following.

10.   **11.**   **12.**

Written In each case, △ABC ≅ △DEF. Find the measure of each side of each triangle.

1.   AB = DE = 6
BC = EF = 15
CA = FD = 10

2. 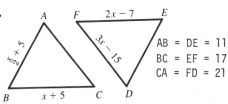   AB = DE = 11
BC = EF = 17
CA = FD = 21

3.   AB = DE = 16
BC = EF = 14
CA = FD = 19

4. 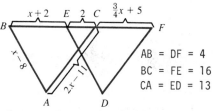   AB = DF = 4
BC = FE = 16
CA = ED = 13

Copy and complete each proof for exercises 5–6. Use the same figure for both.

5. Given: $\overrightarrow{BD} \perp \overline{AC}$, $\overrightarrow{BD}$ bisects ∠ABC.
   Prove: $\overline{AD} \cong \overline{CD}$
   Proof:

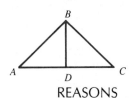

STATEMENTS	REASONS
1. $\overline{BD} \perp \overline{AC}$	1. Given
2. ∠ADB is a right angle. ∠CDB is a right angle.	2. Definition of perpendicular lines
3. ∠ADB ≅ ∠CDB	3. Theorem 4-7
4. $\overrightarrow{BD}$ bisects ∠ABC.	4. Given
5. ∠ABD ≅ ∠CBD	5. Definition of angle bisector

6. $\overline{BD} \cong \overline{BD}$ 6. Theorem 4-5
7. $\triangle ABD \cong \triangle CBD$ 7. ASA
8. $\overline{AD} \cong \overline{CD}$ 8. CPCTC

6. **Given:** $\overline{AB} \cong \overline{CB}$, D is the midpoint of $\overline{AC}$.
   Prove: $\overline{BD} \perp \overline{AC}$
   Proof:

STATEMENTS	REASONS
1. D is the midpoint of $\overline{AC}$.	1. Given
2. $\overline{AD} \cong \overline{CD}$	2. Definition of midpoint
3. $\overline{AB} \cong \overline{CB}$	3. Given
4. $\overline{BD} \cong \overline{BD}$	4. Theorem 4-5
5. $\triangle ABD \cong \triangle CBD$	5. SSS
6. $\angle ADB \cong \angle CDB$	6. CPCTC
7. $\angle ADB$ is _____ to $\angle CDB$. supplementary	7. Supplement Axiom
8. $\angle ADB$ and $\angle CDB$ are <u>right angles</u>	8. If two angles are congruent and supplementary, then each is a right angle.
9. $\overline{BD} \perp \overline{AC}$	9. Definition of perpendicular lines

For answers to exercises 7-18, see the Teacher's Guide.

Write a two-column proof.

7. **Given:** $\overline{ST} \cong \overline{UT}, \angle 3 \cong \angle 4$
   Prove: $\angle 1 \cong \angle 2$
8. **Given:** $\angle 1 \cong \angle 2, \angle 3 \cong \angle 4$
   Prove: $\overline{RS} \cong \overline{RU}$
9. **Given:** $\overline{RS} \cong \overline{RU}, \overline{ST} \cong \overline{UT}$
   Prove: $\angle 1 \cong \angle 2$
10. **Given:** $\angle 1 \cong \angle 2, \overline{RS} \cong \overline{RU}$
    Prove: $\angle S \cong \angle U$

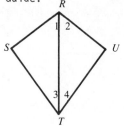
Exercises 7–10

11. **Given:** L is the midpoint of $\overline{KM}$ and $\overline{JN}$.
    Prove: $\overline{JK} \cong \overline{NM}$
12. **Given:** L is the midpoint of $\overline{JN}$, $\angle 2 \cong \angle 6$
    Prove: $\overline{KL} \cong \overline{ML}$
13. **Given:** $\overline{JK} \parallel \overline{NM}, \overline{JK} \cong \overline{NM}$
    Prove: $\overline{JL} \cong \overline{NL}$
14. **Given:** L is the midpoint of $\overline{KM}$ and $\overline{JN}$.
    Prove: $\overline{JK} \parallel \overline{MN}$

Exercises 11–14

15. **Given:** $\overline{BC} \cong \overline{CD}, \angle 2 \cong \angle 4$
    Prove: $\angle D \cong \angle B$
16. **Given:** $\overline{AB} \cong \overline{CD}, \overline{AD} \cong \overline{CB}$
    Prove: $\angle D \cong \angle B$
17. **Given:** $\angle 1 \cong \angle 4, \angle 3 \cong \angle 2$
    Prove: $\overline{AD} \cong \overline{CB}$
18. **Given:** $\overline{AB} \cong \overline{CD}, \overline{AD} \cong \overline{CB}$
    Prove: $\overline{AB} \parallel \overline{CD}$

CHAPTER 5 179

In each case, △ABC ≅ △DEF. Find x, y, AB, DE, BC, and EF.

19. AB = 2x + 2, DE = 9 − y, BC = 5x + 2, EF = 13 − y + x 19. x = 2, y = 3, AB = DE = 6, BC = EF = 12
20. AB = 5x + 2y, DE = 3(6 + x), BC = 3x − y, EF = 1 + x   20. x = $3\frac{1}{3}$, y = $5\frac{2}{3}$, AB = DE = 28, BC = EF = $4\frac{1}{3}$
21. AB = 2(2x + y), DE = 41 − (x + y), BC = 4x + 3, EF = 25 + 3y   21. x = 7, y = 2, AB = DE = 32, BC = EF = 31

Challenge Write a proof for each of the following.
1. Given: $\overline{AB} \cong \overline{CD}$, $\overline{BC} \cong \overline{DA}$, G is the midpoint of $\overline{AC}$.
   Prove: $\overline{EG} \cong \overline{FG}$
2. Given: G is the midpoint of $\overline{DB}$ and $\overline{EF}$.
   Prove: $\overline{AB} \cong \overline{CD}$
3. Given: G is the midpoint of $\overline{AC}$ and $\overline{BD}$.
   Prove: $\overline{EG} \cong \overline{FG}$

For answers to exercises 1-3, see the Teacher's Guide.

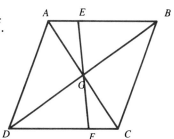

_____ **Mixed Review** _____

In the figure at the right, $\overleftrightarrow{AB} \parallel \overleftrightarrow{CD}$. Find the value of x and m∠4 given the following conditions.

1. m∠2 = 3x − 5, m∠4 = x + 37   x = 21; 58
2. m∠3 = 25, m∠8 = 5x + 2, m∠4 = 2x + 37   x = 20; 77
3. m∠6 = 133, m∠3 = 3x, m∠4 = 3x + 37   x = 16; 85
4. m∠7 = 59, m∠3 = 5x + 12, m∠4 = 4x + 37   x = 8; 69
5. m∠1 = 4x + 41, m∠3 = 2x + 25, m∠4 = 5x + 37   x = 7; 72
6. m∠5 = 6x + 37, m∠3 = 5x − 16, m∠2 = x + 63   x = 8; 71
7. m∠8 = x^2, m∠3 = 4x + 5, m∠2 = 5x + 17   x = 11; 72

8. Suppose $\overline{YD}$ bisects ∠XYZ. If the measure of ∠XYZ is 36 less than three times the measure of ∠XYD, find the measure of ∠XYZ. 36

9. The measure of the supplement of an angle is 6 more than half the measure of the angle. Find the measure of the angle. 116

10. The measures of the angles of a triangle are 4x + 35, 3x + 25, and x^2. Find the measure of each angle of the triangle. 67, 49, 64

In exercises 11–14, tell what additional information you must have in order to use the indicated axiom to prove △ABC ≅ △DEF.

11. ∠B ≅ ∠E. Use SAS. $\overline{AB} \cong \overline{DE}$, $\overline{BC} \cong \overline{EF}$
12. $\overline{AC} \cong \overline{DF}$. Use ASA. ∠A ≅ ∠D, ∠C ≅ ∠F
13. $\overline{AC} \cong \overline{DF}$. Use SSS. $\overline{AB} \cong \overline{DE}$, $\overline{BC} \cong \overline{EF}$
14. ∠B ≅ ∠E. Use ASA. $\overline{AB} \cong \overline{DE}$, ∠A ≅ ∠D or $\overline{BC} \cong \overline{EF}$, ∠C ≅ ∠F

For answer to exercise 15, see the Teacher's Guide.
15. Complete the following proof.
   Given: △ABC with $\overline{AB} \cong \overline{AC}$
   Prove: ∠B ≅ ∠C
   Hint: Think of △ABC as two different triangles, △ABC and △ACB.

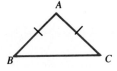

5.4 The AAS Congruence Theorem

Objective: To identify congruent triangles using AAS.

After studying the SSS, SAS, and ASA axioms, a question to be considered is, "Are there AAA, SSA, or AAS congruence principles?"

To investigate the possibility of an AAA principle, look carefully at the two triangles at the right.

Although all three pairs of angles are congruent, the two triangles are clearly not congruent. *Therefore, there cannot be an AAA axiom.*

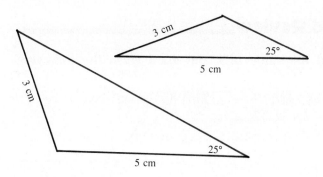

Now consider the two triangles at the left. Each triangle has one 3-cm side, one 5-cm side and one angle that measures 25 *not included between* the given sides.

Clearly, the two triangles are not congruent. There is *no* SSA axiom.

Now, consider the possibility of AAS.

In the two triangles at the right, one angle measures 91, another angle measures 37, and the side opposite the angle whose measure is 37 but not included between the two angles is 3 cm in length.

In this case, it appears that the two triangles are congruent. This is not a chance occurrence. This is a result that can be proved.

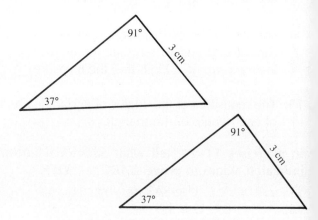

Theorem 5-1
The AAS Theorem

If two angles and a side opposite one of the angles in one triangle are congruent to the corresponding parts of another triangle, then the triangles are congruent.

Example

1 Write a two-column proof of Theorem 5–1.

Given: △ABC and △DEF
∠A ≅ ∠D, ∠B ≅ ∠E
$\overline{AC} \cong \overline{DF}$

Prove: △ABC ≅ △DEF

Proof:

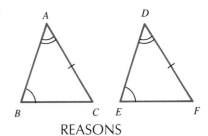

STATEMENTS	REASONS
1. ∠A ≅ ∠D, ∠B ≅ ∠E	1. Given
2. ∠C ≅ ∠F	2. If two angles of one triangle are congruent to two angles of another triangle, then the third pair of angles is congruent.
3. $\overline{AC} \cong \overline{DF}$	3. Given
4. △ABC ≅ △DEF	4. ASA Axiom

Here is a summary of the four ways to prove triangles congruent.

    SSS   SAS   ASA   AAS

Example

2 Write a two-column proof for the following.

Given: ∠S ≅ ∠K, $\overline{SE} \cong \overline{KA}$

Prove: N is the midpoint of $\overline{EA}$.

Proof:

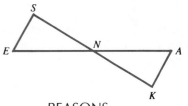

STATEMENTS	REASONS
1. ∠S ≅ ∠K, $\overline{SE} \cong \overline{KA}$	1. Given
2. ∠SNE ≅ ∠KNA	2. If two lines intersect, the vertical angles formed are congruent.
3. △SNE ≅ △KNA	3. AAS
4. $\overline{EN} \cong \overline{AN}$	4. CPCTC
5. N is the midpoint of $\overline{EA}$.	5. Definition of midpoint

Exercises

Max: 1-11, 1-21, Challenge; Avg: 1-11, 1-10, 11-21 odds; Min: 1-11, 1-21 odds

Exploratory Answer the following. For the answers to exercises 1-2., see the Teacher's Guide

1. Explain the difference between ASA and AAS.

2. Is there an SSA principle for establishing congruence of triangles? Explain why or why not.

3. Draw a fully labeled example to illustrate your answer to exercise 2. See students' work.

4. Is there an AAA principle for congruence of triangles? Illustrate your answer with a diagram. No, see students' work.

5. Tara claims that the two right triangles shown in the figure are congruent by AAS. Do you agree? Explain. No, congruent parts do not correspond.

6. For $\triangle ABC$ and $\triangle DEF$, $m\angle A = m\angle D = 30°$; $m\angle C = m\angle F = 50°$, $BC = 2$ cm, $DF = 2$ cm. Is $\triangle ABC \cong \triangle DEF$? Include a diagram in your answer. No, see students' work.

7. Discuss two ways of changing the information in exercise 6 so that your final answer would change. Answers may vary.

In these exercises, two triangles have been marked to show what parts are *given* as congruent. State if the triangles are necessarily congruent and justify your answer.

8. No, congruent parts do not correspond.

9. yes, ASA

10. yes, AAS

11. yes, AAS

Written In each case, determine if the triangles are necessarily congruent. Justify your answers. In each case, include a diagram.

1. $\triangle ABC$ and $\triangle RJK$ with $\overline{AB} \cong \overline{RJ}$, $\angle A \cong \angle R$, and $\angle C \cong \angle K$. yes, AAS

2. no, not enough information
   $\triangle RJK$ and $\triangle DMB$ with $\angle R \cong \angle D$, $\angle J \cong \angle M$, and $\angle K \cong \angle B$.

3. $\triangle DEF$ and $\triangle KLM$ with $\angle D \cong \angle K$, $\angle F \cong \angle M$, $\overline{DF} \cong \overline{KM}$. yes, ASA

4. $\triangle ABD$ and $\triangle CBD$ formed by quadrilateral $ABCD$ and its diagonal $\overline{BD}$ such that $\overline{BD}$ bisects $\angle ABC$.

5. $\triangle AMH$ and $\triangle ATH$ formed by quadrilateral $MATH$ and its diagonal $\overline{AH}$, such that $\overline{AM} \cong \overline{AT}$. no, not enough information

6. $\triangle DRB$ and $\triangle KRJ$ formed by line segments $\overline{JB}$ and $\overline{KD}$ intersecting at R, $\overline{JK}$ and $\overline{DB}$ drawn, and $\overline{RK} \cong \overline{RD}$, $\angle K \cong \angle D$. yes, ASA

7. Ted and Jim have each drawn triangles with sides 4 cm and 7 cm and an angle whose measure is 40 but the two triangles are *not* congruent. Explain. The angle is not included between the sides.

8. Draw two differently shaped triangles that satisfy the conditions of exercise 7. See students' work.

4. no, not enough information

9. Laura says, "If two right triangles have a pair of acute angles congruent, and their hypotenuses congruent, then they must be congruent." Is she right? If she is, *prove* her statement. If she is not, explain why not. yes, AAS

10. Zeb claims that two triangles are congruent if two sides and an angle of one triangle are congruent to the corresponding parts of the other triangle. Do you agree? Explain. No, they are congruent only if the angle is included between the two sides.

Tell whether △SNE and △KNA are congruent under the given conditions. If they *are*, write a full two-column proof.

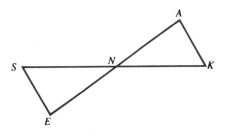

11. ∠A and ∠E are right angles, ∠S ≅ ∠K no
12. ∠A and ∠E are right angles, N is the midpoint of $\overline{SK}$. yes
13. $\overline{AK} \parallel \overline{SE}$, $\overline{SN} \cong \overline{KN}$ yes

For answers to exercises 12-14, see the Teacher's Guide.
Using the figure above, write a two-column proof.
14. **Given:** ∠E ≅ ∠A, N is the midpoint of $\overline{SK}$.
    Prove: △SNE ≅ △KNA

Tell whether or not △ABC and △EDF are congruent under the given conditions. If they are, write a two-column proof.

15. **Given:** ∠A and ∠E are right angles, ∠B ≅ ∠D, ∠ACB ≅ ∠EFD no
16. **Given:** $\overline{AB} \parallel \overline{DE}$, $\overline{AC} \cong \overline{EF}$, ∠ACB ≅ ∠EFD yes
17. **Given:** $\overline{AB} \parallel \overline{DE}$, ∠B ≅ ∠D, $\overline{BC} \cong \overline{DF}$ yes
18. **Given:** $\overline{AB} \parallel \overline{DE}$, $\overline{AC} \cong \overline{EF}$, $\overline{CB} \cong \overline{DF}$ no

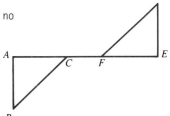

For answers to exercises 16, 17, and 19-21, see the Teacher's Guide.
Write a two-column proof for each of the following.
19. **Given:** ∠4 ≅ ∠3, ∠A ≅ ∠C
    Prove: △ABD ≅ △CDB
20. **Given:** $\overline{DB}$ bisects ∠CDA, ∠A ≅ ∠C
    Prove: △ABD ≅ △CBD
21. **Given:** $\overline{BD}$ bisects ∠ABC, $\overline{AD} \perp \overline{AB}$, $\overline{BC} \perp \overline{DC}$
    Prove: $\overline{AD} \cong \overline{CD}$

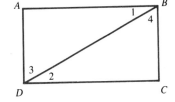

For the answer to the exercise, see the Teacher's Guide.
Challenge Write a proof for the following.
Given: $\overline{TP} \perp \overline{PS}$, $\overline{TP} \parallel \overline{MS}$, $\overline{PQ} \cong \overline{SR}$
∠T ≅ ∠M
Prove: $\overline{TP} \cong \overline{MS}$

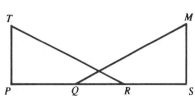

5.5 Triangles That Overlap

Objective: To identify congruent triangles that overlap.

Sometimes it is necessary to show the congruence of two triangles that "overlap." Each of the triangles shares all or part of one or more of its sides or angles with the other triangle. Consider the following examples.

Example

1 Write a plan and a two-column proof for the following.

Given: $\angle B \cong \angle E$
$\angle 1 \cong \angle 2$
$\overline{BD} \cong \overline{EC}$

Prove: $\triangle ABD \cong \triangle AEC$

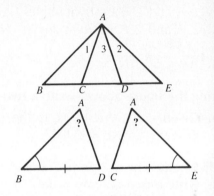

In this case, $\triangle ABD$ and $\triangle AEC$ "partially overlap." They "share" $\angle 3$. Sometimes it is helpful to "take apart" the larger triangle as shown at the right. A new figure is drawn that shows the separate triangles. A plan is developed from these.

Plan: $\angle BAD$ contains $\angle 1$ and $\angle 3$. $\angle CAE$ contains $\angle 3$ and $\angle 2$. If it can be shown that $\angle BAD \cong \angle CAE$, then the proof can be completed by AAS. Remember $\angle 1 \cong \angle 2$ is given.

Proof:

STATEMENTS	REASONS
1. $\angle 1 \cong \angle 2$	1. Given
2. $m\angle 1 = m\angle 2$	2. Definition of congruence
3. $m\angle 3 = m\angle 3$	3. Reflexive Axiom
4. $m\angle 1 + m\angle 3 = m\angle 3 + m\angle 2$	4. Addition Axiom
5. $m\angle 1 + m\angle 3 = m\angle BAD$ $m\angle 3 + m\angle 2 = m\angle EAC$	5. Angle Addition Axiom
6. $m\angle BAD = m\angle EAC$	6. Substitution Axiom
7. $\angle BAD \cong \angle EAC$	7. Definition of congruence
8. $\angle B \cong \angle E$ $\overline{BD} \cong \overline{EC}$	8. Given
9. $\triangle ABD \cong \triangle AEC$	9. AAS Theorem

Sometimes the two triangles are "embedded" in a larger triangle.

Example

2 Write a plan and a two-column proof for the following.

Given: $\angle QPR \cong \angle QRP$
$\overline{QP} \cong \overline{QR}, \overline{QS} \cong \overline{QT}$

Prove: $\triangle SPR \cong \triangle TRP$

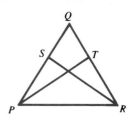

In this case, the two triangles, $\triangle SPR$ and $\triangle TRP$, overlap each other and also share part of the larger triangle, $\triangle QPR$. Angles P and R of the large triangle are also angles of $\triangle SPR$ and $\triangle TRP$. Sides $\overline{SP}$ and $\overline{TR}$ of the smaller triangles are parts of side $\overline{QP}$ and $\overline{QR}$ of the larger triangle.

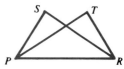

Plan: $\angle QPR \cong \angle QRP$. $\overline{PR} \cong \overline{PR}$ by the Reflexive property. Since $QP = QR$ and $QS = QT$, the Subtraction Axiom can be used to obtain $SP = TR$, or $\overline{SP} \cong \overline{TR}$. Therefore, $\triangle SPR \cong \triangle TRP$ by SAS.

Proof:

STATEMENTS	REASONS
1. $\overline{QP} \cong \overline{QR}, \overline{QS} \cong \overline{QT}$	1. Given
2. $QP = QR, QS = QT$	2. Definition of congruent segments
3. $QP = QS + SP, QR = QT + TR$	3. Segment Addition Axiom
4. $QS + SP = QT + TR$	4. Substitution
5. $SP = TR$	5. Subtraction Axiom
6. $\overline{SP} \cong \overline{TR}$	6. Definition of congruent segments
7. $\angle QPR \cong \angle QRP$	7. Given
8. $\overline{PR} \cong \overline{PR}$	8. Reflexive property
9. $\triangle SRP \cong \triangle TRP$	9. SAS Axiom

Sometimes the situation is more complex and requires careful thinking before proceeding. In the next example, you must first prove two triangles are congruent. Then use corresponding parts of these two triangles to prove a second pair of triangles is congruent.

Example

3 Write a plan and a two-column proof for the following.

Given: $\angle BAE \cong \angle BCE$
$\overline{BD} \perp \overline{AC}$

Prove: $\triangle ABD \cong \triangle CBD$

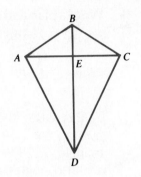

Plan: $\triangle ABD$ and $\triangle CBD$ share a side and an angle with $\triangle ABE$ and $\triangle CBE$. First, show $\triangle ABE \cong \triangle CBE$ by AAS. (Be sure you see how this is done.) Then use corresponding parts of $\triangle ABE$ and $\triangle CBE$ to show $\triangle ABD \cong \triangle CBD$.

Proof:

STATEMENTS	REASONS
1. $\overline{BD} \perp \overline{AC}$	1. Given
2. $\angle BEA$ is a right angle. $\angle BEC$ is a right angle.	2. Definition of perpendicular lines
3. $\angle BEA \cong \angle BEC$	3. All right angles are congruent.
4. $\overline{BE} \cong \overline{BE}$	4. Reflexive property
5. $\angle BAE \cong \angle BCE$	5. Given
6. $\triangle ABE \cong \triangle CBE$	6. AAS Theorem
7. $\angle ABE \cong \angle CBE$ $\overline{AB} \cong \overline{CB}$ ($\angle ABE$ is the same as $\angle ABD$ in $\triangle ABD$. $\angle CBE$ is the same as $\angle CBD$ in $\triangle CBD$.)	7. CPCTC
8. $\overline{BD} \cong \overline{BD}$	8. Reflexive property
9. $\triangle ABD \cong \triangle CBD$	9. SAS Axiom

Exercises

Max: 1-13, 1-24, Challenge; Avg: 1-13, 1-8, 9-23 odds; Min: 1-13 odds, 1-23 odds

Exploratory Name the following.

1. all triangles that contain $\angle A$ △ABC, △ABE, △ADC
2. all triangles that contain $\overline{BC}$ △ABC, △DBC, △EBC, △FBC
3. all triangles that contain $\angle DBC$ △DBC, △ABC
4. all triangles that contain $\angle ECB$ △ECB, △ACB
5. in $\triangle ABE$, sides that include $\angle ABE$ $\overline{AB}$, $\overline{BE}$
6. in $\triangle ACD$, angles that include $\overline{DC}$ $\angle ADC$, $\angle ACD$
7. in $\triangle ECB$, sides that include $\angle C$ $\overline{EC}$, $\overline{CB}$
8. in $\triangle CDB$, sides that include $\angle C$ $\overline{DC}$, $\overline{CB}$

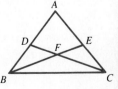

For answers to exercises 9–13, see the Teacher's Guide.

9. all pairs of triangles that appear to be congruent
10. six pairs of corresponding parts of △ABE and △ACD
11. six pairs of corresponding parts of △DCB and △EBC
12. six pairs of corresponding parts of △DFB and △EFC

Answer the following. Use the figure for exercises 1–12.

13. Suppose that △ABE ≅ △ACD and △DCB ≅ △EBC. Give an argument to show that △DFB ≅ △EFC. Be specific.

Written Name all pairs of triangles that have each of the following as corresponding parts.

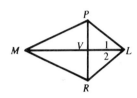

1. $\overline{PM}$, $\overline{RM}$ △MPV, △MRV; △MPL, △MRL
2. ∠PVM, ∠RVM △MPV, △MRV
3. ∠1, ∠2 △PVL, △RVL; △MPL, △MRL
4. $\overline{LR}$, $\overline{LP}$ △MPL, △MRL; △PVL, △RVL

Suppose that ∠1 ≅ ∠2 and $\overline{PR} \perp \overline{ML}$. Which, if any, pairs of triangles must be congruent? Justify your answers.

5. △LVP, △LVR yes; ASA
6. △LPM, △LRM yes; SAS
7. △PVM, △RVM yes; SAS

Supply the reasons for the following proof.

Given: ∠A ≅ ∠D,
∠B ≅ ∠E, $\overline{BF} \cong \overline{CE}$
Prove: △ABC ≅ △DEF

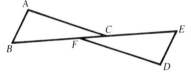

8.

STATEMENTS	REASONS
1. $\overline{BF} \cong \overline{CE}$	1. Given
2. BF = CE	2. Definition of congruent segments
3. FC = FC	3. Reflexive axiom
4. BF + FC = FC + CE	4. Addition axiom
5. BF + FC = BC FC + CE = FE	5. Segment addition axiom
6. BC = FE	6. Substitution
7. $\overline{BC} \cong \overline{FE}$	7. Definition of congruent segments
8. ∠A ≅ ∠D, ∠B ≅ ∠E	8. Given
9. △ABC ≅ △DEF	9. AAS

Write a two-column proof for the following.

For answers to exercises 9–24, see the Teacher's Guide.

9. Given: ∠A ≅ ∠F, ∠B ≅ ∠E, $\overline{BC} \cong \overline{ED}$
   Prove: △ADB ≅ △FCE
10. Given: $\overline{AD} \perp \overline{BE}$, $\overline{FC} \perp \overline{BE}$, $\overline{BC} \cong \overline{DE}$, $\overline{AD} \cong \overline{FC}$ Prove: △ADB ≅ △FCE
11. Given: $\overline{AD} \perp \overline{BE}$, $\overline{FC} \perp \overline{BE}$, $\overline{BC} \cong \overline{DE}$, ∠B ≅ ∠E Prove: AB = FE

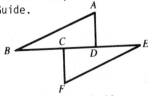

Exercises 9–11

12. **Given:** $\angle 2 \cong \angle 3$, $\overline{SN} \cong \overline{KA}$, $\overline{EN} \cong \overline{EA}$
    Prove: $\triangle SEA \cong \triangle KEN$
13. **Given:** $\triangle ENS \cong \triangle EAK$
    Prove: $\angle SEA \cong \angle KEN$
14. **Given:** $\angle 1 \cong \angle 4$, $\overline{SN} \cong \overline{KA}$, $EN = EA$
    Prove: $\triangle SEA \cong \triangle KEN$
15. **Given:** $\overline{SA} \cong \overline{NK}$, $\angle S \cong \angle K$, $\angle 2 \cong \angle 3$
    Prove: $\angle SEN \cong \angle KEA$

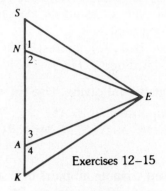
Exercises 12–15

16. **Given:** $\overline{AD} \cong \overline{AE}$, $\angle CDA \cong \angle BEA$
    Prove: $\triangle ADC \cong \triangle AEB$
17. **Given:** $\overline{CD} \perp \overline{AB}$, $\overline{BE} \perp \overline{AC}$, $\overline{AD} \cong \overline{AE}$
    Prove: $\angle ABE \cong \angle ACD$
18. **Given:** $\overline{DB} \cong \overline{EC}$, $\angle DBC \cong \angle ECB$
    Prove: $\overline{DC} \cong \overline{EB}$
19. **Given:** $\overline{AB} \cong \overline{AC}$, $\overline{AD} \cong \overline{AE}$
    Prove: $\angle ABE \cong \angle ACD$
20. **Given:** $\overline{BD} \cong \overline{CE}$, $\overline{AD} \cong \overline{AE}$
    Prove: $\overline{BE} \cong \overline{CD}$

Exercises 16–20

21. **Given:** $\angle DAE \cong \angle DCE$, $\overline{BD} \perp \overline{AC}$
    Prove: $\triangle ABD \cong \triangle CBD$
22. **Given:** $\angle BEA \cong \angle BEC$, $\angle ABE \cong \angle CBE$
    Prove: $AD = CD$
23. **Given:** $\overline{AB} \cong \overline{CB}$, $\angle ABD \cong \angle CBD$
    Prove: $\triangle AED \cong \triangle CED$
24. **Given:** $\overline{BD} \perp \overline{AC}$, $\angle BAC \cong \angle BCA$
    Prove: $\angle DAC \cong \angle DCA$

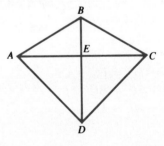

For the answer to the exercise, see the Teacher's Guide.

Challenge Write a proof.
Given: $\angle 1 \cong \angle 2$, $\overline{HB} \cong \overline{TD}$
$\overline{TR} \perp \overline{AD}$, $\overline{HK} \perp \overline{AB}$
Prove: $\triangle HKB \cong \triangle TRD$
$RV = KC$

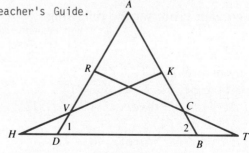

5.6 Isosceles Triangles, Special Segments

Objective: To identify special elements of a triangle and to apply theorems about isosceles triangles.

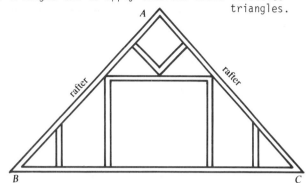

Shown at the right is a cross section of the roof of the Pine City Civic Center. The rafters that support the roof are congruent. That is, $\overline{AB} \cong \overline{AC}$. This means $\triangle ABC$ is an **isosceles triangle**.

It appears that the **base angles**, $\angle B$ and $\angle C$, are also congruent. This is a result that can now be proved for any isosceles triangle.

Before studying isosceles triangles, it will be helpful to name some special segments that can be drawn in any triangle.

You know from Axiom 4–12 that every angle has exactly one bisector. An angle bisector in a triangle is a line segment drawn from a vertex of a triangle to the side opposite, bisecting the angle at the vertex from which it is drawn.

$\overline{AT}$ bisects $\angle A$ if and only if $\angle BAT \cong \angle TAC$.

Two additional definitions of special line segments follow.

Definition of a Median	A **median of a triangle** is a line segment drawn from any vertex to the midpoint of the opposite side.
Definition of an Altitude	An **altitude of a triangle** is a line segment drawn from any vertex and perpendicular to the line containing the opposite side.

Sometimes the altitude is called the *height*.

If M is the *midpoint* of $\overline{BC}$, $\overline{AM}$ is the median from A. If $\overline{AD} \perp \overline{BC}$, then $\overline{AD}$ is the *altitude* from A. Note that sometimes the altitude of a triangle may fall outside the triangle.

Do you think a median could ever fall outside a triangle? You will be asked to provide an answer in the exercises.

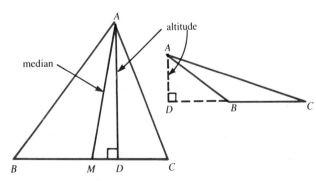

Now consider this important theorem of an isosceles triangle.

Theorem 5-2 If two sides of a triangle are congruent, then the angles opposite those sides are congruent. (Base angles of an isosceles triangle are congruent.)

Example

1 Write a two-column proof of Theorem 5-2.
Given: △ABC with $\overline{AB} \cong \overline{BC}$
Prove: $\angle A \cong \angle C$

Proof:

STATEMENTS	REASONS
1. Draw $\overline{BD}$ so that it bisects $\angle ABC$.	1. An angle has one and only one bisector.
2. $\angle ABD \cong \angle CBD$	2. Definition of angle bisector
3. $\overline{AB} \cong \overline{BC}$	3. Given
4. $\overline{BD} \cong \overline{BD}$	4. Reflexive property
5. △ABD ≅ △CBD	5. SAS Axiom
6. $\angle A \cong \angle C$	6. CPCTC

Is the converse of Theorem 5-2 true? That is, if two angles of a triangle are congruent, are the sides opposite them necessarily congruent? The next theorem provides an answer to this question.

Theorem 5-3
Converse of Theorem 5-2   If two angles of a triangle are congruent, then the sides opposite those angles are congruent.

The proof is left for the exercises.

Exercises

Max: 1-8, 1-20; Avg: 1-8, 1-7, 9-19 odds; Min: 1-8, 1-19 odds

Exploratory Make a diagram and explain each of the following.

1. base and legs of an isosceles triangle
2. base angles of an isosceles triangle
3. vertex angle of an isosceles triangle
4. median of a triangle
5. In △KLM, $\overline{KB}$ is an altitude.
6. $\overline{AQ}$ is an angle bisector in △ARS.

For exercises 1-6, see students' work.

Answer the following.

7. How many medians, altitudes, and angle bisectors can a triangle have? Illustrate your answer. 3 of each; see students' work.

8. Adrienne has just drawn altitude $\overline{KT}$ to side $\overline{LM}$ of $\triangle KLM$ and it has fallen outside the triangle. What kind of triangle is $\triangle KLM$? Explain. Obtuse; the altitudes from the two acute angles will be in the exterior of the triangle.

Written Find the degree measure of each unlabeled angle.

1. 65, 65
2. $78\frac{1}{2}$, $78\frac{1}{2}$
3. 70, 40
4. $37\frac{1}{4}$, $37\frac{1}{4}$, $105\frac{1}{2}$

In each of the following, find the measure of each side of the triangle.

5. 96, 96, 60
6. $3x - 5$ 13, 13, 12
7. $5(x-3)$ 28, 28, 45

For answers to exercises 8-20, see the Teacher's Guide.

Write a two-column proof of each of the following.

8. **Given:** $\overline{AB} \cong \overline{AC}$
   Prove: $\angle 1 \cong \angle 2$
9. **Given:** $\overline{AB} \cong \overline{AC}$, D is the midpoint of $\overline{BC}$, $\angle EDB \cong \angle FDC$
   Prove: $\overline{EB} \cong \overline{FC}$
10. **Given:** $\angle 3 \cong \angle 4$, $\overline{EB} \cong \overline{FC}$
    Prove: $\overline{AE} \cong \overline{AF}$

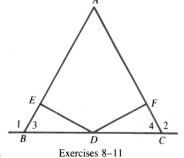

Exercises 8–11

11. **Given:** $\angle BDE \cong \angle CDF$, $\overline{DE} \cong \overline{DF}$, $\angle AED \cong \angle AFD$
    Prove: $\triangle ABC$ is isosceles.
12. **Given:** $\overline{AB} \cong \overline{AC}$, $\overline{BE}$ bisects $\angle CBA$, $\overline{CD}$ bisects $\angle BCA$
    Prove: $\angle 1 \cong \angle 2$
13. **Given:** $\overline{AB} \cong \overline{AC}$, $\overline{BD} \cong \overline{CE}$
    Prove: $\angle 3 \cong \angle 4$
14. **Given:** $\overline{AD} \cong \overline{AE}$, $\overline{BD} \cong \overline{CE}$
    Prove: $\angle CDE \cong \angle BED$
15. **Given:** $\overline{BD} \cong \overline{CE}$, $\overline{BE} \cong \overline{CD}$
    Prove: $\triangle BCF$ is isosceles.

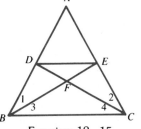

Exercises 12–15

16. If a triangle is equilateral, then it is also equiangular.
17. If a triangle is equiangular, then it is also equilateral.
18. Theorem 5–3
19. Theorem 5–2 by drawing the median from the vertex angle to the base.
20. A triangle is isosceles if it has a median that is also an altitude.

Objective: To prove that two right triangles are congruent.

5.7 Proving Right Triangles Congruent

Very often it is necessary to prove that two right triangles are congruent. Of course, all the methods used to prove triangles congruent apply to right triangles. The following important theorem gives us an additional way of proving two *right* triangles congruent.

Theorem 5–4

> If the hypotenuse and a leg of one right triangle are congruent to the corresponding parts of another right triangle, then the triangles are congruent.

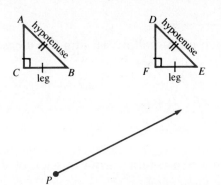

This theorem is often abbreviated by HL, which stands for *hypotenuse leg*.

If $\overline{AB} \cong \overline{DE}$, $\overline{BC} \cong \overline{EF}$, and $\angle C$ and $\angle F$ are right angles, then Theorem 5–4 allows us to conclude that $\triangle ABC \cong \triangle DEF$.

Before proving this theorem, it is necessary to agree on an axiom. Suppose there is a ray with endpoint P. How many points on this ray lie 3 cm from P? If your answer was exactly one point, you already understand Axiom 5–3.

Axiom 5–3

> On a ray, there is exactly one point at a given distance from the endpoint of the ray.

Example

1 Write a plan of proof for Theorem 5–4.

Given: $\triangle ABC$, right $\angle C$
$\triangle DEF$, right $\angle F$
$\overline{AB} \cong \overline{DE}$, $\overline{BC} \cong \overline{EF}$

Prove: $\triangle ABC \cong \triangle DEF$

Plan: By Axiom 5–3, locate G on $\overrightarrow{DF}$ such that $\overline{AC} \cong \overline{GF}$. Since $\overline{AC} \cong \overline{GF}$, $\angle C \cong \angle GFE$ (why?) and $\overline{BC} \cong \overline{EF}$, it follows that $\triangle ABC \cong \triangle GEF$ (SAS). Therefore, $\overline{GE} \cong \overline{AB}$. But $\overline{AB} \cong \overline{DE}$. Thus, by transitivity, $\overline{GE} \cong \overline{DE}$. By Theorem 5–2, $\angle D \cong \angle G$. Since two angles of $\triangle DEF$ are congruent to two angles of $\triangle GEF$, we conclude that $\angle DEF \cong \angle GEF$. Now prove $\triangle DEF \cong \triangle GEF$ by ASA. Since $\triangle ABC \cong \triangle GEF$, $\triangle ABC \cong \triangle DEF$ by transitivity.

Example

2 Write a two-column proof.
Given: $\overline{EB} \perp \overline{AD}$, $\overline{EF} \perp \overline{AG}$, $\overline{EB} \cong \overline{EF}$
Prove: $\overline{AE}$ bisects $\angle DAG$.

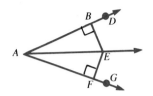

Proof:

STATEMENTS	REASONS
1. $\overline{EB} \perp \overline{AD}$, $\overline{EF} \perp \overline{AG}$	1. Given
2. $\angle EBA$ is a right angle. $\angle EFA$ is a right angle.	2. Definition of perpendicular lines
3. $\triangle EBA$ is a right triangle. $\triangle EFA$ is a right triangle.	3. Definition of a right triangle
4. $\overline{EB} \cong \overline{EF}$	4. Given
5. $\overline{AE} \cong \overline{AE}$	5. Reflexive property
6. $\triangle EBA \cong \triangle EFA$	6. HL Theorem
7. $\angle EAD \cong \angle EAG$	7. CPCTC
8. $\overline{AE}$ bisects $\angle DAG$.	8. Definition of angle bisector

Exercises

Max: 1-10, 1-17; Avg: 1-10, 1-17; Min: 1-4, 5-9 odds, 1-17 odds

Exploratory Draw an isosceles right triangle and label these parts.

1. hypotenuse See students' work.

2. legs See students' work.

Answer the following.

3. Explain HL. See page 192.

4. Restate the Pythagorean Theorem and give a complete example of its use.
See students' work.

Tell whether or not each pair of triangles is congruent. Justify your answers.

5. yes, HL

6. yes, HL

7. yes, SAS

8. No, congruent parts do not correspond.

194 CONGRUENT TRIANGLES

9.
no, not enough information

10.
yes, AAS

Written Answer the following.
1. Can HL be used to prove an SSA Theorem of congruence? Explain. No, HL is only true for right triangles.

Given an altitude to the base of an isosceles triangle, write a two-column proof for the following. For answers to exercises 2-17, see the Teacher's Guide.

2. The altitude bisects the vertex angle.   3. The altitude is also a median.

Write a two-column proof for each of the following.

4. **Given:** $\overline{MD} \perp \overline{AB}$, $\overline{ME} \perp \overline{AC}$, M is the midpoint of $\overline{BC}$, $\overline{DB} \cong \overline{EC}$
   Prove: $\triangle DMB \cong \triangle EMC$

5. **Given:** $\overline{AB} \cong \overline{AC}$, $\overline{MD} \perp \overline{AB}$, $\overline{ME} \perp \overline{AC}$, M is the midpoint of $\overline{BC}$.
   Prove: $\overline{DM} \cong \overline{EM}$

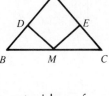

6. Prove that the altitudes drawn to the congruent sides of an isosceles triangle are congruent.

7. **Given:** $\overline{PT} \perp \overline{QS}$, $\overline{RW} \perp \overline{QS}$
   $\overline{QT} \cong \overline{SW}$, $\overline{PS} \cong \overline{QR}$
   Prove: $\triangle PTS \cong \triangle RWQ$

8. **Given:** $\overline{PT} \perp \overline{QS}$, $\overline{RW} \perp \overline{QS}$, $\overline{PT} \cong \overline{RW}$
   $\overline{PS} \cong \overline{RQ}$
   Prove: $\overline{PS} \parallel \overline{QR}$

9. **Given:** $\overline{PS} \parallel \overline{QR}$, $\overline{PT} \parallel \overline{RW}$, $\overline{PT} \cong \overline{RW}$
   Prove: $QT = SW$

10. **Given:** $\overline{PT} \perp \overline{QS}$, $\overline{PT} \parallel \overline{RW}$, $QT = SW$, $PS = RQ$
    Prove: $PT = RW$

Exercises 7–10

11. **Given:** $\overline{RE} \perp \overline{AV}$, $\overline{VL} \perp \overline{AR}$, $\overline{VL} \cong \overline{RE}$
    Prove: $\triangle VER \cong \triangle RLV$

Given $\angle REV \cong \angle REA$, $\angle RLV \cong \angle VLA$, and $\overline{EV} \cong \overline{LR}$, prove the following.

12. $\triangle VER \cong \triangle RLV$ 13. $\triangle VAR$ is isosceles.
14. $\triangle VKR$ is isosceles.   15. $\triangle VKE \cong \triangle RKL$
16. $\triangle VAL \cong \triangle RAE$

Write a two-column proof.

17. **Given:** $\overline{AV} \cong \overline{AR}$, $\overline{EV} \cong \overline{LR}$
    Prove: $\overline{ER} \cong \overline{LV}$

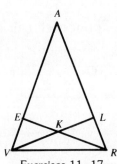
Exercises 11–17

5.8 Inequalities

Objective: To use theorems of inequalities to prove that the measures of angles or sides of triangles are unequal.

Journal

Why do you think it is necessary to study geometry? Give at least two examples.

Previously, we dealt with proving line segments or angles congruent. Sometimes the situation deals with *unequal measures*. For example, consider the statement, *if two sides of a triangle are not congruent, then the angles opposite those sides are not congruent.*

In $\triangle ABC$, $\overline{AB} \not\cong \overline{AC}$. If the above statement is true, then $\angle C \not\cong \angle B$. Look carefully at the figure. It appears that $AB > AC$. Now, consider the measures of $\angle C$ and $\angle B$. It appears that $\angle C$, the angle opposite $\overline{AB}$, has a measure greater than $\angle B$, the angle opposite $\overline{AC}$.

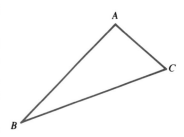

Before trying to prove that this relationship always holds, let us review some ideas about inequalities. Remember $a > b$ means "a is greater than b." A precise definition of $>$ follows.

Definition of Greater Than

Suppose a and b are real numbers. $a > b$ if, and only if, there is a positive real number c such that $a = b + c$. Also, $b < a$ means $a > b$.

Examples

1 Use the definition of $>$ to show $7 > 5$.

Because 2 is a positive real number and $7 = 5 + 2$, it follows from the definition of $>$ that $7 > 5$.

2 Show that $AC > AB$ and $AC > BC$.

From the Segment Addition Axiom, $AC = AB + BC$. By definition of $>$, it follows that $AC > AB$ and $AC > BC$.

3 Show that $m\angle AOC > m\angle AOB$ and $m\angle AOC > m\angle BOC$.

By the Angle Addition Axiom, $m\angle AOC = m\angle AOB + m\angle BOC$. By the definition of $>$, it follows that $m\angle AOC > m\angle AOB$ and $m\angle AOC > m\angle BOC$.

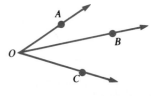

196 CONGRUENT TRIANGLES

It will be helpful to recall some properties of inequalities. The following properties refer to any real numbers a, b, and c.

Trichotomy Property of Inequality

Either $a > b$, $a < b$, or $a = b$.

Transitive Property of Inequality

If $a > b$ and $b > c$, then $a > c$.

Addition Property of Inequality

If $a > b$ then $a + c > b + c$.

Multiplication Properties of Inequality

If $a > b$ and $c > 0$, then $ac > bc$.
If $a > b$ and $c < 0$, then $ac < bc$.

Theorem 5–5

If the lengths of two sides of a triangle are unequal, then measures of the angles opposite those sides are unequal and the greater angle is opposite the greater side.

Example

4 Write a two-column proof.

Given: $\triangle ABC$, $AB > AC$
Prove: $m\angle C > m\angle B$

Proof:

STATEMENTS	REASONS
1. Locate point P on $\overline{AB}$ such that $AP = AC$.	1. On a ray, there is exactly one point at a given distance from the endpoint of the ray.
2. Draw $\overline{CP}$.	2. For any two points, there exists one and only one line containing both of them.
3. $m\angle 1 = m\angle 2$	3. If two sides of a triangle are congruent, then the angles opposite those sides are congruent.
4. $m\angle 2 = m\angle 4 + m\angle 3$	4. Exterior Angle Theorem
5. $m\angle 2 > m\angle 4$	5. Definition of $>$
6. $m\angle 1 > m\angle 4$	6. Substitution Axiom

7. $m\angle BCA = m\angle 1 + m\angle 3$	7. Angle Addition Axiom
8. $m\angle BCA > m\angle 1$	8. Definition of $>$
9. $m\angle BCA > m\angle 4$	9. Transitive Property of $>$

As you might suspect, the converse of Theorem 5–5 is also true.

Theorem 5–6 If two angles of a triangle are unequal in measure, then the sides opposite those angles are unequal in measure and the longer side is opposite the angle of greater measure.

Example

5 Use an indirect proof to prove Theorem 5–6.
Given: $\triangle ABC$, $m\angle B > m\angle C$
Prove: $AC > AB$

Indirect Proof:
Assume $AC \not> AB$. By the trichotomy property, either $AC = AB$ or $AC < AB$.
Case 1: Suppose $AC = AB$. Then $\triangle ABC$ is isosceles and, by Theorem 5–2, $m\angle B = m\angle C$. But this contradicts the given statement that $m\angle B > m\angle C$. Thus, $AC \neq AB$.
Case 2: Suppose $AC < AB$. Consider the angles opposite these two sides, $\angle B$ and $\angle C$. By Theorem 5–5, $m\angle B < m\angle C$. But this contradicts the given statement that $m\angle B > m\angle C$. Thus, $AC \not< AB$.

Cases 1 and 2 show that the original assumption, $AC \not> AB$, leads to a contradiction. Thus, $AC > AB$.

Exercises

Max: 1-10, 1-25; Avg: 1-10, 1-25; Min: 1-10, 1-15, 17-25 odds

Exploratory Tell whether each statement is *always, sometimes,* or *never* true.

1. If $a > b$, then $a + c > b + c$.
2. If $a < b$, then $b < a$.
3. If $AB > EF$, then $AB + BC > BC + EF$.
4. If $a < b$, then $ac < bc$.
5. If $AB > CD$ and $CD > EF$, then $AB > EF$.
6. If $x > y$, then $-x < -y$.
7. If $a \geq b$, then $a - c \geq b - c$.
8. If $x - y > 0$, then $x > y$.
9. If $a < b$ and $c < 0$, then $ac > bc$.
10. If $x \not> y$, then $x < y$.

For answers to exercises 1-10, see the Teacher's Guide.

198 CONGRUENT TRIANGLES

For exercises 1-6, see students' work.
Written Solve each inequality in $\mathcal{R}$ and graph the solution set on a number line.
1. $7 - 2y + 1 \le 29$ $y \ge -10\frac{1}{2}$
2. $5x - (4x - 5) \ge 7$ $x \ge 2$
3. $2(y - 3) - 4 < 10 - 2(y - 3)$ $y < 6\frac{1}{2}$
4. $7m + 9 - 4m \le 51 - 3m$ $m \le 7$
5. $\frac{4y}{3} - \frac{5}{9} \ge \frac{y}{2} + \frac{5}{6}$ $y \ge 1\frac{2}{3}$
6. $2x - 4(x + 2) > -3x + 5$ $x > 13$

Answer each of the following.
7. What is meant by $a > b$? $c < d$? See page 195.
8. $\forall a, b \in R\ [(a > b) \leftrightarrow \exists c \in R\ (a = b + c)]$
8. Write the definition of $a > b$ using logical symbolism.
9. If $AB = CD + EF$, what conclusion can be reached about AB and CD? About AB and EF? AB > CD; AB > EF
10. What is meant by the Trichotomy Property of Inequality? See page 196.
12. $\forall a, b, c \in R\ [(a > b) \wedge (b > c) \rightarrow (a > c)]$
11. What is meant by the Transitive Property of Inequality? See page 196.
12. Rewrite the statement of the Transitive Property given in the text using logical symbolism.
13. Is there a reflexive property of inequality? Explain. no; a = a
14. Is there a symmetric property of inequality? Explain. No, if a > b then b ≯ a.
15. In your own words, describe the logical plan of the proof of Theorem 5–6. Answers may var

Name the angles with the greatest measure and the least measure in each triangle.

16.   ∠W, ∠Y
17.   ∠T, ∠S
18.   ∠C, ∠B

Name the sides of each triangle, in order, from shortest to longest.

19.   YZ, XY, XZ
20. MO, MN, NO
21. PR, PQ, QR

Write a two-column proof. For answers to exercises 22-25, see the Teacher's Guide.
22. **Given:** $BC = DE$
    Prove: $AC > DE$

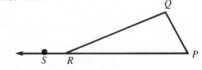

23. **Prove:** An exterior angle of a triangle is greater than either remote interior angle.
24. **Given:** $QR > QP$
    Prove: $m\angle SRQ > m\angle QRP$

25. **Given:** $\triangle ABC$ with right $\angle C$
    Prove: $AB > BC$

5.9 Another Triangle Inequality

Objective: To determine the possible measures of the third side of a triangle given the measures of two sides.

The following theorem expresses an important relationship among the sides of any triangle.

Theorem 5–7 | The sum of the lengths of any two sides of a triangle is greater than the length of the third side.

Example

1 Write a two-column proof.

Given: △ABC
Prove: BA + AC > BC

Proof:

STATEMENTS	REASONS
1. Locate D on $\overrightarrow{CA}$ such that $\overline{DA} \cong \overline{BA}$.	1. On a ray, there is exactly one point at a given distance from the endpoint of the ray.
2. Draw $\overline{BD}$.	2. For any two points, there exists one and only one line containing both of them.
3. ∠1 ≅ ∠2	3. If two sides of a triangle are congruent, then the angles opposite those sides are congruent.
4. m∠1 = m∠2	4. Definition of congruence
5. m∠CBD = m∠2 + m∠3	5. Angle Addition Axiom
6. m∠CBD > m∠2	6. Definition of >
7. m∠CBD > m∠1	7. Substitution Axiom
8. In △BDC, DC > BC.	8. If two angles of a triangle are unequal, then the sides opposite those angles are unequal and the greater side is opposite the greater angle.
9. DC = DA + AC	9. Segment Addition Axiom
10. DA + AC > BC	10. Substitution Axiom
11. DA = BA	11. Definition of congruence
12. BA + AC > BC	12. Substitution Axiom

Example

2 A triangle has two sides whose lengths are 7 cm and 9 cm. Between what two lengths is the length of the third side?

Let x represent the length of the third side. From Theorem 5–7, the length of the third side cannot exceed the sum of the lengths of the two given sides. Therefore, $x < 7 + 9$ or $x < 16$. Also from Theorem 5–7,
$$x + 9 > 7 \quad \text{and} \quad x + 7 > 9$$
$$x > -2 \quad \text{and} \quad x > 2$$
Thus, the length of the third side must be greater than 2 cm, which is the greater of the two values.

Since $x < 16$ and $x > 2$, the length of the third side must be between 2 cm and 16 cm.

Exercises

Max: 1-4, 1-14; Avg: 1-4, 1-14; Min: 1-4, 1-11 odds, 12-14

Exploratory Answer the following.

1. Points M, N, and P are noncollinear. Is it true that $MN < MP + NP$? Why?
   Yes, because M, N, and P form a triangle and Theorem 5-7 applies.

Is the following ever true for any points M, N, and P? Explain your answer.

2. $MN = MP + NP$ Yes, if M, N, and P are collinear and P is between M and N.
3. $MN = NP - MP$ Yes, if M, N, and P are collinear and M is between N and P.
4. $MN > MP + NP$ No

Written Assume all variables represent positive real numbers. Tell whether the following can be measures of the sides of a triangle.

1. 5, 8, 6 yes
2. 6, 6, 10 yes
3. 3, 5, 2 no
4. 8, 4, 14 no
5. x, y, x + y no
6. x, x + 4, x + 6 yes, if x > 2; no, if x ≤ 2
7. 4.48, 3.45, 1.231 yes

The measures of two sides of a triangle are given. Between what two numbers is the measure of the third side?

8. 5, 11 6, 16
9. 4, 3 1, 7
10. 13, 7 6, 20
11. x, y |x - y|, x + y

Answer the following.

12. Two sides of a triangle have lengths 5 cm and 13 cm. Must 5 cm be the length of the shortest side? Must 13 cm be the length of the longest side? yes; no

For answers to exercises 13 and 14, see the Teacher's Guide.

Write a two-column proof for the following.

Given: Equilateral $\triangle ABC$, $\overline{BC}$ is extended to point D.

13. **Prove:** $BD > AD$
14. **Prove:** $AD > AB$

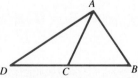

Exercises 13–14

Problem Solving Application: Using Figures

Objective: To recognize information which can and cannot be assumed from a figure, and to use figures in solving problems.

Figures help state given information. They may be used to show any of the following.

Existence — When points, lines, rays, angles, triangles, and so on appear on a figure, you may assume they exist.

Relative Position — When points appear on a line or part of a line, you may assume they are collinear. Also, you may assume betweenness with respect to points on a line and location with respect to interiors and exteriors of angles and figures. Adjacent angles can also be assumed in figures.

Intersection — When lines, rays, or segments appear to intersect, you may assume that they do intersect.

Figures are marked to give further information about congruence and measure. Such information normally is written in the given statements, if the figure is with a proof.

DO NOT ASSUME ANY OF THE FOLLOWING UNLESS A FIGURE IS MARKED.

The properties listed must be given or proven.

congruence
equality of measure
inequality of measure
bisectors or midpoints
perpendiculars or right angles
specific measures

Example

1 Which statements can be assumed from the appearance of the figure shown at the right?

1. V is between Q and T.
2. $\overline{PR}$ bisects $\angle QPS$.
3. $\overline{PT} \cong \overline{TS}$
4. $\overline{PR}$ bisects $\overline{QS}$.
5. $\overline{QR} \cong \overline{RS}$
6. $\overline{QT} \perp \overline{PS}$
7. $\angle QRV$ is a right angle.
8. $\angle QTP$ and $\angle QTS$ are supplementary.

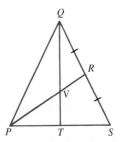

The statements **1, 4, 5,** and **8** can be assumed.

202 CONGRUENT TRIANGLES

Exercises

Max: 1-15, 1-14, Challenge; Avg: 1-15, 1-13; Min: 1-15 odds, 1-7 odds, 9-13

Exploratory Determine whether each of the following may be assumed from a figure.
1. adjacent angles yes
2. existence of a point yes
3. congruent segments no
4. perpendicular lines no
5. equality of distance measure no
6. supplementary adjacent angles yes
7. existence of a ray yes
8. perpendicular rays no
9. existence of a triangle yes
10. collinearity of segments yes
11. intersection of rays yes
12. vertical angles yes
13. intersection of lines yes
14. perpendicular segments no
15. right angles no

Written Determine whether each statement can be assumed from the figure shown.
1. $\angle QVT$ is a right angle. no
2. $\overline{QR} \perp \overline{PV}$ no
3. $\overline{PV}$ bisects $\overline{QS}$. no
4. T is the midpoint of $\overline{PV}$. no
5. $\angle SPT$ and $\angle VQT$ are congruent. no
6. $\overline{PS} \cong \overline{QV}$ no
7. $\overline{PT} \cong \overline{QT}$ yes
8. $PV = TV + PT$ yes

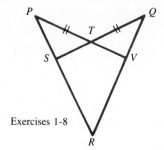

Exercises 1-8

For each exercise, answer
  A if the measure in Column A is greater;
  B if the measure in Column B is greater;
  C if the two measures are equal;
  D if the relationship cannot be determined from the information given.

		Column A	Column B
B	9.	$m\angle 1$	$m\angle 5$
C	10.	$m\angle 8$	$m\angle 3$
D	11.	$m\angle 6$	$m\angle 7$
B	12.	$m\angle 7$	$m\angle 4$
C	13.	$m\angle 1 + m\angle 2$	$m\angle 5$

Exercises 9-13

14. Use graph paper to make several copies of the grid at the right. Shade exactly eight small squares to determine a shaded figure that is congruent to the remaining unshaded figure. Find all six possible shadings.
See students' work.

Challenge

Find four ways to cut a 3 × 3 grid into two congruent halves by following a path of segments whose endpoints are vertices on the grid.   Hint: Your path must pass through the center of the grid.   Answers may vary. See students' work.

Portfolio Suggestion

Review the items in your portfolio. Make a table of contents of the items, noting why each item was chosen. Replace any items that are no longer appropriate.

Performance Assessment

Design a stained glass window. The window should contain two obtuse triangles congruent by SAS, a scalene triangle, an isosceles triangle, and two triangles congruent by AAS. Label the triangles, mark the congruent sides and angles. Tell which triangles are congruent. **See students' work.**

Chapter Summary

1. Two triangles are congruent if and only if there is a correspondence between the vertices such that each side and each angle of one triangle is congruent to the corresponding part of the other triangle. (168)
2. **SSS** Axiom: If three sides of one triangles are congruent to the **corresponding parts** of another triangle, then the triangles are congruent. (171)
3. **SAS** Axiom: If two sides and the included angle of one triangle are congruent to the corresponding parts of another triangle, then the triangles are congruent. (172)
4. **ASA** Axiom: If two angles and the included side of one triangle are congruent to the corresponding parts of another triangle, then the triangles are congruent. (172)
5. Corresponding parts of **congruent triangles** are congruent. (175)
6. Theorem 5–1 (**AAS Theorem**): If two angles and a side opposite one of the angles in one triangle are congruent to the corresponding parts of another triangle, then the triangles are congruent. (180)
7. A **median** of a triangle is a line segment drawn from any vertex to the midpoint of the opposite side. (189)
8. An **altitude** of a triangle is a line segment drawn from any vertex and perpendicular to the line containing the opposite side. (189)
9. Theorem 5–2: **Base angles of an isosceles triangle** are congruent. (190)
10. Theorem 5–3: If two angles of a triangle are congruent, then the sides opposite those angles are congruent. (190)

11. **Theorem 5-4:** If the hypotenuse and a leg of one right triangle are congruent to the corresponding parts of another right triangle, then the triangles are congruent. (192)
12. Suppose a and b are real numbers. $a > b$ if, and only if, there is a positive real number c such that $a = b + c$. Also, $b < a$ means $a > b$. (195)
13. **Trichotomy Property of Inequality:** Either $a > b$, $a < b$, or $a = b$. (196)
14. **Transitive Property of Inequality:** If $a > b$ and $b > c$, then $a > c$. (196)
15. **Addition Property of Inequality:** If $a > b$, then $a + c > b + c$. (196)
16. **Multiplication Properties of Inequality:** If $a > b$ and $c > 0$, then $ac > bc$. If $a > b$ and $c < 0$, then $ac < bc$. (196)
17. **Theorem 5-5:** If the lengths of two sides of a triangle are unequal, then the measures of the angles opposite those sides are unequal, and the greater angle is opposite the greater side. (196)
18. **Theorem 5-6:** If two angles of a triangle are unequal in measure, then the sides opposite those angles are unequal in measure and the longer side is opposite the angle of greater measure. (197)
19. **Theorem 5-7:** The sum of the lengths of any two sides of a triangle is greater than the length of the third side. (199)

Chapter Review

Given $\triangle ABC \cong \triangle UST$.

5.1 1. Name all congruent parts of $\triangle ABC$ and $\triangle UST$.

$\overline{AB} \cong \overline{US}$, $\overline{BC} \cong \overline{ST}$, $\overline{AC} \cong \overline{UT}$, $\angle A \cong \angle U$, $\angle B \cong \angle S$, $\angle C \cong \angle T$

Write a two-column proof for each of the following.

For answers to exercises 2-7 and 9 see the Teacher's Guide.

5.2 2. **Given:** $\overline{ST} \cong \overline{UT}$, $\angle 1 \cong \angle 2$
Prove: $\triangle RST \cong \triangle RUT$

3. **Given:** L is the midpoint of $\overline{KM}$, $\overline{JK} \perp \overline{KM}$, $\overline{NM} \perp \overline{KM}$
Prove: $\triangle KLJ \cong \triangle MLN$

4. **Given:** $\angle 2 \cong \angle 4$, $\overline{JK} \cong \overline{NM}$
Prove: $\triangle KLJ \cong \triangle MLN$

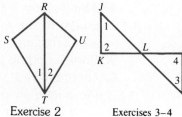

Exercise 2   Exercises 3–4

5.3 5. **Given:** $\overline{AB} \perp \overline{BD}$, $\overline{AF} \perp \overline{FD}$, $\overline{AB} \cong \overline{AF}$, $\angle 1 \cong \angle 4$
Prove: $\overline{AC} \cong \overline{AE}$

6. **Given:** $\overline{AB} \cong \overline{AF}$, $\angle 1 \cong \angle 4$, $\angle 2 \cong \angle 3$
Prove: $\overline{BD} \cong \overline{FD}$

Answer the following.

7. Odella and Rosa plan to use the following method to measure the distance across Pine City Lake. They mark segments as shown on the right so that $m\angle ADB = m\angle ADC$ and $CD = BD$. They conclude that $AC = AB$. Are they right? Explain.

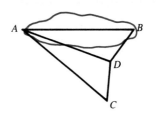

5.4 8. **Given:** $\overline{AB} \perp \overline{BD}, \overline{ED} \perp \overline{BD}$
  C is the midpoint of $\overline{BD}$,
  $\angle A \cong \angle E$.
  Find x and y. $x = 3, y = 2$

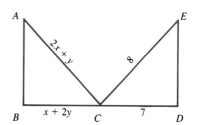

Write a two-column proof for the following.

5.6 9. **Given:** $\overline{AD}$ is the median and altitude of $\triangle ABC$.
  Prove: $\triangle ABC$ is isosceles.

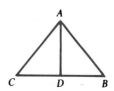

Answers for exercises 10 and 11 may vary.

5.5 and 5.7 Given that D is the midpoint of $\overline{AB}$, $\overline{AB} \cong \overline{AC}$, $\angle AED \cong \angle ECB$, and $\overline{EB} \perp \overline{AC}$, name the following. Justify your answers.

10. four pairs of congruent triangles
11. For one pair of congruent triangles in exercise 10, name six pairs of corresponding parts.

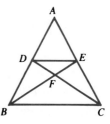

5.8 **Tell whether each statement is *always*, *sometimes*, or *never* true. Justify your answer.**

12. If $a < b$ and $b < c$, then $c > a$. 13. If $a < b$, then $b > a$.
  always; transitive property   always; definition of greater than

5.9 **State whether the following lengths can be sides of a triangle.**

14. 2 cm, 7 cm, 9 cm no 15. 1 cm, 3 cm, 8 cm no 16. 17 cm, 47 cm, 55 cm yes

Write a two-column proof for the following.

17. **Given:** $\triangle ADB$, Equilateral $\triangle ABC$
  Prove: $DB > CA$
  For the answer to exercise 17, see the Teacher's Guide.

Chapter Test

For answers to exercises 1-7, see the Teacher's Guide.

Write a two-column proof for each of the following.

1. Given: $\angle Y \cong \angle V$
   $\angle YZX \cong \angle VZW$
   $\overline{YZ} \cong \overline{VZ}$
   Prove: $\triangle XYZ \cong \triangle WVZ$

Exercise 1

2. Given: $\overline{AB} \cong \overline{CD}, \overline{AD} \cong \overline{CB}$
   Prove: $\overline{AD} \parallel \overline{BC}$
3. Given: $\overline{DA} \perp \overline{AB}, \overline{CB} \perp \overline{CD}$,
   $\overline{AD} \cong \overline{CB}$
   Prove: $m\angle 1 = m\angle 2$
4. Given: $\overline{AB} \parallel \overline{DC}, \overline{AD} \parallel \overline{BC}$
   Prove: $\angle DAB \cong \angle BCD$

Exercises 2–4

5. Given: $\overline{AB} \cong \overline{DC}, \overline{AC} \cong \overline{DB}$
   Prove: $\angle A \cong \angle D$
6. Given: $\overline{AB} \cong \overline{DC}, \angle ABC \cong \angle DCB$
   Prove: $\overline{AC} \cong \overline{DB}$
7. Given: $\angle DBC \cong \angle ACB$,
   $\angle ABC \cong \angle DCB$
   Prove: $\overline{AB} \cong \overline{DC}$

Exercises 5–7

Tell whether each statement is *always, sometimes,* or *never* true. Justify your answer.

8. If $AB + CD = EF$, then $CD > EF$.
   never; $CD < EF$

9. If $a > b$ and $c < 0$, then $\dfrac{a}{c} < \dfrac{b}{c}$.
   always; multiplication property

State whether the following lengths can be sides of a triangle.

10. 3 cm, 8 cm, 11 cm no
11. 3 cm, 5 cm, 7 cm yes
12. 19 cm, 47 c, 55 cm yes

Given D is the midpoint of $\overline{AB}$, $\overline{AB} \cong \overline{AC}$, and $\angle AED \cong \angle ECB$, name the following.

13. a pair of parallel lines $\overline{DE} \parallel \overline{BC}$
14. two medians of a triangle $\overline{CD}, \overline{BE}$
15. an isosceles triangle $\triangle ABC, \triangle ADE$
16. four pairs of congruent triangles
    $\triangle ADC \cong \triangle AEB, \triangle EDC \cong \triangle DEB, \triangle DCB \cong \triangle EBC, \triangle DFB \cong \triangle EFC$
17. Find DB if $AD = 3x - 13$ and $AB = 3x + 13$. 26
18. Find BF if $BF = 2x + 9, EF = 4, EC = 12$, and $FC = 13$. 13
19. Find DC if $DC = x^2, BC = 10x - 24$, and $BE = 10x + 24$. 144

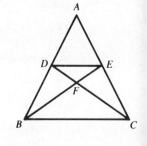

20. The measures of two sides of a triangle are 7 and 10. Between what two numbers is the measure of the third side? 3, 17

CHAPTER 6

Applications of Congruent Triangles

Application in Engineering

On July 4, 1986, the two-year restoration of the Statue of Liberty ended with an elaborate dedication celebration. Concerts, ethnic festivals, fireworks, and the swearing in of 25,000 new American citizens marked the event. The statue was a gift from the people of France given in 1885. Fredric Augutse designed the statue and Gustave Eiffel, the designer of the Eiffel Tower, planned the skeleton of the statue.

The Statue of Liberty is supported by a skeleton made up of iron triangles. **Triangles** are used extensively in engineering because they are rigid. A *rigid* figure will hold its shape. Construct a rectangle from cardboard strips and paper fasteners. Is it rigid? If not, determine a way that it could be made rigid. Write an explanation of how you can make non-rigid figures rigid.
A rectangle is not rigid. One way to make it rigid is to add a cardboard strip diagonal.

Group Project: *Home Economics*

Quilting involves sewing pieces of fabric together to form an interesting and appealing design. Find a book of quilt patterns and study the designs. What kind of geometric figures can you find? Work with your group to design a unique quilt pattern using geometric shapes. Use construction paper cut outs to create a quilt pattern of your own design.

6.1 All About Parallelograms

Objective: To prove and apply theorems about parallelograms.

Knowledge of congruent triangles is used to prove facts about parallelograms. Some of this information will be familiar; other facts will be new.

Definition of a Parallelogram

A **parallelogram** is a quadrilateral with two pairs of parallel sides.

Pictured at the right is parallelogram *ABCD*. You can use "▱*ABCD*" to represent parallelogram *ABCD*.

Does it appear that the opposite sides have the same measure?

Does this happen in *any* parallelogram? Are the opposite angles in this parallelogram congruent? To prove that the opposite sides and opposite angles are congruent, it is possible to show that they are corresponding parts of congruent triangles. But where are the congruent triangles? The following theorem provides an answer.

Theorem 6–1

A diagonal of a parallelogram separates the parallelogram into two congruent triangles.

Example

1 Write a formal proof of Theorem 6–1.

Given: ▱*ABCD*, diagonal $\overline{BD}$
Prove: △*ABD* ≅ △*CDB*
Proof:

STATEMENTS	REASONS
1. ▱*ABCD*, diagonal $\overline{BD}$	1. Given
2. $\overline{AD} \parallel \overline{BC}$, $\overline{AB} \parallel \overline{CD}$	2. Definition of parallelogram
3. ∠1 ≅ ∠2, ∠3 ≅ ∠4	3. If two parallel lines are cut by a transversal, then the alternate interior angles are congruent.
4. $\overline{BD} \cong \overline{BD}$	4. Congruence of segments is reflexive.
5. △*ABD* ≅ △*CDB*	5. ASA Axiom

There are two corollaries of Theorem 6–1.

Theorem 6–2   Opposite sides of a parallelogram are congruent.

Theorem 6–3   Opposite angles of a parallelogram are congruent.

Now, consider ▱ABCD shown here. Does M seem to be the midpoint of both diagonals? You are right if you answered yes.

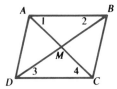

Theorem 6–4   The diagonals of a parallelogram bisect each other.

Before proving a theorem, we should have a plan clearly in mind. The following is a plan for the proof of Theorem 6–4. Show △AMB ≅ △CMD. Then $\overline{AM} \cong \overline{MC}$, and $\overline{DM} \cong \overline{MB}$, because corresponding parts of congruent triangles are congruent. Thus, M is the midpoint of both $\overline{AC}$ and $\overline{BD}$.

Example

2 Using the ▱ABCD above, write a two-column proof of Theorem 6–4.

Given: ▱ABCD, diagonals $\overline{AC}$ and $\overline{BD}$
Prove: $\overline{AC}$ and $\overline{BD}$ bisect each other.
Proof:

STATEMENTS	REASONS
1. ▱ABCD, diagonals $\overline{AC}$ and $\overline{BD}$	1. Given
2. $\overline{AB} \parallel \overline{CD}$	2. Definition of parallelogram
3. ∠1 ≅ ∠4, ∠2 ≅ ∠3	3. If two parallel lines are cut by a transversal, then alternate interior angles are congruent.
4. $\overline{AB} \cong \overline{CD}$	4. Opposite sides of a parallelogram are congruent.
5. △AMB ≅ △CMD	5. ASA Axiom
6. $\overline{AM} \cong \overline{MC}, \overline{DM} \cong \overline{MB}$	6. Corresponding parts of congruent triangles are congruent.
7. M is the midpoint of $\overline{AC}$ and $\overline{BD}$.	7. Definition of midpoint
8. $\overline{AC}$ and $\overline{BD}$ bisect each other.	8. Definition of bisect

Exercises

Max: 1-14, 1-36; Avg: 1-14, 1-19, 21-35 odds; Min: 1-14 odds, 1-35 odds

Exploratory Given ▱ABCD, tell whether each statement is *true* or *false*. Justify your answer.

1. $\overline{AB} \parallel \overline{CD}$ true; definition of a parallelogram
2. $\overline{AB} \cong \overline{BC}$ false; not opposite sides
3. $\triangle ABD \cong \triangle CDB$
4. $\angle A \cong \angle C$ true; Theorem 6-3
5. $\angle ADB \cong \angle CDB$
6. $\angle ABD \cong \angle CDB$ true; Theorem 6-1 and CPCTC
7. $\angle DAB$ and $\angle CBA$ are supplementary.
8. $\angle ADC \cong \angle ABC$

3. true; Theorem 6-1
5. false; not corresponding angles
7. true; Theorem 4-15
8. true; Theorem 6-3

Given ▱RSTU, name the following parts.

9. an angle congruent to $\angle RST$ ∠RUT
10. an angle supplementary to $\angle STU$ ∠RUT, ∠RST
11. a segment congruent to $\overline{RS}$ UT
12. a segment congruent to $\overline{RW}$ TW
13. a segment congruent to $\overline{UW}$ SW
14. a segment congruent to $\overline{ST}$ RU

Written Refer to ▱RSTU above. Find $m\angle URS$, $m\angle RST$, and $m\angle STU$ when $\angle TUR$ has the following degree measure.

1. 47   133, 47, 133
2. 58   122, 58, 122
3. 87   93, 87, 93
4. 78   102, 78, 102
5. 63   117, 63, 117

Find $m\angle URS$ and $m\angle RST$ when $\angle TUR$ and $\angle UTS$ have the following measures.

6. $m\angle TUR = y$, $m\angle UTS = (2y + 60)$ 140, 40
7. $m\angle TUR = (x + 10)$, $m\angle UTS = 3(x - 30)$ 105, 75
8. $m\angle TUR = 2(x - 5)$, $m\angle UTS = (3x - 20)$ 106, 74
9. $m\angle TUR = 4(x - 2)$, $m\angle UTS = 2(x + 14)$ $81\frac{1}{3}$, $98\frac{2}{3}$

Find RS, ST, TU and UR when $\overline{ST}$, $\overline{TU}$, and $\overline{UR}$ have the following measures.

10. $ST = 11y - 10$, $TU = \frac{y}{2} + 40$, $UR = 7y + 30$ 45, 100, 45, 100
11. $ST = 12(y - 4)$, $TU = 4y - 11$, $UR = 7y + 12$ 37, 96, 37, 96
12. $ST = 2(4m - 3) + 8$, $TU = 4m + 8$, $UR = 3(2m + 3)$ 22, 30, 22, 30
13. $ST = x^2$, $TU = 3x^2 - 5x$, $UR = 8x - 16$ 28, 16, 28, 16

Refer to ▱RSTU above. Find RT and US under the given conditions.

14. $RW = 3x - 1$, $WT = 15 - x$, $UW = 3x + 1$, $WS = 4x - 3$ 22, 26
15. $RW = 2x$, $WT = 8 - y$, $UW = x$, $WS = y + 1$ 12, 6
16. $RW = 4x + 2y$, $WT = 17 + x$, $UW = 6x + 2y$, $WS = 16 + y$ $37\frac{1}{3}$, 44
17. $RW = x^2$, $WT = 14x - 49$, $UW = 7(y + 1) - 2(y - 3)$, $WS = 7 - 7(6 - y)$ 98, 266

Describe how these theorems follow directly from Theorem 6–1.

18. Theorem 6–2 CPCTC   **19.** Theorem 6–3 CPCTC

Students may need to review the definitions of parallelogram, rectangle, rhombus, and square before answering exercises 20-31.

Tell whether each statement is *true* or *false*.

20. If quadrilateral *ABCD* is a rectangle, then it is a parallelogram. true
21. If quadrilateral *ABCD* is a rectangle, then it is a square. false
22. If quadrilateral *ABCD* is a rhombus, then it is a square. false
23. If quadrilateral *ABCD* is a trapezoid, then it is a parallelogram. false
24. If quadrilateral *ABCD* is a square, then it is a parallelogram. true
25. If quadrilateral *ABCD* is a rhombus, then it is a rectangle. false

For answers to exercises 26-36, see the Teacher's Guide.

Write the converse of the following statements. Then tell whether each is *true* or *false*.

26. exercise 20   **27.** exercise 21   **28.** exercise 22
29. exercise 23   **30.** exercise 24   **31.** exercise 25

Write a two-column proof for each.

32. Given: ▱*ABCD*, ▱*EFCG*
  Prove: ∠*A* ≅ ∠*FEG*

Exercise 32

33. Given: ▱*ABCD*, *E* is the midpoint of $\overline{DC}$,
  F is the midpoint of $\overline{AB}$.
  Prove: $\overline{AE} \cong \overline{CF}$

Exercise 33

34. Given: ▱*ABEF*, ▱*BCDE*
  Prove: $\overline{AF} \cong \overline{CD}$

Exercise 34

35. Given: ▱*ABCD*,
  $\overline{AE} \perp \overline{DC}$, $\overline{CF} \perp \overline{AB}$
  Prove: $\overline{AE} \cong \overline{CF}$

36. Given: ▱*ABCD*,
  $\overline{DE} \perp \overline{AE}$, $\overline{BF} \perp \overline{FC}$
  Prove: $\overline{AE} \cong \overline{CF}$

Exercises 35–36

6.2 When is a Quadrilateral a Parallelogram?

Objective: To prove when a quadrilateral is a parallelogram.

Suppose we are given a quadrilateral that we suspect is a parallelogram. We would try to prove this fact by showing that both pairs of opposite sides are parallel. In our work with congruent triangles, though, we saw that several ways exist to prove triangles congruent. Perhaps other ways exist to show that a quadrilateral is a parallelogram.

Pictured is a rectangular piece of plywood ABCD. Suppose a point X is located such that AX = 25. Now, pick a point Y on the opposite side and locate point Z such that YZ = 25. Does it appear that quadrilateral AYZX is a parallelogram? That is, if $\overline{AX}$ and $\overline{YZ}$ are both parallel and congruent, must AYZX necessarily be a parallelogram? The next theorem provides an answer.

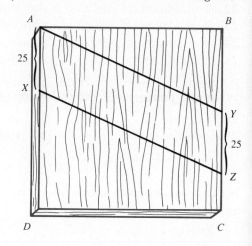

Theorem 6–5

If a quadrilateral has two sides that are both parallel and congruent, then the quadrilateral is a parallelogram.

Example

1 Complete this proof for Theorem 6–5.

Given: Quadrilateral ABCD, with $\overline{AB} \cong \overline{CD}$ and $\overline{AB} \parallel \overline{CD}$

Prove: ABCD is a parallelogram.

Proof:

STATEMENTS	REASONS
1. Draw $\overline{BD}$	1. For any two points, there exists one and only one line containing both of them.
2. $\overline{AB} \parallel \overline{CD}$	2. Given
3. $\angle 1 \cong \angle 2$	3. If two parallel lines are cut by a transversal, then the alternate interior angles are congruent.

4. $\overline{AB} \cong \overline{CD}$	4. Given
5. $\overline{BD} \cong \overline{BD}$	5. Reflexive property
6. $\triangle ABD \cong \triangle CDB$	6. SAS Axiom
7. $\angle 4 \cong \angle 3$	7. Corresponding parts of congruent triangles are congruent.
8. $\overline{BC} \parallel \overline{AD}$	8. If two lines are cut by a transversal and the alternate interior angles are congruent, the lines are parallel.
9. ABCD is a parallelogram.	9. Definition of a parallelogram

Shown at the right are four pieces of wood such that $\overline{AB} \cong \overline{CD}$ and $\overline{EF} \cong \overline{GH}$.

Suppose we fasten the pieces together to form a quadrilateral. Here are some possible arrangements.

Does it appear that each quadrilateral is a parallelogram? The next theorem states that they must be.

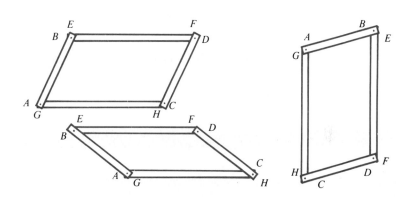

Theorem 6-6 If both pairs of opposite sides of a quadrilateral are congruent, then the quadrilateral is a parallelogram.

Example

2 Complete the proof of Theorem 6-6.

Given: Quadrilateral ABCD with $\overline{AB} \cong \overline{CD}$ and $\overline{BC} \cong \overline{AD}$
Prove: ABCD is a parallelogram.

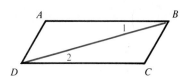

214 APPLICATIONS OF CONGRUENT TRIANGLES

Proof:

STATEMENTS	REASONS
1. Draw $\overline{BD}$.	1. For any two points, there exists one and only one line containing both of them.
2. $\overline{AB} \cong \overline{CD}$, $\overline{BC} \cong \overline{AD}$	2. Given
3. $\overline{BD} \cong \overline{BD}$	3. Any line segment is congruent to itself.
4. $\triangle ABD \cong \triangle CDB$	4. SSS Axiom
5. $\angle 1 \cong \angle 2$	5. Corresponding parts of congruent triangles are congruent.
6. $\overline{AB} \parallel \overline{CD}$	6. If two lines are cut by a transversal and the alternate interior angles are congruent, then the lines are parallel.
7. ABCD is a parallelogram.	7. If a quadrilateral has two sides that are both parallel and congruent, then the quadrilateral is a parallelogram.

Theorem 6–6 is the converse of Theorem 6–2. What about the converse of other theorems proved in that section? Do they provide other ways of proving that a quadrilateral is a parallelogram?

If Theorem 6–1 is placed in if-then form, the result is

If quadrilateral ABCD is a parallelogram, then a diagonal divides ABCD into two congruent triangles.

The converse of this statement is

If a diagonal of quadrilateral ABCD separates it into two congruent triangles, then ABCD is a parallelogram.

Is that statement true?

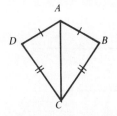

Look carefully at the diagram. Notice that in the figure $\overline{AB} \cong \overline{AD}$ and $\overline{BC} \cong \overline{DC}$. Because $\overline{AC} \cong \overline{AC}$, the SSS axiom tells us that $\triangle ADC \cong \triangle ABC$. Therefore, diagonal $\overline{AC}$ of quadrilateral ABCD separates ABCD into two congruent triangles. However, ABCD is not a parallelogram. Thus, the converse of Theorem 6–1 is *false*.

The search continues for ways of proving a quadrilateral is a parallelogram. This time, put Theorem 6–4 in if-then form.

If a quadrilateral is a parallelogram, then its diagonals bisect each other.

The converse is

If the diagonals of a quadrilateral bisect each other, then the quadrilateral is a parallelogram.

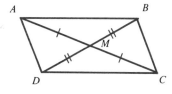

That is, if M is the midpoint of both $\overline{AC}$ and $\overline{BD}$, must $ABCD$ be a parallelogram? The next theorem states that it must be. A plan for this theorem is given in Example 3. The proof is in the exercises.

Theorem 6–7 If the diagonals of a quadrilateral bisect each other, then the quadrilateral is a parallelogram.

Example

3 Here is a plan of proof for Theorem 6–7.

Given: Quadrilateral $ABCD$, $\overline{AC}$ and $\overline{BD}$ bisect each other at M.
Prove: Quadrilateral $ABCD$ is a parallelogram.
Plan: Show $\triangle AMB \cong \triangle CMD$ by SAS. Then show $\overline{AB} \cong \overline{CD}$ and $\overline{AB} \parallel \overline{CD}$. By Theorem 6–5, $ABCD$ is a parallelogram.

Theorem 6–8 If both pairs of opposite angles of a quadrilateral are congruent, then the quadrilateral is a parallelogram.

A quadrilateral may be proved a parallelogram when

1. both pairs of opposite sides are parallel.
2. two sides are both parallel and congruent.
3. both pairs of opposite sides are congruent.
4. the diagonals bisect each other.
5. both pairs of opposite angles are congruent.

216 APPLICATIONS OF CONGRUENT TRIANGLES

Exercises

Max: 1-17, 1-24, Avg: 1-17, 1-15, 17-23 odds; Min: 1-17, 1-23 odds

Exploratory Answer the following. For answers to exercises 1-17, see the Teacher's Guide.

1. Suppose quadrilateral ABCD is a parallelogram. Give as many facts as you can about ⌷ABCD.
2. Explain five ways of showing that a quadrilateral is a parallelogram.

Tell whether each statement is *true* or *false*. Justify your answer.

3. If the opposite sides of a quadrilateral are parallel, then the quadrilateral is a parallelogram.
4. A quadrilateral is a parallelogram if one pair of opposite sides is congruent.
5. If a quadrilateral has one pair of parallel sides, then the quadrilateral is a parallelogram.
6. A quadrilateral is a parallelogram if and only if two sides are both parallel and congruent.
7. A quadrilateral is a parallelogram if both pairs of opposite sides are parallel.
8. The diagonals of a parallelogram bisect each other.
9. If a diagonal of a quadrilateral separates the quadrilateral into two congruent triangles, then the quadrilateral is a parallelogram.
10. A quadrilateral is a parallelogram if its diagonals bisect each other.
11. A quadrilateral is a parallelogram if one pair of opposite angles is congruent.

Tell why quadrilateral ABCD must be a parallelogram under the given conditions. Justify your answer.

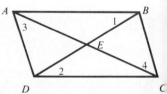

12. Given: $\overline{AB} \parallel \overline{CD}, \overline{AD} \parallel \overline{BC}$
13. Given: $\overline{AD} \cong \overline{BC}, \overline{AD} \parallel \overline{BC}$
14. Given: $\overline{AB} \cong \overline{CD}, \angle 1 \cong \angle 2$
15. Given: $\angle 1 \cong \angle 2, \angle 3 \cong \angle 4$
16. Given: E is the midpoint of $\overline{AC}$ and $\overline{DB}$.
17. Given: $\triangle AED \cong \triangle CEB$

Written Refer to ⌷ABCD. Find the values of x and y for quadrilateral ABCD to be a parallelogram.

1. $AD = 5, AB = 9, DE = 6\frac{1}{2}, BE = x$ $6\frac{1}{2}$
2. $AB = 24, BD = 25, ED = x, DC = y$ $x = 12\frac{1}{2}, y = 24$
3. $AB = 2x + 4, AD = y + 12, DC = 4x - 2, BC = 4 - y$ $x = 3, y = -4$
4. $AB = 3(x + 7), AD = 2(y + 4), DC = 4x + 22, BC = y + 11$ $x = -1, y = 3$
5. $AB = 2x + 1, DC = 5x - 8, AD = 4(y + 6), BC = 2(3y - 5)$ $x = 3, y = 17$
6. $AB = 2(5x + 4), DC = 4(2 + 2x) + 2, AD = 4(3 - 2y), BC = 14 - 6y$ $x = 1, y = -1$
7. $AE = x^2, EC = 2x - 1, DE = y^2, EB = 6y - 9$ $x = 1, y = 3$
8. $AE = x^2 + 2x - 1, EC = 16x - 50, DE = 2y - 4 + y^2, EB = 2y$ $x = 7, y = 2$

For answers to exercises 9-24, see the Teacher's Guide.
Write each theorem named below in if-then form and state its converse. Tell if the converse is *true* or *false*. Justify your answer.

9. Theorem 6–1   10. Theorem 6–2   11. Theorem 6–4   12. Theorem 6–5

Write a two-column proof for each theorem.

13. Theorem 6–7   14. Theorem 6–8

Answer the following.

15. Ted claims that if one pair of opposite sides of a quadrilateral are congruent and the *other* pair of opposite sides are parallel, then the quadrilateral is a parallelogram. Do you agree? Explain.

Write a two-column proof for each.

16. **Given:** ∠1 is supplementary to ∠4,
    $\overline{DC} \cong \overline{AB}$
    Prove: ABCD is a parallelogram.
17. **Given:** $\overline{AD} \cong \overline{CE}$, $\overline{CB} \cong \overline{CE}$,
    ∠1 is supplementary to ∠2.
    Prove: ABCD is a parallelogram.
18. **Given:** ∠1 is supplementary to ∠2,
    ∠DCB ≅ ∠2
    Prove: ABCD is a parallelogram.
19. **Given:** $\overline{AD} \cong \overline{CE}$, $\overline{DC} \cong \overline{AB}$,
    ∠2 ≅ ∠3
    Prove: ABCD is a parallelogram.

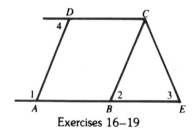
Exercises 16–19

20. **Given:** ▱AFCE, $\overline{DE} \cong \overline{BF}$
    Prove: ABCD is a parallelogram.
21. **Given:** ▱ABCD, E is the midpoint of $\overline{DC}$, F is the midpoint of $\overline{AB}$.
    Prove: AFCE is a parallelogram.
22. **Given:** ▱ABCD, $\overline{AE}$ bisects ∠DAB.
    $\overline{CF}$ bisects ∠DCB.
    Prove: AFCE is a parallelogram.
23. **Given:** ▱ABCD, ∠DAE ≅ ∠BCF
    Prove: AFCE is a parallelogram.

Exercises 20–23

24. **Given:** D is the midpoint of $\overline{AB}$, E is the midpoint of $\overline{AC}$,
    $\overline{DE} \cong \overline{FE}$
    Prove: BCFD is a parallelogram.

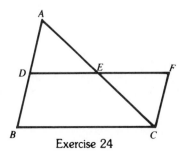
Exercise 24

6.3 Rectangles and Rhombuses

Objective: To prove and apply theorems about rectangles and rhombuses.

Some special parallelograms occur so frequently that they are singled out for special attention. For convenience, the definitions of rectangle and rhombus are given below.

Definition of a Rectangle

> A **rectangle** is a parallelogram with four right angles.

However, only one right angle is necessary to prove that a parallelogram is a rectangle.

Definition of a Rhombus

> A **rhombus** is a parallelogram with four congruent sides.

▱ABCD is a rectangle.   ▱GHJK is a rhombus.

Because rectangles and rhombuses are parallelograms, they possess all the properties of a parallelogram. For example, their diagonals bisect each other. Look carefully at the rectangle shown at the right. It may appear that something else is true about the diagonals. Did you guess the following theorem?

Theorem 6–9

> If a parallelogram is a rectangle, then its diagonals are congruent.

Example

1 Show a plan of proof for Theorem 6–9.

Given: ABCD is a rectangle.
Prove: $\overline{AC} \cong \overline{BD}$

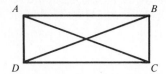

Plan: Show that $\triangle ACD \cong \triangle BDC$.
Then $\overline{AC} \cong \overline{BD}$.
The proof is in the exercises.

Look carefully at the rhombus at the right. Your observations may lead you to Theorems 6-10 and 6-11.

Theorem 6-10 If a parallelogram is a rhombus, then its diagonals are perpendicular.

Example

2 Show a plan of proof for Theorem 6-10.

Given: ▱ABCD is a rhombus.
Prove: $\overline{AC} \perp \overline{BD}$

Plan: Show △ADE ≅ △ABE by SSS. Then ∠1 ≅ ∠2. Therefore, $\overline{AC} \perp \overline{BD}$.
The proof is in the exercises.

Theorem 6-11 If a parallelogram is a rhombus, then each diagonal bisects two opposite angles of the parallelogram.

Example

3 Draw a figure, and write the given statement and the prove statement to use in the proof of Theorem 6-11.

Given: ▱ABCD is a rhombus.
Prove: $\overline{BD}$ bisects ∠ABC.
$\overline{BD}$ bisects ∠ADC.
The proof is in the exercises.

Because a square is both a rhombus and a rectangle, it has all the properties proved for parallelograms, rectangles, and rhombuses.

220 APPLICATIONS OF CONGRUENT TRIANGLES

Exercises

Max: 1-32, 1-22; Avg: 1-32, 1-16, 17-21 odds; Min: 5-26; 1-21 odds

Exploratory Draw an example of each of the following. For exercises 1-4, see students' work

1. parallelogram
2. rectangle
3. rhombus
4. square

Tell whether each statement is *true* or *false*. Justify your answers. For answers to exercises 5-3 see the Teacher's Guid

5. If the diagonals of a parallelogram are congruent, then it is a rectangle.
6. The diagonals of a parallelogram are perpendicular; therefore, it is a rhombus.
7. Each diagonal bisects two opposite angles of a parallelogram; therefore, it is a rhombus.

Tell whether each statement is *true* or *false*. Justify your answer.

8. Every rectangle is a parallelogram.
9. Every square is a rectangle.
10. Every square is a rhombus.
11. Every rhombus is a square.
12. Every rectangle is a rhombus.
13. Every rhombus is a rectangle.
14. The diagonals of a rectangle are perpendicular.
15. Two angles of a rectangle are supplementary.
16. The diagonals of a rhombus bisect each other.
17. The diagonals of a parallelogram are congruent.
18. The diagonals of a rhombus are perpendicular.
19. The diagonals of a rectangle are congruent.
20. A parallelogram that contains a right angle is a rectangle.
21. The diagonals of a rhombus are congruent.
22. Opposite sides of a rectangle are congruent.
23. A diagonal divides a rhombus into two congruent triangles.
24. A diagonal of a parallelogram bisects two opposite angles.
25. A diagonal of a rhombus bisects two opposite angles.
26. If a quadrilateral is a rhombus and a rectangle, then it is a square.

Name as many properties of each figure as you can.

27. parallelogram
28. rectangle
29. rhombus
30. square
31. Name the properties a rhombus must have that a rectangle need not have. Draw a picture to illustrate the properties.
32. Name the properties a rectangle must have that a parallelogram need not have.

Written Refer to rectangle *ABCD* at right.   1. 21, 21, 21, 21   2. 2, 2, 2, 2

1. $AE = 3(4x - 1)$, $BE = 10x + 1$, find AE, BE, DE, and CE.
2. $AE = 2(x - 5)$, $BE = 3x - 16$, find AE, BE, DE, and CE.
3. $AE = 5(10 - 2x)$, $DE = 4(3x - 4)$, find AE, BE, DE, and CE.
4. $AE = 3(4x - 1) - x$, $BE = 4(6 - x) - 12$, find AE, BE, AC, and BD. 8, 8, 16, 16
5. $AE = 2(x - 1) + 3(2x - 2)$, $EC = 4(x + 1) + 2(x - 1)$, find AE, EC, ED, and BE. 32, 32, 32, 32   3. 20, 20, 20, 20
6. $AC = 3(2x + 5) - \frac{1}{2}(2x + 2)$, $BD = \frac{2}{3}(9x + 3) + 5x$, find AC, BD, AE, EC, BE, and ED. 24, 24, 12, 12, 12, 12

For answers to exercises 7-22, see the Teacher's Guide.
Write a two-column proof for each of the following.

7. Theorem 6-9
8. Theorem 6-10
9. Theorem 6-11
10. **Given:** Rectangle ABCD, E is the midpoint of $\overline{BC}$.
    Prove: $\overline{AE} \cong \overline{DE}$
11. **Given:** Rectangle ABCD, $\overline{AE} \cong \overline{DE}$
    Prove: $\angle EDC \cong \angle EAB$
12. **Given:** Rectangle ABCD
    Prove: $\angle BDC \cong \angle ACD$
13. **Given:** Rectangle ABCD
    Prove: $\angle CAD \cong \angle BDA$

Exercises 10–11

Exercises 12–13

State the converse of each theorem named below and decide if it is *true* or *false*. If it is true, write a two-column proof.

14. Theorem 6-9
15. Theorem 6-10
16. Theorem 6-11

Write a two-column proof for each of the following.

17. If the consecutive angles of a quadrilateral are congruent, then the quadrilateral is a rectangle.
18. **Given:** $\overleftrightarrow{EF} \parallel \overleftrightarrow{HG}$, $\overline{AB}$ bisects $\angle EAC$, $\overline{CB}$ bisects $\angle HCA$, $\overline{AD}$ bisects $\angle FAC$, and $\overline{CD}$ bisects $\angle GCA$.
    Prove: Quadrilateral ABCD is a rectangle.

19. **Given:** $\overline{AD} \cong \overline{AB}$, $\overline{AC}$ bisects $\angle DAB$, and $\overline{AE} \cong \overline{EC}$.
    Prove: ABCD is a rhombus.

20. **Given:** $\overline{RT} \parallel \overline{PW}$, $\overline{RT} \parallel \overline{QV}$
    $\overline{TX} \parallel \overline{RP}$
    Q is the midpoint of $\overline{PR}$.
    Prove: V is the midpoint of $\overline{TX}$.
21. **Given:** $\square RTXP$
    V is the midpoint of $\overline{SW}$.
    Prove: $\triangle SVT \cong \triangle WVX$
22. **Given:** $\overline{RT} \parallel \overline{PW}$, $\overline{TX} \parallel \overline{RP}$
    Q is the midpoint of $\overline{PR}$.
    V is the midpoint of $\overline{XT}$.
    Prove: Quadrilateral QRTV is a parallelogram.

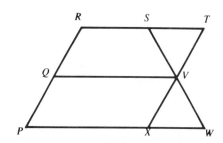

6.4 Trapezoids

Objective: To prove and apply theorems about trapezoids.

To make a frame for a picture, a carpenter usually cuts the corners of the frame on an angle, as shown in the drawing. The result is four quadrilaterals, each of which is a *trapezoid*.

Definition of a Trapezoid

A **trapezoid** is a quadrilateral with exactly one pair of parallel sides.

In the drawing of the frame shown above, *ABCD* is a trapezoid. So is *EFGH* shown at the right.

The parallel sides of a trapezoid are called its *bases*. The other two sides are called its *legs*. A pair of angles that includes one base of the trapezoid are called **base angles**.

In the diagram, $\overline{EF}$ and $\overline{GH}$ are *bases* and $\overline{EH}$ and $\overline{FG}$ are *legs*. One pair of *base angles* is ∠H and ∠G; the other pair is ∠E and ∠F.

Definition of an Isosceles Trapezoid

An **isosceles trapezoid** is a trapezoid with congruent legs.

Do pairs of base angles appear to be congruent? The following theorem answers this question.

Theorem 6–12

The base angles of an isosceles trapezoid are congruent.

Example

1 Write a two-column proof of Theorem 6–12.

Given: Isosceles trapezoid *ABCD* with $\overline{AB} \parallel \overline{DC}$

Prove: ∠DAB ≅ ∠CBA
∠D ≅ ∠C

Proof:

STATEMENTS	REASONS
1. Draw $\overline{AE} \perp \overline{DC}$ and $\overline{BF} \perp \overline{DC}$.	1. Through a point not on a line, there exists one and only one line perpendicular to the given line.
2. $\overline{AE} \parallel \overline{BF}$	2. In a plane, if two lines are perpendicular to the same line, they are parallel.
3. $\overline{AB} \parallel \overline{DC}$, or $\overline{AB} \parallel \overline{EF}$	3. Given
4. ABFE is a parallelogram.	4. Definition of a parallelogram
5. $\overline{AE} \cong \overline{BF}$	5. Opposite sides of a parallelogram are congruent.
6. $\overline{AD} \cong \overline{BC}$	6. Definition of isosceles trapezoid
7. $\triangle AED$ and $\triangle BFC$ are right triangles.	7. Definition of right triangle
8. right $\triangle AED \cong$ right $\triangle BFC$	8. If the hypotenuse and leg of right triangles are congruent, the triangles are congruent.
9. $\angle D \cong \angle C$	9. Corresponding parts of congruent triangles are congruent.
10. $\angle 1 \cong \angle D$, $\angle C \cong \angle 2$	10. If two parallel lines are cut by a transversal, corresponding angles are congruent.
11. $\angle 1 \cong \angle 2$	11. Transitive property
12. $\angle DAB$ is supplementary to $\angle 1$, $\angle CBA$ is supplementary to $\angle 2$.	12. If the exterior sides of two adjacent angles are opposite rays, the angles are supplementary.
13. $\angle DAB \cong \angle CBA$	13. If two angles are supplementary to congruent angles, they are congruent to each other.

Theorem 6–13 The diagonals of an isosceles trapezoid are congruent.

You will be asked to prove this theorem in the exercises.

Exercises

Max: 1-12, 1-18; Avg: 1-12, 1-5, 7-17 odds; Min: 1-12, 1-17 odds

Exploratory Refer to the diagram of isosceles trapezoid ABCD with $\overline{AB} \parallel \overline{CD}$. Name the following.

1. bases $\overline{AB}, \overline{DC}$ ∠D, ∠C; **2.** legs $\overline{AD}, \overline{BC}$
3. a pair of base angles ∠A, ∠B   **4.** a pair of congruent angles
5. a pair of congruent segments $\overline{AD}, \overline{BC}$
                                                                      4. ∠D, ∠C; ∠A, ∠B
6. a pair of supplementary angles ∠B, ∠C; ∠A, ∠D

7. If $m\angle DAB = 4(x + 5)$ and $m\angle CBA = 3(x + 15)$, find $m\angle DAB, m\angle CBA, m\angle ADC$, and $m\angle BCD$. 120, 120, 60, 60
8. If $m\angle DAB = 5(x - 2)$ and $m\angle CBA = 3(x + 4)$, find $m\angle DAB, m\angle CBA, m\angle ADC$, and $m\angle BCD$. 45, 45, 135, 135
9. If $m\angle DAB = 4(x + 5)$ and $m\angle CBA = 2(x + 17) + 4$, find $m\angle DAB, m\angle CBA, m\angle ADC$, and $m\angle BCD$. 56, 56, 124, 124
10. If $m\angle DAB = 5(3x - 1)$ and $m\angle CBA = 5x + 45$, find $m\angle DAB, m\angle CBA, m\angle ADC$, and $m\angle BCD$. 70, 70, 110, 110
11. If $m\angle DAB = 2x + 17$ and $m\angle CBA = 3(2x - 11) + 2$, find $m\angle DAB, m\angle CBA, m\angle ADC$, and $m\angle BCD$. 41, 41, 139, 139
12. If $m\angle ADC$ is 20% of the measure of $\angle DAB$, find $m\angle ADC, m\angle ABC, m\angle BCD$, and $m\angle DAB$. 30, 150, 30, 150

For answers to exercises 1-18, see the Teacher's Guide.

Written Given ABCD, an isosceles trapezoid with $\overline{AB} \parallel \overline{DC}$, prove the following.

1. △ACD ≅ △BDC
2. △AED ≅ △BEC
3. △AEB is isosceles.
4. ∠1 ≅ ∠2

5. When the "given" is a trapezoid, why is it necessary to say which sides are parallel? Give an example of a problem that might occur if this is not done.

Write a two-column proof for the following.

6. Theorem 6–13
7. Given: ABCD is an isosceles trapezoid.
   $\overline{AB} \parallel \overline{DC}, \overline{AE} \perp \overline{DC}, \overline{BF} \perp \overline{DC}$
   Prove: △ADE ≅ △BCF
8. Given: $\overline{AB} \parallel \overline{DC}, \angle D \cong \angle C, \overline{AE} \perp \overline{DC}, \overline{BF} \perp \overline{DC}$
   Prove: ABCD is an isosceles trapezoid.

Exercises 7–8

9. Given: Isosceles trapezoid ABCD, $\overline{DE}$ bisects ∠ADC. $\overline{AB} \parallel \overline{CD}$
   Prove: $\overline{AD} \cong \overline{AE}$

Exercise 9

10. **Given:** Trapezoid ABED with $\overline{EB} \parallel \overline{DA}$, $\overline{BA} \parallel \overline{CD}$, $\angle E \cong \angle 1$.
    Prove: ABED is isosceles.
11. **Given:** ▱ABCD, $\angle E \cong \angle 2$
    Prove: ABED is an isosceles trapezoid.
12. **Given:** ▱ABCD, $\angle E$ is supplementary to $\angle A$.
    Prove: ABED is an isosceles trapezoid.

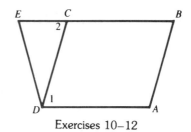

Exercises 10–12

13. **Given:** Isosceles trapezoid ABCD with $\overline{AD} \parallel \overline{BC}$
    Prove: $\overline{AE} \cong \overline{DE}$
14. **Given:** Isosceles trapezoid ABCD with $\overline{AD} \parallel \overline{BC}$
    Prove: $\triangle BEC$ is isosceles.
15. **Given:** $\overline{BE} \cong \overline{CE}$, $\angle EBC \cong \angle EAD$
    Prove: ABCD is an isosceles trapezoid.
16. **Given:** $\overline{AE} \cong \overline{DE}$, $\overline{BC} \parallel \overline{AD}$
    Prove: ABCD is an isosceles trapezoid.

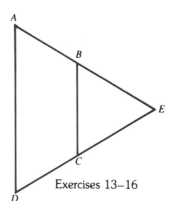

Exercises 13–16

Write each theorem in if-then form. State the converse and decide if it is *true* or *false*. If it is true, write a two-column proof.

17. Theorem 6–12
18. Theorem 6–13

━━━ Mixed Review ━━━

If △MNP is isosceles and has base $\overline{NP}$, find the length of each side of △MNP.

1. MN = 2x + 5, NP = 21, MP = 3x − 7
2. MN = 26, NP = 3x + 1, MP = 5x + 1
3. MN = 4x − 5, NP = 3x, MP = x + 13
4. MN = x^2, NP = 5x − 6, MP = 5x + 6

MN = MP = 29, NP = 21   2. MN = MP = 26, NP = 16   3. MN = MP = 19, NP = 18   4. MN = MP = 36, NP = 24

In exercises 5–8, tell what *one* additional piece of information you must have in order to prove quadrilateral ABCD, with diagonals $\overline{AC}$ and $\overline{BD}$, is a parallelogram.

5. **Given:** $\angle A \cong \angle C$ $\angle B \cong \angle D$
6. **Given:** $\overline{AC}$ bisects $\overline{BD}$. $\overline{BD}$ bisects $\overline{AC}$.
7. **Given:** $\overline{AD} \cong \overline{BC}$ $\overline{AB} \cong \overline{CD}$ or $\overline{AD} \parallel \overline{BC}$
8. **Given:** $\overline{AB} \parallel \overline{CD}$ $\overline{AD} \parallel \overline{BC}$ or $\overline{AB} \cong \overline{CD}$

9. In △DEF, $m\angle D = 39$ and $m\angle E = 71$. Name the sides of △DEF, in order, from shortest to longest. EF, DE, DF

10. In △RST, RS = x + 6, ST = 2x − 3, and RT = 4x − 3. For what values of x will RT be the longest side of △RST?   3 < x < 6

6.5 Some Constructions

Objective: To construct basic geometric figures.

It is possible to construct many geometric figures without making measurements. Two instruments are used, a straightedge and a compass. A **construction** differs from a drawing. Rulers and protractors are used in drawings but may not be used in constructions. Although the edge of a ruler may be used as a straightedge, no markings on a ruler may be used to measure distance. Instead, distances are transferred by a compass.

In this section you will learn how to construct some of the figures with which we have worked. You will also see how to use knowledge about congruent triangles to justify the constructions.

CONSTRUCTION 1: Construct a segment congruent to a given segment.

Given: Segment AB

GIVEN:
$A \longrightarrow B$

Method:

1. Draw a segment clearly longer than $\overline{AB}$. Label one endpoint C.

    1.
    $C \longrightarrow$

2. Place the metal point of the compass on A and open the compass so that the pencil point is at B. The distance between the points of the compass has the measure AB. For convenience, we will call this setting of the compass "radius" AB. This means that a circle could be drawn with center at A through B with radius AB.

    2.

3. Keep the compass open to radius AB. Transfer the metal point to C. Draw an arc that intersects the segment. Label the point of intersection D. Because $CD = AB$, then $\overline{CD} \cong \overline{AB}$.

    3.

CHAPTER 6 227

CONSTRUCTION 2: Construct an angle congruent to a given angle.

Given: ∠A
Method:

1. Draw a ray TR.

2. With A as center, draw an arc that intersects the sides of ∠A in points B and C.

3. With the same radius, draw an arc with center T that intersects $\overrightarrow{TR}$ in point S. The arc should be clearly longer than necessary.

4. With S as center and radius BC, draw an arc that intersects the arc drawn in Step 3 at U.

5. Draw $\overrightarrow{TU}$.
   ∠STU ≅ ∠BAC

   Justification: △BAC ≅ △STU by SSS. Because ∠STU and ∠BAC are corresponding parts, they are congruent.

1.

2.

3.

4.

5.
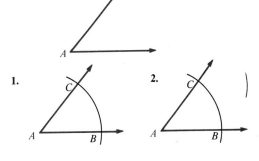

CONSTRUCTION 3: Bisect a given angle.

Given: ∠A
Method:

1. With A as center, draw an arc intersecting the sides of ∠A in B and C.

2. With C as center, draw an arc in the interior of ∠A that has a radius more than half the distance from B to C.

3. Using the same radius, draw an arc with center B that intersects the arc drawn in Step 2 at X.

4. Draw $\overrightarrow{AX}$. $\overrightarrow{AX}$ bisects ∠A.

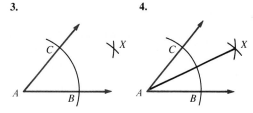

Justification: Draw $\overline{BX}$ and $\overline{XC}$. By construction, $\overline{AB} \cong \overline{AC}$ and $\overline{BX} \cong \overline{CX}$. Since congruence of segments is reflexive, $\overline{AX} \cong \overline{AX}$. Thus, $\triangle ABX \cong \triangle ACX$ by SSS. Since they are corresponding parts, $\angle BAX \cong \angle CAX$.

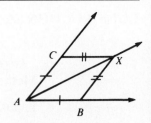

CONSTRUCTION 4: Construct a line perpendicular to a given line at a given point on the line.

Given: Line ℓ and point P

Method:

1. With P as center, draw two arcs with the same radius that intersect ℓ at R and S.

2. With R as center, draw an arc that has radius greater than RP.

3. With center S and the same radius, draw an arc that intersects the one drawn in Step 2 at X.

4. Draw $\overleftrightarrow{XP}$.
  $\overleftrightarrow{XP} \perp \ell$.

Justification: Draw $\overline{XR}$ and $\overline{XS}$. By construction, $\overline{RP} \cong \overline{PS}$ and $\overline{XR} \cong \overline{XS}$. Also, $\overline{XP} \cong \overline{XP}$. Therefore, $\triangle XRP \cong \triangle XSP$ by SSS and it follows that $\angle RPX \cong \angle SPX$. Because these angles are also supplementary, each must be a right angle. Thus, $\overleftrightarrow{XP} \perp \overleftrightarrow{RS}$.

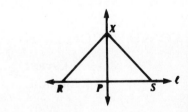

CONSTRUCTION 5: Construct a line perpendicular to a given line from a given point not on the line.

Given: Line ℓ and point P

Method:

1. With P as center, draw an arc that intersects ℓ at A and B.

2. With A as center and radius clearly more than half AB, draw an arc on the side of ℓ opposite P. With center B and the same radius, draw an arc that intersects the one drawn previously at X.

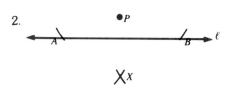

3. Draw $\overleftrightarrow{PX}$.
$\overleftrightarrow{PX} \perp \ell$

Justification: Draw $\overline{PA}$, $\overline{PB}$, $\overline{AX}$ and $\overline{BX}$. Let M be the intersection of $\overleftrightarrow{PX}$ and ℓ. By construction, $\overline{AP} \cong \overline{PB}$ and $\overline{AX} \cong \overline{BX}$. Also, $\overline{XP} \cong \overline{XP}$. Therefore, $\triangle APX \cong \triangle BPX$. Because they are corresponding angles, $\angle APX \cong \angle BPX$. Because $\overline{PM} \cong \overline{PM}$, we have $\triangle APM \cong \triangle BPM$ by SAS.

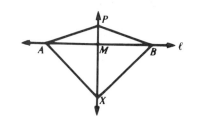

You will be asked to complete the justification in the exercises.

Exercises

Max: 1-3, 1-39, Challenge; Avg: 1-3, 1-17, 19-39 odds; Min: 1-3, 1-11 odds, 12-17, 19-33 odds

Exploratory Answer the following.
1. What is the difference between a construction and a drawing? Rulers and protractors are used in drawings but may not be used in constructions.
2. Do you think constructions are more or less exact than drawings? Explain. Constructions are more exact theoretically.
3. What is meant "by construction" in the "justification"? A figure was constructed to fulfill a requirement.

Written Draw a segment of any convenient measure k. Draw another segment with measure n. Construct segments with the following measures. For exercises 1-33, see students' work.
1. k   2. n   3. k + n   4. 2k   5. 3n   6. 2k + 3n

Draw an angle of any convenient measure m. Construct angles with the following measures.
7. m   8. 2m   9. 3m   10. $\frac{1}{2}m$ 11. $\frac{1}{4}m$

12. Draw an acute angle of convenient measure and construct its bisector.
13. Repeat exercise 12 drawing an obtuse angle.

230 APPLICATIONS OF CONGRUENT TRIANGLES

14. Draw a segment of convenient length and choose a point B on it. Now construct a line through B perpendicular to the segment.
15. Repeat exercise 14 using a different segment.
16. Draw a segment of convenient length. Choose a point B not on the segment. Now construct a line through B perpendicular to the segment.
17. Repeat exercise 14 using a different segment.

Construct an angle that has the following degree measure.

18. 90
19. 45
20. 135
21. $22\frac{1}{2}$
22. $157\frac{1}{2}$
23. $33\frac{3}{4}$
24. 60
25. 30
26. 120
27. 15

28. Construct a right triangle whose legs are congruent to segments s and t.

Make a large copy of the figure shown at the right. Construct a line perpendicular to each of the following lines passing through the given point.

29. $\overleftrightarrow{BP}$, point A
30. $\overrightarrow{BA}$, point A
31. $\overleftrightarrow{BC}$, point P
32. $\overleftrightarrow{BP}$, point C
33. $\overrightarrow{BA}$, point B

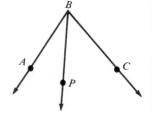

34. Complete the justification of Construction 5.

For exercises 35-39, see students' work.

35. Draw two segments and an angle. Construct a triangle with adjacent sides congruent to the given segments and included angle congruent to the given angle.
36. Draw two acute angles and a segment. Construct a triangle with two angles congruent to the given angles and included side congruent to the given segment.
37. Draw two segments. Construct a right triangle with hypotenuse congruent to the longer segment and one leg congruent to the shorter segment.
38. Draw three segments. Construct a triangle with three sides congruent to the given segments. Is this construction possible for any choice of three segments?
39. Draw two acute angles and a segment. Construct a triangle with two angles congruent to the given angles and a side opposite one of the angles congruent to the given segment. Is this construction possible for any choice of acute angles and segment? no

38. No. The sum of the lengths of any two segments must be greater than the length of the third segment.

Challenge Do some research.

Find out about "Impossible Constructions." Why are they impossible?
Impossible constructions are those that have not been done using only a compass and straightedge.

Objective: To construct basic geometric figures.

6.6 More Constructions

Previously, the bisector of an angle was constructed. Is it possible to find, by construction, the midpoint of a segment? Study Construction 6.

CONSTRUCTION 6: Bisect a given segment.

Given: $\overline{AB}$

Method:

1. With A as center, draw arcs that have the same radius both above and below the segment. The radius should be clearly more than half the distance from A to B.

1.

2. With B as center, and the same radius, draw arcs intersecting the ones drawn in Step 1 at C and D.

3. Draw $\overleftrightarrow{CD}$.
   $\overleftrightarrow{CD}$ bisects $\overline{AB}$.
   You should notice that $\overleftrightarrow{CD}$ is also perpendicular to $\overline{AB}$. For this reason, $\overleftrightarrow{CD}$ is called the perpendicular bisector of $\overline{AB}$.
   Justification of this construction is left for the exercises.

2.

CONSTRUCTION 7: Construct an altitude to a side of a given triangle.

Given: $\triangle ABC$

Method:

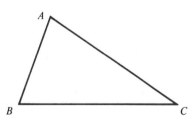

1. With A as center and with radius long enough to draw an arc on $\overline{BC}$, draw the arc on $\overline{BC}$. ($\overline{BC}$ may be extended.) Label the two points D and E where the arc intersects $\overline{BC}$.

2. Using D, then E as centers and the same radius, draw arcs that intersect.

3. Draw a line segment connecting A with the arc intersections. Label the point F where this line intersects $\overline{BC}$.

4. $\overline{AF}$ is the altitude from A to $\overline{BC}$.

232 APPLICATIONS OF CONGRUENT TRIANGLES

CONSTRUCTION 8: Construct a median to a side of a given triangle.

Given: △ABC

Method:
1. With A, then B as centers and using the same radius, draw arcs that intersect above and below $\overline{AB}$.
2. Draw a line through the points of intersection. Label the point D where this line intersects $\overline{AB}$.
3. Draw $\overline{CD}$.

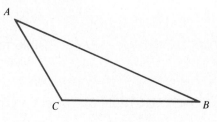

4. $\overline{CD}$ is the median from C to $\overline{AB}$.

CONSTRUCTION 9: Construct a line parallel to a given line through a point not on the line.

Given: line ℓ and point P

Method:

1. Through P draw line m intersecting line ℓ at point S.

2. Use Construction 2 to find point R so that $\angle RPS \cong \angle 1$.

3. Draw $\overleftrightarrow{PR}$.
$\overleftrightarrow{PR} \parallel \ell$

You will be asked to complete the justification in the exercises.

1.

2.

3.

Exercises

Max: 1-5, 1-38, Challenge; Avg: 1-5, 1-37 odds; Min: 1-5, 1-37 odds

Exploratory Answer the following.

1. Explain how Construction 6 can be used to construct the perpendicular bisector of a segment. The bisector that is constructed is also perpendicular to the segment.

2. Explain what is meant by the perpendicular bisector of a segment. It is a line that is perpendicular to and passes through the midpoint of the segment.

CHAPTER 6 233

3. Explain how to bisect a segment. See Construction 6.
4. What is an altitude of a triangle? How many altitudes will a triangle have? Must they all be within the triangle? A segment drawn from any vertex and perpendicular to the line containing the opposite side. 3; no
5. What is a median of a triangle? How many medians will a triangle have? Must they all be within the triangle? A segment drawn from any vertex to the midpoint of the opposite side. 3; yes

Written Complete the following constructions. For exercises 1-23 and 28-38, see students' work.
1. Draw a segment of any convenient length. By construction, find the midpoint of the segment.
2. Draw a segment of any convenient length. Construct the perpendicular bisector of the segment.
3. Given any line segment, explain how you can construct another segment that is one-fourth as long as the given segment.

Draw a line segment of any convenient measure *k*. Construct segments with the following measures.

4. $\dfrac{k}{2}$ 5. $3k$ 6. $\dfrac{k}{4}$ 7. $\dfrac{3}{4}k$

Copy and enlarge each triangle. Construct the altitude from vertex A to $\overline{BC}$.

8.

9.

10.

11.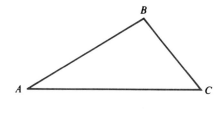

Copy and enlarge the above triangles. For each triangle:
a. construct the median from vertex A to $\overline{BC}$; and
b. construct the angle bisector of $\angle B$.

12. exercise 8   13. exercise 9   14. exercise 10   15. exercise 11

234 APPLICATIONS OF CONGRUENT TRIANGLES

Copy and enlarge the following drawings. Construct a line through B that is parallel to $\overline{AC}$.

16.   17.   18.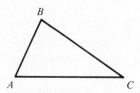

Copy and enlarge the angle shown for each exercise. Construct a line through X that is parallel to each of the following.

19. $\overleftrightarrow{AB}$ 20. $\overleftrightarrow{AC}$ 21. $\overleftrightarrow{BC}$

Copy and enlarge triangle ABC.

22. Construct the midpoint P of $\overline{AC}$.
23. Through P, construct a line parallel to $\overline{BC}$.

Write a justification for each construction.

24. Construction 6   25. Construction 7
26. Construction 8   27. Construction 9

For answers to exercises 24-27, see the Teacher's Guide.

Construct each of the following.

28. Construct a parallelogram that is not a rectangle.
29. Construct a rhombus that contains an angle of 45°.

For each exercise, draw a large triangle. Then construct the following.

30. three medians   31. three altitudes   32. three angle bisectors
33. perpendicular bisectors of each side
34. Repeat exercises 30–33 using a different triangle. For example, if you used an acute triangle, try an obtuse triangle.

Construct each of the following.

35. Draw a segment. Construct an isosceles triangle whose congruent sides are congruent to the given segment.

36. Construct an isosceles right triangle.

37. Draw two segments and an angle. Construct a parallelogram with adjacent sides congruent to the given segments, and included angle congruent to the given angle.

38. Construct a 30°–60°–90° triangle.

See students' work. All of the medians intersect in one point.

Challenge Complete the following.
Draw a large acute triangle and construct all of its medians. What do you notice? Investigate this situation using a ruler and several triangles and try to reach a conclusion.

Objective: To prove and apply a theorem about the segment that joins the midpoints of two sides of a triangle.

6.7 Another Triangle Relationship

Journal

Draw a picture or make a design that contains many of the different geometric shapes you learned about in this chapter. Name the shapes you used.

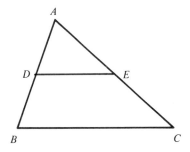

Do the following to discover an important triangle relationship.

Draw any triangle ABC. Find the midpoints of $\overline{AB}$ and $\overline{AC}$. Label the midpoint of $\overline{AB}$ with the letter D and the midpoint of $\overline{AC}$ with the letter E. Draw $\overline{DE}$. Compare the length of $\overline{DE}$ with the length of $\overline{BC}$. Compare $\angle ADE$ and $\angle ABC$. What does this mean about $\overline{DE}$ and $\overline{BC}$? Are you surprised to find $\overline{DE}$ appears to be half as long as $\overline{BC}$? It also seems that $\overline{DE}$ is parallel to $\overline{BC}$. The following theorem shows that this is true in any triangle.

Theorem 6–14

> If a segment joins the midpoints of two sides of a triangle, then it is parallel to the third side and its measure is half that of the third side.

Example

1 Write a two-column proof for the following.

Given: $\triangle ABC$
R is the midpoint of $\overline{AB}$.
S is the midpoint of $\overline{AC}$.

Prove: $\overline{RS} \parallel \overline{BC}$ and $RS = \frac{1}{2}BC$

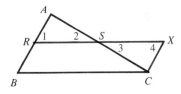

Proof:

STATEMENTS	REASONS
1. Locate X on $\overrightarrow{RS}$ such that $SX = RS$, or $\overline{SX} \cong \overline{RS}$.	1. On a ray, there is exactly one point at a given distance from the endpoint.
2. Draw $\overline{CX}$.	2. For any two points, there exists one and only one line containing both of them.
3. $\angle 2 \cong \angle 3$	3. Vertical angles formed by intersecting lines are congruent.
4. R is the midpoint of $\overline{AB}$. S is the midpoint of $\overline{AC}$.	4. Given

5. $\overline{AS} \cong \overline{SC}$	5. Definition of a midpoint
6. $\triangle ASR \cong \triangle CSX$	6. SAS
7. $\angle 1 \cong \angle 4$	7. Corresponding parts of congruent triangles are congruent.
8. $\overline{AB} \parallel \overline{CX}$	8. If two lines are cut by a transversal and the alternate interior angles are congruent, the lines are parallel.
9. $\overline{AR} \cong \overline{CX}$	9. Corresponding parts of congruent triangles are congruent.
10. $\overline{BR} \cong \overline{AR}$	10. Definition of a midpoint
11. $\overline{BR} \cong \overline{CX}$	11. Transitive property
12. Quadrilateral $RXCB$ is a parallelogram.	12. If a quadrilateral has two sides that are both parallel and congruent, then the quadrilateral is a parallelogram.
13. $\overline{RS} \parallel \overline{BC}$	13. Definition of a parallelogram
14. $RS + SX = RX$	14. Segment Addition Axiom
15. $RS + RS = RX$	15. Substitution Axiom
16. $2RS = RX$	16. Addition
17. $RS = \frac{1}{2}RX$	17. Divide both sides by 2.
18. $\overline{RX} \cong \overline{BC}$, or $RX = BC$	18. Opposite sides of a parallelogram are congruent.
19. $RS = \frac{1}{2}BC$	19. Substitution Axiom

Exercises

Max: 1-8, 1-22, Challenge; Avg: 1-8, 1-17 odds, 18-22; Min: 1-8, 1-21 odds

Exploratory In the figure D and E are midpoints of $\overline{AB}$ and $\overline{AC}$, respectively. Explain the relationship between each of the following pairs.

1. $\overline{DE}$ and $\overline{BC}$ $\overline{DE} \parallel \overline{BC}$
2. DE and BC $DE = \frac{1}{2}BC$
3. $\angle ADE$ and $\angle ABC$ $\angle ADE \cong \angle ABC$
4. $\angle AED$ and $\angle ACB$ $\angle AED \cong \angle ACB$
5. If $DE = 10$, find BC. 20
6. If $BC = 12$, find DE. 6
7. If $DE = 3t + 4$, write BC in terms of t. 6t + 8
8. If $BC = 2x - 2$, write DE in terms of x. x − 1

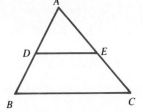

Written In the following, find DE and BC. Use the figure above.

1. $DE = 4x - 10$, $BC = 5x + 1$
   18, 36
2. $DE = 2(x - 10)$, $BC = 3(x - 18) + 29$
   10, 20

3. $DE = 3(x - 2)$, $BC = 5(x - 1)$ 15, 30
4. $DE = 5(x - 3) + 5$, $BC = 4(2x - 3) + 8$ 30, 60
5. $DE = x^2$, $BC = 16(x - 2)$ 16, 32
6. $DE = x^2$, $BC = 24(x - 3)$ 36, 72
7. $DE = x^2$, $BC = 2(5x - 6)$ 4, 8 or 9, 18
8. $DE = x^2$, $BC = 2(11x - 28)$ 16, 32 or 49, 98

In the figure, S, T, and U are midpoints of the sides of the triangle.

9. If $PR = 10$ and $RQ = 17$, find $TU + ST$. $13\frac{1}{2}$
10. If $PR + PQ = 30$, find $TU + SU$. 15

Find the perimeter of △STU, given the following values.

11. $PR = 8$, $RQ = 12$, and $PQ = 18$ 19
12. $PR = 10$, $RQ = 17$, and $PQ = 20$ $23\frac{1}{2}$
13. If $PS = 3x + 5$ and $TU = 21 + x$, find SR. 29
14. If $RU = x^2$ and $ST = 8x - 16$, find UQ. 16
15. If $PQ = x^2$ and $SU = 8(x - 4)$, find PT, TQ, and SU. 32, 32, 32
16. If $PR = y^2$ and $TU = 10(y - 5)$, find PR and TU. 100, 50
17. If $PR = \dfrac{y^2}{2}$ and $TU = -12(12 - y)$, find PR and TU. 288, 144

Answer the following.

18. The segment joining the midpoints of two adjacent sides of a rectangle is 45 cm. Find the length of a diagonal of the rectangle. 90 cm
19. The midpoints of the sides of △ABC are joined to form △DEF. If the perimeter of △DEF is 18 cm, what is the perimeter of △ABC? 36 cm

Write a two-column proof for the following.

20. **Given:** △ABC, D is the midpoint of $\overline{AB}$, E is the midpoint of $\overline{BC}$, and F is the midpoint of $\overline{AC}$. For answers to exercises 20-22, see the Teacher's Guide.
    Prove: The perimeter of △DEF is equal to one-half the perimeter of △ABC.
21. **Given:** △RQS
    $\overline{QR} \cong \overline{QS}$
    T and U are midpoints.
    Prove: △QTU is isosceles.
22. **Given:** △QRS
    U is the midpoint of $\overline{QS}$.
    $\overline{TU} \parallel \overline{RS}$
    Prove: T is the midpoint of $\overline{QR}$.

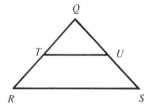

Challenge Write a two-column proof.

Draw any convex quadrilateral that is not a parallelogram. Now connect, in order, the midpoints of each side of the quadrilateral. Prove that the quadrilateral formed is a parallelogram. For the answer to exercise, see the Teacher's Guide.

Problem Solving Application: Indirect Reasoning

Objective: To use indirect reasoning to solve problems.

Sometimes it is difficult to choose the correct answer in a multiple choice test. Often, however, some of the choices can be eliminated. In one form of reasoning, called indirect reasoning, a result is assumed to be true, and a contradiction is then found.

Examples

1 Choose the best answer for $\frac{2}{3} + \frac{3}{4} + \frac{7}{8} = x$.

   a. $x = \frac{4}{5}$ **b.** $x = \frac{11}{24}$ **c.** $x = \frac{55}{24}$ **d.** $x = \frac{17}{12}$

Assume $\frac{4}{5}$ is the solution. Since $\frac{2}{3}, \frac{3}{4},$ and $\frac{7}{8}$ are all greater than $\frac{1}{2}$, their sum is at least $\frac{3}{2}$. Since $\frac{4}{5} < 1$, the assumption is false. The same process eliminates $\frac{11}{24}$. Assume $\frac{17}{12}$ is the solution. Since $\frac{17}{12} < \frac{3}{2}$, the assumption is false. By eliminating possibilities, $\frac{55}{24}$ must be the correct choice.

2 The students in Mr. Block's class tried to guess how many jelly beans were in a large jar. Scott's guess was 3 more than the actual number of jelly beans in the jar. Susan's guess was twice the square of the actual number of jelly beans. Use indirect reasoning to show that Scott and Susan could not have guessed the same number.

Explore   Let $n =$ the actual number of jelly beans in the jar.
Then $n + 3 =$ Scott's guess
and $2n^2 =$ Susan's guess.

Plan   To solve this problem using indirect reasoning, we assume that Scott's guess was the same as Susan's. This assumption leads to the following equation.

Solve

$n + 3 = 2n^2$
$0 = 2n^2 - n - 3$
$0 = (2n - 3)(n + 1)$ Factor.
$2n - 3 = 0$ or $n + 1 = 0$ If $ab = 0$, then $a = 0$ or
$2n = 3$ $n = -1$ $b = 0$.
$n = \frac{3}{2}$

Thus, Scott's guess would be the same as Susan's if the actual number of jelly beans was $\frac{3}{2}$ or -1. Since neither of these results could be the actual number of jelly beans, our assumption that Scott's guess was the same as Susan's is false. Therefore, Scott and Susan could not have guessed the same number. Examine this solution.

Exercises

Max: 1-8; Avg: 1-7; Min: 1-6

Written Use indirect reasoning to choose the best answer for each problem.

1. What is the solution set for $4 - \frac{5}{6} - \frac{7}{16} = x$?

   d a. $\left\{\frac{15}{8}\right\}$ b. $\left\{\frac{59}{24}\right\}$ c. $\left\{\frac{149}{48}\right\}$ d. $\left\{\frac{131}{48}\right\}$

2. What is the solution set for $r^2 + r = 552$?

   b a. $\{-26, 22\}$ b. $\{-24, 23\}$ c. $\{-23, 24\}$ d. $\{-27, 26\}$

3. What is the degree measure of the angle formed by the hands of a clock at 1:30?

   c a. 30   b. 35   c. 135   d. 150

For answers to exercises 4-6, see students' work.

Seth, Sally, Sam, Sarah, Steven, and Sybil each tried to guess the number of jelly beans in the jar in example 2 on page 238. Use indirect reasoning to show that each pair of students given in exercises 4–6 could not have guessed the same number of jelly beans.

4. Seth's guess was 4 more than five times the actual number of jelly beans in the jar. Sally's guess was 3 less than nine times the actual number of jelly beans.

5. Sam's guess was 7 less than four times the actual number of jelly beans in the jar. Sarah's guess was 1 more than four times the actual number of jelly beans.

6. Steven's guess was 2 less than ten times the actual number of jelly beans in the jar. Sybil's guess was the square of the actual number of jelly beans.

Use indirect reasoning and a chart to solve this problem.

7. Tom, Richard, and Sondra are walking down the hall at school. Tom always tells the truth. Sondra never tells the truth. The student walking on the left says, "Tom is in the middle." The student in the middle says, "I am Richard." The student on the right says, "Sondra is in the middle." Name the students from left to right. Richard, Sondra, Tom

8. Mr. Apple, Mr. Pear, Miss Peach, and Mrs. Berry are eating apples, pears, peaches, and berries, although none of them are eating the same fruit as their last name. Neither Mr. Pear nor Miss Peach is eating apples. Mr. Apple is not eating berries, and Miss Peach is not eating pears. By eliminating possibilities, determine who is eating what.

Mr. Apple: pear, Mr. Pear: peach, Miss Peach: berry, Mrs. Berry: apple

Portfolio Suggestion

Select some of your work from this chapter that shows your creativity. Place it in your portfolio.

Performance Assessment   See students' work.

Parallelograms *ABCD* and *EFGH* have all pairs of corresponding sides congruent. Is *ABCD* ≅ *EFGH*? Explain your answer. Include a drawing.

Chapter Summary

1. A **parallelogram** is a **quadrilateral** with two pairs of parallel sides.   (208)
2. Theorem 6–1: A **diagonal** of a parallelogram separates the parallelogram into two congruent triangles.   (208)
3. Theorem 6–2: Opposite sides of a parallelogram are congruent.   (209)
4. Theorem 6–3: Opposite angles of a parallelogram are congruent.   (209)
5. Theorem 6–4: The diagonals of a parallelogram bisect each other.   (209)
6. Theorem 6–5: If a quadrilateral has two sides that are both parallel and congruent, then the quadrilateral is a parallelogram.   (212)
7. Theorem 6–6: If both pairs of opposite sides of a quadrilateral are congruent, then the quadrilateral is a parallelogram.   (213)
8. Theorem 6–7: If the diagonals of a quadrilateral bisect each other, then the quadrilateral is a parallelogram.   (215)
9. Theorem 6–8: If both pairs of opposite angles of a quadrilateral are congruent, then the quadrilateral is a parallelogram.   (215)
10. A **rectangle** is a parallelogram with four right angles.   (218)
11. A **rhombus** is a parallelogram with four congruent sides.   (218)
12. Theorem 6–9: If a parallelogram is a rectangle, then its diagonals are congruent.   (218)
13. Theorem 6–10: If a parallelogram is a rhombus, then its diagonals are perpendicular.   (219)
14. Theorem 6–11: If a parallelogram is a rhombus, then each diagonal bisects two opposite angles of the parallelogram.   (219)
15. A **trapezoid** is a quadrilateral with exactly one pair of parallel sides. **Isosceles trapezoids** have congruent legs.   (222)
16. Theorem 6–12: The **base angles of an isosceles trapezoid** are congruent. (222)
17. Theorem 6–13: The diagonals of an isosceles trapezoid are congruent. (223)
18. Theorem 6–14: If a segment joins the midpoint of two sides of a triangle, then it is parallel to the third side and its measure is half that of the third side.   (235)

Chapter Review

Tell whether each of the following is *sometimes*, *always*, or *never* true.

6.1 1. A parallelogram is a quadrilateral. always 2. A trapezoid is a parallelogram. never
6.3 3. The diagonals of a quadrilateral   4. A parallelogram with two consecu-
6.4   bisect each other. sometimes   tive sides congruent is a rhombus. always

6.1 Quadrilateral *ABCD* is a parallelogram. Find the measures of $\angle DAB$, $\angle ABC$, $\angle BCD$, and $\angle ADC$, given the following.

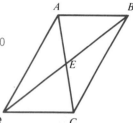

5. $m\angle DAB = y$, $m\angle ABC = 2y - 120$ 100, 80, 100, 80
6. $m\angle DAB = 5x + 18$, $m\angle ABC = x + 60$
7. $m\angle DAB = 3x + 10$, $m\angle BCD = 13(x - 30)$
   6. 103, 77, 103, 77   7. 130, 50, 130, 50

6.1 Find the measures of $\overline{AC}$ and $\overline{BD}$.

8. $AE = 2x - 5$, $EC = \dfrac{3x}{4}$, $BE = 10x - 25$ 6, 30
9. $AE = x^2 - 6$, $EC = 16x - 70$, $BE = 2y(6 - y)$, $ED = 4(y + 4) - y^2$ 116, 32

Write a two-column proof for each of the following. For answers to exercises 10-12, see the Teacher's Guide.

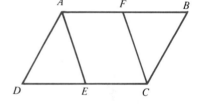

6.2 10. **Given:** *AFCE* is a rhombus, $\triangle AED$ and $\triangle BFC$ are equilateral.
        Prove: *ABCD* is a parallelogram.
    11. **Given:** *ABCD* is a parallelogram. $\overline{DE} \cong \overline{BF}$
        Prove: *AFCE* is a parallelogram.
    12. **Given:** $\angle BAE$ and $\angle AFC$ are supplementary, $\triangle ADE \cong \triangle CBF$
        Prove: *ABCD* is a parallelogram.

6.3 13. If *AFCE* is a rhombus, $AF = 31$, $AE = 4x - 1$, $EC = 7y + 3$, and $CF = 3 - 2z$, find *x*, *y*, and *z*. x = 8, y = 4, z = -14
6.4 14. If *AFCE* is a rhombus, $EC = 40 - 6x$, and $FC = x^2$, find *x* and *AE*. 14. x = -10, AE = 100 or x = 4, AE = 16
    15. If *ADCF* is an isosceles trapezoid, $m\angle ADC = 3x + 8$, and $m\angle DCF = 7x - 138$, find $m\angle ADC$ and $m\angle DAF$. $m\angle ADC = 117.5$, $m\angle DAF = 62.5$
       For exercises 16-18, see students' work.
6.5 16. Construct an angle whose degree measure is 75.
6.6 17. Draw any acute triangle *ABC*. Construct $\triangle QRS \cong \triangle ABC$.
    18. Construct a rhombus having an angle of 30°.
    19. $DE = 2(y - 9)$, $BC = 3(y - 7)$
        $DE = y - 13$, $BC = y^2 - 5(8y - 83)$

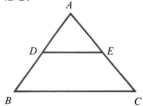

6.7 20. **In the figure, *D* and *E* are midpoints of $\overline{AB}$ and $\overline{AC}$, respectively. Find *BC* and *DE*.**

Chapter Test

Tell whether each of the following is *sometimes*, *never*, or *always* true.

1. A square is a rectangle. always
2. A quadrilateral with four right angles is a rectangle. always
3. A quadrilateral that is separated into congruent triangles by a diagonal is a parallelogram. always
4. The diagonals of a rhombus bisect each other. always
5. An altitude of a triangle lies in the interior of the triangle. sometimes

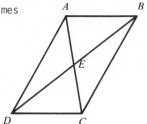

Quadrilateral *ABCD* is a parallelogram. Find the measures of $\angle DAB$, $\angle ABC$, $\angle BCD$, and $\angle CDA$, given the following.

6. $m\angle BCD = x - 86$, $m\angle CDA = x - 94$ 94, 86, 94, 86
7. $m\angle DAB = 3(x + 35)$, $m\angle ABC = 2x + 5$ 147, 33, 147, 33

Find the measures of $\overline{AC}$ and $\overline{BD}$.

8. $AE = 3x - 1$, $EC = x + 13$, $BE = 4y$, $ED = 6y - 36$
   40, 144

For answers to exercises 9–11, see the Teacher's Guide.

Write a two-column proof for each of the following.

9. **Given:** $ABCD$ is an isosceles trapezoid, with $\overline{AB} \parallel \overline{DC}$. $\overline{DB}$ bisects $\angle ADC$.
   Prove: $\overline{AD} \cong \overline{AB}$
10. **Given:** $ABCD$ is an isosceles trapezoid, with $\overline{AB} \parallel \overline{CD}$.
    Prove: $\triangle EAB$ is isosceles.
11. **Given:** $\triangle EDC$, A is the midpoint of $\overline{ED}$, B is the midpoint of $\overline{EC}$.
    Prove: $\angle ABD \cong \angle BDC$

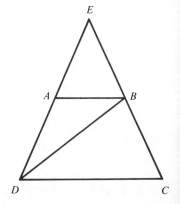

For exercises 12 and 13, see students' work.

12. **Given:** $\overline{DE}$
    Construct: $\triangle ABC$ with all sides congruent to $\overline{DE}$
13. Copy $\angle B$ and $\overline{CB}$. Construct right $\triangle ABC$ with right $\angle C$.

In the figure, D and E are midpoints of $\overline{AB}$ and $\overline{AC}$, respectively. Find BC and DE.

14. $DE = 2x$, $BC = x + 12$ 8, 16
15. $DE = x + 9$, $BC = x^2 - 8x + 43$
    14, 28

CHAPTER 7

Similarity

Application in Architecture

In India, **geometry** is used to expand the Hindu temples in a way that reflects a person's growth. During each stage of development, a body grows but remains **similar** to the earlier form. The temples are built beginning with the altar and then each stage of expansion adds to the structure.

When Hindu temples are expanded, the expansion is a *gnomon.* A gnomon is any figure which when added to another figure leaves the new figure similar to the original. For example, if a temple begins with a square altar, each consecutive step would add a shape that makes the temple into a larger square as shown below.

Find three different ways of expanding an equilateral triangle with a gnomon.

Group Project: *Art*

A Golden Rectangle is a special figure often found Renaissance art, Greek architecture, and in modern art and architecture. Have each person in your group investigate a different art piece or structure that displays a golden rectangle. Then use your findings to create a display containing an explanation and history of the golden rectangle and some examples of its use.

Objective: To apply and solve ratios and proportions.
7.1 Ratio and Proportion

Up to now we have studied geometric figures that have the same size and shape. In many situations, though, figures that have the same *shape* but not the same *size* occur. As you may recall, such figures are said to be similar.

Definition of Similar Polygons

> Two polygons are *similar* if their corresponding angles have equal measure and if the ratios of the measures of corresponding sides are equivalent.

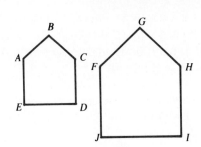

The symbol ~ means "*is similar to.*" In the diagram shown, polygon *ABCDE* ~ polygon *FGHIJ*. As with congruent figures, we agree that corresponding vertices of similar figures are named in the same order. That is, if *ABCDE* ~ *FGHIJ*, then vertex *A* corresponds to vertex *F*, vertex *B* corresponds to vertex *G*, and so on.

It follows from the definition given above that

$$\frac{AB}{FG} = \frac{BC}{GH} = \frac{CD}{HI} = \frac{DE}{IJ} = \frac{EA}{JF}$$

Also $m\angle A = m\angle F$, $m\angle B = m\angle G$, $m\angle C = m\angle H$, $m\angle D = m\angle I$, and $m\angle E = m\angle J$.

Since our work with similar figures involves **ratio** and **proportion**, a review of these ideas will be helpful.

Definition of Ratio

> The ratio of *a* to *b*, where $b \neq 0$, is the quotient $a \div b$ or $\frac{a}{b}$.

The ratio of *a* to *b* can also be written as $a:b$.

Definition of a Proportion

> A proportion is an equality of two ratios.

The following equations are proportions.

$$\frac{4}{5} = \frac{8}{10} \qquad 3\% = 0.03 \qquad \frac{AB}{FG} = \frac{BC}{GH} \qquad \frac{x}{3} = \frac{x+1}{4}$$

In the proportion $\frac{a}{b} = \frac{c}{d}$, *a* is called the *first term*, *b* is called the *second term*, *c* is called the *third term*, and *d* is called the *fourth term*. The numbers a and d are called the **extremes** of the proportion. The numbers b and c are called the **means** of the proportion. When $\frac{a}{b} = \frac{c}{d}$, we say a and c are proportional to b and d. The following two sentences have the same meaning.

The ratios of the measures of corresponding sides of similar polygons are equivalent.

The measures of corresponding sides of similar polygons are proportional.

Since proportions are equations, algebra can be used to prove some important properties of proportions. Suppose we have the following proportion.

$$\frac{a}{b} = \frac{c}{d} \quad \text{where} \quad b \neq 0 \text{ and } d \neq 0$$

Now, multiply each side by bd. This produces

$$\frac{a}{b} \cdot \frac{bd}{1} = \frac{c}{d} \cdot \frac{bd}{1} \quad \text{or} \quad ad = bc.$$

The following property of proportions has just been demonstrated.

Means-Extremes Property

If $\frac{a}{b} = \frac{c}{d}$, then $ad = bc$.

This is often referred to as *cross multiplication*. It may also be stated as follows.

In a proportion, the product of the means is equal to the product of the extremes.

Examples

1 Suppose in the figure below polygon *ABCDE* ~ polygon *RSTUV*. Find *AB* and *BC*.

The ratios of the lengths of corresponding sides are equal.
Therefore,
$$\frac{AB}{RS} = \frac{BC}{ST}$$

or $\frac{k}{2} = \frac{k+4}{5}$

Cross multiplication produces

$$5k = 2(k + 4)$$
$$5k = 2k + 8$$
$$3k = 8$$
$$k = \frac{8}{3} \text{ or } 2\frac{2}{3}$$

Therefore, $AB = 2\frac{2}{3}$ and $BC = k + 4$ or $2\frac{2}{3} + 4$ or $6\frac{2}{3}$.

2 In the figure, $\triangle ABC \sim \triangle DEF$. DE is 65% of AB. Find AB.

Recall that 65% is the ratio $\frac{65}{100}$.

Therefore, the ratio of 7.8 to AB must equal the ratio of 65 to 100.

$$\frac{65}{100} = \frac{7.8}{AB}$$
$$65 \cdot AB = 780$$
$$AB = \frac{780}{65}$$
$$AB = 12$$

Exercises

Max: 1-17, 1-37; Avg: 1-17, 1-9 odds, 10-15, 17-37 odds; Min: 1-9 odds, 10-17, 1-37 odds

Exploratory Suppose $AB = 4$ and $AC = 10$. Find each of the following.

1. $\frac{AB}{AC}$ $\frac{2}{5}$ 2. $\frac{AB}{BC}$ $\frac{2}{3}$ 3. $\frac{BC}{AB}$ $\frac{3}{2}$ 4. $\frac{AC}{AB}$ $\frac{5}{2}$ 5. $\frac{BC}{AC}$ $\frac{3}{5}$ 6. $\frac{AC}{BC}$ $\frac{5}{3}$

In the figure for exercises 1–6, suppose $\frac{AB}{AC} = \frac{3}{5}$. Find each of the following.

7. $\frac{BC}{AB}$ $\frac{2}{3}$ 8. $\frac{BC}{AC}$ $\frac{2}{5}$ 9. $\frac{AC}{AB}$ $\frac{5}{3}$ 10. $\frac{AC}{BC}$ $\frac{5}{2}$

Suppose quadrilateral $ABCD \sim$ quadrilateral $EFGH$.

∠A, ∠E; ∠B, ∠F; ∠C, ∠G; ∠D, ∠H
11. Name the corresponding angles.
12. What is true about the measures of the corresponding angles? They are equal.
13. Name the corresponding sides. $\overline{AB}$, $\overline{EF}$; $\overline{BC}$, $\overline{FG}$; $\overline{CD}$, $\overline{GH}$; $\overline{DA}$, $\overline{HE}$
14. Write three proportions involving the measures of corresponding sides. Answers may vary. Some typical answers are given. $\frac{AB}{EF} = \frac{BC}{FG}$, $\frac{AB}{BC} = \frac{EF}{FG}$, $\frac{AD}{DC} = \frac{EH}{HG}$

Suppose $AB = 5$, $BC = 7$, $AD = 4$, $FG = 3.5$, and $HG = 6$. Find each of the following. Use the figure for exercises 11–14.

15. EF $2\frac{1}{2}$ **16.** DC 12 **17.** EH 2

Written Solve each proportion. Identify the means and extremes of each.

1. $\dfrac{y}{5} = \dfrac{2}{3}$ $y = \dfrac{10}{3}$; 5, 2; y, 3

2. $\dfrac{5}{x} = \dfrac{3}{4}$ $x = \dfrac{20}{3}$; x, 3; 5, 4

3. $\dfrac{5}{3} = \dfrac{2t}{3t}$ $t = \dfrac{2}{5}$; 3, 2; 5, 3t

4. $\dfrac{x}{5} = \dfrac{x+5}{14}$ $x = \dfrac{25}{9}$; 5, x + 5; x, 14

5. $\dfrac{y}{y+1} = 1\dfrac{2}{3}$ $y = \dfrac{5}{2}$; 3 y + 1, 5; y, 3

6. $\dfrac{t-1}{t+1} = \dfrac{-5t}{8}$ $t = \dfrac{3}{13}$; 8 t + 1, -5; t - 1, 8

7. $\dfrac{3+2t}{3t-7} = \dfrac{2t}{5}$ $t = -\dfrac{29}{4}$; 5 3t - 7, 2; 3 + 2t, 5

8. $\dfrac{5}{t} = \dfrac{10-t}{5}$ $t = 5$; t, 10 - t; 5, 5

9. $\dfrac{t}{3} = \dfrac{5}{8-t}$ $t = 3$ or 5; 3, 5; t, 8-t

10. $\dfrac{t-3}{14} = \dfrac{2}{t}$ $t = -4$ or 7; 14, 2; t = 3, t

Answer the following.

11. If $\triangle ABC \cong \triangle DEF$, does it follow that $\triangle ABC \sim \triangle DEF$? Explain. Yes, because their corresponding angles have equal measures.
12. Suppose $\triangle ABC \sim \triangle DEF$. Does it follow that $\triangle ABC \cong \triangle DEF$? Why? No, because all that is known is that the corresponding angles are congruent.

In the figure, $\triangle ABC \sim \triangle ADE$.

13. Name the corresponding vertices. A, A; D, B; E, C
14. Name the corresponding sides. $\overline{AD}$, $\overline{AB}$; $\overline{AE}$, $\overline{AC}$; $\overline{DE}$, $\overline{BC}$
15. Write three proportions involving the corresponding sides of $\triangle ABC$ and $\triangle ADE$.

15. $\dfrac{AD}{AB} = \dfrac{DE}{BC}$; $\dfrac{AD}{AB} = \dfrac{AE}{AC}$; $\dfrac{AE}{AC} = \dfrac{DE}{BC}$

In the following, tell whether all pairs of that figure are similar.

16. rectangles no
17. isosceles triangles no
18. squares yes
19. rhombuses no
20. equilateral triangles yes
21. right triangles no

The measures of two complementary angles are in the following ratios. Find the measure of each angle.

22. 2:3 36, 54
23. 4:5 40, 50
24. 5:1 75, 15
25. 7:8 42, 48

The measures of two supplementary angles are in the following ratios. Find the measure of each angle.

26. 2:7 40, 140
27. 5:7 75, 105
28. 5:13 50, 130
29. 7:3 126, 54

The measures of the angles of a triangle are in the following ratios. Find the measure of each angle.

30. 1:3:5 20, 60, 100
31. 2:2:5 40, 40, 100
32. 1:2:3 30, 60, 90
33. 2:3:5 36, 54, 90

Answer the following.

34. Find the measure of an angle that is 20% of its supplement's measure. 30
35. Find the measure of an angle that is 125% of its complement's measure. 50
36. Find the value of x for which 3x is 80% of 2x + 5. $\dfrac{20}{7}$
37. The sales tax on $140 is $8.40. What is the rate of sales tax? 6%

248 SIMILARITY

Objective: To use properties of proportions to solve problems.

7.2 Some Properties of Proportions

A photographer uses an enlarger to make large prints from snapshot negatives or slides. When a picture is enlarged, the original and an enlargement must be similar just as the three pictures shown at the right. Since the ratios of measures of corresponding sides are equivalent, we can write the proportion

$$\frac{2}{4} = \frac{3}{6}$$

We can write another true proportion using the reciprocals of each ratio.

$$\frac{4}{2} = \frac{6}{3}$$

Is this true of every proportion? Try a few examples. Your results should suggest the following property of proportions, which you will be asked to justify in the exercises. Assume none of the variables is zero.

Reciprocal Property

If $\dfrac{a}{b} = \dfrac{c}{d}$, then $\dfrac{b}{a} = \dfrac{d}{c}$.

Here are three additional properties of proportions, which you will be asked to prove in the exercises.

Interchanging Property

If $\dfrac{a}{b} = \dfrac{c}{d}$, then $\dfrac{a}{c} = \dfrac{b}{d}$.

Addition Property

If $\dfrac{a}{b} = \dfrac{c}{d}$, then $\dfrac{a+b}{b} = \dfrac{c+d}{d}$.

Subtraction Property

If $\dfrac{a}{b} = \dfrac{c}{d}$, then $\dfrac{a-b}{b} = \dfrac{c-d}{d}$.

Referring to the photos shown above, we see that $\dfrac{4}{6} = \dfrac{6}{9}$. In this proportion, the means are each 6. We say that 6 is the **mean proportional** or **geometric mean** between 4 and 9.

Definition of Mean Proportional

If $\dfrac{a}{b} = \dfrac{b}{d}$, then b is called the mean proportional between a and d. We assume a, b, and d are positive.

Example

1 Find the mean proportional between 3 and 27.

Find b such that $\dfrac{3}{b} = \dfrac{b}{27}$. Cross multiplication produces

$b^2 = 81$ or $b = 9$. Reject -9 because b must be positive.

The mean proportional, or geometric mean, between 3 and 27 is 9.

In example 1, we began with $\dfrac{3}{b} = \dfrac{b}{27}$ and then obtained $b^2 = 3 \cdot 27$. As a next step, write $b = \sqrt{3 \cdot 27}$. This illustrates the following property.

Geometric Mean Property

If b is the mean proportional between a and c, then $b = \sqrt{ac}$.

Example

2 Write the geometric mean of 4 and 5 in simplest radical form.

The geometric mean of 4 and 5 is $\sqrt{4 \cdot 5}$ or $2\sqrt{5}$.

In our work with similar figures, we will often be concerned with two or more line segments that are divided proportionally. To see what this means, consider the segments, $\overline{AB}$ and $\overline{CD}$. Suppose point X is on $\overline{AB}$ and point Y is on $\overline{CD}$ such that $\dfrac{AX}{XB} = \dfrac{CY}{YD}$. If this is the case, we say $\overline{AB}$ and $\overline{CD}$ are divided proportionally by X and Y.

250 SIMILARITY

Exercises

Page 236 Max: 1-20, 1-30, Challenge; Avg: 1-6, 7-19 odds, 1-25 odds; Min: 1-19 odds, 1-25 odds

Exploratory Tell whether or not the second equation must be true if the first equation is true. Justify your answer.

1. $\frac{2}{b} = \frac{5}{c}$; $\frac{2}{5} = \frac{b}{c}$ yes; interchanging property

2. $\frac{3}{x} = \frac{y}{7}$; $\frac{x}{3} = \frac{7}{y}$ yes; reciprocal property

3. $\frac{12}{x} = \frac{y}{4}$; $12y = 4x$ no; using the means-extremes property, $xy = 48$

4. $\frac{w}{v} = \frac{5}{8}$; $\frac{w}{5} = \frac{v}{8}$ yes; interchanging property

5. $\frac{4}{11} = \frac{q}{r}$; $\frac{15}{11} = \frac{q + r}{r}$ yes; addition property

6. $\frac{k}{7} = \frac{8}{m}$; $km = 56$ yes; means-extremes property

Complete each of the following.

7. If $\frac{m}{n} = \frac{5}{3}$, then $\frac{n}{m} = $ _?_ $\frac{3}{5}$

8. If $\frac{x}{2} = \frac{y}{5}$, then $5x = $ _?_ $2y$

9. If $\frac{x}{y} = \frac{8}{5}$, then $\frac{x - y}{y} = $ _?_ $\frac{3}{5}$

10. If $\frac{a}{b} = \frac{9}{11}$, then $\frac{a}{9} = $ _?_ $\frac{b}{11}$

11. If $\frac{t}{7} = \frac{p}{q + 1}$, then $\frac{t + 7}{7} = $ _?_ $\frac{p + q + 1}{q + 1}$

12. If $ab = cd$, then $\frac{a}{c} = $ _?_ $\frac{d}{b}$

13. If $\frac{AB}{QR} = \frac{MT}{UV}$, then $\frac{UV}{QR} = $ _?_ $\frac{MT}{AB}$

14. If $3m = 4t$, then $\frac{m}{t} = $ _?_ $\frac{4}{3}$

15. If $\frac{a - c}{c} = \frac{5}{2}$, then $\frac{a}{c} = $ _?_ $\frac{7}{2}$

16. If $\frac{2}{x} = \frac{x}{2}$, then $x = $ _?_ 2

17. If $\frac{b + x}{x} = \frac{10}{3}$, then $\frac{b}{x} = $ _?_ $\frac{7}{3}$

18. If $uv = wx$, then $\frac{w}{v} = $ _?_ $\frac{u}{x}$

19. If $3x = 2y$, then $\frac{x}{y} = $ _?_ $\frac{2}{3}$

20. If $5x - 6y = 0$, then $\frac{x}{y} = $ _?_ $\frac{6}{5}$

Written Find the mean proportional between each of the following. Write results in simplest radical form.

1. 4 and 9 6
2. 2 and 8 4
3. 4 and 16 8
4. 2 and 32 8
5. 4 and 25 10
6. 2 and 25 $5\sqrt{2}$
7. 2 and 10 $2\sqrt{5}$
8. $5\sqrt{2}$ and $6\sqrt{2}$ $2\sqrt{15}$
9. 100 and 200 $100\sqrt{2}$
10. $6\sqrt{8}$ and $5\sqrt{2}$ $2\sqrt{30}$
11. $2\sqrt{18}$ and $3\sqrt{2}$ 6
12. $5\sqrt{32}$ and $9\sqrt{2}$ $6\sqrt{10}$
13. $\frac{1}{3}$ and $\frac{3}{4}$ $\frac{1}{2}$
14. $\frac{2}{3}$ and $\frac{4}{9}$ $\frac{2\sqrt{6}}{9}$
15. r and s $\sqrt{rs}$

Suppose $\overline{AB}$ and $\overline{CD}$ are divided proportionally by X and Y.

16. If $AX = 3$, $XB = 5$, and $CY = 4$, find YD. $\frac{20}{3}$
17. If $CD = 12$, $AX = 3$, and $XB = 4$, find CY and YD. $5\frac{1}{7}$, $6\frac{6}{7}$
18. If $AB = 15$, $CD = 18$, and $YD = 12$, find AX and XB. 5, 10
19. If $AX = \sqrt{2}$, $CY = 5$, and $YD = \sqrt{6}$, find XB. $\frac{2\sqrt{3}}{5}$

In △ABC, $\overline{AB}$ and $\overline{AC}$ are divided proportionally by D and E. Find each of the following.

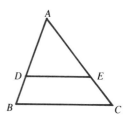

20. EC, if AD = 9, DB = 6, and AE = 12   8
21. DB, if AD = 5, AE = 8, and EC = 16   10
22. AD, if DB = 15, AE = 16, and EC = 20   12
23. AD, if AE = 12, EC = 24, and DB = 6   3
24. AD, if DB = 2, EC = 3, and AC = 7 $\frac{8}{3}$
25. EC, if AB = 27, DB = 15, and AE = 8   10
26. DB, if AB = 24, AE = 15, and AC = 16 $\frac{3}{2}$

Give an argument to justify each of the following properties of proportions.

27. Reciprocal Property
28. Interchanging Property
29. Addition Property
30. Subtraction Property
31. Geometric Mean Property   For answers to exercises 27-31, see the Teacher's Guide.

Challenge   Prove the following property of proportions.

If $\frac{a}{b} = \frac{c}{d} = \frac{e}{f}$, then $\frac{a+c+e}{b+d+f} = \frac{a}{b}$.   Hint: Let $\frac{a}{b} = x$ and show that $\frac{a+c+e}{b+d+f} = x$.

Let $\frac{a}{b} = x$. Then $\frac{c}{d} = x$ and $\frac{e}{f} = x$. Therefore, a = bx, c = dx, and e = fx. By the addition axiom, a + c + e = bx + dx + fx or a + c + e = x(b + d + f). By the division axiom, $\frac{a+c+e}{b+d+f} = x$. Therefore, by substitution, $\frac{a+c+e}{b+d+f} = \frac{a}{b}$.

═══════════ **Mixed Review** ═══════════

Choose the best answer.

1. Which of the following is *not* sufficient to show that △DEF is congruent to △RST?
   c
   a. $\overline{DE} \cong \overline{RS}$, ∠D ≅ ∠R, ∠E ≅ ∠S   b. $\overline{DE} \cong \overline{RS}$, $\overline{DF} \cong \overline{RT}$, $\overline{EF} \cong \overline{ST}$
   c. $\overline{DE} \cong \overline{RS}$, ∠D ≅ ∠R, $\overline{EF} \cong \overline{ST}$ d. $\overline{DE} \cong \overline{RS}$, ∠D ≅ ∠R, $\overline{DF} \cong \overline{RT}$

2. The measures of two sides of a triangle are 21 and 9. Which number *cannot* be the
   b   measure of the third side of this triangle?
   a. 21   b. 9   c. 28   d. 13

3. Which figure *must* have four congruent angles?
   a
   a. rectangle   b. parallelogram   c. rhombus   d. trapezoid

4. Which figure *does not* necessarily have congruent diagonals?
   b
   a. square   b. rhombus   c. rectangle   d. isosceles trapezoid

5. In △ABC, D and E are midpoints of $\overline{AB}$ and $\overline{AC}$, respectively. If DE = 5x + 12 and
   d   BC = x^2, what is the measure of $\overline{BC}$?
   a. 72   b. 16   c. 36   d. 144

6. What is the measure of an angle that is 150% of its supplement's measure?
   d
   a. 72   b. 120   c. 60   d. 108

7. What are all possible values for x if the ratio of 3x to $x^2 + 6$ is 3:5?
   a
   a. 2, 3   b. 1, 6   c. 1, 5   d. 3

Objective: To determine if two triangles are similar and apply the properties of similar triangles.

7.3 Similar Triangles

In our study of congruent triangles, we used shortcuts such as ASA, SAS, and SSS for showing two triangles congruent. The following experiment suggests a shortcut for showing two triangles similar.

1. Draw two line segments, $\overline{AB}$ and $\overline{DE}$, of different lengths, say 6 cm and 9 cm, respectively.
2. Using your protractor, draw an angle whose measure is 30 at D, and an angle whose measure is 75 at E. Label the intersection of the sides of these angles F.
3. Draw angles that measure 30 and 75 at A and B respectively. Label the intersection of their sides C.
4. Compare $m\angle ACB$ to $m\angle DFE$.
5. Carefully measure $\overline{AC}$, $\overline{DF}$, $\overline{BC}$, and $\overline{EF}$.
6. Find $\dfrac{AC}{DF}$, $\dfrac{BC}{EF}$, and $\dfrac{AB}{DE}$. What can be said about $\triangle ABC$ and $\triangle DEF$?

Repeat this experiment using different measurements. You should find that in each case it appears that the two triangles are similar. Because this cannot be proved, we will accept the following axiom.

Axiom 7–1

> If two angles of one triangle are congruent to two angles of another triangle, the triangles are similar.

This axiom is often abbreviated AA, which stands for angle, angle.

Example

1 Prove $\triangle RST \sim \triangle KLM$.

In the figure, $\triangle RST$ and $\triangle KLM$ are right triangles, $m\angle S = 40$ and $m\angle L = 40$. Since all right angles are congruent and $\angle S \cong \angle L$, $\triangle RST \sim \triangle KLM$.

This means $\dfrac{RS}{KL} = \dfrac{ST}{LM} = \dfrac{RT}{KM}$.

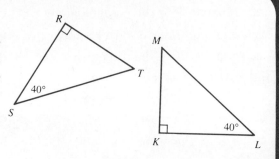

Examples

2 Find the distance across Pine Lake from the dock at *E* to the diving board at *D* if the lengths of $\overline{AB}$, $\overline{BD}$, and $\overline{BC}$ are 100 m, 120 m, and 55 m, respectively.

Since $\angle EDA$ and $\angle CBA$ are both right angles, they are congruent. Also, $\angle A \cong \angle A$. Thus, $\triangle ABC \sim \triangle ADE$ by Axiom 7–1.

Therefore, $\dfrac{DE}{BC} = \dfrac{AD}{AB}$

$\dfrac{DE}{55} = \dfrac{220}{100}$

$DE = 121$

The distance from *D* to *E* is 121 meters.

3 Write a two-column proof.

Given: $\overline{KU} \perp \overline{UT}$
$\overline{MT} \perp \overline{UT}$

Prove: $\dfrac{KS}{US} = \dfrac{MS}{TS}$

Proof:

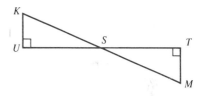

STATEMENTS	REASONS
1. $\overline{KU} \perp \overline{UT}, \overline{MT} \perp \overline{UT}$	1. Given
2. $\angle U$ and $\angle T$ are right angles.	2. Definition of perpendicular lines
3. $\angle U \cong \angle T$	3. Right angles are congruent.
4. $\angle USK \cong \angle TSM$	4. Vertical angles are congruent.
5. $\triangle USK \sim \triangle TSM$	5. AA Axiom
6. $\dfrac{KS}{MS} = \dfrac{US}{TS}$	6. Definition of similar polygons
7. $\dfrac{KS}{US} = \dfrac{MS}{TS}$	7. Interchanging property

In our work with congruence, we proved that congruence of triangles is reflexive, symmetric, and transitive. Similarity of polygons has the same three properties.

254 SIMILARITY

Theorem 7-1 Similarity of polygons is reflexive, symmetric, and transitive.

The proof is left for the exercises.

Exercises

Max: 1-13, 1-22, 1-3; Avg: 1-13, 1-8, 9-21 odds; Min: 1-13 odds, 1-21 odds

Exploratory Tell whether each pair of triangles is similar. Justify your answers.

1. yes; AA

2. yes; AA

3. No, there are not 2 pairs of congruent angles.

4. yes; AA

Refer to the figure at the right. In $\triangle ABC$, $\overline{DE} \parallel \overline{BC}$.

5. Name two similar triangles. Justify your answer.
6. Write three proportions that involve the measures of the sides of two triangles named in exercise 5.
7. If $DE = 5$, $BC = 10$, and $AE = 7$, find AC. 14
8. If $AB = 12$, $AD = 4$, and $AE = 6$, find AC. 18
9. If $AE = 3$, $EC = 7$, and $AD = 5$, find AB. $16\frac{2}{3}$
10. If $AD = 4$, $AE = 5$, and $EC = 10$, find DB. 8
11. If $AD = 7$, $AE = 3$, and $EC = 9$, find DB. 21
12. If $AB = 37$, $BD = 19$, and $AC = 23$, find AE. $11\frac{7}{37}$
13. If $DE = 10$, $BC = 15$, and $AD = 4$, find DB. 2

5. $\triangle ADE \sim \triangle ABC$ by AA since $\angle A \cong \angle A$ and $\angle ADE \cong \angle B$ (or $\angle AED \cong \angle C$).

6. $\frac{AD}{AB} = \frac{DE}{BC}$; $\frac{AD}{AB} = \frac{AE}{AC}$; $\frac{AE}{AC} = \frac{DE}{BC}$

Written In the figure at the right, $\overline{DE} \perp \overline{BC}$, $\overline{BA} \perp \overline{AC}$.

1. Name two similar triangles. Justify your answer. $\triangle BDE \sim \triangle BAC$; AA
2. Write three proportions that involve the measures of the sides of the triangles named in exercise 1.

2. $\frac{BD}{BA} = \frac{DE}{AC}$; $\frac{DE}{AC} = \frac{BE}{BC}$; $\frac{BE}{BC} = \frac{BD}{BA}$

Given $\overline{JK} \parallel \overline{MN}$, **write a two-column proof for each of the following.**

3. Prove: $\triangle JKL \sim \triangle NML$

4. Prove: $\frac{JL}{LN} = \frac{KL}{LM}$

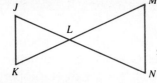

For answers to exercises 3-22, see the Teacher's Guide.

Given △ABC with altitudes $\overline{CD}$ and $\overline{AE}$, write a two-column proof.

5. Prove: △AFD ~ △CFE
6. Prove: $\dfrac{AF}{FC} = \dfrac{DF}{FE}$
7. Prove: △CBD ~ △ABE
8. Prove: $\dfrac{BC}{AB} = \dfrac{DC}{AE}$

Given ▱ABCD, write a two-column proof.

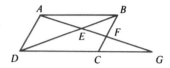

9. Prove: △AED ~ △FEB
10. Prove: △ABF ~ △GCF
11. Prove: △AEB ~ △GED
12. Prove: $\dfrac{AF}{GF} = \dfrac{BF}{CF}$
13. Prove: $\dfrac{AD}{BF} = \dfrac{AE}{EF}$
14. Prove: $EB \cdot GD = ED \cdot AB$

Given △ABC with $\overline{AB} \cong \overline{AC}$, altitudes $\overline{AE}$ and $\overline{CD}$, write a two-column proof.

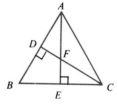

15. Prove: $AC \cdot DB = CB \cdot EC$
16. Prove: $AF \cdot BE = CF \cdot AD$
17. Prove: $AB \cdot DB = CB \cdot EC$
18. Prove: $AC \cdot DB = BE \cdot CB$

Prove the following.

19. If the vertex angles of two isosceles triangles are congruent, then the triangles are similar.
20. Two equilateral triangles are similar.
21. If an acute angle of a right triangle is congruent to an acute angle of another right triangle, the triangles are similar.
22. Theorem 7-1

Challenge   Use proportions to solve the following problems.

1. The mixture for a finish coat of concrete is one part cement to two parts sand. How much cement should be mixed with 15 pounds of sand? $7\frac{1}{2}$ pounds
2. A designated hitter made 8 hits in 9 games. If he continues hitting at that rate, how many hits will he make in 108 games?   96 hits
3. A recipe for preparing material to dye calls for four parts alum to one part washing soda. How much washing soda should be used for 150 grams of alum?   37.5 grams

7.4 Proportions and Triangles

Objective: To prove and apply theorems about proportions and triangles.

On the map of a section of Pine City, notice that $\overline{DE}$ appears to be parallel to $\overline{BC}$. It also appears that $\dfrac{AD}{DB}$ is equal to $\dfrac{AE}{EC}$.

Theorem 7–2 If a line is parallel to one side of a triangle and it intersects the other two sides, then it divides them proportionally.

Example

1 Write a two-column proof.

Given: $\triangle ABC$, $\overleftrightarrow{DE} \parallel \overline{BC}$
Prove: $\dfrac{AD}{DB} = \dfrac{AE}{EC}$

Proof:

STATEMENTS	REASONS
1. $\overleftrightarrow{DE} \parallel \overline{BC}$	1. Given
2. $\angle 1 \cong \angle 3$, $\angle 2 \cong \angle 4$	2. If two parallel lines are cut by a transversal, the corresponding angles are congruent.
3. $\triangle ABC \sim \triangle ADE$	3. AA Axiom
4. $\dfrac{AB}{AD} = \dfrac{AC}{AE}$	4. Definition of similar polygons
5. $\dfrac{AB - AD}{AD} = \dfrac{AC - AE}{AE}$	5. Subtraction Property of Proportions
6. $AD + DB = AB$, $AE + EC = AC$	6. Segment Addition Axiom
7. $DB = AB - AD$, $EC = AC - AE$	7. Subtraction Axiom of Equality
8. $\dfrac{DB}{AD} = \dfrac{EC}{AE}$	8. Substitution
9. $\dfrac{AD}{DB} = \dfrac{AE}{EC}$	9. Reciprocal Property of Proportions

Looking at another section of the map, notice that the streets are parallel and the avenues can be thought of as transversals. The streets appear to divide the avenues proportionally.

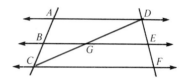

Theorem 7-3 If three parallel lines intersect two transversals, they divide them proportionally.

Example

2 Give a plan of proof.
Given: $\overleftrightarrow{AD} \parallel \overleftrightarrow{BE} \parallel \overleftrightarrow{CF}$
Prove: $\dfrac{AB}{BC} = \dfrac{DE}{EF}$

Plan: Draw $\overline{DC}$. Then use Theorem 7-2 with $\triangle ADC$ and $\triangle DFC$.
The proof is left for the exercises.

This figure could be drawn so that the two transversals intersect on one of the parallel lines or between the given parallel lines.

Another important ratio in similar triangles is expressed in the following.

Theorem 7-4 The perimeters of similar triangles are proportional to the measures of corresponding sides.

The proof is left for the exercises.

Example

3 Find the measure of $\overline{SM}$ in $\triangle RST$ with $\overline{MN} \parallel \overline{RT}$ and $SN = 12$, $NT = 10$, and $SR = 11$.

Let $SM = x$. Then $MR = 11 - x$. Using Theorem 7-2,
$$\dfrac{SM}{MR} = \dfrac{SN}{NT}.$$
$$\dfrac{x}{11-x} = \dfrac{12}{10}$$
$$10x = 132 - 12x$$
$$22x = 132$$
$$x = 6 \quad \text{Therefore, } SM = 6.$$

Because $\triangle SMN \sim \triangle SRT$, an alternative solution is
$$\dfrac{SM}{SR} = \dfrac{SN}{ST} \text{ or } \dfrac{x}{11} = \dfrac{12}{22}.$$
$$x = 6$$

Example

4 The measures of the sides of a triangle are 5, 11, and 12. Find the measure of the longest side of a similar triangle that has a perimeter of 14.

Draw a figure as shown at the right. Using Theorem 7-4,
$$\frac{\text{perimeter of } \triangle ABC}{\text{perimeter of } \triangle DEF} = \frac{12}{x}.$$

$$\frac{28}{14} = \frac{12}{x}$$

$$\frac{2}{1} = \frac{12}{x}$$

$2x = 12$

$x = 6$ The measure of the longest side is 6.

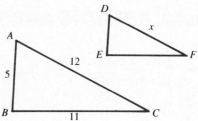

Exercises

Page 244 Max: 1-12, 1-38, Challenge; Avg: 1-12, 1-31 odds, 32-38; Min: 1-11 odds, 1-31 odds

Exploratory In the figure, $\overline{MN} \parallel \overline{KL}$. Complete each of the following. Justify your answers.

1. $\dfrac{JM}{?} = \dfrac{JN}{NL}$ 2. $\dfrac{JK}{MK} = \dfrac{?}{NL}$

3. $\dfrac{JM}{?} = \dfrac{?}{JL}$ 4. $\dfrac{JK - JM}{?} = \dfrac{?}{JN}$

1. MK; Theorem 7-2
2. JL; Theorem 7-2
3. JK, JN; Theorem 7-2
4. JM; JL − JN; Theorem 7-2

Using the given information and the figure, find the following.

5. If JM = 14, MK = 7, and JN = 10, find NL. 5
6. If MK = 3, JM = 9, and NL = 4, find JN. 12
7. If JM = 5, JK = 9, and JL = 18, find JN. 10
8. If JM = 9, MK = 6, and JL = 20, find JN. 12
9. If MN = 5, JN = 4, and JL = 10, find KL. $12\frac{1}{2}$
10. If KL = 10, MK = 2, and JM = 6, find MN. $7\frac{1}{2}$
11. If JM = 6, MN = 8, and KL = 12, find MK. 3
12. If MN = 9, KL = 12, and NL = 2, find JN. 6

Written In the figure, $\overleftrightarrow{PS} \parallel \overleftrightarrow{QT} \parallel \overleftrightarrow{RU}$. Complete each of the following. Justify your answers.

1. $\dfrac{PQ}{?} = \dfrac{ST}{TU}$ 2. $\dfrac{PR}{QR} = \dfrac{SU}{?}$

3. $\dfrac{RQ - PQ}{PQ} = \dfrac{?}{TS}$ 4. $\dfrac{SU}{PR} = \dfrac{?}{QR}$

1. QR; Theorem 7-3
2. TU; Theorem 7-3
3. UT−TS; Theorem 7-3
4. TU; Theorem 7-3

Find the following. Use the figure for exercises 1–4.

5. If $QR = 4$, $UT = 8$, and $PQ = 9$, find TS. 18
6. If $PQ = 3$, $QR = 9$, and $ST = 4$, find UT. 12
7. If $QR = 12$, $ST = 5$, and $TU = 3$, find PQ. 20
8. If $RQ = 4$, $PQ = 9$, and $US = 39$, find UT. 12
9. If $RQ = TS$, $PQ = 4$, and $TU = 9$, find RQ. 6
10. If $RQ = x - 2$, $QP = x + 4$, $UT = x - 3$, and $TS = x + 1$, find x. 5
11. If $TS = x + 5$, $UT = x + 3$, $RQ = x - 2$, and $QP = x - 1$, find x. 7
12. If $QP = x + 1$, $TS = x - 3$, $RQ = x - 1$, and $UT = x - 4$, find x. 7

Solve.

13. Zeb, who is 187.5 cm tall, is standing 9 meters from the base of the light that illuminates the Pine City Tennis Courts. The sign at the bottom of the light says that the light is 6 meters high. How long is Zed's shadow? 4.09 m

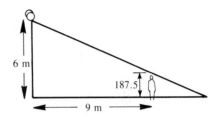

14. The measures of the sides of a triangle are 4, 6, and 8. Find the measure of the shortest side of a similar triangle whose perimeter is 27. 6
15. The measures of the sides of a triangle are 14, 18, and 10. The measure of the shortest side of a similar triangle is 6. Find the perimeter of the smaller triangle. 25.2
16. $\triangle ABC \sim \triangle DEF$, $BC = 10$ and $EF = 16$. If the perimeter of the larger triangle is 8.8 less than twice that of the smaller triangle, find the perimeters of both triangles. 22, 35.2
17. The sum of the perimeters of two similar triangles is 24. The measures of two corresponding sides are as 5 is to 1. Find each perimeter. 4, 20

Write a two-column proof for each. For answers to exercises 18–27, see the Teacher's Guide.

18. Given: $\square ABCD$, $\overline{EG} \parallel \overline{BC}$, $\overline{GF} \parallel \overline{AB}$

    Prove: $\dfrac{BE}{AE} = \dfrac{BF}{FC}$

19. Given: In $\triangle ABC$, $\overline{DE} \parallel \overline{BC}$
    $\overline{AM}$ is a median.

    Prove: $\overline{DQ} \cong \overline{QE}$

Exercise 18

Exercise 19

20. Given: $\angle 1 \cong \angle 2$

    Prove: $\dfrac{SU}{UR} = \dfrac{SV}{VT}$

21. Given: $\angle 2 \cong \angle 3$

    Prove: $RU \cdot WT = US \cdot RW$

22. Given: $\angle 1 \cong \angle 2$
    $\angle 2 \cong \angle 3$

    Prove: $WT \cdot VT = RW \cdot SV$

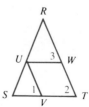

Exercises 20–22

23. Prove Theorem 7–3. Use the plan and diagram in example 2 on page 257.

24. In the diagram in example 2, suppose $\overline{AF}$ is drawn instead of $\overline{CD}$. Prove Theorem 7–3 using this new diagram.

25. Prove Theorem 7–3 for the case where two transversals intersect on one of the given parallel lines.

26. Prove Theorem 7–3 for the case where the two transversals intersect between the given parallel lines.

27. **Prove:** If two triangles are similar, the measures of corresponding altitudes have the same ratio as the measures of corresponding sides.

Solve.

28. $\triangle ABC \sim \triangle DEF$ and $AB:DE = 12:8$. The smaller triangle has an altitude of 5. Find the corresponding altitude of the larger triangle. $7\frac{1}{2}$

29. The corresponding altitudes of two similar triangles have measures of 8 and 14. The perimeter of the first triangle is 30. Find the perimeter of the second triangle. $52\frac{1}{2}$

In $\triangle ABC$, $\overline{DE} \parallel \overline{BC}$ and $\overline{AF} \perp \overline{BC}$.

30. If $BC = 18$, $AF = 20$, and $AG = 5$, find DE. $4\frac{1}{2}$
31. If $BC = 49$, $GF = 7$, and $DE = 22$, find AF. $12\frac{19}{27}$

32. a. Draw $\triangle ABC$ such that $AB = 18$ cm, $AC = 12$ cm, and $BC = 20$ cm. See students' work.
    b. Bisect $\angle A$. Call the intersection of $\overline{BC}$ and the bisector, D. See students' work.
    c. Measure $\overline{BD}$ and $\overline{DC}$ and find $\dfrac{AB}{AC}$ and $\dfrac{BD}{DC}$. $\dfrac{AB}{AC} = \dfrac{3}{2}$, $\dfrac{BD}{DC} = \dfrac{3}{2}$

33. Look at your result in exercise 32. If $\overline{AD}$ bisects $\angle A$, what appears to be true about AB, AC, BD, and DC? $\dfrac{AB}{AC} = \dfrac{BD}{DC}$

34. Your result in exercise 33 may be restated as follows: An angle bisector of a triangle divides the opposite side into segments whose measures are proportional to those of the other two sides. For the answer to exercise 34, see the Teacher's Guide.

To prove this, draw $\overrightarrow{CX} \parallel \overline{AD}$, where X is on $\overrightarrow{BA}$. Show $\triangle ACX$ is isosceles. Now apply Theorem 7–2. Write a complete two-column proof.

In the figure at the right, $\angle 1 \cong \angle 2$. Find each of the following.

35. If $AB = 19$, $AC = 15$, and $DC = 12$, find BD. $15\frac{1}{5}$
36. If $BD = 8$, $BC = 24$, and $AB = 12$, find AC. 24
37. If $AB = 6$, $BC = 19$, and $AC = 9$, find BD. $7\frac{3}{5}$
38. If $AB = 7$, $AC = 9$, and $BC = 15$, find BD and DC. $6\frac{9}{16}$ and $8\frac{7}{16}$

For the answer to the exercise, see the Teacher's Guide.

Challenge

Write a two-column proof of Theorem 7–4.
  Hint: See Challenge exercise on page 251.

Objective: To prove and apply theorems about right triangles and similarity.

7.5 Right Triangles and Similarity

Pictured at the right are three paths from the refreshment stand to Pine Lake. The path represented by $\overline{AC}$ is perpendicular to the path represented by $\overline{CB}$. The path $\overline{CD}$ is perpendicular to $\overline{AB}$. Also, the lengths of $\overline{AD}$ and $\overline{DB}$ are 25 m and 9 m, respectively. Is it possible to find the length of $\overline{CD}$? One plan of reasoning follows. $\triangle ABC$ and $\triangle ADC$ are both right triangles with right angles at C and D. They both contain $\angle A$. Therefore, $\triangle ABC \sim \triangle ACD$ by the AA axiom. In the same manner, $\triangle ABC \sim \triangle CBD$.

It follows that $\triangle ACD \sim \triangle CBD$ by the transitive property of similarity. One pair of corresponding sides is $\overline{AD}$ and $\overline{CD}$. Another pair is $\overline{CD}$ and $\overline{BD}$. Therefore, $\frac{AD}{CD} = \frac{CD}{BD}$ or $\frac{25}{CD} = \frac{CD}{9}$. Cross multiplication produces $CD^2 = 225$. Therefore, $CD = \sqrt{225}$, or the length of $\overline{CD}$ is 15 m.

This problem suggests some interesting relationships in a right triangle. It appears that $\overline{CD}$, which is the altitude to the hypotenuse $\overline{AB}$, separates right $\triangle ABC$ into two triangles that are each similar to $\triangle ABC$. Because similarity is transitive, it follows that these two triangles are similar to each other.

Theorem 7–5 | If the altitude is drawn to the hypotenuse of a right triangle, the two triangles formed are each similar to the original triangle and to each other.

Example

1 Give a plan of proof for the following.

Given: Right $\triangle ABC$, right $\angle C$,
$\overline{CD} \perp \overline{AB}$

Prove:
$\triangle ABC \sim \triangle ACD$
$\triangle ABC \sim \triangle CBD$
$\triangle ACD \sim \triangle CBD$

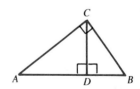

Plan: $\triangle ABC$ and $\triangle ACD$ are right triangles with common angle A. Therefore, $\triangle ABC$ and $\triangle ACD$ are similar. $\triangle ABC$ and $\triangle CBD$ are also right triangles with common angle B. Therefore, $\triangle ABC$ and $\triangle CBD$ are similar. Using the transitive property, $\triangle ACD$ is similar to $\triangle CBD$.

Since $\triangle ACD \sim \triangle CBD$, it follows that $\dfrac{AD}{CD} = \dfrac{CD}{BD}$.

Note that the means of this proportion are the same. Therefore, CD is the *mean proportional*, or *geometric mean*, between AD and BD. The following theorem is a corollary to Theorem 7–5.

Theorem 7–6 The measure of the altitude to the hypotenuse of a right triangle is the mean proportional between the measures of the segments of the hypotenuse.

Example

2 Give a plan of proof for Theorem 7–6.
Given: Right $\triangle ABC$, right $\angle C$, $\overline{CD} \perp \overline{AB}$
Prove: $\dfrac{AD}{CD} = \dfrac{CD}{BD}$
Plan: $\overline{CD}$ is the altitude to the hypotenuse $\overline{AB}$. By Theorem 7–5, it follows that $\triangle ADC \sim \triangle CDB$. Therefore, $\dfrac{AD}{CD} = \dfrac{CD}{BD}$.

Returning to the refreshment stand problem on page 261, is it possible to find the distance from C to A?

Since $\triangle ABC \sim \triangle ACD$, it follows that $\dfrac{AB}{AC} = \dfrac{AC}{AD}$.

Thus, $\dfrac{34}{AC} = \dfrac{AC}{25}$
$(AC)^2 = 850$
$AC = \sqrt{850} \approx 29.2$

That is, the distance from C to A is approximately 29.2 m.

The means of the proportion $\dfrac{AB}{AC} = \dfrac{AC}{AD}$ are the same. Therefore, AC is the mean proportional between AB, the hypotenuse measure and AD, the measure of the segment that is adjacent to $\overline{AC}$. In a like manner, we can show that CB, the measure of the other leg, is the mean proportional between AB and DB.

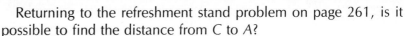

Theorem 7–7 If the altitude is drawn to the hypotenuse in a right triangle, the measure of each leg is the mean proportional between the measure of the hypotenuse and the measure of the segment on the hypotenuse that is adjacent to the leg.

Examples

3 Give a plan of proof for Theorem 7–7.

Given: Right $\triangle ABC$ with right $\angle C$, $\overline{CD} \perp \overline{AB}$

Prove: $\dfrac{AB}{AC} = \dfrac{AC}{AD}$

$\dfrac{AB}{BC} = \dfrac{BC}{DB}$

Plan: Use Theorem 7–5.
The proof is left for the exercises.

4 In $\triangle RST$, $\angle RST$ is a right angle and $\overline{SU} \perp \overline{RT}$. Find RU, if $UT = 9$ and $US = 15$.

By Theorem 7–6, $\dfrac{RU}{US} = \dfrac{US}{UT}$

$\dfrac{RU}{15} = \dfrac{15}{9}$

$9(RU) = 225$

$RU = 25$

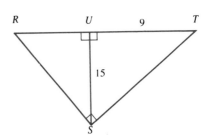

5 Find ST.

By Theorem 7–7, $\dfrac{RT}{ST} = \dfrac{ST}{UT}$

Since $RU = 25$ and $UT = 9$, $RT = 34$.

Therefore, $\dfrac{34}{ST} = \dfrac{ST}{9}$

$(ST)^2 = 306$

$ST = \sqrt{306} \approx 17.5$

Exercises

Max: 1-31, 1-20; Avg: 1-4, 5-31 odds, 1-20; Min: 5-31 odds, 1-19 odds

Exploratory Suppose $\triangle DEF$ has a right angle at D and $\overline{DG} \perp \overline{EF}$. For answers to exercises 1-4, see the Teacher's Guide.

1. Name three pairs of similar triangles.
2. Name the corresponding angles.
3. Name the corresponding sides.
4. Write three proportions involving the measures of the sides of the triangles.

Find the mean proportional of each of the following.
5. 4 and 9 6
6. 2 and 32 8
7. 2 and 8 4
8. 4 and 6 $2\sqrt{6}$
9. $2\sqrt{3}$ and $3\sqrt{3}$ $3\sqrt{2}$

Use the figure for exercises 1–3. Complete each of the following and justify your answer.
10. $\dfrac{EG}{?} = \dfrac{?}{GF}$ DG, DG; Theorem 7-6
11. $\dfrac{?}{ED} = \dfrac{ED}{?}$ EF, EG; Theorem 7-7
12. $\dfrac{?}{DF} = \dfrac{DF}{?}$ EF, GF; Theorem 7-7
13. $DE \cdot DF = EF \cdot \underline{}$ DG; Theorem 7-5
14. $EF \cdot \underline{} = (ED)^2$ EG; Theorem 7-7
15. $\dfrac{EG + GF}{?} = \dfrac{?}{GF}$ DF; Theorem 7-7

Write each of the following in simplest radical form.
16. $\sqrt{8}$ $2\sqrt{2}$
17. $\sqrt{32}$ $4\sqrt{2}$
18. $\sqrt{24}$ $2\sqrt{6}$
19. $\sqrt{200}$ $10\sqrt{2}$
20. $\sqrt{12}$ $2\sqrt{3}$
21. $\sqrt{98}$ $7\sqrt{2}$
22. $3\sqrt{50}$ $15\sqrt{2}$
23. $(3 + \sqrt{2})(5 - \sqrt{2})$ $13 + 2\sqrt{2}$
24. $\sqrt{27} + \sqrt{3}$ $4\sqrt{3}$
25. $2\sqrt{32} - 5\sqrt{8}$ $-2\sqrt{2}$
26. Restate Theorems 7–6 and 7–7 using the term "geometric mean."
For the answer to exercise 26, see the Teacher's Guide.

If $\sqrt{10}$ is the geometric mean between x and the following number, find x.
27. 2 5
28. 5 2
29. 4 $2\frac{1}{2}$
30. 6 $1\frac{2}{3}$
31. 20 $\frac{1}{2}$
32. 0.8 $12\frac{1}{2}$

Written Find the measure of the segment marked x in each right triangle.

1.
2.
3.
4.
5.
6.

Use the figure at the right. Find each of the following.
7. s, if $h = 9$ and $r = 4$ $20\frac{1}{4}$
8. s, if $x = 8$ and $z = 4$ 2
9. h, if $r = 24$ and $s = 6$ 12
10. x, if $s = 4$ and $z = 8$ 16
11. y, if $x = 20$ and $r = 4$ $4\sqrt{5}$
12. h, if $y = 20$ and $x = 50$ $4\sqrt{21}$
13. x, if $z = 8$ and $r = 12$ 16
14. r and s, if $h = 3$ and $s = r + 8$ 1, 9

In right $\triangle ABC$, $\overline{AB}$ is the hypotenuse and $\overline{CD}$ is the altitude to $\overline{AB}$.
15. If $CD = 6$ and DB exceeds AD by 9, find AD and DB. 3, 12
16. If $CD = 4$ and DB exceeds AD by 15, find AD and DB. 1, 16
17. If $CD = 14$ and $AB = 35$, find AD and DB. 7, 28

Write a two-column proof for each. For answers to exercises 18-20, see the Teacher's Guide.
18. Theorem 7–5
19. Theorem 7–6
20. Theorem 7–7

7.6 The Pythagorean Theorem

Objective: To prove and apply the Pythagorean Theorem and its converse.

One of the proofs of the Pythagorean Theorem—perhaps the most famous theorem in all geometry—appears below. It is based on our work with similarity. Over two hundred proofs of this theorem exist.

Theorem 7–8
The Pythagorean Theorem

In a right triangle, the square of the measure of the hypotenuse is equal to the sum of the squares of the measures of the legs.

Example

1 Write a two-column proof.

In the figure, c represents the measure of the hypotenuse and a and b each represent the measure of a leg of right $\triangle ABC$.

Given: Right $\triangle ABC$, right $\angle C$
Prove: $a^2 + b^2 = c^2$
Proof:

STATEMENTS	REASONS
1. Right $\triangle ABC$, right $\angle C$	1. Given
2. Draw the altitude from C to the hypotenuse $\overline{AB}$. Let $AD = x$ and $DB = y$.	2. Through a point not on a given line, there exists one and only one line perpendicular to the given line.
3. $\dfrac{c}{a} = \dfrac{a}{y}$ and $\dfrac{c}{b} = \dfrac{b}{x}$	3. If the altitude is drawn to the hypotenuse of a right triangle, the measure of each leg is the mean proportional between the measure of the hypotenuse and the segment on the hypotenuse adjacent to the leg.
4. $a^2 = cy$ and $b^2 = cx$	4. Means-Extremes property of proportions
5. $a^2 + b^2 = cy + cx$	5. Addition Axiom of Equality
6. $a^2 + b^2 = c(y + x)$	6. Distributive Axiom
7. $a^2 + b^2 = c \cdot c$ or $a^2 + b^2 = c^2$	7. Substitution Axiom

Pythagoras (about 580–500 B.C.), the Greek mathematician and philosopher, may have been the first to prove the theorem. The Babylonians, however, probably knew the theorem and its converse 1000 years earlier.

**Theorem 7–9
Converse of the
Pythagorean Theorem**

If the sum of the squares of the measures of two sides of a triangle is equal to the square of the measure of the third side, then the triangle is a right triangle.

Examples

2 Give a plan of proof.

Given: $\triangle ABC$ with sides of measure a, b, c, where $a^2 + b^2 = c^2$

Prove: $\triangle ABC$ is a right triangle.

Plan: Draw $\overline{DE}$ on line ℓ with measure equal to a. At D, draw line $m \perp \overline{DE}$. Locate point F on m so that $DF = b$. Draw $\overline{FE}$ and call its measure x. Because $\triangle FED$ is a right triangle, $a^2 + b^2 = x^2$. But $a^2 + b^2 = c^2$, so $x^2 = c^2$ or $x = c$. Thus, $\triangle ABC \cong \triangle FED$ by SSS. This means $\angle C \cong \angle D$. Therefore, $\angle C$ must be a right angle.

3 The walls in the Pine City Recreation Center are being covered with wall paneling. The doorway is 0.9 m wide and 2.5 m high. What is the widest rectangular panel that can be taken through the doorway?

The widest panel would be taken through diagonally, as shown. Let x be the measure of that panel. Use the Pythagorean Theorem to find x.

$$(0.9)^2 + (2.5)^2 = x^2$$

Solve for x. $\quad x = \sqrt{7.06}$

$\quad\quad\quad\quad\quad\quad x \approx 2.66$

A width of 2.66 m would be a tight fit.

To allow extra clearance, a narrower panel could be chosen.

Example

4 Sara has three pieces of wood 65 cm, 72 cm, and 97 cm long. Can she connect the pieces end-to-end to form a planter shaped in a right triangle?

$$a^2 + b^2 = c^2$$
$$65^2 + 72^2 \stackrel{?}{=} 97^2 \quad \text{Use Theorem 7–9.}$$
$$4225 + 5184 \stackrel{?}{=} 9409$$
$$9409 = 9409$$

Since $a^2 + b^2 = c^2$, Sara can form a planter shaped in a right triangle using these pieces of wood. A set of three integers x, y, and z that satisfy the equation $x^2 + y^2 = z^2$ is called a Pythagorean Triple. Hence, 65, 72, and 97 is a Pythagorean Triple.

Exercises

Max: 1-14, 1-30, 1-3; Avg: 1-14, 1-29 odds; Min: 1-13 odds, 1-29 odds

Exploratory Find the value of x.

1. 13
2. 8
3. 8
4. 2
5. 6
6. 5

Determine if the three numbers can be the measures of the sides of a right triangle.

7. 8, 15, 17 yes
8. 14, 48, 49 no
9. 12, 35, 37 yes
10. 19, 21, 29 no
11. 3, $\sqrt{2}$, $\sqrt{11}$ yes
12. 4, $4\sqrt{3}$, 8 yes
13. 24, 32, 40 yes
14. 9, 10, 11 no

Written In $\triangle ABC$, $\angle C$ is a right angle. In each case, the measures of two sides are given. Find the measure of the side that is missing.

1. $AC = 21$ and $BC = 20$ 29
2. $AB = 6$ and $AC = 3\sqrt{2}$ $3\sqrt{2}$
3. $AB = 4\sqrt{7}$ and $AC = 3\sqrt{5}$ $\sqrt{67}$

Given are the measures of a rectangle's sides. Find its diagonal measure.

4. 8 and 15 17
5. 6 and 12 $6\sqrt{5}$
6. 7 and 24 25
7. 3.2 and 4 $\sqrt{26.24}$

A square has a side of the given measure. Find the diagonal measure.
8. 1 $\sqrt{2}$
9. 2 $2\sqrt{2}$
10. 16 $16\sqrt{2}$
11. 5.4 $\sqrt{58.32}$

Find the perimeter of a rhombus whose diagonals have the given measures.
12. 8 and 6   20
13. 10 and 24   52
14. 12 and 8 $8\sqrt{13}$
15. 4 and 12 $8\sqrt{10}$

Answer the following.
16. Two adjacent sides of a parallelogram have lengths 21 cm and 28 cm. A diagonal is 35 cm long. Is the parallelogram a rectangle? Explain. Yes, since $21^2 + 28^2 = 35^2$, a right triangle is formed. Therefore, the parallelogram has a right angle and is a rectangle.

Given isosceles $\triangle ABC$ with base $\overline{BC}$. Find the measure of the altitude drawn to $\overline{BC}$ under the following conditions.
17. $AB = 15, BC = 24$ 9
18. $AB = 9, BC = 8\sqrt{2}$ 7
19. $AC = 8, BC = 2\sqrt{3}$ $\sqrt{61}$

Find the measure of an altitude of an equilateral triangle whose sides have the measure given.
20. 8 $4\sqrt{3}$
21. 12 $6\sqrt{3}$
22. $8\sqrt{3}$ 12
23. s $\sqrt{s^2 - (\frac{s}{2})^2}$ or $\frac{s\sqrt{3}}{2}$

Solve.
24. If the perimeter of a square is 32, find the length of its diagonal. $8\sqrt{2}$
25. In $\triangle PQR, PQ = PR = 13$, and $QR = 10$. Find the area of $\triangle PQR$. 60
26. Find the area of an isosceles trapezoid whose legs are 13 and whose bases are 14 and 24. 228
27. In right $\triangle RST$, the measure of the hypotenuse $\overline{ST}$ exceeds the measure of $\overline{SR}$ by 2. If $RT = 10$, find ST. 26
28. In $\triangle ABC, \angle C$ is a right angle. If CB is one more than AC, and AB is two more than AC, find the area of $\triangle ABC$. 6

In the rectangular solid shown at the right, $\overline{AB}$ is a diagonal. Find AB if the solid has the following dimensions.
29. $\ell = 9, w = 12, h = 8$ 17
30. $\ell = 5, w = 12, h = 4$ $\sqrt{185}$

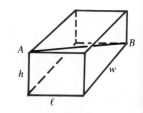

Challenge Solve.

1. A picket fence is to have a gate 42 inches wide. The gate is 54 inches high. Find the length of a diagonal brace for the gate to the nearest inch. 68 in.

2. A plane flies 300 km due north, 400 km due east, and then 500 km due south. How far is the plane from its starting point? State the answer to the nearest kilometer. 447 km

Write a two-column proof.
3. **Given:** $\triangle ABC$, altitudes $\overline{AF}, \overline{BE}$, and $\overline{CD}$ intersect at G.
   Prove: $\dfrac{AE}{EC} \cdot \dfrac{CF}{FB} \cdot \dfrac{BD}{DA} = 1$

   For the answer to exercise 3, see the Teacher's Guide.

7.7 Special Right Triangles

Objective: To use the Pythagorean Theorem to find missing measures of sides of special right triangles.

An isosceles right triangle is often called a 45°–45°–90° triangle. Using the Pythagorean Theorem in the 45°–45°–90° triangle shown at the right,

$$d^2 = x^2 + x^2$$
$$d^2 = 2x^2$$
$$d = \sqrt{2x^2} \quad \text{Reject } -\sqrt{2x^2} \text{ because distance is positive.}$$
$$d = x\sqrt{2}$$

Theorem 7–10

In a 45°–45°–90° triangle, the length of the hypotenuse is $\sqrt{2}$ times the length of a leg.

Example

1 Find the length of the diagonal of a square whose side has a length of 8 m.

The diagonal is the hypotenuse of a 45°–45°–90° triangle. Therefore, the length of the diagonal is $\sqrt{2}$ times 8 or $8\sqrt{2}$ m.

Pictured at the right is equilateral triangle ABC. If altitude $\overline{AD}$ is drawn, two 30°–60°–90° triangles are formed. Let $BD = x$. Then $AB = 2x$. Using the Pythagorean Theorem in $\triangle ABD$,

$$(AD)^2 + x^2 = (2x)^2$$
$$(AD)^2 + x^2 = 4x^2$$
$$(AD)^2 = 3x^2$$
$$AD = x\sqrt{3}$$

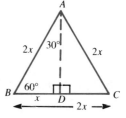

Theorem 7–11

In a 30°–60°–90° triangle, the length of the hypotenuse is two times the length of the side opposite the angle of degree measure 30; the length of the side opposite the angle of degree measure 60 is $\sqrt{3}$ times the length of the other leg.

Example

2 In the figure, find the missing measures.

$\overline{BC}$ is opposite the angle of degree measure 30. Thus, by Theorem 7–11, the hypotenuse measures 16. Because $\overline{AC}$ is opposite the angle of degree measure 60, it measures $8\sqrt{3}$.

Exercises

Max: 1-12, 1-25, 1-3; Avg: 1-12, 1-11 odds, 13-25; Min: 1-11 odds, 1-25 odds

Exploratory Find *x* and *y* in each figure.

1. $x = 5\sqrt{2}$, $y = 5\sqrt{2}$

2. $x = 4$, $y = 4\sqrt{2}$

3. $x = 6\sqrt{2}$, $y = 6$

4. $x = 8$, $y = 8$

5. $x = 8\sqrt{3}$, $y = 16$

6. $x = 9$, $y = 9\sqrt{3}$

7. $x = 2$, $y = 1$ $\sqrt{3}$

8. $x = \frac{20\sqrt{3}}{3}$, $y = \frac{40\sqrt{3}}{3}$

9. $x = 12\sqrt{2}$, $y = 12$

10. $x = 19\sqrt{3}$, $y = 38$

11. $x = \frac{4\sqrt{3}}{3}$, $y = \frac{8\sqrt{3}}{3}$

12. $x = \frac{16\sqrt{3}}{3}$, $y = \frac{32\sqrt{3}}{3}$

Written The length of a leg of an isosceles right triangle is given. Find the length of the hypotenuse in simplest radical form and to the nearest tenth.

1. 6 cm 2. 12 mm 3. 20 ft 4. $3\sqrt{2}$ in. 5. $6\sqrt{3}$ m 6. $2\sqrt{10}$ cm

6. $4\sqrt{5}$ cm, 8.9 cm

1. $6\sqrt{2}$ cm, 8.5 cm 2. $12\sqrt{2}$ mm, 17.0 mm 3. $20\sqrt{2}$ ft, 28.3 ft 4. 6 in., 6 in. 5. $6\sqrt{6}$ m, 14.7 m

The length of each side of an equilateral triangle is given. Find the length of an altitude.

7. 4 in. 8. 10 cm 9. 14 mm 10. 20 m 11. 15 cm 12. 23 ft

$2\sqrt{3}$ in. $5\sqrt{3}$ cm $7\sqrt{3}$ mm $10\sqrt{3}$ m $\frac{15\sqrt{3}}{2}$ cm $\frac{23\sqrt{3}}{2}$ ft

Refer to the figure at the right.

13. If $AD = 8$, find AB, BD, DC, BC, and AC. 16, $8\sqrt{3}$, 24, $16\sqrt{3}$, 32
14. If $AB = 4$, find AC, AD, DC, BD, and BC. 8, 2, 6, $2\sqrt{3}$, $4\sqrt{3}$
15. If $AC = 12$, find AB, AD, DC, BD, and BC. 6, 3, 9, $3\sqrt{3}$, $6\sqrt{3}$
16. If $BD = 2\sqrt{3}$, find AD, AB, AC, DC, and BC. 2, 4, 8, 6, $4\sqrt{3}$

CHAPTER 7 271

17. The base angles of an isosceles trapezoid each measure 45. The bases have lengths of 11 m and 21 m. Find the length of the altitude of the trapezoid. 5

18. An angle of a rhombus measures 120. If the longer diagonal has length of 18 in., find the length of the shorter diagonal. $6\sqrt{3}$

19. The perimeter of an equilateral triangle is $21\sqrt{3}$ cm. Find the length of the altitude of the triangle. $\frac{21}{2}$

Find the area of each polygon.

20.

21.

22.

23.

24.

25.

Challenge Find the area of the shaded region.

1.
36.48

2.
61.37

3.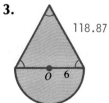
118.87

7.8 More Construction

Objective: To construct geometric figures based on similarity.

The study of geometric constructions continues with the first two constructions based on similarity.

CONSTRUCTION 10: Construct a triangle similar to a given triangle on a given base.

Given: $\triangle ABC$, base $\overline{DE}$

Method:
1. Use Construction 2 to construct $\angle EDT \cong \angle BAC$.
2. Using Construction 2 again, construct $\angle DES \cong \angle ABC$.
3. $\overrightarrow{DT}$ and $\overrightarrow{ES}$ will intersect in a point. Label the point F. $\triangle DEF \sim \triangle ABC$

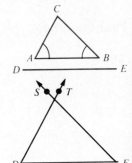

Justification: Since $\angle A \cong \angle D$ and $\angle B \cong \angle E$, the AA Axiom tells us that $\triangle ABC \sim \triangle DEF$.

CONSTRUCTION 11: Divide a given segment into a given number of congruent segments.

Given: $\overline{AB}$ to be divided into three congruent segments.

Method:
1. Draw $\overrightarrow{AC}$ so as not to contain $\overline{AB}$.
2. With a convenient radius and center A, mark point X on $\overrightarrow{AC}$.
3. Now, locate Y and Z on $\overrightarrow{AC}$ so that $\overline{XY} \cong \overline{AX}$ and $\overline{YZ} \cong \overline{AX}$.
4. Draw $\overline{BZ}$.
5. Use Construction 9 to construct a line through X that is parallel to $\overline{BZ}$. Its intersection with $\overline{AB}$ is R.
6. Use Construction 9 again to construct a line through Y parallel to $\overline{BZ}$. Its intersection with $\overline{AB}$ is S.

Thus, $\overline{AR} \cong \overline{RS} \cong \overline{SB}$.

Justification: $\overleftrightarrow{AB}$ and $\overleftrightarrow{AZ}$ are transversals that intersect three parallel lines, $\overleftrightarrow{RX}$, $\overleftrightarrow{SY}$, and $\overleftrightarrow{BZ}$. By Theorem 7-2, $\overleftrightarrow{AB}$ and $\overleftrightarrow{AZ}$ cut off segments whose measures are proportional. Since $\overline{AX} \cong \overline{XY} \cong \overline{YZ}$, $\overline{AR} \cong \overline{RS} \cong \overline{SB}$.

There are many constructions that involve circles. One is presented on the next page. First it will be helpful to introduce a new definition.

Definition of a Circumscribed Circle or an Inscribed Polygon

A circle is **circumscribed** about a polygon if the circle passes through each vertex of the polygon. The polygon is said to be **inscribed** in the circle.

In each case below, circle O is circumscribed about a polygon.

  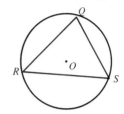

Journal

What do you like and dislike about doing geometric constructions? Give specific examples.

Can a circle be circumscribed about any triangle?

CONSTRUCTION 12: Construct a circle circumscribed about a given triangle.

Given: △ABC

Method:

1. Use Construction 6 to construct $\overleftrightarrow{DE}$, the perpendicular bisector of $\overline{AB}$.
2. Now construct the perpendicular bisector of $\overline{AC}$. Call it $\overleftrightarrow{FG}$.
3. Call the intersection of $\overleftrightarrow{DE}$ and $\overleftrightarrow{FG}$, point O.
4. Using O as center, construct a circle with radius $\overline{OC}$.

The circle with center O and radius $\overline{OC}$ passes through points C, A, and B, and thus is circumscribed about △ABC.

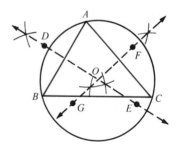

Justification: Let the intersection of $\overleftrightarrow{DE}$ and $\overline{AB}$ be H. Let the intersection of $\overleftrightarrow{FG}$ and $\overline{AC}$ be J.
Therefore, H and J are midpoints. Draw $\overline{AO}$, $\overline{OB}$, and $\overline{OC}$. △BHO ≅ △AHO and △AOJ ≅ △COJ by SAS. Thus, $\overline{OB} \cong \overline{OA} \cong \overline{OC}$. A circle with center O and radius $\overline{OC}$ must pass through points A and B.

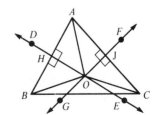

Exercises

Max: 1-5, 1-20; Avg: 1-5, 1-10, 11-19 odds; Min: 1-5, 1-10, 11-19 odds

Exploratory Answer the following.

1. Suppose △DEF is a given triangle and $\overline{GH}$ is a given line segment. Explain how to construct a triangle with base $\overline{GH}$ similar to △DEF. Construct ∠G ≅ ∠D and ∠H ≅ ∠F. The triangle formed is similar to △DEF.
2. Draw △ABC. Then draw a line segment $\overline{DE}$ that is clearly longer than $\overline{AB}$. Using $\overline{DE}$ as a base, construct a triangle similar to △ABC. Construct ∠D ≅ ∠A and ∠E ≅ ∠B. The triangle formed is similar to △ABC.

For exercises 3-5, see students' work.

Draw △ABC. Construct △DEF ~ △ABC so that each of the following is true.

3. $DE = 2(AB)$
4. $DE = 3(AB)$
5. $EF = 4(BC)$

For exercises 1-16, see students' work.

Written Copy the figure. Then perform the construction in each exercise.

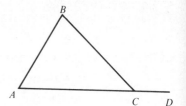

1. Construct a triangle with base $\overline{AD}$ that is similar to △ABC.
2. Construct a triangle not congruent to the one in exercise 1 that also has base $\overline{AD}$ and is similar to △ABC.

For each of the following, draw a line segment. Then, by construction, divide it into the following number of congruent segments.

3. 3
4. 5

Draw a line segment.

5. By construction, separate it into four congruent segments.
6. Repeat exercise 5 using a different method of construction.

Draw $\overline{AB}$. Divide it into two segments whose measures have the ratios given.

7. 1:2
8. 2:3
9. 3:4

Copy the figure shown at the right.

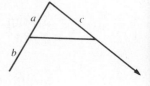

10. Construct a segment of measure x such that $\frac{a}{b} = \frac{c}{x}$ and justify the construction.

Construct each of the following. Then circumscribe a circle about it.

11. an equilateral triangle
12. a scalene right triangle
13. an isosceles right triangle
14. an obtuse triangle
15. a square
16. a rectangle

For answers to exercises 17-20, see the Teacher's Guide.

Justify the construction in each of the following.

17. exercise 11
18. exercise 12
19. exercise 13
20. exercise 14

7.9 Relations

Objective: To identify and write relations.

The idea of a **relation** appears frequently in all branches of mathematics as well as in nonmathematical situations. Before defining a relation, it will be helpful to review the idea of an ordered pair. Recall that pairs such as (2, 3) and (4, 5), in which order counts, are called **ordered pairs.** Each member of an ordered pair is called a **coordinate.** In the ordered pair (3, 2), 3 is the first coordinate and 2 is the second coordinate.

Definition of a Relation

> A relation is a set of ordered pairs.

The sets T, S, and C are relations.

$T = \{(3, 1), (1, 3), (4, 5)\}$
$S = \{(0, 0), (1, 1), (4, 2), (9, 3)\}$
$C = \{(\text{Montgomery, Alabama}), (\text{Juneau, Alaska}), \ldots,$
   $(\text{Cheyenne, Wyoming})\}$

Note that the elements of each relation are ordered pairs. If the ordered pair (a, b) is in the relation R, we may write either

$$(a, b) \in R \quad \text{or} \quad aRb$$

The symbol $x\cancel{R}y$ means (x, y) is not in the relation R.

Usually there is a rule that tells how the two coordinates of each ordered pair in a relation are related to each other. For instance, look carefully at relation S above. The first coordinate of each pair is the square of the second coordinate. That is,

0 is the square of 0,
1 is the square of 1,
4 is the square of 2,
and 9 is the square of 3.

Because this relationship exists between the coordinates of each pair in the relation S, we can use the rule *is the square of* to refer to the relation S. Now, look at relation C. Here, C can be described by the rule *is the capital of*.

Very often we deal with a relation on some specific set. For example, suppose set $A = \{3, 4, 5\}$. Suppose further that the relation L on set A is described by the rule, *is less than*. To find the elements of L, use the numbers from set A to form every possible ordered pair for which the first coordinate *is less than* the second coordinate. The results are $3L4$, $3L5$, and $4L5$. That is,

$$L = \{(3, 4), (3, 5), (4, 5)\}.$$

In this case we could refer to L more simply as the familiar relation "<" and write 3 < 4, 3 < 5, and 4 < 5.

Example

1 Give some examples of relations.
  a. Suppose the relation S *is the square of* is defined on the set of integers. Then, $S = \{(0, 0), (1, 1), (1, -1), (4, 2), (4, -2), (9, 3), (9, -3), \ldots\}$. Note that aSb if and only if $a = b^2$.
  b. Suppose T is the set of all possible triangles in a given plane. Suppose the relation R *is congruent to* is defined on T. Then xRy if and only if x and y are each triangles and $\triangle x \cong \triangle y$.
  c. Let M be the set of all lines in a plane. Suppose the relation R *is parallel to* is defined on M. Therefore, xRy if and only if x and y are each lines and $x \parallel y$.

Exercises

Max: 1-19, 1-14, 1-6; Avg: 1-19, 1-14,; Min: 1-19 odds, 1-3, 7-13 odds

Exploratory Let $M = \{(-1, 3), (2, 5), (3, 3), (-4, 6)\}$. Tell whether each of the following is *true* or *false*.

1. M is a relation. true
2. $(-1, 3) \in M$ true
3. $5M2$ false
4. $-4M6$ true
5. $6M(-4)$ false
6. $3M3$ true
7. $3M(-1)$ false
8. $2M5$ true

Let $A = \{2, 3, 4, 9, 16\}$. Write each relation as a set of ordered pairs.

9. is less than
10. is greater than
11. is equal to
12. is the square of
13. is the double of
14. is a factor of
15. is the positive square root of
16. is one less than

For answers to exercises 9-16, see the Teacher's Guide.

Let T be the set of all triangles in the figure shown.
17. Name the elements of T. T = {△ABC, △ABD, △ACD}
18. Name the elements of T that are similar to each other. △ABC, △ABD, △ACD
19. Suppose S is the relation is similar to. Write the elements in S.
S = {(△ABC, △ABC), (△ABC, △DBA), (△ABC, △DAC),(△ABD, △ABD),(△ABD, △CBA), (△ABD, △CAD), (△ACD, △ACD), (△ACD, △BCA), △ACD, △BAD)}

Written Given ▱ABCD with diagonals $\overline{AC}$ and $\overline{BD}$, let T be the set of all triangles in the figure. For answers to exercises 1-3, see the Teacher's Guide.

1. Write the elements of T.
2. Name the elements of T that are congruent to each other.
3. Suppose C is the relation is congruent to. Write the elements of C.

Use the figure for exericses 1–3. Let $L = \{\overline{AB}, \overline{BC}, \overline{CD}, \overline{AD}, \overline{AC}, \overline{BD}\}$.
4. Name the elements of L that are parallel to each other. $\overline{AB}, \overline{CD}; \overline{AD}, \overline{BC}$
5. Suppose P is the relation *is parallel to*. Write the elements of P. $P = \{(\overline{AB}, \overline{CD}), (\overline{CD}, \overline{AB}), (\overline{AD}, \overline{BC}), (\overline{BC}, \overline{AD})\}$

The diagram at the right is part of the Smith's family tree. The tree shows that Sarah had four children: Joe, Zeb, Mary, and Nancy. Joe had a daughter, Elsie. Zeb had two daughters, Flora and Clara. Mary had no children. Nancy had two children, Brian and Laurie. Use the first letters of each name to write the ordered pairs in each relation.

For answers to exercises 6-14, see the Teacher's Guide.

6. is a grandmother of
7. is the father of
8. is the mother of
9. is a sister of
10. is a brother of
11. is an aunt of
12. is an uncle of
13. is a cousin of

Let $S = \{0, 1, 2, 3, 4, 5, 6, 7\}$. Let R be a relation on S such that xRy if and only if x and y both have the same remainder when divided by 2.

14. Write R as a set of ordered pairs.

Challenge Let $S = \{$all students in your math class$\}$. Write each relation S as a set of ordered pairs. For exercises 1-6, see students' work.

1. is taller than
2. sits next to
3. has the same color eyes as
4. lives within one mile of
5. is a sister of
6. is the father of

Mathematical Excursions — Pythagorean Triples

Three integers, a, b, and c, that satisfy the equation $a^2 + b^2 = c^2$ are called a **Pythagorean triple**. Common examples are 3, 4, 5 and 5, 12, 13. *Primitive* Pythagorean triples are, pairwise, relatively prime. The positive integers x, y, z constitute a primitive Pythagorean triple *if and only if* there exist two relatively prime positive integers s and t with $s > t$ and one number is odd and the other even, such that $x = s^2 - t^2$, $y = 2st$, and $z = s^2 + t^2$.

Exercise Make a table of all primitive Pythagorean triples with $z < 100$.
For the answer to the exercise, see the Teacher's Guide.

7.10 Equivalence Relations

Objective: To identify a relation as reflexive, symmetric, or transitive, or as an equivalence relation.

Every segment is congruent to itself. This can be expressed as the congruence of segments is reflexive. The relation $\cong$ on the set of all segments is a **reflexive relation**. A more general definition of a reflexive relation follows.

Definition of Reflexive Relation

Let R be a relation on set A. Then R is said to be reflexive if and only if, for every a in A, the ordered pair (a, a) is in R.

Example

1 Give some examples of reflexive relations and relations that are not reflexive.

Reflexive

a. The relation *is congruent to* on the set of all triangles in a plane is a reflexive relation. Every triangle in a plane is congruent to itself.

b. The relation *is equal to* on the set of real numbers is reflexive because every real number equals itself.

Not Reflexive

a. The relation *is less than* on the set of real numbers is not reflexive. No matter which real number x is chosen, it is never true that $x < x$.

b. The relation $R = \{(1, 1), (2, 1)\}$ defined on $A\{1, 2\}$ is *not* reflexive because $(2, 2)$ is not in R.

If $\triangle ABC \cong \triangle DEF$, then $\triangle DEF \cong \triangle ABC$. The relation $\cong$ on the set of all triangles is a **symmetric relation**.

Definition of Symmetric Relation

Let R be a relation on set A. For all x and y in A, R is said to be symmetric if, for any (x, y) in R, (y, x) is also in R.

Example

2 Give some examples of symmetric relations and relations that are not symmetric.

Symmetric
a. The relation *has the same color eyes as* on the set of students in your math class is symmetric. Whenever x has the same color eyes as y, it follows that y has the same color eyes as x.
b. The relation *is similar to* on the set of all polygons is symmetric. For all polygons x and y, whenever $x \sim y$ it follows that $y \sim x$.

Not Symmetric
a. The relation *is less than* on the set of real numbers is not symmetric. For real numbers x and y, if $x < y$, it is never true that $y < x$.

Recall that for all segments $\overline{AB}$, $\overline{CD}$, and $\overline{EF}$, if $\overline{AB} \cong \overline{CD}$ and $\overline{CD} \cong \overline{EF}$, then $\overline{AB} \cong \overline{EF}$. The relation $\cong$ on the set of all segments is a **transitive relation**.

Definition of Transitive Relation

> Let R be a relation on a set A. For all a, b, and c in A, R is said to be transitive if, whenever (a, b) is in R and (b, c) is in R, (a, c) is in R.

Example

3 Give examples of transitive relations and relations that are not transitive.

Transitive
a. The relation *lives on the same block* on the set of all students in your school is a transitive relation. If x lives on the same block as y and y lives on the same block as z, then x must live on the same block as z.
b. The relations *is congruent to* and *is similar to* on the set of all triangles in a plane are transitive relations.

Not Transitive
a. The relation $N = \{(1, 2), (2, 1)\}$ on the set $\{1, 2\}$ is not transitive. Because (2, 1) is in N and (1, 2) is in N, the ordered pair (2, 2) would have to be in N for N to be transitive. Since (2, 2) is not in N, N is not transitive.

Notice that some relations are reflexive, symmetric, and transitive. Such relations are called **equivalence relations**.

Definition of Equivalence Relation

An equivalence relation is a relation that is reflexive, symmetric, and transitive.

Example

4 Give examples of equivalence relations.

a. The relations *is congruent to* and *is similar to* on a set of triangles are equivalence relations since they are reflexive, symmetric, and transitive.
b. The relation *is equal to* on the set of real numbers is an equivalence relation.
c. The relation *is parallel to* on the set of all lines in a plane is an equivalence relation.

Exercises

Max: 1-10, 1-8; Avg: 1-10, 1-8; Min: 1-10, 1-8

Exploratory Let $S = \{2, 3, 4\}$. Suppose $R = \{(2, 2), (2, 3), (3, 4), (3, 3), (4, 4), (3, 2)\}$. Tell whether each statement is *true* or *false*. Justify your answer.

1. R is a relation on S.
2. $2R3$
3. $4R3$
4. R is reflexive.
5. R is symmetric.
6. R is transitive.
7. R is an equivalence relation. For answers to exercises 1-7, see the Teacher's Guide.

Let $S = \{1, 5, 6\}$. Each of the following describes a relation on S. State whether each relation is *reflexive*, *symmetric*, or *transitive*.

8. $R = \{(1, 1), (1, 5), (1, 6), (5, 1), (5, 6)\}$ none
9. $T = \{(1, 1), (5, 6), (6, 5), (1, 5), (6, 1)\}$ none
10. $U = \{(1, 5), (5, 6), (1, 6)\}$ transitive

Written Each of the following describes a relation on the set of whole numbers. For each relation tell whether it is *reflexive*, *symmetric*, *transitive*, or an *equivalence relation*.

1. $<$ transitive
2. $=$ equivalence relation
3. $\leq$ reflexive, transitive
4. is a factor of reflexive, transitive
5. $>$ transitive
6. $\geq$ reflexive, transitive

Let $S = \{$all people in the United States$\}$. Is each relation on S an equivalence relation? Justify your answer. For answers to exercises 7 and 8, see the Teacher's Guide.

7. lives in the same state as
8. has the same weight as

Problem Solving Application: Perimeter And Area

Objective: To solve perimeter and area problems using equations.
Sometimes we need to determine the perimeter or the area of a polygon in order to solve a problem.

Example

1 The Pine City Garden Club wants to build a fence around its rectangular rose garden. The length of the garden is 10 ft more than its width. A 3-foot wide sidewalk surrounds the garden. If the area of the sidewalk is 396 ft², how much fencing is needed to enclose the garden?

Explore Let x represent the width of the garden. Then $x + 10$ represents the length of the garden, $x + 6$ represents the width of the garden and sidewalk, and $x + 16$ represents the length of the garden and sidewalk.

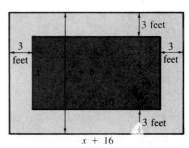

In order to find out how much fencing is needed to enclose the rose garden, we must find the perimeter of the garden.

Plan To determine the dimensions of the garden, write and solve an equation.

area of garden and sidewalk − area of garden = area of sidewalk

$$(x + 16)(x + 6) - (x + 10)x = 396$$

Solve
$$x^2 + 22x + 96 - (x^2 + 10x) = 396$$
$$x^2 + 22x + 96 - x^2 - 10x = 396$$
$$12x = 300$$
$$x = 25$$

The width is 25 ft and the length is 25 + 10 or 35 ft. The perimeter of the garden is 25 + 35 + 25 + 35 or 120 ft. Thus, 120 ft of fencing is needed to enclose the rose garden.

Examine
$$\text{area of garden} + \text{area of sidewalk} = \text{length of garden and sidewalk} \times \text{width of garden and sidewalk}$$

$$25 \times 35 + 396 \stackrel{?}{=} (35 + 6) \times (25 + 6)$$
$$875 + 396 \stackrel{?}{=} 41 \times 31$$
$$1271 = 1271$$

The solution is correct.

Exercises

Max: 1-12, Challenge; Avg: 1-12; Min: 1-9, 11

Written Solve each problem.

1. The length of a rectangle is 4 feet more than twice the width. The perimeter is 116 feet. Find the dimensions of the rectangle. 18 ft by 40 ft

2. The perimeter of a football field is 1040 feet. The length of the field is 120 feet less than 3 times the width. What are the dimensions of the field? 160 ft by 360 ft

3. In a certain isosceles triangle, the third side is 17 cm shorter than either of the congruent sides. If the perimeter is 91 cm, what is the length of the third side? 19 cm

4. The measures of the sides of a triangle are 4x, 2x + 6, and 7x − 9. If the perimeter of the triangle is 62 inches, find the lengths of the three sides. 16., 20., 26 in.

5. To get a square photograph to fit into a square frame, Lynette had to trim a 1-inch strip from each side of the photo. In all, she trimmed off 40 square inches. What were the original dimensions of the photograph? 11 in. by 11 in.

6. A rectangular garden is 5 feet longer than twice its width. It has a 3-foot wide sidewalk on one end and on a side adjacent to that end. The area of the sidewalk is 213 square feet. Find the dimensions of the garden. 21 ft by 47

7. The second side of a triangle is twice the length of the first. The third side is 3 cm less than the second side. What are the lengths of the sides if the perimeter of the triangle is 37 cm? 8 cm, 16 cm, 13 cm

8. The three sides of a triangle have measures that are consecutive odd numbers. What are the lengths of the sides if the perimeter is 87 m? 27 m, 29 m, 31

9. A trapezoid has an area of 162 m² and a height of 12 m. One base is 6 m shorter than twice the length of the other base. Find the length of each base. Use $A = \frac{1}{2}h(b_1 + b_2)$. 7 m

10. A trapezoid has an area of 81 ft² and a height of 9 ft. One base is 14 ft shorter than 3 times the length of the other base. Find the length of each base. 8 ft Use $A = \frac{1}{2}h(b_1 + b_2)$.

11. The length of a rectangle is 20 yards greater than the width. If the length is decreased by 5 yards and the width is increased by 4 yards, the area remains unchanged. Find the original dimensions of the rectangle. 60 yd by 80 yd

12. The length of a rectangle is 7 cm less than twice its width. If the length is increased by 11 cm and the width is decreased by 6 cm, the area decreases by 40 square centimeters. Find the original dimensions of the rectangle. 16 cm by 25 cm

Challenge

Mr. Herrera has a concrete sidewalk built on three sides of his yard as shown at the right. The yard measures 24 by 42 feet. The longer walk is 3 feet wide. The price of the concrete is $22 per square yard, and the total bill is $902. What is the width of the walk on the two remaining sides? 4.5 ft

Portfolio Suggestion

Select some of your work from this chapter that shows how you used a calculator or computer. Place it in your portfolio.

Performance Assessment

Engineers frequently solve problems using scale drawings. The truss shown at the right is drawn to scale. If 1 mm = 0.75 ft, find the dimensions of the truss. **Sample answers: AC = 34.5, BC = 18, AB = 26.25**

Chapter Summary

1. Two polygons are **similar** if their corresponding angles have equal measure and if the ratios of the measures of corresponding sides are equal. (244)
2. The **ratio** of a to b, where $b \neq 0$, is the quotient $a \div b$ or $\frac{a}{b}$. (244)
3. A **proportion** is an equality of two ratios. (244)
4. **Means-Extremes Property:** If $\frac{a}{b} = \frac{c}{d}$, then $ad = bc$. (245)
5. Reciprocal Property: If $\frac{a}{b} = \frac{c}{d}$, then $\frac{b}{a} = \frac{d}{c}$. (248)
6. Interchanging Property: If $\frac{a}{b} = \frac{c}{d}$, then $\frac{a}{c} = \frac{b}{d}$. (248)
7. Addition Property: If $\frac{a}{b} = \frac{c}{d}$, then $\frac{a+b}{b} = \frac{c+d}{d}$. (248)
8. Subtraction Property: If $\frac{a}{b} = \frac{c}{d}$, then $\frac{a-b}{b} = \frac{c-d}{d}$. (248)
9. If $\frac{a}{b} = \frac{b}{d}$, then b is called the **mean proportional** between a and d. (We assume a, b, and d are positive.) (249)
10. **Geometric mean:** If b is the mean proportional between a and c, then $b = \sqrt{ac}$. (249)
11. Axiom 7–1: If two angles of one triangle are congruent to two angles of another triangle, the triangles are similar. (252)

12. Theorem 7–1: Similarity of polygons is reflexive, symmetric, and transitive. (254)
13. Theorem 7–2: If a line is parallel to one side of a triangle and it intersects the other two sides, then it divides them proportionally. (256)
14. Theorem 7–3: If three parallel lines intersect two transversals, they divide them proportionally. (257)
15. Theorem 7–4: The perimeters of similar triangles are proportional to the measures of corresponding sides. (257)
16. Theorem 7–5: If the altitude is drawn to the hypotenuse of a right triangle, the two triangles formed are each similar to the original triangle and to each other. (261)
17. Theorem 7–6: The measure of the altitude to the hypotenuse of a right triangle is the mean proportional between the measures of the segments of the hypotenuse. (262)
18. Theorem 7–7: If the altitude is drawn to the hypotenuse in a right triangle, the measure of each leg is the mean proportional between the measure of the hypotenuse and the measure of the segment on the hypotenuse that is adjacent to the leg. (262)
19. Theorem 7–8: The **Pythagorean Theorem:** In a right triangle, the square of the measure of the hypotenuse is equal to the sum of the squares of the measures of the legs. (265)
20. Theorem 7–9: Converse of the Pythagorean Theorem: If the sum of the squares of the measures of two sides of a triangle is equal to the square of the measure of the third side, then the triangle is a right triangle. (266)
21. Theorem 7–10: In a $45° - 45° - 90°$ triangle, the length of the hypotenuse is $\sqrt{2}$ times the length of a leg. (269)
22. Theorem 7–11: In a $30° - 60° - 90°$ triangle, the length of the hypotenuse is two times the length of the side opposite the angle of degree measure 30; the length of the side opposite the angle of degree measure 60 is $\sqrt{3}$ times the length of the other leg. (269)
23. A circle is **circumscribed** about a polygon if the circle passes through each vertex of the polygon. The polygon is said to be **inscribed** in the circle. (273)
24. A **relation** is a set of **ordered pairs.** (275)
25. Let R be a relation on set A. Then R is said to be **reflexive** if and only if, for every a in A, the ordered pair (a, a) is in R. (278)
26. Let R be a relation on set A. For all x and y in A, R is said to be **symmetric** if, for any (x, y) in R, (y, x) is also in R. (278)
27. Let R be a relation on set A. For all a, b, and c in A, R is said to be **transitive** if, whenever (a, b) is in R and (b, c) is in R, (a, c) is in R. (279)
28. An **equivalence relation** is a relation that is reflexive, symmetric, and transitive. (280)

Chapter Review

7.1 Solve the proportions.

1. $\dfrac{t}{4} = \dfrac{8}{11}$ $\dfrac{32}{11}$
2. $\dfrac{m+38}{50} = \dfrac{m+3}{15}$ 12
3. $\dfrac{x}{9} = \dfrac{1}{x-8}$ 9, -1
4. The measures of two complementary angles are in the ratio of 4:5. Find the measure of each angle. 40, 50

7.2 and 7.3 Refer to the figure. Write a proportion and solve.

5. If $\overline{UV} \parallel \overline{ST}$, $RS = 18$, $US = 12$, and $RT = 15$, find VT. 10

7.4 Find the following measures.

6. Two triangles are similar and the sides of the first triangle measure 3, 4, and 5. If the shortest side of the second triangle measures 15, how long are the other two sides? 20, 25

7.5 In right $\triangle FGH$ with right $\angle H$, $\overline{HJ} \perp \overline{FG}$. Complete the following.

7. $\dfrac{?}{GH} = \dfrac{GH}{?}$ FG, JG
8. $\dfrac{GJ}{HJ} = \dfrac{?}{FJ}$ HJ
9. If $FJ = 3$, $JG = 14$, find JH. $\sqrt{42}$

10. If $FG = 18$, $JG = 12$, find FH. $6\sqrt{3}$
11. If $FH = 5$, $FJ = 3$, find HG. $6\tfrac{2}{3}$

7.6 Find the measure of the side of an equilateral triangle whose altitude has the following measure.

12. $5\sqrt{3}$ 10
13. $\sqrt{3}$ 2
14. $8\sqrt{3}$ 16

7.7 Solve.

15. Sam decides to take a shortcut to school by walking diagonally across a square field that is 180 yards on a side. How many yards shorter is the shortcut than the usual route? 105.44 yd

7.8 Use the following ratios to construct triangles similar to $\triangle ABC$. For exercises 16-18, see students' work.

16. 2:1
17. 3:1
18. 4:3

7.9 Let $S = \{0, 1, 2, 3, 4, 5, 6, 7, 8\}$. Let R be a relation on S such that xRy if and only if x and y both have the same remainder when divided by 3.

19. Write R as a set of ordered pairs. $R = \{(0,0), (0,3), (0,6), (3,3), (3,0), (3,6), (6,6), (6,0), (6,3), (1,1), (1,4), (1,7), (4,4), (4,1), (4,7), (7,7), (7,1), (7,4), (2,2), (2,5), (2,8), (5,5), (5,2), (5,8), (8,8), (8,2), (8,5)\}$

7.10 Is each relation an equivalence relation on the given set? Explain.

20. $\cong$ on {all triangles in a plane}
Yes, because congruence of triangles is reflexive, symmetric, and transitive.

21. $\parallel$ on {all lines in a plane}
Yes, because is parallel to is reflexive, symmetric, and transitive.

Chapter Test

Solve the proportions.

1. $\dfrac{x}{6} = \dfrac{24}{18}$ 8

2. $\dfrac{x}{5} = \dfrac{x+10}{10}$ 10

3. $\dfrac{2x+1}{3x-4} = \dfrac{1}{3} - \dfrac{7}{3}$

4. The measures of two supplementary angles are in the ratio 2:3. Find the measure of each angle. 72, 108

5. Find the measure of an angle that is 20% of the measure of its complement. 15

Refer to the figures. Write a proportion and solve. 36, 64, 96

6. If $\overline{UV} \parallel \overline{RT}$, $RS = 54$, $UR = 18$, and $VT = 32$, find US, SV, and ST.

7. If $\triangle RST \sim \triangle UVW$, find UW when $RT = 8$, $RS = 5$, and $UV = 2$. $3\tfrac{1}{5}$

Exercise 6

Exercise 7

Find the following measures.

8. A triangle has sides that measure 42, 36, and 18. Find the perimeter of a similar triangle whose longest side is 7. 16

9. Two triangles are similar and the sides of the first triangle measure 17, 21, and 30. If the shortest side of the second triangle measures 34, how long are the other two sides? 42, 60

In right $\triangle XYZ$ with right $\angle Y$, $\overline{YW} \perp \overline{XZ}$. Complete the following.

10. $\dfrac{?}{XY} = \dfrac{XY}{?}$ XZ, XW

11. $\dfrac{XW}{YW} = \dfrac{?}{WZ}$ YW

12. If $XW = 4$ and $WZ = 9$, find YW. 6
13. If $XZ = 20$ and $XW = 15$, find YZ. 10
14. If $XW = 4$ and $WZ = 16$, find WY. 8

Write a two-column proof.

15. **Given:** In $\triangle RST$, $\overline{UW} \parallel \overline{ST}$, $\overline{UV} \parallel \overline{RT}$
    Prove: $RW \cdot SV = VT \cdot WT$

For the answer to exercise 15, see the Teacher's Guide.

Find the measure of a side of each polygon.

16. The diagonal of a square is 14 cm long. What is the length of each side of the square? $7\sqrt{2}$ cm

17. The altitude of an equilateral triangle is $7\sqrt{3}$ cm. What is the length of each side of the triangle? 14 cm

Construct $\triangle ABC \sim \triangle RST$ shown above so that each ratio is the ratio of AB to RS.

18. 3:2

19. 4:1

20. 1:2

For exercises 18-20, see students' work.

CHAPTER 8
Geometry with Coordinates

Application in Oceanography

Earth's oceans are the biggest unexplored territories remaining today. Oceanographers use mathematics, chemistry, physics, and biology to study every aspect of the seas. They uncover mysteries and sometimes find new sources of food, energy, medicines, or mineral resources.

The atmospheric pressure on divers increases as they descend into the ocean. Use the chart at the right to write a **linear equation** that approximates the relationship between depth in feet and pressure in pounds per square inch. Then graph the equation on a coordinate plane.
**Sample answer: $y = 0.445x + 14.7$;
See students' graphs.**

Depth (in feet)	Pressure (in lb/in²)
0	14.7
600	269
1200	536
3000	1338
7200	3208
18,000	8019

Group Project: *Economics*

Business executives estimate the cost and revenue of a product before it is produced to determine if they can expect to make a reasonable profit. Have your group choose a product to market. Determine the fixed and variable costs of producing your product. Write an equation for the cost per unit of production. Then determine a price and write an equation for the revenue that will be generated. How much product do you propose making? What will the level of profit be? Present your findings in a business plan which includes charts and graphs that illustrate your points.

8.1 Naming Points in a Plane

Objective: To name and locate points on a plane.

To set up a **coordinate system** in a plane, select two number lines, one horizontal and one vertical. The point where they intersect is called the **origin**. The horizontal number line is usually called the **x-axis**. The vertical number line is usually called the **y-axis**. The four regions of a plane formed by the two axes are called **quadrants**. Each is numbered as shown in the figure at the right.

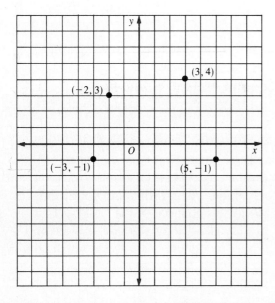

Directions can be given to any point in the plane. Begin at the origin and tell how many units to move in a horizontal direction. Then, tell how many units to move in a vertical direction. These movements are symbolized by an **ordered pair**. For example, the ordered pair (3, 4) names the point three units to the right of, and four units above, the origin. Some other points are named in the figure at the left.

Each number of the ordered pair is called a **coordinate**. The first coordinate of the ordered pair is called the **x-coordinate**, or **abscissa**. The second coordinate is called the **y-coordinate**, or **ordinate**. A point chosen at random is named simply as (x, y).

Finding the point named by an ordered pair is called plotting, or **graphing,** the point. The point itself is the graph of the ordered pair.

If you pick any point in the plane, you can name it with an ordered pair of real numbers. Conversely, if you pick any ordered pair of real numbers, you can find a point in the plane that corresponds to that pair. The system of intersecting perpendicular lines that allows every point to be named as described above is called a **rectangular coordinate system.**

Example

1 Graph the triangle whose vertices are $A(-2, -1)$, $B(1, 4)$, and $C(6, -1)$. Then, find its area.

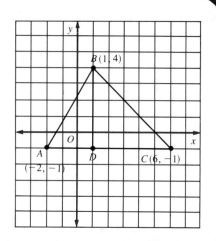

The triangle is shown at the right. By counting, we see that $AC = 8$. The altitude $\overline{BD}$ has a length of 5 units. Using the formula $A = \frac{1}{2}bh$, we have the following.

$$A = \frac{1}{2} \cdot 8 \cdot 5 \quad \text{or} \quad 20$$

The area is 20 square units.

Exercises

Max: 1-14, 1-50; Avg: 1-14, 1-20, 21-49 odds; Min: 1-9 odds, 10-14, 1-37 odds

Exploratory Name the ordered pair for each point on the graph.

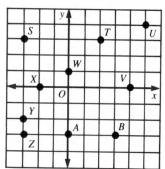

1. S (-3,3)
2. T (2,3)
3. U (5,4)
4. V (4,0)
5. W (0,1)
6. X (-2,0)
7. Y (-3,-2)
8. Z (-3,-3)
9. A (0,-3)
10. B (3,-3)
11. Name the abscissa of B. 3
12. Name the ordinate of X. 0
13. What is true about the abscissa of every point on the y-axis? It is zero.
14. What is true about the ordinate of every point on the x-axis? It is zero.

For the answers to exercises 1-12, see the Teacher's Guide.

Written Locate each point on a coordinate plane. Name them using the letter given.

1. $A(2, 3)$
2. $B(3, 4)$
3. $C(-2, 5)$
4. $D(5, -2)$
5. $E(0, 5)$
6. $F(5, 0)$
7. $G(-3, -4)$
8. $H(0, -3)$
9. $I\left(-1\frac{1}{2}, 3\frac{1}{2}\right)$
10. $J(-10, 5)$
11. $K(0, 2.5)$
12. $L(-\sqrt{3}, \sqrt{2})$

Name three points in each quadrant. Answers may vary.

13. I
14. II
15. III
16. IV

If x and y satisfy the following conditions, tell in what quadrant (x, y) lies.

17. $x > 0 \wedge y > 0$ I
18. $x > 0 \wedge y < 0$ IV
19. $x < 0 \wedge y > 0$ II
20. $x < 0 \wedge y < 0$ III

For the graphs of exercises 21-28, see students' work.

Graph the following pairs of points. Find the distance between them.

21. $(0, 1), (0, 2)$ 1
22. $(6, 2), (6, -3)$ 5
23. $(5, 2), (5, 7)$ 5
24. $(-7, 4), (-7, -2)$ 6
25. $(-5, 4), (6, 4)$ 11
26. $(-10, -3), (20, -3)$ 30
27. $(0, 3), (4, 0)$ 5
28. $(5, 0), (0, 12)$ 13

For the graphs of exercises 29-34, see students' work.

Graph the triangle determined by these vertices. Then, find its area.

29. $A(0, 0), B(4, 0), C(1, 3)$ 6
30. $A(0, 0), B(0, 6), C(5, 0)$ 15
31. $A(-1, 0), B(7, 0), C(0, 4)$ 16
32. $A(3, 1), B(4, 1), C(4, 2)$ 0.5
33. $A(-1, -3), B(-1, 5), C(4, 1)$ 20
34. $A(-1, -2), B(-1, 2), C(-3, -4)$ 4

If the three points are the vertices of a rectangle, find the fourth vertex.

35. $(0, 0), (0, 5), (5, 0)$ (5,5)
36. $(0, 0), (-3, 0), (0, 3)$ (-3,3)
37. $(1, 5), (1, -6), (-3, 5)$ (-3,-6)
38. $(-2, 6), (4, 6), (4, -2)$ (-2,-2)
39. $(-4, -3), (5, -3), (-4, 2)$ (5,2)
40. $(m, n) (m, p), (q, p)$ (q,n)

For the answers to exercises 41-50, see students' work.

Graph a set of four points that satisfies each of the following conditions.

41. The x-coordinate is 3.
42. The y-coordinate is 3.
43. The x-coordinate is -2.
44. The x-coordinate is less than 5.
45. The y-coordinate equals the x-coordinate.
46. The x-coordinate plus the y-coordinate equals 4.
47. The x-coordinate minus the y-coordinate equals 4.
48. The x-coordinate minus the y-coordinate is less than 4.
49. The x-coordinate minus the y-coordinate is more than 2 and less than or equal to 4.
50. The y-coordinate minus the x-coordinate equals 5.

8.2 Graphing Linear Equations

Objective: To graph equations and inequalities.

Look carefully at each of the following ordered pairs.

$$(1, 3) \quad (2, 2) \quad (5, -1) \quad (4, 0) \quad (-2, 6)$$

In each case, the sum of the x-coordinate and y-coordinate is 4. That is, $x + y = 4$.

The graph of all points satisfying the equation $x + y = 4$ appears at the right. Note that the points lie on a straight line. If you select any ordered pair that satisfies the equation $x + y = 4$, the point represented by the orderd pair will lie on this line.

Calculator Hint

You can learn how to graph a linear equation on a graphing calculator in Activity 2 on pages A3-A4.

In general, the graph of any equation in the form $Ax + By = C$ is a straight line. An equation in the form $Ax + By = C$ is called a **linear equation in two variables.**

Definition of a Linear Equation in Two Variables

An equation is a linear equation in two variables if, and only if, it can be written in the form $Ax + By = C$, where A, B, and C are real numbers and A and B are not both 0.

Example

1 Graph $y = 2x - 5$.

Because the equation $y = 2x - 5$ is equivalent to $y - 2x = -5$, its graph is a straight line. We can make a table of values as shown. The equation is graphed at the right.

The x-coordinate of the point where the line crosses the x-axis is called the **x-intercept.** The y-coordinate of the point where the line crosses the y-axis is called the **y-intercept.**

Table of Values

x	y
−1	−7
0	−5
$\frac{1}{2}$	−4
1	−3
2	−1
$\frac{5}{2}$	0

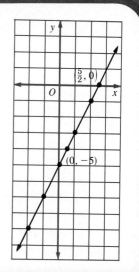

Examples

2 Graph $y = x^2$.

The table of values shown at the right lists some ordered pairs that satisfy $y = x^2$. Note that the graph of these points does not lie on a straight line. The curve pictured here, called a *parabola*, appears often in many applications of mathematics. It will appear again in Chapter 10.

Table of Values

x	y
0	0
$\frac{1}{2}$	$\frac{1}{4}$
$-\frac{1}{2}$	$\frac{1}{4}$
1	1
-1	1
2	4
-2	4
3	9
-3	9

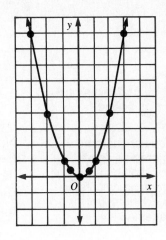

3 Find the values of x and y, if such values exist, that satisfy the equations $x + y = 3$ and $2x + y = 4$.

The two equations $x + y = 3$ and $2x + y = 4$ together are called a **system of equations.**

To solve this system of equations, graph each line on the same set of axes. The point of intersection appears to be (1, 2). Check this value in both equations as follows.

$$x + y = 3 \qquad 2x + y = 4$$
$$1 + 2 \stackrel{?}{=} 3 \qquad 2(1) + 2 \stackrel{?}{=} 4$$
$$3 = 3 \qquad 2 + 2 \stackrel{?}{=} 4$$
$$4 = 4$$

The solution of the system is (1, 2).

4 Graph $2x + y > 2$.

First, graph $2x + y = 2$. This line is shown as a dashed line in the figure at the right. The coordinates of all points in the half-plane that lie above the line satisfy $2x + y > 2$. Therefore, the graph of $2x + y > 2$ is the shaded half-plane above the dashed line.

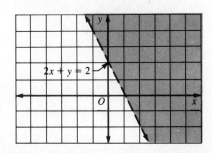

Exercises

Max: 1-13, 1-38; Avg: 1-13, 1-37 odds; Min: 1-13 odds, 1-37 odds

Exploratory State whether each equation is a linear equation in two variables. If it is, write the equation in the form $Ax + By = C$.

yes; 1. $2x + y = 5$ $2x + y = 5$ 2. $x - y = -6$ yes, $x - y = -6$ 3. $y = 2x$ yes; $2x - y = 0$

yes; 4. $x = -y + 4$ $x + y = 4$ 5. $y = |x|$ no   6. $x^2 + y = 2$ no

no 7. $y = x^2 + 1$ yes; 8. $y = -\frac{2}{3}x + 5\frac{2}{3}$ $\frac{2}{3}x + y = 5$ 9. $y = x^3$ no

State which, if any, of the three given points lie on the graph of the given equation.

10. $y = x + 1$ (1, 1), (1, 0), (0, 1)   (0,1)
11. $x + y = 5$ (3, 2), (−1, 6), (0, −5)   (3,2),(-1,6)
12. $y = x^2$ (1, 1), $\left(\frac{1}{2}, \frac{1}{4}\right)$, (4, 2)   (1,1),($\frac{1}{2},\frac{1}{4}$)
13. $x = |y|$ (−1, 1), (1, −1), (5, −5)   (1,-1),(5,-5)

For the answers to exercises 1-18, see the Teacher's Guide and students' work.

Written Graph each of the following.

1. $y = 2x$
2. $y = -3x$
3. $y = x$
4. $x = -3y$
5. $x = 2$
6. $x = -4$
7. $y = 0$
8. $x + y = 6$
9. $x - y = 4$
10. $y = x^2 + 3$
11. $y = x^2 - 4$
12. $y < -2$
13. $x \geq -3$
14. $y \geq 2x$
15. $y + 5 \geq x$
16. $x + 2y \geq 4$
17. $5x - y \geq 16$
18. $|x| + |y| = 4$

Find the x-intercept and y-intercept of the graph of each equation.

19. $x - 3y = 9$ 9; -3   20. $2x + 5y = 10$ 5; -2   21. $y = -6x + 3$ $\frac{1}{2}$; 3
22. $x = -y - 5$ -5; -5   23. $y = -2x$ 0; 0   24. $y = x^2 - 2$ $\sqrt{2}, -\sqrt{2}$; -2

Given each equation and point, find the value of k so that the point lies on the graph of the given equation. For the graphs of exercises 25-38, see students' work.

Sample: $x + y = 5$; (2, k)

If (2, k) lies on the graph of $x + y = 5$, then $2 + k = 5$ or $k = 3$.

25. $x + 2 = y$; (4, k)   6   26. $x - y = 2$; (k, 3)   5
27. $y = 2x + 1$; (−2, k)   -3   28. $y = x^2$; (k, 9)   3, -3
29. $y = |x|$; (−3, k)   3   30. $x = y^2$; (25, k)   5, -5
31. $2x - y = 5$; (4, k)   3   32. $3x + y = 4$; (4, k)   -8

Solve each system of equations by graphing.

33. $x + y = 5$
    $x - y = -1$ {(2, 3)}

34. $y = 4$
    $x - y = -3$ {(1, 4)}

35. $x - y = -3$
    $y = 4x$ {(1, 4)}

36. $x + y = 1$
    $y = 3x + 5$ {(-1, 2)}

37. $x - y = 1$
    $x = 2y + 3$ {(-1, -2)}

38. $2x + y = 7$
    $x - y = 5$ {(4, -1)}

8.3 Algebraic Solutions

Objective: To solve systems of equations using the substitution method and the elimination method.

It is not always practical to solve a system of equations by graphing. Consider the system of equations $x + 2y = 8$ and $y = 2x$. Can you find the solution to this system by graphing?

Systems of equations are most often solved by algebraic methods. Two such methods are the **substitution method** and the **elimination method**.

Examples

1 Use the substitution method to solve the system of equations $x + y = 16$ and $x = 3y$.

The second equation gives a value for x in terms of y. Substitute this value, $3y$, for x in the first equation.

$$x + y = 16$$
$$3y + y = 16$$

Substitute $3y$ for x.

The resulting equation has only one variable, y. Solve the equation.

$$3y + y = 16$$
$$4y = 16$$
$$y = 4$$

Now, find x by substituting 4 for y in the second equation, $x = 3y$.

$$x = 3y$$
$$x = 3(4)$$
$$x = 12$$

Therefore, the solution is $(12, 4)$.

2 Use the elimination method to solve the system of equations $x + 5y = 20.5$ and $-x + 3y = 13.5$.

Add the second equation to the first equation.

$$x + 5y = 20.5$$
$$-x + 3y = 13.5$$
$$8y = 34$$
$$y = 4.25$$

The variable x is eliminated.

Now substitute 4.25 for y in $x + 5y = 20.5$ to find the value of x.

$$x + 5y = 20.5$$
$$x + 5(4.25) = 20.5$$
$$x = -0.75$$

Because $x = -0.75$ and $y = 4.25$, the solution is $(-0.75, 4.25)$.

Example

3 Use the elimination method to solve the system of equations $3x - 2y = -1$ and $2x + 5y = 12$.

In this case, adding or subtracting the two equations will not eliminate a variable. However, suppose both sides of the first equation are multiplied by 2, and both sides of the second equation are multiplied by -3. Then, the system can be solved by adding the equations.

$3x - 2y = -1$ Multiply by 2. $6x - 4y = -2$ Notice that the coeffici-
$2x + 5y = 12$ Multiply by -3. $-6x - 15y = -36$ ents of x are additive inverses.

Now, add to eliminate x. Then, solve for y.

$$\begin{aligned} 6x - 4y &= -2 \\ -6x - 15y &= -36 \\ \hline -19y &= -38 \\ y &= 2 \end{aligned}$$ The variable x is eliminated.

Finally, substitute 2 for y in the second equation. Then, solve for x.

$$\begin{aligned} 2x + 5y &= 12 \\ 2x + 5(2) &= 12 \\ 2x + 10 &= 12 \\ 2x &= 2 \\ x &= 1 \end{aligned}$$

Because $x = 1$ and $y = 2$, the solution is $(1, 2)$.

Exercises

Max: 1-12, 1-58; Avg: 1-11 odds, 1-57 odds; Min: 1-11 odds, 1-35 odds, 47-57 odds

Exploratory Solve each equation for y in terms of x.

1. $y + 3 = x$ $y = x - 3$
2. $y - 6 = 3x$ $y = 3x + 6$
3. $-2x + y = \frac{3}{4}$ $y = 2x + \frac{3}{4}$
4. $2y + 3x = -3$ $y = -\frac{3}{2}x - \frac{3}{2}$
5. $-y = 4x + 2$ $y = -4x - 2$
6. $3x - 2y = 8$ $y = \frac{3}{2}x - 4$

State whether adding or subtracting the two equations will eliminate a variable.

7. $2x + y = 8$ 8. $2x + y = 7$ 9. $\frac{1}{2}x + \frac{1}{4}y = 1$
    adding   neither   adding
   $x - y = 2$ $3x - 2y = 7$ $\frac{1}{3}x - \frac{1}{4}y = 6$

10. $x + 3y = 4$
    $-x + 2y = 1$ adding

11. $3x + 4y = 8$
    $3x + y = 5$ subtracting

12. $x + 2y = 1$
    $2x + y = 1$ neither

Written Solve each system of equations by the substitution method.

1. $y = 8$
   $7x = 1 - y$ (-1, 8)

2. $y = x - 1$
   $4x - y = 19$ (6, 5)

3. $y = x - 4$
   $2x + y = 5$ (3, -1)

4. $x - y = -5$
   $x + y = 25$ (10, 15)

5. $x = y + 10$
   $2y = x - 6$ (14, 4)

6. $3x + 4y = -7$
   $2x + y = -3$ (-1, -1)

7. $9x + y = 20$
   $3x + 3y = 12$ (2, 2)

8. $x + 2y = 5$
   $2x + y = 7$ (3, 1)

9. $y = -2x + 4$
   $x = -2y + 4$ ($\frac{4}{3}$, $\frac{4}{3}$)

10. $2x - y = 7$
    $\frac{3}{4}x - \frac{1}{2}y = 3$ (2, -3)

11. $3x + 5y = 2x$
    $x + 3y = y$ (0, 0)

12. $x + \frac{1}{2}y = 4$
    $\frac{1}{3}x - y = -1$ (3, 2)

Solve each system of equations by the elimination method.

13. $x + y = 7$
    $x - y = 9$ (8, -1)

14. $3x + y = 13$
    $2x - y = 2$ (3, 4)

15. $2x - 3y = -9$
    $-2x - 2y = -6$ (0, 3)

16. $-5 = 4x + 3y$
    $3 = -4x - 2y$ ($\frac{1}{4}$, -2)

17. $2x + 3y = 6$
    $2x - 5y = 22$ (6, -2)

18. $3x + 4y = -7$
    $2x + y = -3$ (-1, -1)

19. $x - y = 6$
    $2x + 3y = 7$ (5, -1)

20. $2x - 5y = 1$
    $3x - 4y = -2$ (-2, -1)

21. $2x + 2y = 8$
    $5x - 3y = 4$ (2, 2)

22. $2x + y = 7$
    $3x - 2y = 7$ (3, 1)

23. $-4x + 3y = -1$
    $8x + 6y = 10$ ($\frac{3}{4}$, $\frac{2}{3}$)

24. $3x + 4y = -25$
    $2x - 3y = 6$ (-3, -4)

Use either the elimination or the substitution method to solve each system of equations.

25. $4x + 3y = 6$
    $-2x + 6y = 7$ ($\frac{1}{2}$, $\frac{4}{3}$)

26. $x + 2y = 400$
    $x - 100 = y$ (200, 100)

27. $x + 5y = 20.25$
    $x + 3y = 13.05$ (2.25, 3.6)

28. $3x - 2y = 7$
    $2x + 5y = 9$ ($\frac{53}{19}$, $\frac{13}{19}$)

29. $4x + 2y = 10.5$
    $2x + 3y = 10.75$ (1.25, 2.75)

30. $x + y = 40$
    $0.2x + 0.45y = 10.5$ (30, 10)

31. $y = x - 4$
    $2x + y = 5$ (3, -1)

32. $y = \frac{1}{2}x + 5$
    $3x + 2y = 2$ (-2, 4)

33. $2x + 3y = 13$
    $2x - 3y = -17$ (-1, 5)

34. $2x + y = 4$
    $6x + 2y = 9$ ($\frac{1}{2}$, 3)

35. $x + y = 20$
    $x - y = 6$ (13, 7)

36. $x + 2y = 6$
    $y = \frac{1}{2}x - 3$ (6, 0)

37. $x - y = 3$
    $2x + y = 4$ ($\frac{7}{3}$, $-\frac{2}{3}$)

38. $2x + y = 7$
    $x - y = 5$ (4, -1)

39. $x - y = -3$
    $y = 4x$ (1, 4)

40. $3x + 10y = 2$
    $x - 2y = 6$ (4, -1)

41. $x + 6y = -1$
    $0.3y - 0.2x = 1.2$ (-5, $\frac{2}{3}$)

42. $y - 6x = 1$
    $5x + y = 12$ (1, 7)

43. $y = 6 - 2x$
    $5x + 3y = -14$ (32, -58)

44. $12x = 48$
    $2x + 2y = 24$ (4, 8)

45. $4x - y = -2$
    $5x + y = 6.5$ (0.5, 4)

For each problem, define two variables. Then use a system of equations to solve the problem.

46. The sum of two numbers is 45. Their difference is 7. Find the numbers. 19, 26
47. The sum of two numbers is 48. Their difference is 24. Find the numbers. 12, 36
48. The length of a rectangle is 5 times its width. The perimeter is 24 cm. Find the dimensions of the rectangle. 2 cm by 10 cm
49. Lindsey is twice as old as Kyung. The sum of their ages is 51 years. Find their ages. Lindsey: 34, Kyung: 17
50. A father is three times as old as his daughter. In 10 years, the father will be twice as old as his daughter. Find their present ages. father: 30, daughter: 10
51. A 140-meter rope is cut into 2 pieces. One piece is four times as long as the other. How long is each piece of rope? 28 m, 112 m
52. The perimeter of a rectangle is 88 cm. Five times its width exceeds twice its length by 3 cm. Find the dimensions of the rectangle. 13 cm by 31 cm
53. Aaron has a total of 14 quarters and dimes. The total value of his coins is $2.15. How many quarters and how many dimes does he have? 5 quarters, 9 dimes
54. Bonnie sold 30 pears for a total of $7.50. She sold small ones for 20¢ each and large ones for 35¢ each. How many of each kind did she sell? 10 large, 20 small
55. The ratio of the length of a rectangle to its width is 8:5. The perimeter of the rectangle is 286 mm. What are the dimensions of the rectangle? 55 mm by 88 mm
56. In parallelogram $ABCD$, $AB = 3x + 1$, $BC = 2x - 1$, $CD = 5y - 3$, and $AD = 3y - 2$. Find AB, BC, CD, and AD.
57. In parallelogram $PQRS$, $m\angle P = 7x$, $m\angle Q = 3y - 5$, and $m\angle R = 11y + 3$. Find $m\angle P, m\angle Q, m\angle R,$ and $m\angle S$.
58. The perimeter of an isosceles trapezoid is 75 cm. The measures of the congruent legs are $8x - 7$ and $3y + 2$. The measures of the bases are $6x$ and $5y - 2$. What are the lengths of the sides of the isosceles trapezoid? 58. 17 cm, 18 cm, 17 cm, 23 cm

56. $AB = CD = 22$, $BC = AD = 13$
57. $m\angle P = m\angle R = 146$, $m\angle Q = m\angle S = 34$

Mixed Review

For each triangle, find the values of x and y.

1.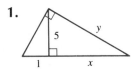
   $x = 25$, $y = 5\sqrt{26}$

2.
   $x = 12$, $y = 6\sqrt{5}$

3.
   $x = 15$, $y = 9$

In the figure at the right, $\overline{UV} \parallel \overline{RT}$ and $\overline{UW} \parallel \overline{ST}$. Solve each problem.

4. If $RS = 27$, $UR = 18$, and $VT = 20$, find ST. 30
5. If $RT = 18$, $RS = 9$, $RU = x + 1$, and $RW = 3x - 1$, find RU and RW. 4, 8
6. If $SU = 2x$, $SR = x + 6$, $SV = x + 5$, $VT = 4$, and $RW = 4$, find RU and RT. 3, 12

7. Find the perimeter of a square whose diagonal is 32 cm long. $64\sqrt{2}$ cm
8. Find the perimeter of an equilateral triangle whose altitude is 27 m long. $54\sqrt{3}$ m

8.4 Distance Between Two Points

Objective: To find the distance between two points and to apply the distance formula.

It is not hard to find the distance between Alice's house at point A and Clara's house at point C. Simply count the number of units and find that $AC = 4$. Similarly, we have no trouble finding that the distance from Clara's house to Beth's house at point B is 3 units.

But how far is it from Alice's house to Beth's house?

Solve this problem by using the Pythagorean Theorem. Clearly, triangle ABC is a right triangle with right angle at C, legs $\overline{AC}$ and $\overline{BC}$, and hypotenuse $\overline{AB}$. Therefore,

$$(AB)^2 = (AC)^2 + (BC)^2$$
$$= 4^2 + 3^2$$
$$= 16 + 9$$
$$= 25$$
$$AB = \sqrt{25} \text{ or } 5 \qquad -5 \text{ is rejected because distance is positive.}$$

The distance from A to B is 5 units.

Example

1 **Find the distance between $D(2, 3)$ and $E(7, 9)$.**

Form right triangle DEF by drawing the horizontal segment DF and the vertical segment EF. The x-coordinate of F is the same as the x-coordinate of E. The y-coordinate of F is the same as the y-coordinate of D. Thus, F has coordinates $(7, 3)$.

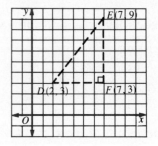

The measure of $\overline{DF}$ is 5 and the measure of $\overline{EF}$ is 6. How were these measures obtained?

Using the Pythagorean Theorem in right $\triangle DEF$, we have the following.

$$\begin{aligned}(DE)^2 &= (DF)^2 + (EF)^2 \\ &= (7-2)^2 + (9-3)^2 \\ &= 5^2 + 6^2 \\ &= 25 + 36 \\ &= 61 \\ DE &= \sqrt{61}\end{aligned}$$

The distance is $\sqrt{61}$ units.

Example 1 suggests the possibility of a general formula for finding the distance between two points.

Consider two general points, A and B. Suppose A has coordinates (x_1, y_1) and B has coordinates (x_2, y_2). The coordinates of C must be (x_2, y_1).

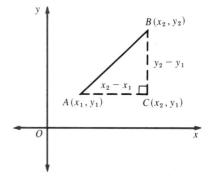

The distance from A to C is $x_2 - x_1$.
The distance from B to C is $y_2 - y_1$.

Using the Pythagorean Theorem in right $\triangle ABC$, we obtain the following.

$$\begin{aligned}(AB)^2 &= (AC)^2 + (BC)^2 \\ &= (x_2 - x_1)^2 + (y_2 - y_1)^2 \\ AB &= \sqrt{(x_2 - x_1)^2 + (y_2 - y_1)^2}\end{aligned}$$

The figure shows points A and B in Quadrant 1. The procedure used above to find AB can be repeated for any two points with coordinates (x_1, y_1) and (x_2, y_2).

Theorem 8–1
Distance Formula

If point A has coordinates (x_1, y_1) and point B has coordinates (x_2, y_2), then the distance from A to B is given by the following formula.

$$AB = \sqrt{(x_2 - x_1)^2 + (y_2 - y_1)^2}$$

Examples

2 Find the distance between $(-5, 4)$ and $(6, 3)$.

Let A be $(-5, 4)$. Then $x_1 = -5$ and $y_1 = 4$.
Let B be $(6, 3)$. Then $x_2 = 6$ and $y_2 = 3$.

$$\begin{align} AB &= \sqrt{(x_2 - x_1)^2 + (y_2 - y_1)^2} \\ &= \sqrt{(6-(-5)^2) + (3-4)^2} \\ &= \sqrt{(6+5)^2 + (-1)^2} \\ &= \sqrt{121 + 1} \\ AB &= \sqrt{122} \end{align}$$

The distance is $\sqrt{122}$ units.

3 Show that the triangle with vertices $(-4, 6)$ $(3, 10)$, and $(-1, 3)$ is isosceles.

Let A be $(-4, 6)$, B be $(3, 10)$, and C be $(-1, 3)$. Find AB, AC, and BC. If two of these are equal, then $\triangle ABC$ is isosceles.

$AB = \sqrt{(3-(-4))^2 + (10-6)^2}$
$AB = \sqrt{65}$
$AC = \sqrt{(-1-(-4))^2 + (3-6)^2}$
$AC = \sqrt{18}$ or $3\sqrt{2}$
$BC = \sqrt{(-1-3)^2 + (3-10)^2}$
$BC = \sqrt{65}$

Because $AB = BC$, $\triangle ABC$ is isosceles.

Calculator Hint

You can enter the expression as it is written on a graphing calculator and press ENTER to find an approximate square root. Then you can press x^2 to find the exact number under the radical.

Exercises

Max: 1-12, 1-40, 1-6; Avg: 1-12; 1-33 odds, 34, 35-39 odds; Min: 1-11 odds 1-27 odds

Exploratory Explain how the Pythagorean Theorem can be used to find the distance between each pair of points. For the answers to exercises 1-3, see the Teacher's Guide.

1. $(2, 1)$, $(5, 9)$
2. $(3, 1)$, $(6, 5)$
3. $(1, 1)$, $(5, 8)$

For the graphs of exercises 4-12, see students' work.
Graph each pair of points and find the distance between them.

4. $(0, 0)$, $(0, 7)$ 7
5. $(3, 2)$, $(3, 7)$ 5
6. $(-2, 5)$, $(4, 5)$ 6
7. $(9, 8)$, $(1, 2)$ 10
8. $(1, 4)$, $(4, 8)$ 5
9. $(-7, 8)$, $(1, 2)$ 10
10. $(-1, -2)$, $(2, -6)$ 5
11. $(0, 0)$, $(-3, 4)$ 5
12. $(9, -7)$, $(13, -5)$ $2\sqrt{5}$

Written Find the distance between each pair of points.

1. $(1, 2), (3, 6)$ $2\sqrt{5}$
2. $(2, 4), (5, 9)$ $\sqrt{34}$
3. $(1, 6), (4, 3)$ $3\sqrt{2}$
4. $(1, 2), (5, -6)$ $4\sqrt{5}$
5. $(3, 7), (-2, 4)$ $\sqrt{34}$
6. $(5, -2), (1, 8)$ $2\sqrt{29}$
7. $(-4, 6), (3, 5)$ $5\sqrt{2}$
8. $(-4, -2), (-1, 6)$ $\sqrt{73}$
9. $(-7, -4), (6, -5)$ $\sqrt{170}$
10. $(-8, -1), (3, -11)$ $\sqrt{221}$
11. $\left(-\frac{1}{2}, 1\frac{1}{2}\right), \left(-\frac{3}{4}, -\frac{1}{2}\right)$ $\frac{1}{4}\sqrt{65}$
12. $(3.2, -4.6), (-5.8, -2.7)$ $\sqrt{84.61}$

Explain.

13. Show that the argument given for Theorem 8–1 holds regardless of the quadrants in which the endpoints of the segment lie. For the answer to exercise 13, see the Teacher's Guide.

For exercises 14-21, measures of the triangle's sides are given.
Show that a triangle with the following vertices is isosceles.

14. $(0, 0), (3, 4), (6, 0)$ 5, 5, 6
15. $(0, 1), (-4, 4), (-8, 1)$ 5, 5, 8
16. $(0, 2), (4, 0), (0, -3)$ $2\sqrt{5}$, 5, 5
17. $(5, -1), (11, -1), (8, 4)$ 6, $\sqrt{34}$, $\sqrt{34}$
18. $(1, 4), (4, 1), (7, 7)$ $3\sqrt{5}$, $3\sqrt{2}$, $3\sqrt{5}$
19. $(-3, 0), (-1, 4), (1, -2)$ $2\sqrt{5}$, $2\sqrt{10}$, $2\sqrt{5}$
20. $(2, 4), (9, 3), (5, 0)$ $5\sqrt{2}$, 5, 5
21. $(-7, 1), (-3, 3), (-5, 7)$ $2\sqrt{10}$, $2\sqrt{5}$, $2\sqrt{5}$

For each set of points, show that quadrilateral ABCD is a parallelogram.

22. $A(0, 0), B(5, 0), C(7, 3), D(2, 3)$ AB = CD = 5, AD = BC = $\sqrt{13}$
23. $A(0, 4), B(5, 7), C(5, 0), D(0, -3)$ AB = CD = $\sqrt{34}$, AD = BC = 7
24. $A(-5, 1), B(0, 2), C(-3, 6), D(-8, 5)$ AB = CD = $\sqrt{26}$, AD = BC = 5
25. $A(1, -4), B(7, 1), C(11, -2), D(5, -7)$ AB = CD = $\sqrt{61}$, AD = BC = 5
26. $A(-3, -4), B(-5, 2), C(2, 4), D(4, -2)$ AB = CD = $2\sqrt{10}$, AD = BC = $\sqrt{53}$
27. $A(-1, 1), B(2, 5), C(5, 1), D(2, -3)$ AB = CD = 5, AD = BC = 5

Find the length of each diagonal for the following exercises.

28. exercise 22
29. exercise 23
30. exercise 24
31. exercise 25
32. exercise 26
33. exercise 27

28. AC = $\sqrt{58}$, BD = $3\sqrt{2}$
29. AC = $\sqrt{41}$, BD = $5\sqrt{5}$
30. AC = $\sqrt{29}$, BD = $\sqrt{73}$
31. AC = $2\sqrt{26}$, BD = $2\sqrt{17}$
32. AC = $\sqrt{89}$, BD = $\sqrt{97}$
33. AC = 6, BD = 8

34. Show that the parallelogram of exercise 27 is a rhombus.
AB = BC = CD = DA = 5 By definition, the parallelogram is a rhombus.

Determine whether or not a triangle with the following vertices is a right triangle.

35. $(0, 0), (3, 3), (5, 1)$ yes
36. $(0, 0), (2, 4), (6, 2)$ yes
37. $(0, 0), (1, 4), (4, 3)$ no
38. $(0, 3), (2, 0), (-3, 2)$ no
39. $(-1, 1), (-3, 4), (2, 3)$ yes
40. $(-4, -2), (-1, 3), (0, 3)$ no

Challenge Write an equation for the set of points in the plane equidistant from each of the following pairs of points.

1. $(1, 1), (4, 3)$ 6x + 4y = 23
2. $(2, 2), (3, 5)$ x + 3y = 13
3. $(-1, 2), (3, 5)$ 8x + 6y = 29
4. $(-2, 3), (3, -4)$ 5x - 7y = 6
5. $(-5, -2), (3, 6)$ x + y = 1
6. $(6, -4), (3, 8)$ 2x - 8y = -7

302 GEOMETRY WITH COORDINATES

Objective: To find the midpoint of a line segment.

8.5 The Midpoint Formula

In the preceding section, we found a formula for the distance between any two points (x_1, y_1) and (x_2, y_2). In this section, a formula for finding the midpoint of a line segment will be used.

Consider $\overline{AB}$ shown at the right. By counting, you can find that its midpoint is $(4, -5)$. In this case, it is clear that the y-coordinate of the midpoint *has* to be -5, because $\overline{AB}$ is a horizontal segment. But how is the x-coordinate, 4, obtained?

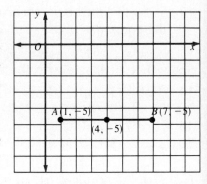

Suppose we take the average of the x-coordinates of A and B. We obtain the following.

$$\frac{1+7}{2} = \frac{8}{2} \text{ or } 4$$

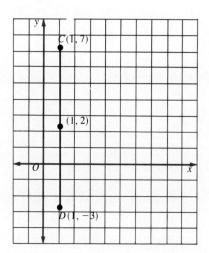

Try this procedure again, this time with a vertical segment. Consider $\overline{CD}$ shown in the diagram. By counting, we see that the midpoint is $(1, 2)$. Because $\overline{CD}$ is a vertical segment, the x-coordinate of the midpoint is 1. The average of the y-coordinates of C and D is found below.

$$\frac{7+(-3)}{2} = \frac{4}{2} \text{ or } 2$$

It appears that the average of the y-coordinates gives the y-coordinate of the midpoint.

Now the coordinates of the midpoint of a segment that is neither horizontal nor vertical can be found.

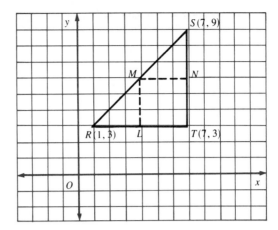

Suppose we are given $\overline{RS}$, as shown. Suppose also that M is the midpoint of $\overline{RS}$, and that we want to find its coordinates.

Let $\overline{MN}$ be parallel to $\overline{RT}$. By Theorem 7–2, $\overline{MN}$ divides both sides of $\triangle RST$ proportionally. Thus, N is the midpoint of $\overline{ST}$. Because $\overline{ST}$ is a vertical segment, the coordinates of N are $\left(7, \dfrac{9+3}{2}\right)$ or $(7, 6)$. But M has the same y-coordinate as N. Therefore, the y-coordinate of M is $\dfrac{9+3}{2}$, or 6. That is, the y-coordinate of M is the average of the y-coordinates of R and S.

Now draw $\overline{ML}$ parallel to $\overline{ST}$. Using Theorem 7–2 again, we conclude that L is the midpoint of horizontal $\overline{RT}$. Therefore, L has coordinates $\left(\dfrac{1+7}{2}, 3\right)$ or $(4, 3)$. But the x-coordinate of M is the same as the x-coordinate of L. Therefore, M has x-coordinate $\dfrac{1+7}{2}$, the average of the x-coordinates of R and S.

Thus, M has coordinates $\left(\dfrac{1+7}{2}, \dfrac{3+9}{2}\right)$ or $(4, 6)$. Because this argument may be repeated for any pair of points (x_1, y_1) and (x_2, y_2), the following theorem results.

Theorem 8–2
The Midpoint Formula

If A has coordinates (x_1, y_1) and B has coordinates (x_2, y_2), then the midpoint M of $\overline{AB}$ has coordinates $\left(\dfrac{x_1 + x_2}{2}, \dfrac{y_1 + y_2}{2}\right).$

Example

1 Find the coordinates of the midpoint of the segment whose endpoints are $(-3, 5)$ and $(4, -7)$.

Let A be the point having coordinates $(-3, 5)$ and B be the point having coordinates $(4, -7)$.

The coordinates of the midpoint, M, are $\left(\dfrac{-3+4}{2}, \dfrac{5+(-7)}{2}\right)$, or $\left(\dfrac{1}{2}, -1\right)$.

Thus, the coordinates of the midpoint of the segment are $\left(\dfrac{1}{2}, -1\right)$.

Exercises

Max: 1-12, 1-42, 1-5; Avg: 1-11 odds, 1-31 odds, 33-42; Min: 1-11 odds, 1-31 odds

Exploratory Find the coordinates of the midpoint of the segment joining each pair of points.

1. $(0, 0), (4, 6)$ (2,3)
2. $(2, 4), (6, 10)$ (4,7)
3. $(5, 2), (3, 8)$ (4,5)
4. $(4, 6), (2, 8)$ (3,7)
5. $(2, 5), (4, 9)$ (3,7)
6. $(4, -2), (6, 8)$ (5,3)
7. $(-7, 2), (3, 12)$ (-2,7)
8. $(-9, 10), (-5, 2)$ (-7,6)
9. $(7, -11), (8, -5)$ $(7\frac{1}{2}, -8)$
10. $(-4, -2), (9, -9)$ $(2\frac{1}{2}, -5\frac{1}{2})$
11. $(-2, 10), (1, -12)$ $(-\frac{1}{2}, -1)$
12. $(-4, -3), (-8, -1)$ (-6,-2)

Written Find the coordinates of the midpoint of the segment joining each pair of points.

1. $(-17, 23), (9, -30)$ $(-4, -3\frac{1}{2})$
2. $\left(-3\dfrac{1}{4}, 5\dfrac{1}{8}\right), \left(2\dfrac{1}{2}, -6\dfrac{3}{4}\right)$ $(-\frac{3}{8}, -\frac{13}{16})$
3. $\left(\dfrac{1}{4}, \dfrac{2}{3}\right), \left(\dfrac{3}{4}, -\dfrac{1}{3}\right)$ $(\frac{1}{2}, \frac{1}{6})$
4. $\left(-\dfrac{1}{2}, 2\right), \left(4, -\dfrac{1}{4}\right)$ $(1\frac{3}{4}, \frac{7}{8})$
5. $(2.3, 1.5), (0.1, 0.3)$ (1.2, 0.9)
6. $(-1.2, 3.4), (-2.4, -3.4)$ (-1.8, 0)
7. $(\sqrt{2}, \sqrt{5}), (-3\sqrt{2}, 7\sqrt{5})$ $(-\sqrt{2}, 4\sqrt{5})$
8. $(2 + \sqrt{3}, 4 + \sqrt{5}), (3 + 3\sqrt{3}, 2 + 3\sqrt{5})$ $(\frac{5}{2} + 2\sqrt{3}, 3 + 2\sqrt{5})$

For exercises 9–18, M is the midpoint of $\overline{AB}$. If A and M have the given coordinates, find the coordinates of B.

9. $A(4, 2), M(2, 1)$ (0,0)
10. $A(5, 3), M(6, 4)$ (7,5)
11. $A(4, 7), M(9, 4)$ (14,1)
12. $A(-1, 5), M(3, 6)$ (7,7)
13. $A(2, -3), M(3, -4)$ (4,-5)
14. $A(3, -4), M(5, -3)$ (7,-2)
15. $A(-2, -5), M(2, -2)$ (6,1)
16. $A(-7, 4), M(-5, -2)$ (-3,-8)
17. $A(-11, -9), M(-7, 7)$ (-3,23)
18. $A(-18, -2), M\left(-3\dfrac{1}{2}, -2\dfrac{1}{2}\right)$ (11,-3)

Find the center of circle P with diameter $\overline{AB}$, given the coordinates of A and B.

19. $A(-1, 4)$, $B(3, -6)$ (1,-1)
20. $A(-4, -6)$, $B(-8, 2)$ (-6,-2)
21. $A(1, 19)$, $B(-7, -5)$ (-3,7)
22. $A(c, d)$, $B(3, f)$ $\frac{c+3}{2}, \frac{d+f}{2}$

If points A, B, and C are the vertices of a triangle, find the midpoint of each side.

23. $A(0, 0)$, $B(4, 6)$, $C(8, 2)$ (2,3),(4,1),(6,4)
24. $A(0, 4)$, $B(2, 6)$, $C(4, -2)$ (1,5),(2,1),(3,2)
25. $A(-1, 5)$, $B(2, 7)$, $C(3, -6)$
26. $A(-2, 8)$, $B(3, 5)$, $C(7, -4)$ $(\frac{1}{2},6\frac{1}{2}),(2\frac{1}{2},2),(5,\frac{1}{2})$
27. $A(5, -2)$, $B(0, -4)$, $C(9, -11)$
28. $A(-11, 12)$, $B(-3, -8)$, $C(5, -12)$
$(\frac{1}{2},6),(1,-\frac{1}{2}),(2\frac{1}{2},\frac{1}{2})$ $(2\frac{1}{2},-3),(7,-6\frac{1}{2}),(4\frac{1}{2},-7\frac{1}{2})$ (-7,2), (-3,0),(1,-10)

For quadrilateral ABCD, determine whether the diagonals of ABCD bisect each other.

29. $A(8, 6)$, $B(5, 5)$, $C(4, 2)$, $D(7, 3)$ yes
30. $A(-2, 6)$, $B(2, 11)$, $C(3, 8)$, $D(-1, 3)$ yes
31. $A(11, 6)$, $B(1, -2,)$, $C(-2, 4)$, $D(3, 8)$ no
32. $A(8, -2)$, $B(3, -5)$, $C(-3, 5)$, $D(2, 8)$ yes

For the answers to exercises 33-36, see the Teacher's Guide.
In exercises 33–36, do the following.

a. Draw the graph of $\triangle ABC$.
b. Find midpoint D of $\overline{AB}$.
c. Find midpoint E of $\overline{AC}$.
d. Find the measure of $\overline{DE}$.
e. Find the measure of $\overline{BC}$.
f. Compare your answers in **d** and **e**.

33. $A(0, 0)$, $B(2, 6)$, $C(6, 2)$
34. $A(0, 4)$, $B(0, -6)$, $C(6, -8)$
35. $A(-4, 0)$, $B(2, 6)$, $C(2, -4)$
36. $A(-6, 1)$, $B(-4, 7)$, $C(2, 10)$

37. Look at your results in exercises 33–36. Draw a conclusion. The measure of a line segment that connects the midpoints of two sides of a triangle is half the measure of the third side.

For the answers to exercises 38-41, see the Teacher's Guide.
In exercises 38–41, do the following.

a. Graph right $\triangle ABC$.
b. Find the coordinates of the midpoint of the hypotenuse.
c. Find the measure of the median to the hypotenuse.
d. Find the measure of the hypotenuse.
e. Compare your answers to **c** and **d**.

38. $A(0, 0)$, $B(0, 3)$, $C(4, 0)$
39. $A(0, 0)$, $B(6, 0)$, $C(0, 2)$
40. $A(-2, -1)$, $B(-2, -5)$, $C(2, -5)$
41. $A(0, 0)$, $B(3, 3)$, $C(6, 0)$

42. Look at your results in exercises 38–41. Draw a conclusion. In a right triangle, the measure of the median to the hypotenuse is half the measure of the hypotenuse.

Challenge Suppose A and C have the following coordinates. If point D lies on $\overline{AC}$ and is one-third of the way from A to C, find the coordinates of D.

1. $A(0, 6)$, $C(9, 12)$ (3,8)
2. $A(0, 2)$, $C(12, 20)$ (4,8)
3. $A(0, -3)$, $C(15, 9)$ (5,1)
4. $A(1, 1)$, $C(7, 10)$ (3,4)

See the Teacher's Guide.
For any two points A and B, prove the following.

5. **Given:** $A(x_1, y_1)$ and $B(x_2, y_2)$
   Prove: Point P with coordinates $\left(\dfrac{bx_1 + ax_2}{a+b}, \dfrac{by_1 + ay_2}{a+b}\right)$ divides $\overline{AB}$ such that $AP : PB = a : b$.

CHAPTER 8 305

306 GEOMETRY WITH COORDINATES

Objective: To find the slope of a line.

8.6 Slope

You are already familiar with the idea of the steepness of a line. For example, the solid line in the graph is clearly "steeper" than the dashed line.

A precise description of the "steepness" of a line can be given. You can find the ratio of the vertical change to the horizontal change as you move between any two points on the line. This ratio is called the **slope** of the line.

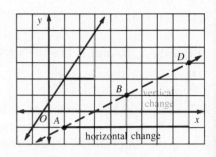

The slope of a line segment is the slope of the line that contains the segment.

$$\text{slope} = \frac{\text{vertical change}}{\text{horizontal change}}$$

Examples

1 Find the slope of the dashed line in the graph above.

The horizontal change from point A to point B is 4. The vertical change is 2. Therefore, the slope is $\frac{2}{4}$, or $\frac{1}{2}$. Note that it does not matter which two points on the line are selected. Moving from A to D, the horizontal change is 8 and the vertical change is 4. The slope is $\frac{4}{8}$, or $\frac{1}{2}$.

2 Find the slope of $\overleftrightarrow{AB}$.

The horizontal change is 6 − 2, or 4 units. The vertical change is 8 − 3, or 5 units. Therefore, the slope is $\frac{5}{4}$. Note that the horizontal change between the two points is the difference of the x-coordinates. Similarly, the vertical change is the difference of the y-coordinates.

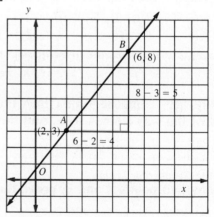

| Definition of Slope | If A has coordinates (x_1, y_1) and B has coordinates (x_2, y_2), and $x_1 \neq x_2$, then the slope m of $\overleftrightarrow{AB}$ is given by the following. $$m = \frac{y_2 - y_1}{x_2 - x_1}$$ |

Example

3 **Find the slope of the line containing $(-5, 7)$ and $(2, -3)$.**

Let $x_1 = -5$, $y_1 = 7$, $x_2 = 2$, and $y_2 = -3$. Therefore, the slope m of the line is as follows.

$$m = \frac{y_2 - y_1}{x_2 - x_1} = \frac{-3 - 7}{2 - (-5)} \text{ or } \frac{-10}{2 + 5} \text{ or } \frac{-10}{7}$$

In the figure, the dashed line has slope $-\frac{3}{2}$. The solid line has slope $\frac{2}{3}$. In general, a line that is *higher* on the right than the left has *positive* slope. A line that is *higher* on the left has *negative* slope. Can a line have slope 0? The next example provides an answer to this question.

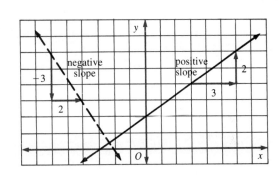

Example

4 **Find the slope of the line shown in the diagram.**

Choose two points on the line, $(-1, 3)$ and $(4, 3)$. Now, use the formula for m.

$$m = \frac{3 - 3}{4 + 1} \text{ or } \frac{0}{5} \text{ or } 0$$

The slope of the line is 0.

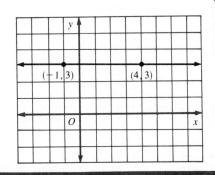

Whenever a line is horizontal, the y-coordinates of all points on the line are equal. Therefore, the numerator of the formula for m, $y_2 - y_1$, will be 0. Hence, a horizontal line has slope 0.

Example

5 Find the slope of the line shown in the figure.

Choose two points $(2, -1)$ and $(2, 5)$.

$$m = \frac{y_2 - y_1}{x_2 - x_1} = \frac{5 + 1}{2 - 2} \quad \text{or} \quad \frac{6}{0}$$

Because division by 0 is not defined, we conclude that the vertical line has no slope.

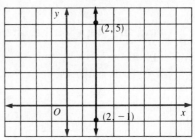

The argument in example 5 can be applied to all vertical lines. Hence, any vertical line has no slope.

Exercises

Max: 1-14, 1-42; Avg: 1-14, 1-39 odds, 40-42; Min: 1-13 odds, 1-39 odds

Exploratory For each line, state whether the slope is *positive, negative, zero,* or *there is no slope*.

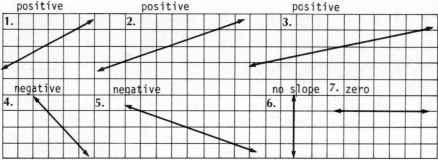

Find the slope of each line.
- **8.** exercise 1 $1\frac{1}{2}$
- **9.** exercise 2 $\frac{1}{3}$
- **10.** exercise 3 $\frac{1}{5}$
- **11.** exercise 4 -1
- **12.** exercise 5 $-\frac{1}{3}$
- **13.** exercise 6 no slope
- **14.** exercise 7 0

For the answers to exercises 1-15, see students' work.

Written Draw a line having the given slope. Use graph paper. Remember to write each slope as a fraction.

1. 2
2. −2
3. 3
4. −1
5. 1
6. $\frac{1}{4}$
7. $-\frac{1}{4}$
8. $\frac{2}{3}$
9. $-\frac{3}{2}$
10. $-\frac{2}{9}$
11. 40%
12. $37\frac{1}{2}$%
13. 0
14. −1.6
15. 0.625

Find the slope of $\overleftrightarrow{AB}$.

16. $A(2, 3), B(4, 5)$ 1
17. $A(2, 7), B(-3, 6)$ $\frac{1}{5}$
18. $A(5, 3), B(4, 0)$ 3
19. $A(6, -1), B(4, -1)$ 0
20. $A(-5, 11), B(-5, -7)$ no slope
21. $A(a, b), B(c, d)$ $\frac{d-b}{c-a}$
22. $A(a, b), B(b - a, a - b)$ $\frac{a-2b}{b-2a}$ See students' work.
23. $A(e + f, e - f), B(e - f, e + f)$ −1

Find two points on each line. Then, find the slope of the line.

24. $x + y = 4$ −1
25. $2x + y = 3$ −2
26. $2x + 4y = 6$ $-\frac{1}{2}$
27. $x = y$ 1
28. $x - 2y = 8$ $\frac{1}{2}$
29. $x = 7$ no slope
30. $y + 3 = 0$ 0
31. $3x - 8 = 0$ no slope
32. $y = \sqrt{2}$ 0

Find the missing coordinate so that the line passing through the two points has the given slope.

33. $(3, 5), (5, y)$; slope $\frac{1}{2}$ 6
34. $(4, y), (6, 4)$; slope $\frac{5}{2}$ −1
35. $(x, 3), (7, 5)$; slope −2 8
36. $(-3, 4), (x, 2)$; slope 4 $-3\frac{1}{2}$
37. $(-9, y), (2, -3)$; slope 0 −3
38. $(x, -8), (-3, 15)$; no slope −3
39. $(-2, k - 1), (k - 1, k)$; slope $-\frac{1}{3}$ K = -4

Graph $\triangle ABC$ given points $A(-2, -4), B(1, 2)$, and $C(5, -1)$. Find the slope of each.

40. median to $\overline{AC}$ -9
41. median to $\overline{AB}$ 0
42. median to $\overline{BC}$ $\frac{9}{10}$

───────── **Mixed Review** ─────────

Solve each system of equations.

1. $x + y = 10$
   $x - y = 14$ (12, -2)
2. $3x - 5y = 7$
   $-2x + y = -9$ $(5\frac{3}{7}, 1\frac{6}{7})$
3. $6x + 13y = -23$
   $4x + 5y = 3$ (7, -5)

For graph to exercise 8, see students' work

4. Determine if the triangle with vertices (1, 3), (8, 2), and (0, −4) is isosceles, right, neither, or both. both
5. The midpoint of $\overline{AB}$ is (5, −4). If the coordinates of A are (−2, 11), what are the coordinates of B? (12, -19)
6. Determine the point on the graph of $3x + 4y = 13$ whose x-coordinate is five less than its y-coordinate. (-1, 4)
7. In $\triangle ABC$, $\angle C$ is a right angle, and the length of $\overline{BC}$ exceeds the length of $\overline{AC}$ by 7 cm. If $AB = 17$, find AC and BC. 8, 15
8. Graph $\triangle ABC$ given $A(1, -3), B(1, 3)$, and $C(-3, 7)$. Then find its area. 12
9. Find the value of k so that (4, 9) lies on the graph of $kx - 2y = k + 3$. 7
10. At Burger Heaven, 2 cheeseburgers and a milkshake cost $3.77. A cheeseburger and 3 milkshakes cost $4.36. Find the cost of one cheeseburger and of one milkshake. cheeseburger: $1.39; milkshake: $0.99

8.7 Parallel and Perpendicular Lines

Objective: To determine if lines are parallel, perpendicular, or neither by comparing their slopes

The lines shown in the figure are all parallel. Find the slope of each line by selecting any two points on each line. The result should suggest the next theorem.

Theorem 8–3 If two nonvertical lines are parallel, they have the same slope.

Consider the converse of Theorem 8-3. That is, if two or more nonvertical lines have the same slope, must they necessarily be parallel? The next theorem states that they must be.

Theorem 8–4 If two nonvertical lines have the same slope, they are parallel.

Example

1 Show that the quadrilateral with vertices $A(0, 0)$, $B(3, 4)$, $C(8, 4)$, and $D(5, 0)$ is a rhombus.

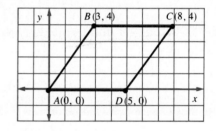

First, show that ABCD is a parallelogram by showing that opposite sides are parallel.

$$\text{slope of } \overline{AB} = \frac{4}{3}$$

$$\text{slope of } \overline{CD} = \frac{4-0}{8-5} \text{ or } \frac{4}{3}$$

Therefore, $\overline{AB} \parallel \overline{CD}$.

Because $\overline{BC}$ and $\overline{AD}$ are both horizontal segments, they are parallel. Therefore, quadrilateral ABCD is a parallelogram.

Now, show that two adjacent sides are congruent.

$AB = \sqrt{(4-0)^2 + (3-0)^2}$ or $\sqrt{16 + 9}$ or $\sqrt{25}$
$AB = 5$
$AD = 5$

Therefore, quadrilateral ABCD is a rhombus.

We know from example 1 that $\square ABCD$ is a rhombus. Hence, the diagonals, $\overline{AC}$ and $\overline{BD}$, are perpendicular.

The slope of $\overline{AC}$ is $\frac{4}{8}$, or $\frac{1}{2}$. The slope of $\overline{BD}$ is $\frac{4-0}{3-5}$ or $\frac{4}{-2}$ or -2. Note that the two slopes, $\frac{1}{2}$ and $-\frac{2}{1}$, are *negative reciprocals* of each other.

The product of the slopes is -1. This happens for any two non-vertical perpendicular lines.

Theorem 8-5

Two nonvertical lines are perpendicular if and only if their slopes, m_1 and m_2, are negative reciprocals of each other. That is, if and only if the following is true.

$$m_1 = \frac{-1}{m_2} \quad \text{or} \quad m_1 m_2 = -1$$

Example

2 Show that the triangle with vertices $R(-1, 2)$, $S(4, -3)$, and $T(-2, -1)$ is a right triangle.

Find the slope of each side of $\triangle RST$. If two sides have slopes whose product is -1, we can conclude that the sides are perpendicular, or that $\triangle RST$ is a right triangle.

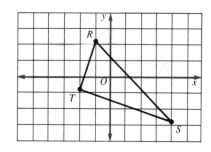

slope of $\overline{RS} = \frac{-3-2}{4-(-1)}$ or $\frac{-5}{4+1}$ or -1

slope of $\overline{ST} = \frac{-1-(-3)}{-2-4}$ or $\frac{-1+3}{-6}$ or $-\frac{1}{3}$

slope of $\overline{RT} = \frac{-1-2}{-2-(-1)}$ or $\frac{-3}{-2+1}$ or 3

Because $-\frac{1}{3} \cdot 3 = -1$, it follows that $\overline{ST} \perp \overline{RT}$. Therefore, $\triangle RST$ is a right triangle.

Exercises

Max: 1-12, 1-50, 1-4; Avg: 1-12, 1-21 odds, 22, 24, 27-49 odds;
Min: 5-11 odds, 1-21 odds, 22, 24, 27-45 odds

Exploratory State what must be true about the slopes of nonvertical lines ℓ and m under the following conditions.

1. ℓ and m are parallel. They are the same.
2. ℓ and m are perpendicular. They are negative reciprocals.
3. ℓ is horizontal. It is zero.
4. ℓ and m intersect. They are not the same.

Suppose line ℓ_1 has slope m_1 and line ℓ_2 has slope m_2. Tell whether ℓ_1 and ℓ_2 are *parallel, perpendicular,* or *neither.*

5. $m_1 = \frac{1}{2}$, $m_2 = 2$ neither
6. $m_1 = -\frac{1}{5}$, $m_2 = 5$ perpendicular
7. $m_1 = 4$, $m_2 = 4$ parallel
8. $m_1 = 0.1$, $m_2 = -10$ perpendicular
9. $m_1 = \frac{3}{5}$, $m_2 = 0.6$ parallel
10. $m_1 = \frac{2}{3}$, $m_2 = -\frac{3}{2}$ perpendicular
11. $m_1 = 0$, $m_2 = 0$ parallel
12. ℓ_1 has no slope, $m_2 = 0$ perpendicular

Written Tell whether $\overrightarrow{AB} \perp \overrightarrow{CD}$, $\overrightarrow{AB} \parallel \overrightarrow{CD}$, or *neither.*

1. $A(3, 5)$, $B(2, 7)$, $C(5, 9)$, $D(4, 11)$ parallel
2. $A(-1, 4)$, $B(2, 6)$, $C(-5, -4)$, $D(-8, -6)$ parallel
3. $A(2, -3)$, $B(4, -5)$, $C(-4, -6)$, $D(10, 8)$ perpendicular
4. $A(-3, 9)$, $B(-1, -7)$, $C(-2, -3)$, $D(-18, -5)$ perpendicular
5. $A(-2, 10)$, $B(1, -5)$, $C(4, 2)$, $D(6, 12)$ neither
6. $A(-12, 18)$, $B(-3, 7)$, $C(-13, -13)$, $D(2, -4)$ neither
7. $A(-6, 15)$, $B(4, -19)$, $C(3, 5)$, $D(8, -12)$ parallel
8. $A(-7, -11)$, $B(6, -10)$, $C(-17, 10)$, $D(-4, 9)$ neither
9. $A(-9, 14)$, $B(-2, 19)$, $C(26, -6)$, $D(16, 8)$ perpendicular
10. $A(-2, 18)$, $B(0, -13)$, $C(24, 21)$, $D(26, -10)$ parallel

Determine if $\triangle ABC$ is a right triangle.

11. $A(-1, 3)$, $B(6, 5)$, $C(4, 8)$ no
12. $A(2, -3)$, $B(-3, 7)$, $C(-7, 2)$ no
13. $A(5, -1)$, $B(4, -6)$, $C(2, -3)$ yes
14. $A(4, -9)$, $B(2, 7)$, $C(20, -7)$ yes
15. $A(-2, 5)$, $B(1, 6)$, $C(2, 9)$ no
16. $A(4, 9)$, $B(1, -3)$, $C(4, -3)$ yes

Classify quadrilateral ABCD as a *parallelogram, rhombus, square,* or *other.*

17. $A(4, 2)$, $B(7, 3)$, $C(8, 6)$, $D(5, 5)$ rhombus
18. $A(-1, 3)$, $B(-2, 6)$, $C(2, 11)$, $D(3, 8)$ parallelogram
19. $A(-4, -3)$, $B(-1, -1)$, $C(1, -4)$, $D(-2, -6)$ square
20. $A(-2, 4)$, $B(3, 8)$, $C(11, 6)$, $D(1, -2)$ other (trapezoid)
21. $A(-3, 5)$, $B(2, 8)$, $C(8, -2)$, $D(3, -5)$ other (rectangle)

Determine if the diagonals of quadrilateral ABCD are perpendicular.

22. exercise 17 yes
23. exercise 18 no
24. exercise 19 yes
25. exercise 20 no
26. exercise 21 no

Find the slope of each side of △ABC.

27. $A(-1, 4)$, $B(2, 5)$, $C(3, -7)$
28. $A(2, -4)$, $B(-3, 2)$, $C(1, 2)$
29. $A(-3, 7)$, $B(5, -2)$, $C(-1, 6)$
30. $A(10, 2)$, $B(-7, 5)$, $C(-4, -9)$

27. $\overline{AB}$: $\frac{1}{3}$; $\overline{AC}$: $-\frac{11}{4}$; $\overline{BC}$: -12
28. $\overline{AB}$: $-\frac{6}{5}$; $\overline{AC}$: -6; $\overline{BC}$: 0
29. $\overline{AB}$: $-\frac{9}{8}$; $\overline{AC}$: $-\frac{1}{2}$; $\overline{BC}$: $-\frac{4}{3}$
30. $\overline{AB}$: $-\frac{3}{17}$; $\overline{AC}$: $\frac{11}{14}$; $\overline{BC}$: $-\frac{14}{3}$

Find the slope of each altitude of △ABC in exercises 27–30.

31. exercise 27 -3; $\frac{4}{11}$; $\frac{1}{12}$
32. exercise 28 $\frac{5}{6}$; $\frac{1}{6}$; no slope
33. exercise 29 $\frac{8}{9}$; 2; $\frac{3}{4}$
34. exercise 30 $\frac{17}{3}$; $-\frac{14}{11}$; $\frac{3}{14}$

Given points $A(2, 3)$, $B(4, 5)$, and $C(-6, -2)$, find the value of k that makes $\overleftrightarrow{AB} \parallel \overleftrightarrow{CD}$.

35. $D(2, k)$ 6
36. $D(k, 4)$ 0
37. $D(-3, k)$ 1
38. $D(2k, k + 2)$ -2

Find the value of k that makes $\overleftrightarrow{AB} \perp \overleftrightarrow{CD}$ in exercises 35–38.

39. exercise 35 -10
40. exercise 36 -12
41. exercise 37 -5
42. exercise 38 $-\frac{10}{3}$

State whether the graph of each equation is parallel to the graph of $y = \frac{1}{3}x + 2$.

43. $y = 3x + 2$ no
44. $3y + x = 5$ no
45. $-x + 3y - 4 = 0$ yes

Classify quadrilateral ABCD completely as possible. Justify your answers.

46. $A(-2, 2)$, $B(-2, 5)$, $C(0, 2)$, $D(1, 5)$ trapezoid
47. $A(2, 4)$, $B(5, 2)$, $C(4, 6)$, $D(7, 4)$ parallelogram
48. $A(3, 1)$, $B(8, 1)$, $C(12, 4)$, $D(7, 4)$ rhombus
49. $A(4, 3)$, $B(8, 10)$, $C(15, 6)$, $D(11, -1)$ square
50. $A(-6, -2)$, $B(-3, -4)$, $C(2, -1)$, $D(8, -5)$ trapezoid

For the answers to exercises 1-4, see the Teacher's Guide.

Challenge Prove.

1. Theorem 8-3
2. Theorem 8-4
3. Theorem 8-5

Mathematical Excursions

Age of Diophantus

The solution to the following riddle is the age of the ancient Greek mathematician Diophantus at the time of his death.

His youth lasted one-sixth of his life. He grew a beard after one-twelfth more. He married after one-seventh more. He had a son five years later. His son lived half as long as his father. Diophantus died four years after his son died.

84 years old

Objective: To write equations for nonvertical lines.

8.8 Equations of Lines

You know that the equation of a straight line can be written in the form $Ax + By = C$, where A and B are not both 0.

The equation of any nonvertical line can be written as

$$y = mx + b.$$

In this form, y stands alone on one side of the equation, m is the coefficient of x, and b is some number.

Example

1 Rewrite $2x + y = 4$ in $y = mx + b$ form.

Subtract $2x$ from both sides to obtain the following.

$y = -2x + 4$ In this case, $m = -2$ and $b = 4$.

An equation in the form $y = mx + b$ has special significance. To see what is special about this form, consider the graph of $y = -2x + 4$ at the right.

From the graph, we see that the slope of the line $y = -2x + 4$ is -2. We also see that the graph crosses the y-axis at the point $(0, 4)$. That is, the y-intercept of the line $y = -2x + 4$ is 4.

x	y
-2	8
-1	6
0	4
1	2
2	0

When the equation of a nonvertical line is written in the form $y = mx + b$,

1. m is the slope of the line.
2. b is the y-intercept of the line.

Examples

2 Find the slope and y-intercept of the line $3x - 2y = 4$. Then graph the line.

Rewrite $3x - 2y = 4$ in $y = mx + b$ form.
$3x - 2y = 4$
$-2y = -3x + 4$
$y = \frac{3}{2}x - 2$

Now m, the slope, is $\frac{3}{2}$. The y-intercept, b, is -2. This information is used to draw the graph of the line. Begin at the point $(0, -2)$. Move two units to the right. Then go three units up. The point reached must be on the line. Draw the line through the two points using a straightedge.

3 **Find the slope and y-intercept of the line $x = -5$.**

The line $x = -5$ is a vertical line. Therefore, its equation cannot be expressed in the form $y = mx + b$. Previously, we learned that vertical lines have no slope. Because the line does not cross the y-axis, it has no y-intercept.

4 **Write an equation of a line with y-intercept 5 that is perpendicular to the line $2x + y = 3$.**

The equation of the required line can be written in the form $y = mx + 5$. First, find m. Because the line is perpendicular to the line $2x + y = 3$, its slope is the negative reciprocal of the slope of $2x + y = 3$.
The equation $2x + y = 3$ is equivalent to $y = -2x + 3$. The slope is -2. Therefore, the line we seek has slope $\frac{1}{2}$. Its equation is $y = \frac{1}{2}x + 5$.

5 **Write an equation of a line that passes through the point $(3, 4)$ and has slope $-\frac{2}{3}$.**

Since the slope is $-\frac{2}{3}$, the equation of the line has the form $y = -\frac{2}{3}x + b$. To find b, use the fact that the line contains the point $(3, 4)$. Substituting $x = 3$ and $y = 4$ in the equation above yields the following.

$$4 = -\frac{2}{3}(3) + b$$
$$4 = -2 + b$$
$$6 = b$$

An equation of the line is

$$y = -\frac{2}{3}x + 6.$$

Here is another method of solution for example 5.
Let $P(x, y)$ be any point on the line. Using the formula for the slope, m, we have the following.

$$\frac{y - 4}{x - 3} = m$$

$$\frac{y - 4}{x - 3} = -\frac{2}{3} \qquad \text{The slope of the line is } -\frac{2}{3}.$$

$$y - 4 = -\frac{2}{3}(x - 3) \qquad \text{Multiply each side by } x - 3.$$

An equation of the line passing through (x_1, y_1) with slope m is $y - y_1 = m(x - x_1)$.

Point-Slope Equation of a Line

The equation $y - y_1 = m(x - x_1)$ is called the **point-slope form** of an equation of a line.

Slope-Intercept Equation of a Line

The equation $y = mx + b$ is called the **slope-intercept form** of an equation of a line.

Example

6 Write an equation of the line through (2, 6) and (1, −5).

Method I Find the slope of the line.

$$m = \frac{-5 - 6}{1 - 2} \quad \text{or} \quad \frac{-11}{-1} \quad \text{or} \quad 11 \qquad y = 11x + b$$

Now, find b using the coordinates of point (2, 6). Substituting $x = 2$ and $y = 6$ yields the following.

$$6 = 11(2) + b$$
$$-16 = b$$

What is the result if (1, −5) is used rather than (2, 6)?

An equation for the line is $y = 11x - 16$.

Method II An equation for the line can also be written in point-slope form, $y - y_1 = m(x - x_1)$, where (x_1, y_1) is a point on the line. Find $m = 11$ by the procedure used in **Method I**. Then, substitute the coordinates of point (1, −5) for x_1 and y_1.

$$y + 5 = 11(x - 1)$$
$$y + 5 = 11x - 11$$
$$y = 11x - 16$$

What is the result if (2, 6) is used rather than (1, −5)?

Example

7 △ABC has vertices A(2, 0), B(8, 6), and C(1, 5). If $\overline{CD}$ is an altitude to side $\overline{AB}$, find the coordinates of D.

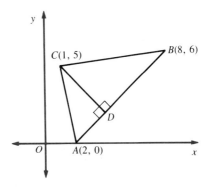

Plan: Find an equation for $\overleftrightarrow{AB}$. Because $\overline{CD} \perp \overline{AB}$, the slope of $\overleftrightarrow{CD}$ is the negative reciprocal of the slope of $\overleftrightarrow{AB}$. Find an equation for $\overleftrightarrow{CD}$. Solve the two equations simultaneously to find the coordinates of D. You will be asked to complete the details as an exercise.

Exercises

Max: 1-10, 1-70; Avg: 1-10, 1-5, 10-14, 19-61 odds; Min: 1-9 odds, 1-5, 10-14, 19-61 odds

Exploratory Rewrite each equation in slope-intercept form.
1. $x + y = 3$ $y = -x + 3$
2. $x + y = -6$ $y = -x - 6$
3. $-x + y = 1$ $y = x + 1$
4. $-3x + y = 5$ $y = 3x + 5$
5. $y = 6$ $y = 6$
6. $2y = 8 - x$ $y = -\frac{1}{2}x + 4$
7. $3y - \frac{1}{2}x = 1$ $y = \frac{1}{6}x + \frac{1}{3}$
8. $\frac{2}{3}x = 5y + 7$ $y = \frac{2}{15}x - \frac{7}{5}$
9. $-5y - 15x = 4$ $y = -3x - \frac{4}{5}$
10. Explain why equations of vertical lines cannot be written in slope-intercept form.
Vertical lines have no slope and no y - intercept.

Written Find the slope and y-intercept of the graph of each equation.
1. $y = x + 2$ 1; 2
2. $y = 2x - 5$ 2; -5
3. $y = -3x + 1$ -3; 1
4. $y = -\frac{1}{3}x - 1$ $-\frac{1}{3}$; -1
5. $y = \frac{3}{5}x + 2$ $\frac{3}{5}$; 2
6. $x + y = 3$ -1; 3
7. $x - y = 4$ 1; -4
8. $3x - 4y = 1$ $\frac{3}{4}$; $-\frac{1}{4}$
9. $x - \frac{1}{2}y = 2$ 2; -4

For the answers to exercises 10-18, see the Teacher's Guide.
Graph each equation using the slope and y-intercept found in exercises 1-9.
10. exercise 1
11. exercise 2
12. exercise 3
13. exercise 4
14. exercise 5
15. exercise 6
16. exercise 7
17. exercise 8
18. exercise 9

Write an equation of the line that has the given slope and y-intercept.
19. slope, -2; y-intercept, 5 $y = -2x + 5$
20. slope, 0; y-intercept, -2 $y = -2$
21. slope, $\frac{1}{2}$; y-intercept, -3 $y = \frac{1}{2}x - 3$
22. slope, $-\frac{3}{5}$; y-intercept, $\frac{1}{2}$ $y = -\frac{3}{5}x + \frac{1}{2}$

Find the slope and one point contained on each line. Points may vary

23. $y - 1 = \frac{1}{2}(x - 3)$ $\frac{1}{2}, (3,1)$
24. $y - 3 = 2(x - 5)$ $2, (5,3)$
25. $y + 3 = -\frac{3}{5}(x + 5)$ $-\frac{3}{5}, (-5,-3)$
26. $y = -\frac{3}{4}(x + 2)$ $-\frac{3}{4}, (-2,0)$
27. $y + 1 = x + 6$ $1, (-6,-1)$
28. $y + 3 = -2(x - 7)$ $-2, (7,-3)$

Write an equation of the line that passes through P and has slope m.

29. $P(1, 5); m = 2$ $y = 2x + 3$
30. $P(-2, 3); m = -3$ $y = -3x - 3$
31. $P(5, 0); m = -2$ $y = -2x + 10$
32. $P(3, -4); m = \frac{1}{2}$ $y = \frac{1}{2}x - \frac{11}{2}$
33. $P(4, -7); m = -\frac{2}{3}$ $y = -\frac{2}{3}x - \frac{13}{3}$
34. $P(-5, -8); m = -\frac{3}{4}$ $y = -\frac{3}{4}x - \frac{47}{4}$

Write an equation of the line that passes through the given pair of points.

35. $(1, 2), (2, 3)$ $y = x + 1$
36. $(2, 13), (8, 13)$ $y = 13$
37. $(0, 0), (4, -2)$ $y = -\frac{1}{2}x$
38. $(0, 1), (1, 5)$ $y = 4x + 1$
39. $(4, 6), (1, 5)$ $y = \frac{1}{3}x + \frac{14}{3}$
40. $(0, 0), (3, 1)$ $y = \frac{1}{3}x$
41. $(3, 3), (-5, -5)$ $y = x$
42. $(8, 6), (-8, -6)$ $y = \frac{3}{4}x$
43. $(3, 1), (100, 1)$ $y = 1$
44. $(0, 0), (x_1, y_1)$ $y = \frac{y_1 x}{x_1}$
45. $(x_1, 0), (x_2, 0)$ $y = 0$
46. $(x_1, y_1), (2, 2)$ $y = \frac{y_1 - 2}{x_1 - 2}(x - 2) + 2$

State whether each pair of lines is *parallel, perpendicular, or neither.*

47. $y = 2x - 3; y = 2x + 5$ parallel
48. $y - x = 5; x - y = -2$ parallel
49. $x + y = 3; x - y = -2$ perpendicular
50. $5x = 4y - 6; 5y - 4x = 3$ neither
51. $x + 2y = 6; y = -\frac{1}{2}x + 3$ parallel
52. $y = \frac{2}{3}x - 6; y = -\frac{3}{2}x + 7$ perpendicular

Find an equation of the line through the given point and parallel to the given line.

53. $(2, 3); y = x + 5$ $y = x + 1$
54. $(1, -2); y = -2x + 7$ $y = -2x$
55. $(-5, -4); 2x + 3y = -1$ $y = -\frac{2}{3}x - \frac{22}{3}$
56. $(0, -5); -2y + 5x = -7$ $y = \frac{5}{2}x - 5$
57. $(2, -1); y = -\frac{1}{2}x + 2$ $y = -\frac{1}{2}x$
58. $(-3, 2); 2x - 3y = 6$ $y = \frac{2}{3}x + 4$

Find an equation of the line through the given point and perpendicular to the given line.

59. exercise 53 $y = -x + 5$
60. exercise 54 $y = \frac{1}{2}x - \frac{5}{2}$
61. exercise 55 $y = \frac{3}{2}x + \frac{7}{2}$
62. exercise 56 $y = -\frac{2}{5}x - 5$

For each $\triangle ABC$ in exercises 63–68, do the following.

a. Find the midpoint D of $\overline{BC}$.
b. Find an equation of $\overleftrightarrow{AD}$.
c. Find the slope of the altitude $\overline{AE}$ to side $\overline{BC}$.
d. Find an equation of $\overleftrightarrow{AE}$.
   For the answers to exercises 63-68, see the Teacher's Guide.
e. Find an equation of $\overleftrightarrow{BC}$.
f. Find the coordinates of point E.

63. $A(-1, 7), B(0, 0), C(6, 6)$
64. $A(-3, 4), B(0, 0), C(2, 4)$
65. $A(7, 6), B(2, 1), C(11, 4)$
66. $A(2, 4), B(-8, 9), C(1, -3)$
67. $A(1, 3), B(-4, 2), C(6, 1)$
68. $A(1, 1), B(-4, 5), C(7, 3)$

Refer to example 7 on page 317.

69. Find the coordinates of D by completing the steps in example 7. $(4,2)$

Objective: To find the area of polygons graphed on a coordinate plane.

8.9 Area Using Coordinates

In this section, the area of some familiar polygons that are graphed on a coordinate plane will be found.

Examples

1 Find the area of the parallelogram whose vertices are $A(-3, -2)$, $B(2, -2)$, $C(4, 1)$, and $D(-1, 1)$.

Parallelogram $ABCD$ is shown here. To find the area of a parallelogram, the lengths of a side and the altitude drawn to that side are needed.

Let $\overline{AB}$ be the base. Then $\overline{DE}$ is the altitude to $\overline{AB}$. By counting, we see that $AB = 5$ and $DE = 3$.

$$A = bh$$
$$= 5 \cdot 3 \quad \text{or} \quad 15$$

The area of $\square ABCD$ is 15 square units.

2 Find the area of the triangle whose vertices are $A(-4, -6)$, $B(4, 2)$, and $C(0, 4)$.

Draw $\triangle ABC$, as shown at the right. Then, "enclose" $\triangle ABC$ in a rectangle as shown.

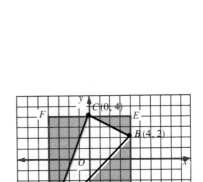

Area of $\triangle ABC =$
Area of rectangle $ADEF$ $-$ Area of $\triangle ACF$ $-$ Area of $\triangle CEB$ $-$ Area of $\triangle ABD$

Area of rectangle $ADEF = AD \cdot DE$ area of rectangle = length · width
$= 8 \cdot 10$
$= 80$

Because △ACF, △CEB, and △ABD are right triangles, we know that the area of each triangle is half the product of the lengths of the two legs.

Area of △ACF $= \frac{1}{2} AF \cdot FC$ or $\frac{1}{2}(10)(4)$ or 20

Area of △CEB $= \frac{1}{2} CE \cdot EB$ or $\frac{1}{2}(4)(2)$ or 4

Area of △ABD $= \frac{1}{2} AD \cdot DB$ or $\frac{1}{2}(8)(8)$ or 32

Area of △ABC $= 80 - 20 - 4 - 32$
$= 24$ The area of △ABC is 24 square units.

Example 3 shows another method for finding the area of △ABC.

Example

3 Find the area of △ABC from example 2 by finding the length of a side and the length of the altitude drawn to that side.

Begin by drawing △ABC as shown.

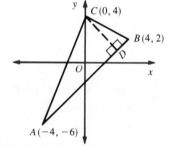

Let $\overline{AB}$ be the base and let $\overline{CD}$ be the altitude from C to $\overline{AB}$. Use the distance formula to find AB.

$AB = \sqrt{(-4-4)^2 + (-6-2)^2}$
$AB = \sqrt{64 + 64}$ or $\sqrt{128}$ or $8\sqrt{2}$

To find CD, first find the coordinates of D. Because D is on $\overleftrightarrow{AB}$ and $\overleftrightarrow{CD}$, the equations for both $\overleftrightarrow{AB}$ and $\overleftrightarrow{CD}$ can be solved simultaneously to find the coordinates of D.

Find an equation for $\overleftrightarrow{AB}$.

$$\text{slope of } \overleftrightarrow{AB} = \frac{-6-2}{-4-4} \text{ or } \frac{-8}{-8} \text{ or } 1$$

Therefore, an equation for $\overleftrightarrow{AB}$ has the form $y = x + b$.

To find b, use the fact that $B(4, 2)$ is on the line.

$$2 = 4 + b$$
$$-2 = b$$

Therefore, an equation of $\overleftrightarrow{AB}$ is $y = x - 2$.

Now, find an equation for $\overleftrightarrow{CD}$. Because $\overleftrightarrow{CD} \perp \overleftrightarrow{AB}$, their slopes are negative reciprocals. Therefore, the slope of $\overleftrightarrow{CD}$ is -1. An equation for $\overleftrightarrow{CD}$ has the following form.

$$y = -x + b$$

Because the coordinates of $C(0, 4)$ satisfy the equation,

$$4 = 0 + b \quad \text{or} \quad b = 4.$$

Therefore, an equation of $\overleftrightarrow{CD}$ is $y = -x + 4$.
Now, solve the equations for $\overleftrightarrow{AB}$ and $\overleftrightarrow{CD}$ simultaneously.

$$y = x - 2$$
$$y = -x + 4$$
$$2y = 2$$
$$y = 1$$

Because $y = x - 2$, $1 = x - 2$, or $x = 3$.
Thus, D has coordinates $(3, 1)$.

Now, find CD.

$$\begin{aligned} CD &= \sqrt{(3 - 0)^2 + (1 - 4)^2} \\ &= \sqrt{9 + 9} \quad \text{or} \quad \sqrt{18} \\ &= 3\sqrt{2} \end{aligned}$$

For $\triangle ABC$, $\begin{aligned} A &= \frac{1}{2} AB \cdot CD \\ &= \frac{1}{2}(8\sqrt{2})(3\sqrt{2}) \quad \text{or} \quad \frac{1}{2}(48) \\ &= 24 \end{aligned}$

The area of $\triangle ABC$ is 24 square units.

Exercises

Max: 1-8, 1-54, 1-7; Avg: 1-8, 1-10, 17-32, 37-53 odds; Min: 1-7 odds, 1-26

Exploratory Show that the quadrilateral with the following vertices is a rectangle.

1. (0, 0), (0, 5), (7, 0), (7, 5)
2. (0, 0), (4, 4), (7, 1), (3, −3)
3. (−3, 1), (−1, 5), (5, 2), (3, −2)
4. (−5, −2), (−6, 2), (3, 0), (2, 4)

For exercises 1-4, quadrilaterals are proved rectangular by using slopes to show opposite sides are parallel and adjacent sides are perpendicular.

Find the area of each rectangle.

5. exercise 1 35
6. exercise 2 24
7. exercise 3 30
8. exercise 4 34

Written Show that the quadrilateral with the following vertices is a parallelogram.

1. (0, −4), (0, 3), (5, 0), (5, 7)
2. (−2, 6), (4, 6), (−5, 2), (1, 2)
3. (3, −4), (−2, −3), (2, −10), (7, −11)
4. (−5, −4), (6, −4), (3, 4), (−8, 4)
5. (−7, 5), (2, 5), (6, −4), (−3, −4)

For exercises 1-5, pairs of opposite sides have the same slope, proving that the quadrilaterals are parallelograms.

Find the area of each parallelogram in the following.

6. exercise 1 35
7. exercise 2 24
8. exercise 3 31
9. exercise 4 88
10. exercise 5 81

Find the area of a triangle with the following vertices.

11. (0, 0), (0, 4), (8, 0) 16
12. (0, 0), (18, 0), (2, 10) 90
13. (1, 3), (7, 3), (3, 9) 18
14. (3, −3), (5, −3), (−1, −10) 7
15. (−3, 9), (−3, −6), (3, 2) 45
16. (0, 0), (−4, 4), (2, 2) 8

Show that the quadrilateral with the following vertices is a trapezoid.

17. (0, 0), (2, 3), (10, 3), (11, 0)
18. (0, 0), (3, 4), (8, 4), (10, 0)
19. (−2, 2), (2, 7), (9, 7), (14, 2)
20. (−1, 3), (8, 8), (8, −9), (−1, −2)
21. (−5, −3), (1, 1), (4, 6), (−8, −2)

For exercises 17-21, there is one pair of parallel, noncongruent sides, proving that the quadrilaterals are trapezoids.

Find the area of each trapezoid in the following.

22. exercise 17 $28\frac{1}{2}$
23. exercise 18 30
24. exercise 19 $57\frac{1}{2}$
25. exercise 20 99
26. exercise 21 27

Find the area of a triangle with the following vertices.

27. (1, 1), (4, 5), (9, 3) 13
28. (2, 3), (6, 7), (16, 5) 24
29. (−6, −4), (0, 6), (4, 2) 32
30. (−8, −4), (1, 4), (6, 2) 29
31. (−4, 2), (1, 8), (6, 3) $27\frac{1}{2}$
32. (12, 10), (5, −4), (1, 4) 56
33. (3, −5), (−1, 6), (8, 2) $41\frac{1}{2}$
34. (16, 6), (2, −3), (5, 8) $63\frac{1}{2}$
35. (−2, −10), (5, −4), (7, 3) $18\frac{1}{2}$
36. (−16, −8), (−9, −4), (−2, −6) 21

Complete.

37. Show that quadrilateral ABCD is a trapezoid.
    Use slopes to show $\overline{AB} \parallel \overline{DC}$ but $\overline{AD} \not\parallel \overline{BC}$.

Find the area of each of the following.

38. rectangle GFEA 20
39. △CDG $\frac{1}{2}$
40. △BCF $1\frac{1}{2}$
41. △ABE 8
42. trapezoid ABCD 10

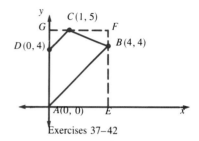

Exercises 37–42

Copy the figure for exercises 37–42. Draw $\overline{BD}$. Find the area of each of the following.

43. △ABD 8
44. △DBC 2
45. trapezoid ABCD 10

Find the area of the trapezoid with the following vertices.

46. (0, 0), (0, 5), (1, 6), (4, 5) 12
47. (0, 0), (0, 6), (1, 8), (3, 6) 12
48. (0, 0), (0, 8), (2, 9), (8, 4) 40
49. (0, 0), (5, 5), (15, 10), (5, 0) $37\frac{1}{2}$
50. (0, 0), (3, 3), (8, 5), (3, 0) 12

Find the area of a quadrilateral with the following vertices.

51. (−1, 0), (−4, 4), (2, 13), (5, 4) $58\frac{1}{2}$
52. (0, 3), (2, 0), (8, 0), (4, 5) 23
53. (−2, 3), (3, 10), (8, 7), (13, 5) 45
54. (−3, −5), (0, 4), (1, 2), (7, −2) $36\frac{1}{2}$

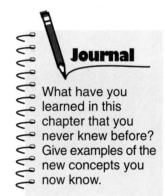

Journal

What have you learned in this chapter that you never knew before? Give examples of the new concepts you now know.

Challenge Refer to the figure for exercises 37–42. Suppose $\overline{DH}$ is the altitude to side $\overline{AB}$. Find each of the following.

1. an equation for $\overleftrightarrow{AB}$ y = x
2. an equation for $\overleftrightarrow{DH}$ y = -x + 4
3. the coordinates of H   (2, 2)
4. the length of $\overline{DH}$ $2\sqrt{2}$
5. the length of $\overline{AB}$ $4\sqrt{2}$
6. the length of $\overline{DC}$ $\sqrt{2}$
7. Using the results of the above exercises and the formula for the area of a trapezoid, find the area of ABCD. 10

Problem Solving Application: Graphs and Estimation

Objective: To solve problems using graphs and estimation.

Some problems can be solved by drawing graphs and then estimating values from the graphs. This strategy is particularly useful when several values can be determined from the same graph.

Example

1 Water freezes at 0°C or 32°F and boils at 100°C or 212°F. Estimate the Celsius equivalent for a Fahrenheit temperature of 100°.

Explore The temperatures 0°C and 32°F are equivalent. The temperatures 100°C and 212°F are also equivalent. A Celsius temperature approximately equivalent to 100°F must be found.

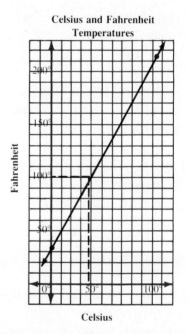

Plan To solve this problem, make a graph with Celsius temperatures on the horizontal axis and Fahrenheit temperatures on the vertical axis. The coordinates of two points on the graph are (0°, 32°) and (100°, 212°). Since this relationship is linear, the line through these two points can be used to estimate other equivalent temperatures.

Solve To estimate the Celsius equivalent temperature for 100°F, locate 100° on the Fahrenheit scale. Follow the value horizontally to the line and then drop vertically to the Celsius scale. The corresponding Celsius temperature appears to be about 38°. Therefore, 100°F is approximately 38°C.

Examine Since 100°F and 38°C are both slightly less than halfway between the freezing temperature and boiling temperature on each scale, the estimate seems reasonable.

Exercises

Max: 1-18; Avg: 1-18; Min: 1-10, 11-17 odds

Written Use the graph on page 324 to estimate the Celsius equivalent for each Fahrenheit temperature.

1. 60°F   15°C
2. 140°F   60°C
3. 90°F   32°C
4. 125°F   51°C

Use the graph on page 324 to estimate the Fahrenheit equivalent for each Celsius temperature.

5. 80°C   176°F
6. 50°C   122°F
7. 10°C   50°F
8. 65°C   149°F

9. Write an equation to represent the relationship between Fahrenheit and Celsius temperature. $F = \frac{9}{5}C + 32$ or $C = \frac{5}{9}(F - 32)$

10. At what temperature are the Celsius and Fahrenheit equivalents the same?   40°

The graph at the right shows the relationship between time and the distance traveled by sound through water. Use this graph to solve each problem.

11. The sonar on a ship finds that it takes sound waves 3.5 seconds to reach the ocean floor. Estimate the depth of the ocean at this point.   about 5 km

12. The sonar on another ship finds that it takes sound waves 6 seconds to reach the ocean floor. Estimate the depth of the ocean at this point.   about 9 km

13. Estimate how many seconds it would take sound waves to reach a depth of 4 km.   about 3 seconds

14. Estimate how many seconds it would take sound waves to reach a depth of 6.5 km.   about 4.5 seconds

15. Write an equation to represent the relationship between time and distance traveled by sound through water. $d = \frac{3}{2}t$ or $t = \frac{2}{3}d$

Use a graph to solve each problem. Assume that the rate is constant in each problem.

16. David cuts lawns during the summer to earn extra money. After working for $2\frac{1}{2}$ hours, David has earned $10. After working for an additional $1\frac{1}{2}$ hours, David has earned a total of $16. How much is David paid per hour? How much will David earn if he works a total of 6 hours? $4 per hour; $24

17. Larry consulted a chart to see if he was the average weight for his height. The chart gave the weight 130 pounds for a person 60 inches tall. The average weight for a person 66 inches tall was 143 pounds. Larry is 6 feet in height. What should Larry's weight be?   *Hint: Change feet to inches.*   156 pounds

18. When Maria arrives at work there are 50 filled and addressed envelopes on her desk. After she fills and addresses envelopes for 2 hours, she has a total of 190 envelopes ready to mail. How many completed envelopes did Maria have after her first hour at work? How many completed envelopes will Maria have after she works for $3\frac{1}{2}$ hours?

120 envelopes, 295 envelopes

Portfolio Suggestion

Select a homework item or class notes that shows something new you learned in this chapter and place it in your portfolio.

Performance Assessment   See students' graphs.

Draw a line ℓ through $(-1, 2)$ and $(3, 5)$ on a coordinate plane. $\frac{3}{4}$
a. Find the slope of the line.
b. Draw line m through $(-2, 3)$ parallel to line ℓ. Explain how you know it is parallel.   **Same slope as line ℓ**
c. Draw a line n through $(-2, 3)$ perpendicular to ℓ. Explain how you know it is perpendicular.   **Slope is negative reciprocal of slope of ℓ.**

Chapter Summary

1. A **linear equation** in two variables is an equation that can be written in the form $Ax + By = C$ where A, B, and C are real numbers and A and B are not both zero. The graph of linear equation in two variables is a straight line. (291)
2. A **system of equations** is two or more equations that contain the same variables. The solution to a system of equations satisfies each of the equations. (294)
3. Theorem 8–1: The distance between two points (x_1, y_1) and (x_2, y_2) is given by the formula $\sqrt{(x_2 - x_1)^2 + (y_2 - y_1)^2}$. (299)
4. Theorem 8–2: The midpoint of a segment having endpoints (x_1, y_1) and (x_2, y_2) has coordinates $\left(\dfrac{x_1 + x_2}{2}, \dfrac{y_1 + y_2}{2}\right)$. (303)
5. If a line contains the points (x_1, y_1) and (x_2, y_2), then the **slope** of the line is $\dfrac{y_2 - y_1}{x_2 - x_1}$. (307)
6. Theorems 8–3 and 8–4: Two nonvertical lines have the same slope if, and only if, they are parallel and vice versa. (310)
7. Theorem 8–5: Two nonvertical lines are perpendicular if, and only if, their slopes are negative reciprocals of one another. (311)
8. The equation $y - y_1 = m(x - x_1)$ is called the **point-slope form** of an equation of a line. (316)
9. The equation $y = mx + b$ is called the **slope-intercept form** of an equation of a line. (316)

Chapter Review

For the answers to exercises 1-10, see the Teacher's Guide.

8.1 Graph each of the following on a coordinate plane.
1. $A(3, 2)$
2. $B(-1, 4)$
3. $C(0, 5)$
4. $D(-3, 0)$

8.2 Graph each of the following on a coordinate plane.
5. $y = 3x + 2$
6. $y = 2x - 3$
7. $x - y = 1$
8. $2y = x + 4$
9. $3x = y - 6$
10. $3x + 2y = 0$

8.3 Solve each system of equations.
11. $x + y = 6$
    $x - y = 2$ (4,2)
12. $x + y = 4$
    $2x + 3y = 9$ (3,1)
13. $4x + y = 9$
    $3x - 2y = 4$ (2,1)
14. $2y - 3x = 0$
    $x - y + 2 = 0$ (4,6)
15. $2x + 3y = 8$
    $x - y = 2$ ($\frac{14}{5}, \frac{4}{5}$)
16. $6x - 4y = -6$
    $3x + y = 3$ ($\frac{1}{3}, 2$)

8.4 Find the distance between each pair of points.
17. $(1, 2), (4, 6)$ 5
18. $(5, -1), (11, 7)$ 10
19. $(-1, 2), (5, 10)$ 10
20. $(-8, -4), (-3, 8)$ 13
21. $(-4, 2), (4, 17)$ 17
22. $(5, 4), (-3, 8)$ $4\sqrt{5}$

8.5 Find the midpoint of the segment joining each pair of points.
23. $(5, 1), (9, 3)$ (7,2)
24. $(6, 5), (8, -3)$ (7,1)
25. $(2, 0), (0, 12)$ (1,6)
26. $(9, -4), (1, 14)$ (5,5)
27. $(5, -6), (-7, 3)$ $(-1, -\frac{3}{2})$
28. $(2, 5), (11, 2)$ $(\frac{13}{2}, \frac{7}{2})$

8.6 Find the slope of the line segment joining each pair of points.
29. $(3, 1), (5, 4)$ $\frac{3}{2}$
30. $(8, 0), (7, -2)$ 2
31. $(6, 1), (6, 8)$ no slope
32. $(0, 0), (5, -2)$ $-\frac{2}{5}$
33. $(6, 4), (-6, -4)$ $\frac{2}{3}$
34. $(0, 0), \left(\frac{1}{3}, \frac{1}{6}\right)$ $\frac{1}{2}$

8.7 Determine whether $\overleftrightarrow{AB}$ and $\overleftrightarrow{CD}$ are *parallel, perpendicular,* or *neither.*
35. $A(1, 2), B(3, 3), C(0, 1), D(-1, 2)$ neither
36. $A(-3, 4), B(-1, 8), C(0, 0), D(-2, 1)$ perpendicular
37. $A(-2, -5), B(2, 4), C(3, -2), D(7, 7)$ parallel
38. $A(1, 1), B(3, 4), C(1, 1), D(4, -1)$ perpendicular

8.8 Write an equation of the line passing through each pair of points. Write the equation in slope-intercept form. 39. $y = 2x - 5$ 40. $y = x - 7$ 41. $y = \frac{1}{5}x + \frac{7}{5}$
39. $(3, 1), (2, -1)$
40. $(5, -2), (3, -4)$
41. $(-2, 1), (3, 2)$
42. $(0, 1), (-5, 2)$ 42. $y = -\frac{1}{5}x + 1$
43. $(-4, 3), (6, -5)$ 43. $y = -\frac{4}{5}x - \frac{1}{5}$
44. $(8, 0), (2, 5)$ 44. $y = -\frac{5}{6}x + \frac{20}{3}$

8.9 Find the area of a triangle with the following vertices.
45. $A(4, 3), B(0, 5), C(5, 5)$ 5
46. $A(3, 6), B(4, 9), C(0, 7)$ 5
47. $A(-2, 3), B(3, -2), C(5, 5)$ $22\frac{1}{2}$
48. $A(0, 0), B(5, 0), C(3, 4)$ 10
49. $A(2, 0), B(2, 4), C(10, 0)$ 16
50. $A(0, 3), B(18, 3), C(2, 13)$ 90

Chapter Test

Graph each equation. For the answers to exercises 1-6, see the Teacher's Guide.
1. $y = 3x$
2. $y + 4 = 0$
3. $y = 2x + 4$
4. $y + \frac{1}{3}x = 9$
5. $y - 2 = \frac{1}{4}(x + 7)$
6. $y = x^2 + 4$

Determine whether △ABC is *isosceles, right, neither,* or *both*.
7. $A(-4, 4)$, $B(-9, 1)$, $C(-4, -2)$ isosceles
8. $A(2, 3)$, $B(4, 6)$, $C(9, 3)$ neither
9. $A(-2, -2)$, $B(-5, -5)$, $C(-1, -9)$ right
10. $A(2, 1)$, $B(6, 5)$, $C(8, -1)$ isosceles

Find the midpoint of each side of the triangle in exercises 7–10. For answers to exercises 11-14, see the Teacher's Guide.
11. exercise 7
12. exercise 8
13. exercise 9
14. exercise 10

Solve each system of equations.
15. $x = 4y$ $(14\frac{2}{3}, 4\frac{1}{3})$
    $2x - 5y = 11$
16. $x - y = 6$
    $x + y = 2$
    $(4, -2)$
17. $3x + 2y = 18$
    $2x - y = -2$
    $(2, 6)$
18. $7x + 4y = -6$
    $-2x + 5y = 57$
    $(-6, 9)$

Write an equation for the line that satisfies the following conditions.
19. slope is $-\frac{1}{3}$; y-intercept is 5 $y = -\frac{1}{3}x + 5$
20. slope is 3; y-intercept is -2 $y = 3x - 2$
21. contains $(2, 2)$; slope is 1 $y = x$
22. contains $(3, 2)$ and $(6, 1)$ $y = -\frac{1}{3}x + 3$

Determine whether $\overleftrightarrow{AB}$ and $\overleftrightarrow{CD}$ are *parallel, perpendicular,* or *neither*.
23. $A(-2, 2)$, $B(-2, 5)$, $C(0, 2)$, $D(1, 5)$ neither
24. $A(2, 4)$, $B(5, 2)$, $C(4, 6)$, $D(7, 4)$ parallel
25. $A(3, 1)$, $B(9, 1)$, $C(14, 1)$, $D(14, 5)$ perpendicular
26. $A(4, 3)$, $B(8, 10)$, $C(0, -4)$, $D(-7, 0)$ perpendicular

Given points $A(5, 8)$, $B(13, 2)$, and $C(6, 1)$, find the value of k so that $\overleftrightarrow{AB} \perp \overleftrightarrow{CD}$.
27. $D(2, k)$ $-4\frac{1}{3}$
28. $D(k, -2)$ $3\frac{3}{4}$
29. $D(2k, k)$ $4\frac{1}{5}$
30. $D(k, k + 9)$ 48

Determine whether quadrilateral ABCD is a parallelogram.
31. $A(-6, 2)$, $B(-2, 5)$, $C(2, 2)$, $D(-2, -1)$ yes
32. $A(3, -4)$, $B(7, -1)$, $C(10, -5)$, $D(6, -8)$ yes
33. $A(1, 7)$, $B(7, 10)$, $C(9, 6)$, $D(3, 3)$ yes
34. $A(-3, 7)$, $B(8, 7)$, $C(10, 4)$, $D(-5, 4)$ no

Find the area of quadrilateral ABCD in each of the following.
35. exercise 31 24
36. exercise 32 25
37. exercise 33 30
38. exercise 34 39

39. The sum of two numbers is 57. Their difference is 11. Find the numbers. 23, 34

40. The perimeter of a rectangle is 86 m. Twice its width exceeds its length by 2 m. Find the width of the rectangle. 15 m

CHAPTER 9

Using Coordinates

Application in Astronomy

On April 25, 1990, the Hubble Space Telescope was deployed from the cargo bay of the space shuttle *Discovery*. In the first pictures sent back to Earth, the stars had diffused halos of light around them. These fuzzy images were caused when the light from the inner and outer edges of the curved mirror were not brought into focus at the optical receptor. Scientists concluded that the curvature of the mirror is not quite correct at its outer edges. New optic instruments that compensate for the incorrect curvature were installed, and the telescope is now functioning properly.

The curved mirror of the Hubble Space Telescope is intended to be shaped like a **parabola**. Follow the steps below to draw a parabola.

1. Draw a line and a point off of the line on a piece of unlined or waxed paper.
2. Fold the paper so that the point falls somewhere on the line. Make a crease. Continue making folds in different places so that the point always falls somewhere on the line.
3. When enough lines have been folded, the shape of the parabola will become visible. Use a pencil to draw a smooth curve.

What is true about the points on each of the lines you folded? Compare this method of drawing a parabola to the definition on page 351. **See students' work.**

Individual Project: *Engineering*

The Gateway Arch in St. Louis, Missouri is the tallest monument in the United States. The 630-foot arch was constructed to honor the westward expansion of the United States. Investigate the curve of the arch. What is the name of its shape? Why was the shape of the arch chosen for the monument? How was the arch constructed? Make a model of the arch. Include the data that you found.

9.1 Using Coordinates to Prove Theorems

Objective: To use coordinates to prove theorems involving quadrilaterals.

You are already familiar with one type of proof in coordinate geometry. This type involves specific numerical coordinates such as (4, 2) and (0, −8).

Example

1 Show that the quadrilateral with vertices $A(4, 9)$, $B(6, 12)$, $C(5, 8)$, and $D(3, 5)$ is a parallelogram.

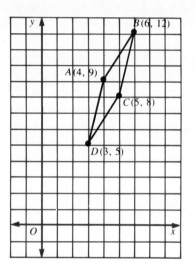

We have to show that the opposite sides of $ABCD$ are parallel.

$$\text{slope of } \overline{AB} = \frac{12 - 9}{6 - 4}$$
$$= \frac{3}{2}$$

$$\text{slope of } \overline{DC} = \frac{8 - 5}{5 - 3}$$
$$= \frac{3}{2}$$

Since the slope of $\overline{AB}$ equals the slope of $\overline{DC}$, $\overline{AB} \parallel \overline{DC}$.

$$\text{slope of } \overline{AD} = \frac{5 - 9}{3 - 4} \qquad \text{slope of } \overline{BC} = \frac{8 - 12}{5 - 6}$$
$$= \frac{-4}{-1} \text{ or } 4 \qquad\qquad = \frac{-4}{-1} \text{ or } 4$$

Since the slope of $\overline{AD}$ equals the slope of $\overline{BC}$, $\overline{AD} \parallel \overline{BC}$. Since the opposite sides of quadrilateral $ABCD$ are parallel, $ABCD$ is a parallelogram.

In another type of coordinate proof, variables are used for coordinates. Since the coordinates are not specific numerical values, geometric theorems can be proved.

The placement of a geometric figure in a coordinate plane and the resulting vertex labels are important. For example, suppose we are to prove a theorem about a rectangle. The figures shown below give three different ways of arranging the coordinate axes relative to a rectangle. Which do you think is the most convenient way?

Figure 1

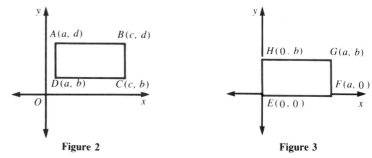

Figure 2   **Figure 3**

The choice of axes in figure 1 does not take advantage of the fact that the opposite sides of rectangle *RSTU* are congruent or that each angle is a right angle.

The choice of axes in figure 2 is an improvement over that of figure 1. However, if the coordinates of rectangle *ABCD* had to be used in, say, the distance formula, the computations would be tedious.

Figure 3 clearly has the most convenient choice. Any computation involving the coordinates of rectangle *EFGH* will have many zeros.

Example

2   **Given:** Rectangle *EFGH*
       Prove: $\overline{EG} \cong \overline{FH}$ (The diagonals of a rectangle are congruent.)

Choose axes as shown. Then use the distance formula to find *EG* and *FH*.

$$EG = \sqrt{(a-0)^2 + (b-0)^2}$$
$$= \sqrt{a^2 + b^2}$$

$$FH = \sqrt{(0-a)^2 + (b-0)^2}$$
$$= \sqrt{a^2 + b^2}$$

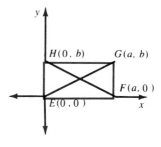

Since *EG* = *FH*, we can conclude that $\overline{EG} \cong \overline{FH}$.

Example 3 demonstrates a general method of positioning and labeling a parallelogram in the coordinate plane.

Examples

3 Show that the quadrilateral with vertices $A(0, 0)$, $B(b, c)$, $C(a + b, c)$, and $D(a, 0)$ is a parallelogram.

Show that the opposite sides of quadrilateral $ABCD$ are parallel by showing that they have the same slope.

slope of $\overline{AB} = \dfrac{c - 0}{b - 0}$ or $\dfrac{c}{b}$

slope of $\overline{DC} = \dfrac{c - 0}{a + b - a}$ or $\dfrac{c}{b}$

slope of $\overline{BC} = \dfrac{c - c}{a + b - b} = \dfrac{0}{a}$ or 0

slope of $\overline{AD} = \dfrac{0 - 0}{a - 0} = \dfrac{0}{a}$ or 0

Since $\overline{AB}$ and $\overline{DC}$ have equal slopes, $\overline{AB} \parallel \overline{DC}$. Likewise, $\overline{BC}$ and $\overline{AD}$ have equal slopes and $\overline{BC} \parallel \overline{AD}$. We conclude that quadrilateral $ABCD$ is a parallelogram.

4 Show that the diagonals of a parallelogram bisect each other.

We will show that the intersection of the two diagonals is also the midpoint of each. Therefore, each diagonal bisects the other.

1. Position and label parallelogram $ABCD$ as shown at the right.
2. The diagonal $\overline{AC}$ has midpoint $\left(\dfrac{a + b}{2}, \dfrac{c}{2}\right)$.
3. The diagonal $\overline{BD}$ has midpoint $\left(\dfrac{a + b}{2}, \dfrac{c}{2}\right)$.

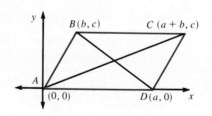

4. Since these midpoints are the same point, it is the intersection of $\overline{AC}$ and $\overline{BD}$. We conclude that the diagonals bisect each other.

Exercises

Max: 1-15, 1-22, 1-2; Avg: 1-15, 1-22; Min: 1-15 odds, 1-9 odds, 10-22

Exploratory Find the missing coordinates for each of the following.

1. C(4,4), D(0,4)

square

2. A(-3,0), C(3,6), D(-3,6)

square with OA = OB

3. B(8,0), D(0,5)

rectangle

4. C(11,5)

parallelogram

5. B(7,0), D(4,5)

parallelogram

6. C(2, 6)

isosceles trapezoid

7.
parallelogram

8.
parallelogram

9.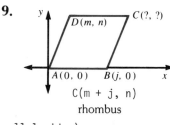
rhombus

Parallelograms can be justified by two pairs of equal slopes (parallel sides).

In each exercise, four points are given. Classify the quadrilateral formed by connecting these points as a *parallelogram*, a *trapezoid*, or *neither*. Justify your answer.

10. (0, 0), (3, 3), (9, 3), (6, 0) Parallelogram
11. (1, 0), (6, 5), (9, 4), (7, −1) neither
12. (−1, 3), (1, 2), (−1, −3), (−3, −2)
13. (−1, −2), (2, 5), (6, 3), (5, −4) neither
14. (−1, −2), (2, 1), (5, 0), (5, −4) trapezoid
15. (−3, 4), (−2, −1), (3, 5), (7, −2) neither
12. parallelogram

Trapezoids are justified by one pair of equal slopes.

Written Find the missing coordinates for each of the following.

1. A(-b,0), C(b,2b), D(-b,2b)

square with OA = OB

2. C(-a,0), D(0,-a)

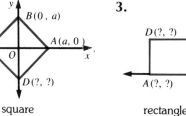
square

3. A(-f,0), B(f,0), D(-f,g)

rectangle with OA = OB

Show that the quadrilateral with the following vertices is a parallelogram.

4. (0, 0), (5, 0), (5, 5), (0, 5)
5. (0, 0), (4, 0), (4, 3), (0, 3)

Parallelograms can be justified by two pairs of equal slopes (parallel sides).

6. (0, 0), (a, 0), (a, a), (0, a)
7. (0, 0), (a, 0), (a, b), (0, b)
8. A(0, 0), B(r, 0), C(r + s, t), D(s, t)
9. R(0, 0), S(a, b), T(a, b + c), U(0, c)

Answer the following.

10. Suppose square WXYZ has vertices W(0, 0) and X(2a, 0). What are the coordinates of Y and Z? (Do not introduce any new variables.) Y(2a,2a), Z(0, 2a)
11. Suppose rectangle PQRS has vertices P(0, 0) and R(2a, 2b). What are the coordinates of Q and S? (Do not introduce any new coordinates.) Q(2a, 0), S(0, 2b)
12. Suppose isosceles trapezoid ABCD with bases $\overline{AB}$ and $\overline{CD}$ has vertices A(0, 0), B(2a, 0), and D(2b, 2c). What are the coordinates of C? C(2a - 2b, 2c)
13. Suppose rectangle JKLM has vertices J(2, 2), K(10, 2), L(10, 6), and M(2, 6). For point P(8, 8), is it true that $(PJ)^2 + (PL)^2 = (PK)^2 + (PM)^2$? yes

Rhombus ABCD has vertices A(0, 0), B(b, c), C(a + b, c), and D(a, 0) where a, b, and c are positive numbers. For the answers to exercises 14 and 17, see the Teacher's Guide.

14. Show that $c^2 = a^2 - b^2$.
15. Find the slope of $\overline{AC}$. $\frac{c}{a + b}$
16. Find the slope of $\overline{BD}$. $\frac{c}{b - a}$
17. Show that $\overline{AC} \perp \overline{BD}$.

Justify each step in the following proof.

18. **Prove:** If two sides of a quadrilateral are congruent and parallel, then the quadrilateral is a parallelogram.

    18c. Definition of slope, Definition of inverse

    a. Assume that $\overline{BC} \cong \overline{AD}$ and $\overline{BC} \parallel \overline{AD}$. Given
    b. The slope of $\overline{AD} = 0$. So the slope of $\overline{BC}$ must be 0. Theorems 8-3 and 8-4
    c. Therefore, $\frac{f - c}{e - b} = 0$. This means $f = c$.
    d. Since AD = a, BC = a as well. Thus, e - b = a, or e = a + b. Substitution; Distance formula
    e. The coordinates of C are (a + b, c). Substitution
    f. The slope of $\overline{CD}$ is, therefore, $\frac{0 - c}{a - (a + b)} = \frac{-c}{-b}$ or $\frac{c}{b}$. Definition of slope
    g. The slope of $\overline{AB}$ is $\frac{c - 0}{b - 0} = \frac{c}{b}$. Definition of slope
    h. Thus, $\overline{AB} \parallel \overline{CD}$ and the theorem is proved. Theorems 8-3 and 8-4

Use coordinate geometry to prove each of the following.

19. The diagonals of a rectangle are congruent.
20. The opposite sides of a parallelogram are congruent.
21. The diagonals of a square are perpendicular.
22. The diagonals of a rhombus are perpendicular.

For answers to exercises 19-22, see the Teacher's Guide.

Challenge Prove the following.

1. If two lines are both perpendicular to the line $y = ax$, $(a \neq 0)$, then the lines are parallel.
2. If the diagonals of a quadrilateral bisect each other, then the quadrilateral is a parallelogram.

For answers to exercises 1 and 2, see the Teacher's Guide.

9.2 More Coordinate Proofs

Objective: To use coordinates to prove theorems involving triangles.

Figure 1

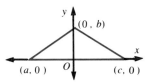

Figure 2

There are several ways of arranging coordinate axes relative to a given triangle.

With the choice of axes in figure 1, one of the vertices has co-ordinates (0, 0). This will simplify computations involving slope or distance.

Sometimes we will want to choose the coordinate system shown in figure 2. This choice of axes is especially useful when working with the altitude of the triangle. We know that the y-axis is perpendicular to the x-axis. There is no reason we cannot place our axes over the triangle to take advantage of that fact.

Example

1 Prove that the line segment joining the midpoints of two sides of a triangle is parallel to the third side.

We will prove this theorem using two methods.

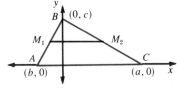

Method 1

The midpoint M_1 of $\overline{AB}$ has coordinates $\left(\frac{b}{2}, \frac{c}{2}\right)$. The midpoint M_2 of $\overline{BC}$ is $\left(\frac{a+b}{2}, \frac{c}{2}\right)$. The slope of the line containing these midpoints is

$$\frac{\frac{c}{2} - \frac{c}{2}}{\frac{a+b}{2} - \frac{b}{2}} \quad \text{or} \quad 0.$$

Since the x-axis also has a slope of 0, we conclude that $\overline{M_1 M_2}$ is parallel to $\overline{AC}$. Thus, the theorem is proved.

Method 2

The midpoint M_1 of $\overline{AB}$ is $\left(\frac{b}{2}, \frac{c}{2}\right)$.

The midpoint M_2 of $\overline{BC}$ is $\left(\frac{a}{2}, \frac{c}{2}\right)$.

The slope of $\overline{M_1 M_2}$ is $\dfrac{\frac{c}{2} - \frac{c}{2}}{\frac{a}{2} - \frac{b}{2}}$ or 0.

Since $a \neq b$, the denominator cannot also be 0.

Since the x-axis also has a slope of 0, we conclude that $\overline{M_1 M_2}$ is parallel to $\overline{AC}$. Thus, the theorem is proved.

The next example gives an even simpler method of proving the theorem in example 1. In this method, we use the fact that *arbitrary* values may be assigned to the coordinates of the vertices of the triangle. Thus, we may prefer to call the coordinates of the triangle $(0, 0)$, $(2a, 0)$, and $(2b, 2c)$ instead of $(0, 0)$, $(a, 0)$, and (b, c).

Examples

2 **Prove the theorem of example 1 by a third method.**

Method 3

Label the vertices of $\triangle ABC$ as shown in the figure.

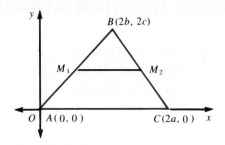

The midpoint M_1 of $\overline{AB}$ is (b, c). The midpoint M_2 of $\overline{BC}$ is $(a + b, c)$.

The slope of $\overline{M_1 M_2}$ is $\dfrac{c - c}{a + b - b}$ or 0.

The rest of the proof is the same as in example 1.

The "trick" of labeling used in Method 3 is often quite useful when proving theorems dealing with midpoints. Can you see why?

3 **In the figure, $\triangle QRS$ is equilateral. Find the coordinates of Q and R.**

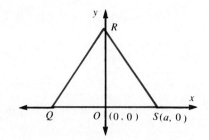

Since $\overline{RO} \perp \overline{QS}$ and $\triangle QRS$ is equilateral, $OQ = OS$.

Therefore, Q has coordinates $(-a, 0)$.

$\triangle ORS$ is a 30°–60°–90° triangle. Since $OS = a$ and $\overline{OR}$ is opposite the 60° angle, $OR = a\sqrt{3}$. Therefore, the coordinates of R are $(0, a\sqrt{3})$.

Explain how to use the Pythagorean Theorem to verify that the coordinates of Q and R are correct.

Exercises

Max: 1-10, 1-22, 1-4; Avg: 1-10, 1-22; Min: 1-9 odds, 1-22

Exploratory Find the missing coordinates for each of the following.

1.
isosceles right triangle

2.
isosceles triangle

3.
equilateral triangle

4.
$B(2a,0)$, $C(2a,b)$, $D(0, b)$
rectangle

5.
$C(2a + \frac{1}{2}c, w)$
parallelogram

6.
$C(4z,d)$
parallelogram

In each exercise, the three given points are the vertices of a triangle. Classify the triangles as *isosceles*, *right*, *both*, or *neither*.

7. (0, 0), (5, 2), (3, 7) both
8. (4, 5), (4, 8), (7, 5) both
9. (1, 6), (−2, 3), (5, 4) neither
10. (3, 5), (6, 9), (10, 6) both

Written For $\triangle PQR$, find the following.

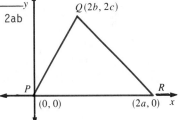

1. PQ $2\sqrt{b^2 + c^2}$
2. PR $2a$
3. QR $2\sqrt{a^2 + b^2 + c^2 - 2ab}$
4. the midpoint of $\overline{PQ}$ (b, c)
5. the midpoint of $\overline{PR}$ $(a, 0)$
6. the midpoint of $\overline{QR}$ $(a + b, c)$

For answers to exercises 7 and 8, see the Teacher's Guide.
Triangle ABC has vertices $A(-3, 5)$, $B(1, 7)$, and $C(4, 3)$.
E is the midpoint of $\overline{AB}$ and F is the midpoint of $\overline{BC}$.

7. Show that $\overline{EF} \parallel \overline{AC}$.
8. Show that $EF = \frac{1}{2}AC$.

For $\triangle KLM$, find the following. 10. $\sqrt{b^2 + c^2 + d^2 - 2bd}$

9. KL $\sqrt{b^2 + c^2}$
10. LM
11. KM d
12. the midpoints of each side
13. the measure of the median to $\overline{KM}$ 13. $\sqrt{b^2 + c^2 + \frac{d^2}{4} - db}$
14. How would a change in coordinates for the vertices of triangle KLM make the computation in exercise 13 simpler? Using coordinates like $L(2b, 2c)$ and $M(2d, 0)$ would eliminate fractions.

12. midpoint of $\overline{KL}$, $(\frac{b}{2}, \frac{c}{2})$; midpoint of $\overline{LM}$, $(\frac{b+d}{2}, \frac{c}{2})$; midpoint of $\overline{KM}$, $(\frac{d}{2}, 0)$

Justify each step in the following proof.

15. Prove: The measure of the segment joining the midpoints of two sides of a triangle is one-half the measure of the third side.

a. The midpoint M_1 of $\overline{BA}$ is $\left(\dfrac{b}{2}, \dfrac{c}{2}\right)$. Midpoint formula

The midpoint M_2 of $\overline{BC}$ is $\left(\dfrac{a+b}{2}, \dfrac{c}{2}\right)$.

b. The measure of $\overline{AC}$ is a. Distance formula

c. The measure of $\overline{M_1 M_2}$ is $\sqrt{\left(\dfrac{a+b}{2} - \dfrac{b}{2}\right)^2 + \left(\dfrac{c}{2} - \dfrac{c}{2}\right)^2}$. Distance formula

d. This expression equals $\sqrt{\left(\dfrac{a}{2} + \dfrac{b}{2} - \dfrac{b}{2}\right)^2 + 0^2}$. Simplifying

e. This equals $\sqrt{\left(\dfrac{a}{2}\right)^2}$ or $\sqrt{\dfrac{a^2}{4}}$. Simplifying

f. This is equal to $\dfrac{1}{2}a$. Simplifying

g. The theorem is proved. Substitution

For answers to exercises 16-22, see the Teacher's Guide.

Answer the following.

16. Why does the argument in exercise 15 prove the case for all three pairs of midpoints?
17. Reprove the theorem in exercise 15, but use different coordinates for the vertices of the triangle.
18. If you had to prove a theorem about an isosceles triangle, how would you set up your coordinate system? Discuss different alternatives.

Use coordinate geometry to prove the following.

19. The midpoint of the hypotenuse of a right triangle is equidistant from all the vertices of the triangle.
20. The triangle formed by connecting the midpoints of the sides of an isosceles triangle is also isosceles.
21. The median drawn to the base of an isosceles triangle is perpendicular to the base of the triangle.
22. The medians drawn to the congruent sides of an isosceles triangle are congruent.

For answers to exercises 1-4, see the Teacher's Guide.

Challenge Use coordinate geometry to prove each of the following.

1. The measure of the median of a trapezoid is equal to one-half the sum of the measure of the bases. The median of a trapezoid is the segment joining the midpoints of the nonparallel sides.
2. The segments joining the midpoints of consecutive sides of a quadrilateral form a parallelogram.
3. The segments joining the midpoints of consecutive sides of an isosceles trapezoid form a rhombus.
4. **Given:** $\square ABCD$
   Prove: $(AB)^2 + (BC)^2 + (CD)^2 + (AD)^2 = (AC)^2 + (BD)^2$

Objective: To describe and draw a locus of points that satisfies certain conditions.

9.3 Locus

In the spring of 1979, an accident at a nuclear power plant in Pennsylvania caused authorities to consider a possible evacuation of everyone living within five miles of the plant. Government officials and journalists illustrated this region by drawing a circle with center at the power plant and a radius of five miles.

We can see that this **circle** is the set of all the points, and only those points, that are five miles from its center.

A special term is used to describe a situation such as this one.

Definition of Locus

> The set of all those points, and only those points, that satisfy a certain condition is called a **locus**. The plural of locus is loci (pronounced *LOW-sigh*).

In the diagram above, the locus of points five units from a given point is a circle with a radius of five units. Unless otherwise specified, we will assume loci are in a plane.

Examples

1 Describe the locus of points equidistant from two given points *A* and *B*.

First, draw two points *A* and *B*.
Then, locate several points that are equidistant from *A* and *B*.
The locus is the perpendicular bisector of $\overline{AB}$.

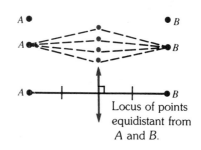

Locus of points equidistant from *A* and *B*.

2 Find the locus of points 2 inches from a given line.

Let the line be line ℓ. As the figure shows, there is a set of points (a line) lying on each side of line ℓ that satisfies the given condition. The locus is two parallel lines.

Examples

3 **Find the locus of points equidistant from two parallel lines.**

The figure at the right shows the situation. No point in the gray-shaded area can be in the locus because every point in that area is clearly farther from one line than the other. We conclude that the required locus is the "line up the middle." It is a line parallel to both of the given lines and at a distance halfway between them.

4 **Find the locus of points equidistant from two intersecting lines.**

Let the two given intersecting lines be ℓ and m. The locus is the pair of lines n and s. Each line bisects an angle formed by the intersecting lines.

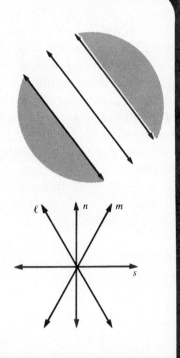

In the preceding examples, it is assumed we are dealing with a locus in a plane. A locus may differ depending on whether the points under consideration are in a plane or in space.

Example

5 **Find the locus of all points in *space* that are 5 cm from a given point.**

Let P be the given point. The locus of points in space that are 5 cm from P is a sphere with center P and a radius of 5 cm.

Exercises

Max: 1-5, 1-17, 1-7; Avg: 1-5, 1-17; Min: 1-5, 1-17 odds

For answers to exercises 1-5, see the Teacher's Guide.

Exploratory Draw and describe the locus for each of the following.

1. points 3 cm from a given point P
2. points 5 cm from a given line k
3. points equidistant from two given points P and Q
4. points equidistant from two parallel lines m and n
5. points equidistant from two intersecting lines r and s

For answers to exercises 1-17, see the Teacher's Guide

Written Draw and describe the locus for each of the following.

1. points 4 cm from a given line
2. points in the interior of an angle that are equidistant from the sides of the angle
3. points equidistant from the four vertices of a square
4. points equidistant from two parallel lines that are 10 cm apart
5. points equidistant from two points 10 cm apart
6. points 4 cm from a circle with a radius of 6 cm
7. points equidistant from two opposite vertices of a square
8. points in the interior of square $ABCD$ that are equidistant from $\overline{AB}$ and $\overline{BC}$
9. all points in the exterior of a 5-inch square that are 2 inches from a side of the square
10. points in *space* 8 cm from a given point
11. points in *space* 5.5 cm from a given point
12. points in *space* equidistant from two parallel planes
13. points in *space* equidistant from two intersecting planes
14. points in *space* 4 cm from a given line

Describe each of the following loci.

15. The path of a swimmer who is swimming in such a way as to always be equally distant from two anchored floats.
16. The path of the center of a bicycle wheel as the wheel moves down a smooth road.
17. A microscopically small nurd-bug walks on the groove of a recording of Beethoven's Fifth Symphony from the outermost groove to the innermost. Describe the bug's path.

For answers to exercises 1-7, see the Teacher's Guide.

Challenge Describe each of the following loci in *space*. Try to draw a picture of the situation.

1. points 4 cm from point A
2. points 4 cm from line C
3. points equidistant from a pair of parallel lines 4 cm apart
4. points 4 cm from plane $\mathcal{A}$

Describe each of the following loci.

5. A small circle with center C rolls along the outside of a larger circle with center Q. What is the path of point C? Make a drawing.
6. Answer exercise 5, if the small circle rolls inside the larger circle.
7. Try to describe the path of a point on the circumference of the smaller circle in exercises 5 and 6.

9.4 Locus, Coordinates, and Circles

Objective: To write an equation of a locus of points that satisfies certain conditions.

If we are asked to describe a locus in a coordinate plane, it is often possible to find an equation of the locus.

Examples

1 Graph the locus of points that are 5 units from the line with equation $y = 3$. Then, find an equation of the locus.

a. The locus of points 5 units from the given line is the two parallel lines shown at the right.

b. One line is 5 units above the line $y = 3$. Its equation is $y = 8$. The other line is 5 units below $y = 3$. Its equation is $y = -2$.

2 Graph the locus of points 5 units from the origin. Then, find an equation of the locus.

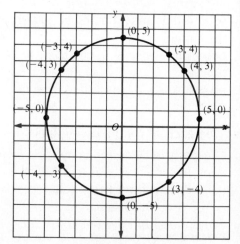

a. This locus is shown at the right. It is a circle with a radius of 5 units and center at (0, 0). Some of the points it passes through are (0, 5), (5, 0), (0, −5), and (−5, 0).

In addition, recall that $3^2 + 4^2 = 5^2$ and that distance is always positive. We see that the circle also passes through (3, 4), (4, 3), (−3, 4), (3, −4), and so on.

b. To find the general equation of all the points on the circle, pick a general point on the circle and call it (x, y). Then use the Pythagorean Theorem.

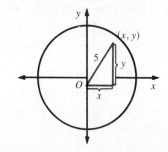

The point (x, y) will be one vertex of a right triangle with a hypotenuse 5 units long and legs x units and y units long. For all such points, $x^2 + y^2 = 5^2$ or 25.

Calculator Hint

You can learn how to use a graphing calculator to graph a circle in Activity 3 on page A5.

In example 2, we saw that the **equation of a circle** with center at the origin and radius 5 is $x^2 + y^2 = 5^2$. The reasoning used in example 2 can be repeated for any circle with center at the origin and radius r.

> The equation of any circle with center at the origin and radius r is $x^2 + y^2 = r^2$.

Example

3 Describe the graph of the equation $x^2 + y^2 = 13$.

The graph of the equation $x^2 + y^2 = 13$ is a circle with center at the origin and a radius of $\sqrt{13}$ units.

Often circles have centers that are not at the origin.

Example

4 Draw the locus of points 4 units from the point (2, 1). Then, find an equation for the locus.

a. The locus is drawn at the right. It is a circle with center at (2, 1) and a radius of 4 units.

b. To find its equation, let a point of the circle be (x, y). Now use the distance formula (Theorem 8–1). The measure of the distance from (x, y) to $(2, 1)$ is
$$\sqrt{(x - 2)^2 + (y - 1)^2}.$$
But this distance equals 4 units. An equation of our circle is
$$\sqrt{(x - 2)^2 + (y - 1)^2} = 4.$$
A less complicated form is obtained by squaring both sides.
$$(x - 2)^2 + (y - 1)^2 = 16$$

Equation of a Circle with Center (h, k) and Radius r

The reasoning used in example 4 can be repeated for any circle with center at (h, k) and radius r.

The equation of any circle with center at (h, k) and radius r is $(x - h)^2 + (y - k)^2 = r^2$.

Examples

5 Describe the graph of the equation $(x + 3)^2 + (y - 14)^2 = 121$.

The equation can be written as $(x - (-3))^2 + (y - 14)^2 = 11^2$. Therefore, its graph is a circle with center at $(-3, 14)$ and a radius of $\sqrt{121}$ or 11 units.

6 Draw the locus of points equidistant from the points $A(1, 4)$ and $B(7, 2)$. Then, find an equation of the locus.

a. The locus of the points equidistant from points A and B is the perpendicular bisector of $\overline{AB}$. It is drawn at the right.

b. *Step 1* If the line is a bisector, it must pass through the midpoint of the segment. Use the Midpoint Formula (Theorem 8–2). This midpoint is $(4, 3)$.

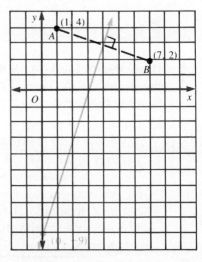

Step 2 If the line is perpendicular to $\overline{AB}$, then its slope must be the negative reciprocal of the slope of $\overline{AB}$, which is $\dfrac{2-4}{7-1}$ or $-\dfrac{1}{3}$. Therefore, the slope of the locus must be 3.

Step 3 The required line passes through $(4, 3)$ and has slope 3. Using the point-slope form of the equation of a line, we have

$$y - y_1 = m(x - x_1)$$
$$y - 3 = 3(x - 4) \quad m = 3 \text{ and } (x_1, y_1) = (4, 3)$$
$$y - 3 = 3x - 12$$
$$y = 3x - 9$$

The equation of the locus is $y = 3x - 9$.

Exercises

Max: 1-10, 1-55, 1-8; Avg: 1-10, 1-55 odds; Min: 1-10, 1-55 odds

Exploratory Find an equation of the locus of each of the following. Use graph paper.
1. points 4 units to the right of the y-axis $x = 4$
2. points 6 units to the left of the y-axis $x = -6$
3. points 10 units above the x-axis $y = 10$
4. points 7 units below the x-axis $y = -7$
5. points 7 units to the right of the line $x = 8$ $x = 15$
6. points 3 units from the x-axis $y = 3$ and $y = -3$
7. points 5 units from the y-axis $x = 5$ and $x = -5$
8. points 2 units from the line $y = 4$ $y = 6$ and $y = 2$
9. points 4 units from the line $x = 8$ $x = 12$ and $x = 4$
10. points 6 units from the line $y = -4$ $y = 2$ and $y = -10$

For answers to exercises 1-6, see the Teacher's Guide.

Written Describe fully the graph of each equation.
1. $x^2 + y^2 = 1$
2. $x^2 + y^2 = 9$
3. $x^2 + y^2 = 4$
4. $x^2 + y^2 = 10$
5. $x^2 + y^2 = 12$
6. $2x^2 + 2y^2 = 50$

Write the equation of the locus of points that are the following distances from the origin.
7. 3 units $x^2 + y^2 = 9$
8. 8 units $x^2 + y^2 = 64$
9. 10 units $x^2 + y^2 = 100$
10. 0.4 units $x^2 + y^2 = 0.16$
11. $2\sqrt{5}$ units $x^2 + y^2 = 20$
12. r units $x^2 + y^2 = r^2$

Which of the following points lie on the locus of points 5 units from the origin?
13. $(-5, 0)$ yes
14. $(0, 25)$ no
15. $\left(12\frac{1}{2}, 0\right)$ no
16. $(3, -4)$ yes
17. $(1, 24)$ no
18. $(-4, -3)$ yes
19. $(2\sqrt{6}, 1)$ yes
20. $(-2\sqrt{5}, \sqrt{5})$ yes
21. $(2, \sqrt{21})$ yes

For the answers to exercises 22-37, see the Teacher's Guide.
Describe fully (center and radius) the circle given by each equation.
22. $(x - 2)^2 + (y - 4)^2 = 25$
23. $(x - 6)^2 + (y - 3)^2 = 64$
24. $(x + 5)^2 + (y - 3)^2 = 121$
25. $(x - 3)^2 + (y + 8)^2 = 5$
26. $(x + 6)^2 + (y + 10)^2 = 18$
27. $x^2 + (y + 7)^2 = 8$

Write an equation of the circle for each of the following.
28. center, (5, 5); radius, 1 unit
29. center, (4, 6); radius, 9 units
30. center, (4, 0); radius, 4 units
31. center, (1, −2); radius, 13 units
32. center, (6, 0); radius, $\sqrt{6}$ units
33. center, (−4, −6); radius, $\sqrt{11}$ units

Find the circumference and area of each circle in exercises 30–33.
34. exercise 30
35. exercise 31
36. exercise 32
37. exercise 33

Find an equation of the locus of points equidistant from each pair of lines. Use graph paper.
38. $x = 2$ and $x = 6$ $x = 4$
39. $x = -5$ and $x = 7$ $x = 1$
40. $y = 4$ and $y = -8$ $y = -2$
41. $x = -2$ and $x = 9$ $x = 3.5$
42. $y = x$ and $y = x + 2$ $y = x + 1$
43. $y = 2x - 3$ and $y = 2x + 5$ $y = 2x + 1$

Find an equation of the locus of points equidistant from each pair of points. Use graph paper.

44. (1, 0) and (7, 0) $x = 4$
45. (−3, 0) and (1, 0) $x = -1$
46. (0, 2) and (0, 6) $y = 4$
47. (3, 5) and (3, 7) $y = 6$
48. (8, 4) and (10, 4) $x = 9$
49. (−3, 5) and (−3, −7) $y = -1$

For each of the following pairs of points:
a. plot points A and B on graph paper;
b. find the midpoint of $\overline{AB}$;
c. find the slope of $\overline{AB}$; and See student graphs for part a of exercises 50-55.
d. find an equation for the locus of points equidistant from A and B.

50. $A(2, 4)$ and $B(0, 6)$
51. $A(-3, 3)$ and $B(1, 5)$
52. $A(-5, 4)$ and $B(1, 0)$
53. $A(-1, 3)$ and $B(5, 5)$
54. $A(-1, 1)$ and $B(7, -1)$
55. $A(2, 7)$ and $B(4, 3)$

50. b. (1,5); c. -1; d. y = x + 4

Challenge Write an equation of the circle for each of the following.

1. the circle passing through the points (1, −2), (5, 4), and (10, 5)
   $(x - 9)^2 + (y + 3)^2 = 65$
2. the circle circumscribed about the triangle whose vertices are (−2, 3), (5, 2), and (6, −1) $(x - 1)^2 + (y + 1)^2 = 25$

Describe a transformation such that the circle given by each equation is the image of $x^2 + y^2 = 9$ for that transformation.

3. $(x - 1)^2 + y^2 = 9$ $T_{1, 0}$
4. $x^2 + (y + 2)^2 = 9$ $T_{0, -2}$
5. $(x + 3)^2 + (y - 4)^2 = 9$ $T_{-3, 4}$
6. $x^2 + y^2 = 36$ $D_{0, 2}$
7. $x^2 + y^2 = 144$ $D_{0, 4}$
8. $x^2 + y^2 = 63$ $D_{0, \sqrt{7}}$

51. b. (-1,4); c. $\frac{1}{2}$; d. y = -2x + 2
52. b. (-2,2); c. $-\frac{2}{3}$; d. y = $\frac{3}{2}$x + 5
53. b. (2,4); c. $\frac{1}{3}$; d. y = -3x + 10
54. b. (3,0); c. $-\frac{1}{4}$; d. y = 4x - 12
55. b. (3,5); c. -2; d. y = -2x + 11

Mathematical Excursions

Ellipses

A circle can be considered as a special case of a more general curve called an *ellipse*. A circle is defined in terms of a given point and a given distance. An ellipse is defined in terms of *two* given points and *two* distances. The following instructions show one way to draw an ellipse.

Use a piece of string about 25 cm long. Put two tacks through a piece of paper from the back, about 10 cm apart. Loop the string around the tacks. Place a pencil in the loop. Keep the string tight and draw around the tacks.

Repeat this procedure using other lengths of string and distances between the tacks.

The points where the tacks are placed are called the *foci* (plural of focus) of the ellipse. Ellipses can be defined in terms of their foci.

Definition of Ellipse An ellipse is the locus of all points in the plane such that the sum of the distances from two given points, called the foci, is constant.

CHAPTER 9

9.5 Some Locus Problems

Objective: To find a locus of points that satisfies two different conditions simultaneously.

Sometimes we need to find the locus of points satisfying *two different conditions at once*. In such cases, we are dealing with **compound loci**.

Examples

1 **Point A is on line ℓ. Find the locus of points 3 cm from ℓ and also 5 cm from A.**

First, draw the locus of points 3 cm from line ℓ. This is the gray pair of parallel lines shown at the right.

Next, draw the locus of points 5 cm from point A. This is the circle shown at the right in color.

Notice that the circle and the gray lines intersect in four points. Therefore, the required locus is the set of four points B, C, D, and E.

2 **Point C is 7 cm from line m. Find the locus of points 5 cm from line m and also 3 cm from point C.**

The situation is shown at the right. The points in gray satisfy the first condition and those in color satisfy the second.

The locus we want is the set of two points, A and B, that satisfies both conditions.

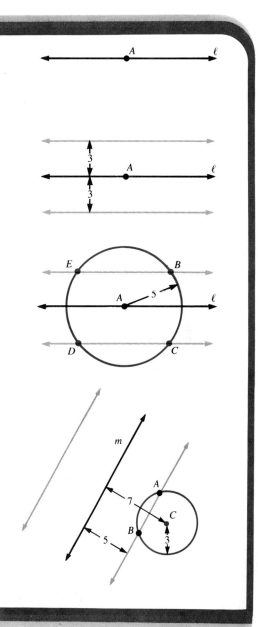

Examples

3 Find the locus of points 3 units from the line with equation $x = 5$ and 2 units from the line with equation $y = 4$.

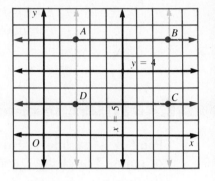

The situation is shown at the right. The gray lines satisfy the first condition. The lines in color satisfy the second condition.

The compound loci is the set of four points $A(2, 6)$, $B(8, 6)$, $C(8, 2)$, and $D(2, 2)$.

4 Find the locus of points equidistant from three noncollinear points A, B, and C.

Locate points A, B, and C, as shown at the right. Then reason as follows.

a. The locus must be all or part of line ℓ, the perpendicular bisector of $\overline{AB}$, since all the points on that line are equidistant from A and B.

b. Furthermore, the locus must be all or part of line m, the perpendicular bisector of $\overline{BC}$, since all the points on that line are equidistant from B and C.

c. Since the required locus must satisfy both of the preceding conditions, we conclude that the intersection point of these two lines, point E, is the required locus.

The following theorem is a general statement of example 4.

Theorem 9–1

The locus of points equidistant from three noncollinear points A, B, and C is the intersection of the perpendicular bisectors of $\overline{AB}$ and $\overline{BC}$.

Exercises

Max: 1-8, 1-29, 1-2; Avg: 1-8, 1-29; Min: 1-8, 1-29 odds

For answers to exercises 1-8, see the Teacher's Guide.

Exploratory Draw a diagram and find the locus of each of the following.

1. points 4 cm from a given line and also 6 cm from a given point on the line
2. points 5 cm from a given line and also equidistant from two points on the line
3. points 3 cm from a given line and also 3 cm from a given point on the line
4. points equidistant from two given points R and S and also 5 cm from $\overline{RS}$.

Assume point P is 8 cm from line ℓ. Draw a diagram and find the locus of each of the following.

5. points 5 cm from P and 4 cm from ℓ
6. points 3 cm from ℓ and 5 cm from P
7. points 11 cm from P and 3 cm from ℓ
8. points 6 cm from ℓ and 15 cm from P

For answers to exercises 1-10, see the Teacher's Guide.

Written Answer each of the following. Draw a diagram for each exercise.

1. Two parallel lines ℓ and m are 4 cm apart. Point A is on line ℓ. Find the locus of points 5 cm from A and 2 cm from ℓ and m.
2. Points A and B are 6 cm apart. Find the locus of points 7 cm from A and equidistant from A and B.
3. Points Q and R are 15 cm apart. Find the locus of points 13 cm from Q and 5 cm from R.
4. Find the locus of points equidistant from the sides of $\angle ABC$ and 5 cm from B.
5. Points A and B are 1 cm apart. Find the locus of points 2 cm from both A and B.
6. Points A and B are on line ℓ. Find the locus of points 4 cm from ℓ and equidistant from A and B.
7. Triangle ABC is acute. Find the locus of points equidistant from $\overrightarrow{AB}$ and $\overrightarrow{AC}$ and equidistant from A and C.
8. Points R and S are on line ℓ. Point T is not on ℓ. Find the locus of points equidistant from R, S, and T.
9. Points P, Q, and R are collinear. Explain why the locus of points equidistant from P, Q, and R contains no points.
10. Find the locus of points equidistant from the sides of rhombus ABCD.

14. (-1,-1), (7,-1), (-1,9), (7,9)
16. (3,4), (3,-4), (-3,4), (-3, -4)

For exercises 11-21, see students' work.

Draw the locus of points that satisfies the following conditions. Use graph paper.

11. points equidistant from both the x-axis and y-axis y = x, y = -x
12. points 5 units from the x-axis and 5 units from the y-axis
13. points 3 units from the line with equation x = 5 and 3 units from the line with equation y = 5 (2,2), (8,2), (2,8), (8,8)
14. points 4 units from the line with equation x = 3 and 5 units from the line with equation y = 4
15. points 5 units from the origin and 3 units from the y-axis
16. points 5 units from the origin and 4 units from the x-axis
17. points 10 units from the origin and 6 units from the y-axis (6,8), (6,-8), (-6,8), (-6,-8)
18. points 10 units from the origin and 8 units from the x-axis (6,8), (6,-8), (-6,8), (-6,-8)

15. (3,4), (3,-4), (-3,4), (-3,-4)

12. (5,5), (5,-5), (-5,5), (-5,-5)

350 USING COORDINATES

19. points 1 unit from the line with equation $x = 2$ and 2 units from the origin $(1,\sqrt{3}), (1,-\sqrt{3})$
20. points 4 units from the origin and 2 units from the line with equation $y = 6$ (0
21. points equidistant from the graphs of $x^2 + y^2 = 36$ and $x^2 + y^2 = 4$ $x^2 + y^2 = 16$

Determine the number of points that satisfy each condition for each the following.

22. equidistant from both the x-axis and y-axis and also 4 units from the origin 4
23. 2 cm from a given line and 5 cm from a point in the line 4
24. equidistant from two parallel lines and also equidistant from two points on one of the lines 1
25. equidistant from two parallel lines that are 6 cm apart and also 3 cm from a point on one of the lines 1
26. equidistant from two parallel lines that are 6 cm apart and also 5 cm from a point on one of the lines 2
27. 3 cm from a given line and 2 cm from a point on the line 0
28. equidistant from (6, 5) and (6, 1) and also on the graph of $(x - 6)^2 + (y - 5)^2 = 9$ 2
29. 3 units from the y-axis and also on the graph of $(x - 1)^2 + (y - 3)^2 = 16$ 2

Challenge

1. Describe as completely as you can the locus of points equidistant from the points $A(2, 1)$, $B(8, 1)$, and $C(6, 5)$. (5,2), the intersection of the perpendicular bisectors of the sides of △ABC.
2. Find the coordinates of all points that are 5 units from the point (5, 2) and also 3 units from the line with equation $y = 2$. (9,5), (1,5), (1,-1), (9,-1)

Mixed Review

Find the area of each triangle having the given vertices.

1. (0, 0), (4, −4), (3, 3) 12
2. (1, 4), (4, 1), (−4, −6) $22\frac{1}{2}$

Describe the locus of each of the following.

3. all points less than 3 cm from a given point The interior of the circle with center at the given point and a radius of 3 cm.
4. all points in *space* equidistant from two given points The plane that contains all the perpendicular bisectors of the line segment joining the two points.
5. Write an equation for the locus of points 11 units from the point $(-1, 4)$. $(x + 1)^2 + (y - 4)^2 = 121$
6. Write an equation for the locus of points equidistant from the points $(3, -4)$ and $(4, 3)$. $y = -\frac{1}{7}x$
7. Given points $A(4k, 3k - 4)$, $B(-5, -4)$, $C(1, 3)$, and $D(7, 5)$, find the value of k so that $\overleftrightarrow{AB} \parallel \overleftrightarrow{CD}$. 1
8. Given points $A(4k, 3k - 4)$, $B(-5, -4)$, $C(1, 3)$, and $D(7, 5)$, find the value of k so that $\overleftrightarrow{AB} \perp \overleftrightarrow{CD}$. -1
9. Find an equation for the line through (5, 2) and parallel to the line $3x + 4y = 1$. $3x + 4y = 23$
10. Find the value of k that makes the line $kx - 3y = 5$ perpendicular to the line $(k + 1)x + 4y = 7$. -4 or 3

9.6 A Special Locus: Parabola

Objective: To write an equation of a parabola.

In the figure at the right, we are given point F and line ℓ. P_1 is a point equidistant from point F and line ℓ.

Now, suppose we wish to find the locus of *all* points equidistant from F and ℓ. Shown at the right are the points P_1, P_2, P_3, and so on. Each of these points is equidistant from F and ℓ.

The smooth curved line formed by all points of this locus is shown at the right. This curved line is called a *parabola*.

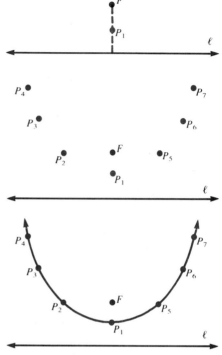

Definition of Parabola

The locus of points equidistant from a given point and a given line is called a **parabola**. The line is the **directrix** and the point is the **focus** of the parabola.

Example

1 Sketch the parabola with focus (0, 3) and directrix $y = 1$.

Graph the focus, (0, 3), and the line $y = 1$. The distance between the focus and directrix is 2 units. One point on the parabola is located at (0, 2). The points (2, 3) and (−2, 3) are each located 2 units from the focus and directrix. Sketch the parabola by drawing a smooth curve (shown in color) connecting points located at (0, 2), (2, 3), and (−2, 3).

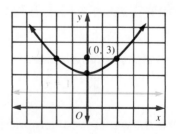

Example

2 Find an equation of the parabola with directrix $y = -2$ and focus $(2, 1)$.

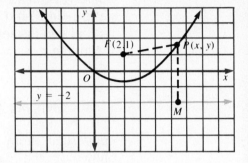

Let $P(x, y)$ be any point on the parabola. By the definition of a parabola, $PM = PF$.

Point P is y units up from the x-axis. It is also another 2 units from the graph of $y = -2$. Therefore, PM is $y + 2$ units in length.

Use the distance formula to find the distance from $P(x, y)$ to $F(2, 1)$.

$$PM = PF$$
$$y + 2 = \sqrt{(x-2)^2 + (y-1)^2} \qquad PM = y + 2; PF = \sqrt{(x-2)^2 + (y-1)^2}$$
$$(y+2)^2 = (x-2)^2 + (y-1)^2 \qquad \text{Square both sides.}$$
$$y^2 + 4y + 4 = x^2 - 4x + 4 + y^2 - 2y + 1 \qquad \text{Square each binomial.}$$
$$6y = x^2 - 4x + 1 \qquad \text{Simplify.}$$
$$y = \frac{x^2}{6} - \frac{4x}{6} + \frac{1}{6} \qquad \text{Divide both sides by 6.}$$
$$y = \frac{1}{6}x^2 - \frac{2}{3}x + \frac{1}{6} \qquad \text{Write } \frac{x^2}{6} \text{ as } \frac{1}{6}x^2 \text{ and simplify } \frac{4}{6} \text{ as } \frac{2}{3}.$$

An equation of the parabola is $y = \frac{1}{6}x^2 - \frac{2}{3}x + \frac{1}{6}$.

Exercises

Max: 1-10, 1-20, 1-4; Avg: 1-10, 1-20; Min: 1-9 odds, 1-20

Exploratory Draw a line. Choose a point that is each of the following distances from the line. Then carefully sketch the locus of points equidistant from the line and the point.

1. 2 cm
2. 1 cm
3. 1 in.
4. 3 in.

Sketch the parabola determined by each focus and directrix.

5. $(3, 1), y = -2$
6. $(1, 1), y = -1$
7. $(2, -2), y = 2$
8. $(-3, 0), y = -4$
9. $(1, 0), x = -1$
10. $(2, 3), x = -1$

For exercises 1-10, see students' work.

Written Refer to the figure at the right.
1. Explain why the distance from a point (x, y) on the parabola to the directrix y = −1 is y + 1. (x,y) is y + 1 units from the x-axis.
2. Write an expression for the distance from a point (x, y) on the parabola to the focus (3, 2).
3. Show that an equation of the parabola is $y = \frac{1}{6}x^2 - x + 2$. For answers to exercises 3 and 4, see the Teacher's Guide.
4. Show that the point (12, 14) lies on the parabola.

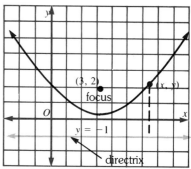

For exercises 5-10, see the Teacher's Guide.   2. $\sqrt{(x-3)^2 + (y-2)^2}$
Find an equation of the parabola with each pair of foci and directrices. Include a sketch.

5. (0, 1), y = −1   6. (1, 1), y = −2   7. (4, 1), y = −1
8. (3, 3), y = 1   9. (1, 2), y = −3   10. (−2, −3), y = 0

Refer to the figure at the right.
11. Explain why the distance from a point (x, y) on the parabola to the directrix x = −2 is x + 2. (x,y) is x + 2 units from the y-axis.
12. Write an expression for the distance from a point of the parabola to the focus (2, 0). $\sqrt{(x-2)^2 + y^2}$
13. Show that the equation of the parabola is $x = \frac{1}{8}y^2$.

For the answer to exercise 13, see the Teacher's Guide.

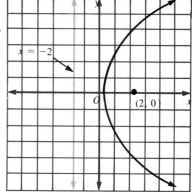

For exercises 14-19, see the Teacher's Guide.
Find an equation of the parabola with each pair of foci and directrices. Include a sketch.

14. (1, 0), x = −1   15. (2, 1), x = −2   16. (1, 3), x = −1
17. (4, 4), x = 3   18. (1, 1), x = 0   19. (−1, 2), x = 4

20. Draw several diagrams to show that two parabolas can intersect in one, two, three, or four points. Interpret this fact in terms of compound loci. See students' work.

For answers to exercises 1 and 2, see the Teacher's Guide.
Challenge Answer the following.

1. Show that the equation of a parabola with focus (0, p) and directrix y = −p is $y = \frac{1}{4p}x^2$.
2. Show that the equation of a parabola with focus (p, 0) and directrix x = −p is $y^2 = 4px$ $\left(\text{or } x = \frac{1}{4p}y^2\right)$. For exercises 3 and 4, see students' work.
3. Describe the parabola whose equation is $y^2 = 20x$. (That is, give its focus and directrix.) Include a sketch. focus, (5,0); directrix, x = −5
4. Answer exercise 3 for equations $y^2 = -11x$ and $y^2 = 23x$.
focus, $(-2\frac{3}{4}, 0)$; directrix, $x = 2\frac{3}{4}$ and
focus, $(5\frac{3}{4}, 0)$; directrix, $x = -5\frac{3}{4}$

9.7 Coordinates in Space

Objective: To name a point in three-dimensional space using three coordinates.

Your work with coordinate geometry so far has dealt with a single, two-dimensional plane. By choosing two perpendicular axes and a unit length on each, you set up a system in which you could name any point in the plane. Then you were able to consider lines and their equations, parallel and perpendicular lines, and so forth.

Now we are going to investigate how to broaden our outlook from all the points in a single plane to all the points in **three-dimensional space**. Just as a plane contains an infinite number of lines, so space contains an infinite number of planes.

> **Journal**
> Describe the locations of ten objects in your classroom using coordinates in space.

A plane contains many lines.

Three-dimensional space contains many planes.

Think of sitting in a classroom with a coordinate system covering the entire front chalkboard. You hold up a pencil. You could give the location of the point of the pencil in terms of the coordinate system on the chalkboard, but you would have to give a third number.

For example, you might say, "Go to the point (20, 5) on the chalkboard. Then come out 30 units into the classroom." Then, (20, 5, 30) would be the coordinates of the pencil point.

A student in the classroom on the other side of the chalkboard might be at (20, 5 −30). Someone in your own classroom, but on the other side of the room, on your left, might be at (−20, 5, 30).

Our investigation suggests that naming a point in three-dimensional space requires three coordinates. In terms of your classroom, with the origin at the center of the chalkboard, the first coordinate is "how much to the left or right." The second coordinate is "how much up or down" and the third is "how much in front of the board or in back of it." The first two coordinates should be familiar to you. We will call them the x-coordinate and the y-coordinate. The third coordinate is the z-coordinate.

Shown below are the x-axis, y-axis, and z-axis. These three axes intersect at the origin.

Since two intersecting lines determine a unique plane, each pair of coordinate axes determines a coordinate plane. Thus, we have the xy-coordinate plane, the xz-coordinate plane, and the yz-coordinate plane.

Each of the three coordinate planes is perpendicular to each of the others. The three planes together divide space into eight regions. Each region is called an **octant**. In one of the octants, all three of the coordinates are positive. This region is called the First Octant. Usually, number names are not given to the others.

At the right is a figure of the First Octant, where all coordinates are positive. The "general" point is called (x, y, z). A point in the xy-coordinate plane has 0 as its z-coordinate since 0 is the number of units moved in z direction to reach it.

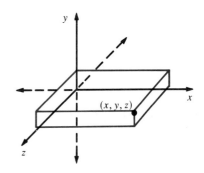

Also, a point in the xz-plane has a y-coordinate of 0. A point in the yz-plane has an x-coordinate of 0.

Examples

1 Describe the location of the point $A(3, 0, -5)$.

It is three units in the positive x direction, zero units in the y direction, and five units in the negative z direction, "back into the paper."

2 Give the coordinates of points B, C, D, and E in the figure.

The coordinates are $B(0, 1, 4)$, $C(-3, 1, 4)$, $D(0, 0, 4)$, and $E(-3, 0, 4)$.

In example 2, rectangle $CBDE$ lies completely in its own plane. This plane is four units in the positive z direction from the xy-plane and is parallel to it. Some other points in this plane are $F(4, 0, 4)$, $G(4, -1, 4)$, and $H(5, -1, 4)$.

All of these points are four units in the positive z direction from the xy-plane. So all points in this plane are of the form $(a, b, 4)$ where a and b are any real numbers. Therefore, the equation $z = 4$ describes the plane four units in the positive z direction from the xy-plane and parallel to it.

Example

3 Write an equation for the plane two units in the positive x direction from the yz-plane.

Every point in this plane must have first coordinate 2. The other two coordinates can be any real numbers. The equation is $x = 2$.

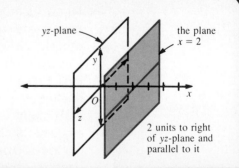

Exercises

Max: 1-8, 1-14; Avg: 1-8, 1-14; Min: 1-8, 1-14

For answers to exercises 1-8, see the Teacher's Guide.

Exploratory Tell whether each of the following is *always, sometimes,* or *never* true. Explain your answers and sketch the situation, if possible.

1. If two planes intersect, they must intersect in a line.
2. If two planes are each parallel to a third plane, then they are parallel to each other.
3. If two lines are parallel, then there must be a unique plane containing both of them.
4. If two lines in space are not parallel, then they must intersect.
5. If two lines intersect, then there must be a unique plane containing both of them.
6. Given three noncollinear points, D, E, and F, there must be exactly one plane containing all three of them.
7. Given four noncollinear points D, E, F, and G, there must be exactly one plane containing all four of them.
8. If a line is not in a plane and is not parallel to the plane, then it intersects the plane in exactly one point.

Written A box of Crunchy-Wunchies has dimensions 3 in. by 8 in. by 12 in. high. Such a box is placed in a corner as shown in the figure. Let the corner be the origin of a coordinate system with 1-inch units on all three axes.

1. Give the coordinates of the four corners of the top of the box. (0,12,0), (0,12,3), (8,12,0), (8,12,3)
2. Give the coordinates of the four bottom corners of the box. (0,0,0), (8,0,0), (8,0,3), (0,0,3)

Give the equation for each plane of the following parts of the box.

3. the front $z = 3$
4. the top $y = 12$
5. the two side panels $x = 0, x = 8$
6. Is the point (5, −7, 0) in the same plane as the back of the box? yes
7. Name three points in the same plane as the bottom of the box.
8. The top of the box is pictured in this book as a parallelogram. Why?
9. If the diagonals of the front of the box are drawn, at what point will they intersect? Are these diagonals congruent? How do you know?
10. What is the length of each diagonal of the top of the box? $\sqrt{73}$ or about 8.54

Tell if each statement about three-dimensional space is *true* or *false.* Justify your answers as carefully as you can. For answers to exercises 7-9 and 11-14, see the Teacher's Guide.

11. Three distinct planes cannot intersect in a straight line.
12. For any real number k, the point (0, k, 0) is on the y-axis.
13. If a point lies in one of the coordinate planes, then one of its coordinates must be zero.
14. The locus of points equidistant from the origin and the point (2, 0, 0) is the plane x = 1.

Problem Solving Application: Using Inverse Variation

Objective: To solve problems using inverse variation.

ℓ	1 m	2 m	3 m	4 m	6 m
w	36 m	18 m	12 m	9 m	6 m

The area of a rectangle is 36 m². The table at the left shows some possible values for its length (ℓ) and width (w). Notice that as the length of the rectangle increases, its width decreases.

The relationship between the length and the width of this rectangle is given by the equation $\ell w = 36$. Such an equation is called an **inverse variation.** We say that ℓ varies inversely as w, or ℓ and w vary inversely.

Inverse Variation

> An inverse variation is described by an equation of the form $xy = k$, where $k \neq 0$. The variable k is called the *constant of variation.*

Sometimes an inverse variation is written in the form $y = \frac{k}{x}$.

Example

1 Hope's trip takes 3 hours driving at 50 miles per hour (mph). How long will the return trip take if she drives at 45 miles per hour?

Explore The equation that relates distance (d), time (t), and rate (r) is $d = rt$. Since the distance is constant, this is an inverse variation problem. As the rate decreases, the time increases.

Plan First find the distance of Hope's trip using the rate (50 mph) and the time (3 hours). Then use this distance and the return rate (45 mph) to find the time for her return trip.

Solve
$d = rt$
$d = (50 \text{ mph}) \times (3 \text{ hours})$
$d = 150$ miles   Hope traveled 150 miles one way on her trip.

$t = \frac{d}{r}$ Use the $y = \frac{k}{x}$ form of inverse variation.

$t = \frac{150 \text{ miles}}{45 \text{ mph}}$

$t = 3\frac{1}{3}$ hours   or   3 hours and 20 minutes

The return trip will take 3 hours and 20 minutes.

Examine Since this is an inverse variation, as the rate decreases, the time should increase. The answer seems reasonable.

Exercises

Max: 1-8, 1-14; Avg: 1-8, 1-14; Min: 1-7 odds, 1-8, 9-13 odds

Exploratory State whether each equation represents an inverse variation. Write yes or no. For each inverse variation, state the constant of variation.

1. $ab = 5$ yes, 5
2. $x = 5$ no
3. $c = 5d$ no
4. $mn = -3$ yes, -3
5. $\dfrac{50}{y} = x$ yes, 50
6. $\dfrac{w}{5} = z$ no
7. $q = \dfrac{-1}{r}$ yes, -1
8. $a = \dfrac{7}{b}$ yes, 7

Written Solve each problem. Assume y varies inversely as x.

1. If $y = 24$ when $x = 8$, find y when $x = 6$. 32
2. If $y = 6$ when $x = 2$, find y when $x = 5$. 2.2
3. If $y = 2$ when $x = 8$, find x when $y = \frac{2}{3}$. 24
4. If $y = \frac{1}{3}$ when $x = 5$, find x when $y = \frac{1}{4}$. $6\frac{2}{3}$
5. If $y = -8$ when $x = 2$, find y when $x = 7$. $-2\frac{2}{7}$
6. If $y = 7$ when $x = -3$, find x when $y = 4$. $-5\frac{1}{4}$
7. Ruth's trip takes 4 hours driving at 80 km/h. How long will the return trip take if she drives at 75 km/h? 4 h 16 min
8. Bert's trip takes 6 hours driving at 50 mph. At what rate must he drive to complete the return trip in 5.5 hours? about 55 mph
9. If y^2 and x vary inversely, and $y = 4$ when $x = 2$, find x when $y = 8$. $\frac{1}{2}$
10. If y^2 and x vary inversely, and $y = 6$ when $x = 3$, find y when $x = 9$. $2\sqrt{3}$

Boyle's law states that the volume of a gas (V) varies inversely with the applied pressure (P). This relationship is shown by the formula $P_1V_1 = P_2V_2$. Use this formula to solve each problem.

11. A helium-filled balloon has a volume of 16 m³ when the applied pressure is 1 atmosphere. What is the volume of the balloon when the applied pressure is 0.75 atmospheres? $21\frac{1}{3}$ m³
12. The pressure acting on 8.0 m³ is 20 atmospheres. The pressure is reduced until the volume of the gas is 20 m³. What is the new pressure acting on the gas? 8 atmospheres

In sound and harmonics, the frequency of a vibrating string varies inversely with the length of the string. Use this information to solve each problem.

13. A 10-inch violin string vibrates at a frequency of 512 cycles/second. Find the frequency of an 8-inch string. 640 cycles/sec
14. A 36-inch piano string vibrates at a frequency of 480 cycles/second. Find the length of a string that vibrates at a frequency of 720 cycles/second. 24 in.

Portfolio Suggestion

Select an item from your homework or classwork from this chapter that you feel shows your best work and place it in your portfolio. Explain why you selected it.

Performance Assessment

A treasure map indicates that a treasure chest is buried 10 feet from a fence and 30 feet from an oak tree. There is more than one oak tree in the vicinity.
parallel lines
a. Describe the locus of points 10 feet from a straight fence. **to fence**
b. Describe the locus of points 30 feet from a tree. **a circle about the tree**
c. Draw the loci that satisfy the conditions given on the map for three different locations of the tree. **See students' work.**

Chapter Summary

1. In coordinate proofs, choose axes relative to the polygon that will result in simpler computations involving the coordinates. Variable coordinates such as (a, b) can be used. Arbitrary values may be assigned to coordinates to further simplify proofs of theorems.   (330)
2. The set of all those points, and only those points, that satisfy a certain condition is called a **locus**.   (339)
3. The equation of any **circle** with center at the origin and radius r is $x^2 + y^2 = r^2$.   (343)
4. The equation of any circle with center at (h, k) and radius r is $(x - h)^2 + (y - k)^2 = r^2$.   (344)
5. **Compound loci** involve a locus of points satisfying two different conditions at once.   (347)
6. Theorem 9-1: The locus of points equidistant from three noncollinear points A, B, and C is the intersection of the perpendicular bisectors of $\overline{AB}$ and $\overline{BC}$.   (348)
7. The locus of points equidistant from a given point and a given line is called a **parabola**. The line is the **directrix**, and the point is the **focus** of the parabola.   (351)
8. The x-, y-, and z-coordinate planes in **three-dimensional space** are perpendicular to each other, intersect at the origin, and divide space into eight regions, each of which is called an **octant**.   (354, 355)

Chapter Review

Parallelograms can be justified by two pairs of equal slopes (parallel sides).
9.1 Show that the quadrilateral with the following vertices is a parallelogram.
  1. $A(0, 0)$, $B(6, 0)$, $C(9, 6)$, $D(3, 6)$
  2. $T(h, k)$, $U(i + h, k)$, $V(i, 0)$, $W(0, 0)$

For answers to exercises 3, 9–13, and 21–22, see the Teacher's Guide.
Use coordinate geometry to prove the following.
  3. The diagonals of an isosceles trapezoid are congruent.

9.2 For $\triangle ABC$, find the following.

  4. AB a   5. BC $\sqrt{a^2 + b^2}$ 6. CA b
  7. the midpoint of $\overline{BC}$ $(\frac{a}{2}, \frac{b}{2})$
  8. the measure of the median to $\overline{BC}$ $\frac{1}{2}\sqrt{a^2 + b^2}$

  9. Show that $\triangle ABC$ is a right triangle.

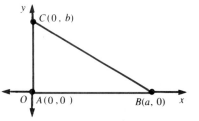

9.3 Draw a diagram and describe the locus of each of the following.
  10. points 8 cm from the origin
  11. points equidistant from two points 2 cm apart
  12. points 10 cm from a circle with a diameter of 20 cm
  13. points in *space* 5 cm from a given point

9.4 Which points lie on the locus of points 10 units from the origin?
  14. $(10, 0)$ yes   15. $(-6, -8)$ yes   16. $(5, 5)$ no

Describe fully (center and radius) the circle given by each equation.
  17. $(x - 1)^2 + (y - 3)^2 = 16$ (1,3), 4 units
  18. $(x + 5)^2 + (y - 8)^2 = 100$ (-5,8), 10 units
  19. Find an equation for the locus of points equidistant from $A(4, 4)$ and $B(-2, 12)$. $3x - 4y = -29$
  20. Find an equation for the locus of points equidistant from the lines $y = x - 1$ and $y = x + 5$. $y = x + 2$

9.5
  21. Two parallel lines, r and s, are 5 cm apart. Point A is on r. Find the locus of points 6 cm from A and 2.5 cm from r and s.
  22. Points C and D are 10 cm apart. Find the locus of points 8 cm from C and 6 cm from D.

9.6 Find an equation of the parabola with each focus and directrix.
  23. $(0, 0)$, $y = -2$ $y = \frac{1}{4}x^2 - 1$
  24. $(5, 3)$, $y = -1$ $y = \frac{1}{8}x^2 - \frac{5}{4}x + 4\frac{1}{8}$
  25. $(-3, -2)$, $x = 0$ $x = -\frac{1}{6}y^2 - \frac{2}{3}y - 2\frac{1}{6}$

9.7 Tell if each statement about three-dimensional space is *true* or *false*.
  26. For any real numbers m and n, the point $(0, m, n)$ lies in the xy-plane.   false
  27. Three distinct planes cannot intersect in two distinct lines. false

Chapter Test

Find the coordinates of R for each of the following. Do not introduce new letters.

1.
parallelogram
$(h+m, n)$

2.
isosceles triangle
$(2f, 0)$

3.
isosceles trapezoid
$(h-m, n)$

For answers to exercises 4 and 10-16, see the Teacher's Guide.

4. Show that the triangle whose vertices are (3, 2), (20, 2), and (11, 17) is an isosceles triangle.

Quadrilateral *ABCD* is a trapezoid with base $\overline{AB}$ longer than base $\overline{DC}$. The vertices are $A(7, -3)$, $B(2k, 2)$, $C(k, 5)$, and $D(3, 2)$.

5. Express the slope of $\overline{AB}$ in terms of k. $\frac{5}{2k-7}$
6. Express the slope of $\overline{DC}$ in terms of k. $\frac{3}{k-3}$
7. Write an equation that can be used to solve for k. $\frac{5}{2k-7} = \frac{3}{k-3}$
8. Solve the equation written for exercise 7. $k = 6$
9. Find an equation of the line that passes through points *B* and *D*. $y = 2$
10. Use coordinate geometry to prove that the diagonals of a square are perpendicular.

Draw a diagram and describe the locus of each of the following.

11. points 5 cm from a given point *B*
12. points equidistant from two points *A* and *B*
13. points 4 units from the point (2, 5)
14. points equidistant from two parallel lines
15. points 3 cm from a given line ℓ
16. points equidistant from two intersecting lines

Write an equation or equations of the locus of each of the following.

17. points 3 units from the y-axis $x = 3$, $x = -3$
18. points 9 units from the origin $x^2 + y^2 = 81$
19. points equidistant from the points, (5, 3) and (5, 7) $y = 5$
20. points 4 units from the point $(-3, 4)$ $(x + 3)^2 + (y - 4)^2 = 16$

21. Find the coordinates of the center of the circle with equation $(x - 6)^2 + (y - 4)^2 = 64$. (6, 4)
22. Find the radius of the circle in exercise 21. 8 units
23. How many points are 3 cm from a given line ℓ and 4 cm from a point *P* on ℓ. 4
24. Find an equation of the locus of points equidistant from $A(-2, 4)$ and $B(2, 6)$. $y = -2x +$
25. Sketch on graph paper the locus of points equidistant from the point (0, 4) and the x-axis.
26. Sketch the parabola with focus (4, 2) and directrix $y = -2$. For exercises 25-26, see students' work.

CHAPTER 10

Solving Quadratic Equations

Application in Sports

"Take me out to the ballgame...." Baseball is a favorite American pastime, but physics is probably not on your mind as you root for your favorite team. But did you know that the path of a baseball can be described by a **quadratic equation**? The path that an object follows when it is thrown, hit, or dropped, is called its *trajectory*. The trajectory of a baseball is a **parabola**.

Juan Gonzalez, outfielder for the Texas Rangers, comes to bat with the bases loaded and the Rangers down 4 to 1 against the New York Yankees. The count on Gonzalez is 3 and 2, and the pitcher throws a smoking fastball about waist high. Gonzalez connects and the ball sails toward center field. As Gonzalez rounds the bases, the ball flies straight at the 420-foot marker in center field on the 15-foot high wall. The equation for the path of the baseball is $y = -0.00081x^2 + 0.36x + 3$, where x is the horizontal distance and y is the vertical distance. Will the ball clear the fence for a home run?

No, the height of the ball will be 11.36 feet off the ground when it reaches the fence.

Group Project: *Physics*

Construct a Packard apparatus like the one shown at the right, or borrow one from a physics teacher. Place a piece of carbon paper on top of a piece of graph paper on the board. Have one group member roll a heavy steel marble down the ramp while another uses a stop watch to record the time it takes for the marble to roll to the bottom. Observe the path on the graph paper. Research the paths of falling objects in a physics book. Can you find the equation for the height of the steel ball based on the time elapsed? Write a report about your findings.

10.1 Multiplying and Factoring Polynomials

Objective: To multiply and factor polynomials.

Before beginning a study of quadratic equations, it will be helpful to review your knowledge of multiplying and factoring polynomials.

Recall that a *monomial* is a polynomial with one term. A polynomial with two different terms is a *binomial*, and a polynomial with three different terms is a *trinomial*. In a monomial such as $5x^3$, the 5 is the numerical coefficient, the x is a variable, and the 3 is the exponent of the variable.

Examples

1 Find $(4x^3)(3x^2y)$.

Rearrange the factors with the numerical coefficients first and then each variable.

$$(4x^3)(3x^2y) = 4 \cdot 3 \cdot x^3 \cdot x^2 \cdot y$$
$$= 12 \cdot x^5 \cdot y$$
$$= 12x^5y$$

Caution: $x^3 \cdot x^2 = x^5$, not x^6. When multiplying like variables, *add* the exponents.

2 Find $2c^2(5c - 1)$.

Use the Distributive Law and then simplify.

$$2c^2(5c - 1) = 2c^2(5c) + 2c^2(-1)$$
$$= 10c^3 + (-2c^2)$$
$$= 10c^3 - 2c^2$$

The Distributive Law is also used to find the product of two binomials.

Product of Two Binomials

$$(a + b)(c + d) = a(c + d) + b(c + d)$$
$$= ac + ad + bc + bd$$

Often two of the terms can be combined to simplify the expression.

Examples

3 Find $(2x - 3)(x + 5)$.

Use the rule for the product of two binomials.

$$(2x - 3)(x + 5) = 2x(x + 5) - 3(x + 5)$$
$$= 2x^2 + \underbrace{10x - 3x} - 15 \quad \text{Combine the like terms.}$$
$$= 2x^2 + 7x - 15$$

4 Find $(x + 7)(x - 7)$.

$$(x + 7)(x - 7) = x^2 - 7x + 7x - 49 \quad \text{The sum of the like terms is 0.}$$
$$= x^2 - 49$$

You may recall that products such as those in example 4 have a special name.

Difference of Two Squares

$$(a + b)(a - b) = a^2 - b^2$$

In example 4, $a = x$ and $b = 7$, yielding $x^2 - 7^2$ or $x^2 - 49$.

The "opposite" of multiplying is factoring. When you multiply $5 \cdot 2$, the result is 10; when you **factor** 10, the result is $5 \cdot 2$. A polynomial may seem to have more than one factorization. However, we can eliminate all other possibilities by seeking the most completely factored form.

Example

5 Factor $30x^2 - 6x$.

First search for the greatest monomial factor common to both terms (GCF). Then, use the Distributive Law.

$$30x^2 - 6x = (6x)(5x) - (6x)(1)$$
$$= 6x(5x - 1)$$

Note that while $6(5x^2 - x)$ is a factorization of $30x^2 - 6x$, it is not the most complete factorization. There still remains a monomial factor, x, common to both $5x^2$ and x.

Examples

6 **Factor $x^2 - 4x - 12$.**

The first term of each binomial must be x since their product is x^2.

$$x^2 - 4x - 12 = (x + \Box)(x + \Box)$$

The product of these two numbers must be -12. Their sum must be -4.

The numbers are -6 and 2. So,

$$x^2 - 4x - 12 = (x - 6)(x + 2).$$

7 **Factor $4x^2 - 25$.**

Recall that this is the special case of $a^2 - b^2 = (a + b)(a - b)$. Here, $a^2 = 4x^2$ and $b^2 = 25$. Therefore, $a = 2x$ and $b = 5$.

$$4x^2 - 25 = (2x + 5)(2x - 5)$$

8 **Factor $2x^2 + 13x + 6$.**

This example needs more investigation than was needed in example 6 to find the correct binomials.

First consider the factors for the first term. Our only choices are x and $2x$. Therefore,

$$2x^2 + 13x + 6 = (2x + \Box)(x + \Box).$$

The choices for the last term are $1 \cdot 6$ and $2 \cdot 3$. After some experimentation with the placement of these factors, we see that only one correct factorization exists.

$$2x^2 + 13x + 6 = (2x + 1)(x + 6)$$

Because factoring is the opposite of multiplying, any factorization can be checked by multiplying.

$$(2x + 1)(x + 6) \stackrel{?}{=} 2x^2 + 13x + 6$$
$$(2x + 1)(x + 6) = 2x^2 + 12x + x + 6$$
$$= 2x^2 + 13x + 6 \quad \text{The factorization is correct.}$$

Example

9 Factor $2x^2y + 10xy + 12y$.

Notice that there is a common factor of $2y$. First factor the polynomial using the common factor and the Distributive Law.

$$2x^2y + 10xy + 12y = 2y(x^2 + 5x + 6)$$

The remaining trinomial can also be factored as $(x + 2)(x + 3)$. So, the complete factorization requires two steps.

$$2x^2y + 10xy + 12y = 2y(x + 2)(x + 3)$$

Exercises

Max: 1-21, 1-68, 1-10; Avg: 1-21, 1-61 odds, 64-68; Min: 1-21 odds, 1-61 odds

Exploratory Explain the following terms. For answers to exercises 1-5, see the Teacher's Guide.

1. binomial
2. trinomial
3. difference of two squares
4. complete factorization
5. Distributive Law

Multiply.

6. $3x(4x)$ $12x^2$
7. $2x^2(4x)$ $8x^3$
8. $4y^3(-5x^2y)$ $-20x^2y^4$
9. $(-2x^2y)\left(-\dfrac{1}{2}xy^3z\right)$ x^3y^4z
10. $7(x - 2)$ $7x - 14$
11. $-3(x + 4)$ $-3x - 12$
12. $-5(2 - x)$ $-10 + 5x$
13. $x^2(x - 1)$ $x^3 - x^2$

Name the greatest common factor for each pair of terms.

14. $2x, 6x$ $2x$
15. $2x^2, 3x^2$ x^2
16. $5xy, 10xy$ $5xy$
17. $3xy, 6x^2$ $3x$
18. $14c^2, 35c^3$ $7c^2$
19. $100x^2z, 50xz$ $50xz$
20. $9x^2, 3xy^2$ $3x$
21. $6yz^2, 8y^2z$ $2yz$

Written Find each product. For answers to exercises 1-68, see the Teacher's Guide.

1. $8(c - 2)$
2. $5x(x - 1)$
3. $2y(y - 4y^2)$
4. $x^3(x^2 - 4y)$
5. $4x^2(y^2 + xy)$
6. $3x^2y(5xy^2 + 6x^2z^2)$
7. $(x + 2)(x + 5)$
8. $(x - 2)(x + 4)$
9. $(x - 9)(x + 9)$
10. $(c + 9)(c - 1)$
11. $(x + 5)(x - 5)$
12. $(x - 13)(x - 5)$
13. $(2w + 5)(w + 7)$
14. $(n - 6)(3n + 4)$
15. $(4x + 1)(2x - 1)$
16. $(6x - 1)(6x + 1)$
17. $(3y + 4)(2y - 5)$
18. $(4 - 3y)(2 + y)$

Factor each polynomial by finding the greatest common factor.

19. $8c - 16$
20. $4c - 4d$
21. $5x^2 - 5x$
22. $4x^2 + 8y^2 - 16z^2$
23. $2xy + 2xy^2$
24. $3y^2 - 9y - 3y^3$
25. $13xyz - 26xy$
26. $-5x^2y + 10x^2$
27. $xy^3z + 3x^2y$
28. $abc^2 + ab^2c - abc$
29. $100x^2 - 100x^3$
30. $-xy + 6x^2y^2$

Factor each polynomial as a product of binomials.

31. $x^2 - 4$
32. $x^2 + 4x - 5$
33. $m^2 + 2m + 1$
34. $x^2 - 5x + 6$
35. $x^2 - 81$
36. $b^2 - 0.01$
37. $x^2 - 2x - 15$
38. $x^2 + 20x - 21$
39. $4c^2 - 1$
40. $x^2 + x - 30$
41. $x^2 - 5x - 24$
42. $y^2 + 17y + 60$
43. $2c^2 + 11c + 5$
44. $3n^2 + 10n + 8$
45. $5x^2 + 19x - 4$
46. $2y^2 - y - 3$
47. $2x^2 - 11x + 12$
48. $8x^2 - 6x - 9$

Factor each polynomial completely. A second factorization may be necessary.

49. $9y^2 - 9$
50. $3c^2 - 60c - 63$
51. $x^2 - 2x$
52. $2x^2 - 10x + 12$
53. $3b^2 - 48$
54. $14x^2 - 28x$
55. $-2x^2 + 2x + 30$
56. $30x^2 - 15x$
57. $h^3 - h^2$
58. $x^3 - x^5$
59. $2m^6 - 2m^4$
60. $6w^2 + 3w - 9$
61. $4d^2x - 4dx - 24x$
62. $x^4 - 1$
63. $y^6 - y^4$

Explain each of the following.

64. What is the relationship between $a^2 - 9$ and $9 - a^2$?
65. When $a^2 - b^2$ and $b^2 - a^2$ are factored, what is the relationship between the two sets of factors?
66. What is the relationship between factoring and the Distributive Law?
67. How can the Distributive Law be used to show $(x + r)(x + s) = x^2 + (r + s)x + rs$?
68. How can the sum of two squares, $a^2 + b^2$, be factored?

1. $(3x - 5)(4x + 7)$ 2. $(5x^2 + 4)(12x^2 - 5)$ 3. $(x - 1)(x + 1)(x^2 + 1)(x^4 + 1)(x^8 + 1)$

Challenge Factor completely.

1. $12x^2 + x - 35$
2. $60x^4 + 23x^2 - 20$
3. $x^{16} - 1$
4. $r^4 - r^2s - 2s^2$
5. $x^4 - 18x^2 + 81$
6. $2a^4 - 32b^4$

4. $(r^2 + s)(r^2 - 2s)$ 5. $(x - 3)^2(x + 3)^2$ 6. $2(a - 2b)(a + 2b)(a^2 + 4b^2)$

Some polynomials can be factored by grouping the terms to find a common binomial factor. For example, $xy + xb + ay + ab = x(y + b) + a(y + b)$. The common factor is $y + b$. The factored form is $(y + b)(x + a)$. Factor each of the following by grouping to find the common binomial factor.

7. $r^2 + rs - rt - st$ $(r - t)(r + s)$
8. $x^2 + xby + ax + aby$ $(x + a)(x + by)$
9. $3xy + 9xz + 2y + 6z$ $(3x + 2)(y + 3z)$
10. $z^3 + 4z^2 - 3z - 12$ $(z^2 - 3)(z + 4)$

10.2 Solving Quadratics by Factoring

Objective: To solve quadratic equations by factoring.
Here are some examples of **quadratic equations**.

1. $x^2 - 5x + 4 = 0$ 2. $y^2 - 3y = 38$ 3. $4x^2 = -4x - 1$

In a quadratic equation, a cannot equal zero.

The standard form of a quadratic equation is $ax^2 + bx + c = 0$, where a is the *coefficient of* x^2, b is the *coefficient of* x, and c is the *constant term*. In the examples above, equation **1** is already in standard form. Equation **2** is equivalent to $y^2 - 3y - 38 = 0$, and equation **3** is equivalent to $4x^2 + 4x + 1 = 0$.

You already know that the method of solving quadratic equations by factoring consists of the following steps.

Step 1: Rewrite the given quadratic equation in standard form.
Step 2: Factor the nonzero side of the equation.
Step 3: Set each factor equal to zero and solve.

Step 3 is an application of the Zero Product Property:
For any numbers a and b, $ab = 0$ if and only if $a = 0$ or $b = 0$.

Examples

1 Solve $y^2 - 3y = 28$.

Step 1: Write the quadratic equation in standard form.
Step 2: Factor the nonzero side.
Step 3: Set each factor equal to zero and solve.

$y^2 - 3y - 28 = 0$
$(y - 7)(y + 4) = 0$
$y - 7 = 0$ or $y + 4 = 0$
$y = 7$ $y = -4$

The solution set is $\{-4, 7\}$.

2 Write a quadratic equation whose solution set is $\{-4, 7\}$.

Use the steps from example 1 in reverse.

$x = -4$ or $x = 7$ } Step 3
$x + 4 = 0$ or $x - 7 = 0$
$(x + 4)(x - 7) = 0$ Step 2
$x^2 - 3x - 28 = 0$ Step 1

Example

3 Solve $x^2 - 3x = 0$.

Step 1: $x^2 - 3x = 0$ Notice that $c = 0$.
Step 2: $x(x - 3) = 0$
Step 3: $x = 0$ or $x - 3 = 0$
                   $x = 3$

The solution set is $\{0, 3\}$.

Some quadratic equations have only one solution or *root*. This happens when the polynomial involved is a perfect square trinomial. For example, $x^2 + 10x + 25$ is a *perfect square trinomial* because it factors into $(x + 5)(x + 5)$ or $(x + 5)^2$.

Example

4 Solve $x^2 + 10x = -25$.

Step 1: $x^2 + 10x + 25 = 0$
Step 2: $(x + 5)(x + 5) = 0$
Step 3: $x + 5 = 0$ or $x + 5 = 0$
         $x = -5$ $x = -5$ The equation has two equal roots.

The solution set is $\{-5\}$.

In the equations studied so far, the coefficient of x^2, a, has been 1. Quadratic equations with $a \neq 1$ are sometimes more challenging to solve.

Examples

5 Solve $4c^2 - 1 = 0$.

There are two methods of solving this equation. Each uses the special form $a^2 - b^2 = (a + b)(a - b)$.

Method 1

$4c^2 - 1 = 0$
$(2c + 1)(2c - 1) = 0$
$2c + 1 = 0$ or $2c - 1 = 0$
$\quad 2c = -1 \qquad\qquad 2c = 1$
$\quad\; c = -\dfrac{1}{2} \qquad\qquad c = \dfrac{1}{2}$

Method 2

$4c^2 - 1 = 0$
$c^2 - \dfrac{1}{4} = 0 \quad$ Divide both sides by 4.
$\left(c + \dfrac{1}{2}\right)\left(c - \dfrac{1}{2}\right) = 0$
$c + \dfrac{1}{2} = 0$ or $c - \dfrac{1}{2} = 0$
$\quad c = -\dfrac{1}{2} \qquad\qquad c = \dfrac{1}{2}$

The solution set is $\left\{-\dfrac{1}{2}, \dfrac{1}{2}\right\}$.

6 Solve $3x^2 = 14x + 5$.

$3x^2 = 14x + 5$
$3x^2 - 14x - 5 = 0$
$(3x + 1)(x - 5) = 0$
$3x + 1 = 0$ or $x - 5 = 0$
$\quad 3x = -1 \qquad\qquad x = 5$
$\quad\; x = -\dfrac{1}{3}$

The solution set is $\left\{-\dfrac{1}{3}, 5\right\}$.

7 Solve $-x^2 + x = -20$.

$-x^2 + x = -20$
$-x^2 + x + 20 = 0$
$x^2 - x - 20 = 0 \quad$ Multiply both sides by -1 so that the
$(x - 5)(x + 4) = 0 \quad$ coefficient of x^2, a, is positive.
$x - 5 = 0$ or $x + 4 = 0$
$\quad x = 5 \qquad\qquad x = -4 \quad$ The solution set is $\{-4, 5\}$.

Exercises

Max: 1-26, 1-57; Avg: 1-26, 1-57 odds; Min: 1-25 odds, 1-57 odds

Exploratory State the conditions under which each quadratic equation will be true. Then, state the solution set. For answers to exercises 1-18, see the Teacher's Guide.

1. $x(x - 6) = 0$
2. $m(m + 3) = 0$
3. $2y(y + 4) = 0$
4. $a(4a + 1) = 0$
5. $(x + 1)(x + 5) = 0$
6. $(b + 4)(b - 2) = 0$
7. $(c - 5)(c - 3) = 0$
8. $(r - 6)(r + 4) = 0$
9. $(2y + 1)(y - 2) = 0$
10. $(4z - 1)(z + 2) = 0$
11. $(3x - 4)(4x - 3) = 0$
12. $(4y + 3)(2y - 2) = 0$

Write each quadratic equation in standard form.

13. $3x^2 - 3x = 1$
14. $x^2 - 13 = 12x$
15. $x^2 = -5x - 1$
16. $-4 = -4x - x^2$
17. $5x = -x^2 + 13$
18. $24 = -3x^2 + 10x$

The following quadratic equations are in standard form, $ax^2 + bx + c = 0$. Name a, b, and c.

19. $3x^2 - 5x - 1 = 0$ 3, -5, -1
20. $2x^2 + 4x + 2 = 0$ 2, 4, 2
21. $4x^2 - 1 = 0$ 4, 0, -1

In the following, choose the expression that is the difference of two squares.

22. a. $x^2 + y^2$ b. $2x^2 - y^2$ c. $4x^2 - y^2$ c
23. a. $16 - 4c^2$ b. $16 + 4c^2$ c. $16 - 4c^3$ a

In the following, choose the perfect square trinomial.

24. a. $x^2 + 4$ b. $x^2 - 4$ c. $x^2 + 4x + 4$ c
25. a. $m^2 - m + 2$ b. $m^2 - 2m + 4$ c. $m^2 - 2m + 1$ c

Answer the following.

26. Find a value of k that makes $x^2 + 8x + k$ a perfect square trinomial. 16

Written Solve each equation by factoring.

1. $x^2 + 5x + 4 = 0$ {-4,-1}
2. $b^2 - 11b + 30 = 0$ {5, 6}
3. $a^2 - 7a = 0$ {0, 7}
4. $h^2 - h = 0$ {0, 1}
5. $y^2 + 7y + 10 = 0$ {-5, -2}
6. $x^2 - 5x - 24 = 0$ {-3, 8}
7. $m^2 - 16 = 0$ {-4, 4}
8. $y^2 - 9 = 0$ {-3, 3}
9. $p^2 - 5p - 50 = 0$ {-5, 10}
10. $r^2 - 3r - 18 = 0$ {-3, 6}
11. $z^2 + 4z + 4 = 0$ {-2}
12. $w^2 + 10w + 25 = 0$ {-5}
13. $2m^2 - m = 0$ {0, $\frac{1}{2}$}
14. $5x^2 - x = 0$ {0, $\frac{1}{5}$}
15. $t^2 - 8t = 9$ {-1, 9}
16. $c^2 + 6c = -5$ {-5, -1}
17. $y^2 + 6y = -9$ {-3}
18. $z^2 + 8z = -16$ {-4}
19. $x^2 = 8x + 20$ {-2, 10}
20. $w^2 = -12w - 32$ {-8, -4}
21. $m = -m^2 + 20$ {-5, 4}
22. $-16x = -x^2 - 60$ {6, 10}
23. $-30 = -7p - p^2$ {-10, 3}
24. $48 = 14x - x^2$ {6, 8}
25. $3x^2 + 5x + 2 = 0$ {$-\frac{2}{3}$, -1}
26. $2y^2 + 11y + 5 = 0$ {-5, $-\frac{1}{2}$}
27. $2c^2 - 7c + 6 = 0$ {$\frac{3}{2}$, 2}
28. $2t^2 + 9t + 4 = 0$ {-4, $-\frac{1}{2}$}
29. $4b^2 - 17b + 4 = 0$ {$\frac{1}{4}$, 4}
30. $6x^2 + 4x - 10 = 0$ {$-\frac{5}{3}$, 1}
31. $2m^2 - 11m = 21$ {$-\frac{3}{2}$, 7}
32. $3y^2 + 13y = 10$ {-5, $\frac{2}{3}$}
33. $r = 14 - 3r^2$ {$-\frac{7}{3}$, 2}
34. $-23t = 6 - 4t^2$ {$-\frac{1}{4}$, 6}
35. $3m^2 - 147 = 0$ {-7, 7}
36. $5t^2 - 5 = 0$ {-1, 1}
37. $3x^2 + 18x + 15 = 0$ {-5, -1}
38. $4z^2 + 16z - 48 = 0$ {-6, 2}
39. $-y^2 + 7y = 12$ {3, 4}
40. $-p^2 + p = -30$ {-5, 6}
41. $-x^2 = 32 - 12x$ {4, 8}
42. $-a^2 = -5a - 24$ {-3, 8}

Solve each equation by the two methods shown in example 5.

43. $9c^2 - 1 = 0$ {$-\frac{1}{3}$, $\frac{1}{3}$}
44. $16x^2 - 1 = 0$ {$-\frac{1}{4}$, $\frac{1}{4}$}
45. $16y^2 - 4 = 0$ {$-\frac{1}{2}$, $\frac{1}{2}$}

For the answers to exercises 46-57, see the Teacher's Guide.

Write quadratic equations whose solution sets are the following.

46. {1, -7}
47. {-3, -1}
48. {10, 90}
49. {1, 2}
50. {$\frac{1}{2}$, 2}
51. {$-\frac{1}{2}$, 5}
52. {$\frac{1}{3}$, $\frac{1}{2}$}
53. {$\frac{1}{4}$, $\frac{1}{6}$}
54. {$-\frac{2}{3}$, $\frac{4}{5}$}
55. {$-\frac{1}{6}$, $-\frac{1}{2}$}
56. {6}
57. {-1}

10.3 Graphing $y = ax^2 + bx + c$

Objective: To identify and graph parabolas.

Equations of the form $y = ax^2 + bx + c$ can be graphed on a set of coordinate axes. We will begin with the most basic case.

Examples

1 Graph $y = x^2$.

In this case, the y-coordinate must be the square of the x-coordinate. Carefully make a table of values. Note that, unlike graphing a line, a single y-coordinate is paired with *two* different x-coordinates.

x	y
0	0
1	1
−1	1
2	4
−2	4
3	9
−3	9

Calculator Hint

You can learn how to graph a quadratic equation on a graphing calculator in Activity 4 on pages A6-A7.

2 Graph $y = x^2 + 3$.

To obtain each y value, first square the value of x and then add 3.

x	y
0	3
1	4
−1	4
2	7
−2	7
3	12
−3	12

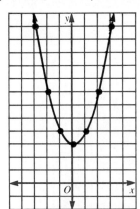

Notice that the graph lies 3 units above the graph of $y = x^2$. (See example 1.)

Graphs of equations of the form $y = ax^2 + bx + c$ are called **parabolas**. The coefficient of x^2, a, determines how "narrow" or "wide" the parabola will be.

Example

3 Graph (a) $y = 2x^2$ and (b) $y = \frac{1}{2}x^2$.

(a)

x	y
0	0
1	2
−1	2
2	8
−2	8

(b)

x	y
0	0
1	$\frac{1}{2}$
−1	$\frac{1}{2}$
2	2
−2	2

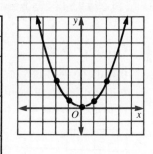

If the coefficient of x^2 is negative, the parabola will open downward. (It will "frown" instead of "smile".)

Example

4 Graph $y = -x^2$.

x	y
0	0
1	−1
−1	−1
2	−4
−2	−4

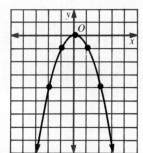

You have seen enough parabolas by now to agree that every one is **symmetric** about some line. A parabola can be "folded" exactly in half at this line. Furthermore, the turning point, or **vertex**, of the parabola, whether a *maximum* or a *minimum*, must lie on this line. This line is called the **axis of symmetry**.

Some experimentation should convince you that the following statement is true.

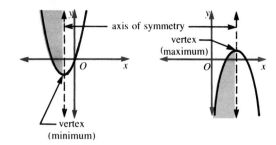

Equation of the Axis of Symmetry of a Parabola

For the graph of $y = ax^2 + bx + c$, the axis of symmetry is $x = -\dfrac{b}{2a}$.

If the coefficient of x^2, a, is 1, then the axis of symmetry is $x = -\dfrac{b}{2}$. The axis of symmetry for a graph where $b = 0$ is the y-axis.

Examples

5 Find the axis of symmetry for the graph of $y = x^2 - 3x - 4$.

Since the coefficient of x^2 is 1, use $x = -\dfrac{b}{2}$.

$x = -\left(\dfrac{-3}{2}\right)$ or $\dfrac{3}{2}$. The axis of symmetry is $x = \dfrac{3}{2}$.

6 Graph $y = x^2 - 3x - 4$.

The axis of symmetry provides a "starting place" for the graph. Since the vertex is on the axis of symmetry, find the value of y corresponding to $x = \dfrac{3}{2}$. Then, plot several points on either side of the axis.

x	y
$\dfrac{3}{2}$	$-\dfrac{25}{4}$
1	-6
2	-6
0	-4
3	-4

Notice that the pairs of values chosen for x (1 and 2, 0 and 3) are the same distance from the axis of symmetry.

The vertex is $\left(\dfrac{3}{2}, -\dfrac{25}{4}\right)$.

Exercises

Max: 1-16, 1-29, 1-3; Avg: 1-16, 1-29; Min: 1-16, 1-27 odds

Exploratory Describe or explain the following.

1. axis of symmetry
2. $y = ax^2 + bx + c$
3. $x = -\dfrac{b}{2a}$

Answer the following. For answers to exercises 1-7, see the Teacher's Guide.

4. Why does the graph of $y = -x^2$ open downward?
5. In general, under what circumstances will the graph of $y = ax^2 + bx + c$ open upward? Open downward?
6. How do the graphs of $y = x^2$ and $y = x^2 + 5$ compare?
7. What is the axis of symmetry for the graphs of the equations in exercise 6?

Name the axis of symmetry for the following.

8. $y = x^2 + 4x + 5$ $x = -2$
9. $y = x^2 - 2x + 7$ $x = 1$
10. $y = 2x^2 - 8x + 11$ $x = 2$

State whether the following equations describe parabolas.

11. $y = 3x$ no
12. $x + y = 4$ no
13. $y = x^2 + 4$ yes
14. $2x + 3y = 6$ no
15. $x - 2y = 5$ no
16. $x^2 - y = 5$ yes

For answers to exercises 1-12, see the Teacher's Guide.

Written Graph the following equations.

1. $y = x^2 + 2$
2. $y = x^2 - 5$
3. $y = x^2 - 4x + 1$
4. $y = x^2 - 2x - 4$
5. $y = x^2 + 4x$
6. $y = x^2 - 2x - 15$
7. $y = -x^2 + 2$
8. $y = -x^2 - 9$
9. $y = 3x^2$
10. $y = 4x^2$
11. $y = 2x^2 + 2x - 3$
12. $y = \dfrac{1}{2}x^2 + 1$

Name the axis of symmetry and the vertex for the graphs of the following.

13. $y = x^2$ $x = 0, (0,0)$
14. $y = 2x^2$ $x = 0, (0,0)$
15. $y = x^2 + 4x - 9$ $x = -2, (-2,-13)$
16. $y = x^2 + 2x$ $x = -1, (-1,-1)$
17. $y = -x^2 + 4x - 1$ $x = 2, (2,3)$
18. $y = -x^2 + 1$ $x = 0, (0,1)$
19. $y = 2x^2 - 6x$ $x = \dfrac{3}{2}, (\dfrac{3}{2}, -\dfrac{9}{2})$
20. $y = 2x^2 + x + 3$ $x = -\dfrac{1}{4}, (-\dfrac{1}{4}, \dfrac{2}{8})$
21. $y = 3x^2 - x + 1$ $x = \dfrac{1}{6}, (\dfrac{1}{6}, \dfrac{11}{12})$

Graph the following equations by finding the axis of symmetry, the vertex, and three points on either side of the axis of symmetry. For answers to exercises 22-29, see the Teacher's Guide.

22. $y = x^2 - 2x$
23. $y = x^2 - 4x + 3$
24. $y = x^2 - 6x + 1$
25. $y = x^2 + 6x - 5$
26. $y = -x^2 + 2x$
27. $y = -x^2 + 2x - 3$

Answer the following.

28. Graph $x = y^2$. What is its axis of symmetry?
29. Describe the graph of $x = y^2 + 1$. Check your answer by graphing.

Challenge Complete the following. For answers to exercises 1-3, see the Teacher's Guide.

1. Graph $y = \sqrt{x}$.
2. Graph $y = x^3$.
3. Graph $y = x^2 + 2$ and $x + y = 8$ on the same set of axes. What do you notice about the graphs?

10.4 Solving Quadratic Equations by Graphing

Objective: To solve quadratic equations by graphing.

Quadratic equations can be solved by graphing, if the graphs are carefully drawn.

Experiment 1

a. Solve the quadratic equation, $x^2 - 3x - 4 = 0$, by factoring.

b. Graph the associated equation, $y = x^2 - 3x - 4$.

Solution by factoring

$$x^2 - 3x - 4 = 0$$
$$(x + 1)(x - 4) = 0$$
$$x + 1 = 0 \text{ or } x - 4 = 0$$
$$x = -1 \qquad\qquad x = 4$$

The solution set is $\{-1, 4\}$.

Graph of $y = x^2 - 3x - 4$

Experiment 2

a. Solve the quadratic equation, $x^2 + 6x + 9 = 0$, by factoring.

b. Graph the associated equation, $y = x^2 + 6x + 9$.

Solution by factoring

$$x^2 + 6x + 9 = 0$$
$$(x + 3)(x + 3) = 0$$
$$x + 3 = 0 \text{ or } x + 3 = 0$$
$$x = -3 \qquad\qquad x = -3$$

The solution set is $\{-3\}$.

Graph of $y = x^2 + 6x + 9$

Calculator Hint

You can learn how to solve a quadratic equation on a graphing calculator in Activity 4 on pages A6-A7.

In the experiments, the solutions of the quadratic equations are the same as the x-coordinates of the points where the associated parabolas cross the x-axis. Will this always work?

We know that the equation of the x-axis is $y = 0$. When we graph $y = x^2 - 3x - 4$, the points at which the parabola crosses the x-axis are where $y = 0$, and therefore are also the points at which $x^2 - 3x - 4 = 0$. Thus, the x-coordinates of these points are the values that make $x^2 - 3x - 4 = 0$. Therefore, they are roots of the equation $x^2 - 3x - 4 = 0$.

Examples

1 Solve $x^2 - 2x - 3 = 0$ by graphing.

x	y
1	-4
0	-3
2	-3
-1	0
3	0
-2	5
4	5

In this case we can check our work by factoring.

$$x^2 - 2x - 3 = 0$$
$$(x + 1)(x - 3) = 0$$
$$x + 1 = 0 \quad \text{or} \quad x - 3 = 0$$
$$x = -1 \qquad\qquad x = 3$$

The solution set is $\{-1, 3\}$.

2 a. Solve $x^2 - 3x - 5 = 0$ by factoring.
b. Graph the associated equation, $y = x^2 - 3x - 5$.

The trinomial $x^2 - 3x - 5$ is not factorable. However, the graph of the associated equation crosses the x-axis.

x	y
$\frac{3}{2}$	$-7\frac{1}{4}$
1	-7
2	-7
0	-5
3	-5
-1	-1
4	-1

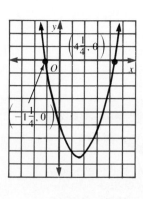

Our conclusion is that the quadratic equation does have roots even though they cannot be found by factoring. Precise roots cannot be found by graphing either, but a fairly good approximation is possible if the graph is carefully drawn. In this case, the solution set can be approximated to be $\left\{-1\frac{1}{4}, 4\frac{1}{4}\right\}$.

Example

3 Solve $x^2 - x + 5 = 0$ by graphing.

x	y
$\frac{1}{2}$	$4\frac{3}{4}$
0	5
1	5
-1	7
2	7
-2	11
3	11

In this case, the parabola does *not* cross the x-axis. Therefore, the equation $x^2 - x + 5 = 0$ has no real roots.

Here are two points to remember about solving quadratic equations by graphing.

1. Even if a quadratic cannot be solved by factoring, it *may* have real roots. These roots can be *estimated* by a carefully drawn graph.
2. If the graph of a quadratic, $y = ax^2 + bx + c$, does not cross the x-axis, then the associated quadratic equation, $ax^2 + bx + c = 0$, does *not* have real roots.

Exercises

Max: 1-10, 1-30, 1-2; Avg: 1-10, 1-29 odds; Min: 5-9 odds, 1-29 odds

Exploratory Answer the following. For answers to exercises 1-4, see the Teacher's Guide.

1. What are the steps used in solving a quadratic equation by graphing?
2. If $ax^2 + bx + c$ is not factorable, can the graph of $y = ax^2 + bx + c$ cross the x-axis? Explain.

380 SOLVING QUADRATIC EQUATIONS

3. Under what circumstances will the graph of a quadratic equation touch the x-axis at exactly one point?

4. Which equation has a graph that touches the x-axis at exactly one point?
   a. $y = x^2 + 4$ b. $y = x^2 + x + 1$
   c. $y = x^2 + 4x + 4$

State the solution set of each quadratic equation whose associated parabola is graphed below.

5. {-1,1}

6. {-3,1}

7. ∅

8. {0,2}

9. {2}

10. {-1½, 1½}

For graphs to exercises 1-30, see students' work.

Written Solve each equation by graphing. Then check the solutions by factoring.

1. $x^2 - 4 = 0$ {-2, 2}
2. $x^2 - 16 = 0$ {-4, 4}
3. $x^2 - 3x - 4 = 0$ {-1, 4}
4. $x^2 + 7x + 10 = 0$ {-5, -2}
5. $x^2 + 3x - 4 = 0$ {-4, 1}
6. $x^2 + 9x + 18 = 0$ {-6, -3}
7. $x^2 - 5x + 6 = 0$ {2, 3}
8. $x^2 + 10x + 16 = 0$ {-8, -2}
9. $4x^2 - 1 = 0$ {-½, ½}
10. $2x^2 - 18 = 0$ {-3,3}
11. $x^2 + 10x + 25 = 0$ {-5}
12. $x^2 + 6x + 9 = 0$ {-3}
13. $x^2 - 7x = -6$ {1,6}
14. $x^2 - 5x = 24$ {-3, 8}
15. $x^2 - 8x = -16$ {4}

Solve each equation by graphing. Estimate the roots as carefully as possible. If the equation appears to have no solution, explain why.

16. $x^2 - 2 = 0$ {-1⅖, 1⅖}
17. $x^2 - 5 = 0$ {-2⅕, 2⅕}
18. $x^2 + 3 = 0$ ∅
19. $x^2 - x - 7 = 0$ {-2 1/10, 3 1/10}
20. $x^2 + 4x = 1$ {-4⅕, ⅕}
21. $2x^2 - 1 = 0$ {-7/10, 7/10}
22. $x^2 - x + 4 = 0$ ∅
23. $x^2 - 2x = 5$ {-1⅖, 3⅖}
24. $2x^2 + x + 3 = 0$ ∅
25. $2x^2 + 3x - 1 = 0$ {-7/4, ¼}
26. $3x^2 + 4x - 1 = 0$ {-1½, ⅕}
27. $3x^2 + 5x + 1 = 0$ {-1⅖, -⅕}
28. $2x^2 - 3x + 5 = 0$ ∅
29. $5x^2 + 2x + 6 = 0$ ∅
30. $4x^2 - x + 3 = 0$ ∅

Challenge Answer the following.

1. For what value of k will the graph of $y = x^2 + 26x + k$ intersect the x-axis at exactly one point? Explain. 169

2. Graph $y = x^3 + 1$. What do you notice about the graph? Answers may vary.
   For graph to exercise 2, see the Teacher's Guid

10.5 Completing the Square

Objective: To solve quadratic equations by making perfect square trinomials.

Solving quadratic equations graphically can give a good approximation of the roots of the equation, even if the quadratic is *not* factorable.

However, the exact roots of a quadratic equation are often needed. A method called **completing the square** can be used to find such roots. This method is based on perfect square trinomials and square roots.

i. $x^2 + 6x + 9$ is a perfect square trinomial because it factors into $(x + 3)(x + 3) = (x + 3)^2$.

ii. $x^2 + 5x + \frac{25}{4}$ is a perfect square trinomial. It factors into $\left(x + \frac{5}{2}\right)^2$.

iii. $4x^2 + 4x + 1$ is a perfect square trinomial. It factors into $(2x + 1)^2$.

iv. $x^2 - 36$ is *not* a perfect square trinomial. It *is* factorable, but not into two identical binomials.

Examples

1 Find the value of k that will make $x^2 + 10x + k$ a perfect square trinomial.

Recall that $(x + a)^2 = x^2 + 2ax + a^2$. In this case, $10x$ is the "$2ax$" term. Thus, $2a = 10$ or $a = 5$. Since the last term must be a^2, $k = a^2 = 5^2 = 25$.

2 Find the value of k that will make $x^2 + kx + 49$ a perfect square trinomial.

In this case, $a^2 = 49$, so $a = 7$ or $a = -7$. So for $k = 2a$, $k = 14$ or $k = -14$.

In general, to find the constant term, c, of a perfect square trinomial of the form $x^2 + bx + c$, take half of b, and then square the result.

Example

3 Find the value of m that will make $x^2 + 7x + m$ a perfect square trinomial.

In this case, $b = 7$. Therefore, $m = \left(\frac{7}{2}\right)^2 = \frac{49}{4}$.

Recall that to solve an equation such as $y^2 = 13$, the square root of both sides is found. Thus, $y = \sqrt{13}$ or $y = -\sqrt{13}$. This same concept can be used to solve quadratic equations such as $(x + 3)^2 = 5$.

$$(x + 3)^2 = 5$$
$$x + 3 = \sqrt{5} \text{ or } -\sqrt{5}$$
$$x + 3 = \pm\sqrt{5} \qquad \pm\sqrt{5} \text{ means ``}\sqrt{5} \text{ or } -\sqrt{5}.\text{''}$$
$$x = -3 \pm \sqrt{5} \qquad -3 \pm \sqrt{5} \text{ means ``}-3 + \sqrt{5} \text{ or } -3 - \sqrt{5}.\text{''}$$

The solution set is $\{-3 + \sqrt{5}, -3 - \sqrt{5}\}$.

The following examples use completing the square.

Examples

4 Solve $x^2 + 6x + 2 = 0$.

Note that if $x^2 + 6x + 2$ was a perfect square trinomial, we could proceed as shown above. However, it is not a perfect square trinomial, but it can be changed to one.

$$x^2 + 6x + 2 = 0$$
$$x^2 + 6x = -2 \qquad \text{Subtract 2 from both sides.}$$
$$x^2 + 6x + 9 = -2 + 9 \qquad \left(\frac{6}{2}\right)^2 \text{ or 9 will make the left side a perfect square trinomial. Therefore, add 9 to both sides.}$$
$$(x + 3)^2 = 7 \qquad \text{Factor and simplify.}$$
$$x + 3 = \pm\sqrt{7} \qquad \text{Find the square root of both sides.}$$
$$x = -3 \pm \sqrt{7} \qquad \text{Subtract 3 from both sides.}$$

The solution set is $\{-3 - \sqrt{7}, -3 + \sqrt{7}\}$.

5 Solve $x^2 - 4x + 10 = 0$.

$$x^2 - 4x + 10 = 0$$
$$x^2 - 4x = -10 \qquad \text{Subtract 10 from both sides.}$$
$$x^2 - 4x + 4 = -10 + 4 \qquad \text{Add } \left(\frac{-4}{2}\right)^2 \text{ or 4 to both sides.}$$
$$(x - 2)^2 = -6 \qquad \text{Factor and simplify.}$$
$$x - 2 = \pm\sqrt{-6} \qquad \text{Find the square root of both sides.}$$

Stop. The square root of a negative number is meaningless. Therefore, the quadratic equation has no real solutions.

Example

6 Solve $2x^2 + 8x + 7 = 0$.

If the leading coefficient is not 1, *divide by that coefficient first*. Then, proceed as before. At the end of a solution, a radical expression may have to be simplified.

$2x^2 + 8x + 7 = 0$

$x^2 + 4x + \dfrac{7}{2} = 0$ Divide both sides by 2 so that the coefficient of x^2 is 1.

$x^2 + 4x = -\dfrac{7}{2}$ Subtract $\dfrac{7}{2}$ from both sides.

$x^2 + 4x + 4 = -\dfrac{7}{2} + 4$ Add $\left(\dfrac{4}{2}\right)^2$ or 4 to both sides.

$(x + 2)^2 = \dfrac{1}{2}$ Factor and simplify.

$x + 2 = \pm\sqrt{\dfrac{1}{2}}$ Find the square root of both sides.

$x = -2 \pm \sqrt{\dfrac{1}{2}}$ Subtract 2 from both sides.

$x = -2 \pm \dfrac{\sqrt{2}}{2}$ $\sqrt{\dfrac{1}{2}} = \dfrac{\sqrt{1}}{\sqrt{2}} \cdot \dfrac{\sqrt{2}}{\sqrt{2}} = \dfrac{\sqrt{2}}{2}$

The solution set is $\left\{-2 - \dfrac{\sqrt{2}}{2},\ -2 + \dfrac{\sqrt{2}}{2}\right\}$.

Exercises

Max: 1-7, 1-51, 1-4; Avg: 1-7, 1-51 odds; Min: 1-7, 1-35 odds

Exploratory Answer the following. For answers to exercises 1-7, see the Teacher's Guide

1. What shortcomings does the graphical solution to quadratic equations have? Why is it necessary to have an algebraic method?

2. Does $\sqrt{5}$ name a positive number, a negative number, both, or neither? Explain.

3. Solve the equation $x^2 = 121$ without subtracting 121 from both sides. Express the answer in radical form, then simplify.

4. Explain the use of the $\pm$ symbol. Use $\pm\sqrt{7}$ as an example.

SOLVING QUADRATIC EQUATIONS

5. Explain how to find a value of c that will make $x^2 + mx + c$ a perfect square trinomial.

6. Explain how to find a value of m that will make $x^2 + mx + c$ a perfect square trinomial.

Give reasons for each step in the following solution of $x^2 + 8x + 3 = 0$.

7. a. $x^2 + 8x + 3 = 0$
    b. $x^2 + 8x = -3$
    c. $x^2 + 8x + 16 = -3 + 16$
    d. $(x + 4)^2 = 13$
    e. $x + 4 = \pm\sqrt{13}$
    f. $x = -4 \pm \sqrt{13}$
    g. The solution set is $\{-4 - \sqrt{13}, -4 + \sqrt{13}\}$.

Written Solve each equation by finding the square root of both sides.

1. $(x + 1)^2 = 2$ $\{-1 \pm \sqrt{2}\}$
2. $(x + 3)^2 = 7$ $\{-3 \pm \sqrt{7}\}$
3. $(x - 5)^2 = 13$ $\{5 \pm \sqrt{13}\}$
4. $(x + 10)^2 = 3$ $\{-10 \pm \sqrt{3}\}$
5. $(x + 6)^2 = 6$ $\{-6 \pm \sqrt{6}\}$
6. $(x - 2)^2 = 14$ $\{2 \pm \sqrt{14}\}$
7. $(x - 1)^2 = \frac{1}{2}$ $\{\frac{2 \pm \sqrt{2}}{2}\}$
8. $(x + 2)^2 = \frac{1}{3}$ $\{\frac{-6 \pm \sqrt{3}}{3}\}$
9. $\left(x + \frac{3}{4}\right)^2 = \frac{2}{3}$ $\{\frac{-9 \pm 4\sqrt{6}}{12}\}$
10. $\left(x - \frac{3}{2}\right)^2 = \frac{1}{2}$ $\{\frac{3 \pm \sqrt{2}}{2}\}$
11. $\left(x + \frac{1}{3}\right)^2 = \frac{3}{2}$ $\{\frac{-2 \pm 3\sqrt{6}}{6}\}$
12. $\left(x - \frac{1}{4}\right)^2 = \frac{1}{6}$ $\{\frac{3 \pm 2\sqrt{6}}{12}\}$

Find the value of c that will make each expression a perfect square trinomial.

13. $x^2 + 6x + c$ 9
14. $x^2 - 14x + c$ 49
15. $x^2 + 26x + c$ 169
16. $x^2 + x + c$ $\frac{1}{4}$
17. $x^2 + 3x + c$ $\frac{9}{4}$
18. $9x^2 + 24x + c$ 16
19. $x^2 + cx + 25$ ± 10
20. $x^2 + cx + 64$ ± 16
21. $x^2 + cx + 121$ ± 22

Solve each equation by completing the square.

22. $x^2 - 2x = 24$ $\{-4, 6\}$
23. $x^2 + 4x = 96$ $\{-12, 8\}$
24. $x^2 + 3x = 88$ $\{-11, 8\}$
25. $x^2 - 3x = 10$ $\{-2, 5\}$
26. $x^2 - 8x + 15 = 0$ $\{3, 5\}$
27. $x^2 - 6x + 8 = 0$ $\{2, 4\}$
28. $x^2 + 8x - 20 = 0$ $\{-10; 2\}$
29. $x^2 + 2x - 48 = 0$ $\{-8, 6\}$
30. $x^2 - 7x + 12 = 0$ $\{3, 4\}$
31. $x^2 - 10x + 21 = 0$ $\{3, 7\}$
32. $x^2 + 8x - 84 = 0$ $\{-14, 6\}$
33. $x^2 + 3x - 180 = 0$ $\{-15, 12\}$
34. $x^2 + 3x - 40 = 0$ $\{-8, 5\}$
35. $x^2 + 12x + 4 = 0$
36. $x^2 - 8x + 14 = 0$ $\{4 \pm \sqrt{2}\}$
37. $x^2 + 5x - 8 = 0$
38. $x^2 - 5x - 10 = 0$
39. $x^2 - 3x + 1 = 0$
40. $4x^2 + 19x - 5 = 0$ $\{-5, \frac{1}{4}\}$
41. $6x^2 + 7x - 3 = 0$ $\{-\frac{3}{2}, \frac{1}{3}\}$
42. $3x^2 - 14x + 8 = 0$ $\{\frac{2}{3}, 4\}$
43. $2x^2 + 11x - 21 = 0$
44. $12x^2 - 17x - 5 = 0$
45. $3x^2 + 4x - 15 = 0$
46. $6x^2 + 2x + 3 = 0$ ∅
47. $2x^2 - 3x + 4 = 0$ ∅
48. $4x^2 - x + 3 = 0$ ∅
49. $6x^2 + 14x + 10 = 0$ ∅
50. $5x^2 - 7x + 2 = 0$ $\{\frac{2}{5}, 1\}$
51. $3x^2 - 6x + 10 = 0$ ∅

For answers to exercises 1-4, see the Teacher's Guide.

35. $\{-6 \pm 4\sqrt{2}\}$
37. $\{\frac{-5 \pm \sqrt{57}}{2}\}$
38. $\{\frac{5 \pm \sqrt{65}}{2}\}$
39. $\{\frac{3 \pm \sqrt{5}}{2}\}$
43. $\{-7, \frac{3}{2}\}$
44. $\{-\frac{1}{4}, \frac{5}{3}\}$
45. $\{-3, \frac{5}{3}\}$

Challenge Describe and graph the following equations.

1. $x^2 - 6x + 9 + y^2 + 10y + 25 = 49$
2. $x^2 + 6x + 7 + y^2 + 4y + 1 = 4$
3. $x^2 + y^2 + 4x + 6y - 3 = 0$
4. $x^2 + y^2 + 10x - 14y - 50 = 0$

10.6 The Quadratic Formula

Objective: To solve quadratic equations by using the quadratic formula.

Our work with algebraic solutions of quadratic equations has led to the conclusion that in all cases where a solution exists, we can find it by completing the square.

This means that there is always some way of manipulating a, b, and c—the coefficients of x^2, x, and the constant in the equation $ax^2 + bx + c = 0$—to find the value of x in terms of a, b, and c.

Such manipulations have been completed for the general case, resulting in a *formula* that can be used in every individual case. This formula is called the **quadratic formula**. It is a *very* important mathematical tool.

The following shows how the method of completing the square is used to derive the quadratic formula. On the right, the formula is derived in the general case. On the left, a specific case is given to help you follow along.

Specific

$$3x^2 + 11x + 5 = 0$$

$$x^2 + \frac{11}{3}x + \frac{5}{3} = 0$$

$$x^2 + \frac{11}{3}x = -\frac{5}{3}$$

$$x^2 + \frac{11}{3}x + \left(\frac{11}{6}\right)^2 = \left(\frac{11}{6}\right)^2 - \frac{5}{3}$$

$$\left(x + \frac{11}{6}\right)^2 = \frac{121}{36} - \frac{60}{36}$$

$$\left(x + \frac{11}{6}\right)^2 = \frac{61}{36}$$

$$x + \frac{11}{6} = \pm\sqrt{\frac{61}{36}}$$

$$x = -\frac{11}{6} \pm \sqrt{\frac{61}{36}}$$

$$x = -\frac{11}{6} \pm \frac{\sqrt{61}}{6}$$

$$x = \frac{-11 \pm \sqrt{61}}{6}$$

General

$$ax^2 + bx + c = 0$$

$$x^2 + \frac{b}{a}x + \frac{c}{a} = 0$$

$$x^2 + \frac{b}{a}x = -\frac{c}{a}$$

$$x^2 + \frac{b}{a}x + \left(\frac{b}{2a}\right)^2 = \left(\frac{b}{2a}\right)^2 - \frac{c}{a}$$

$$\left(x + \frac{b}{2a}\right)^2 = \frac{b^2}{4a^2} - \frac{4ac}{4a^2}$$

$$\left(x + \frac{b}{2a}\right)^2 = \frac{b^2 - 4ac}{4a^2}$$

$$x + \frac{b}{2a} = \pm\sqrt{\frac{b^2 - 4ac}{4a^2}}$$

$$x = -\frac{b}{2a} \pm \sqrt{\frac{b^2 - 4ac}{4a^2}}$$

$$x = -\frac{b}{2a} \pm \frac{\sqrt{b^2 - 4ac}}{2a}$$

$$x = \frac{-b \pm \sqrt{b^2 - 4ac}}{2a}$$

Quadratic Formula

If $ax^2 + bx + c = 0$ and $a \ne 0$, then $x = \dfrac{-b \pm \sqrt{b^2 - 4ac}}{2a}$.

Examples

1 Solve $x^2 + 3x - 5 = 0$ using the quadratic formula.

$$x = \frac{-b \pm \sqrt{b^2 - 4ac}}{2a}$$

$$= \frac{-3 \pm \sqrt{3^2 - 4(1)(-5)}}{2(1)} \quad \text{Substitute the following values into the formula: } a = 1, b = 3, c = -5.$$

$$= \frac{-3 \pm \sqrt{9 - (-20)}}{2}$$

$$= \frac{-3 \pm \sqrt{29}}{2} \quad \text{The solution set is } \left\{\frac{-3 - \sqrt{29}}{3}, \frac{-3 + \sqrt{29}}{3}\right\}.$$

2 Solve $x^2 + 3x + 11 = 0$ using the quadratic formula.

$$x = \frac{-b \pm \sqrt{b^2 - 4ac}}{2a}$$

$$= \frac{-3 \pm \sqrt{(3)^2 - 4(1)(11)}}{2(1)} \quad a = 1, b = 3, c = 11$$

$$= \frac{-3 \pm \sqrt{-35}}{2}$$

Stop. Because the square root of a negative number cannot be found in $\mathcal{R}$, the equation has no real roots.

The expression $b^2 - 4ac$ is called the **discriminant** of a quadratic equation. As shown in example 2, if $b^2 - 4ac$ is negative, the equation has no real roots.

Example

3 Find the value of the discriminant for $2x^2 + 4x + 9 = 0$. Tell if the equation has real roots.

$$b^2 - 4ac = (4)^2 - 4(2)(9) \quad a = 2, b = 4, c = 9$$
$$= -56$$

Because the value of the discriminant is negative, the equation has no real roots.

What can be concluded if $b^2 - 4ac$ is exactly equal to zero? You will be asked to check your answer in the exercises.

Another relationship exists between a, b, and c, and the roots of a quadratic equation.

Sum and Product of Roots

> 1. The sum of the roots of $ax^2 + bx + c = 0$ is $-\dfrac{b}{a}$.
> 2. The product of the roots of $ax^2 + bx + c = 0$ is $\dfrac{c}{a}$.

In the exercises, you will be asked to show that these statements are true.

Example

4 Find the sum and the product of the roots of $3x^2 + 24x - 1 = 0$.

We have $a = 3$, $b = 24$, and $c = -1$.

The sum of the roots is $-\dfrac{b}{a} = -\dfrac{24}{3} = -8$.

The product of the roots is $\dfrac{c}{a} = \dfrac{-1}{3} = -\dfrac{1}{3}$.

Exercises

Max: 1-23, 1-32, 1-3; Avg: 1-23, 1-31 odds; Min: 5-23 odds, 1-31 odds

Exploratory Answer the following. For answers to exercises 1-23, see the Teacher's Guide.

1. What is the quadratic formula? Explain what a, b, and c represent.
2. What is the discriminant of a quadratic equation?
3. What information does the discriminant give you? Give an example.

Write each equation in standard form, $ax^2 + bx + c = 0$. Name a, b, and c.

4. $1 = -4x^2 + 4x$
5. $2x^2 - 1 = 5x$
6. $18x^2 = 1$
7. $1 - x^2 = 12x - 1$
8. $8x = 3 - x^2$
9. $x = -x - x^2$

Find the value of the discriminant for each equation. State whether the equation has real roots.

10. $x^2 - x + 1 = 0$
11. $x^2 - 3x + 4 = 0$
12. $2x^2 + x - 1 = 0$
13. $4x^2 + 4x + 1 = 0$
14. $2x^2 - x - 3 = 0$
15. $2x^2 = 3 - 4x^2$

Answer the following.

16. What is true about the roots of a quadratic equation if its discriminant is zero?
17. What is true about the roots of a quadratic equation if its discriminant is a perfect square?

Find the sum and the product of the roots of each equation.

18. $x^2 - 5x - 7 = 0$
19. $2x^2 - 13x + 2 = 0$
20. $2x^2 + x - 1 = 0$
21. $3x^2 - 4x = -5$
22. $5x^2 - 1 = 0$
23. $-x^2 = 4x - 20$

For answers to exercises 1-32, see the Teacher's Guide.

Written Solve each equation using the quadratic formula.

1. $x^2 - x - 30 = 0$
2. $x^2 + 10x + 16 = 0$
3. $x^2 + 2x - 15 = 0$
4. $x^2 + 13x + 42 = 0$
5. $x^2 - 10x + 24 = 0$
6. $x^2 + 5x - 24 = 0$
7. $x^2 - 5x + 4 = 0$
8. $5x^2 - x - 4 = 0$
9. $3x^2 - 7x - 20 = 0$
10. $4x^2 - 11x - 3 = 0$
11. $6x^2 - x - 15 = 0$
12. $24x^2 - 14x - 5 = 0$
13. $14x^2 + 33x - 5 = 0$
14. $6x^2 + 19x + 15 = 0$
15. $20x^2 + 3x - 2 = 0$
16. $15x^2 + 34x + 15 = 0$
17. $24x^2 - 2x = 15$
18. $3x^2 + 5x = 28$
19. $2x^2 - x - 14 = 0$
20. $4x^2 - 9x + 4 = 0$
21. $13x^2 - 16x - 4 = 0$
22. $2x^2 - 5x + 2 = 0$
23. $x^2 - 13x = 0$
24. $3x^2 - 12x = 0$
25. $7x^2 - 7x - 5 = 0$
26. $20x^2 + 20x - 1 = 0$
27. $7x^2 + 20x - 32 = 0$
28. $6x^2 - 5x - 6 = 0$
29. $8x^2 = 60$
30. $12x^2 - 11x = 3$

State whether the following equations have real roots.

31. $0.01x^2 - 1.020x + 0.001 = 0$
32. $1.34x^2 - 1.1x + 1.02 = 0$

Challenge Complete the following. For answers to exercises 1-3, see the Teacher's Guide.

1. The following is an alternative derivation of the quadratic formula. Give reasons for each step.

    a. $ax^2 + bx + c = 0$
    b. $ax^2 + bx = -c$
    c. $4a^2x^2 + 4abx = -4ac$
    d. $4a^2x^2 + 4abx + b^2 = b^2 - 4ac$
    e. $(2ax + b)^2 = b^2 - 4ac$
    f. $2ax + b = \pm\sqrt{b^2 - 4ac}$
    g. $2ax = -b \pm \sqrt{b^2 - 4ac}$
    h. $x = \dfrac{-b \pm \sqrt{b^2 - 4ac}}{2a}$

2. Show that the sum of the roots of a quadratic equation of the form $ax^2 + bx + c = 0$ is $-\dfrac{b}{a}$. Hint: Use the quadratic formula.

3. Show that the product of the roots of a quadratic equation of the form $ax^2 + bx + c = 0$ is $\dfrac{c}{a}$.

10.7 Problem Solving with Quadratic Equations

Objective: To solve problems using quadratic equations

Many kinds of problems can be described or represented by a quadratic equation. The solutions for these problems can be determined by using the methods for solving quadratic equations.

Example

1 The area of a rectangle is 28 in². The width is 3 inches less than the length. Find both dimensions.

Explore Let x represent the length of the rectangle.
Then $x - 3$ represents the width of the rectangle.

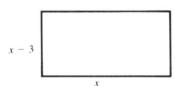

It is usually helpful to make a drawing and label it as completely as possible.

Plan The area of the rectangle is equal to the product of its length and width.

length	times	width	is equal to	area
x	×	$(x - 3)$	=	28

Solve
$$x(x - 3) = 28$$
$$x^2 - 3x = 28 \quad \text{Multiply.}$$
$$x^2 - 3x - 28 = 0 \quad \text{Write in standard form.}$$
$$(x - 7)(x + 4) = 0 \quad \text{Factor.}$$
$$x - 7 = 0 \text{ or } x + 4 = 0 \quad \text{Set each factor equal to zero.}$$
$$x = 7 \qquad\qquad x = -4 \quad \text{Solve.}$$

The length of the rectangle must be 7 inches since -4 inches is not a reasonable answer. Length cannot be negative.
The width of the rectangle is $7 - 3$ or 4 inches.

Examine The width of the rectangle is 4 inches, which is 3 inches less than its length, 7 inches. Since the area of the rectangle is (7 in.) × (4 in.) or 28 in², the solution is correct.

Examples

2 Two numbers differ by 2. The sum of their squares is 130. Find the numbers.

Explore   Let x represent the greater number.
Then $x - 2$ represents the lesser number.

Plan

The square of the greater number plus the square of the lesser number is 130.

$$x^2 + (x - 2)^2 = 130$$

Solve
$$x^2 + (x - 2)^2 = 130$$
$$x^2 + x^2 - 4x + 4 = 130$$
$$2x^2 - 4x - 126 = 0$$
$$x^2 - 2x - 63 = 0$$
$$(x - 9)(x + 7) = 0$$
$$x - 9 = 0 \text{ or } x + 7 = 0$$
$$x = 9 \qquad\qquad x = -7$$

For $x = 9$, the two numbers are 9 and $9 - 2$ or 7.
For $x = -7$, the two numbers are -7 and $-7 - 2$ or -9.

Examine   Since $9^2 + 7^2 = 81 + 49$ or 130 and $(-7)^2 + (-9)^2 = 49 + 81$ or 130, both solutions are correct.

3 The measure of the hypotenuse of a right triangle is 30. The measure of one leg is 3 times the measure of the other. Find the measures of both legs.

Explore   Let x represent the measure of the shorter leg.
Then $3x$ represents the measure of the longer leg.

Plan   Use the Pythagorean Theorem to write an equation.

Solve
$$a^2 + b^2 = c^2$$
$$x^2 + (3x)^2 = 30^2$$
$$x^2 + 9x^2 = 900$$
$$10x^2 = 900$$
$$x^2 = 90$$
$$x = \pm\sqrt{90}$$
$$x = \pm 3\sqrt{10}$$

You could also use the Quadratic Formula to solve this equation.

The measure of the shorter leg must be $3\sqrt{10}$, since $-3\sqrt{10}$ is not a reasonable solution. The measure of the longer leg is $3(3\sqrt{10})$ or $9\sqrt{10}$.   Examine this solution.

Examples

4 In right triangle *KLM*, the measure of altitude $\overline{MD}$ that is drawn to hypotenuse $\overline{KL}$ is 10. If *KD* is 21 less than *DL*, find *KL*.

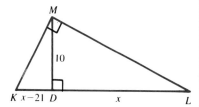

Explore Let x represent the measure of $\overline{DL}$.
Then $x - 21$ represents the measure of $\overline{KD}$.

Plan To solve this problem, we must use Theorem 7–6, which states that the measure of the altitude of the hypotenuse of a right triangle is the mean proportional between the measures of the segments of the hypotenuse.

$$\frac{x - 21}{10} = \frac{10}{x} \qquad \frac{KD}{MD} = \frac{MD}{DL}$$

Solve
$$x(x - 21) = 100$$
$$x^2 - 21x = 100$$
$$x^2 - 21x - 100 = 0$$
$$(x - 25)(x + 4) = 0$$
$$x = 25 \quad \text{or} \quad x = -4$$

If $DL = 25$, then $KD = 25 - 21$ or 4. Since $KL = KD + DL$, we have $KL = 25 + 4$ or 29. Examine this solution.

5 A rectangular picture is 18 inches by 10 inches. It is placed in a frame of uniform width. If the area of the picture and the frame is 384 in^2, find the width of the frame.

Explore Let x represent the width of the frame.
Then $18 + 2x$ represents the length and $10 + 2x$ represents the width of the picture and the frame.

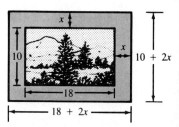

Plan $(18 + 2x)(10 + 2x) = 384$ length × width = area

Solve
$$180 + 36x + 20x + 4x^2 = 384$$
$$4x^2 + 56x - 204 = 0$$
$$x^2 + 14x - 51 = 0$$
$$(x + 17)(x - 3) = 0$$
$$x = -17 \quad \text{or} \quad x = 3$$

The width of the frame must be 3 inches since −17 inches is not a reasonable solution.

Examine The length of the picture and frame is 18 + 2(3) or 24 inches. The width of the picture and frame is 10 + 2(3) or 16 inches. Since (24 in.) × (16 in.) = 384 in², the solution is correct.

Exercises

Max: 1-8, 1-27; Avg: 1-8, 1-27; Min: 1-7 odds, 1-27 odds

Exploratory For each problem, define a variable. Then write a quadratic equation to represent the problem. Do *not* solve. For answers to exercises 1-8, see the Teacher's Guide.

1. The area of a rectangle is 24 in². The length is 5 in. greater than the width. Find both dimensions.
2. The area of a rectangle is 90 cm². The length is 1 cm more than the width. Find both dimensions.
3. The area of a rectangle is 18 m². The length is 1 m more than 4 times the width. Find both dimensions.
4. The area of a rectangle is 12 cm². The measures of the sides are in the ratio of 3:16. Find both dimensions.
5. A number is 2 larger than another number. Their product is 35. Find the numbers.
6. Two numbers differ by 3. The sum of their squares is 45. Find the numbers.
7. Two numbers differ by 5. Their product is 24. Find the numbers.
8. Find two consecutive integers whose product is 72.

<u>1</u>. 3 in. by 8 in.   <u>2</u>. 9 cm by 10 cm   <u>3</u>. 2 m by 9 m   <u>4</u>. 1.5 cm by 8 cm

Written Solve the equation and answer the problem in each Exploratory Exercise.

1. exercise 1
2. exercise 2
3. exercise 3
4. exercise 4
5. exercise 5
6. exercise 6
7. exercise 7
8. exercise 8

<u>5</u>. -7, -5; 5, 7   <u>6</u>. -6, -3; 3, 6   <u>7</u>. -8, -3; 3, 8   <u>8</u>. -9, -8; 8, 9

Solve each problem.

9. A number is twice another number. Their product is 98. Find the numbers. -14, -7; 7, 14
10. A number is 3 more than twice another number. Their product is 65. Find the numbers. -10, -6.5; 5, 13
11. If the product of two consecutive integers is decreased by 20 times the greater integer, the result is 442. Find the integers. -14, -13; 33, 34
12. The area of $\triangle DEF$ is 30 m². The measure of altitude $\overline{DR}$ drawn to $\overline{EF}$ is 7 m more than EF. Find DR and EF. DR = 12, EF = 5
13. The area of $\triangle RST$ is 48 in². The measure of altitude $\overline{RD}$ drawn to $\overline{ST}$ is 4 in. less than ST. Find RD and ST. RD = 8, ST = 12
14. The area of $\triangle RSW$ is 36 cm². The measure of altitude $\overline{RQ}$ drawn to $\overline{SW}$ is one-half SW. Find RQ and SW. RQ = 6, SW =
15. The area of $\triangle RTU$ is 45 in². The measure of $\overline{RU}$ is 2.5 times the measure of altitude $\overline{TM}$ drawn to $\overline{RU}$. Find TM and RU. TM = 6, RU = 15
16. The hypotenuse of a right triangle is 15 cm long. The length of one leg is 3 cm more than the length of the other. Find the lengths of both legs. 9, 12

17. A rectangular picture is 9 inches by 14 inches. It is placed in a frame of uniform width that has an area of 78 in². Find the width of the frame. $1\frac{1}{2}$ in.

18. A rectangular garden 25 ft by 50 ft is increased on all sides by the same amount. Its area increases 400 ft². How much does each dimension increase? 5 ft

19. A rectangular flower bed in Pine City Park is 20 ft by 28 ft. It is surrounded by a walk of uniform width. If the area of the flower bed and walk is 1008 ft², find the width of the walk. 4 ft

20. The length of a rectangular garden is 6 ft more than its width. A walkway 3 ft wide surrounds the outside of the garden. The area of the walkway is 288 ft². Find the dimensions of the garden.

21. A rectangular park is 30 m long by 20 m wide. Plans were made to double the area by adding a strip at one end and another of the same width on one side. Find the width of the strips. 10 m

22. The Hillside Garden Club wants to double the area of its rectangular display of roses. If it is now 6 m by 4 m, by what equal amount must each dimension be increased? 2 m

20. 18 ft by 24 ft

For each figure, find the value of x.

23. 2.52

24. 5.94

25. 1.38

26. 8

27. 27

Mixed Review

Name the axis of symmetry and the vertex for the graph of each equation.

1. $y = -x^2 - 2$
   $x = 0$, $(0, -2)$

2. $y = x^2 + 8x - 6$
   $x = -4$, $(-4, -22)$

3. $y = 3x^2 - 4x + 4$
   $x = \frac{2}{3}$, $(\frac{2}{3}, \frac{8}{3})$

Solve each equation.

4. $x^2 - 9x + 20 = 0$ $\{4, 5\}$
5. $2y^2 - 3y - 2 = 0$ $\{-\frac{1}{2}, 2\}$
6. $20r^2 + 11r = 3$ $\{-\frac{3}{4}, \frac{1}{5}\}$
7. $a^2 + 10a + 1 = 0$ $\{-5 \pm 2\sqrt{6}\}$
8. $n^2 - 8n = 2$ $\{4 \pm 3\sqrt{2}\}$
9. $9k = -5k^2 - 3$ $\{\frac{-9 \pm \sqrt{21}}{10}\}$

10. Factor $x^4y^2 - 64x^2y^2$. $x^2y^2(x-8)(x+8)$
11. Solve $2x^2 + 3x = -6$ by graphing. ∅
12. Find the locus of points equidistant from $(-1, -1)$, $(3, 7)$, and $(7, -1)$. $(3, 2)$
13. Find an equation for the parabola with focus $(3, 1)$ and directrix $y = -1$.
14. Find the value of k that will make $4x^2 + kx + 49$ a perfect square trinomial. ± 28
15. Write an equation for the plane 3 units in the negative z direction from the xy-plane. $z = -3$

13. $y = \frac{1}{4}x^2 - \frac{3}{2}x + \frac{9}{4}$

10.8 Systems of Equations

Objective: To solve systems of equations graphically or algebraically.

Systems of linear equations can be solved by graphing. For example, the graphs of $y = 3x$ and $x + y = 4$ are shown at the right. The graphs intersect at $(1, 3)$. Therefore, $(1, 3)$ is the solution of the system of equations $y = 3x$ and $x + y = 4$.

A system with one linear equation and one quadratic equation is called a **quadratic-linear system**. As with systems of linear equations, quadratic-linear systems can be solved by graphing. To *solve a system* means to find all the values of x and y that make both equations true at the same time.

Examples

1 Solve the following system by graphing.

$$y = x^2$$
$$y = 2x + 3$$

The parabola and the line are graphed on the same set of axes. There are two intersection points, $(-1, 1)$ and $(3, 9)$. Therefore, the solution set is $\{(-1, 1), (3, 9)\}$.

2 Solve the following system by graphing.

$$y = x^2 + 3$$
$$x - 2y = 2$$

The graphs do not intersect. There are no values of x and y that satisfy both equations at the same time. The solution set is $\emptyset$.

Systems of equations whose graphs are circles and lines can be solved by graphing.

Example

3 Solve the following system by graphing.
$$x^2 + y^2 = 25$$
$$y = x - 7$$

The graph of the first equation is a circle centered at the origin with a radius of 5. The graph of the second equation is a line with a slope of 1 and a y-intercept of -7. The graphs intersect at $(4, -3)$ and $(3, -4)$.
The solution set is $\{(4, -3), (3, -4)\}$.

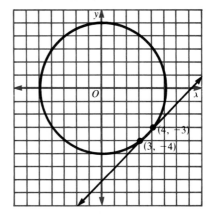

Systems of linear equations can be solved *algebraically, without graphing*. The same is true of quadratic-linear systems.

Example

4 Solve the following system algebraically.
$$y = 2x^2 + 6x + 5$$
$$y = x + 3$$

If both of the expressions on the right equal y, then they must be equal to each other.

$$x + 3 = 2x^2 + 6x + 5$$
$$0 = 2x^2 + 5x + 2$$
$$0 = (2x + 1)(x + 2)$$
$$2x + 1 = 0 \quad \text{or} \quad x + 2 = 0$$
$$x = -\frac{1}{2} \qquad x = -2$$

Substituting these values of x in the original equations, we obtain $y = \frac{5}{2}$ and $y = 1$. The solution set is $\left\{\left(-\frac{1}{2}, \frac{5}{2}\right), (-2, 1)\right\}$.

Exercises

Max: 1-5, 1-27, 1-7; Avg: 1-5, 1-27 odds; Min: 1-5, 1-27 odds

Exploratory Answer the following. For answers to exercises 1-5, see the Teacher's Guide.

1. What does it mean to "solve a system" of equations?
2. What is the maximum number of points in which a line can intersect a parabola? Give an example.
3. What does it mean if the graphs of a quadratic equation and a linear equation do not intersect? Give an example.
4. What is meant by "quadratic-linear system?"
5. In general, if a quadratic-linear system has no solution, what will happen if you try to solve the system algebraically?

For graphs to exercises 1-15, see students' work.

Written Solve each system by graphing. Check each solution in the original equations.

1. $y = x^2$
   $y = 2x$ {(0,0),(2,4)}

2. $y = x^2$
   $y = -3x$ {(-3,9),(0,0)}

3. $y = \frac{1}{2}x^2$
   $x = y$ {(0,0),(2,2)}

4. $y = x^2 + 4$
   $x + y = 4$ {(-1,5),(0,4)}

5. $y = x^2 - 9$
   $y + x = -7$ {(-2,-5),(1,-8)}

6. $y = x^2 - 6x + 2$
   $x + y = 2$ {(0,2),(5,-3)}

7. $y = x^2 - 6x + 2$
   $0 = x - y - 4$ {(1,-3),(6,2)}

8. $y = x^2 + x + 6$
   $y = \frac{1}{2}x + 1$ ∅

9. $y = x^2 + 2x - 1$
   $y = \frac{1}{2}x - 1$ {(-1½,-1¾),(0,-1)}

10. $y = x^2 - 4x + 9$
    $2x - y + 1 = 0$ {(2,5),(4,9)}

11. $x - y = 3$
    $y = x^2 + 1$ ∅

12. $y = x^2 - 4x + 7$
    $x + y = 2$ ∅

13. $x^2 + y^2 = 4$
    $x + y = 4$ ∅

14. $x^2 + y^2 = 9$
    $y - x = 1$ {(-2⅗,-1⅗),(1⅗,2½)}

15. $x^2 + y^2 = 3$
    $y = x - 5$ ∅

Solve the following systems of equations algebraically. For the answers to exercises 16-27, see the Teacher Guide.

16. exercise 1
17. exercise 2
18. exercise 3
19. exercise 4
20. exercise 5
21. exercise 6
22. exercise 7
23. exercise 8
24. exercise 9
25. exercise 10
26. exercise 11
27. exercise 12

Challenge Estimate the roots of the following systems by graphing. Then check your work algebraically. For the answers to exercises 1-3, see the Teacher's Guide.

1. $2x + 3y = 6$
   $y = x^2 - 4x + 4$

2. $y = 2x^2 + x - 1$
   $4x + 2y = 8$

3. $y = x^2 - 4x + 3$
   $-x + 4y = 4$

4. no, if 0 is substituted for x, we get 0 = 12; no

Consider the equation $xy = 12$. Answer the following.

4. Is there a value of y that corresponds to x = 0? Why or why not? Is there a value of x that corresponds to y = 0?
5. Can the graph of $xy = 12$ ever enter Quadrants II or IV? Explain. no; x and y must always have the same sign.
6. Make a table of values for $xy = 12$ and graph the equation carefully. Does it cross either axis? See the Teacher's Guide
7. Solve the following system.
   $xy = 12$ (-4,-3),(2,6)
   $2y = 3x + 6$

Mathematical Excursions

THE GOLDEN RECTANGLE

The ancient Greeks very highly prized a certain kind of rectangle, called a **golden rectangle**, because of its shape. This rectangle occurs in much of their art. The front of the famous Parthenon contains golden rectangles.

Golden rectangles have interesting mathematical properties.

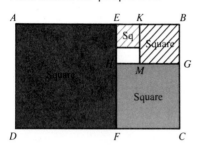

For example, if you remove a square from one end of a golden rectangle, a rectangle that is similar to the original is left. As you can see from the figure at the left, this process can go on forever.

rectangle $EBCF \sim$ rectangle $ABCD$
rectangle $EBGH \sim$ rectangle $EBCF$
rectangle $HEKM \sim$ rectangle $EBGH$

⋮

Golden rectangles appear in many places in nature. The photograph shown at the left is of a chambered nautilus shell. Notice that the spiral in the shell closely follows the pattern in the rectangle shown above.

Another mathematical property of golden rectangles is the ratio of the measures of their sides. This ratio is the same for all golden rectangles and is about 1 to 1.618. This ratio is called the **golden ratio**.

For answers to exercises 1 and 2, see the Teacher's Guide.

Exercises Solve each problem.

1. Write a proportion for the golden rectangle shown at the right.
2. Cross multiply the proportion to get a quadratic equation. Then use the quadratic formula to check that the golden ratio is 1 to 1.618.
   Hint: Let $a = 1$, $b = y$, and $c = -y^2$. The ratio you are looking for is $\dfrac{x}{y}$.

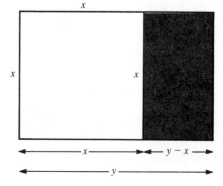

10.9 Simplifying Rational Expressions

Objective: To simplify rational expressions.

Fractions whose numerator and denominator are polynomials are called *rational expressions*. A rational expression can be simplified in the same manner as a fraction. Recall that to simplify a fraction, first factor the numerator and denominator, and then eliminate common factors.

$$\frac{24}{42} = \frac{2 \cdot 2 \cdot 2 \cdot 3}{2 \cdot 3 \cdot 7} \quad \text{Factor.}$$
$$= \frac{2 \cdot 2}{7} \quad \text{Eliminate common factors.}$$
$$= \frac{4}{7}$$

When simplifying rational expressions, remember that division by zero is undefined. Therefore, zero cannot be used as a denominator.

Examples

1 Simplify $\dfrac{42y^2}{27xy}$.

$$\frac{42y}{27xy} = \frac{2 \cdot 3 \cdot 7 \cdot y \cdot y}{3 \cdot 3 \cdot 3 \cdot x \cdot y}$$
$$= \frac{2 \cdot 7 \cdot y}{3 \cdot 3 \cdot x} \quad \begin{array}{l} x \neq 0 \\ \text{and} \\ y \neq 0. \end{array}$$
$$= \frac{14y}{9x}$$

2 Simplify $\dfrac{5y^2 + 10xy}{5y}$.

$$\frac{5y^2 + 10xy}{5y} = \frac{5y^2}{5y} + \frac{10xy}{5y} \quad y \neq 0.$$
$$= \frac{5 \cdot y \cdot y}{5 \cdot y} + \frac{2 \cdot 5 \cdot x \cdot y}{5 \cdot y}$$
$$= y + 2x$$

3 Simplify $\dfrac{2y - 6}{y^2 - y - 6}$.

$$\frac{2y - 6}{y^2 - y - 6} = \frac{2(y - 3)}{(y + 2)(y - 3)}$$
$$= \frac{2}{y + 2} \quad \text{Why are 3 and } -2 \text{ excluded values?}$$

4 Simplify $\dfrac{x^2 + 4x + 3}{x^2 + 2x + 1}$.

$$\frac{x^2 + 4x + 3}{x^2 + 2x + 1} = \frac{(x + 3)(x + 1)}{(x + 1)(x + 1)}$$
$$= \frac{(x + 3)}{(x + 1)} \quad \text{What values of } x \text{ must be excluded?}$$

Exercises

Max: 1-21, 1-30; Avg: 1-21 odds, 1-30; Min: 1-21 odds, 1-29 odds

Exploratory Factor.

1. 28   2(2)(7)
2. 18xy $2(3^2)xy$
3. $z^2 - z$ z(z - 1)
4. $2a^2 + 2ab$ 2a(a + b)
5. $3x^2 + 15x$ 3x(x + 5)
6. $4 - x^2$ (2 - x)(2 + x)
7. $t^2 - 25$ (t - 5)(t + 5)
8. $q^2 - 8q + 15$ (q - 5)(q - 3)
9. $4n^2 + 13n - 12$ (4n - 3)(n + 4)

Find the greatest common factor of each numerator and denominator.

10. $\dfrac{28}{42}$ 7
11. $\dfrac{48a}{64a^2}$ 16a
12. $\dfrac{15x^3}{5xm}$ 5 x   5x
13. $\dfrac{x - 5}{(x - 5)(x + 5)}$ x - 5
14. $\dfrac{(x + 2)(x - 1)}{x - 1}$ x - 1
15. $\dfrac{18(c - 4)}{30(c - 4)}$ 6(c - 4)
16. $\dfrac{4n - 16n^2}{4n}$ 4n
17. $\dfrac{x - 4}{4 - x}$ x - 4
18. $\dfrac{9x - 27y}{18}$ 9
19. $\dfrac{28y + 7xy}{7y}$ 7y
20. $\dfrac{2ab + 4a^2}{2a}$ 2a
21. $\dfrac{y^2 + 7y}{3(y + 7)}$ y + 7

For answers to exercises 1-30, see the Teacher's Guide.

Written Simplify. Assume that no denominator is equal to zero.

1. $\dfrac{3}{6a}$
2. $\dfrac{4p}{2p^2}$
3. $\dfrac{25mn^2}{-5mn}$
4. $\dfrac{2x + 8y}{2}$
5. $\dfrac{15q - 30q^2}{15q}$
6. $\dfrac{24w^2 - 6wz}{3w}$
7. $\dfrac{(x + 1)^2}{x + 1}$
8. $\dfrac{2(y - 1)^2}{y^2 - 1}$
9. $\dfrac{4x + 16}{x + 4}$
10. $\dfrac{k^2 + 3k}{k^3 + 3k^2}$
11. $\dfrac{6b - 3}{1 - 4b^2}$
12. $\dfrac{4x + 16}{x^3 + 4x^2}$
13. $\dfrac{2 - c}{c^2 - 4}$
14. $\dfrac{n^2 + 8n + 16}{n^2 - 16}$
15. $\dfrac{4 - 4x^2}{2 + 4x + 2x^2}$
16. $\dfrac{21y^2 + 3xy}{3y}$
17. $\dfrac{p^2 - q^2}{p^2 + 2pq + q^2}$
18. $\dfrac{x^2 + 7x + 12}{x^2 + 9x + 20}$
19. $\dfrac{5x^4 - 15x^5}{5 - 17x + 6x^2}$
20. $\dfrac{p^2 - 9}{p^2 - 4p - 21}$
21. $\dfrac{4x - 20}{x^2 - 2x - 15}$
22. $\dfrac{t^2 + 2t + 1}{t^2 - 2t - 3}$
23. $\dfrac{j^3k + j^2k^2}{j^3k - j^2k^2}$
24. $\dfrac{4x^2y - 4xy^2}{18x^3 + 12x^2y}$
25. $\dfrac{9 - a^2}{a^2 - a - 6}$
26. $\dfrac{25 - x^2}{x^2 + x - 30}$
27. $\dfrac{-x^2 + 6x - 9}{x^2 - 6x + 9}$
28. $\dfrac{a^2 - 2a + 1}{-a^2 + 2a - 1}$
29. $\dfrac{4y^2 + 7y - 2}{8y^2 + 15y - 2}$
30. $\dfrac{4x^2 - 6x - 4}{2x^2 - 8x + 8}$

10.10 Multiplying and Dividing Rational Expressions

Objective: To multiply and divide rational expressions.

The rules that apply to rational numbers also apply to rational expressions. To multiply rational expressions, multiply the numerators and multiply the denominators.

$$\frac{3}{7} \cdot \frac{2}{5} = \frac{3 \cdot 2}{7 \cdot 5} \qquad \frac{3}{8} \cdot \frac{x}{y} = \frac{3 \cdot x}{8 \cdot y}$$

$$= \frac{6}{35} \qquad\qquad\qquad = \frac{3x}{8y}$$

Multiplying Rational Expressions

> If b and d are not zero, then $\frac{a}{b} \cdot \frac{c}{d} = \frac{ac}{bd}$.

Exercises

1 Find $\frac{3a}{4b} \cdot \frac{12b}{21a}$ and simplify.

$$\frac{3a}{4b} \cdot \frac{12b}{21a} = \frac{3a \cdot 12b}{4b \cdot 21a} \qquad a \neq 0 \text{ and } b \neq 0.$$

$$= \frac{36ab}{84ab}$$

$$= \frac{3}{7}$$

2 Find $\frac{3x - 1}{2x + 1} \cdot \frac{4x^2 - 1}{9x^2 - 1}$ and simplify.

$$\frac{3x - 1}{2x + 1} \cdot \frac{4x^2 - 1}{9x^2 - 1} = \frac{3x - 1}{2x + 1} \cdot \frac{(2x + 1)(2x - 1)}{(3x - 1)(3x + 1)}$$

$$= \frac{(3x - 1)(2x + 1)(2x - 1)}{(2x + 1)(3x - 1)(3x + 1)} \qquad \text{What values of } x \text{ must be excluded?}$$

$$= \frac{2x - 1}{3x + 1}$$

It will be assumed from this point on that all values of the variable that result in a denominator of zero will be excluded.

Dividing by a rational number is the same as multiplying by its multiplicative inverse. To divide rational expressions, you also multiply by the multiplicative inverse.

$$\frac{2}{7} \div \frac{3}{8} = \frac{2}{7} \cdot \frac{8}{3} \text{ or } \frac{16}{21} \qquad \frac{8x}{5y} \div \frac{3y}{2x} = \frac{8x}{5y} \cdot \frac{2x}{3y} \text{ or } \frac{16x}{15y}$$

Dividing Rational Expressions

If b, c, and d are not zero, then $\dfrac{a}{b} \div \dfrac{c}{d} = \dfrac{a}{b} \cdot \dfrac{d}{c}$ or $\dfrac{ad}{bc}$.

Examples

3 Find $\dfrac{3ab}{4c} \div \dfrac{6a^2}{5b^2}$ and simplify.

$$\dfrac{3ab}{4c} \div \dfrac{6a^2}{5b^2} = \dfrac{3ab}{4c} \cdot \dfrac{5b^2}{6a^2} \qquad \text{The multiplicative inverse of } \dfrac{6a^2}{5b^2} \text{ is } \dfrac{5b^2}{6a^2}.$$

$$= \dfrac{15ab^3}{24a^2c}$$

$$= \dfrac{5b^3}{8ac} \qquad \text{What values of the variables are excluded?}$$

4 Find $\dfrac{(x+y)^2}{x-y} \div \dfrac{x^2-y^2}{(x-y)^2}$ and simplify.

$$\dfrac{(x+y)^2}{x-y} \div \dfrac{x^2-y^2}{(x-y)^2} = \dfrac{(x+y)^2}{x-y} \cdot \dfrac{(x-y)^2}{x^2-y^2}$$

$$= \dfrac{(x+y)(x+y)}{x-y} \cdot \dfrac{(x-y)(x-y)}{(x+y)(x-y)}$$

$$= x + y \qquad \text{Why must } x = y \text{ and } x = -y \text{ be excluded?}$$

5 Find $\dfrac{d^2+3d+2}{d^2+7d+12} \div \dfrac{d^2+9d+14}{d^2+10d+21}$ and simplify.

$$\dfrac{d^2+3d+2}{d^2+7d+12} \div \dfrac{d^2+9d+14}{d^2+10d+21} = \dfrac{d^2+3d+2}{d^2+7d+12} \cdot \dfrac{d^2+10d+21}{d^2+9d+14}$$

$$= \dfrac{(d+1)(d+2)}{(d+4)(d+3)} \cdot \dfrac{(d+3)(d+7)}{(d+2)(d+7)}$$

$$= \dfrac{d+1}{d+4}$$

Exercises

Max: 1-12, 1-24; Avg: 1-11 odds, 1-24; Min: 1-11 odds, 1-23 odds

Exploratory Find each product or quotient and simplify.

1. $\dfrac{2}{3} \cdot \dfrac{3}{4}$ $\dfrac{1}{2}$
2. $\dfrac{7}{8} \cdot \left(-\dfrac{4}{5}\right)$ $-\dfrac{7}{10}$
3. $\dfrac{3}{a} \cdot \dfrac{2b}{9}$ $\dfrac{2b}{3a}$
4. $\dfrac{x-1}{x+1} \cdot \dfrac{3}{2x}$ $\dfrac{3x-3}{2x^2+2x}$
5. $\dfrac{2}{3} \div \dfrac{3}{4}$ $\dfrac{8}{9}$
6. $\dfrac{-7}{16} \div \dfrac{2}{3}$ $-\dfrac{21}{32}$
7. $\dfrac{5}{x} \div \dfrac{y}{x^2}$ $\dfrac{5x}{y}$
8. $\dfrac{x-1}{x+1} \div (x-1)$ $\dfrac{1}{x+1}$

Write each quotient as a product.

9. $\dfrac{x}{2y} \div \dfrac{-x}{4y}$ $\dfrac{x}{2y} \cdot \dfrac{4y}{-x}$
10. $\dfrac{3m}{m+1} \div (m-5)$ $\dfrac{3m}{m+1} \cdot \dfrac{1}{m-5}$
11. $5k \div 2\dfrac{1}{4}$ $5k \cdot \dfrac{4}{9}$
12. $\dfrac{p^2-4}{3} \div \dfrac{p+2}{2}$ $\dfrac{p^2-4}{3} \cdot \dfrac{2}{p+2}$

For answers to exercises 1-24, see the Teacher's Guide.

Written Find each product or quotient and simplify.

1. $\dfrac{9x^2}{2} \cdot \dfrac{15x}{4}$
2. $\dfrac{8}{15m} \cdot \dfrac{m^2}{2}$
3. $\dfrac{-3a}{10b} \cdot \dfrac{20b}{-24a}$
4. $\dfrac{3n^2}{-2y} \div \dfrac{6n^2}{4y^2}$
5. $\dfrac{3x}{-4y^3} \div \dfrac{-32}{15x^2y}$
6. $\dfrac{a^2-b^2}{b-a} \cdot \dfrac{1}{a}$
7. $\dfrac{(x-1)^2}{y} \div \dfrac{x}{x^2-1}$
8. $\dfrac{a^2-b^2}{b-a} \div \dfrac{1}{a}$
9. $\dfrac{1}{x} \cdot \dfrac{x^2+x}{x-1}$
10. $\dfrac{2b+4}{3b+9c} \cdot \dfrac{9b+3c}{b+2}$
11. $\dfrac{2-x}{x} \div \dfrac{x-2}{2x-4}$
12. $\dfrac{x^2-y^2}{2} \div \dfrac{x-y}{4}$
13. $\dfrac{4x+60}{3x} \cdot \dfrac{9b}{3x+30}$
14. $\dfrac{2x^2-4x}{3xy-9y} \cdot \dfrac{3(xy-3y^2)}{4x^3+4x^2}$
15. $\dfrac{y^2-2y+1}{x+4} \div \dfrac{y-1}{8x+32}$
16. $\dfrac{a^3b-2a^2b^2}{3ab^2+3b^3} \div \dfrac{a^2-8ab+12b^2}{2a^2+4ab+2b^2}$
17. $\dfrac{2x^2-7x-4}{x^2-7x+12} \div \dfrac{4x^2-1}{2x+6}$
18. $\dfrac{2x^2-18}{x^2-6x+9} \div \dfrac{3x+6}{x^2+2x}$
19. $\dfrac{x^2-4x-12}{x-4} \cdot \dfrac{x^2-x-12}{x+12}$
20. $\dfrac{y^2+5y+6}{y^2+4y+4} \cdot \dfrac{y^2+3y-10}{y^2+8y+15}$
21. $\dfrac{a^3-b^3}{a+b} \cdot \dfrac{a^2-b^2}{a^2+ab+b^2}$
22. $\dfrac{y^2-y-12}{y+12} \div \dfrac{y^2-4y-12}{y-4}$
23. $\dfrac{w^2-11w+24}{w^2-18w+80} \div \dfrac{w^2-15w+50}{w^2-9w+20}$
24. $\dfrac{2x^2+x-15}{4x^2+2x-30} \cdot \dfrac{6x^2-8x+2}{3x^2+8x-3}$

For exercise 21, you may want to remind students of the difference of cubes formula: $x^3 - y^3 = (x-y)(x^2 + xy + y^2)$.

10.11 Adding and Subtracting Rational Expressions

Objective: To add and subtract rational expressions

> **Journal**
> How have you used fractions in your everyday life? Give examples.

Rational numbers or rational expressions that have a common denominator are added in the same manner.

$$\frac{3}{11} + \frac{4}{11} = \frac{3+4}{11} \text{ or } \frac{7}{11} \qquad \frac{3}{a} + \frac{4}{a} = \frac{3+4}{a} \text{ or } \frac{7}{a}$$

To add rational numbers with different denominators, find equivalent rational numbers with common denominators and then add. Use the least common denominator (LCD) to determine equivalent rational numbers.

One way to find the least common denominator is by first factoring each denominator into its prime factors. For example, suppose you wish to add $\frac{5}{18}$ and $\frac{7}{24}$. The least common denominator of each fraction contains the prime factors of each denominator the greatest number of times that it appears.

> The greatest number of times 3 appears is twice. The greatest number of times 2 appears is 3 times.

prime factors of 18: $2 \cdot 3^2$ prime factors of 24: $2^3 \cdot 3$

Thus, the LCD is $2^3 \cdot 3^2$ or 72.

Now, rename $\frac{5}{18}$ and $\frac{7}{24}$ using the LCD and then add.

$$\frac{5}{18} + \frac{7}{24} = \frac{5 \cdot 4}{18 \cdot 4} + \frac{7 \cdot 3}{24 \cdot 3}$$
$$= \frac{20}{72} + \frac{21}{72} \text{ or } \frac{41}{72}$$

Example

1 Find $\frac{2}{xy} + \frac{y}{x}$ and simplify.

$$\frac{2}{xy} + \frac{y}{x} = \frac{2}{xy} + \frac{y \cdot y}{x \cdot y} \qquad \text{The LCD is } xy.$$
$$= \frac{2}{xy} + \frac{y^2}{xy}$$
$$= \frac{2 + y^2}{xy} \qquad \text{Add the numerators.}$$

Examples

2 Find $\dfrac{2}{y-2} + \dfrac{3}{1-y}$ and simplify.

$$\dfrac{2}{y-2} + \dfrac{3}{1-y} = \dfrac{2(1-y)}{(y-2)(1-y)} + \dfrac{3(y-2)}{(1-y)(y-2)} \qquad \text{The LCD is } (y-2)(1-y).$$

$$= \dfrac{2-2y}{(y-2)(1-y)} + \dfrac{3y-6}{(1-y)(y-2)}$$

$$= \dfrac{2-2y+3y-6}{(y-2)(1-y)} \qquad \text{Add the numerators.}$$

$$= \dfrac{y-4}{-y^2+3y-2} \qquad \text{Simplify.}$$

The same procedure can be used to subtract rational expressions. Rename the rational expressions using the LCD and subtract the numerators.

Example

3 Find $\dfrac{x}{x-1} - \dfrac{2}{x^2-1}$ and simplify.

$$\dfrac{x}{x-1} - \dfrac{2}{x^2-1} = \dfrac{x(x+1)}{(x-1)(x+1)} - \dfrac{2}{(x-1)(x+1)} \qquad \text{The LCD is } (x-1)(x+1) \text{ or } x^2-1.$$

$$= \dfrac{x^2+x}{(x-1)(x+1)} - \dfrac{2}{(x-1)(x+1)}$$

$$= \dfrac{x^2+x-2}{(x-1)(x+1)} \qquad \text{Subtract the numerators.}$$

$$= \dfrac{(x+2)(x-1)}{(x-1)(x+1)} \qquad \text{Factor.}$$

$$= \dfrac{x+2}{x+1}$$

Exercises

Max: 1-17, 1-28, 1-6; Avg: 1-17 odds, 1-28; Min: 1-17 odds, 1-27 odds

Exploratory State the least common denominator for each pair of fractions.

1. $\dfrac{5}{8}, \dfrac{7}{12}$ 24
2. $\dfrac{3}{5a}, \dfrac{2}{15}$ 15a
3. $\dfrac{4}{7a}, \dfrac{8}{21b}$ 21ab
4. $\dfrac{7}{x}, \dfrac{5}{y}$ xy
5. $\dfrac{3}{20m^2}, \dfrac{1}{24mn^3}$ $120m^2n^3$
6. $\dfrac{6}{k^3}, \dfrac{5}{kj}k^3j$
7. $\dfrac{7}{x-3}, \dfrac{5}{3-x}$ $x-3$
8. $\dfrac{x+5}{3x+6}, \dfrac{x+3}{x-2}$ $(3x+6)(x-2)$

Copy and complete.

9. $\dfrac{2}{5} = \dfrac{?}{15a}$ 6a
10. $\dfrac{3}{20x} = \dfrac{?}{60xy}$ 9y
11. $\dfrac{t}{t+w} = \dfrac{?}{t(t+w)}$ t^2
12. $\dfrac{x}{x+y} = \dfrac{?}{x^2+xy}$ x^2
13. $\dfrac{x}{x+1} = \dfrac{?}{x^2-1}$ $x^2 - x$
14. $\dfrac{3}{a+2} = \dfrac{?}{a^2+4a+4}$ $3a+6$
15. $\dfrac{y-1}{y+3} = \dfrac{?}{2y^2+5y-3}$ $2y^2 - 3y + 1$
16. $\dfrac{2n}{m} = \dfrac{?}{m(n+1)}$ $2n^2 + 2n$
17. $\dfrac{4}{x-2} = \dfrac{?}{2-x}$ -4

For answers to exercises 1-28, see the Teacher's Guide.

Written Find each sum or difference and simplify.

1. $\dfrac{3}{8} + \dfrac{5}{12}$
2. $\dfrac{3}{4} - \dfrac{1}{8}$
3. $\dfrac{3}{a} - \dfrac{4}{b}$
4. $\dfrac{2}{x} + \dfrac{5}{7}$
5. $\dfrac{3x}{4} + \dfrac{2x}{7}$
6. $\dfrac{5}{2y} - \dfrac{3}{6y}$
7. $\dfrac{7}{3n} - \dfrac{5}{6n^2}$
8. $\dfrac{7}{x} + \dfrac{x+1}{2x}$
9. $\dfrac{5k}{7x} + \dfrac{3k}{21x^2}$
10. $\dfrac{a}{a+b} - \dfrac{b}{a+b}$
11. $\dfrac{x^2}{x-5} - \dfrac{3}{5-x}$
12. $\dfrac{m+n}{3} + \dfrac{3m-2n}{12}$
13. $\dfrac{2x+y}{3} + \dfrac{x-3y}{18}$
14. $\dfrac{1}{8x} - \dfrac{1}{x-1}$
15. $\dfrac{x+2}{x-2} + \dfrac{x-2}{x+2}$
16. $\dfrac{y}{y+6} + \dfrac{7}{y^2-36}$
17. $\dfrac{7}{x^2-7x+12} - \dfrac{5}{x-3}$
18. $\dfrac{7}{b^2-16} - \dfrac{3}{4-b}$
19. $\dfrac{3}{y^2-2y} + \dfrac{y}{y^2-4}$
20. $\dfrac{3}{4-b} + \dfrac{1}{b+4}$
21. $\dfrac{7}{4x^2-1} - \dfrac{3}{2x+1}$
22. $\dfrac{12a}{a^2-4} + \dfrac{4}{a+2}$
23. $\dfrac{2x+1}{(x-1)^2} - \dfrac{x-2}{x-1}$
24. $\dfrac{x-y}{x^2+2xy+y^2} - \dfrac{x+y}{x-y}$
25. $\dfrac{a-2}{a^2+4a+4} + \dfrac{a+2}{a-2}$
26. $\dfrac{3m}{m^2+3m+2} - \dfrac{3m-6}{m^2+4m+4}$
27. $\dfrac{4a}{6a^2-a-2} - \dfrac{5a+1}{2-3a}$
28. $\dfrac{2x+1}{(x-1)^2} + \dfrac{x-2}{(1-x)(x+4)}$

Challenge Solve each equation.

Sample: $\dfrac{5}{x} + \dfrac{1}{2} = 3 \implies 2x\left(\dfrac{5}{x}\right) + 2x\left(\dfrac{1}{2}\right) = 2x(3)$

$10 + x = 6x$
$10 = 5x$
$x = 2$

1. $\dfrac{1}{3} + \dfrac{1}{x} = \dfrac{7}{12}$ {4}

2. $\dfrac{2}{x} - \dfrac{1}{7} = \dfrac{13}{21}$ $\{\dfrac{21}{8}\}$

3. $\dfrac{3}{x} + \dfrac{1}{5} = 1$ $\{\dfrac{15}{4}\}$

4. $\dfrac{3}{5x} + \dfrac{7}{2x} = 1$ $\{\dfrac{41}{10}\}$

5. $\dfrac{3}{2x} - \dfrac{2x}{x+1} = -2$ $\{-\dfrac{3}{7}\}$

6. $\dfrac{4x}{3x-2} + \dfrac{2x}{3x+2} = 2$ {-2}

7. If 3 is added to the reciprocal of a number, the result is 2 times the reciprocal of the number. Find the number. $\dfrac{1}{2}$

Mixed Review

Choose the best answer.

1. Which equation has real roots?
   c a. $x^2 - 5x = -9$ b. $2x^2 + 8x = -9$ c. $2x^2 - 9x = -9$ d. $2x^2 + 6x = -5$

2. A rectangular photo is 8 in. by 10 in. It is placed in a frame of uniform width that has an
   b area of 63 in². What equation can be used to find the width, w, of the frame?
   a. $(8 + w)(10 + w) = 80 + 63$ b. $(8 + 2w)(10 + 2w) = 80 + 63$
   c. $(8 + 2w)(10 + 2w) = 63$ d. $(8 + 2w)(10 + 2w) = 80$

3. The solution set is ∅ for which of the following systems of equations?
   d a. $x - 3y = 3$ b. $4x + y = 6$ c. $x + y = 3$ d. $3x - 3y = 1$
       $y = x^2 - 1$ $x^2 + y^2 = 4$ $x^2 + y^2 = 5$ $y = x^2$

4. Which rational expression is the simplest form of $\dfrac{y^2 - 16}{y^2 - 36} \div \dfrac{y^2 + 2y - 24}{y^2 - 2y - 24}$?
   a
   a. $\dfrac{y^2 + 8y + 16}{y^2 + 12y + 36}$ b. $\dfrac{y^2 - 8y + 16}{y^2 - 12y + 36}$ c. $\dfrac{y^2 + 10y + 24}{y^2 - 10y + 24}$ d. $\dfrac{y^2 - 10y + 24}{y^2 + 10y + 24}$

5. For which value(s) of k will $x^2 + kx + (k + 3)$ be a perfect square trinomial?
   b a. $k = 6$ b. $k = 6$ or $k = -2$ c. $k = -6$ or $k = 2$ d. no values exist

6. The graph of which equation touches the x-axis at exactly one point?
   d a. $y = x^2 - 2$ b. $y = x^2 - 2x$ c. $y = x^2 - 2x + 2$ d. $y = x^2 - 2x + 1$

Portfolio Suggestion

Select a topic from this chapter that shows how you used a calculator or computer. Place it in your portfolio.

Performance Assessment

Mr. Babb is cutting two 8-foot timbers to outline a garden as shown at the right. Write a quadratic equation to represent the situation. Solve the equation graphically and then by factoring. Where should he make his cut if the area is 15 square feet? **See students' work; x = 3 or 5**

Chapter Summary

1. The standard form of a **quadratic equation** is $ax^2 + bx + c = 0$, where a is the coefficient of x^2, b is the coefficient of x, and c is the constant term. (369)
2. A **perfect square trinomial** is a polynomial that factors into a binomial squared. An equation in this form has two equal roots. (370)
4. Parabolas are graphs of equations of the form $y = ax^2 + bx + c$. All parabolas have an **axis of symmetry** and a **vertex.** The equation of the axis of symmetry for the graph of $y = ax^2 + bx + c$ is $x = -\dfrac{b}{2a}$. (374)
5. A quadratic equation of the form $ax^2 + bx + c = 0$ can be solved by graphing the equation $y = ax^2 + bx + c$. (377)
6. Quadratic equations can be solved using a method called **completing the square.** (381)
7. **Quadratic formula:** If $ax^2 + bx + c = 0$ and $a \neq 0$, then
$$x = \frac{-b \pm \sqrt{b^2 - 4ac}}{2a}.$$ (385)
8. The expression $b^2 - 4ac$ is called the **discriminant** of a quadratic equation. If it is negative, the equation has no real roots. (386)

9. The sum of the roots of $ax^2 + bx + c = 0$ is $-\frac{b}{a}$. The prouct of the roots of $ax^2 + bx + c = 0$ is $\frac{c}{a}$. (387)

10. A system of equations with one linear equation and one quadratic equation is called a **quadratic-linear system**. (394)

11. Fractions whose numerators and denominators are polynomials are called **rational expressions**. (398)

12. If b and d are not 0, then $\frac{a}{b} \cdot \frac{c}{d} = \frac{ac}{bd}$. (400)

13. If b, c, and d are not 0, then $\frac{a}{b} \div \frac{c}{d} = \frac{a}{b} \cdot \frac{d}{c}$ or $\frac{ad}{bc}$. (401)

14. To add or subtract rational expressions with unlike denominators, first rename the expressions so the denominators are alike. Then add or subtract these equivalent expressions. (403)

Chapter Review

10.1 Factor each polynomial completely.
1. $x^2 - x - 12$ $(x - 4)(x + 3)$
2. $y^2 - 5y + 4$ $(y - 4)(y - 1)$
3. $2a^2 + 16a + 14$ $2(a + 7)(a + 1)$
4. $3b^2 - 9b - 12$ $3(b - 4)(b + 1)$

10.2 Solve each equation by factoring.
5. $x^2 + 4x + 4 = 0$ $\{-2\}$
6. $m^2 - 11m + 30 = 0$ $\{5, 6\}$
7. $y^2 - 1 = 0$ $\{-1, 1\}$
8. $6r^2 + 13r + 6 = 0$ $\{-\frac{3}{2}, -\frac{2}{3}\}$
9. $4x^2 = x + 3$ $\{-\frac{3}{4}, 1\}$
10. $2z = 6 - 4z^2$ $\{-\frac{3}{2}, 1\}$

For answers to exercises 11-14, see the Teacher's Guide.

10.3 Graph the following equations.
11. $y = x^2 + 2$
12. $y = x^2 + 4x + 3$
13. $y = 4x^2$
14. $y = 2x^2 + 4x$

For graphs to exercises 15-18, see students' work

10.4 Solve each equation by graphing. Then check the solutions by factoring.
15. $x^2 - 9 = 0$ $\{-3, 3\}$
16. $x^2 + 2x - 15 = 0$ $\{-5, 3\}$
17. $x^2 - 5x = 0$ $\{0, 5\}$
18. $2x^2 + 4x + 10 = 0$ $\emptyset$

10.5 Solve each equation by completing the square.
19. $x^2 + 2x = 15$ $\{-5, 3\}$
20. $x^2 + 4x - 1 = 0$ $\{-2 \pm \sqrt{5}\}$
21. $x^2 + 3x + 9 = 0$ $\emptyset$
22. $2x^2 - x - 5 = 0$ $\{\frac{1 \pm \sqrt{41}}{4}\}$

10.6 Solve each equation using the quadratic formula.
23. $x^2 + 4x - 60 = 0$ $\{-10, 6\}$
24. $x^2 + 3x + 1 = 0$ $\{\frac{-3 \pm \sqrt{5}}{2}\}$
25. $4x^2 + 4x + 1 = 0$ $\{-\frac{1}{2}\}$
26. $5x^2 + 4x - 13 = 0$ $\{\frac{-2 \pm \sqrt{69}}{5}\}$

10.7 Solve each problem.

27. The area of a rectangle is 45 cm². The length is 9 cm less than 8 times the width. Find both dimensions.   3 cm by 15 cm

28. Two numbers differ by 6. Their product is 160. Find the numbers.   -16, -10; 10, 16

29. The square of a number exceeds 11 times the number by 312. Find the number.   24; -13

30. A rectangular lawn is 24 feet by 32 feet. A sidewalk will be constructed along the outside edges of all four sides. If the area of the lawn and the sidewalk is 1073 square feet, what will be the width of the walk?   2.5 ft

For graphs to exercises 31-34, see students' work.

10.8 Solve the following systems of equations first by graphing and then algebraically.

31. $y = x^2$
    $y - 2x = 0$ {(0, 0), (2, 4)}

32. $y = x^2 - 4x + 9$
    $y = x + 5$ {(1, 6), (4, 9)}

33. $y = x^2 - 6x + 6$
    $4y - 4 = 0$ {(1, 1), (5, 1)}

34. $y = x^2 + 2x - 3$
    $y + x = -3$ {(0, -3), (-3, 0)}

For answers to exercises 35-50, see the Teacher's Guide.

10.9 Simplify each expression.

35. $\dfrac{4a^2b^3c^4 + 13ab - 12a^4b^2c}{2abc}$

36. $\dfrac{k + 3}{4k^2 + 7k - 15}$

37. $\dfrac{x^2 + 10x + 21}{x^3 + x^2 - 42x}$

38. $\dfrac{x^2 - x - 56}{x^2 + x - 42}$

10.10 Find each product or quotient and simplify.

39. $\dfrac{5x^2y}{8ab} \cdot \dfrac{12a^2b}{25x}$

40. $\dfrac{a^2 - b^2}{6b} \div \dfrac{a + b}{36b^2}$

41. $\dfrac{x^2 + x - 12}{x + 2} \cdot \dfrac{x + 4}{x^2 - x - 6}$

42. $\dfrac{3x^2 + 5x - 28}{x^2 - 3x - 28} \cdot \dfrac{x^2 - 8x + 7}{3x - 7}$

43. $\dfrac{7a^2b}{x^2 + x - 30} \div \dfrac{3a}{x^2 + 15x + 54}$

44. $\dfrac{m^2 + 4m - 21}{m^2 + 8m + 15} \div \dfrac{m^2 - 9}{m^2 + 12m + 35}$

10.11 Find each sum or difference and simplify.

45. $\dfrac{2x}{x - 3} - \dfrac{6}{x - 3}$

46. $\dfrac{x}{x^2 - 1} + \dfrac{1}{x^2 - 1}$

47. $\dfrac{5a}{3x} - \dfrac{2}{4x^2y}$

48. $\dfrac{x}{x + 3} - \dfrac{5}{x - 2}$

49. $\dfrac{2x + 3}{x^2 - 4} + \dfrac{6}{x + 2}$

50. $\dfrac{2}{y^2 - 4y - 5} + \dfrac{5}{y^2 - 2y - 15}$

Chapter Test

Solve each equation by factoring.
1. $x^2 + 6x + 9 = 0$ {-3}
2. $y^2 - y - 12 = 0$ {-3, 4}
3. $m^2 - 16 = 0$ {-4, 4}
4. $3a^2 - a - 2 = 0$ {$-\frac{2}{3}$, 1}
5. $4x^2 - 8x = 5$ {$-\frac{1}{2}$, $\frac{5}{2}$}
6. $-8r = -6 - 2r^2$ {1, 3}

Graph the following equations.
7. $y = x^2 - 4$
8. $y = x^2 - 3x - 4$
9. $y = 5x^2$
10. $y = -x^2 + 4x$

Solve each equation by graphing.
11. $x^2 - 4 = 0$
12. $x^2 + x - 6 = 0$
13. $x^2 + 6x = 0$
14. $x^2 + 4x + 5 = 0$

Solve each equation by completing the square.
15. $x^2 + 2x = 3$
16. $x^2 + 4x - 5 = 0$
17. $x^2 + 5x + 9 = 0$
18. $3x^2 - 2x - 3 = 0$

Solve each equation using the quadratic formula.
19. $x^2 + 3x - 10 = 0$
20. $x^2 - 5x + 6 = 0$
21. $2x^2 + x + 5 = 0$
22. $x^2 - 4x + 4 = 0$

23. The area of a rectangle is 108 in². The length is 3 times the width. Find both dimensions.
24. One number is 3 more than another number. Their product is 180. Find the numbers.
25. A rectangular photograph is 14 inches by 20 inches. It is placed in a frame of uniform width that has an area of 111 in². Find the width of the frame. 1.5 in.

For graphs to exercises 26 and 27, see students' work.

26. $y = x^2 + 1$
    $y + 2x = 0$ {(-1, 2)}
27. $y = 2x^2 + 4x + 5$
    $y = x + 3$ ∅

Simplify each expression.
28. $\dfrac{21x^2y}{28ax}$ $\dfrac{3xy}{4a}$
29. $\dfrac{7x^2 - 28}{5x^3 - 20x}$ $\dfrac{7}{5x}$
30. $\dfrac{2x^2 - 5x - 3}{x^2 + 2x - 15}$ $\dfrac{2x + 1}{x + 5}$

For answers to exercises 32-35, see the Teacher's Guide.

Perform the indicated operations and simplify.
31. $\dfrac{2x}{x+7} + \dfrac{14}{x+7}$ 2
32. $\dfrac{2x}{x+12} - \dfrac{4}{x+4}$
33. $\dfrac{3x-8}{x+4} + \dfrac{9}{x+1}$
34. $\dfrac{x^2 + 4x - 32}{x+5} \cdot \dfrac{x-3}{x^2 - 7x + 12}$
35. $\dfrac{4x^2 + 11x + 6}{x^2 - x - 6} \div \dfrac{x^2 + 8x + 16}{x^2 + x - 12}$

CHAPTER 11

Probability and Combinatorics

Application in Ecology

Maintaining the delicate balance of the ecosystems is necessary for life to continue and thrive on Earth. Environmental scientists collect **data** from an environment and monitor the effects that different practices have on the environment. Seemingly small changes can have a drastic impact on the future of the planet.

In the beginning of a long environmental study of a proposed hydroelectric plant, marine biologists caught and tagged 25 fish in Antrium Lake. They wished to estimate the total fish population, so ten groups of 50 fish were caught and counted. The numbers of fish in each trial are recorded in the chart at the right. Find the average number of tagged fish caught per trial. Determine the **probability** of catching a tagged fish. Then use a proportion to estimate the number of fish in Antrium Lake.   **133 fish**

Trial Number	Tagged Fish Caught	Total Fish Caught
1	7	50
2	8	50
3	4	50
4	12	50
5	15	50
6	7	50
7	11	50
8	9	50
9	13	50
10	8	50

Individual Project: *Criminology*

Often it is difficult to prove a legal point "beyond a shadow of a doubt". Probability is used to argue legal cases in several ways. The use of an expert witness, blood typing, and evaluation of circumstantial evidence all involve probability. Research the legal uses of probability in a courtroom. How might a knowledge of probability help an attorney prepare a case? Are there any situations where you think a juror could be misled by someone using probability in a courtroom? Present a mock trial that uses probability to your class.

11.1 Introduction

Objective: To find the outcome set of a situation and the probability that an event will occur.

Perhaps the most important idea in the study of probability is that the **probability of an event E** is equal to the number of ways E can occur divided by the total number of possible outcomes. This statement is written

$$\text{probability of } E = \frac{\text{number of ways } E \text{ can occur}}{\text{total number of outcomes possible}}.$$

More compactly, it becomes

$$P(E) = \frac{n(E)}{n(S)}.$$

$n(E)$ means the number of ways E can occur and $n(S)$ means the number of outcomes possible.

Sometimes, S represents a set of outcomes and S is called the **outcome set.** Then, $n(S)$ names the number of elements in set S. Below are some probability situations and their outcome sets.

Remind students that these probabilities are computed under the assumption that the outcomes are all equally likely.

1. For tossing a coin, the outcome set, S, is {heads, tails}. This can be shortened to {H, T}. In this case $n(S) = 2$.
2. For tossing one die, $S = \{1, 2, 3, 4, 5, 6\}$ and $n(S) = 6$.
3. For picking one state from the United States,
   $S = \{\text{Alabama, Alaska, ..., Wyoming}\}$ and $n(S) = 50$.

Diagrams of outcome sets are often useful. Below are diagrams of the outcome sets for tossing two coins and for tossing two dice.

Calculator Hint

You can learn how to use a graphing calculator program to simulate rolling a die in Activity 5 on page A8.

Figure 1

Figure 2

Some event E must necessarily be a subset of the outcome set S. After all, the outcomes of the set E must be among the total number of outcomes possible.

Example

1 **Two coins are tossed. Find *P(2 heads)*.**

Look at figure 1. Because $S = \{(H, H), (H, T), (T, H), (T, T)\}$ and $E = \{(H, H)\}$, E is a subset of S. Also, $n(S) = 4$ and $n(E) = 1$. Therefore, $P(2 \text{ heads}) = \frac{1}{4}$.

Sometimes the desired event is called a *success*. In example 1, the probability of success is $\frac{1}{4}$.

Example

2 **Two fair dice are tossed. Find *P(sum of 7 or 11)*.**

Look at the outcome set of figure 2. The successes are (1, 6), (2, 5), (3, 4), (4, 3), (5, 2), (6, 1), (6, 5), and (5, 6). There are eight of them, so $n(E) = 8$. Because $n(S) = 36$, it follows that $P(\text{sum of 7 or 11}) = \frac{8}{36}$ or $\frac{2}{9}$.

Consider an event that *cannot* occur.

$$P(\text{impossible event}) = 0$$

Because E is an event that is impossible, E must be the empty set, $\emptyset$, and $P(\emptyset) = 0$.

If an event must occur, then every outcome is also a success. In other words, whatever S is, $n(S) = n(E)$.

Therefore, $P(E) = \frac{n(E)}{n(S)} = \frac{n(S)}{n(S)}$ or 1.

$$P(\text{certain event}) = 1$$

No event can ever have a probability greater than 1 or less than zero.

Basic Probability Concepts

1. If S is an outcome set and E is an event, then $P(E) = \dfrac{n(E)}{n(S)}$.
2. If E is impossible, then $P(E) = 0$.
3. If E is certain, then $P(E) = 1$.
4. For every event E, $P(E)$ must be equal to or greater than zero and equal to or less than one. In other words, for every event E, $0 \leq P(E) \leq 1$.

Exercises

Max: 1-10, 1-32, 1-5; Avg: 1-10, 1-25 odds, 26-32; Min: 1-9 odds, 1-25 odds

Exploratory Write the outcome set for each situation.

1. picking a letter from the first ten letters of the alphabet {A,B,C,D,E,F,G,H,I,J}
2. picking a letter from the word TOPICAL {T,O,P,I,C,A,L}
3. picking a color from red, yellow, and blue {red, yellow, blue}
4. picking a letter from the last five letters of the alphabet {V,W,X,Y,Z}
5. picking a whole number that is less than 15 {0,1,2,3,...,14}
6. picking an even number between 3 and 11 {4,6,8,10}

Use the outcome set shown in figure 1 on page 412 to find each probability when two fair coins are tossed.

7. P(2 heads) $\frac{1}{4}$
8. P(one head and one tail) $\frac{1}{2}$
9. P(both the same side up) $\frac{1}{2}$
10. P(at least one head) $\frac{3}{4}$

Written Use figure 2 on page 412 to find each probability when two fair dice are tossed.

1. P(sum of 3) $\frac{1}{18}$
2. P(sum of 4) $\frac{1}{12}$
3. P(sum of 5) $\frac{1}{9}$
4. P(sum of 9) $\frac{1}{9}$
5. P(sum of 11 or 12) $\frac{1}{12}$
6. P(sum of 8 or 10) $\frac{2}{9}$
7. P(sum less than 7) $\frac{5}{12}$
8. P(double or sum of 11) $\frac{2}{9}$
9. P(double or sum of 12)
10. P(odd number sum) $\frac{1}{2}$
11. P(sum less than or equal to 5) $\frac{5}{18}$

Answer the following.

12. Give three examples of impossible events involving the tossing of two dice. Answers may vary.
13. Give three examples of certain events involving the tossing of two dice. Answers may vary.
14. Explain what is known about E if $P(E)$ is closer to 0 than to 1. P(E) is less than $\frac{1}{2}$.
15. Explain what is known about E if $P(E)$ is closer to 1 than to 0. P(E) is greater than

A letter is drawn at random from A, B, E, Q, X, R, T, and L. Find each probability.

16. $P(A)$ $\frac{1}{8}$
17. P(X or R) $\frac{1}{4}$
18. P(vowel) $\frac{1}{4}$
19. P(L or <u>not</u> X) $\frac{7}{8}$
20. P(E or a letter from RAT) $\frac{1}{2}$

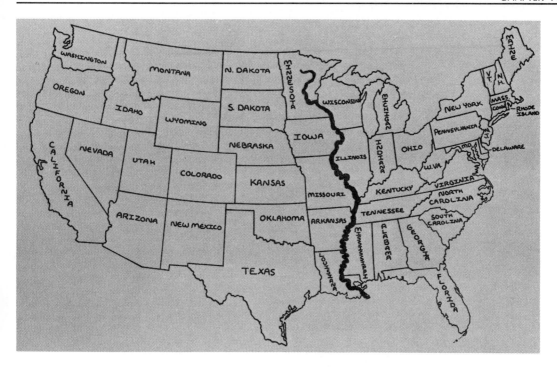

A state is chosen at random from the 50 states. Find each probability.

21. P(on the Pacific coast of the continental U.S.) $\frac{3}{50}$
22. P(part of its boundary is on the Mississippi River) $\frac{1}{5}$
23. P(begins with the letter M) $\frac{4}{25}$
24. P(has three United States senators) 0
25. P(has at least one representative in the House of Representatives) 1

Complete the following.

26. Two tetrahedral (four-sided) dice are tossed. Each has one of the numbers 1, 2, 3, and 4 on its faces. Draw the outcome set for this experiment.   Landing down is an outcome.
{(1,1),(1,2),(1,3),(1,4),(2,1),(2,2),(2,3),(2,4),(3,1),(3,2),(3,3),(3,4),(4,1),(4,2),(4,3),(4,4)}

For the experiment in exercise 26, find each probability.

27. P(sum is 2) $\frac{1}{16}$
28. P(sum is prime) $\frac{9}{16}$
29. P(sum is 9) 0
30. P(sum is 3 or 6) $\frac{5}{16}$
31. P(sum is odd) $\frac{1}{2}$
32. P(sum is less than 12) 1

Challenge   Answer the following.

1. Recall the definition of an ordered pair. Now define ordered triple and ordered quadruple. three or four numbers in which the order is specified
2. Suppose *three* four-sided dice are tossed. Use ordered triples to list the outcome set. For the answer to exercise 2, see the Teacher's Guide.

For the experiment in exercise 2, find each probability.

3. P(sum is 3) $\frac{1}{64}$
4. P(sum is 3 or 12) $\frac{1}{32}$
5. P(sum is <u>not</u> 11) $\frac{61}{64}$

11.2 "Or" Statements

Objective: To find the probability of mutually exclusive or inclusive events.

In the last section you found the probability of rolling a 7 or an 11 with two dice. This probability was $\frac{8}{36}$. The idea of "or" did not cause a problem here. You could just *count* the successful outcomes.

Another method of finding the probability of rolling a 7 or an 11 is to find the probability of rolling a 7 and the probability of rolling an 11. Then, add the two probabilities. Because $P(7) = \frac{6}{36}$ and $P(11) = \frac{2}{36}$, $P(7 \text{ or } 11) = \frac{6}{36} + \frac{2}{36}$ or $\frac{8}{36}$. But note carefully that the events "seven" and "eleven" have no elements in common.

Consider the probability of rolling a 6 or a "double." In this case, one outcome, (3, 3), exists that satisfies both events. If $P(6)$ is found to be $\frac{5}{36}$ and $P(double)$ is found to be $\frac{6}{36}$ and these two probabilities are added together, one element would be counted twice. The true probability of *(6 or double)* is $\left(\frac{5}{36} + \frac{6}{36}\right) - \frac{1}{36}$ or $\frac{10}{36}$.

Examples

1 A card is drawn from a standard deck of 52. Find *P(king or queen)*.

A deck has four kings and four queens—eight successes in all. No card can be a king *and* a queen, so *no* successes have been counted twice.

$$P(\text{king or queen}) = \frac{4}{52} + \frac{4}{52} = \frac{8}{52} \text{ or } \frac{2}{13}$$

2 A card is drawn from a standard deck of 52. Find *P(king or face card)*.

The set of kings is a subset of the set of face cards.

$$P(\text{king or face card}) = \left(\frac{4}{52} + \frac{12}{52}\right) - \frac{4}{52} = \frac{12}{52} \text{ or } \frac{3}{13}$$

Examples 1 and 2 illustrate the following rule.

Rule for P(A or B)

$$P(A \text{ or } B) = P(A) + P(B) - P(A \text{ and } B)$$

If A and B have no elements in common, then $P(A \text{ and } B) = 0$ and the rule becomes $P(A \text{ or } B) = P(A) + P(B)$.

Examples

3 The members of the Pine City Board of Supervisors are listed below, together with the sex and political affiliation of each.

Name	Sex	Political Affiliation
1. K. Adams	F	Independent
2. M. Barnard	F	Republican
3. S. Cohen	M	Democrat
4. D. Davidson	M	Democrat
5. C. Eastman	M	Conservative
6. S. Falk	M	Republican
7. B. Garber	F	Liberal
8. J. Halloran	M	Independent
9. M. Ianelli	F	Conservative
10. R. Jarmyn	F	Republican

A supervisor is selected at random to go out and get coffee and donuts. Find *P(not a Republican)*.

Count the non-Republicans. $P(\text{not a Republican}) = \frac{7}{10}$.

4 Refer to example 3. Find *P(woman or Republican)*.

Five women sit on the board. There are three Republicans. Because two of the women are Republicans,

$$P(\text{woman or Republican}) = \left(\frac{5}{10} + \frac{3}{10}\right) - \frac{2}{10} = \frac{6}{10} \text{ or } \frac{3}{5}.$$

5 Refer to example 3. Find *P(Democrat or Conservative)*.

In this case, *no* person satisfies both conditions.

$$P(\text{Democrat or Conservative}) = \frac{2}{10} + \frac{2}{10} = \frac{4}{10} \text{ or } \frac{2}{5}$$

Exercises

Max: 1-10, 1-24; Avg: 1-10, 1-24; Min: 1-10, 1-23 odds

Exploratory For exercises 1–10, name the occurrences that would be counted twice if someone simply added $P(A)$ and $P(B)$ in an attempt to find $P(A \text{ or } B)$.

A letter is chosen at random from the word SHARP.

1. A is "choosing a vowel;" B is "choosing a letter from the word RAT." choosing an A

A state is chosen at random from the United States.

2. A is "choosing a state beginning with M;" B is "choosing one of the five largest states." choosing Montana
3. A is "choosing a state that borders Canada;" B is "choosing a state that begins with M." choosing Montana, Minnesota, Michigan, or Maine

Two fair dice are tossed.

4. A is "tossing an even number;" B is "tossing a number greater than 5." tossing a 6
5. A is "tossing a double;" B is "tossing an 8." tossing a (4,4)

A card is drawn from a standard deck.

6. A is "choosing a red card;" B is "choosing a king." choosing a red king
7. A is "choosing a club;" B is "choosing a face card." choosing a club that is a face card
8. A is "choosing a queen;" B is "choosing a face card." choosing a queen
9. A is "choosing a heart;" B is "choosing a red card." choosing a heart
10. A is "choosing a heart or diamond;" B is "choosing a heart or diamond." choosing a heart or diamond

Written A card is drawn from a standard deck. Find each probability.

1. $P(ace)$ $\frac{1}{13}$
2. $P(ace \text{ or } two)$ $\frac{2}{13}$
3. $P(ace \text{ or } heart)$ $\frac{4}{13}$
4. $P(heart \text{ or } club)$ $\frac{1}{2}$
5. $P(king \text{ or } black \text{ } card)$ $\frac{7}{13}$
6. $P(seven \text{ of } clubs \text{ or } red \text{ } card)$ $\frac{27}{52}$
7. $P(face \text{ } card \text{ or } red \text{ } card)$ $\frac{8}{13}$
8. $P(diamond \text{ or } red \text{ } card)$ $\frac{1}{2}$

Two fair dice are tossed. Find each probability.

9. $P(double)$ $\frac{1}{6}$
10. $P(double \text{ or } 10)$ $\frac{2}{9}$
11. $P(double \text{ or } 11)$ $\frac{2}{9}$
12. $P(8 \text{ or } 9)$ $\frac{1}{4}$
13. $P(prime \text{ } number \text{ or } two)$ $\frac{5}{12}$
14. $P(less \text{ } than \text{ } 5 \text{ or } greater \text{ } than \text{ } 3)$ 1
15. $P(even \text{ } number \text{ or } double)$ $\frac{1}{2}$
16. $P(even \text{ } number \text{ or } odd \text{ } number)$ 1

For exercises 17–24, refer to the Pine City Board of Supervisors whose names are given on page 417. One member of the board is selected at random to be recording secretary. Find each probability.

17. $P(member \text{ of the } board)$ 1
18. $P(woman)$ $\frac{1}{2}$
19. $P(woman \text{ or } Independent)$ $\frac{3}{5}$
20. $P(man \text{ or } Democrat)$ $\frac{1}{2}$
21. $P(Conservative \text{ or } Republican)$ $\frac{1}{2}$
22. $P(man \text{ or } woman)$ 1
23. $P(Independent \text{ or } Conservative)$ $\frac{2}{5}$
24. $P(woman \text{ or } Liberal)$ $\frac{1}{2}$

11.3 Applications of the Counting Principle

Objective: To use the Counting Principle and to draw tree diagrams.

The **Counting Principle** is another extremely important idea of probability.

Counting Principle

> Suppose one activity can occur in any of *m* ways. Another can occur in any of *n* ways. The total number of ways both activities can occur is given by the product *mn*.

The following version of the Counting Principle is valid if and only if the events are **independent.** When two events are independent, the occurrence of one event does *not* affect the occurrence of the other.

Counting Principle (Alternate Version)

> Suppose the probability of one event *E* is *r* and the probability of another event *F* is *s*. If the events are independent, then the probability of *E* and *F* both occurring is the product *rs*.

Examples

1 Three different airlines fly from Alphaville to Betaville. Five different airlines fly from Betaville to Gammaville. How many different ways can a traveler book flights for a journey from Alphaville to Gammaville via Betaville?

Use the Counting Principle directly. The answer is 3 · 5 or 15.

2 If Epsilon Airlines is one of those that flies from Alphaville to Betaville, and Iota Airways flies from Betaville to Gammaville, what is the probability that a traveler will make the trip by these two airlines if the choice is made at random?

Use the *Alternate Version of the Counting Principle.*
Because $P(Epsilon)$ is $\frac{1}{3}$ and $P(Iota) = \frac{1}{5}$, $P(Epsilon \text{ and } Iota)$ is $\frac{1}{3} \cdot \frac{1}{5}$ or $\frac{1}{15}$.

Example

3 Two dice are tossed twice. Find *P(doubles the first time, different doubles the second).* Which version of the Counting Principle will you use?

The probability of a double the first time is $\frac{1}{6}$ and a different double the second time is $\frac{5}{36}$. The required probability is $\frac{1}{6} \cdot \frac{5}{36}$ or $\frac{5}{216}$.

A device frequently used to analyze probability problems is the **tree diagram.**

Examples

4 An urn contains 3 red marbles and 5 blue marbles. Gwen selects a marble at random from the urn, tosses it away, and selects another one. What is the probability that both marbles selected are red?

The tree diagram at the right shows the situation. If the first marble is red, then seven marbles remain in the urn, two of which are red. This means that P(red on second pick) is $\frac{2}{7}$. Using the Counting Principle, P(red on first, red on second) = $\frac{3}{8} \cdot \frac{2}{7} = \frac{6}{56}$ or $\frac{3}{28}$.

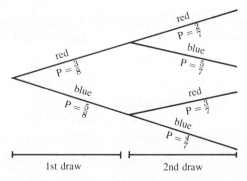

5 Refer to example 4. What is the probability that both marbles selected are the same color?

Use the diagram to find P(both marbles the same color).

P(both same color) = P(both red) + P(both blue) = $\frac{3}{28} + \frac{10}{28} = \frac{13}{28}$

Examples 4 and 5 show an urn problem *without replacement*. If Gwen replaces the marble each time, then the probability of red on every pick is $\frac{3}{8}$. Example 6 is similar to this situation.

Example

6 A coin has been "loaded" so that $P(head) = \frac{2}{3}$. If the coin is tossed three times, what is the probability of obtaining three heads?

At each toss, $P(H) = \frac{2}{3}$. Thus, $P(3 \text{ heads}) = \frac{2}{3} \cdot \frac{2}{3} \cdot \frac{2}{3} = \left(\frac{2}{3}\right)^3 = \frac{8}{27}$.

Exercises

Max: 1-10, 1-40; Avg: 1-10, 1-26, 27-39 odds; Min: 1-10, 1-39 odds

Exploratory Use the Counting Principle (either version) to answer the following.

1. Gwen has five skirts and four blouses. How many different skirt-blouse outfits can she choose to wear on any given day? 20

2. At the Burger Shoppe, a customer can order a burger rare, medium, or well done. It can be plain, or have one of these toppings: onions, relish, or mayonnaise. How many different styles of burgers can be ordered? 12

3. Four ferryboats make the crossing between point A and point B. How many different ways can a traveler make a round trip? 16

4. Using the ferryboats in exercise 3, how many different ways can a traveler make a round trip, but return on a different ferryboat from the one he went on? 12

5. Jack will fail French with probability $\frac{1}{3}$. He will fail chemistry with probability $\frac{1}{4}$. What is the probability he will fail both subjects? $\frac{1}{12}$

6. Use the information in exercise 5. What is the probability that Jack will pass French and fail chemistry? Assume that if he does not fail, he passes. $\frac{1}{6}$

7. What is the probability that a student will answer two 3-choice multiple choice questions correctly if the student is forced to guess at both? $\frac{1}{9}$
8. What is the probability that the student will get the first one right and the second one wrong, using the information in exercise 7? $\frac{2}{9}$
9. A coin is "loaded" so that P(heads) = $\frac{3}{5}$. If it is tossed twice, what is the probability of two heads? $\frac{9}{25}$
10. In exercise 9, what is P(head first, tail second)? $\frac{6}{25}$

Written Two fair dice are tossed twice. Find each probability.

1. P(doubles twice) $\frac{1}{36}$
2. P(7 the first time, 8 the second) $\frac{5}{216}$
3. P(10 both times) $\frac{1}{144}$
4. P(perfect square both times) $\frac{49}{1296}$
5. P(double the first time, 2 the second) $\frac{1}{216}$

A bowl contains 4 yellow marbles and 2 red ones. Nancy draws out a marble, sets it aside, then draws another. Complete the following. For the answer to exercise 6, see the Teacher's Gu

6. Draw a completely labeled tree diagram for this situation.
7. Find P(both yellow). $\frac{2}{5}$
8. Find P(both red). $\frac{1}{15}$
9. Find P(both same color). $\frac{7}{15}$
10. Find P(different colors). $\frac{8}{15}$
11. Find P(both yellow) if the first marble drawn is *replaced* in the bowl. $\frac{4}{9}$

An urn contains 3 green, 4 red, and 2 white marbles. A marble is drawn, *not* replaced, and another drawn. Complete the following. For the answer to exercise 12, see the Teacher's Guide

12. Draw a complete tree diagram for this situation.
13. Find P(red the first time, white the second). $\frac{1}{9}$
14. Find P(same color twice). $\frac{5}{18}$
15. Find P(neither white). $\frac{7}{12}$

Answer the following. For the answer to exercise 16, see the Teacher's Guide.

16. Draw a tree diagram for example 6 on page 421.
17. Find P(4 heads) for the coin in example 6. $\frac{16}{81}$
18. Find P(6 heads) for the coin in example 6. $\frac{64}{729}$
19. Find P(5 heads) for the coin in example 6. $\frac{32}{243}$
20. Use ordered triples to write all the possibilities, by sex, for a three-child family. For example, if the oldest and youngest are boys, and the middle child is a girl, this can be written as (B, G, B). {(B,B,B),(B,B,G),(B,G,B),(B,G,G),(G,B,B),(G,B,G),(G,G,B),(G,G,G)}

For a three-child family, find each probability.

21. P(all the children are girls) $\frac{1}{8}$
22. P(all the children are the same sex) $\frac{1}{4}$
23. P(the two youngest are girls) $\frac{1}{4}$
24. P(the oldest is a boy) $\frac{1}{2}$
25. P(the oldest and the youngest are the same sex) $\frac{1}{2}$
26. P(the oldest is a boy and the youngest is a girl) $\frac{1}{4}$

For each shrimp, lobster, or chicken dinner in a restaurant, you have a choice of soup or salad. With shrimp you may have hash browns or a baked potato. With lobster you may have rice or hash browns. With chicken you may have rice, hash browns, or a baked potato. If all combinations are equally likely to be ordered, find the probability of an order containing each of the following.

27. shrimp $\frac{2}{7}$
28. salad $\frac{1}{2}$
29. baked potato $\frac{2}{7}$
30. rice $\frac{2}{7}$
31. lobster and hashbrowns $\frac{1}{7}$
32. soup and hashbrowns $\frac{3}{14}$
33. shrimp and rice 0
34. chicken, salad, and rice $\frac{1}{14}$

35. A committee of 2 is to be selected from a group of 6 men and 3 women. What is the probability that the 2 selected are women? $\frac{1}{12}$

36. A bag contains 5 red, 3 white, and 8 blue marbles. Three are selected in sequence without replacement. What is the probability of selecting a red, a white, and a blue, in that order? $\frac{1}{28}$

37. For the bag in exercise 36, what is the probability if each marble is replaced after it is drawn? $\frac{15}{512}$

38. For the bag in exercise 36, what is the probability of selecting a red, a white, and a blue, in any order? $\frac{3}{14}$

39. Uri has 6 blue and 4 black socks in a drawer. One dark morning, he pulls out 2 socks. What is the probability that he has a matching pair of socks? $\frac{7}{15}$

40. Deanna guesses on all 5 questions on a true-false test. What is the probability that she answered exactly 4 questions correctly? $\frac{5}{32}$

Mixed Review

Perform the indicated operations and simplify.

1. $\dfrac{x^2 - 2x - 15}{2x^2 - 7x - 15} \cdot \dfrac{4x^2 - 4x - 15}{2x^2 + x - 15}$ 1

2. $\dfrac{k + 2}{m^2 + 4m + 4} \div \dfrac{4k + 8}{m + 4}$ $\dfrac{m + 4}{4(m + 2)^2}$

3. $\dfrac{15n}{5n + 3} + \dfrac{9}{5n + 3}$ 3

4. $\dfrac{25}{5 - g} - \dfrac{g^2}{5 - g}$ $5 + g$

5. $\dfrac{c}{c^2 - 4c} - \dfrac{5c}{c - 4}$ $\dfrac{1 - 5c}{c - 4}$

6. $\dfrac{m - 1}{m + 1} + \dfrac{4}{2m + 5}$ $\dfrac{2m^2 + 7m - 1}{2m^2 + 7m + 5}$

Two dice are tossed. Find each probability.

7. P(sum greater than 6) $\frac{7}{12}$
8. P(sum divisible by 3) $\frac{1}{3}$
9. P(sum is even or less than 7) $\frac{2}{3}$
10. P(sum is odd or prime) $\frac{19}{36}$

11. A box contains red, blue, and green marbles. There are 5 more red marbles than blue and three times as many green marbles as blue. If you choose a marble at random, the probability that it is green is $\frac{2}{5}$. How many marbles are in the box? 15

12. Suppose you choose a marble from the box in exercise 11, set it aside, then draw another. Find P(red the first time, blue the second). $\frac{1}{15}$

11.4 Permutations

Objective: To use permutations to count possible orders of objects.

Someone has suggested that if six chimpanzees are put at six typewriters, sooner or later one of them will type out the play, *Hamlet*. There is a possibility—however slight—that a chimpanzee's random selection of thousands of letters will turn out to be *Hamlet*. However, not many persons would want to wait around for it to happen.

Consider a similar, but less time-consuming problem. Suppose the four letters, A, H, M, and T are placed in a hat and drawn out one by one. What is the probability that MATH will be spelled?

To answer this question, first decide how many possibilities exist altogether. In other words, how many different ways can four symbols be arranged? To solve this problem, proceed as follows.

There are 4 different ways to choose the first symbol. After that, 3 different choices are left for the second. In the same manner, 2 different choices are left for the third position, and finally only 1 choice is left for the last position. You can use the Counting Principle to conclude that

$$4 \cdot 3 \cdot 2 \cdot 1 \quad \text{or} \quad 24$$

different arrangements of the four symbols are possible.

The probability of picking the letters out of the hat in the order M, A, T, H is $\frac{1}{24}$.

M, A, T, H is an arrangement of the four letters. You will recall that another term for arrangement is **permutation**.

Example

1 Find the number of permutations of 5 letters.

Use an argument similar to the one in the MATH example. The answer is $5 \cdot 4 \cdot 3 \cdot 2 \cdot 1$ or 120.

Examples

2 Find the number of permutations of the letters A, E, X, Q, B, and R if the first letter must be A.

If the permutation must begin with A, then the first place is fixed. The other five letters must be arranged in the five remaining places.

$1 \cdot 5 \cdot 4 \cdot 3 \cdot 2 \cdot 1 = 120$ There are 120 permutations.

3 Find the number of permutations of the letters A, E, X, Q, B, and R if the first and last letters must be vowels.

In this case, place the vowels first. The first place can be filled in either of two ways. Once this has been done, the last place can be filled in one way. Then fill in the other four places.

$2 \cdot 4 \cdot 3 \cdot 2 \cdot 1 \cdot 1 = 48$ There are 48 permutations.

A symbol for $5 \cdot 4 \cdot 3 \cdot 2 \cdot 1$ is $5!$.   This is read "5 factorial."

Definition of n Factorial

For any positive integer n,
$n! = n(n - 1)(n - 2) \cdot \ldots \cdot 1$

A symbol for "the number of permutations of 7 things taken seven at a time" is $_7P_7$. From our previous work, we know that $_7P_7 = 7 \cdot 6 \cdot 5 \cdot 4 \cdot 3 \cdot 2 \cdot 1$ or $7!$.

$_nP_n$

The symbol, $_nP_n$, names the number of permutations of n things taken n at a time.

$_nP_n = n!$

Example

4 How many ways can the 100 United States senators seat themselves in 100 seats, if there are no restrictions?

The answer is $_{100}P_{100}$, which equals $100!$.   We will not compute this number at this time. It contains over 150 digits.

Exercises

Max: 1-8, 1-20; Avg: 1-8, 1-20; Min: 1-8, 1-19 odds

Exploratory Find the value of each of the following.

1. 5! 120
2. (5 − 2)! 6
3. 7! 5040
4. 8! − 3 40,317
5. $_6P_6$ 720
6. $_3P_3$ 6
7. $_1P_1$ 1
8. $_4P_4$ 24

Written Give the number of permutations of the letters in each word.

1. CLOP 24
2. SLOPE 120
3. QUICKLY 5040

Give the number of permutations of the letters in PARIS for each situation.

4. There are *no* restrictions. 120
5. The first letter must be P. 24
6. The first letter must be a vowel. 48
7. The first letter *cannot* be a vowel. 72
8. The letter R must be in the middle place. 24
9. The last two letters must be RP in that order. 6

Give the number of permutations of the letters in QUICKLY for each situation.

10. The last letter must be Y. 720
11. The last letter *cannot* be Y. 4320
12. The last two letters must be C and K in either order. 240
13. The letter Q must be first and the letter Y must be last. 120

Answer the following.

14. In how many ways can three club officers arrange themselves as president, vice-president, and treasurer? 6

15. How many ways can five persons be seated in a row of five seats if one of them, the guest of honor, must be in the center? 24

16. What expression names the number of ways 135 passengers on a jet plane can arrange themselves in 135 unreserved seats? $_{135}P_{135}$ = 135!

17. Jason says that $n! \div n = (n - 1)!$. Is he right? yes

18. Professor Stone has these seven books to be placed on a shelf: *Shakespeare's Tragedies, Shakespeare's Histories, Shakespeare's Comedies,* Plays of Marlowe, Poems of Spenser, Poems of Milton, Works of Chaucer. In how many different ways can the books be placed on the shelf if the three Shakespeare volumes must be together in any order? 720

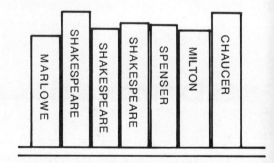

19. How many ways can six books be arranged if the first and last must always be the same two books? 24

20. How many ways can students in your classroom row be arranged using all the available desks in the row?
Answers may vary.

11.5 $_nP_r$

Objective: To use permutations to count possible orders of objects that do not include all available objects.

There are some permutation problems in which *not* all the available symbols are used. For example, how many four-letter arrangements can be made of the letters in HAMLET? To solve this problem, proceed as follows.

There are four places to fill. The first may be filled by any one of 6 letters. Once this has been done, the second place can be filled by any one of the 5 remaining letters. Proceed in this manner until all four places are filled. You can use the Counting Principle to conclude that there are

$$6 \cdot 5 \cdot 4 \cdot 3 \quad \text{or} \quad 360$$

different possible arrangements, or permutations, of six letters taken four at a time.

A symbol for "the number of permutations of n things taken r at a time" is $_nP_r$. Thus, $_6P_4$ equals 360.

Example

1 Find $_7P_5$.

There are five places to fill, with any of seven symbols.

$$_7P_5 = 7 \cdot 6 \cdot 5 \cdot 4 \cdot 3 = 2520$$

What is the general formula for $_nP_r$? Using the information in example 1,

$$_7P_5 = 7 \cdot 6 \cdot 5 \cdot 4 \cdot 3$$
$$= \frac{7 \cdot 6 \cdot 5 \cdot 4 \cdot 3 \cdot 2 \cdot 1}{2 \cdot 1}.$$

Note that the 2 in the denominator comes from 7 − 5.

Also,

$$_8P_3 = 8 \cdot 7 \cdot 6$$
$$= \frac{8 \cdot 7 \cdot 6 \cdot 5 \cdot 4 \cdot 3 \cdot 2 \cdot 1}{5 \cdot 4 \cdot 3 \cdot 2 \cdot 1}$$

Note that the 5 in the denominator comes from 8 − 3.

In general, $_nP_r = \dfrac{n!}{(n-r)!}$.

Another convenient version of this formula is given below.

$$_nP_r = n(n-1)(n-2)(n-3) \cdot \ldots \cdot (n-r+1)$$

The following summarizes $_nP_r$.

$_nP_r$

> The symbol $_nP_r$ names the number of permutations of n things taken r at a time.
>
> $$_nP_r = \frac{n!}{(n-r)!}$$
>
> $$_nP_r = n(n-1)(n-2)(n-3) \cdot \ldots \cdot (n-r+1)$$

Examples

2 Find $_9P_4$.

Use the first formula as follows.

$$_9P_4 = \frac{9!}{(9-4)!} = \frac{9!}{5!} = \frac{9 \cdot 8 \cdot 7 \cdot 6 \cdot 5 \cdot 4 \cdot 3 \cdot 2 \cdot 1}{5 \cdot 4 \cdot 3 \cdot 2 \cdot 1}$$

$$= 9 \cdot 8 \cdot 7 \cdot 6 = 3024$$

Use the second statement of the formula, $n = 9$, $r = 4$, and $n - r + 1 = 9 - 4 + 1 = 6$. This gives $9 \cdot 8 \cdot 7 \cdot 6$ or 3024 directly.

3 Sea-going vessels use different flags or arrangements of flags to send messages to other vessels. If ten different flags are available, and if every message consists of three different flags, how many different messages are possible?

This problem is equivalent to "Find the number of permutations of ten symbols taken three at a time." Using the second formula, $_{10}P_3 = 10 \cdot 9 \cdot 8 = 720$.

4 In example 3, how many messages are possible if the top flag must be one of three specific flags?

The first flag can be chosen in any of three ways. For the second and third, there are nine and eight possibilities respectively. The answer is $3 \cdot 9 \cdot 8$ or 216.

Exercises

Max: 1-12, 1-34; Avg: 1-12, 1-33 odds; Min: 1-11 odds, 1-33 odds

Exploratory Find the value of each of the following. 1,814,400

1. $_5P_3$ 60
2. $_6P_2$ 30
3. $_4P_1$ 4
4. $_{10}P_8$
5. $_7P_5$ 2520
6. $_7P_6$ 5040
7. $_{100}P_3$ 970,200
8. $_{75}P_4$
29,170,800

State how many three letter "words" can be made from the letters in each word. Assume a "word" is any arrangement of three different letters.

9. CAT 6
10. GOAT 24
11. CLONE 60
12. HUXLEY 120

Written In geometry, polygons are usually named by placing letters of the alphabet at their vertices. The letters S, N, A, K, and E are available. Tell how many different ways each of the following figures can be labeled if there are no other restrictions. 120

1. triangle 60
2. quadrilateral 120
3. pentagon

Tell how many four letter "words" can be made from the letters X, B, T, L, R, V, and A for each situation.

4. There are *no* restrictions. 840
5. The first letter must be A. 120
6. The third letter must be A. 120
7. The last two letters must be R or T in either order. 40

Tell how many five letter "words" can be made from E, Q, B, X, R, T, L, A, and V for each situation.

8. There are *no* restrictions. 15,120
9. The first letter must be A and the last must be T. 210
10. The first letter *cannot* be a vowel. 11,760
11. The last two letters must be vowels. 420
12. The letters A, X, L, and E must be used. 600
13. The letter X *cannot* be used. 6720

Answer the following.

14. Look at example 3 on page 428. How many messages are possible if nine flags are available? 504
15. Using the same flags in exercise 14, how many messages are possible if one particular flag must be at the top? 56
16. How many points *not* on the line $x = y$ can be plotted using ordered pairs whose coordinates are drawn from 3, -1, $\frac{1}{4}$, and 4? 12
17. In some states, license plates consist of two letters followed by four digits. Letters and numbers may be repeated. How many different license plates are possible if letters I and O may *not* be used, but there are *no* other restrictions? 5,760,000
18. Repeat exercise 17 for license plates consisting of three letters followed by three digits, with no restrictions on the letters used. 17,576,000

19. In exercise 17, how many different license plates are possible if the first digit *cannot* be zero? 5,184,000

20. In exercise 17, how many different license plates are possible if the letters must be different? 5,520,000

Tell how many three-digit, three-letter license plates are possible for each of the following situations.

21. There are *no* restrictions. 17,576,000
22. The letters must be different. 15,600,000
23. The letter O *cannot* be used. 15,625,000
24. The three digits *cannot* be 000. 12,812,904

Answer the following.

25. How many two-letter, five-digit telephone numbers can be made if the first digit cannot be zero? 60,840,000

26. How many seven-digit phone numbers are possible if the first three numbers must be 033? 10,000

27. Look at the Social Security card shown at the right. How many different Social Security numbers are possible? (Assume *no* restrictions.) 10^9

28. A combination lock has 60 numbers around its dial. How many different combinations are possible if every combination has three numbers? 216,000

29. How many numbers can be formed from the digits 3, 5, 6, and 9? Every digit is used once. 205,320

30. How many numbers greater than 3000 but less than 7000 can be formed from the digits 2, 3, 5, and 8? Every digit is used once. 12

31. How many even numbers between 3000 and 7000 can be formed from the digits 2, 3, 5, and 8 if each digit is used once? 8

32. From all the four digit numbers that can be formed from 2, 3, 5, and 7, using each digit once, one is chosen at random. What is the probability that it is divisible by 25? $\frac{1}{12}$

33. Of all the five digit numbers that can be formed from 1, 2, 3, 5, and 7, using each digit once, one is chosen at random. What is the probability that it is divisible by 25? $\frac{1}{20}$

34. On the planet Glorfum II, the alphabet has 76 letters. What expression represents the number of 19 letter "words" that can be formed from 19 different letters? $_{76}P_{19} = \frac{76!}{(76-19)!}$

11.6 Permutations with Repetitions

Objective: To use permutations to count possible orders of objects when some objects are identical.

You have seen that the number of permutations of six things taken all at a time is 6! or 720.

But what if *not all of the six* things are different? For example, find the number of arrangements, or permutations, of the letters in BOBBLE. In this case, there are six symbols of which three—the B's—are the same.

Assume that the B's in BOBBLE are different from each other in some way—say, by color. List some of the different arrangements of the six letters.

B O B B L E   B O B B L E   B O B B L E
B O B B L E   B O B B L E   B O B B L E

B B B O L E   B B B O L E   B B B O L E
B B B O L E   B B B O L E   B B B O L E

Of course, there would be many others.

But the B's in the original problem are *not* colored. Therefore when the color is omitted, the first six arrangements are the same. So are the next six. Every final arrangement will represent six arrangements that use color. Why does this occur? Because three different symbols—in this case, three different colored B's—can be arranged in 3! or 6 different ways. This means that the final answer to the BOBBLE problem will be the number of arrangements that were determined using colored B's divided by 6. The number of different permutations of 6 items, with 3 of the items identical, is $\frac{720}{6}$ or 120.

In general, the number of permutations of n things, with r of the things identical, is $\frac{n!}{r!}$.

Example

1 Find the number of permutations of the letters Q, B, B, R, X, T, and V.

There are seven letters and two of the letters are identical.

The answer is $\frac{7!}{2!} = \frac{7 \cdot 6 \cdot 5 \cdot 4 \cdot 3 \cdot 2 \cdot 1}{2 \cdot 1}$ or 2520.

432 PROBABILITY AND COMBINATORICS

A famous problem in the study of permutations is to find how many arrangements exist of the letters in MISSISSIPPI. But, be careful. This problem is tough because there are r things of one kind (I's), s things of another kind (S's), and still t things of a third kind (P's).

Permutations with Repetitions

> The number of permutations of n things with r things identical, s things identical, and t things identical is $$\frac{n!}{r!\,s!\,t!}.$$

Example

2 Find the number of permutations of the letters in MISSISSIPPI.

In the word MISSISSIPPI, three sets of identical letters can be found. There are four S's, four I's, and two P's.

Use the formula to find the answer. There are $\dfrac{11!}{4!\,4!\,2!}$ or 34,650 permutations.

Exercises

Max: 1-8, 1-25, 1-3; Avg: 1-8, 1-25; Min: 1-8, 1-25 odds

Exploratory Find the number of permutations of the letters in each of the following.

1. BUBBLE 120
2. GOBBLE 360
3. POP 3
4. CLAMMY 360
5. LOOK 12
6. HAPPY 60
7. SORRY 60
8. WRITTEN 2520

Written Find the number of permutations of the letters in each of the following.

1. TROPICAL 40,320
2. MUMMY 20
3. QUADRATIC 181,440
4. BANANA 60
5. XXXXXX 1
6. LONDON 180
7. MADRID 360
8. GIGGLING 840
9. PARSNIPS 10,080
10. ARREARS 420
11. VOODOO 30
12. CLOCKMAKER
13. COMMITTEE 45,360
14. BASKETBALL 453,600
15. BOOKKEEPER 151,200
16. PARALLEL 3360

12. 907,200

17. Don has 5 pennies, 3 nickels, and 4 dimes. The coins of each denomination are indistinguishable. How many ways can he arrange the coins in a row? 27,720

18. Jill has 8 quarters, 5 dimes, 3 nickels, and a penny. The coins of each denomination are indistinguishable. How many ways can she place the coins in a straight line? 12,252,240

19. How many ways can 4 nickels and 5 dimes be distributed among 9 children if each is to receive one coin? 126

20. How many 6-digit numbers can be made using the digits from 833,284? 180

21. Ten scores received on a test were 82, 91, 75, 83, 91, 64, 83, 77, 91, and 75. In how many different orders might they be recorded? 151,200

22. There are 3 identical red flags and 5 identical white flags that are used to send signals. All 8 flags must be used. How many signals can be given? 56

23. How many numbers between 3000 and 7000 can be formed from the digits 2, 3, 3, and 8 if each digit is used once? 6

24. How many odd numbers between 3000 and 7000 can be formed from the digits 2, 3, 3, and 8 if each digit is used once? 2

25. From all the numbers that can be formed from the digits 2, 5, 5, 7, using each digit once, one is chosen at random. What is the probability it is divisible by 25? $\frac{1}{3}$

Challenge

1. A certain word has n letters, 3 of which are identical. If there are 20 permutations of the letters of this word, find n. 5

2. A certain word has 8 letters, r of which are identical. If there are 56 permutations of the letters of this word, find r. 6

3. To win a math contest, Lois must guess how many marbles are in a box. She knows that there are 2 red marbles and some number of white marbles in the box. She also knows that there are 15 permutations of the marbles. What number should Lois guess? 6

Mathematical Excursions

Puzzle

Three dimes and two quarters are arranged as shown at the right.

The object is to rearrange the coins in the least number of moves so that they appear as follows.

The rules for moving the coins are:
1. With each move, a dime and an adjacent quarter are moved as a whole.
2. During a move, the dime and quarter may not be interchanged.

11.7 Picking a Committee

Objective: To determine if the number of groupings of objects is found using a combination or a permutation.

Pine City Board of Supervisors		
K. Adams	D. Davidson	J. Halloran
M. Barnard	C. Eastman	M. Ianelli
S. Cohen	S. Falk	R. Jarmyn
	B. Garber	

Journal

Give a real-life example using permutations and one using combinations.

The Pine City Board of Supervisors has ten members. Suppose the board must choose a subcommittee of three persons to represent Pine City before the State Legislature. How many ways can this be done?

Some might be tempted to say that the answer is $_{10}P_3$. However, permutations are arrangements. The arrangement

    Adams   Falk   Ianelli

may be different from the arrangement

    Adams   Ianelli   Falk

but both arrangements *name the same committee!* The three persons—Adams, Falk, and Ianelli—form the *same subcommittee despite the order in which they are chosen.* Using $_{10}P_3$ gives an answer that is much too large.

You are no longer dealing with a permutation. In this problem, order does *not* count. This is similar to finding a subset of a given set. For example, the set {a, b, c} is the same subset of the set of letters of the alphabet as the set {c, a, b}. The elements contained in the subset is the important issue, not their order of appearance. This kind of set is called a **combination**.

A combination can be thought of as a set, or as a committee, in which *order does not count*. This is in contrast to permutations, or arrangements, in which order does count.

Sometimes the word *selection* is used to describe a combination.

Examples

1 **Decide if picking three students to plan the prom is a combination or permutation situation.**

Three students are chosen to form a committee and the order is *not* significant. This is a combination situation.

2 **Decide if picking a prom chairperson, a decoration chairperson, and a ticket chairperson is a combination or permutation situation.**

Three students are chosen, but for three different jobs. Order is significant. This is a permutation situation.

3 **Consider the vowels A, E, I, O, and U. Find the number of combinations of the five vowels.**

There are $5! = 120$ different permutations of these letters. However, they all name the same combination. This suggests that there is only one combination of five things taken five at a time.

The symbol for combination is C.

$_nC_n$

> The symbol $_nC_n$ names the number of combinations of n things taken n at a time.
> $$_nC_n = 1$$

This value of $_nC_n$ is reasonable because if you have to choose a committee of n persons, and you have n from which to choose, there is obviously only one way you can make your selection. Furthermore, $_nC_0 = 1$. After all, if no one is chosen from n persons, there is only one way the choice can be made.

$_nC_r$ In general, the symbol, $_nC_r$, names the number of combinations of n things taken r at a time.

The Pine City Board problem is really asking you to find $_{10}C_3$.

Exercises

Max: 1-6, 1-15; Avg: 1-6, 1-15; Min: 1-6, 1-11 odds, 12-15

Exploratory Find the value of each of the following.

1. $_3C_3$ 1
2. $_7C_7$ 1
3. $_1C_1$ 1
4. $_7C_0$ 1
5. $_{135}C_{135}$ 1
6. $_{19}C_0$ 1

Written Decide if each of the following is a combination or permutation situation.

1. selecting three persons to go out and get pizza for everybody combination
2. selecting three persons to be editor-in-chief, photo editor, and feature editor for the school yearbook permutation
3. seating six persons in six seats combination
4. finding the number of subsets of a given set combination
5. finding the number of orders in which five prize-winning tickets can be pulled from a hat if all the prizes are alike combination
6. same as exercise 5, except the prizes are different permutation
7. choosing a committee from the faculty of a school combination
8. dealing five cards to a poker player combination
9. picking three ferryboats from a fleet of ten for a special charter trip combination
10. choosing four kinds of sandwiches to serve at the picnic combination
11. picking the batting order from a list of fifteen players permutation

Answer the following. For the answers to exercises 12-15, see the Teacher's Guide.

12. What do you think the value of $_9C_1$ is? Why?
13. Explain the meaning of the symbol "$_nC_n$."
14. Explain the meaning of the symbol "$_4C_4$." What is its value? Why is this a reasonable answer?
15. Why is it reasonable that $_nC_0$ should equal 1?

11.8 $_nC_r$

Objective: To use combinations to count possible groupings of objects.

BALLOT
Vote For Three
☐ AARON
☐ BELINDA
☐ CHRIS
☐ DAVE

How many committees of three persons can be selected from four persons—**A**aron, **B**elinda, **C**hris, and **D**ave? The answer to this problem is the value of $_4C_3$. Call the persons **A, B, C,** and **D**.

Consider $_4P_3$. Following are the permutations of the symbols **A, B, C,** and **D** taken three at a time.

A B C	B C D	A B D	A C D
B A C	C B D	B A D	C A D
A C B	B D C	A D B	A D C
B C A	C D B	B D A	C D A
C A B	D C B	D A B	D C A
C B A	D B C	D B A	D A C

Calculator Hint

If you highlight PRB on the MATH menu on the TI-82 graphing calculator, you will find functions to calculate $_nP_r$, $_nC_r$, and factorials.

Now, note that *all the arrangements in the first column are the same committee.* The same is true of each of the other columns. To find $_4C_3$, divide the total number of *arrangements* of the four symbols taken three at a time, by the number of ways the three chosen symbols can be arranged among themselves. This last number is 3! or 6. Therefore,

$$_4C_3 = \frac{_4P_3}{3!} = \frac{24}{6} = 4.$$

If there are four persons—**A, B, C, D**—there are four possible three-person committees—**A, B, C; B, C, D; A, B, D; A, C, D.**

Consider $_5C_3$. Now there are $_5P_3$ different arrangements. But the three elements can be arranged among themselves in $_3P_3 = 3!$ or 6 ways. Hence, $_5C_3 = \frac{_5P_3}{3!} = \frac{60}{6}$ or 10.

Here is a general rule.

Value of $_nC_r$

To find the number of combinations of *n* things taken *r* at a time, divide the number of permutations of *n* things taken *r* at a time by *r*!.

$$_nC_r = \frac{_nP_r}{r!}$$

Because $_nP_r = \frac{n!}{(n-r)!}$, the rule can also be written

$$_nC_r = \frac{n!}{r!(n-r)!}.$$

Examples

1 Find how many combinations of four things can be selected from nine.

Use the last rule.

$$_9C_4 = \frac{9!}{4!5!} = \frac{9 \cdot \overset{2}{\cancel{8}} \cdot 7 \cdot \cancel{6} \cdot \cancel{5} \cdot 4 \cdot \cancel{3} \cdot \cancel{2} \cdot \cancel{1}}{\cancel{4} \cdot \cancel{3} \cdot \cancel{2} \cdot 1 \cdot \cancel{5} \cdot \cancel{4} \cdot \cancel{3} \cdot \cancel{2} \cdot \cancel{1}} = 126$$

There are 126 combinations.

2 Find the number of subsets of five elements in a set with seven elements in it.

Recall that with sets, order does *not* count. So this is a combination problem. Use the second rule to find the value of $_7C_5$.

$$_7C_5 = \frac{7!}{5!2!} = \frac{7 \cdot 6}{2 \cdot 1} = 21$$

There are 21 subsets.

3 Find the number of subcommittees of three in the ten-member Pine City Board of Supervisors.

The answer is $_{10}C_3$, which equals 120. Check this one.

4 In the card game, bridge, every player is dealt a hand of 13 cards. Find how many different hands there are in bridge.

In this case, the order in which the cards are dealt is *not* important. Therefore, this is a combination problem.

$$_{52}C_{13} = \frac{52!}{13!39!}$$

The value of $\frac{52!}{13!39!}$ is 635,013,559,600.

Exercises

Max: 1-14, 1-33, 1-6; Avg: 1-5, 7-13 odds, 1-15, 17-33 odds; Min: 1-13 odds, 1-33 odds

Exploratory List the combinations of three elements that can be chosen from each of the following.

1. R, S, T, W

R,S,T; R,S,W; R,T,W; S,T,W

2. R, S, T, W, Z

R,S,T; R,S,W; R,S,Z; R,T,W; R,T,Z;
R,W,Z; S,T,W; S,T,Z;
S,W,Z; T,W,Z

3. S, T, W

S,T,W

Answer the following. For the answers to exercises 4-5, see the Teacher's Guide.

4. List all the two-ingredient desserts that can be prepared from a quantity of apples, oranges, dates, grapes, and strawberries.
5. List the combinations of four elements that can be chosen from A, B, C, D, E, and F.

Find the value of each of the following.

6. $_5C_3$ 10
7. $_6C_4$ 15
8. $_7C_4$ 35
9. $_8C_5$ 56
10. $_8C_2$ 28
11. $_9C_4$ 126
12. $_9C_5$ 126
13. $_6C_5$ 6
14. $_{12}C_{11}$ 12

Written Tell how many committees of three persons can be chosen from each of the following.

1. six persons 20
2. ten persons 120
3. fifteen persons 455
4. forty persons 9880

Answer the following. Refer to page 434.

5. In how many ways can the Pine City Board of Supervisors send a committee of three to the capital if Ianelli must be on the committee? 36
6. In how many ways can the Pine City Board of Supervisors send a committee of three to the capital if both Ianelli and Falk must be on the committee? 8

Set B has seven elements. Answer the following.

7. How many subsets of one element does it have? 7
8. How many subsets of three elements does it have? 35
9. How many subsets of five elements does it have? 21
10. How many subsets does it have altogether? Be careful. 128

Tell how many ways two of Beethoven's nine symphonies can be chosen for a concert program for each situation. The order is *not* important.

11. There are *no* restrictions. 36
12. The Ninth Symphony *cannot* be chosen. 28
13. The Second Symphony must be chosen. 8
14. The Ninth Symphony *cannot* be paired with the Third, Sixth, or Seventh. 33
15. From a list of 12 books, how many groups of 5 books can be selected? 792
16. How many baseball teams of 9 members can be formed from 14 players? 2002
17. Suppose there are 9 points on a circle. How many different 4-sided polygons can be formed by joining any 4 of these points? 126
18. There are 85 telephones at Kennedy High School. How many 2-way connections can be made among the school telephones? 3570
19. How many different groups of 25 people can be formed from 27 people? 351
20. From a deck of 52 cards, how many different 4-card hands exist? 270,725
21. Suppose there are 8 points in a plane, no 3 of which are collinear. How many distinct triangles could be formed with these points as vertices? 56

From a group of 8 men and 10 women, a committee of 5 is to be selected. In how many ways can the committee be selected if it must be comprised as follows.

22. All are men. 56
23. All are women. 252
24. There are 4 women and 1 man. 1680
25. There are 3 men and 2 women. 2520

26. From a deck of 52 playing cards, how many different 5-card hands can have 5 cards of the same suit? 5148
27. From a deck of 52 playing cards, how many different 4-card hands can have each card from a different suit? 28,561
28. How many softball teams can be formed from 15 players if only 3 pitch while the others play the remaining 8 positions? 1485
29. Answer exercise 28 if, in addition, only 2 of the players can catch. 720

Find the total number of diagonals that can be drawn in each polygon.
30. hexagon 9   **31.** decagon 35   **32.** 13-gon 65   **33.** n-gon $\frac{n}{2}(n-3)$

Challenge Find each of the following pairs of values. What do you notice? They are equa
1. $_5C_2$; $_5C_3$ 10; 10
2. $_{10}C_8$; $_{10}C_2$ 45; 45
3. $_{11}C_4$; $_{11}C_7$ 330; 330

4. Give an argument to show that for $n \in \mathcal{N}$ and $r \leq n$, $_nC_r = {}_nC_{n-r}$.
5. For any n, find the value of $_nC_1$. n
For answers to exercises 4 and 6, see the Teacher's Guide
6. For any n, find the value of $_nC_{n-1}$. Justify your answer in two ways.

Mixed Review

There are 5 fudgesicles and 8 popsicles in the freezer. If 2 are selected at random, find the probability of each selection.

1. 2 fudgesicles $\frac{5}{39}$ **2.** 2 popsicles $\frac{14}{39}$ **3.** 1 fudgesicle and 1 popsicle $\frac{20}{39}$

Give the number of permutations of the letters in STUDYING for each situation.
4. The last letter must be G. 5040
5. The last letter cannot be G. 35,280
6. The first letter must be a vowel. 10,080
7. The middle two letters must be S and T in either order. 1440

8. Five algebra and four geometry books are to be placed on a shelf. How many ways can they be arranged if all the algebra books are together? 14,400
9. There are 1 blue, 1 red, and 4 green books on a shelf. How many ways can they be arranged if the red book and the blue book are separated? 480
10. How many ways can the letters from the word TELEVISION be arranged? 907,200
11. How many 5-digit even numbers can be made using the digits from 83,384?
12. An urn contains blue and gold marbles. The probability of drawing one blue marble from the urn is $\frac{2}{5}$. If 4 blue marbles are removed from the urn, then the probability of drawing one blue marble is $\frac{1}{4}$. How many of each type of marble are in the urn?
8 blue, 12 gold

Objective: To apply various principles of probability and combinatorics.

11.9 Applications

The following examples use various principles of probability and combinatorics.

Examples

1 A committee has five Republicans and four Democrats. In how many ways can a subcommittee of four persons be selected if the subcommittee must contain two Republicans and two Democrats?

The two Republicans can be chosen from the five on the committee in $_5C_2$ ways. The two Democrats can be chosen from the four on the committee in any of $_4C_2$ ways. Now, for every choice of two Republicans, there is a different choice of two Democrats. Use the Counting Principle to conclude that the number of ways the subcommittee can be selected is

$$_5C_2 \cdot {_4C_2} = 10 \cdot 6 \text{ or } 60.$$

2 If, in example 1, the subcommittee is chosen *at random*, what is the probability that it will contain two Republicans and two Democrats?

There are 60 ways of choosing two Republicans and two Democrats. Altogether, there are $_9C_4 = 126$ different ways of choosing the subcommittee of four. Using $P(E) = \dfrac{n(E)}{n(S)}$,

$$P(2 \text{ Republicans, 2 Democrats}) = \frac{60}{126} \text{ or } \frac{10}{21}.$$

3 All the taxis operated by the Dick'n'Doug Taxicab Company seat a maximum of four persons. A party of six persons must therefore use two cabs. In how many different ways can a party of six group themselves?

They can divide themselves either three-three or four-two. Now, once three persons are selected for one cab, the three who go in the other cab are *automatically determined*. Such a three-three grouping can be made in any of $_6C_3$ ways. The value of $_6C_3$ is 20. If they choose a four-two split, this can be done in any of $_6C_4$ ways. The value of $_6C_4$ is 15 ways. The group can therefore separate itself in any of 20 + 15 or 35 ways.

Example

4 Suppose there are five points in a plane, as shown in the diagram at the right. No three of the points can be contained in any one line. How many different line segments are there connecting the points?

The segment *AB* is the same as the segment *BA*—that is, order does *not* count. This is a *combination problem*. Any two of the five points determine a segment. The answer is $_5C_2$ or 10.

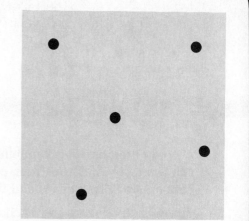

A card is drawn from a standard deck. The probability of picking a king is $\frac{4}{52}$ or $\frac{1}{13}$. The probability of *not* picking a king is $\frac{48}{52}$ or $\frac{12}{13}$.

$$P(king) = \frac{1}{13}$$

$$P(not\ king) = \frac{12}{13}$$

Notice that

$$\frac{1}{13} = 1 - \frac{12}{13}$$

and

$$P(king) = 1 - P(not\ king).$$

P(E) and P(not E)   In general, $P(E) = 1 - P(not\ E)$.

Likewise, $P(not\ E) = 1 - P(E)$.

Examples

5 If six fair coins are tossed, what is the probability that *at least* one lands heads up?

"At least one" means "exactly one, or exactly two, or exactly three, and so on, up to exactly six." Another way of thinking about "at least one head" is to consider it as *"not all tails."* Use the fact that $P(E) = 1 - P(\text{not } E)$. Now, $P(\text{all tails in six tosses})$ is $\left(\frac{1}{2}\right)^6$ or $\frac{1}{64}$. The probability of tossing at least one head is $1 - \frac{1}{64}$ or $\frac{63}{64}$.

6 The planets Merglorf and Venglorf have been at war for centuries. At the Merglorf spy-training school, ten agents are being trained for a secret mission. What the Merglorfians do not know is that two of the ten are actually Venglorf counterspies. If five of the agents being trained are chosen *at random* for a secret mission, what is the probability that at least one of them will be a Venglorf counterspy?

The number of ways the five can be chosen from ten is

$$_{10}C_5 = \frac{10 \cdot 9 \cdot 8 \cdot 7 \cdot 6}{5 \cdot 4 \cdot 3 \cdot 2 \cdot 1} = 252.$$

What is the event, *"not 'at least one counterspy'?"* The answer is "all genuine spies" and the number of ways five agents can be chosen from the eight genuine spies is $_8C_5$ or 56. This means that the probability of picking all genuine spies is $\frac{56}{252}$ and the probability of choosing at least one counterspy is $1 - \frac{56}{252} = \frac{196}{252}$ or $\frac{7}{9}$.

Exercises

Max: 1-8, 1-46, 1-7; Avg: 1-8, 1-13, 15-45 odds; Min: 1-7 odds, 1-13, 15-45 odds

Exploratory For each value of $P(E)$, find the value of $P(\text{not } E)$.

1. $P(E) = \frac{1}{3}$ $\frac{2}{3}$

2. $P(E) = \frac{1}{2}$ $\frac{1}{2}$

3. $P(E) = \frac{3}{11}$ $\frac{8}{11}$

4. $P(E) = \frac{14}{45}$ $\frac{31}{45}$

5. $P(E) = \frac{21}{64}$ $\frac{43}{64}$

6. $P(E) = \frac{5}{102}$ $\frac{97}{102}$

7. $P(E) = 0$ 1

8. $P(E) = 1$ 0

Written Helen has four Latin books and seven Greek books. Tell how many ways she can make each selection.

1. five books, no restrictions 462
2. two Latin books, two Greek books 126
3. two Latin books, three Greek books 210
4. one Latin book, four Greek books 140

Answer each question about Helen's books.

5. One of Helen's Greek books was given to her by her sister. If she chooses five books *at random,* what is the probability she will choose that book? $\frac{5}{11}$
6. If her selection is three books of each language, what is the probability the book in exercise 5 will be selected? $\frac{3}{7}$

From a group of six men and three women, tell how many committees can be formed consisting of the following.

7. three men 20
8. any three persons 84
9. a man and a woman 18
10. two men and a woman 45
11. all women (any size committee) 7

Answer each question about the committees.

12. How many committees of three are possible if one of the women, Dr. Snow, must be on the committee? 28
13. If the committee of three is chosen at random, what is the probability Dr. Snow will be on the committee? $\frac{1}{28}$

Tell how many ways a group of ten persons can regroup themselves into each of the following groups.

14. seven and three 120
15. five and five 252
16. four and six 210

Tell how many line segments can be drawn using the following numbers of points in a plane. Assume *no* three points lie on the same line.

17. 4 6
18. 3 3
19. 6 15
20. 8 28
21. 12 66

Tell how many triangles can be drawn using the following numbers of points in a plane. Assume *no* three points lie on the same line.

22. 5 10
23. 8 56
24. 10 120
25. 12 220
26. 24 2,024

Find the probability of getting at least one tail if the following numbers of fair coins are tossed.

27. 3 $\frac{7}{8}$
28. 5 $\frac{31}{32}$
29. 6 $\frac{63}{64}$
30. 8 $\frac{255}{256}$
31. 10 $\frac{1023}{1024}$

Answer the following.

32. What is the probability in exercise 28 if the coin has been "loaded" so that $P(tails) = \frac{2}{3}$? $\frac{242}{243}$
33. What is the probability of getting at least one 6 when three fair dice are tossed? four fair dice? n fair dice?
    $\frac{91}{216}$; $\frac{671}{1296}$ $1 - (\frac{5}{6})^n$

34. What is the probability that a five-child family has at least one girl? $\frac{31}{32}$

35. What is the probability that a seven-child family has at least one girl? $\frac{127}{128}$

Rework example 6 on page 443 for the following information.

36. 12 agents being trained; 5 selected for the mission; 2 counterspies $\frac{15}{22}$

37. 12 agents being trained; 4 selected for the mission; 2 counterspies $\frac{19}{33}$

38. 9 agents being trained; 3 selected for the mission; 1 counterspy $\frac{1}{3}$

A bag contains 4 red, 6 white, and 9 blue marbles. Five are selected at random. Find the probability of each selection.

39. All the marbles are white. $\frac{1}{1938}$

40. All the marbles are blue. $\frac{7}{646}$

41. All the marbles are red. 0

42. Two are red, 2 are white, and 1 is blue. $\frac{45}{646}$

43. Two must be blue. $\frac{120}{323}$

44. Two are 1 color and 3 are another color. $\frac{78}{323}$

45. Michele is a lawyer, likes Spanish books, and collects coins. Her library consists of eight law books, seven Spanish books, and three coin books. For her summer reading she plans to pack five law books, three Spanish books, and two coin books. How many ways can she do this? 5,880

46. Look at exercise 45. Suppose Michele has eight Spanish books. How many ways can Michele choose the books for her summer reading? 9408

Challenge Answer the following.

1. How many ways can six persons sit around a circular table? Note that the two orders in the diagram are the same. Hint: Suppose you "broke the circle and stretched it out to form a linear arrangement." 120

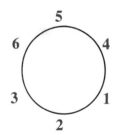

2. How many ways can seven persons sit around a circular table? 720

3. A light with a parallel circuit will go on if any one, or more, of its switches are closed. Suppose a light has a parallel circuit with 3 switches. How many combinations are there of open and closed switches that will permit the light to go on? 7

4. Repeat exercise 3 for a light that has a parallel circuit with 8 switches. 255

5. Repeat exercise 3 for a light that has a parallel circuit with n switches.
$${}_nC_n + {}_nC_{n-1} + {}_nC_{n-2} + \cdots + {}_nC_2 + {}_nC_1 = 2^n - 1$$

A student guessed at every question on a four-choice multiple choice test with q questions. Answer the following.

6. What is the probability he answered them all correctly? $(\frac{1}{4})^q$

7. What is the probability that he answered one-half of the questions correctly? $(\frac{1}{4})^{\frac{q}{2}}$

Mathematical Excursions

DIAGONALS

A diagonal of a polygon is a line drawn from any vertex to any nonadjacent vertex.

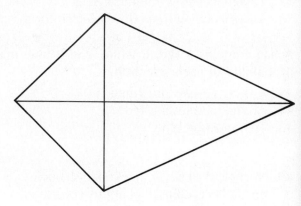

Thus, a quadrilateral has two diagonals.

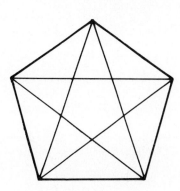

A pentagon has five diagonals.

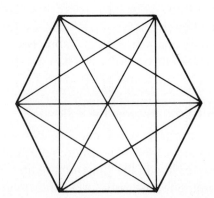

A hexagon has nine diagonals.

The problem of finding the number of diagonals of a polygon is similar to the problem of finding the number of line segments determined by a certain number of points.

Exercises Answer the following.

1. How many diagonals does a triangle have? 0
2. How many diagonals does a heptagon (a seven-sided polygon) have? 14
3. How many diagonals does an octagon have? 20
4. How many diagonals does a nonagon (a nine-sided polygon) have? 27
5. How many diagonals does a decagon (a ten-sided polygon) have? 35
6. How many diagonals does a polygon with n sides have? $\dfrac{n(n-3)}{2}$

Problem Solving Application: List Possibilities

Objective: To solve problems by listing possibilities.

Certain problems can be solved by determining the number of items that satisfy the conditions of the problem. These problems can often be solved by *listing possibilities*.

Example

1 Leshlock Construction Co. is building a new apartment complex that contains 1037 units. How many single digits will be needed for the numbers on the apartment doors in the complex?

Explore Some of the apartment numbers will require only one digit. Others will require two digits, three digits, or four digits. We must determine how many total digits are needed to make the 1037 apartment numbers.

Plan To solve this problem, find the numbers that will require only one digit. Then determine how many digits will be needed for all of these numbers. Repeat this process for the numbers that will require two, three, and four digits. The total number of digits needed for the apartment doors can now be determined from this list by adding these results.

Solve

Apartment Numbers	Number of Apartments	Number of Digits		
1 – 9	9	9 × 1	or	9
10 – 99	90	90 × 2	or	180
100 – 999	900	900 × 3	or	2700
1000 – 1037	38	38 × 4	or	152
				3041 Add.

A total of 3041 single digits will be needed for the numbers on the apartment doors in the complex.

Examine Most of the 1037 doors will require three digits for the apartment number. Thus, approximately 1000 × 3 or 3000 digits will be needed for all of the apartments numbers. The answer seems reasonable.

Exercises

Max: 1-25; Avg: 1-13, 15-25 odds; Min: 1-25 odds

Written Refer to example 1 on page 447. Determine how many of each digit will be needed for the 1037 apartment doors in the complex.

1. 0   241
2. 1   352
3. 2   314
4. 3   312
5. 5   304
6. 6   304
7. 7   304
8. 8   303
9. 9   303

10. Five computers are to be directly connected to one another using circuits. How many circuits will be needed to make all of these connections?   10

11. A vending machine dispenses products that each cost 50¢. It accepts quarters, dimes, and nickels only. How many different combinations of coins must the machine be programmed to accept?

12. Four people sit down at a table for a game of cards. How many different seating arrangements are there? If the two people that sit across the table from each other are partners, how many different sets of partners are possible?   6; 3

13. The two co-captains from BK High School meet with the three co-captains from EC High School for the coin toss. If each co-captain shakes hands with the referee and each opposing co-captain, how many hand shakes occur?

14. In darts, you can receive any of the following point totals on one throw: 1, double 1 (2), triple 1 (3), 2, double 2 (4), triple 2 (6), and so on up to 20, double 20 (40), and triple 20 (60). You can also receive either 25 or 50 points for hitting the appropriate circle on the board. If Jim Hardthrower scored 30 points on 3 throws, what combination of point totals could he have?   412

15. To determine total bases in softball, 4 bases are given for a home run, 3 for a triple, 2 for a double, and 1 for a single. If Casey Atthebat had 12 total bases in 5 at bats, what combinations of hits could she have?   11

The Leaning Tower of Pizza will add any combination of the following items to their cheese pizza: pepperoni, sausage, mushrooms, onions, olives, or hot peppers. How many of each type of pizza are possible? (Cheese does not count as an item.)

16. pizza with two items on it   15
17. pizza with three items on it   21
18. pizza with at least three items on it   42
19. pizza with sausage on it   32
20. pizza *without* onion on it   32
21. pizza *without* meat on it   16
22. How many different pizzas can be ordered at the Leaning Tower of Pizza?   64

Dawn, Dawne, Don, Donn, and Lester are to be seated in a row of five chairs. Find the probability of each seating arrangement.

23. Lester sits next to Dawne. $\frac{2}{5}$
24. Someone sits between Don and Dawn.
25. Don and Donn do not sit next to each other. $\frac{3}{5}$

Portfolio Suggestion

Place your favorite word problem from this chapter in your portfolio with a note explaining why it is your favorite. Be sure to include your solution.

Performance Assessment

Women's socks are to be displayed along an aisle of a department store.
a. If there are 3 styles, 5 colors, and 3 sizes of socks, how many different arrangements would be possible? **45**
b. Not all of the arrangements in part a make good sense as it may cause a customer some confusion in trying to locate a particular pair of socks. Describe a poor arrangement and why it is poor. **See students' work.**

Chapter Summary

1. If S is an **outcome** and E is an **event,** then $P(E) = \frac{n(E)}{n(S)}$. If E is impossible, then $P(E) = 0$. If E is certain, then $P(E) = 1$. (412, 413)
2. For every event E, $0 \leq P(E) \leq 1$. (414)
3. For two events A and B, $P(A \text{ or } B) = P(A) + P(B) - P(A \text{ and } B)$. (417)
4. **Counting Principle:** Suppose one activity can occur in any of m ways. Another can occur in any of n ways. The total number of ways both activities can occur is given by the product mn. (419)
5. **Counting Principle Alternate Version:** Suppose the probability of one event E is $r (0 \leq r \leq 1)$. The probability of another event F is s $(0 \leq s \leq 1)$. Then the probability of E and F both occurring is the product rs. (419)
6. A **tree diagram** shows the number of possible outcomes. (420)
7. For any positive integer n, $n! = n(n - 1)(n - 2) \cdot \ldots \cdot 1$. (425)
8. $_nP_n$ names the number of **permutations** of n things taken n at a time. $_nP_n = n!$. (425)

9. $_nP_n$ names the number of permutations of n things taken r at a time

$$_nP_r = \frac{n!}{(n-r)!} \text{ or } n(n-1)(n-2)(n-3)\cdots(n-r+1). \quad (428)$$

10. The number of permutations of n things with r things identical, s things identical, and t things identical is $\frac{n!}{r!s!t!}$. (432)

11. $_nC_n$ names the number of **combinations** of n things taken n at a time. $_nC_n = 1$ and $_nC_0 = 1$. (435)

12. $_nC_r$ names the number of combinations of n things taken r at a time. To find the number of combinations of n things taken r at a time, divide the number of permutations of n things taken r at a time by $r!$. That is,

$_nC_r = \frac{_nP_r}{r!}$. Because $_nP_r = \frac{n!}{(n-r)!}$, the rule can also be written

$$_nC_r = \frac{n!}{r!(n-r)!} \quad (437)$$

13. In general, $P(E) = 1 - P(\text{not } E)$. (442)

Chapter Review

11.1 Complete the following.   {(1,H),(1,T),(2,H),(2,T),(3,H),(3,T),(4,H),(4,T)}
1. Draw the outcome set for the experiment of tossing a four-sided die and a coin.

Use the outcome set drawn in exercise 1 to find each probability.
2. $P(3, H)$ $\frac{1}{8}$
3. $P(\text{prime number}, T)$ $\frac{1}{4}$
4. $P(\text{even number}, H)$

11.2 A card is drawn from a standard deck of 52 cards. Find each probability.
5. $P(\text{black card})$ $\frac{1}{2}$
6. $P(\text{king or ace})$ $\frac{2}{13}$
7. $P(\text{ace, red card})$ $\frac{7}{13}$
8. $P(\text{black five or face card})$ $\frac{7}{26}$

11.3 Use the Counting Principle (either version) to answer each of the following.
9. The luncheon special at the Burger Shoppe includes sandwich, salad, and beverage. There are five sandwiches, four salads, and three beverages. How many different luncheon specials can be ordered? 60
10. The probability that Ron will make the football team is $\frac{1}{4}$. The probability that he will make the track team is $\frac{3}{5}$. What is the probability Ron will make both teams? $\frac{3}{20}$

11.4 Find the value of each of the following.

11. 3!   6
12. 4!   24
13. $_5P_5$ 120
14. $_8P_8$ 40,320

Give the number of permutations of the letters in SHARP for each situation.

15. There are *no* restrictions. 120
16. The first letter must be S. 24
17. The first letter must be a vowel. 24
18. The first two letters must be *SH* in that order. 6

11.5 Find the value of each of the following.

19. $_5P_4$ 120
20. $_{13}P_3$ 1,716
21. $_{50}P_2$ 2450
22. $_9P_4$ 3024

Tell how many four-letter "words" can be made from the letters A, C, E, P, I, K, and M for each situation.

23. There are *no* restrictions. 840
24. The first letter must be P or K. 240
25. The letter E *cannot* be used. 360
26. The first letter *cannot* be a vowel. 480

11.6 Find the number of permutations of the letters in each of the following.

27. RAMPAGE   2520
28. PEPPER   60
29. BALLOON   1260
30. PRESS ROOM   45,360

11.7 Find the value of each of the following.

31. $_4C_4$ 1
32. $_6C_0$ 1
33. $_9C_9$ 1
34. $_4C_0$ 1

Decide if each of the following is a combination or permutation situation.

35. picking seven runners for a cross-country meet   combination
36. picking president, vice president, secretary, and treasurer of the senior class   permutation

11.8 Find the value of each of the following.

37. $_4C_3$ 4
38. $_9C_4$ 126
39. $_{10}C_3$ 120
40. $_{20}C_{17}$ 1140

Tell how many ways a selection of five books can be made from each of the following.

41. 12 books   792
42. 5 books   1
43. 14 books   2002
44. 20 books   15,504

11.9 Answer the following.

45. Greg has 3 Latin, 5 Greek, and 5 Hebrew books. How many ways can he select two books of each language?   300
46. A Donald 'n' Elden Limousine holds five persons. How many ways can a group of eight persons ride in two limousines?   126
47. There are seven points in a plane such that no three of the points lie on the same line. How many different line segments connect the points?   21
48. If four fair coins are tossed, what is the probability that at least one lands tails up? $\frac{15}{16}$

Chapter Test

Find the value of each of the following.

1. 6! 720
2. 8! 40,320
3. $_7P_7$ 5040
4. $_{120}P_7$ 720
5. $_9C_9$ 1
6. $_7C_0$ 1
7. $_8C_7$ 8
8. $_{12}C_4$ 495

Answer the following.

9. A letter is drawn *at random* from A, B, C, D, E, and F. What is the probability the letter is a vowel? $\frac{1}{3}$

10. A state is chosen *at random* from the United States. What is the probability that the state is one of the five smallest states? $\frac{1}{10}$

11. A card is drawn from a standard 52-card deck. What is the probability the card will be the ace of diamonds or a club? $\frac{7}{26}$

12. A card is drawn from a standard 52-card deck. What is the probability the card will be a face card or a diamond? $\frac{1}{2}$

13. Darrell is buying a new car. He has three decisions to make. He can choose a 4-cylinder or a 6-cylinder engine, a standard or an automatic transmission, and a red, blue, or green exterior. How many choices does he have? 12

14. What is the probability that Darcy will answer two 5-choice multiple choice questions correctly if she is forced to guess at both? $\frac{1}{25}$

15. How many permutations can be formed using all the letters in FAMILY? 720

16. How many permutations can be formed using all the letters in HOUSE if the first letter must be a vowel? 48

17. How many five letter "words" can be made from the letters in THURSDAY? 6720

18. How many three-letter "words" can be made from the letters in CAMEL if the first letter must be M? 12

19. How many permutations can be formed using the letters in BENZENE? 420

20. A set has eight elements. How many subsets of four elements does it have? ?

21. A group has six men and four women. How many committees of two men and three women can be selected? 60

22. A Dick'n'Doug Deluxe Limousine holds five persons. How many ways can a group of seven persons ride in two limousines? 56

23. There are nine points in a plane such that no three of the points lie on the same line. How many different line segments connect the points? 36

24. A coin is "loaded" so that $P(tails) = \frac{3}{4}$. What is the probability of getting at least one head if the coin is tossed three times? $\frac{37}{64}$

25. From a group of four Republicans, three Democrats, and two Independents, a committee of four is to be selected *at random*. What is the probability that the committee will contain two Democrats, one Republican, and one Independent? $\frac{4}{21}$

CHAPTER 12

Transformation Geometry

Application in Music

Jazz, rap, classical, rock and roll—there are all kinds of music. Whatever your favorite kind of music is, it can be represented mathematically. Both the music itself and the methods for producing and recording music are dependent on mathematics.

Musical composers refer to **transposing** a piece of music from one key to another. When music is transposed, each note is moved, or **translated**, the same distance up or down the musical scale. The music below is from Mozart's *Clarinet Quintet in A Major, K*. The music is shown in the key of C major.

Copy the musical staff below. Translate the music into the key of A major. The first note is shown for you.

Individual Project: Art

Dutch artist M.C. Escher is famous for his art involving impossibilities and *tessellations*. A tessellation is a pattern that covers a plane with repeated shapes so that there are no empty spaces. Find some examples of Escher's use of tessellation. Then create your own piece of art using a tessellation. Draw your tessellation on a poster to display in the classroom.

Objective: To identify mappings, one-to-one mappings, and to draw arrow diagrams.

12.1 Mappings

The chart at the right lists students' names and test grades. Notice that each name is paired with exactly one grade. This is an example of a **mapping** from the set of names to the set of grades.

Student	Grade
Keith	73
Susan	85
James	65
Rosalind	73
Carlos	92

Definition of Mapping

A mapping pairs each member of one set with exactly one member of the same, or another set. The first set is called the **domain**. The second set is called the **range**.

A mapping is also called a <u>function</u>.

In the mapping described by the chart above, the domain is {Keith, Susan, James, Rosalind, Carlos}. The range is {73, 85, 65, 92}.

Another way to represent a mapping is to make an arrow diagram.

"Keith → 73" is read "Keith maps to 73."

Domain		Range
Keith	→	73
Susan	→	85
James	→	65
Rosalind		
Carlos	→	92

Consider the following mapping.

Domain		Range
−3	→	9
2	→	4
3		
4	→	16

Notice that each number in the domain is assigned its square in the range.
$x \rightarrow x^2$

Lowercase letters such as f, g, or h are used to name mappings. For example, the mapping shown above in the arrow diagram can be written as follows.

$$f: x \longrightarrow x^2 \quad \text{or} \quad x \xrightarrow{f} x^2$$

It is read "f is the function or mapping that maps x to x^2." In this mapping, 2 maps to 4. We write

$$f: 2 \longrightarrow 4 \quad \text{or} \quad 2 \xrightarrow{f} 4$$

The **image** of 2 is 4 and 2 is the **preimage** of 4. Preimages are always in the domain of a mapping. Images are always in the range.

Example

1 Find the images of −1, 0, and 1 if $h: x \longrightarrow x^2 + x + 1$.

$h: -1 \longrightarrow (-1)^2 + (-1) + 1$ or $h: -1 \longrightarrow 1$
$h: 0 \longrightarrow 0^2 + 0 + 1$ or $h: 0 \longrightarrow 1$
$h: 1 \longrightarrow 1^2 + 1 + 1$ or $h: 1 \longrightarrow 3$

The image of −1 and 0 is 1 and the image of 1 is 3.

Domain		Range
1	$\xrightarrow{f}$	5
2	$\xrightarrow{f}$	3
3	$\xrightarrow{f}$	18

one-to-one

Domain		Range
4	$\xrightarrow{g}$	a
5	$\xrightarrow{g}$	b
6	$\xrightarrow{g}$	

<u>not</u> one-to-one

Consider the mappings *f* and *g* shown at the left. In mapping *f*, each member of the range has exactly one preimage. In mapping *g*, *b* has two preimages, 5 and 6.

Definition of One-to-One Mapping

A mapping is a **one-to-one mapping** if and only if each member of the range has exactly one preimage.

Example

2 The domain is $\mathcal{Z}$. Is the mapping represented by $y = |x|$ a one-to-one mapping?

Consider two members of the domain, −1 and 1.

$$-1 \longrightarrow 1 \quad \text{and} \quad 1 \longrightarrow 1$$

The image 1 has two preimages. Thus, $y = |x|$ does not represent a one-to-one mapping if the domain is $\mathcal{Z}$.   What if the domain is $\mathcal{W}$?

	g	
1	$\longrightarrow$	2
2	$\longrightarrow$	4
3	$\longrightarrow$	6
4	$\longrightarrow$	8

	f	
2	$\longrightarrow$	4
4	$\longrightarrow$	16
6	$\longrightarrow$	36
8	$\longrightarrow$	64

In many applications of mappings, it is necessary to combine the action of two mappings to produce a new mapping. For example, suppose we have the mappings $f: x \longrightarrow x^2$ and $g: x \longrightarrow 2x$, shown to the left.

The situation pictured suggests that it is possible to combine the actions of *f* and *g* as shown on the next page.

```
           g        f
    1 ────▶ 2 ────▶ 4
    2 ────▶ 4 ────▶ 16
    3 ────▶ 6 ────▶ 36
    4 ────▶ 8 ────▶ 64
```

```
  f ∘ g
1 ────▶ 4
2 ────▶ 16
3 ────▶ 36
4 ────▶ 64
```

By combining f and g in this way, we produce a new mapping with domain $\{1, 2, 3, 4\}$ and range $\{4, 16, 36, 64\}$. This mapping is called the **composite mapping** of f and g. The composite of f and g is symbolized "$f \circ g$," which is read, "f following g." To the left is a diagram for $f \circ g$.

Example

3 If f and g are the mappings shown below, draw an arrow diagram for $f \circ g$.

```
      f              g
0 ────▶ 5      3 ────▶ 0
1 ────▶ 6      4 ────▶ 1
2 ────▶ 7      5 ────▶ 2
```

$f \circ g$ means "f following g." So, we perform a mapping of g first, and then a mapping of f.

```
      g        f                 f ∘ g
3 ────▶ 0 ────▶ 5          3 ────▶ 5
4 ────▶ 1 ────▶ 6          4 ────▶ 6
5 ────▶ 2 ────▶ 7          5 ────▶ 7
```

Exercises

Max: 1-10, 1-40; Avg: 1-10, 1-10, 11-39 odds; Min: 1-9 odds, 1-10, 11-39 odds

Exploratory State whether each of the following represents a mapping. If so, state the domain and range of the mapping. For the answers to exercises 1-6, see the Teacher's Guide.

1.
Degrees Celsius	Degrees Fahrenheit
100	212
35	95
20	68
0	32

2.
Cost of Telegram	Number of Words
$5.00	1, 2, 3, or 4
$10.00	5, 6, 7, or 8

3.
Table	Students
1	John
	Ted
	Rosemary
2	Denny
	Laura

4.
City	Population
Kenton	22,348
McGuffey	4,225
Alger	4,225

5. 5 ⟶ 8
9 ⟶ 6
17 ⟶

6. 8 ⟶ x
13 ⟶ y
14 ⟶ z

State whether each of the following represents a one-to-one mapping.

7. 2 ⟶ 3
4 ⟶ 5   yes
6 ⟶ 7

8. a ⟶ 1
b ⟶   no
c ⟶ 11

9.
x	-3	4	5	8	9
y	0	1	3	0	1

no

10. Suppose h is a one-to-one mapping. If its domain has 21 members, how many members are in its range?   21

Written Use the arrow diagram below for exercises 1–10.

1. Write the domain of f.   {-3, -2, -1, 0, 1, 2}
2. Write the range of f.   {-10, -5, 0, -1}

Find the image of each of the following.

3. 0   -1   4. 1   0   5. -2   -5   6. -1   0

Find all preimages of each of the following.

7. -10   -3   8. -5   -2, 2   9. 0   -1, 1   10. -1   0

If $g: x \longrightarrow 3x - 5$, find the image of each of the following.

11. 0   -5   12. 1   -2   13. 2   1   14. $\frac{1}{3}$ -4   15. -2   -11   16. -3   -14

The domain is $\{-3, -2, -1, 0, 1, 2, 3\}$. Draw an arrow diagram for each of the following mappings and then state the range.   For the answers to exercises 17-22, see the Teacher's Guide.

17. $g: x \longrightarrow 2x$
18. $x \xrightarrow{f} x - 5$
19. $m: x \longrightarrow x^2$
20. $t: x \longrightarrow x^2 - x + 1$
21. $x \xrightarrow{s} x - 2$
22. $f: x \longrightarrow |x| - 2$

Exercises 23–40 refer to the mappings f and g at the right.

f		g	
3 ⟶ 7		7 ⟶ 3	
4 ⟶ 9		13 ⟶ 6	
5 ⟶ 11		11 ⟶ 4	
6 ⟶ 13		9 ⟶ 5	

Find each of the following.

23. 3 $\xrightarrow{f}$ _?_   7
24. 9 $\xrightarrow{g}$ _?_   5
25. 5 $\xrightarrow{f}$ _?_   11
26. 11 $\xrightarrow{g}$ _?_   4

For the answers to exercises 27 and 28, see the Teacher's Guide.

27. Construct an arrow diagram for $f \circ g$.
28. Construct an arrow diagram for $g \circ f$.

Find each of the following.

29. domain of $f \circ g$ {7, 9, 11, 13}
30. domain of $g \circ f$ {3, 4, 5, 6}
31. range of $f \circ g$ {7, 9, 11, 13}
32. range of $g \circ f$ {3, 4, 5, 6}
33. $g \circ f: 3 \to$ _?_   3
34. $g \circ f: 4 \to$ _?_   5
35. $g \circ f: 5 \to$ _?_   4
36. $g \circ f: 6 \to$ _?_   6
37. $f \circ g: 7 \to$ _?_   7
38. $f \circ g: 9 \to$ _?_   11
39. $f \circ g: 11 \to$ _?_   9
40. $f \circ g: 13 \to$ _?_   13

Objective: To find the image of a figure over a line of reflection.

12.2 Line Reflections

The domain and range of some mappings is the set of all points in a plane. These mappings are used in geometry.

Definition of Transformation

> A **transformation** is a one-to-one mapping whose domain and range are the set of all points in the plane.

When water in a lake is perfectly calm, often mirror-like reflections can be seen of objects on the shore. This situation suggests the idea of a **line reflection**.

In the figure below, point A is flipped, or reflected over line ℓ onto point A'. We read A' as A prime. We say that A' is the image of A and B' is the image of B. A is the preimage of A' and B is the preimage of B'. Notice that point C, which is on line ℓ, is its own image.

Line ℓ is called the **line of reflection**. Notice that line ℓ is perpendicular to $\overline{AA'}$ and $\overline{BB'}$ and bisects each segment. In other words, ℓ is the perpendicular bisector of the segment joining a point and its image.

A line reflection is a one-to-one mapping because each point maps to a unique image. Therefore, a line reflection is a transformation.

Definition of Line Reflection

> A reflection in line ℓ (or reflection over line ℓ) is a transformation that maps each point P onto a point P' in one of the following ways.
> 1. If P is on line ℓ, then the image of P is P.
> 2. If P is not on line ℓ, then ℓ is the perpendicular bisector of $\overline{PP'}$.

The symbol $A \xrightarrow{R_\ell} A'$ means that A' is the image of point A for a reflection over line ℓ.

Examples

1 If $\triangle ABC$ is flipped, or reflected over line n, its image is $\triangle A'B'C'$.

Hence, $\triangle ABC \xrightarrow{R_n} \triangle A'B'C'$.

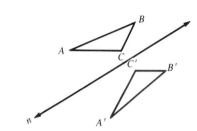

2 Check that each point on the figure satisfies the requirements.

The flag figures in color and black are images when you reflect in line ℓ.

3 Find $P \xrightarrow{R_\ell} P'$ using construction.

Use Construction 5 from Chapter 6.

1. With P as center, draw an arc that intersects ℓ in two points, X and Y.
2. With X as center and with radius $\overline{XP}$, draw an arc as shown.
3. With Y as center and with radius $\overline{XP}$, draw an arc that intersects the previous one at P'.

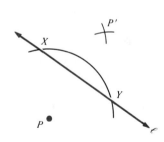

Because $\underline{\ell}$ is the perpendicular bisector of $\overline{PP'}$, point P' is the image of P.

Look carefully at the line reflection over line ℓ shown at the right. It appears that segment AB and its image, $\overline{A'B'}$, have the same length. That is, the distance from A to B is the same as the distance from A' to B'. (Check this using a ruler.) This observation can be summarized as follows.

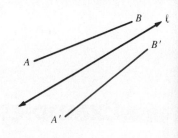

Journal

What other ways are mirrors used besides looking at your reflection?

Line reflections preserve distance.

Can you think of any other property of a figure that seems to be preserved by a line reflection? Are there any properties of figures that are not preserved? You will be asked to think about these questions in the exercises that follow.

Max: 1-8, 1-28; Avg: 1-8, 1-28; Min: 1-7 odds, 1-27 odds

Exercises

For the answers to exercises 1-8, see students' work.

Exploratory Copy each figure onto your paper. Then carefully sketch the image of each figure for R_m. Label all image points using prime notations; that is A, A'; B, B'; and so on.

Written Name the images of each of the following for line reflection R_ℓ.

1. point R point R
2. point L point U
3. point T point M
4. $\overline{RH}$ $\overline{RK}$
5. $\overline{SF}$ $\overline{NQ}$
6. $\angle USF$ $\angle LNQ$
7. quadrilateral $FSUH$ quadrilateral QNLK

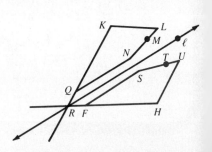

CHAPTER 12 461

For the answer to exercise 8, see the Teacher's Guide.
8. Explain how to find the image of point P for R_ℓ using a ruler and protractor. Justify your answer.

Copy the figures onto your paper and enlarge them. Use a ruler and protractor to find the images for a reflection over line *t*. For the answers to exercises 9-17, see student's work.

9.   10.   11.   12.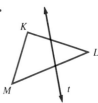

For each figure in exercises 9–12, copy and enlarge the figure and then use only a compass and straightedge to construct the image.

13. exercise 9   14. exercise 10   15. exercise 11   16. exercise 12

17. Find the image of points A, B, C, and D for R_ℓ.

For the answers to exercises 18-20, see the Teacher's Guide.
Explain how your results in exercise 17 suggest that each of the following is true.
18. Line reflections preserve collinearity.   19. Line reflections preserve betweenness.

20. Do line reflections preserve angle measure? Explain.
21. Do line reflections preserve congruence? Explain. yes, see 20
22. In a transformation, does each point have exactly one preimage? yes
23. Try to find a property of geometric figures not preserved by line reflections. (Hint: Think of a line reflection of a clock.) orientation
24. Suppose $A \xrightarrow{R_n} A'$, where A and A' are not the same point. How is n related to $\overline{AA'}$?
    n is the perpendicular bisector of $\overline{AA'}$. See the definition of Line Reflection.

Study the first figure and the four patterns that follow. Then answer each question.

   a.   b.   c.   d.

25. Which pattern is a reflection of the first figure over line k? b
26. Which pattern is a reflection of the first figure over line ℓ? d
27. Which pattern is a reflection of the first figure over line m? a
28. Which pattern is a reflection of the first figure over line n? c

Objective: To find the line of symmetry in a figure.
12.3 Line Symmetry

Nature and art provide many interesting examples of shapes that are their own images after a line reflection. For example, if you reflect the butterfly over line *m*, each point on the butterfly has an image that is also on the butterfly. That is, the butterfly is its own image when it is reflected over line *m*. Such a figure has **line symmetry** in line *m* or is symmetrical with respect to line *m*.

Definition of Line Symmetry

If a figure is its own image after a line reflection over line *m*, then the figure is said to have line symmetry about line *m*, or to be symmetrical with respect to line *m*. Line *m* is called a **line of symmetry**.

Examples

1 **Find the line of symmetry.**

Isosceles trapezoid *ABCD* is symmetrical with respect to line ℓ. It is *not*, however, symmetrical about line *k*. For example, if point *D* is reflected over line *k*, its image is clearly *not* on the trapezoid.

2 **Find the line of symmetry.**

The word M O M has line symmetry with respect to vertical line *m* through O. Can you think of other words that have line symmetry?

Example

3 Can a figure have more than one line of symmetry?
Consider these figures.

The rectangle has two lines of symmetry. The square has four lines of symmetry. The M O M–W O W figure has two lines of symmetry.

Exercises

Max: 1-8, 1-10;   Avg: 1-8, 1-10;   Min: 1-8, 1-9

Exploratory Copy each figure and sketch the lines of symmetry. If there are none, write *none*.

1. 2. none **DAD** 3.   4. none

5. none 6.   7.   8.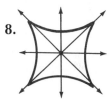

Written Answer the following. For the answers to exercises 1 and 2, see students' work.

1. Draw a triangle that has exactly one line of symmetry.
2. Draw a triangle that has exactly three lines of symmetry.
3. Can a triangle have exactly two lines of symmetry? Explain. For the answer to exercise 3, see the Teacher's Guide.

Consider all the letters of the alphabet.

A B C D E F G H I J K L M
N O P Q R S T U V W X Y Z

4. Name the letters with a vertical line of symmetry. A, H, I, M, O, T, U, V, W, X, Y
5. Name the letters with a horizontal line of symmetry. B, C, D, E, H, I, K, O, X
6. Name the letters with lines of symmetry other than vertical or horizontal. O
7. Some words have horizontal lines of symmetry, for example, B̶O̶O̶K̶. Find five other words that have horizontal line symmetry. Answers may vary.

Copy each pattern. Shade squares so that the patterns have the given line(s) of symmetry.

8. vertical line of symmetry

9. two diagonal lines of symmetry

10. vertical *and* horizontal lines of symmetry

For exercises 8-10, answers may vary. See student's work.

Mixed Review

Suppose the pattern squares shown above are dartboards. What is the probability that when a dart lands on the board shown in each exercise, it lands on the shaded part of the pattern?

1. exercise 8 $\frac{5}{16}$

2. exercise 9 $\frac{1}{4}$

3. exercise 10 $\frac{3}{16}$

Study the first figure and the four patterns that follow in each row. Then determine which of the patterns, if any, represent a line reflection of the first figure.

4. a, b
 a. b. c. d.

5. b
 a. b. c. d.

6. a, b
 a. b. c. d.

7. Find the midpoint of the segment whose endpoints are (5, −4) and (1, 6).
(3, 1)

8. Find the equation of the perpendicular bisector of the segment whose endpoints are (−5, −4) and (1, −6).
$y = 3x - 11$

9. A group has five men and seven women. How many committees of three men and three women can be selected? 350

10. If $g: x \longrightarrow x^2 - 6x + 4$, find the preimage of −1. Is g a one-to-one mapping? Explain.
5 or 1; no; −1 has more than one preimage.

12.4 Line Reflections in the Coordinate Plane

Objective: To find the image of a figure over a line of reflection in the coordinate plane.

You already know how to find the image of a point for a given line reflection.

Suppose the given line of reflection is the y-axis of the coordinate plane. Look carefully at the figure below. When A is reflected over the y-axis, its image is $A'(-3, 4)$. Why?

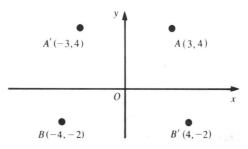

What is the image of $B(-4, -2)$ for a reflection over the y-axis? If you said $B'(4, -2)$, you are correct. In each case, it appears that the y-coordinate of the original point and its image are the same. The x-coordinate of the image is the additive inverse of the original x-coordinate. This rule is expressed as follows.

Reflection over the y-axis

If point $P(x, y)$ is reflected over the y-axis, its image is $P'(-x, y)$.

$$(x, y) \xrightarrow{R_{y\text{-axis}}} (-x, y)$$

Example

1 Given points $A(-1, 2)$, $B(-4, 1)$ and $C(-3, -2)$, find the image of $\triangle ABC$ for $R_{y\text{-axis}}$.

$A(-1, 2) \xrightarrow{R_{y\text{-axis}}} A'(1, 2)$

$B(-4, 1) \xrightarrow{R_{y\text{-axis}}} B'(4, 1)$

$C(-3, -2) \xrightarrow{R_{y\text{-axis}}} C'(3, -2)$

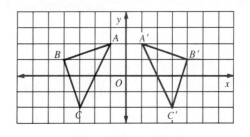

The next example shows a reflection over other lines.

Example

2 Reflect $\triangle ABC$ of example 1 over the line $y = 4$.

By counting, we obtain the following images of A, B, and C for a reflection over the line $y = 4$.

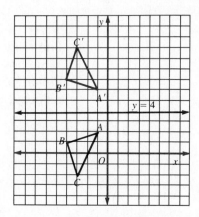

$$A(-1, 2) \xrightarrow{R_{y=4}} A'(-1, 6)$$
$$B(-4, 1) \xrightarrow{R_{y=4}} B'(-4, 7)$$
$$C(-3, -2) \xrightarrow{R_{y=4}} C'(-3, 10)$$

We can find a rule for reflection over the line $y = 4$. Consider this argument. Let $A(x, y)$ be any point. Because $y = 4$ is a horizontal line, A' must be on the same vertical line as A. That is, A' has the same x-coordinate as A. Therefore,

$$(x, y) \xrightarrow{R_{y=4}} (x, \quad).$$

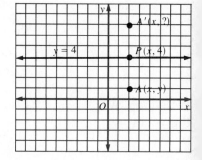

Now, find the y-coordinate. The line $y = 4$ is the perpendicular bisector of $\overline{AA'}$. Therefore, point P is the midpoint of $\overline{AA'}$. The coordinates of P are $(x, 4)$. If y' is the y-coordinate of A', then the midpoint formula tells us the following.

$$\frac{y' + y}{2} = 4, \quad \text{or} \quad \begin{aligned} y' + y &= 8, \\ y' &= 8 - y \end{aligned}$$

Thus, our rule may be completed as follows.

$$(x, y) \xrightarrow{R_{y=4}} (x, 8 - y)$$

A general rule for the reflection over a line of the form $y = b$ is as follows.

Reflection over a Horizontal Line

If point $P(x, y)$ is reflected over the line $y = b$, its image is $P'(x, 2b - y)$.

Exercises

Max: 1-15, 1-46, 1-6; Avg: 1-15 odds, 1-12, 13-37 odds, 38-46; Min: 1-15 odds, 1-12, 13-37 odds

Exploratory Find the image of each of the following points for a reflection over the y-axis.

1. (2, 3)   (-2, 3)
2. (−4, 6)   (4, 6)
3. $\left(2\frac{1}{2}, 7\right)$ $(-2\frac{1}{2}, 7)$
4. (0, 6)   (0, 6)
5. (−4, −4)   (4, -4)
6. $\left(53, -5\frac{1}{2}\right)$ $(-53, -5\frac{1}{2})$
7. (3, −5)   (-3, -5)
8. (−6, 0)   (6, 0)
9. $\left(7, -3\frac{1}{2}\right)$ $(-7, -3\frac{1}{2})$

Tell which of the following curves appear to be symmetric with respect to the y-axis, the x-axis, both, or neither.

10.
x-axis

11.
neither

12.
neither

13.
x-axis

14.
both

15.
both

Written Use graph paper. Find the image of each of these points for a reflection over the x-axis.

1. (1, 3)   (1, -3)
2. (0, 5)   (0, -5)
3. (−3, 4)   (-3, -4)
4. (5, −3)   (5, 3)
5. $\left(3\frac{1}{2}, 3\frac{1}{2}\right)$ $(3\frac{1}{2}, -3\frac{1}{2})$
6. (5, 0)   (5, 0)

7. Use exercises 1–6 to complete the following rule for finding the image of a point for a reflection over the x-axis.

$(x, y) \xrightarrow{R_{x\text{-axis}}} \underline{\quad ? \quad}$ (x, -y)

Use graph paper. Find the image of each of these points for a reflection over the line y = x.

8. (2, 1)   (1, 2)
9. (4, 0)   (0, 4)
10. (−5, −1)   (-1, -5)
11. (2, 7)   (7, 2)

12. Use exercises 8–11 to complete the following rule for finding the image of a point for a reflection over the line y = x.

$(x, y) \xrightarrow{R_{y=x}} \underline{\quad ? \quad}$ (y, x)

For the answers to exercises 13-24, see the Teacher's Guide.

For each pair of points A and B, find the images, A' and B' for $R_{y\text{-axis}}$. Then use the distance formula to find AB and A'B'.

13. $A(4, 2)$ and $B(1, 5)$
14. $A(0, 3)$ and $B(3, -1)$
15. $A(-3, 4)$ and $B(5, 6)$
16. $A(-2, 3)$ and $B(5, 1)$

Repeat exercises 13–16 for $R_{x\text{-axis}}$.

17. exercise 13
18. exercise 14
19. exercise 15
20. exercise 16

Repeat exercises 13–16 for a reflection over the line $y = x$.

21. exercise 13
22. exercise 14
23. exercise 15
24. exercise 16

Use graph paper. Find the image of each point for a reflection over the line $x = 4$.

25. $(1, 4)$ (7,4)
26. $(-2, 3)$ (10,3)
27. $(-2, -4)$ (10,-4)
28. $(5, -3)$ (3,-3)
29. Complete this rule: $(x, y) \xrightarrow{R_{x=4}}$ ___?___ (8 - x, y)

Find the coordinates of the images of $A(3, -2)$ and $B(-2, 5)$ for a reflection over a line with the following equations.

30. $x = -3$
    A'(-9,-2), B'(-4,5)
31. $y = -2$
    A'(3,-2), B'(-2,-9)
32. $x = -2$
    A'(-7,-2), B'(-2,5)
33. $y = 10$
    A'(3,22), B'(-2,

Find a rule of the form $(x, y) \longrightarrow$ ___?___ for each reflection in exercises 30–33.

34. exercise 30 (x,y)→(-6 - x, y)
35. exercise 31 (x,y)→(x, -4 - y)
36. exercise 32 (x,y)→(-4 - x, y)
37. exercise 33 (x,y)→(x, 10 - y)

Solve. Use graph paper for exercise 38. For the answers to exercises 38 and 39, see the Teacher's Guide.

38. Graph the quadrilateral with vertices $A(2, 1)$, $B(5, -2)$, $C(9, 4)$, and $D(6, 7)$.
39. Find the images of A, B, C, and D for $R_{y\text{-axis}}$.
40. Find the slopes of $\overleftrightarrow{AB}$ and $\overleftrightarrow{CD}$. -1, -1
41. Find the slopes of $\overleftrightarrow{A'B'}$ and $\overleftrightarrow{C'D'}$. 1, 1
42. Do line reflections preserve parallelism? Explain. Yes, if two lines are parallel and their slopes are equal, the slopes of their images are equal and the images are parallel.

On graph paper, draw a figure that satisfies each condition, if possible.

43. symmetric about the y-axis but not the x-axis
44. symmetric about the y-axis and the x-axis
45. symmetric about the line $y = x$, but neither of the two coordinate axes
46. symmetric about the line $x = 3$ and the line $y = 4$

For the answers to exercises 43-46, see students' work.

Challenge Find, if possible, a line of symmetry of the graph of each equation. If there is no line of symmetry, write *none*.

1. $y = x^2 + 5$ y-axis
2. $y = |x|$ y-axis
3. $y = x^2 + 4x - 1$ x = -2
4. $y = x$ y = -x
5. $|y| = |x|$ x-axis, y-axis, y = x, y = -x
6. $|x| + |y| = 4$ x-axis, y-axis, y = x, y = -x

Objective: To find the image of a figure for a translation.

12.5 Translations

Calculator Hint

You can learn how to use a graphing calculator to draw geometric figures and their translated images in Activity 6 on page A9.

We have seen that line reflections are a way of assigning a unique image point to every point in the plane. In this section, another way of assigning image points will be investigated.

Consider "shifting" or "sliding" $\triangle ABC$ one inch in the direction shown below.

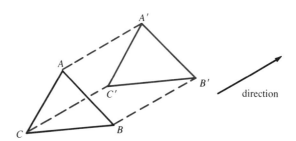

In this case, A is assigned A', B is assigned B', and C is assigned C'. This method of assigning images is a **translation**. In a translation, the segments connecting each point and its image are parallel and congruent.

Reflections and translations are two different ways of assigning images to points in the plane.

Definition of Translation

> A translation is a transformation that maps every point in the plane to its image by moving each point the same distance in the same direction.

Example

1 Each figure below has been translated 2 cm in the direction shown.

Example

2 For each figure, describe a transformation under which one triangle appears to be the image of the other.

Figure 1

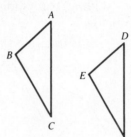

Figure 2

In figure 1, △ABD and △ACD appear to be images of each other for a line reflection over line AD.

In figure 2, consider the translation that takes point A to point D. With this translation, △DEF appears to be the image of △ABC.

It is also possible to consider translations in the coordinate plane. In the diagram, every point is moved three units to the right and then two units up. Here are some points and their images.

$$A(0, 0) \longrightarrow A'(3, 2)$$
$$B(1, 5) \longrightarrow B'(4, 7)$$
$$C(-5, -2) \longrightarrow C'(-2, 0)$$

The rule of this translation can be expressed in the following way.

$$(x, y) \longrightarrow (x + 3, y + 2)$$

If the symbol $T_{3,2}$ is used to symbolize this translation, then we can write

$$(x, y) \xrightarrow{T_{3,2}} (x + 3, y + 2).$$

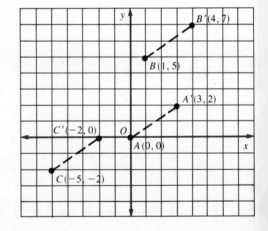

Rule for Translation

If $P(x, y)$ is transformed by translation $T_{a,b}$, then

$$P(x, y) \xrightarrow{T_{a,b}} P'(x + a, y + b).$$

Examples

3 Given points $A(3, 2)$, $B(-1, 4)$, and $C(1, -2)$, find the image of $\triangle ABC$ for $T_{-6, -2}$.

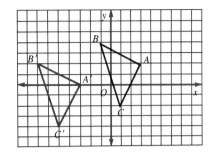

The rule for $T_{-6, -2}$ is
$(x, y) \xrightarrow{T_{-6, -2}} (x - 6, y - 2)$.

Thus, $A(3, 2) \longrightarrow A'(-3, 0)$,

$B(-1, 4) \longrightarrow B'(-7, 2)$, and

$C(1, -2) \longrightarrow C'(-5, -4)$.

4 Given points $A(-5, 2)$ and $B(3, -1)$ and translation $T_{1, 4}$, show that $AB = A'B'$.

The rule for $T_{1, 4}$ is $(x, y) \xrightarrow{T_{1, 4}} (x + 1, y + 4)$.

Therefore, $A(-5, 2) \longrightarrow A'(-4, 6)$ and $B(3, -1) \longrightarrow B'(4, 3)$.

Using the distance formula gives these results.

$AB = \sqrt{(-5 - 3)^2 + [2 - (-1)]^2}$ $\qquad A'B' = \sqrt{(-4 - 4)^2 + (6 - 3)^2}$
$\quad = \sqrt{64 + 9}$ $\qquad\qquad\qquad\qquad\qquad = \sqrt{64 + 9}$
$\quad = \sqrt{73}$ $\qquad\qquad\qquad\qquad\qquad\qquad = \sqrt{73}$

We see that $AB = A'B'$.

Max: 1-13, 1-39; Avg: 1-13, 1-31 odds, 32-39; Min: 1-13 odds, 1-31 odds

Exercises

For the answers to exercises 1-6, see the Teacher's Guide.

Exploratory Use graph paper. Plot the points $A(-3, -1)$, $B(0, 5)$, $C(4, 2)$, and $D(1, -3)$. Find their images for each of the following rules.

1. $(x, y) \longrightarrow (x, -y)$
2. $(x, y) \longrightarrow (x^2, y)$
3. $(x, y) \longrightarrow (x - 7, y)$
4. $(x, y) \longrightarrow (x, y + 6)$
5. $(x, y) \longrightarrow (y, x)$
6. $(x, y) \longrightarrow (|x|, |y|)$

Which of the rules in exercises 1-6 are translations?

7. exercise 1 no
8. exercise 2 no
9. exercise 3 yes
10. exercise 4 yes
11. exercise 5 no
12. exercise 6 no

Copy the figure at the right.

13. For a certain translation, A' is the image of A. Draw the the images of B and C.

For the answer to exercise 13, see students' work.

Written In the figure, D is the image of E for a translation. Name the image of each of the following.

1. $\overline{JK}$
2. $\overline{HI}$
3. $\overline{GH}$ $\overline{FI}$
4. $\angle HGF$ $\angle IFE$
5. $\angle INP$ $\angle JMN$
6. $\overline{NI}$ $\overline{JM}$

In the figure for exercises 1–6, suppose F is the image of J for a translation. Name the image of each of the following.

7. N H
8. K E
9. J F
10. $\overline{NM}$ $\overline{HI}$
11. $\overline{JM}$ $\overline{FI}$
12. $\angle KLM$ $\angle EJI$
13. $\angle INM$ $\angle GHI$
14. $\overleftrightarrow{KL}$ $\overleftrightarrow{EJ}$

Find the image of $(-2, 9)$ for each transformation.

15. $T_{2,3}$ (0,12)
16. $T_{-5,6}$ (-7,15)
17. $T_{-4,-8}$ (-6,1)
18. $T_{-2,-3}$ (-4,6)
19. $R_{y\text{-axis}}$ (2,9)
20. $R_{x\text{-axis}}$ (-2,-9)
21. $R_{y=x}$ (9,-2)
22. $R_{x=3}$ (8,9)

Find the coordinate rule of a translation for which B is the image of A.

23. $A(1, 1)$; $B(2, 4)$ $T_{1,3}$
24. $A(1, -3)$; $B(-3, -1)$ $T_{-4,2}$
25. $A(5, -12)$; $B(0, 0)$ $T_{-5,12}$
26. $A(3\sqrt{2}, 1)$; $B(\sqrt{2}, -1)$ $T_{-2\sqrt{2},-2}$

Find the images of the points $A(1, 3)$ and $B(5, 1)$ for $T_{4,-1}$. Then justify each.

27. $AB = A'B'$ AB = 2√5, A'B' = 2√5
28. $\overleftrightarrow{AB} \parallel \overleftrightarrow{A'B'}$ slope of $\overleftrightarrow{AB}$ = $-\frac{1}{2}$, slope of $\overleftrightarrow{A'B'}$ =
29. $AA' = BB'$ AA' = √17, BB' = √17
30. $ABB'A'$ is a parallelogram.
31. $\overline{AB'}$ and $\overline{A'B}$ bisect each other.   because AB = A'B' and $\overleftrightarrow{AB} \parallel$
    31. The diagonals of a parallelogram bisect each other.

Graph the triangle with vertices $A(-2, 0)$, $B(1, 5)$, and $C(3, -1)$. Then complete the following exercises. For the answers to exercises 32 and 33, see the Teacher's Guide.

32. Graph the image of $\triangle ABC$ for $T_{-2,3}$. Label the image $\triangle A'B'C'$.
33. Graph the image of $\triangle A'B'C'$ for $T_{-3,4}$. Label the image $\triangle A''B''C''$.
34. To what point is (x, y) finally assigned for $T_{-3,4}$ following $T_{-2,3}$? $(x - 5, y + 7)$
35. Name as many properties as you can that are preserved by translations.
36. Can you think of a property preserved by translation but not by line reflection? Explain.

For the answers to exercises 35–39, see the Teacher's Guide.

Graph the triangle with vertices $A(-5, 3)$, $B(-4, -2)$, and $C(-1, 1)$. Then complete the following exercises.

37. Find the image of $\triangle ABC$ for a reflection over the y-axis.
38. Reflect $\triangle A'B'C'$ over the line $x = 7$. Label the image $A''B''C''$.
39. Compare $\triangle ABC$ and $\triangle A''B''C''$. Do you think there is a single transformation that takes $\triangle ABC$ to $\triangle A''B''C''$? Explain.

Objective: To find the image of a figure for a rotation.

12.6 Rotations

The triskelions pictured at the left suggest another kind of transformation called a **rotation**.

Definition of Rotation

A rotation is a transformation that maps every point in the plane to its image by rotating the plane around a fixed point. The fixed point is called the *center of rotation* and is its own image.

Example

1 In the figure, suppose $\triangle ABC$ is our original triangle. With center Q, rotate the plane through an angle of 110° counterclockwise.

$\triangle A'B'C'$ is the image of $\triangle ABC$.

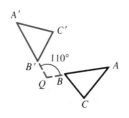

The symbol "$\text{Rot}_{Q, 110°}$" stands for a rotation with center Q through an angle of 110° *counterclockwise*. We write

$$\triangle ABC \xrightarrow{\text{Rot}_{Q, 110°}} \triangle A'B'C'.$$

For a rotation with center Q through an angle of 110° *clockwise*, we write

$$\triangle ABC \xrightarrow{\text{Rot}_{Q, -110°}} \triangle A'B'C'.$$

Examples

2 Find the image of segment AB for a rotation of 75° clockwise with center O.

First, find the image of A.
1. Draw $\overline{OA}$.
2. Using a protractor, draw an angle measuring 75° with vertex O, and sides $\overrightarrow{OA}$ and $\overrightarrow{OX}$.
3. Using a compass, locate A' on $\overrightarrow{OX}$ such that $OA' = OA$. A' is the image of A.

Repeat steps 1–3 to find the image of B.
Thus,

$$\overline{AB} \xrightarrow{\text{Rot}_{O,\,-75°}} \overline{A'B'}$$

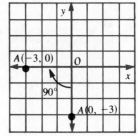

3 In the coordinate plane, find the image of $A(0, -3)$ for a clockwise rotation of 90° about the origin.

From the figure notice that
$$A(0, -3) \xrightarrow{\text{Rot}_{O,\,-90°}} A'(-3, 0).$$
Later you will be asked to complete this general rule for $\text{Rot}_{O,\,-90°}$.

$$(x, y) \xrightarrow{\text{Rot}_{O,\,-90°}} \underline{\quad ? \quad}$$

Some figures have line symmetry, others have **rotational symmetry**.

Definition of Rotational Symmetry

A figure has rotational symmetry if it is the image of itself under some rotation. The angle of rotation must be greater than 0° and less than 360°.

Example

4 The figure at the right has rotational symmetry about point O.

If you rotate the figure about O through 120° clockwise, or counterclockwise, it is its own image. It may also be rotated through an angle of 240° either clockwise or counterclockwise.

Exercises

Max: 1-10, 1-25; Avg: 1-10, 1-11, 13-25 odds; Min: 1-10, 1-11, 13-25 odds

Exploratory Explain each symbol.
1. $Rot_{C, 30°}$ a rotation with center C through an angle of 30° counterclockwise
2. $Rot_{C, -30°}$ a rotation with center C through an angle of 30° clockwise

The figure at the right shows a rotation of △KTP. Name each of the following.

3. the center of rotation V
4. the image of K M
5. Estimate the measure of the angle of rotation. Answers may vary.
6. Use a protractor to measure the angle of rotation. How close was your estimate in exercise 5? 66°; Answers may vary.

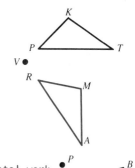

Copy and enlarge the figure at the right. Draw the image of $\overline{AB}$ for each rotation. For the answers to exercises 7-10, see students' work.
7. $Rot_{P, 90°}$
8. $Rot_{P, -90°}$
9. $Rot_{P, 60°}$
10. $Rot_{P, -240°}$

Written Find the image of each point for $Rot_{O, 90°}$. Use graph paper.
1. (4, 0) (0,4)
2. (0, 5) (-5,0)
3. (2, 2) (-2,2)
4. (−3, 3) (-3,-3)

Complete the following.
5. Complete this rule for a rotation with center at the origin through an angle 90° counterclockwise. $(x, y) \longrightarrow \underline{\quad ? \quad}$ (-y,x)
6. Explain why $Rot_{P, 180°}$ is equivalent to $Rot_{P, -180°}$. $Rot_{P,180°}$ and $Rot_{P,-180°}$ are both half rotations.
7. Complete this rule for $Rot_{O, 180°}$ in the coordinate plane. $(x, y) \longrightarrow \underline{\quad ? \quad}$ (-x,-y)

Copy and enlarge the figure. Draw the image for each of the following. For the answers to exercises 8-11, see students' work.
8. $Rot_{P, 90°}$
9. $Rot_{P, -75°}$
10. $R_{\overleftrightarrow{PB}}$
11. $R_{\overleftrightarrow{AB}}$

Name the figures that have rotational symmetry. Justify your answer.

12. A no
13. B no
14. C no
15. H yes
16. K no
17. N yes
18. O yes

19. S yes
20. X yes
21. △ yes
22. no
23. yes
24. yes
25. yes

CHAPTER 12 475

12.7 Half-Turns or Point Reflections

Objective: To find the image of a figure for a half-turn (point reflection).

Rotations of 180° about a point occur frequently, so these are given a special name.

Definition of Half-turn

> A rotation of 180° about a point P is called a **half-turn** with center P. (A half-turn is also called a **point reflection** or a reflection in a point.) The symbol $\text{Rot}_{P,\,180°}$ represents a half-turn about point P.

Example

1 Find the image of $\overline{AB}$ for a half-turn with center P.

First, find the image of A for a 180° rotation about P. Then, find the image of B. Segment $\overline{A'B'}$ is the image of $\overline{AB}$ for a half-turn with center P. In symbols,
$$\overline{AB} \xrightarrow{\text{Rot}_{P,\,180°}} \overline{A'B'}.$$

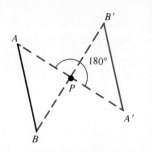

Look carefully at how A' and B' were obtained in example 1. Note that point P is the midpoint of $\overline{AA'}$ and $\overline{BB'}$. A half-turn with center P (a reflection in point P) is a transformation that maps every point A to a point A' such that P is the midpoint of $\overline{AA'}$.

Example

2 Find the image of $\triangle ABC$ for a half-turn with center P.

Find the images of A, B, and C. First, draw ray $\overrightarrow{AP}$. Using a compass, locate A' such that P is the midpoint of $\overline{AA'}$. In the same way, locate points B' and C'. $\triangle A'B'C'$ is the image of $\triangle ABC$ for a half-turn with center P.

Consider the problem of finding the image of a point for a half-turn whose center is the origin of the coordinate plane.

Examples

3 Find the image of $A(2, 3)$ for a half-turn about the origin.

Locate A' such that O is the midpoint of $\overline{AA'}$. Thus,
$$A(2, 3) \xrightarrow{\text{Rot}_{O, 180°}} A'(-2, -3).$$

4 Find the image of $B(-3, 5)$ for a half-turn about the origin.

Locate point B' such that O is the midpoint of $\overline{BB'}$. Thus,
$$B(-3, 5) \xrightarrow{\text{Rot}_{O, 180°}} B'(3, -5).$$

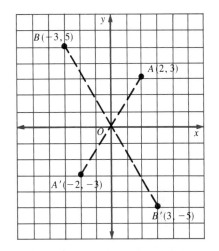

Rule for Half-turn about the Origin

If $P(x,y)$ is transformed by half-turn $\text{Rot}_{O, 180°}$, then
$$P(x, y) \xrightarrow{\text{Rot}_{O, 180°}} P'(-x, -y).$$

Example

5 Find the image of a triangle with vertices $A(2, 1)$, $B(6, 3)$, and $C(3, 5)$ for a half-turn about the origin.

Use the rule $(x, y) \xrightarrow{\text{Rot}_{O, 180°}} (-x, -y)$.

$A(2, 1) \longrightarrow A'(-2, -1)$

$B(6, 3) \longrightarrow B'(-6, -3)$

$C(3, 5) \longrightarrow C'(-3, -5)$

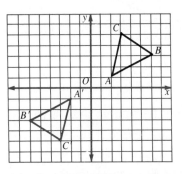

If a figure is its own image for some half-turn with center P, then it is said to have **point symmetry** with respect to P.

Exercises

Max: 1-8, 1-33; Avg: 1-8, 1-31; Min: 1-7 odds, 1-13, 15-31 odds

Exploratory Find the image of each of the following points for a half-turn whose center is the origin of the coordinate plane.

1. (2, 3) (-2,-3)
2. (−4, 5) (4,-5)
3. (−5, 0) (5,0)
4. $\left(\frac{1}{2}, -\frac{1}{4}\right)$ $(-\frac{1}{2}, \frac{1}{4})$

Find the image of (1, 3) for a half-turn whose center is each point. Use graph paper.

5. (0, −1) (-1,-5)
6. (−1, 0) (-3,-3)
7. (−2, 5) (-5, 7)
8. (1, 3) (1,3

Written Copy and enlarge the figure. Then use a compass and ruler to find the image of △PQR for a half-turn for each of the following centers. For the answers to exercises 1-3, see students' work.

1. P
2. Q
3. C

For the points A(−1, 5) and B(−3, −4), find their images for a half-turn whose center is the origin. Then justify each of the following.

4. $\overline{AB} \cong \overline{A'B'}$ AB = √85, A'B' = √85
5. $\overleftrightarrow{AB} \parallel \overleftrightarrow{A'B'}$ slope of $\overleftrightarrow{AB} = \frac{9}{2}$, slope of $\overleftrightarrow{A'B'} = \frac{9}{2}$
6. $\overleftrightarrow{AB'} \parallel \overleftrightarrow{A'B}$ slope of $\overleftrightarrow{AB'} = -\frac{1}{4}$, slope of $\overleftrightarrow{A'B} = -\frac{1}{4}$

For each figure, describe a transformation for which one triangle is the image of the other.

7. translation
8. reflection
9. rotation
10. rotation
11. rotation
12. reflection

Complete.

13. Sketch a figure that has point symmetry but does not have line symmetry. Answers may vary.

Name the figures that have point symmetry. Name the ones that have line symmetry.

14. parallelogram point
15. rhombus point, line
16. isosceles trapezoid line

17. **A** line
18. **S** point
19. line
20. ◯ point, line
21. ⬠ point, line
22.   point

If B is the image of A under a reflection in point P, find the coordinates of P.

23. A(0, 0); B(6, 4) P(3,2)
24. A(−8, 6); B(0, 0) P(-4,3)
25. A(7, −6); B(−1, −2) P(3,-4)
26. A(1, 11); B(8, −12) P($4\frac{1}{2}$,-$\frac{1}{2}$)

Find the image of (2, −5) under the reflection in the given point.

27. (0, 0) (-2,5)
28. (5, 6) (8,17)
29. (−4, 1) (-10,7)
30. (x, y) (2x − 2, 2y + 5)

Copy each pattern. Shade squares so that the patterns have point symmetry with respect to the point at the center of the pattern.

31.
32.
33.

For exercises 31-33, answers may vary. See students' work.

Mixed Review

Use the Greek letters shown below for Exercises 1–5.

Δ Φ Γ ϑ Λ Π Θ Σ Ω

1. Which letters have a vertical line of symmetry? Δ, Φ, Λ, Π, Θ, Ω
2. Which letters have a horizontal line of symmetry? Φ, Θ, Σ
3. Which letters have both vertical and horizontal lines of symmetry? Φ, Θ
4. Which letters have no lines of symmetry? Γ, ϑ
5. Which letters have point symmetry? Δ, Φ, Θ

Find the image of (−3, −4) under each transformation.

6. $T_{2,3}$ (-1,-1)
7. $T_{-4,-3}$ (-7,-7)
8. $R_{y\text{-axis}}$ (3,-4)
9. $R_{x=1}$ (5,-4)
10. $Rot_{O, 90°}$ (4,-3)
11. $Rot_{O, -90°}$ (-4,3)

Find the preimage of (2, −1) under each transformation.

12. $T_{-1, 6}$ (3,-7)
13. $T_{5, -4}$ (-3,3)
14. $R_{x\text{-axis}}$ (2,1)
15. $R_{y = x}$ (-1,2)
16. $R_{y = -4}$ (2,-7)
17. $Rot_{O, 180°}$ (-2,1)

18. Find the coordinate rule of a translation for which (10, −4) is the image of (2, −2). $T_{8, -2}$
19. If (3, 7) is the image of (−5, 7) for a reflection over line m, find the equation of line m. x = -1
20. Urns X and Y contain red and blue disks. Urn X has 7 more blue disks than red disks. Urn Y has the same number of blue disks as Urn X, but four times as many red disks. The probability of choosing a blue disk from Urn X is the same as the probability of choosing a red disk from Urn Y. How many of each type of disk are in each urn?
Urn X: 7 red, 14 blue; Urn Y: 28 red, 14 blue

12.8 Composing Transformations

Objective: To find the image a figure for a composition of transformations.

So far we have considered several different kinds of transformations of the plane: line reflections, translations, and rotations. In this section we investigate the effect of applying one transformation after another on a figure.

Consider $\triangle ABC$ in the figure. First, reflect $\triangle ABC$ over line ℓ, producing $\triangle A'B'C'$. Now, reflect $\triangle A'B'C'$ over line m, producing $\triangle A''B''C''$. This combined action of two transformations is called a **composition**. In this case, the composition of line reflections R_m and R_ℓ "takes" $\triangle ABC$ into $\triangle A''B''C''$. The notation for "R_m following R_ℓ" is "$R_m \circ R_\ell$." This means first apply R_ℓ, and then R_m.

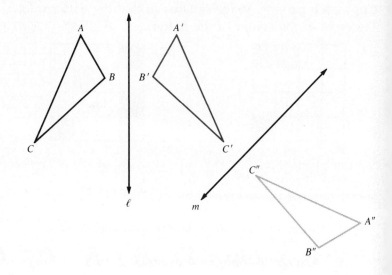

That is,

$$\triangle ABC \xrightarrow{R_\ell} \triangle A'B'C' \xrightarrow{R_m} \triangle A''B''C''$$

or

$$\triangle ABC \xrightarrow{R_m \circ R_\ell} \triangle A''B''C''.$$

$R_m \circ R_\ell$ is called the composite of R_m and R_ℓ.

Example

1 The figure at the right shows a clockwise rotation of $\triangle K'L'M'$ 60° about Q following a translation of $\triangle KLM$ 3 cm in the given direction.

$\triangle KLM \rightarrow \triangle K''L''M''$.

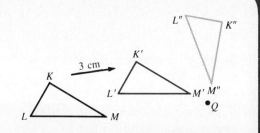

Examples

2 Given a triangle with vertices $A(-5, 2)$, $B(-1, 4)$, and $C(-3, 7)$, find its image for
$$T_{6, -5} \circ R_{y\text{-axis}}.$$

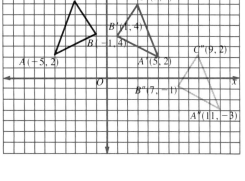

First use the rule,
$$(x, y) \xrightarrow{R_{y\text{-axis}}} (-x, y).$$

$A(-5, 2) \longrightarrow A'(5, 2)$

$B(-1, 4) \longrightarrow B'(1, 4)$

$C(-3, 7) \longrightarrow C'(3, 7)$

Then use the rule, $(x, y) \xrightarrow{T_{6, -5}} (x + 6, y - 5)$.

$A'(5, 2) \longrightarrow A''(11, -3)$

$B'(1, 4) \longrightarrow B''(7, -1)$

$C'(3, 7) \longrightarrow C''(9, 2)$ Thus, $\triangle ABC \xrightarrow{T_{6, -5} \circ R_{y\text{-axis}}} \triangle A''B''C''$.

3 Using the triangle of example 2, find the image of $\triangle ABC$ for $R_{y\text{-axis}} \circ T_{6, -5}$.

This time, begin with the rule $(x, y) \xrightarrow{T_{6, -5}} (x + 6, y - 5)$.

$A(-5, 2) \longrightarrow A'(1, -3)$

$B(-1, 4) \longrightarrow B'(5, -1)$

$C(-3, 7) \longrightarrow C'(3, 2)$

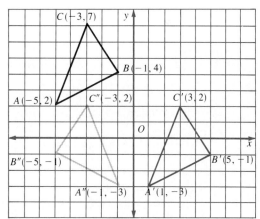

Then use the rule,
$$(x, y) \xrightarrow{R_{y\text{-axis}}} (-x, y).$$

$A'(1, -3) \longrightarrow A''(-1, -3)$

$B'(5, -1) \longrightarrow B''(-5, -1)$

$C'(3, 2) \longrightarrow C''(-3, 2)$

Thus, $\triangle ABC \xrightarrow{R_{y\text{-axis}} \circ T_{6, -5}} \triangle A''B''C''$.

Examples 2 and 3 show that composition of transformation is not a commutative operation.

In our work with line reflections, translations, and rotations, we saw that a segment AB and its image, segment $A'B'$, have the same length. That is, each of these transformations preserves distance.

Definition of Isometry

An **isometry** is a transformation that preserves distance.

Line reflections, translations, and rotations are examples of isometries.

Look carefully at examples 1, 2, and 3. Each one has the composite of two isometries. Compare the lengths AB, AC, and BC with the lengths $A''B''$, $A''C''$, and $B''C''$. The results should lead to the conclusion that the composite of any number of isometries is an isometry.

Exercises

Max: 1-10, 1-42, Challenge; Avg: 1-10, 1-23 odds, 24-42; Min: 1-10, 1-23 odds, 25-29

Exploratory Explain each symbol. For the answers to exercises 1-3, see the Teacher's Guide.

1. $R_m \circ R_\ell$
2. $\text{Rot}_{P, 90°} \circ R_m$
3. $T_{2, 3} \circ \text{Rot}_{O, 180°}$

Copy and enlarge the figure at the right. Find the image of $\triangle ABC$ for each of the following.

4. $R_m \circ R_\ell$
5. $R_\ell \circ R_m$
6. $R_\ell \circ R_\ell$
7. $R_m \circ R_m$

For the answers to exercises 4-7, see students' work.

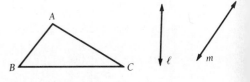

Answer the following. For the answers to exercises 8-10, see the Teacher's Guide.

8. Is $T_{a, b} \circ T_{c, d}$ equivalent to $T_{c, d} \circ T_{a, b}$? Explain.
9. If R_ℓ and R_m are line reflections, is $R_\ell \circ R_m$ equivalent to $R_m \circ R_\ell$? Explain.
10. Is composition of rotations with the same center commutative? Explain.

For the answers to exercises 1-12, see the Teacher's Guide.

Written Graph the triangle with the vertices $A(-6, 4)$, $B(-3, 2)$, and $C(-1, -2)$. Then find the image of $\triangle ABC$ for the following.

1. $R_{y\text{-axis}} \circ R_{x\text{-axis}}$
2. $R_{x\text{-axis}} \circ R_{y\text{-axis}}$
3. $\text{Rot}_{O, 180°} \circ R_{y\text{-axis}}$
4. $T_{-2, 3} \circ T_{4, 2}$
5. $T_{4, 2} \circ T_{-2, 3}$
6. $R_{x\text{-axis}} \circ T_{-2, 4}$
7. $T_{-2, 4} \circ R_{x\text{-axis}}$
8. $\text{Rot}_{O, 180°} \circ T_{-2, 5}$
9. $T_{-2, 5} \circ \text{Rot}_{O, 180°}$
10. $T_{0, 7} \circ R_{y\text{-axis}}$
11. $R_{y\text{-axis}} \circ T_{0, 7}$
12. $\text{Rot}_{O, 180°} \circ \text{Rot}_{O, 180°}$

CHAPTER 12   483

Find the length of $\overline{AB}$ and $\overline{A'B'}$ for each. For exercises 13-24, AB = A'B' = $\sqrt{13}$
13. exercise 1
14. exercise 2
15. exercise 3
16. exercise 4
17. exercise 5
18. exercise 6
19. exercise 7
20. exercise 8
21. exercise 9
22. exercise 10
23. exercise 11
24. exercise 12

Sketch an example to show that each is an isometry. For the answers to exercises 25-28, see students' work.
25. a line reflection following another line reflection
26. a line reflection following a translation
27. a line reflection following a half-turn
28. a rotation following a translation
29. Explain why the composite of any number of isometries must be an isometry.
Because each isometry preserves distance, the composition of any number of isometries will preserve distance.

Copy the figure at the right.

30. Find the image of $\triangle KLM$ under R_ℓ. Label it $\triangle K'L'M'$. See students' work.
31. Do the letters of $\triangle KLM$ "run" clockwise or counterclockwise? counterclockwise
32. Do the letters of $\triangle K'L'M'$ "run" clockwise or counterclockwise? clockwise

The clockwise or counterclockwise property of points in a geometric figure is called its *order* or *orientation*. An isometry that preserves order is called a *direct isometry*. Which of the following are direct isometries?

33. line reflections no
34. translations yes
35. rotations yes
36. half-turns yes

Another kind of transformation is called a *glide reflection*. This is defined as the composite of a line reflection and a translation in a direction that is parallel to the line of reflection.

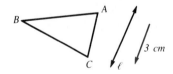

37. Copy and enlarge the figure at the right. Construct the image of $\triangle ABC$ under the glide reflection consisting of R_ℓ and a translation 3 cm in the indicated direction. See students' work.
38. Is a glide reflection an isometry? Explain.
39. Is a glide reflection a direct isometry? Explain.
40. Is the composite of any number of parallel line reflections a direct isometry? Explain.
41. Is the composite of any number of translations a direct isometry? Explain.
42. Is the composite of any number of rotations with a given center a direct isometry? Explain.
For the answers to exercises 38-42, see the Teacher's Guide.

Challenge Answer the following. See the Teacher's Guide.
Let ℓ be a given line and P a point not on line ℓ. Let G be a glide reflection with respect to line ℓ. Suppose P' is the image of P under G. Show that ℓ bisects $\overline{PP'}$.

12.9 A Mathematical System

Objective: To relate the composition of transformations to a mathematical system.

You are familiar with three kinds of symmetry: line symmetry, rotational symmetry, and point symmetry. You also have worked with the operation of composition of transformations. In this section, these ideas lead to another familiar idea—a mathematical system.

To begin, consider equilateral triangle ABC shown at the right. From previous work, we know that an equilateral triangle has several symmetries. The symmetries are listed and illustrated in the following table. Lines ℓ_1, ℓ_2, and ℓ_3 are angle bisectors that intersect at point O.

Transformation	Symbol	Original Position	Final Position
Line Reflection over ℓ_1	R_{ℓ_1}		
Line Reflection over ℓ_2	R_{ℓ_2}		
Line Reflection over ℓ_3	R_{ℓ_3}		
Clockwise Rotation of 120° about point O	$Rot_{O,\ -120°}$		
Clockwise Rotation of 240° about point O	$Rot_{O,\ -240°}$		

Example

1 What happens to the position of $\triangle ABC$ if we first apply $\text{Rot}_{O, -120°}$, and then $\text{Rot}_{O, -240°}$?

The composite $\text{Rot}_{O, -240°} \circ \text{Rot}_{O, -120°}$ is shown below. Do $\text{Rot}_{O, -120°}$ first.

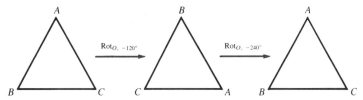

We get the triangle back in its original position. Thus, the composition of $\text{Rot}_{O, -120°}$ and $\text{Rot}_{O, -240°}$ has the same effect as leaving the original triangle fixed.

Definition of Identity Transformation

The transformation that assigns every point to itself is called the **identity transformation**. The symbol $T_{0, 0}$ represents the identity transformation.

Example

2 Is there a single transformation of $\triangle ABC$ that has the same effect as $\text{Rot}_{O, -120°} \circ R_{\ell_1}$?

The composite $\text{Rot}_{O, -120°} \circ R_{\ell_1}$ is shown below.

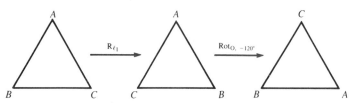

Now look at the table of transformations on page 484. R_{ℓ_2} "moves" $\triangle ABC$ into the same position as the final one shown above. That is, the composition of $\text{Rot}_{O, -120°}$ and R_{ℓ_1} is R_{ℓ_2}. We write

$$\text{Rot}_{O, -120°} \circ R_{\ell_1} = R_{\ell_2}.$$

Example

3 **Determine $\text{Rot}_{O,\,-240°} \circ R_{\ell_3}$.**

First perform R_{ℓ_3} and then $\text{Rot}_{O,\,-240°}$ as shown at the right.

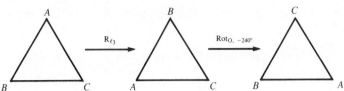

Then, observe the action of R_{ℓ_2}.

Thus, $\text{Rot}_{O,\,-240°} \circ R_{\ell_3} = R_{\ell_2}$.

The results of the examples can be summarized conveniently in a table. Each entry is obtained by first performing the transformation at the top, and then the transformation at the left. You will be asked to complete this table in the exercises.

$\circ$	$T_{0,0}$	R_{ℓ_1}	R_{ℓ_2}	R_{ℓ_3}	$\text{Rot}_{O,\,-120°}$	$\text{Rot}_{O,\,-240°}$
$T_{0,0}$						
R_{ℓ_1}						
R_{ℓ_2}						
R_{ℓ_3}						
$\text{Rot}_{O,\,-120°}$			R_{ℓ_2}			
$\text{Rot}_{O,\,-240°}$				R_{ℓ_2}	$T_{0,0}$	

Exercises

Max: 1-7, 1-27; Avg: 1-7, 1-27; Min: 1-7, 1-11 odds, 13-27

Exploratory Explain why each composite is equivalent to the identity transformation. Include a sketch. For the answers to exercises 1-7, see the Teacher's Guide.

1. a line reflection over line ℓ following a line reflection over line ℓ
2. a half-turn with center P following a half-turn with center P
3. a translation $T_{-a,\,-b}$ following translation $T_{a,\,b}$
4. a clockwise rotation with center P through an angle of 120° following a clockwise rotation with center P through an angle of 240°

5. Is there a translation that is equivalent to the identity transformation? Explain.
6. Is there a rotation that is equivalent to the identity transformation? Explain.
7. Is there a line reflection that is equivalent to the identity transformation? Explain.

For the answers to exercises 1-12, see students' work.

Written For each of the following, draw a sketch of $\triangle ABC$ as it appears at the right. Then draw its image for each of the following. (R_{ℓ_1}, R_{ℓ_2}, R_{ℓ_3}, $Rot_{O,\,-120°}$, and $Rot_{O,\,-240°}$ are defined in the text.)

1. $T_{0,\,0}$
2. R_{ℓ_2}
3. R_{ℓ_3}
4. $Rot_{O,\,-120°}$
4. $Rot_{O,\,-240°}$
6. $T_{0,\,0} \circ Rot_{O,\,-120°}$
7. $Rot_{O,\,-120°} \circ T_{0,\,0}$
8. $R_{\ell_1} \circ Rot_{O,\,-120°}$
9. $R_{\ell_1} \circ R_{\ell_2}$
10. $R_{\ell_2} \circ R_{\ell_1}$
11. $R_{\ell_3} \circ R_{\ell_3}$
12. $Rot_{O,\,-120°} \circ R_{\ell_3}$

For the answer to exercise 13, see the Teacher's Guide.

13. Copy and complete the table on page 486.
14. How does the table show that the operation of composition, as it is defined on the set $\{T_{0,\,0},\,R_{\ell_1},\,R_{\ell_2},\,R_{\ell_3},\,Rot_{O,\,-120°},\,Rot_{O,\,-240°}\}$ is a closed operation? For each pair of elements of the set, the operation of composition takes it back into an element of the set.

What single transformation has the same effect as each composite transformation?

15. $R_{\ell_1} \circ (R_{\ell_2} \circ R_{\ell_3})$ R_{ℓ_2}
16. $(R_{\ell_1} \circ R_{\ell_2}) \circ R_{\ell_3}$ R_{ℓ_2}
19. Associative property
17. $Rot_{O,\,-120°} \circ (R_{\ell_1} \circ R_{\ell_2})$ I
18. $(Rot_{O,\,-120°} \circ R_{\ell_1}) \circ R_{\ell_2}$ I

19. What property of the operation "$\circ$" is suggested by your results in exercises 15–18?
20. Does the system defined by the table on page 486 have an identity element? Explain.

20. Yes, because for every element a of the set $a \circ I = I \circ a = a$

What single transformation has the same effect as each composite transformation?

21. $R_{\ell_2} \circ R_{\ell_2}$ I
22. $Rot_{O,\,-120°} \circ Rot_{O,\,-240°}$ I
23. $R_{\ell_3} \circ R_{\ell_3}$ I

24. What is meant by saying that R_{ℓ_2} is its own inverse? R_2 is such that $R_2 \circ R_2 = I$.
25. Does every element in $\{T_{0,\,0},\,R_{\ell_1},\,R_{\ell_2},\,R_{\ell_3},\,Rot_{O,\,-120°},\,Rot_{O,\,-240°}\}$ have an inverse with respect to the operation of composition? Explain.
26. Does the set $\{T_{0,\,0},\,R_{\ell_1},\,R_{\ell_2},\,R_{\ell_3},\,Rot_{O,\,-120°},\,Rot_{O,\,-240°}\}$ together with the operation "$\circ$" form a group? A commutative group? Explain.
27. Does the system in exercise 26 look familiar? Where have you seen it before? (Hint: Recall your work in Chapter 2.) For the answers to exercises 25-27, see the Teacher's Guide.

12.10 Dilations

Objective: To find the image of a figure for a dilation.

The transformations so far have all been isometries. That is, they all preserve distance. Some transformations are *not* isometries.

Suppose a plane is made of an elastic material that can be stretched or shrunk at will. In the figure at the left, a stretching of the plane, with point O held fixed, might produce △OP'Q' from △OPQ. Note that every point in the plane is assigned a unique image as required by a transformation.

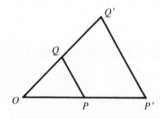

In the figure, P' is assigned to P, Q' is assigned to Q, and O is assigned to O. Note also that P' is on ray OP and Q' is on ray OQ. In this case, OQ' = 2 · OQ and OP' = 2 · OP. This transformation is an example of a **dilation** with center O and **scale factor** 2.

Example

1 Find the image of △ABC for a dilation with center O and scale factor 3.

First, draw rays $\overrightarrow{OA}$, $\overrightarrow{OB}$, and $\overrightarrow{OC}$. Then locate A', B', and C' such that OA' = 3 · OA, OB' = 3 · OB, and OC' = 3 · OC.

The scale factor of a dilation with center O can be negative.

Example

2 Find the image of △ABC for a dilation with center O and scale factor −2.

Locate A', B', and C' on the rays opposite $\overrightarrow{OA}$, $\overrightarrow{OB}$, and $\overrightarrow{OC}$ such that OA' = 2 · OA, OB' = 2 · OB, and OC' = 2 · OC.

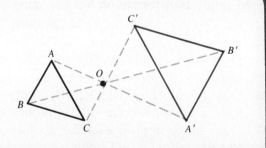

Images of figures after a dilation retain some characteristics of the original figure. But one important feature is not retained—size. Dilations do not preserve distance. The symbol $D_{O,k}$ represents the dilation having scale factor k and whose center is the origin on the coordinate plane.

Example

3 Find the image of $A(-1, 3)$ for $D_{O,2}$.

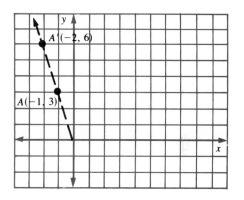

Draw ray OA. Locate A' such that $OA' = 2 \cdot OA$. It appears that the coordinates of A' are $(-2, 6)$. To check this, note that the slopes of $\overleftrightarrow{OA}$ and $\overleftrightarrow{AA'}$ are equal. This means O, A, and A' are collinear.

Using the distance formula, find OA and OA'.

$OA = \sqrt{1 + 9} = \sqrt{10}$
$OA' = \sqrt{4 + 36} = \sqrt{40} = 2\sqrt{10}$

We see that $OA' = 2 \cdot OA$. Thus, $A(-1, 3) \xrightarrow{D_{O,2}} A'(-2, 6)$.

Rule of Dilation with Center at the Origin

If $P(x, y)$ is transformed by dilation $D_{O,k}$, then
$$P(x, y) \xrightarrow{D_{O,k}} P'(kx, ky).$$

Example

4 Find the image of the triangle with vertices $A(1, 1)$, $B(1, 4)$, and $C(5, 1)$ for $D_{O,3}$.

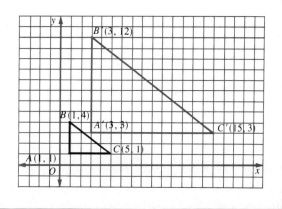

Use the rule,
$(x, y) \xrightarrow{D_{O,3}} (3x, 3y)$.
Thus, $A(1, 1) \longrightarrow A'(3, 3)$
$B(1, 4) \longrightarrow B'(3, 12)$
$C(5, 1) \longrightarrow C'(15, 3)$.
$\triangle A'B'C'$ is the image.

You can see that although distance is not preserved by dilations, ratio of distance is preserved. Since $A'B' = 3 \cdot AB$, $A'C' = 3 \cdot AC$, and $B'C' = 3 \cdot BC$, we have $\dfrac{AB}{A'B'} = \dfrac{AC}{A'C'} = \dfrac{BC}{B'C'}$. This means that $\triangle ABC \sim \triangle A'B'C'$.

Exercises

Max: 1-10, 1-20; Avg: 1-10, 1-20; Min: 1-9 odds, 1-20

Exploratory Copy $\triangle ABC$. Use a ruler to draw its image for each dilation.

1. center O, scale factor 2
2. center O, scale factor $\dfrac{1}{2}$
3. center O, scale factor -1
4. center A, scale factor 3
5. center B, scale factor $\dfrac{1}{2}$
6. center C, scale factor -1

For the answers to exercises 1-10, see students' work.

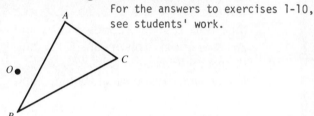

A triangle has vertices $(0, 8)$, $(-5, 9)$, and $(-3, 2)$. Draw the figure on graph paper. Find its image for each given dilation.

7. $D_{O, 2}$
8. $D_{O, -1}$
9. $D_{O, -3}$
10. $D_{O, \frac{3}{4}}$

Written For a dilation with center P and scale factor k, A' is the image of A, and B' is the image of B. Find k for the given conditions.

1. $PA = 4$ and $PA' = 12$ 3 or -3
2. $PB = 16$ and $PB' = 8$ $\dfrac{1}{2}$ or $-\dfrac{1}{2}$
3. $PA' = 4$ and $AA' = 16$ $\dfrac{1}{5}$ or $-\dfrac{1}{3}$
4. $AB = 4$ and $A'B' = 24$ 6 or -6

Find the image of the point $(4, -3)$ for each composition.

5. $D_{O, 2} \circ D_{O, 4}$ (32,-24)
6. $D_{O, -3} \circ D_{O, 2}$ (-24,18)
7. $D_{O, -2} \circ D_{O, 4}$ (-32,24)
8. State a rule for finding the image of a point (x, y) for $D_{O, k} \circ D_{O, \ell}$. Explain your rule.
9. Can a dilation ever be an isometry? Explain.   For the answers to exercises 8-10,
10. Show that a dilation with scale factor -1 is a half-turn.   see the Teacher's Guide.

Which properties are preserved by dilations? Justify your answer. Include a sketch.

11. collinearity yes
12. angle measure yes
13. distance no
14. ratio of distances yes
15. parallelism of lines yes
16. perpendicularity of lines yes

Let p be "Transformation t is an isometry;" let r be "Transformation t is a dilation with scale factor $k = 1$;" let s be "Transformation t does not preserve distance." Which of the following are true? Justify your answers. For the answers to exercises 17-20, see the Teacher's Guide.

17. $p \longrightarrow s$
18. $r \longrightarrow s$
19. $\sim s \longrightarrow r$
20. $\sim r \longrightarrow \sim s$

Problem Solving Application: Using Transformations

Objective: To solve problems by using reflections.

Transformations can often be used to find the shortest path between two points when there are restrictions that make a direct path impossible.

Examples

1 The towns of Ada and Bluffton are located on the same side of a major highway. Two roads are to be built to allow access for the towns to the highway. Where should the highway interchange be built so that the two roads are as short as possible?

Explore If the towns were on opposite sides of the highway, then the intersection of the segment joining the towns and the highway would be the site for the interchange. We can use a reflection to "place" the cities on opposite sides of the highway.

Plan Draw a figure to represent the situation. We must find point X on line n so that the path from A to X and the path from X to B $(AX + XB)$ is the shortest path.

Solve Find the reflection of B with respect to line n. Label this point B'. Then draw $\overline{AB'}$. Point X, the intersection of $\overline{AB'}$ and line n, is the location of the interchange.

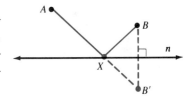

Examine Since $\overline{AB'}$ is the shortest path from A to B' and $\overline{XB} \cong \overline{XB'}$, $AX + XB$ is the shortest distance. Thus, this choice of point X produces the shortest path.

2 For the miniature golf hole shown at the right, determine a path that could be used to score a hole-in-one.

Explore The path $\overline{BH}$ will not work since it intersects a wall. By using a reflection, we can determine a path where the golf ball hits a wall and goes into the hole.

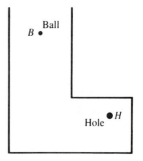

Plan If a golf ball does not have much spin, it will bounce off a wall in such a way that the two angles formed by the path of the ball and the wall will be congruent. We must find point X on wall $\overline{EF}$ so that $\angle BXE \cong \angle HXF$.

Solve Find the reflection of H with respect to $\overline{EF}$. Label this point H'. Then draw $\overline{BH'}$. Point X is the intersection of $\overline{BH'}$ and $\overline{EF}$. The shot should be aimed at this point in order to score a hole-in-one.

Examine Since $\angle BXE$ and $\angle FXH'$ are vertical angles, $\angle BXE \cong \angle FXH'$. Since reflections preserve angle measure, $\angle FHX' \cong \angle FXH$. Thus, $\angle BXE \cong \angle FXH$, and the solution is correct.

Exercises

Max: 1-5; Avg: 1, 3-5; Min: 1-5 odds
Written Copy each figure. Then use the results of example 1 to solve each problem.

1. Gilboa and Rawson are located 13 miles apart on the same side of a highway. Gilboa is 10 miles from the highway, and Rawson is 15 miles from the highway. Where should a highway interchange be built so that the two roads connecting the towns to the interchange are as short as possible?
2. Determine the total length of the two roads in exercise 1. about 27.7 miles

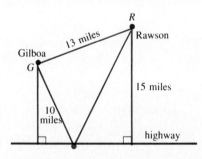

Copy each miniature golf hole. Then use one or more reflections to determine a path that could be used to score a hole-in-one for each hole.

3.   4.   5.

Portfolio Suggestion

Select one of the assignments from this chapter that you found especially challenging and place it in your portfolio.

Performance Assessment

Draw △GHI and two lines *m* and *n* that intersect at an angle of 30°. Use a composite of reflections with respect to *m* and *n* to produce a rotation image of △GHI. **See students' work.**

Chapter Summary

1. A **mapping** pairs each member of one set called the **domain** with exactly one member of the same, or another, set called the **range.** The **preimages** of a mapping are in the domain. The **images** are in the range.   (454)
2. A mapping is a **one–to–one mapping** if, and only if, each image has exactly one preimage.   (455)
3. The **composite** of two mappings, $f \circ g$, is a mapping in which an image of mapping g is taken of a preimage for mapping f.   (456)
4. A **transformation** is a one–to–one mapping whose domain and range are the set of all points in the plane.   (458)
5. A **reflection** in line ℓ (or reflection over line ℓ) is a transformation that maps each point P onto a point P' in one of the following ways.   (458)
   a. If P in on line ℓ, then the image of P is P.
   b. If P is not on line ℓ, then ℓ is the perpendicular bisector of $\overline{PP'}$.
6. If a figure is its own image after a line reflection over line m, then the figure is said to have **line symmetry** about line m, or to be symmetrical with respect to line m. Line m is called a **line of symmetry.**   (462)
7. If point $P(x, y)$ is reflected over the y-axis, its image is $P'(-x, y)$.   (465)
8. If point $P(x, y)$ is reflected over the line $y = b$, its image is $P'(x, 2b - y)$. (466)
9. A **translation** is a transformation that maps every point in the plane to its image by moving each point the same distance in the same direction.   (469)

494 TRANSFORMATION GEOMETRY

10. If $P(x, y)$ is transformed by translation $T_{a, b}$, then its image is $P'(x + a, y + b)$. (470)
11. A **rotation** is a transformation that maps every point in the plane to its image by rotating the plane around a fixed point. The fixed point is called the center of rotation and is its own image. (473)
12. A figure has **rotational symmetry** if it is the image of itself under some rotation. The angle of rotation must be greater than 0° and less than 360°. (474)
13. A rotation of 180° about a point P is called a **half-turn** with center P. A half-turn with center P is a transformation that maps every point A to a point A' such that P is the midpoint of $\overline{AA'}$. If $P(x, y)$ is transformed by a half-turn $\text{Rot}_{O, 180°}$, then its image is $P'(-x, -y)$. (476)
14. If a figure is its own image for some half-turn with center P, then it has point symmetry with respect to P. (477)
15. An **isometry** is a transformation that preserves distance. (482)
16. The transformation that assigns every point to itself is called the **identity transformation,** symbolized by $T_{0, 0}$. (485)
17. A **dilation** is a transformation of the plane that changes the size of a figure, but not its shape. If $P(x, y)$ is transformed by a dilation $D_{O, k}$, then its image is $P'(kx, ky)$. (488)

Chapter Review

12.1 State whether each represents a mapping. If so, state the domain and range.

1. $1 \longrightarrow a$ yes
   $2 \longrightarrow b$ domain = {1,2,3}
   $3 \longrightarrow c$ range = {a,b}

2. $x \longrightarrow 1$ yes
   $y \longrightarrow 2$ domain = {x,y,z}
   $z \longrightarrow 1$ range = {1,2}

3. $A \longrightarrow 1$
   $B \longrightarrow 2$ no
   $C \longrightarrow 3$

Find the image for each of the following given the mapping and the preimage.

4. $f: x \longrightarrow 2x + 1; \ x = 7$ 15
5. $g: x \longrightarrow (x - 1)^2; \ x = -2$ 9
6. $h: x \longrightarrow x^2 + x + 1; \ x = -1$ 1
7. $j: x \longrightarrow \dfrac{1}{x}; \ x = \dfrac{1}{2}$ 2

Use the mappings from exercises 4–7 to find the image of each of the following given the composite and the preimage.

8. $f \circ g; \ x = 2$ 3
9. $g \circ f; \ x = 1$ 4
10. $g \circ j; \ x = \dfrac{1}{3}$ 4

For the answers to exercises 11 and 12, see students' work.

12.2 Copy the following. Then draw the image of each for a reflection over ℓ.

11.   12.

12.3 For each of the following, determine whether ℓ is a line of symmetry.

13. no   14. yes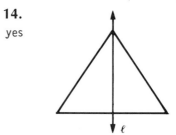

12.4 Complete the following.

15. $(2, 3) \xrightarrow{R_{\text{x-axis}}} \underline{\ ?\ }$ $(2,-3)$ 16. $(-3, 2) \xrightarrow{R_{\text{x-axis}}} \underline{\ ?\ }$ $(-3,-2)$

17. $(-1, 4) \xrightarrow{R_{\text{y-axis}}} \underline{\ ?\ }$ $(1,4)$ 18. $(5, 0) \xrightarrow{R_{\text{y-axis}}} \underline{\ ?\ }$ $(-5,0)$

12.5 Complete the following.

19. $(1, 7) \xrightarrow{T_{1, 4}} \underline{\ ?\ }$ $(2,11)$ 20. $(2, -3) \xrightarrow{T_{3, 2}} \underline{\ ?\ }$ $(5,-1)$

21. $(5, 4) \xrightarrow{T_{-1, 3}} \underline{\ ?\ }$ $(4,7)$ 22. $(-4, 6) \xrightarrow{T_{-2, -1}} \underline{\ ?\ }$ $(-6,5)$

12.6 Complete the following.

23. $(1, 3) \xrightarrow{\text{Rot}_{O, 90°}} \underline{\ ?\ }$ $(-3,1)$ 24. $(2, 5) \xrightarrow{\text{Rot}_{O, -90°}} \underline{\ ?\ }$ $(5,-2)$

25. $(4, -1) \xrightarrow{\text{Rot}_{O, -90°}} \underline{\ ?\ }$ $(-1,-4)$ 26. $(-3, 0) \xrightarrow{\text{Rot}_{O, -90°}} \underline{\ ?\ }$ $(0,3)$

12.7 Find the image for each of the following for a half-turn whose center is the origin of the coordinate plane.

27. $(3, 4)$ $(-3,-4)$ 28. $(-2, 5)$ $(2,-5)$ 29. $(-1, -1)$ $(1,1)$

12.8 Find the image for each of the following given the composite and the preimage.

30. $(5, 6)$; $R_{\text{y-axis}} \circ R_{\text{x-axis}}$ $(-5,-6)$ 31. $(2, 0)$; $R_{\text{x-axis}} \circ R_{\text{y-axis}}$ $(-2,0)$

32. $(1, 4)$; $R_{\text{x-axis}} \circ T_{2, 4}$ $(3,-8)$ 33. $(3, 6)$; $\text{Rot}_{O, 90°} \circ T_{-1, 3}$ $(-9,2)$

12.9 Find the following. Use the table on page 486.

34. $R_{\ell_1} \circ R_{\ell_2} \text{Rot}_{O, -240°}$ 35. $R_{\ell_2} \circ \text{Rot}_{O, -120°}$ R_{ℓ_1} 36. $\text{Rot}_{O, -120°} \circ \text{Rot}_{O, -120°}$ $\text{Rot}_{O, -240°}$

12.10 Complete the following.

37. $(3, 7) \xrightarrow{D_{O, 2}} \underline{\ ?\ }$ $(6,14)$ 38. $(-2, 4) \xrightarrow{D_{O, 3}} \underline{\ ?\ }$ $(-6,12)$

Chapter Test

State whether each of the following is *always, sometimes,* or *never* true.

1. Transformations preserve distance. sometimes
2. Line reflections preserve collinearity. always
3. A figure has at least one line of symmetry. sometimes
4. A dilation is an isometry. sometimes
5. A figure with rotational symmetry also has line symmetry. sometimes
6. A figure with point symmetry has rotational symmetry. always

State whether each figure has *line symmetry, rotational symmetry, neither,* or *both*.

7. both
8. both HIH
9. HAH line symmetry
10. both

Find the image of (4, −2) for each of the following.

11. $R_{y\text{-axis}}$ (-4,-2)
12. $R_{y=x}$ (-2,4)
13. $R_{x=3}$ (2,-2)
14. $T_{2,3}$ (6,1)
15. $T_{4,-8}$ (8,-10)
16. $\text{Rot}_{O, 90°}$ (2,4)
17. $D_{O,2}$ (8,-4)
18. $D_{O,-3}$ (-12,6)

Graph the triangle with vertices A(2, 1), B(5, −2), and C(9, 4). Then, graph the image of △ABC for each of the following. For the answers to exercises 19-26, see the Teacher's Guide.

19. $R_{y\text{-axis}}$
20. $R_{y=4}$
21. $T_{-2,3}$
22. $D_{O,2}$
23. $\text{Rot}_{O,90°}$
24. $R_{x\text{-axis}} \circ T_{4,-2}$
25. $\text{Rot}_{O,-90°} \circ D_{O,-3}$
26. $T_{1,1} \circ \text{Rot}_{O,180°}$

For the answers to exercises 27-30, see students' work.

Label three noncollinear points A, B, and C. Draw each of the following.

27. the image of A for $R_{\overline{BC}}$
28. the image of A for $\text{Rot}_{C,90°}$
29. the image of A for a translation that takes B to C
30. the image of $\overline{AB}$ for a half-turn with center C

For the answers to exercises 31-38, see the Teacher's Guide.

Name the transformation for which △A'B'C' is the image of △ABC.

31.
32.
33.
34.
35.
36.
37.
38.

39. Find the coordinate rule of a translation for which (−3, 6) is the image of (4, 1). $T_{-7,5}$
40. If (3, −4) is the image of (−1, −12) under a reflection in point P, find the coordinates of point P. P(1,-8)

CHAPTER 13

Introduction to Trigonometry

Application in Meteorology

Meteorologists who work at airports keep close tabs on the weather to help ensure that airplanes can fly safely. Visibility is key to a pilot as he or she is flying, so meteorologists must make sure that the cloud ceiling is high enough to fly. The cloud ceiling is the lowest altitude at which solid cloud is present. If the cloud ceiling is below a certain level, usually about 61 meters, meteorologists advise pilots to avoid taking off or landing.

One way that meteorologists can determine the cloud ceiling at night is to shine a searchlight that is located a fixed distance from their office vertically on the clouds. They measure the *angle of elevation* to the spot produced on the clouds. Trigonometry can then be used to determine the cloud ceiling. A searchlight 200 meters from the weather office produces a spot on the clouds that has an angle of elevation of 35°. How high is the cloud ceiling?   **140 meters**

Group Project: Surveying

The ancient Egyptians are credited with developing the first tools for surveying land. Each year, the Nile River would flood its banks and wash away the boundary markings for the fields at the river's edge. The people had to find a way to replace the markers accurately. One of the tools found the Egyptians used was a rope with thirteen knots that divide the rope into twelve equal parts. Make a model of this tool. Investigate how the tool could be used to approximate a right angle. Use a protractor to measure the right angle you make with your tool. Is it accurate? Explain your findings in a poster or one-page report.

13.1 The Right Triangle

Objective: To use the Pythagorean Theorem and the properties of similarity to solve problems.

A branch of mathematics known as **trigonometry** is based upon the properties of the right triangle.

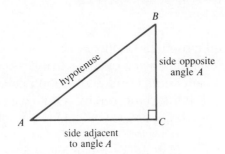

In a right triangle, the side opposite the right angle is called the **hypotenuse**. The other two sides are called the **legs**. A leg may be referred to as an **adjacent side** or an **opposite side**, in reference to a particular angle of the right triangle.

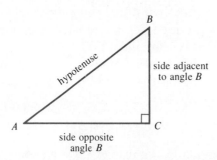

One important aspect of right triangles is the concept of similar right triangles.

Example

1 △ABC and △KLM are similar right triangles. If AC = 4, KM = 1, and AB = 6, find KL, LM, and BC.

The following relationships hold for the sides of the two triangles.

$$\frac{AB}{KL} = \frac{AC}{KM} = \frac{BC}{LM}$$

In addition, the following relationships exist.

$$\frac{AB}{AC} = \frac{KL}{KM} \qquad \frac{AB}{BC} = \frac{KL}{LM} \qquad \frac{AC}{BC} = \frac{KM}{LM}$$

Use the Pythagorean Theorem to find BC.

$$(AB)^2 = (AC)^2 + (BC)^2$$
$$36 = 16 + (BC)^2$$
$$20 = (BC)^2$$
$$\sqrt{20} = BC$$
$$2\sqrt{5} = BC$$

The measure of $\overline{BC}$ is $2\sqrt{5}$.

Use the appropriate proportion to find the measures of $\overline{KL}$ and $\overline{LM}$.

$$\frac{AC}{KM} = \frac{BC}{LM}$$
$$\frac{4}{1} = \frac{2\sqrt{5}}{LM}$$
$$4(LM) = 2\sqrt{5}$$
$$LM = \frac{1}{2}\sqrt{5}$$

The measure of $\overline{LM}$ is $\frac{1}{2}\sqrt{5}$.

$$\frac{AC}{KM} = \frac{AB}{KL}$$
$$\frac{4}{1} = \frac{6}{KL}$$
$$4(KL) = 6$$
$$KL = \frac{3}{2}$$

The measure of $\overline{KL}$ is $\frac{3}{2}$.

Exercises

Max: 1-14, 1-20; Avg: 1-14, 1-20; Min: 1-13 odds, 1-19 odds

Exploratory Given $\triangle PQR$ with a right angle at Q, name the following parts of the triangle.

1. hypotenuse $\overline{PR}$
2. legs $\overline{PQ}, \overline{QR}$
3. side adjacent to $\angle R$ $\overline{QR}$
4. side opposite $\angle R$ $\overline{PQ}$
5. side adjacent to $\angle P$ $\overline{PQ}$
6. side opposite $\angle P$ $\overline{QR}$

In $\triangle ABC$, $m\angle C = 90$. Use the Pythagorean Theorem and the given measures of two sides of the triangle to find the measure of the third side.

7. $AB = 25$, $BC = 20$ 15
8. $AC = 8$, $AB = 17$ 15
9. $AC = 8$, $BC = 8\sqrt{3}$ 16
10. $AC = 10$, $AB = 26$ 24
11. $AB = 4$, $AC = 3$ $\sqrt{7}$
12. $AC = BC = 3$ $3\sqrt{2}$

Explain how to find the length of $\overline{AB}$ in each of the following.

13.

Form right $\triangle ABC$ with the right angle at C. Measure $\overline{AC}$ and $\overline{BC}$. Then use the Pythagorean Theorem ($AB = \sqrt{AC^2 + BC^2}$).

14.

It is not possible to easily solve this problem using the Pythagorean Theorem.

500 INTRODUCTION TO TRIGONOMETRY

Written Complete the following proportions for the similar triangles below.

1. $\dfrac{AC}{KM} = \dfrac{AB}{?}$ KL

2. $\dfrac{AC}{CB} = \dfrac{KM}{?}$ ML

3. $\dfrac{?}{AB} = \dfrac{ML}{CB}$ KL

4. $\dfrac{RS}{?} = \dfrac{TS}{CB}$ AB

5. $\dfrac{KL}{ML} = \dfrac{?}{TS}$ RS

6. $\dfrac{AC}{CB} = \dfrac{?}{TS}$ RT

7. $\dfrac{AB}{RS} = \dfrac{AC}{?}$ RT

8. $\dfrac{?}{KM} = \dfrac{TS}{ML}$ RT

9. $\dfrac{2(AC)}{2(RT)} = \dfrac{AB}{?}$ RS

10. $\dfrac{RS}{TS} = \dfrac{?}{ML}$ KL

11. $\dfrac{8(CB)}{8(AB)} = \dfrac{5(TS)}{?\ 5(RS)}$

12. $\dfrac{k(AC)}{k(AB)} = \dfrac{?\ k(KM)}{k(KL)}$

In each of the following, the two triangles are similar. Find the measure of the indicated segment.

13.

14.

15.

16.

17.

18.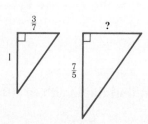

Solve.

19. In the figure $\overline{LQ} \parallel \overline{MR} \parallel \overline{NS} \parallel \overline{PT}$. Show that $\triangle KLQ \sim \triangle KMR \sim \triangle KNS \sim \triangle KPT$.

△KLQ ~ △KMR ~ △KNS ~ △KPT because when parallel lines are cut by a transversal, corresponding angles are congruent. Also, ∠K ≅ ∠K.

20. In the figure below, the vertical lines are each perpendicular to line m. Give an argument to show that the vertical lines are parallel to each other.

By Theorem 4-11, any two lines are parallel. By repeatedly applying this theorem, all of the lines are parallel.

Objective: To use the tangent ratio to find the measures of the angles of a right triangle.

13.2 The Tangent Ratio

The figure below shows several right triangles, all sharing angle A.

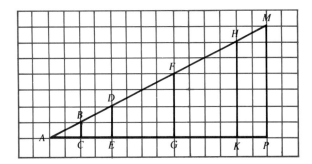

Because $m\angle A = m\angle A$, and all the right angles have the same measure, all the triangles are similar. Therefore, by the AA axiom, we have the following.

$$\triangle ABC \sim \triangle ADE \sim \triangle AFG \sim \triangle AHK \sim \triangle AMP$$

This means that the following ratios are equal.

$$\frac{BC}{AC} = \frac{DE}{AE} = \frac{FG}{AG} = \frac{HK}{AK} = \frac{MP}{AP}$$

By measuring the sides of the triangles on the grid, the following ratio for an angle with the measure of angle A can be found.

$$\frac{\text{the measure of the side opposite } \angle A}{\text{the measure of the side adjacent to } \angle A} \approx \frac{1}{2}$$

This idea can be generalized for any right triangle, by making the following definition.

Journal

Look up the word *tangent*. List the different definitions that you find.

Definition of Tangent

> For any acute angle A in a right triangle, the ratio of the measure of the side opposite angle A to the measure of the side adjacent to angle A is called the **tangent of angle A**.

The symbol for the tangent of angle A is tan A.

In the case discussed above, $\tan A \approx \frac{1}{2}$. Because $m\angle A$ is approximately 27, $\tan 27° = \frac{1}{2}$.

The symbol = will be used instead of ≈ for trigonometric ratios, even though the ratios are approximations.

Examples

1 **Find tan 35°.**

Draw a right triangle with an angle measuring 35.

There are many sizes of right triangles that have a 35° angle.

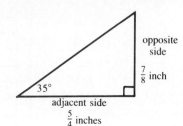

Identify the opposite and adjacent sides. Measure them. Write the tangent ratio.

$$\tan 35° = \frac{\text{measure of side opposite the 35° angle}}{\text{measure of side adjacent to the 35° angle}}$$

$$\tan 35° = \frac{\frac{7}{8}}{\frac{5}{4}} \quad \text{or} \quad \frac{28}{40}$$

As a decimal, tan 35° is approximately 0.7.

2 **Given tan B = 2, find $m \angle B$.**

Draw a right triangle with the opposite side of $\angle B$ twice as long as the adjacent side. Then, using a protractor, measure $\angle B$.

The measure of $\angle B$ is approximately 63.

You can also use a calculator to find the tangent of an angle.

Instead of drawing, measuring, and then dividing to find the tangent of an angle, a table of tangent values can be used. The table consists of each whole number of degrees between 0° and 90° and their tangent values accurate to four decimal places. This table can be found on page 528.

Examples

3 Find tan 64°.

Locate 64° in the column labeled *Angle*.

Angle	sin	cos	tan
61°	0.8746	0.4848	1.8040
62°	0.8829	0.4695	1.8807
63°	0.8910	0.4340	1.9626
64°	0.8988	0.9384	2.0503
65°	0.9063	0.4226	2.1445
66°	0.9135	0.4067	2.2460

Locate the value for tan 64° in the column labeled *tan*.

Using a calculator:

ENTER: 64 [tan]
DISPLAY: 2.05030384

An approximate value for tan 64° is 2.0503.

4 Given tan A = 0.27, find $m\angle A$ to the nearest degree.

Look in the column labeled *tan*.

Angle	sin	cos	tan
12°	0.2079	0.9781	0.2126
13°	0.2250	0.9744	0.2309
14°	0.2419	0.9703	0.2493
15°	0.2588	0.9659	0.2679
16°	0.2756	0.9613	0.2867
17°	0.2924	0.9563	0.3057

Think of 0.27 as 0.2700.
The value of 0.2700 is between 0.2679 and 0.2867.
It is closer to 0.2679.

Locate the measure of the angle that corresponds to 0.2679. The measure of $\angle A$ is about 15.

Using a calculator: **ENTER:** 0.27 [tan^{-1}]
DISPLAY: 15.1095751

Exercises

Max: 1-12, 1-32, 1-2; Avg: 1-12, 1-5, 7-31 odds; Min: 1-12, 1-31 odds

Exploratory Use the figure to answer the following exercises.

1. Name all the right triangles.
∠EFQ, ∠GHQ, ∠KLQ, ∠MRQ, ∠STQ, ∠VWQ

2. Give a reason for all the right triangles being similar. Since ∠Q ≅ ∠Q, and all the right angles are congruent, all the right triangles are similar by AA.

504 INTRODUCTION TO TRIGONOMETRY

Complete each proportion using the figure in the previous exercise.

3. $\dfrac{GH}{QH} = \dfrac{?}{QT}$ ST

4. $\dfrac{MR}{?} = \dfrac{VW}{QW}$ QR

5. $\dfrac{?}{QF} = \dfrac{GH}{QH}$ EF

6. $\dfrac{?}{QL} = \dfrac{EF}{QF}$ KL

7. $\dfrac{MR}{QR} = \dfrac{?}{QH}$ GH

8. $\dfrac{EF}{QF} = \dfrac{VW}{?}$ QW

9. Complete the proportion $\dfrac{GH}{QH} = \dfrac{?}{QW}$.
Check your answer by finding the number of units in GH, QH, QW, and your answer, and then dividing. VW; GH = 2, QH = 8, QW = 20, VW = 5. Since $\dfrac{2}{8} = \dfrac{5}{20}$, the answer checks.

10. Find the length of $\overline{ST}$ and $\overline{QT}$ in units. Use these numbers to find the ratio $\dfrac{ST}{QT}$. $\dfrac{1}{4}$

11. Define tangent ratio. See page 465.

12. Find the tangent of angle Q. 0.25

Written For each $m\angle A$ given below, use a protractor to draw a right triangle ABC. Identify the sides opposite and adjacent to $\angle A$. Then, by measuring and dividing, find the approximate value of tan A. For exercises 1-5, see students' work. Student answers should be close to the answers given.

1. 45 1
2. 30 0.5774
3. 60 1.7321
4. 76 4.0108
5. 17 0.3057

For each right triangle, name the side opposite and the side adjacent to the indicated angle.

6. △ECG, ∠G
7. △ADG, ∠G
8. △DKG, ∠G
9. △ABE, ∠A
10. △ABE, ∠E
11. △DKG, ∠D
12. △AKD, ∠A
13. △AKD, ∠D
14. △ADG, ∠A

For answers to exercises 6-14, see the Teacher's Guide.

For each value for tan B, draw a right triangle ABC on graph paper.

15. $\dfrac{1}{2}$
16. 3
17. $\dfrac{2}{5}$
18. $\dfrac{7}{3}$
19. 0.2
20. 3.5000

For exercises 15-20, see students' work.

Use the table on page 528 or a calculator to find the tangent of each angle.

21. 25° 0.4663
22. 30° 0.5774
23. 3° 0.0524
24. 85° 11.4301
25. 61° 1.8040
26. $52\dfrac{1°}{2}$ 1.3032

Find the measures of the angles having the following tangents. State your answer to the nearest degree. Use the table on page 528 or a calculator.

27. 0.4450 24
28. 0.3100 17
29. 1.4 54
30. $\dfrac{3}{5}$ 31
31. $\dfrac{3}{2}$ 56
32. 25.0 88

Challenge

1. Draw a right triangle ABC with a right angle at C. Show whether or not $\tan A \cdot \tan B = 1$. Since $\tan A = \dfrac{BC}{CA}$ and $\tan B = \dfrac{CA}{BC}$, then $\tan A \cdot \tan B = \dfrac{BC}{CA} \cdot \dfrac{CA}{BC} = 1$.

2. Generalize your answer in exercise 1 for any right triangle, if possible. By the definition of tangent, the tangents the acute angles are reciprocals and product of reciprocals is 1.

Objective: To use the tangent ratio to find missing measures of sides in a right triangle.

13.3 Using the Tangent Ratio

Surveyors use a variety of tools to obtain measures. Once some measures are known, the principles of geometry and trigonometry can be used to find other measures.

The tangent ratio can be used to find missing measures in right triangles.

Example

1 A surveyor is 15 meters from a rock formation near the planned location of a new highway. The formation makes an angle of 90° with the ground, as shown. If the surveyor uses surveying equipment, she can find the angle at which the top of the formation is sighted. Suppose the angle measures 58°. Find the height of the rock formation.

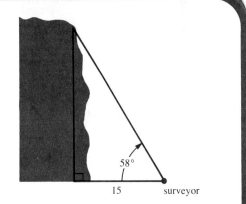

$$\tan 58° = \frac{\text{measure of side opposite the 58° angle}}{\text{measure of side adjacent to the 58° angle}}$$

$\tan 58° = \dfrac{x}{15}$ The value for tan 58° can be found in the table on page 528 or by using a calculator.

$1.6003 = \dfrac{x}{15}$

$x = 24.005$

The rock formation is about 24 meters high.

The measure of a leg of a right triangle can be found using the tangent of either acute angle.

Example

2 Find the length of $\overline{KM}$ in $\triangle KLM$ by using two different angles.

a. $\tan 40° = \dfrac{33}{x}$

$0.8391 = \dfrac{33}{x}$

$0.8391x = 33$ Multiply both sides by x.

$x = 39.328$ Divide both sides by 0.8391.

b. Because $m\angle K = 40$, $m\angle L = 50$.

$\tan 50° = \dfrac{x}{33}$

$1.1918 = \dfrac{x}{33}$ Multiply both sides by 33.

$x = 39.33$

Sometimes one tangent ratio is more convenient to use than another. Which of these two solutions do you find easier?

The length of $\overline{KM}$ is about 39.33 cm to the nearest hundredth.

The measure of an angle can also be found without measuring.

Example

3 Find the measure of $\angle R$ in $\triangle RST$.

$\tan R = \dfrac{33}{70}$

$\tan R = 0.4714$

Use the table on page 528 or a calculator.

The measure of $\angle R$ is 25, to the nearest degree.

CHAPTER 13 507

Exercises

Max: 1-7, 1-58, 1; Avg: 1-7, 1-21 odds, 24-58 evens; Min: 1-7, 1-21 odds, 24-46 evens

Exploratory For each measure of angle *B*, draw a right triangle. For exercises 1-5, see students' work.

1. 35   2. 70   3. 10   4. 80   5. 20

6. Use the measure-divide method to find tan *B* in exercise 1. about 0.7

7. Use the table on page 528 or a calculator to find tan *B* in exercise 1. Compare this result to the one in exercise 6. 0.7002

For exercises 1-12, approximate answers are given to the nearest hundredth.

Written For each triangle, find the measure of the indicated side. Use the table on page 528 or a calculator.

1.

2.

3.

4.

5.

6.

7.

8.

9.

10.

11.

12.

For each value for tan *B*, find $m\angle B$, to the nearest degree. Use the table on page 528 or a calculator.

13. $\frac{3}{10}$ 17
14. $\frac{5}{6}$ 40
15. 0.6667 34
16. 1.3333 53
17. 0.8 39

508 INTRODUCTION TO TRIGONOMETRY

For each triangle, find tan A. For exercises 20 and 22, answers are rounded to 4 decimal places.

18. 0.6
19. $\frac{7}{8}$
20. 0.2679
21. 2.7
22. 0.5188

For exercises 23–58, angle measures are rounded to the nearest degree and lengths are rounded to two decimal places.

Use the table on page 528 or a calculator to find $m\angle A$ for exercises 18–22.

23. exercise 18 31
24. exercise 19 41
25. exercise 20 15
26. exercise 21 70
27. exercise 22 27

Use the tangent ratio, the Pythagorean Theorem, or both to find the missing measure in each triangle. If a solution is not possible, explain why.

28.
29. 52.16
30. 16.07
31.

32. 37
33. not a right triangle
34.
35. 4

Find the measure of the angle that the diagonal of a rectangle forms with the longer side of the rectangle if the rectangle's sides have these measures.

36. 3, 7   23
37. 1, 12   5
38. 2, 2.5   39
39. $\sqrt{2}, \sqrt{3}$ 39

40. The measure of the side opposite $\angle L$ in right triangle LMP is $1\frac{1}{2}$ times the measure of the side adjacent to $\angle L$. Find tan L and $m\angle L$.
tan∠L = 1.5, m∠L = 56

41. The lengths of the diagonals of a rhombus are 8 meters and 3 meters. Find the measure of the angle that the longer diagonal makes with a side of the rhombus. 21

42. The area of a rhombus is 15 in² and the length of the shorter diagonal is 3 in. Find the measure of a side of the rhombus. Give the measure of the angle that the longer diagonal makes with a side.
5.22 in., 17

43. Find the area of the parallelogram shown at the right.
119.29 cm²

44. Find the height of the flagpole below. 11.55 m

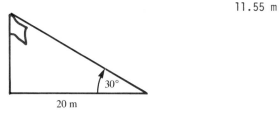

45. Find the height of the flagpole below if a 5-foot person views it. 16.55 ft

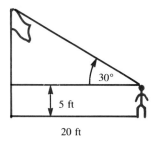

46. The line of sight of Ken Engel, 1000 feet away, makes an angle of 20° with the ground when he sights the top of the water tower. Find the height of the tower. 364 ft

47. If Ken Engel's eyes are five feet above the ground, find the height of the tower in exercise 46. 369 ft

The following exercises refer to the terms defined below.

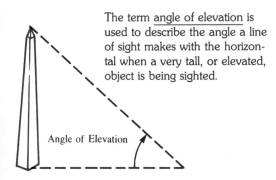

The term <u>angle of elevation</u> is used to describe the angle a line of sight makes with the horizontal when a very tall, or elevated, object is being sighted.

An <u>angle of depression</u> is the angle formed by the line of sight and the horizontal when an object beneath the viewer is being sighted.

48. A person at the top of a lighthouse sights a boat in the water. The angle of depression is 50°. If the lighthouse is 70 feet high, find the distance from the foot of the lighthouse to the boat. 58.73 ft

49. Refer to exercise 48. Find the distance from the foot of the lighthouse to the boat, if the angle of depression is 60° and the lighthouse is 80 feet tall. 46.19 ft

50. Refer to exercise 48. Find the distance from the foot of the lighthouse to the boat if the angle of depression is 45° and the lighthouse is 20 meters high. 20 m

51. A person in a boat sights the top of a lighthouse at an angle of elevation of 70°. If the lighthouse is 40 feet tall, find the distance from the foot of the lighthouse to the boat. 14.56 ft

52. Refer to exercise 51. Find the distance from the foot of the lighthouse to the boat if the angle of elevation is 28°. 75.23 ft

53. Mei and her sister are in Conifer Park, flying their kites. When Mei's kite is directly over her sister, standing 120 feet from Mei, the line of the kite string makes an angle of 48° with the ground. Find the height of the kite. 133.27 ft

54. The path of the cable car connecting Peak Wood and Peak Stone rises 20 meters for every 100 meters of horizontal distance. Find the measure of the angle the cable makes with the horizontal. 11°

55. A surveyor is 100 meters from a building. The angle of elevation to the top of the building is 23°. The surveyor's instrument is 1.55 meters above the ground. Find the height of the building. 44 ft

56. A surveyor is 100 meters from a bridge. The angle of elevation to the top of the bridge is 35°. The surveyor's instrument is 1.45 meters above the ground. Find the height of the bridge. 71.47 m

57. In a parking garage, each level is 22 feet apart. Each ramp to a level is 122 feet long. Find the measure of the angle of elevation for each ramp. 10°

58. To secure a 500-meter radio tower against high winds, guy wires are attached to a ring on the tower. The ring is 5 meters from the top. The wires form a 15° angle with the tower. Find the distance from the tower to the guy wire anchor in the ground. 132.63 m

Challenge   Solve.

1. Before Apollo 11 LEM descended to the surface of the moon, it made one orbit at a distance of three miles from the surface. At one point in its orbit, the onboard guidance system measured the angles of depression to the near and far edges of a huge crater. The angles measured 25 and 18. Find the distance across the crater.   about 2.8 miles

Mixed Review

Find the image of $(2, -3)$ for each composite transformation.

1. $R_{\text{x-axis}} \circ T_{3, 2}$ (5, 1)
2. $D_{O, 3} \circ \text{Rot}_{O, 90°}$ (9, 6)
3. $\text{Rot}_{O, 180°} \circ T_{-5, 8}$ (3, -5)

Find the preimage of $(-4, -8)$ for each composite transformation.

4. $\text{Rot}_{O, -90°} \circ R_{\text{y-axis}}$ (-8, -4)
5. $T_{4, -4} \circ D_{O, 3}$ $(-\frac{8}{3}, -\frac{4}{3})$
6. $D_{O, -4} \circ R_{y = x}$ (2, 1)

Use the figure at the right for Exercises 7–10. For Exercises 9–10, use the table on page 528 or a calculator.

7. If $BG = 12$, $BD = 15$, and $GH = 8$, find BE. $5\sqrt{5}$
8. If $AC = 9$, $DE = 6$, and $AB = 15$, find EC.   4
9. If $m\angle A = 50$ and $BH = 4$, find GH.   about 3.36
10. If $AB = 10\sqrt{2}$ and $AC = 6\sqrt{5}$, find $m\angle B$.   about 72°

11. A committee of 5 people is to be formed from a group of 5 men and 8 women. What is the probability that the committee will have at least one woman? $\frac{1286}{1287}$

13.4 The Sine and Cosine Ratios

Objective: To find the sine, cosine, or tangent of angles in a right triangle.

Two other right triangle ratios are the **sine** and **cosine** ratios.

The sine of angle A is the ratio

$$\frac{\text{measure of side opposite angle } A}{\text{measure of hypotenuse}}.$$

The cosine of angle A is the ratio

$$\frac{\text{measure of side adjacent to angle } A}{\text{measure of hypotenuse}}.$$

The symbol for the sine of angle A is $\sin A$.
The symbol for the cosine of angle A is $\cos A$.
The tangent, sine, and cosine ratios can be stated as follows.

Trigonometric Ratios

$$\tan A = \frac{\text{side opposite}}{\text{side adjacent}} \qquad \sin A = \frac{\text{side opposite}}{\text{hypotenuse}}$$

$$\cos A = \frac{\text{side adjacent}}{\text{hypotenuse}}$$

Example

1 Find the sine, cosine, and tangent of angles A and B in the figure at the right.

$\sin A = \dfrac{\text{side opposite}}{\text{hypotenuse}}$ or $\dfrac{3}{5}$ $\qquad \sin B = \dfrac{4}{5}$

$\cos A = \dfrac{\text{side adjacent}}{\text{hypotenuse}}$ or $\dfrac{4}{5}$ $\qquad \cos B = \dfrac{3}{5}$

$\tan A = \dfrac{\text{side opposite}}{\text{side adjacent}}$ or $\dfrac{3}{4}$ $\qquad \tan B = \dfrac{4}{3}$

Examples

2 **Find the measure of angle A in example 1.**

There are several methods of finding $m\angle A$. The sine ratio will be used here.

$$\sin A = \frac{3}{5}$$
$$\sin A = 0.6000$$

Use the table on page 528 or a calculator to find $m \angle A$.
Look in the column labeled *sin*.

Angle	sin	cos	tan
35°	0.5736	0.8192	0.7002
36°	0.5878	0.8090	0.7265
37°	0.6018	0.7986	0.7536
38°	0.6157	0.7880	0.7813
39°	0.6293	0.7771	0.8098

The value of 0.6000 is between 0.5878 and 0.6018. It is closer to 0.6018.

Locate the measure of the angle that corresponds to the column that contains *0.6018*.

The measure of $\angle A$ is 37, to the nearest degree.

Using a calculator:

ENTER: 0.6 $\boxed{\sin^{-1}}$
DISPLAY: 36.8698977

3 **The distance across Pine County Swamp cannot be measured directly. Use trigonometry to find the distance.**

Use either tan P or sin P to solve for x. Sin P is used below.

$$\sin P = \frac{x}{1500}$$
$$\sin 68° = \frac{x}{1500}$$
$$0.9272 = \frac{x}{1500}$$
$$1390.8 = x$$

Use the table on page 528 or a calculator to find sin 68°.

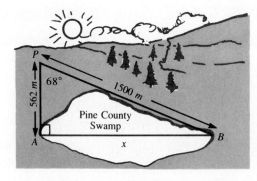

The distance across Pine County Swamp is 1390.8 meters.

Using a calculator: **ENTER:** 68 $\boxed{\sin}$
DISPLAY: 0.92718385

Example

4 Suppose that for maximum safety, a ladder should be placed against a wall at a 75° angle with the ground. If the ladder is 14 feet long, how far from the wall should the foot of the ladder be placed?

$$\cos 75° = \frac{x}{14} \quad \text{The cosine ratio is used. Why?}$$

$$0.2588 = \frac{x}{14}$$

$$x = 3.6232$$

For maximum safety, the ladder should be placed slightly more than $3\frac{1}{2}$ feet from the wall.

Exercises

Max: 1-8, 1-77, 1-4; Avg: 1-8, 1-77 odds; Min: 1-8, 1-59 odds

Exploratory Use the figure at the right to name the following.
1. Name the hypotenuse in △RST. $\overline{RS}$
2. Name the side opposite ∠R. $\overline{ST}$
3. Name the side adjacent to ∠R. $\overline{RT}$

Use △RST to complete each ratio.

4. $\sin R = \frac{ST}{?}$ RS
5. $\cos R = \frac{?}{RS}$ RT
6. $\sin S = \frac{RT}{?}$ RS
7. $\cos S = \frac{?}{?}$ $\frac{ST}{RS}$

8. Draw a right triangle whose acute angles measure 30 and 60. Use a ruler to find values for sin 30°, sin 60°, and cos 60°. See students' work. sin 30° = $\frac{1}{2}$, sin 60° = 0.8667, cos 60° = $\frac{1}{2}$

For exercises 1-12, see students' work. Students answers should be close to the answers given.
Written By drawing, measuring, and dividing, find approximations for the following.

1. sin 40° 0.6428
2. sin 70° 0.9397
3. cos 40° 0.7660
4. cos 70° 0.3420
5. sin 45° 0.7071
6. cos 45° 0.7071
7. sin 10° 0.1736
8. cos 10° 0.9848
9. sin 80° 0.9848
10. cos 80° 0.1736
11. sin 58° 0.8480
12. cos 58° 0.5299

Use the table on page 528 or a calculator to find approximations for the trigonometric ratios in each of the following.

13. exercise 1 0.6428
14. exercise 2 0.9397
15. exercise 3 0.7660
16. exercise 4 0.3420
17. exercise 5 0.7071
18. exercise 6 0.7071
19. exercise 7 0.1736
20. exercise 8 0.9848
21. exercise 9 0.9848
22. exercise 10 0.1736
23. exercise 11 0.8480
24. exercise 12 0.5299

Use the table on page 528 or a calculator to find the value of each of the following.

25. $\sin 7°$ 0.1219
26. $\cos 53°$ 0.6018
27. $\cos 15°$ 0.9659
28. $\sin 55°$ 0.8192

Find approximate values for each of the following.

29. $\sin 22\frac{1°}{2}$ 0.3827
30. $\sin 71\frac{1°}{2}$ 0.9483
31. $\cos 33\frac{1°}{2}$ 0.8339
32. $\sin 42\frac{1°}{2}$ 0.6756

Use the Pythagorean Theorem and your knowledge of trigonometry to find the value of each of the following for the triangle at the right.

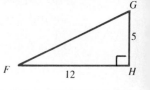

33. $\tan F$ $\frac{5}{12}$
34. $\tan G$ $\frac{12}{5}$
35. $\sin G$ $\frac{12}{13}$
36. $\cos G$ $\frac{5}{13}$
37. $\sin F$ $\frac{5}{13}$
38. $\cos F$ $\frac{12}{13}$
39. $m\angle F$ 23
40. $m\angle G$ 67

Use rectangle ABCD to find the values of the following.

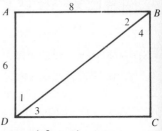

41. measure of $\overline{BD}$ 10
42. $\sin \angle 1$ $\frac{4}{5}$
43. $\cos \angle 1$ $\frac{3}{5}$
44. $\sin \angle 2$ $\frac{3}{5}$
45. $\cos \angle 2$ $\frac{4}{5}$
46. $\sin \angle 3$ $\frac{3}{5}$
47. $\cos \angle 3$ $\frac{4}{5}$
48. $\sin \angle 4$ $\frac{4}{5}$
49. $\cos \angle 4$ $\frac{3}{5}$
50. $m\angle 1$ 53
51. $m\angle 2$ 37
52. $m\angle 3$ 37
53. $m\angle 4$ 53

For exercises 54–77, angle measures are rounded to the nearest degree and lengths are rounded to two decimal places.

Use any method to find the missing measure in each triangle.

54.

55.

56.

57.

58. 1643.25

59. 2.93

60.   61.   62.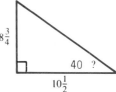

Find the distance across the following swamps. The measures are in meters.

63.   64.   65.

Suppose that for maximum safety, a ladder should be placed against a wall at a 75° angle with the ground. How far from the wall should the foot of the ladder be placed when the ladder has each length?

66. 14 ft 3.62 ft   67. 16 ft 4.14 ft   68. 20 ft 5.18 ft   69. 5 m 1.30 m   70. 100 ft 25.88 ft

Solve.

71. Given $\triangle ABC$, $AB = BC$, $m\angle A = 28$, $AC = 18$, find AB, BC, and the measure of altitude $\overline{BD}$. AB = BC = 10.19, BD = 4.79

72. A kite is held by a taut string pegged to the ground. If the string is 40 feet long and makes a 33° angle with the ground, how high is the kite? 21.78 ft

73. A metal wire fixed to the ground braces a 20-foot utility pole. The wire is 17 feet long and reaches 15 feet up the pole. What is the measure of the angle the wire makes with the ground? 62

74. A ground-to-air missile is test-fired at a 15° angle with the ground. It hits a target 8000 feet from the ground. How far does the missile travel? (Assume the missile follows a straight path.) 30,911.90 ft

75. Given rectangle $ABCD$, $AB = 30$, $m\angle CBD = 44$, find BD, BC, and the area and perimeter of $ABCD$. BD = 43.19, BC = 31.07 perimeter = 122.14, area = 932.1

76. Find the measure of the acute angles in a right triangle whose sides measure 15 cm, 20 cm, and 25 cm. 53, 37

77. A cross-sectional diagram of a house is shown at the right. Find the measure of the angle the rafters make with the horizontal beam. 39

Challenge Answer the following.

1. The symbol $\sin^2 A$ means $(\sin A)^2$. Use the table to find $\sin^2 30°$. 0.25

2. Use the table to find the value of $\cos^2 30°$. 0.75

3. Find the sine and cosine of an acute angle and its complement. Do this for three more acute angles and their complements. What conclusions can be made? the sine of an angle is the same as the cosine of its complement, and vice versa

4. Verify the following statement. The tangent of an angle can get greater and greater without bound, but the sine and cosine can never be greater than one. For the answer to exercise 4, see the Teacher's Guide.

Mathematical Excursions

MATHEMATICS IN SPACE

Many elements go into the planning of space exploration, such as the design and construction of the launch vehicle and spacecraft. The launch into space, insertion into orbit, change of orbit, tracking, and landing at a specified target, are all selected before the spacecraft ever leaves earth. All these events require mathematical computations.

Suppose two of NASA's tracking stations are located near the equator: one in Ethiopia at 40° E longitude, the other in Ecuador at 78° W longitude. Assume both stations, represented by A and B in the figure at the right, are on the equator and that the radius is 3960 miles.

Find the distance between the two stations on a straight line through the earth. The angular distance between the two meridians of longitude is 78° + 40° = 118°.

To find the distance, consider right triangle DCB, and $2CB = AB$.

$$\sin 59° = \frac{CB}{3960}$$
$$CB = 3960(\sin 59°)$$
$$CB = 3960(0.8572)$$
Hence, $2CB = 2(3960)(0.8572)$
$$AB = 6789 \text{ miles}$$

The distance between the two stations is approximately 6789 miles.

Exercise Solve.

If the circumference of the earth is 24,900 miles, find the distance along the surface of the earth between the two tracking stations. $8161\frac{2}{3}$ miles

13.5 The Unit Circle

Objective: To find the values of the sine or cosine of an angle using a unit circle.

Another aspect of trigonometry depends on the use of the unit circle. A **unit circle** is a circle whose radius has a length of one unit and whose center is the origin.

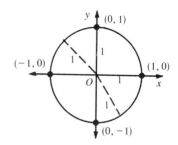

Consider a unit circle on the labeled set of axes as shown to the right.

Two rays extend from the center to form an angle with measure a. One ray is fixed along the positive x-axis. It is called the **initial side** of the angle. The other ray can rotate about the center. It is called the **terminal side** of the angle. Such an angle is said to be in **standard position**.

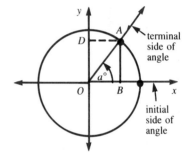

$\triangle AOB$ is a right triangle whose hypotenuse is $\overline{OA}$. Because $\overline{OA}$ is also a radius in the unit circle, it is one unit long. Therefore,

$$\sin a = \frac{\text{side opposite}}{\text{hypotenuse}} = \frac{AB}{1} \quad \text{or} \quad AB.$$

The measure of $\overline{AB}$ is the y-coordinate of point A. A is the point at which the terminal side of the angle intersects the unit circle. The sine of an angle can be defined using this idea.

Definition of Sine

Let a be the measure of an angle in standard position. Sin a is the y-coordinate of the intersection of the terminal side with the unit circle.

The cosine of an angle can be defined in a similar manner.

$$\cos a = \frac{\text{side adjacent}}{\text{hypotenuse}} = \frac{OB}{1} \quad \text{or} \quad OB$$

The measure of $\overline{OB}$ equals the measure of $\overline{DA}$. The measure of $\overline{DA}$ is the x-coordinate of point A. Therefore, the measure of $\overline{OB}$ is also equal to the x-coordinate of point A. The cosine of an angle can be defined using this idea.

Definition of Cosine

Cos a is the x-coordinate of the intersection of the terminal side with the unit circle.

Because sin a and cos a are determined according to their definitions, the first coordinate of point A is cos a and the second coordinate is sin a. It is important to note that acute angles are not the only angles that have sines and cosines.

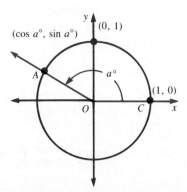

∠AOC is an obtuse angle. Sin a and cos a are determined precisely according to the new definition. The first coordinate of point A is cos a and the second coordinate is sin a.

When the terminal side of the angle falls in Quadrant III or IV, the measure of the angle is greater than 180, but less than 360. Such an angle is called a **reflex angle**. ∠BOC falls in Quadrant IV.

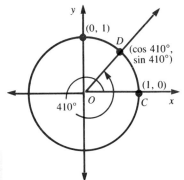

The terminal side of ∠DOC has made more than one complete counterclockwise revolution. Such angles have measures greater than 360. In this case, the angle measure is 410. From the figure, we see that the terminal side of 410° is in the same position that the terminal side of a 50° angle would be. Therefore, sin 410° = sin 50°, and cos 410° = cos 50°.

Calculator Hint

You can use a scientific calculator to find the value of the trigonometric function of any angle, regardless of size. Make sure the calculator is in the correct mode for the angle measure you are using.

The terminal side of an angle can also rotate clockwise around the circle.

In such a case, the angle has a negative measure. ∠COE has a measure of −30, but the terminal side is in the same position as the terminal side in a 330° angle. Therefore, sin (−30°) = sin 330°, and cos (−30°) = cos 330°.

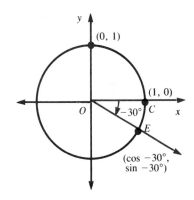

Exercises

Max: 1-4, 1-30, 1-6; Avg: 1-4, 1-19 odds, 21-30; Min: 1-4, 1-29 odds

Exploratory Answer the following.   See students' work.

1. Define unit circle. a circle with a radius of 1 unit
2. Draw a unit circle. Include an angle with a measure between 0 and 90.
3. Draw a unit circle. Include an angle with a measure between 180 and 270. See students' work.
4. Explain the method of finding the sine and cosine of a given angle using your drawings in exercises 2 and 3. Answers will vary.

Written Find the approximate value of each trigonometric ratio. Use unit circles drawn on graph paper. Let 10 squares equal one unit.

1. sin 30° 0.5
2. sin 20° 0.3420
3. cos 60° 0.5
4. sin 150° 0.5
5. sin 200° -0.3420
6. cos 120° -0.5
7. cos 200° -0.9397
8. sin 420° 0.8660
9. sin (−60°) -0.8660
10. cos (−60°) 0.5
11. cos (−200°) -0.9397
12. sin (−1000°) 0.9848
13. sin 90° 1
14. cos 90° 0
15. sin 180° 0
16. cos 180° -1
17. sin 270° -1
18. cos 270° 0
19. sin 360° 0
20. cos 360° 1

520 INTRODUCTION TO TRIGONOMETRY

For exercises 21-28, answers may vary. Typical answers are given.

Give the measure of two angles that have the same sine as each angle whose measure is listed below.

21. 40   140, 400
22. 210   330, 570
23. 300   240, 660
24. −60   240, 300

Give a negative measure of an angle that has the same sine as the angle listed in the following exercises.

25. exercise 1   −330°
26. exercise 5   −160°
27. exercise 8   −300°
28. exercise 9   −120°

Answer the following.

29. At what point will the terminal ray of an angle intersect the circle centered at the origin but with radius whose length is 3 units? if the angle has measure a, (3 cos a, 3 sin a)

30. $\sin^2 a$ means $(\sin a)^2$. Show that $\sin^2 a + \cos^2 a = 1$.
For any angle of measure a on the unit circle, sin a and cos a are legs of a right triangle whose hypotenuse is 1. By the Pythagorean Theorem, $\sin^2 a + \cos^2 a = 1$.

Challenge   Complete.

1. Copy and complete the following table.

x	0°	30°	45°	60°	90°	120°	135°	150°	180°
sin x	0	0.5	0.7071	0.8660	1	0.8660	0.7071	0.5	0

x	210°	225°	240°	270°	300°	315°	330°	360°
sin x	-0.5	-0.7071	-0.8660	-1	-0.8660	-0.7071	-0.5	0

For answers to exercises 2 and 4, see the Teacher's Guide.

2. Draw a pair of coordinate axes. Let every box on the x-axis represent 15°. Use the completed table from exercise 1 to graph the relation y = sin x from x = 0° to x = 360°.

3. Copy and complete the following table.

x	0°	30°	45°	60°	90°	120°	135°	150°	180°
cos x	1	0.8660	0.7071	0.5	0	-0.5	-0.7071	-0.8660	-1

x	210°	225°	240°	270°	300°	315°	330°	360°
cos x	-0.8660	-0.7071	-0.5	0	0.5	0.7071	0.8660	1

4. Draw a pair of coordinate axes. Let every box on the x-axis represent 15°. Use the completed table from exercise 3 to graph the relation y = cos x from x = 0° to x = 360°.

5. Compare your sin and cos graphs. How are they similar? How do they differ?
The graph of cosine is the graph of sine shifted 90°.

6. What do you think would happen to your graphs if they were each extended to x = 720°? The graphs repeat in periodic cycles of 360°.

Problem Solving Application: Using Right Triangles

Objective: To solve problems using right triangle ratios and the Pythagorean Theorem.

Right triangle trigonometry can sometimes be used to solve problems. Often, problems are solved by using more than one right triangle ratio or by using the Pythagorean Theorem along with one of the right triangle ratios.

Example

1 The base of a television antenna and two points on the ground are in a straight line. The two points are 100 feet apart. From the two points, the angles of elevation to the top of the antenna are 30° and 20°. Find the height of the antenna to the nearest foot.

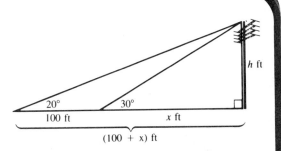

Explore Let h = height of the antenna.
Let x = distance from closer point to the base of the antenna.
Then $100 + x$ = distance from farther point to the base of the antenna.

Plan The tangent ratio can be used to obtain two equations.

$$\tan 30° = \frac{h}{x} \qquad \tan 20° = \frac{h}{100 + x}$$

We can solve each of these equations for h in terms of x, and then use substitution to solve for x.

Solve

$$\tan 30° = \frac{h}{x} \qquad\qquad \tan 20° = \frac{h}{100 + x}$$

$x(\tan 30°) = h \qquad\quad (100 + x)(\tan 20°) = h$ The values for
$x(0.5774) = h \qquad\quad (100 + x)(0.3640) = h$ $\tan 30°$ and
$0.5774x = h \qquad\qquad\ \ 0.3640x + 36.4 = h$ $\tan 20°$ can be
found in the
$\qquad\qquad 0.5774x = 0.3640x + 36.4$ table on page 528
$\qquad\qquad 0.2134x = 36.4$ or by using a
$\qquad\qquad\qquad x \approx 170.6$ calculator.

$\qquad 0.5774(170.6) \approx h$ Substitute for x in
$\qquad\qquad\qquad h \approx 98.5$ the equation above.

The height of the antenna is about 99 feet.   Examine this solution.

Exercises

Max: 1-11; Avg: 1-10; Min: 1-5, 7, 9

Written Solve each problem. Round all lengths to the nearest unit and all angle measures to the nearest degree.

1. An observer at the top of a lighthouse sights two sailboats in line due west of the lighthouse. From the observer, the angles of depression to the sailboats are 21° and 19°. How far above sea level are the observer's eyes if the sailboats are 100 feet apart at the time of the sighting? 334 ft

2. The pilot of a plane flying 2000 meters above sea level observes two ships in line due east of the plane. From the pilot, the angles of depression to the ships are 42° and 30°. How far apart are the two ships at the time of the sighting? 1243 m

3. To find the height of a mountain peak, points A and B are located on a plain in line due north of the peak. The angles of elevation are measured from each point. The angle at A is 37°, and the angle at B is 21°. The distance from A to B is 720 meters. How high is the peak above the level of the plain? 470 m

4. The angle of elevation to the top of a building from a point on the ground is 39°. From a point 50 feet closer to the building, the angle of elevation is 45°. What is the height of the building? 213 f

5. A ladder is braced by a 4-foot rod placed at a right angle to the ladder. One end of the rod is 3 feet up from the base of the ladder. The other end of the rod is 7 feet from the base of the wall on which the ladder is resting. How far is the top of the ladder from the base of the wall? 16 ft

6. A wire is strung from the base of one antenna to the top of a shorter antenna and then back to the top of the first antenna. The angle of elevation from the top of the shorter antenna to the top of the other antenna is 42°. How long is the wire if the shorter antenna is 5 meters high and the antennas are 12 meters apart? 31 m

7. In △ABC, the measure of ∠A is 15, and ∠B is an obtuse angle. The length of altitude $\overline{CX}$, which is drawn to the line that contains side $\overline{AB}$, is 15 cm. The length of $\overline{BX}$ is 8 cm. Find the perimeter of △ABC. 123 cm

8. A radio antenna sits atop a building. From a point 200 feet from the base of the building, the angle of elevation to the top of the antenna is 80°. The angle of elevation from the same point to the bottom of the antenna is 75°. How tall is the antenna? 388 ft

9. A pendulum 50 cm long is moved 40° from the vertical. How far did the tip of the pendulum rise? 12 cm

10. A ship sails due north from port for 90 km, then 40 km east, and then 70 km north. How far is the ship from port? 16!

11. In square QRST, M is the midpoint of $\overline{QR}$, N is the midpoint of $\overline{QM}$, and P is the midpoint of $\overline{MR}$. If $\overline{MT}$, $\overline{NT}$, and $\overline{PT}$ are each drawn, find the measures of ∠QMT, ∠QNT, and ∠QPT. m∠QMT = 64, m∠QNT = 76, m∠QPT = 53

Portfolio Suggestion

Review the items in your portfolio. Make a table of contents of the items, noting why each item was chosen. Replace any items that are no longer appropriate.

Performance Assessment

$A(-3, -\sqrt{3})$ lies on the terminal side of an angle in standard position. Give the degree measure of three angles that fit this description. Tell how to find the cosine of these angles and then do so. **Sample answers: 240°, 600°, 960°; cos θ = −0.5**

Chapter Summary

1. The **hypotenuse** is the side opposite the right angle in a right triangle. (498)
2. A **leg** of a right triangle can either be an adjacent side or an opposite side. (498)
3. For any acute angle A in a right triangle, the ratio of the measure of the side opposite angle A to the measure of the side adjacent to angle A is called the **tangent** of angle A. The symbol for the tangent of angle A is tan A. (501)
4. Trigonometric Ratios: $\tan A = \dfrac{\text{side opposite}}{\text{side adjacent}}$,

   $\sin A = \dfrac{\text{side opposite}}{\text{hypotenuse}}$, $\cos A = \dfrac{\text{side adjacent}}{\text{hypotenuse}}$. (511)
5. A **unit circle** is a circle whose radius has a length of one unit and whose center is the origin. (517)
6. A fixed ray along the positive x-axis is called the **initial side** of an angle. (517)
7. The ray of an angle that can rotate about the center of a unit circle is called the **terminal side** of the angle. (517)
8. Let a be the measure of an angle in **standard position**. Sin a is the y-coordinate of the intersection of the terminal side with the unit circle. (517)

9. Cos a is the x-coordinate of the intersection of the terminal side with the unit circle. (518)

 Chapter Review

13.1 Complete the following proportions for the figure at the right.

1. $\dfrac{DE}{BE} = \dfrac{?}{BC}$ AC
2. $\dfrac{BH}{BE} = \dfrac{GH}{?}$ DE
3. $\dfrac{AB}{CB} = \dfrac{DB}{?}$ EB
4. $\dfrac{AC}{BC} = \dfrac{?}{BH}$ GH

In each of the following, the two triangles are similar. Find the missing measure.

5. ? $2\dfrac{1}{2}$

6. ?

7. ?

For exercises 8–11, see students' work.

13.2 For each $m \angle A$ given below, use a protractor to draw a right triangle ABC.

8. 20
9. 60
10. 35
11. 56

For each right triangle, name the side opposite and the side adjacent to the indicated angle. $\overline{CD}$, $\overline{BD}$

$\overline{AF}$, $\overline{BF}$ 12. △ABF, ∠B   13. △BCD, ∠B   14. △FBE, ∠E $\overline{BF}$, $\overline{EF}$
$\overline{ED}$, $\overline{BD}$ 15. △EDB, ∠B   16. △BEC, ∠E   17. △ABE, ∠A $\overline{BE}$, $\overline{BA}$
                                         $\overline{BC}$, $\overline{BE}$

Use the table on page 528 or a calculator to find the tangent for the angle whose measure is given below.

18. 60° 1.7321
19. 30° 0.5774
20. 5° 0.0875
21. $36\dfrac{1}{2}°$ 0.7400

Find the measure of the angle having each tangent, to the nearest degree.

22. 1.1918 50
23. 0.1755 10
24. 3.7562 75
25. $\dfrac{4}{5}$ 39

13.3 For each triangle, find the measure of the indicated side. Use the table on page 528 or a calculator.

26. 　　27. 　　28.

29. Suppose the top of a lighthouse 152 feet tall can be seen from a boat. The angle of elevation is 9°. Find the distance from the foot of the lighthouse to the boat. 959.60 ft

30. From an airplane flying 7000 feet above the ground, the angle of depression to the base of the control tower is 19°. Find the distance from a point on the ground directly beneath the airplane to the control tower. 20,331.11 ft

For exercises 26-30 and 43-45, angle measures are rounded to the nearest degree and lengths are rounded to two decimal places.

13.4 Use the table on page 528 or a calculator to find each of the following.

31. cos 7° 0.9925　　32. sin 53° 0.7986　　33. sin 15° 0.2588　　34. cos 45° 0.7071

For the triangle at the right, find each of the following.

35. sin B $\frac{4\sqrt{41}}{41}$　36. sin C $\frac{5\sqrt{41}}{41}$　37. cos B $\frac{5\sqrt{41}}{41}$　38. cos C $\frac{4\sqrt{41}}{41}$
39. tan B $\frac{4}{5}$　40. tan C $\frac{5}{4}$　41. m∠B 39　42. m∠C 51

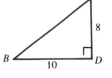

Use any method to find the missing measure in each triangle.

43. 　　44. 　　45.

For answers to exercises 46, 47, and 49, see the Teacher's Guide.

13.5 Answer the following.

46. Give an argument to show that sin 45° = cos 45° = $\frac{\sqrt{2}}{2}$.

47. Find the sine and cosine of 90°, 180°, 270°, and 360° angles.

48. At what point will the terminal ray of an angle intersect the circle centered at the origin but with a radius whose length is two units? For angle a, (2 cos a, 2 sin a).

49. Given an argument to show that tan a = $\frac{\sin a}{\cos a}$.

Find each of the following without referring to the table.

50. sin 30° 0.5　　51. cos 60° 0.5　　52. cos 120° -0.5　　53. sin 90° 1
54. cos 180° -1　　55. sin 45° 0.7071　　56. sin 210° -0.5　　57. cos 240° -0.5
58. sin (−30°) -0.5　　59. cos 225° -0.7071　　60. sin 270° -1　　61. cos (−135°) -0.7071

Chapter Test

Complete the following proportions for the similar triangles at the right.

1. $\dfrac{FG}{AG} = \dfrac{?}{AC} BC$ 2. $\dfrac{AE}{AG} = \dfrac{DE}{?} FG$

3. $\dfrac{AF}{AG} = \dfrac{AD}{?} AE$ 4. $\dfrac{DE}{AE} = \dfrac{BC}{?} AC$

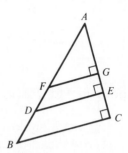

Find the tangent, sine, and cosine of $\angle A$ in each of the following.

5. $\dfrac{3}{4}, \dfrac{3}{5}, \dfrac{4}{5}$

6. $\dfrac{5}{12}, \dfrac{5}{13}, \dfrac{12}{13}$

7. $\dfrac{4}{3}, \dfrac{4}{5}, \dfrac{3}{5}$

8. $1, \dfrac{\sqrt{2}}{2}, \dfrac{\sqrt{2}}{2}$

Find the tan a, sin a, and cos a for the following values of a. Use the table on page 528 or a calculator.

9. 33° 0.6494 / 0.5446 / 0.8387

10. 58° 1.6003 / 0.8480 / 0.5299

11. 89° 57.2900 / 0.9998 / 0.0175

12. 17° 0.3057 / 0.2924 / 0.9563

Find the value of a in each of the following. Use the table on page 528 or a calculator.

13. tan a = 0.8693 41

14. sin a = 0.3420 20

15. cos a = 0.2924 73

Solve. For exercises 16–19, angle measures are rounded to the nearest degree and lengths are rounded to two decimal places.

16. The height of the Empire State Building is 1250 feet. Find the angle of elevation of the top from a point 1.4 miles away. 10

17. A rectangle is 30 cm wide. The diagonal makes an angle of 36° with the longer side. Find the length of the diagonal and the longer side. 60 cm, 51.96

18. In a parking garage, each level is 20 feet apart. The ramp from one level to the next is 130 feet long. Find the measure of the angle of elevation for the ramp. 9°

19. A tree is broken by the wind. The top touches the ground 13 meters from the base. It makes an angle of 29° with the ground. How tall was the tree before it was broken? 22.07 m

20. Give the measures of two angles that have the same cosine as a −40° angle. Answers may vary. Typical answers are 40° and 320°.

Squares and Approximate Square Roots

n	n^2	$\sqrt{n}$	n	n^2	$\sqrt{n}$
1	1	1.000	51	2601	7.141
2	4	1.414	52	2704	7.211
3	9	1.732	53	2809	7.280
4	16	2.000	54	2916	7.348
5	25	2.236	55	3025	7.416
6	36	2.449	56	3136	7.483
7	49	2.646	57	3249	7.550
8	64	2.828	58	3364	7.616
9	81	3.000	59	3481	7.681
10	100	3.162	60	3600	7.746
11	121	3.317	61	3721	7.810
12	144	3.464	62	3844	7.874
13	169	3.606	63	3969	7.937
14	196	3.742	64	4096	8.000
15	225	3.873	65	4225	8.062
16	256	4.000	66	4356	8.124
17	289	4.123	67	4489	8.185
18	324	4.243	68	4624	8.246
19	361	4.359	69	4761	8.307
20	400	4.472	70	4900	8.367
21	441	4.583	71	5041	8.426
22	484	4.690	72	5184	8.485
23	529	4.796	73	5329	8.544
24	576	4.899	74	5476	8.602
25	625	5.000	75	5625	8.660
26	676	5.099	76	5776	8.718
27	729	5.196	77	5929	8.775
28	784	5.292	78	6084	8.832
29	841	5.385	79	6241	8.888
30	900	5.477	80	6400	8.944
31	961	5.568	81	6561	9.000
32	1024	5.657	82	6724	9.055
33	1089	5.745	83	6889	9.110
34	1156	5.831	84	7056	9.165
35	1225	5.916	85	7225	9.220
36	1296	6.000	86	7396	9.274
37	1369	6.083	87	7569	9.327
38	1444	6.164	88	7744	9.381
39	1521	6.245	89	7921	9.434
40	1600	6.325	90	8100	9.487
41	1681	6.403	91	8281	9.539
42	1764	6.481	92	8464	9.592
43	1849	6.557	93	8649	9.644
44	1936	6.633	94	8836	9.695
45	2025	6.708	95	9025	9.747
46	2116	6.782	96	9216	9.798
47	2209	6.856	97	9409	9.849
48	2304	6.928	98	9604	9.899
49	2401	7.000	99	9801	9.950
50	2500	7.071	100	10000	10.000

Trigonometric Ratios

Angle	sin	cos	tan	Angle	sin	cos	tan
0°	0.0000	1.0000	0.0000	45°	0.7071	0.7071	1.0000
1°	0.0175	0.9998	0.0175	46°	0.7193	0.6947	1.0355
2°	0.0349	0.9994	0.0349	47°	0.7314	0.6820	1.0724
3°	0.0523	0.9986	0.0524	48°	0.7431	0.6691	1.1106
4°	0.0698	0.9976	0.0699	49°	0.7547	0.6561	1.1504
5°	0.0872	0.9962	0.0875	50°	0.7660	0.6428	1.1918
6°	0.1045	0.9945	0.1051	51°	0.7771	0.6293	1.2349
7°	0.1219	0.9925	0.1228	52°	0.7880	0.6157	1.2799
8°	0.1392	0.9903	0.1405	53°	0.7986	0.6018	1.3270
9°	0.1564	0.9877	0.1584	54°	0.8090	0.5878	1.3764
10°	0.1736	0.9848	0.1763	55°	0.8192	0.5736	1.4281
11°	0.1908	0.9816	0.1944	56°	0.8290	0.5592	1.4826
12°	0.2079	0.9781	0.2126	57°	0.8387	0.5446	1.5399
13°	0.2250	0.9744	0.2309	58°	0.8480	0.5299	1.6003
14°	0.2419	0.9703	0.2493	59°	0.8572	0.5150	1.6643
15°	0.2588	0.9659	0.2679	60°	0.8660	0.5000	1.7321
16°	0.2756	0.9613	0.2867	61°	0.8746	0.4848	1.8040
17°	0.2924	0.9563	0.3057	62°	0.8829	0.4695	1.8807
18°	0.3090	0.9511	0.3249	63°	0.8910	0.4540	1.9626
19°	0.3256	0.9455	0.3443	64°	0.8988	0.4384	2.0503
20°	0.3420	0.9397	0.3640	65°	0.9063	0.4226	2.1445
21°	0.3584	0.9336	0.3839	66°	0.9135	0.4067	2.2460
22°	0.3746	0.9272	0.4040	67°	0.9205	0.3907	2.3559
23°	0.3907	0.9205	0.4245	68°	0.9272	0.3746	2.4751
24°	0.4067	0.9135	0.4452	69°	0.9336	0.3584	2.6051
25°	0.4226	0.9063	0.4663	70°	0.9397	0.3420	2.7475
26°	0.4384	0.8988	0.4877	71°	0.9455	0.3256	2.9042
27°	0.4540	0.8910	0.5095	72°	0.9511	0.3090	3.0777
28°	0.4695	0.8829	0.5317	73°	0.9563	0.2924	3.2709
29°	0.4848	0.8746	0.5543	74°	0.9613	0.2756	3.4874
30°	0.5000	0.8660	0.5774	75°	0.9659	0.2588	3.7321
31°	0.5150	0.8572	0.6009	76°	0.9703	0.2419	4.0108
32°	0.5299	0.8480	0.6249	77°	0.9744	0.2250	4.3315
33°	0.5446	0.8387	0.6494	78°	0.9781	0.2079	4.7046
34°	0.5592	0.8290	0.6745	79°	0.9816	0.1908	5.1446
35°	0.5736	0.8192	0.7002	80°	0.9848	0.1736	5.6713
36°	0.5878	0.8090	0.7265	81°	0.9877	0.1564	6.3138
37°	0.6018	0.7986	0.7536	82°	0.9903	0.1392	7.1154
38°	0.6157	0.7880	0.7813	83°	0.9925	0.1219	8.1443
39°	0.6293	0.7771	0.8098	84°	0.9945	0.1045	9.5144
40°	0.6428	0.7660	0.8391	85°	0.9962	0.0872	11.4301
41°	0.6561	0.7547	0.8693	86°	0.9976	0.0698	14.3007
42°	0.6691	0.7431	0.9004	87°	0.9986	0.0523	19.0811
43°	0.6820	0.7314	0.9325	88°	0.9994	0.0349	28.6363
44°	0.6947	0.7193	0.9657	89°	0.9998	0.0175	57.2900
45°	0.7071	0.7071	1.0000	90°	1.0000	0.0000	∞

Symbols

$=$	is equal to	π	pi
$\neq$	is not equal to	$\lvert a \rvert$	absolute value of a
$>$	is greater than	$\sqrt{}$	principal square root
$<$	is less than	$a:b$	ratio of a to b
$\geq$	is greater than or equal to	$\{\ \}$	set
$\leq$	is less than or equal to	$\in$	is a member of
$\approx$	is approximately equal to	$\cup$	union
$*$	an operation in a group or a field	$\cap$	intersection
$\cdot$	times	$\wedge$	conjunction
$-$	negative	$\vee$	disjunction
$+$	positive	$\rightarrow$	conditional; is mapped onto
$\pm$	positive or negative	$\leftrightarrow$	biconditional
$\overleftrightarrow{AB}$	line containing points A and B	$n!$	n factorial
$\overrightarrow{AB}$	ray with endpoint A passing through B	$x \xrightarrow{f} x^2$	f is the function that maps x to x^2
$\overline{AB}$	line segment with endpoints A and B	$f \circ g$	f following g
AB	measure of AB	R_ℓ	reflection over line ℓ
$\overset{\frown}{AB}$	arc with endpoints A and B	$T_{2,3}$	translation 2 units to the right and 3 units up
$m\overset{\frown}{AB}$	measure of arc AB		
$\angle$	angle	$\text{Rot}_{A,65°}$	rotation with center A through an angle of 65° counterclockwise
$m\angle A$	measure of angle A	$D_{P,2}$	dilation with center P and a scale factor of 2
$°$	degree		
$\triangle$	triangle	$\mathcal{N}$	set of natural numbers
$\square$	parallelogram	$\mathcal{W}$	set of whole numbers
$\odot P$	circle with center P	$\mathcal{Z}$	set of integers
$\cong$	is congruent to	$\mathcal{Q}$	set of rational numbers
$\sim$	is similar to; negation	$\mathcal{R}$	set of real numbers
$\parallel$	is parallel to		
$\perp$	is perpendicular to		

Formulas

Distance Formula	$\sqrt{(x_2 - x_1)^2 + (y_2 - y_1)^2}$
Midpoint Formula	$\left(\dfrac{x_1 + x_2}{2}, \dfrac{y_1 + y_2}{2}\right)$
Slope of a Line	$m = \dfrac{y_2 - y_1}{x_2 - x_1}$
Slope-Intercept Form of a Linear Equation	$y = mx + b$
Point-Slope Form of a Linear Equation	$y - y_1 = m(x - x_1)$
Equation of a Circle with Center (h, k) and Radius r	$(x - h)^2 + (y - k)^2 = r^2$
Equation of a Parabola	$y = ax^2 + bx + c$
Axis of Symmetry of a Parabola	$x = \dfrac{-b}{2a}$
Quadratic Formula	For $ax^2 + bx + c = 0$, $x = \dfrac{-b \pm \sqrt{b^2 - 4ac}}{2a}$.
Sum of the Measures of the Interior Angles of a n-gon	$(n - 2)180$
Pythagorean Theorem	$a^2 + b^2 = c^2$
Area of a Parallelogram	$A = bh$
Area of a Triangle	$A = \dfrac{1}{2}bh$
Area of a Rectangle	$A = bh$
Area of a Trapezoid	$A = \dfrac{(b_1 + b_2)h}{2}$
Factorials	$n! = n(n - 1)(n - 2) \cdot \ldots \cdot 1$
Probability	$P(E) = \dfrac{n(E)}{n(S)}$ $P(A \text{ and } B) = P(A) \cdot P(B)$ $P(A \text{ or } B) = P(A) + P(B) - P(A \text{ and } B)$
Permutations	$_nP_n = n!$ or $\dfrac{n!}{r!}$ (for repetitions) $_nP_r = \dfrac{n!}{(n - r)!}$
Combinations	$_nC_n = 1$ $_nC_r = \dfrac{n!}{r!(n - r)!}$
Trigonometric Functions	$\sin A = \dfrac{\text{side opposite}}{\text{hypotenuse}}$ $\cos A = \dfrac{\text{side adjacent}}{\text{hypotenuse}}$ $\tan A = \dfrac{\text{side opposite}}{\text{side adjacent}}$

Glossary

abscissa (288) The x-coordinate of an ordered pair.

adjacent angles (123) Two angles that have a common side, the same vertex, and no interior points in common.

altitude of a triangle (189) A line segment drawn from any vertex perpendicular to the opposite side.

angle (123) The union of two rays that have the same endpoint.

angle bisector (129) A ray that is the common side of two adjacent angles having the same measure.

angle of depression (509) The angle formed by the line of sight and the horizontal when an object beneath the viewer is being sighted.

antecedent (6) The part of a compound statement logically associated with "if."

associative (58) Let S be a set. The operation $*$ is associative on S if for every a, b, and c in S, $a * (b * c) = (a * b) * c$.

axiom (71, 125) A statement that is accepted as true without proof.

base angles of an isosceles triangle (189) A pair of angles that includes the base of an isosceles triangle.

base angles of a trapezoid (222) A pair of angles that includes one base of a trapezoid.

between (122) Point B is between point A and point C if points A, B, and C lie on the same line and $AB + BC = AC$.

bi-conditional (7) A statement formed by the conjunction of the conditionals $p \rightarrow q$ and $q \rightarrow p$.

bisector (126) A line, ray, line segment, or plane that intersects a line segment at its midpoint.

circle (339) The set of all points in a plane a given distance from a point in the plane called the center.

circumscribed (273) A polygon is circumscribed about a circle if its sides are tangent to the circle.

closure (57) Let S be a set. Then S is closed under an operation $*$ if, for every a and b in S, $a * b$ is an element of S.

collinear (122) Two or more points are collinear if and only if they lie on the same line.

combination (434) A subset of a set in which order does not matter.

commutative (58) Let S be a set. The operation $*$ is commutative on S if, for every a and b in S, $a * b = b * a$.

complementary (129) Two angles are complementary if and only if the sum of their degree measures is 90.

composite mapping (456) The result of performing two or more mappings in a prescribed order.

composition (480) The effect of applying one transformation after another on a figure.

compound loci (347) The locus of points that satisfy two different conditions at once.

conclusion (10) A statement that follows from the premises.

conditional (6) A compound statement formed by joining two statements with the words *if* . . . , *then*.

congruent angles (128) Angles having the same degree measure.

congruent line segments (126) Line segments having the same measure.

congruent triangles (168) Two triangles with correspondence between the vertices such that each side and each angle of one triangle is the same measure as the corresponding part of the other triangle.

conjunct (2) Each statement that forms a conjunction.

conjunction (2) Compound statement formed by joining two statements with the word *and*.

connective (2) A word or phrase that connects two or more simple statements to form a compound statement.

consequent (6) The part of a compound sentence associated with "then."

construction (226) The process of producing a figure that will satisfy certain given conditions, using only a ruler or compass.

contrapositive (17) A conditional in which the antecedent and consequent are both reversed and negated.

converse (17) A conditional formed by reversing the order of the antecedent and the consequent.

convex polygon (157) A polygon in which any line containing a side of the polygon does not contain a point interior to the polygon.

coordinate (288) The number that corresponds to a point on the number line.

coplanar (122) Two or more points contained in the same plane.

corollary (155) A theorem that follows directly from another theorem.

cosine (511) The ratio between the measure of the side adjacent to an acute angle in a right triangle and the measure of the hypotenuse.

counting principle (419) Suppose one activity can occur in any of *m* ways. Another can occur in any of *n* ways. The total number of ways both activities can occur is given by the product *mn*.

De Morgan Laws (25) **1.** $\sim(p \wedge q) \leftrightarrow (\sim p \vee \sim q)$, **2.** $\sim(p \vee q) \leftrightarrow (\sim p \wedge \sim q)$.

dilation (488) A transformation in which the size of a plane figure is altered based on a center and a scale factor.

directrix (351) The line from which the locus of points in a parabola are equidistant.

discriminant (386) The expression $b^2 - 4ac$ is called the discriminant of a quadratic equation.

disjunct (2) Each statement that forms a disjunction.

disjunction (2) A compound statement formed by joining two statements with the word *or*.

Distributive Law (88) For all *a*, *b*, and *c*, in a given field, $a(b + c) = ab + ac$.

domain (4) The set of all possible replacements for the placeholder or variable in an open sentence.

domain (454) The first set in a mapping.

endpoint (122) Either of two points at the end of a line segment, or the point at the end of a ray.

equivalence (8) A set of statements that are always either both true or both false.

equivalence relation (280) A relation that is reflexive, symmetric, and transitive.

existential quantifier (31) A quantifier that tells something about only part of a population.

extremes (245) In the proportion $\frac{a}{b} = \frac{c}{d}$, the terms *a* and *d* are the extremes.

factorial (425) The expression *n!* (*n* factorial) is the product of all the numbers from 1 to *n* for any positive integer *n*.

field (93) A field is a two-fold operational system that satisfies these requirements: $(F, +)$ is a commutative group, $(F/\{0\}, \cdot)$ is a commutative group, and $\cdot$ distributes over $+$.

focus (351) A given point from which the locus of points in a parabola are equidistant.

formal proof (10) A method used to demonstrate that a given set of premises leads to a certain conclusion.

geometric mean (249) If b is the mean proportional between a and c, then $b^2 = ac$.

glide reflection (483) The composite of a line reflection and a translation in a direction that is parallel to the line of reflection.

group (65) A group is a set G together with an operation $*$ that satisfies these requirements: $(G,*)$ is an operational system, the operation $*$ is associative on G, there is an identity element for $*$ in G, and every element in G has an inverse with respect to $*$ in G.

half-plane (124) Part of a plane on one side of a straight line drawn in the plane.

half-turn (476) A rotation of 180° about point P with center P.

hypotenuse (498) The side opposite the right angle in a right triangle.

identity element (61) The element in a set for which operating with it leaves the other elements unchanged.

identity transformation (485) The transformation that assigns every point to itself.

image (454) If A is mapped to A', then A' is called the image of A.

indirect proof (151) A method of proving theorems by assuming the opposite of the conclusion and proving that the opposite contradicts a given set of conditions.

initial side (517) The ray of a unit circle that is fixed along the positive x-axis.

inscribed (273) A polygon is inscribed in a circle if the circle contains the vertices of the polygon.

inverse (17) A conditional formed by keeping the antecedent and consequent in the same places, but negating them both.

inverse (61) a_{inv} is the inverse of a if the combination of a and a_{inv} is the identity.

isometry (482) A transformation that preserves distance.

isosceles trapezoid (222) A trapezoid with congruent legs.

isosceles triangle (189) A triangle with at least two sides that have the same measure.

Law of the Contrapositive (18) A conditional and its contrapositive are equivalent.

Law of Detachment (13) If $p \to q$ is a true conditional and p is true, then q is true.

Law of Disjunctive Inference (10) If a disjunction is true and one of its disjuncts is false, then the other disjunct is true.

Law of Double Negation (8) The equivalence $p \leftrightarrow (\sim p)$.

Law of Syllogism (20) If $p \to q$ and $q \to r$ are true, then $p \to r$ is true.

leg (498) **1.** Any side of a right triangle not opposite the right angle. **2.** Either of the two nonparallel sides of a trapezoid.

linear equation in two variables (291) An equation that can be written in the form $Ax + By = C$, where A, B, and C are any number and A and B are not both 0.

line of reflection (458) The line ℓ in **line reflection** below.

line of symmetry (462) The line m in **line symmetry** below.

line reflection (458) A reflection in line ℓ is a transformation that maps each point P onto a point P' in one of the following ways. **1.** If P is on line ℓ, then the image of P is P. **2.** If P is not on line ℓ, then ℓ is the perpendicular bisector of PP'.

line segment (122) Part of a line that consists of two endpoints and all the points between them.

line symmetry (462) A plane figure is said to have line symmetry if a line m can be drawn through it so that the figure on one side is a reflection of the figure on the opposite side.

locus (339) The set of all those points, and only those points, that satisfy a certain condition.

mapping (454) A relation that pairs each member of one set with exactly one member of the same or another set.

mean proportional (249) In the proportion $\frac{a}{b} = \frac{b}{d}$, b is the mean proportional between a and d.

means (245) In the proportion $\frac{a}{b} = \frac{c}{d}$, the terms b and c are the means.

median of a triangle (189) A line segment drawn from any vertex to the midpoint of the opposite side.

midpoint (126) A point that lies between two points such that the two line segments formed are congruent.

negation (2) If a statement is represented by p, then *not p* is the negation of that statement.

noncollinear (122) Points that are not contained on the same line.

noncoplanar (122) Two or more points (or lines) that are not contained in the same plane.

octant (355) The regions into which three coordinate planes divide three-dimensional space.

one-to-one mapping (455) A mapping in which each member of the range has exactly one preimage.

open sentence (4) A sentence containing a placeholder(s) to be replaced in order to determine if the sentence is true or false.

operation on a set (56) Let S be a set. Then $*$ is an operation on set S if, for every a and b in S, there exists a single element c, also in S, such that $a * b = c$.

opposite rays (123) Two rays that lie in opposite directions from a common endpoint.

ordered pair (275, 288) A pair of numbers in which the order is specified. An ordered pair is used to locate points in a plane.

order relation (111) A system in which the rationals, reals, and integers are ordered.

ordinate (288) The y-coordinate of an ordered pair.

origin (288) The point of intersection of the two axes of the coordinate plane.

outcome set (412) A set of all the possible outcomes in a probability experiment.

parabola (351, 374) The locus of points equidistant from a given point and a given line. Its equation is in the form of $y = ax^2 + bx + c$.

parallel lines (143) Two distinct lines that lie in the same plane and do not intersect.

parallelogram (208) A quadrilateral with two pairs of parallel sides.

perpendicular bisector (140) A line, line segment, or ray that is perpendicular to the line segment at the midpoint of the line segment.

perpendicular lines (139) Two lines that intersect to form right angles.

point-slope form (316) The point-slope form of an equation of a line is $y - y_1 = m(x - x_1)$. The point on the line is (x, y) and the slope is m.

point symmetry (477) A figure is said to have point symmetry if it is its own image for some half-turn with center P.

point reflection (476) See **half-turn.**

postulates (125) Statements that are accepted as true.

preimage (454) If A is mapped to A', then A is called the preimage of A'.

premise (10) A statement that is assumed true.

probability of an event (412) The ratio between the number of ways an event can occur and the total number of possible outcomes.

proportion (244) An equality between ratios.

quadrant (288) One of the four regions into which the coordinate plane is separated by the axes.

quadratic equation (369) The standard form of a quadratic equation is $ax^2 + bx + c = 0$.

quadratic formula (385) If $ax^2 + bx + c = 0$ and $a \neq 0$, then $x = -b \pm \dfrac{\sqrt{b^2 - 4ac}}{2a}$.

quadratic-linear system (394) A system with one linear equation and one quadratic equation.

quantifier (30) A word or group of words that acts to quantify a statement.

range (454) The second set in a mapping.

ratio (244) A comparison of two numbers by division.

ray (123) Part of a line consisting of one endpoint and all the points on one side of the endpoint.

rectangle (218) A parallelogram with four right angles.

rectangular coordinate system (289) A system of intersecting perpendicular lines that allows every point to be named on a plane.

reflex angle (518) An angle whose degree measure is greater than 180 but less than 360.

reflexive relation (278) A relation $\overline{R}$ on set A in which for every a in A, the ordered pair (a, a) is in R.

regular polygon (158) A polygon that is both equiangular and equilateral.

relation (275) A set of ordered pairs

rhombus (218) A parallelogram with all sides having the same measure.

right angle (129) An angle whose degree measure is 90.

rotation (473) A transformation that maps every point in the plane to its image by rotating the plane around a fixed point.

rotational symmetry (474) A plane figure is said to have rotational symmetry if it can be rotated through a certain angle and be the image of itself.

scale factor (488) A multiple by which a measure is increased or decreased.

skew lines (143) Two lines that do not lie in the same plane and do not intersect.

similar polygons (244) Two polygons whose corresponding angles have the same measure and whose corresponding sides have measures in proportion.

sine (511) The ratio between the measure of the side opposite an acute angle in a right triangle and the measure of the hypotenuse.

slope (306) The ratio of the change in y to the corresponding change in x in a line on the coordinate plane.

slope-intercept form (316) The slope-intercept form of the equation of a line is $y = mx + b$. The slope of the line is m and the y-intercept is b.

solution set (4) The set of all replacements from the domain that make an open sentence true.

statement (2) Any declarative sentence that is either true or false, but not both.

supplementary (129) Two angles are supplementary if and only if the sum of their degree measures is 180.

symmetric relation (278) A relation R on set A in which for all x and y in A, for any (x, y) in R, (y, x) is also in R.

system of equations (292) Two equations that have the same variables and a common solution set.

tangent (501) The ratio of the measure of the side opposite an acute angle in a right triangle to the measure of the side adjacent to that angle.

terminal side (517) The ray of a unit circle that rotates about the center.

theorem (71, 125) A mathematical statement that can be proved.

transformation (458) A one-to-one mapping whose domain and range are the set of all points in the plane.

transitive relation (279) A relation R on set A in which for all a, b, and c in R, whenever (a, b) is in R and (b, c) is in R, (a, c) is in R.

translation (469) A transformation that maps every point in the plane to its image by moving each point the same distance in the same direction.

transversal (143) A line that intersects two other lines in the same plane in two different points.

trapezoid (222) A quadrilateral with exactly one pair of parallel sides.

tree diagram (420) A diagram used to show the total number of possible outcomes in a probability experiment.

trigonometry (498) A branch of mathematics based on the properties of the right triangle.

truth value (2) The truth or falsity of a statement.

two-fold operational system (92) A mathematical system that uses two operations.

undefined term (122) A term whose meaning is accepted without formal definition. Point, line, and plane are undefined terms in geometry.

unit circle (517) A circle on the coordinate plane whose radius has a length of one unit and whose center is the origin.

universal quantifier (30) A quantifier that tells about an entire population.

vertex (123, 375) 1. The common endpoint of the two rays (or segments) that form an angle. 2. The turning point of a parabola.

vertical angles (123) Two nonadjacent angles formed by two intersecting lines.

x-axis (288) The horizontal number line on the coordinate plane.

x-coordinate (288) The first number in an ordered pair.

x-intercept (291) The x-coordinate of the point at which the graph of an equation crosses the x-axis.

y-axis (288) The vertical number line on the coordinate plane.

y-coordinate (288) The second number in an ordered pair.

y-intercept (291) The y-coordinate of the point at which the graph of an equation crosses the y-axis.

Selected Answers

CHAPTER 1 WORKING WITH LOGIC

Page 5 Lesson 1.1
Exploratory 1. true 3. true 5. true
Written 1. 0.5 is rational and -8 is an integer. true 3. 0.5 is rational or 15 is not a multiple of 3. true 5. 0.5 is not rational and 15 is not a multiple of 3. false 7. 0.5 is not rational or -8 is an integer. true 9. false
11. true 13. true 15. false 25. never
27. sometimes 29. {7} 31. {-3} 33. {-16}
35. {6} 37. {$-1, 0, 1, 2, \ldots$}
39. {$-3, -2, -1, 0, \ldots$}

Page 9 Lesson 1.2
Exploratory 1. antecedent: If $2 + 3 = 4$; consequent: then $3 + 3 = 6$; true
3. antecedent: If $-2 < -7$; consequent: then $0.15 < 0.14$; true 5. antecedent: If $-2 - (-5) = -7$; consequent: then Jane Austen is the author of Macbeth; true 7. antecedent: If $x^2 + 9 = (x + 3)^2$; consequent: then $\frac{2}{x}$ is a monomial; true **Written:** 1. If Labor Day is in November, then 2 is a prime number. true 3. If Labor Day is not in November, then $3y + 1$ is a binomial. true 5. Labor Day is in November if and only if 2 is a prime number. false 7. 2 is not a prime number if and only if Labor Day is in November. true 9. false
11. true 13. true 15. $p \to q, \sim p \lor q$
17. $p \lor \sim q, q \to p$ 19. The answers are equivalent. 21. There is a heavy turnout on Election Day and Brullman does not win.

Pages 11–12 Lesson 1.3
Exploratory 1. t 3. q 5. e 7. no conclusion 9. Sally dated Felix last weekend.
11. $q, \sim(\sim p) \leftrightarrow p$ **Written** 1. Given; Given; Law of disjunctive inference, 2, 3 5. No conclusion. A disjunction may have two true disjuncts. If one is true, the other may be true or false. Therefore, although $\sim B$ is true, C may be true or false. 7. No conclusion. A disjunction may have two true disjuncts. If one is true, the other may be true or false. Therefore, even though it is true that Rose R is red, Violet V may or may not be blue.
11. $a > 2$ 13. $m < 0$

Pages 15–16 Lesson 1.4
Exploratory 1. q 3. m 5. no conclusion
7. r 9. Law of disjunctive inference
Written 1. Given; Given; $\sim A$; B, Law of detachment, 1, 3 3. Given; Given; Law of detachment, 1, 2; Definition of conjunction, 3; Given; Law of disjunctive inference, 4, 5

9.
STATEMENTS	REASONS
1. $Q \lor R$	1. Given
2. $\sim R$	2. Given
3. Q	3. Law of disjunctive inference
4. $Q \to T$	4. Given
5. T	5. Law of detachment

15. No conclusion. Although Crunchy-Wunchies are made from oats, they may or may not be healthful.

Pages 18–19 Lesson 1.5
Exploratory 1. $w \to r, \sim r \to \sim w, \sim w \to \sim r$
3. $\sim t \to \sim s, s \to t, t \to s$ 5. If $\angle A \cong \angle B$, then $\angle A$ and $\angle B$ are both right angles. If $\angle A$ and $\angle B$ are not right angles, then $\angle A \not\cong \angle B$. If $\angle A \cong \angle B$, then $\angle A$ and $\angle B$ are not right angles. 7. not equivalent 9. not equivalent
Written 3. $A \to N; \sim N \to \sim A$; Given; Law of detachment, 2, 3; $\sim A \to W$; Law of detachment, 4, 5

5.
STATEMENTS	REASONS
1. $\sim C \lor \sim F$	1. Given
2. C	2. Given
3. $\sim F$	3. Law of disjunctive inference
4. $\sim B \to F$	4. Given
5. $\sim F \to B$	5. Law of the contrapositive
6. B	6. Law of detachment

Pages 21–23 Lesson 1.6
Exploratory 1. $p \to r$ 3. $p \to m$ 5. If Flora is good at chess, then she gets good grades.
7. His feet are tired. Law of detachment.
Mixed Review 1. false 2. true 3. true
4. true 5. true 6. false 7. yes 8. no
9. yes 10. yes

Pages 26–27   Lesson 1.7
Exploratory   1. $a \neq b$ and $c \neq d$ **3.** $c \geq d$ or $d \geq e$ **5.** m is not positive and $m \neq -1$ **7.** Binky is not good at Latin or Mary is not good at French.   **9.** Otto is not going to be class treasurer and Gwen is not.
Written   1. $\sim r \vee \sim s$ **3.** $p \vee q$ **5.** $\sim r \vee t$ **7.** $(\sim r \wedge s) \vee \sim t$ **9.** $(p \wedge \sim q) \vee (\sim t \wedge s)$ **15.** Given; DeMorgan Laws, 1; Given; Law of disjunctive inference, 2, 3; Given; Given; Law of syllogism, 5, 6; Law of detachment, 4, 7

17.

STATEMENTS	REASONS
1. $\sim(\alpha \wedge \beta)$	1. Given
2. $\sim\alpha \vee \sim\beta$	2. DeMorgan Laws
3. α	3. Given
4. β	4. Law of disjunctive inference

23. If $r \neq 0$ and $r \leq 4$, then $r \geq n$.   **25.** If Ted is not wearing his tweeds, then Marsha is not going to the Big Dance or Sally is.
27. $(c \vee \sim d) \rightarrow (\sim t \wedge w)$

Page 32   Lesson 1.8
Exploratory   1. not quantified   **3.** universally quantified, all   **5.** not quantified   **7.** existentially quantified, at least one
Written   1. Some whole numbers are less than zero. false   **3.** All rational numbers are integers. false   **5.** Some integers are composite. true   **7.** $\exists x \in Z (x^2 > 0)$, true   **9.** $\exists x \in Z$ (x is even), true   **11.** All multiples of 3 are rational. true   **13.** All multiples of 6 are multiples of 3. true   **15.** Some multiples of 6 are multiples of 3 and multiples of 4. true

Page 35   Lesson 1.9
Exploratory   1. c, This choice states that there is at least one x that is not green.   **3.** No students take Latin.   **5.** Some birds do not fly.   **7.** No history teacher plays the flute.
Written   1. Some daffodils are not yellow. all, some   **3.** No Canadians were born in Kansas. at least one, no   **5.** All American citizens were born in the United States. some, all   **7.** Some rhombuses are not parallelograms. every, some   **9.** All integers are even. some, all   **11.** There does not exist an x in Q such that $x^2 < 1$. there exists, there does not exist   **13.** Some roses are not red or violets are not blue.   **15.** All elves were not jolly or some trolls were not mean.   **17.** Some persons in this class are not tall or smart.   **19.** $\exists x \in T$ (x is not rational)   **21.** $\forall z \in Z$, (z is not even)   **23.** $\exists x \in S$ ($x \notin T$)   **25.** $\exists x \in S$ ($x \notin F$)

Page 37   Problem Solving Application
1. Ed and Cari, Fred and Teri, Ted and Mary   **3.** Angie: carpenter; Brenda: cab driver; Carol: lawyer; Darlene: cook   **5.** Marty Hood, red; Morey Frame, blue; Mervin Enjin, black; Murray Tyre, tan; Mike Chassie, white

Pages 40–41   Chapter Review
1. 2 is a prime number and Van Gogh painted the Mona Lisa. false   **3.** A square is a parallelogram or 2 is not a prime number. true   **9.** {5}   **11.** {−18}   **13.** {. . . , −5, −4, −3}   **15.** If a square is not a parallelogram, then 2 is a prime number. true   **17.** 2 is a prime number if and only if Van Gogh did not paint the Mona Lisa. true   **27.** $t \rightarrow s$, $\sim s \rightarrow \sim t$, $\sim t \rightarrow \sim s$ **29.** $\sim B \rightarrow \sim A$, $A \rightarrow B$, $B \rightarrow A$

31.

STATEMENTS	REASONS
1. $S \rightarrow \sim T$	1. Given
2. $T \rightarrow \sim S$	2. Law of the contrapositive
3. T	3. Given
4. $\sim S$	4. Law of detachment

39. $a \vee b$ **41.** $\sim x \wedge (y \vee z)$ **45.** For every integer y, $y \neq -5$. false   **47.** For every whole number a, a is an integer. true   **49.** $\forall t \in Q (t + 1 > t)$, true   **51.** $\forall x \in W$ (x is prime), false   **53.** Some choir members play the bassoon. none, some   **55.** No teachers have red hair. at least one, no

CHAPTER 2   MATHEMATICAL SYSTEMS

Page 47   Lesson 2.1
Exploratory   1. 9:00   **3.** 8:00   **5.** 6:00   **7.** 4:00   **9.** 1:00   **11.** 5:00   **Written   3.** 2   **5.** 0   **7.** 3   **9.** 4   **11.** 2   **13.** 3   **15.** 4   **17.** 2   **19.** 0   **21.** 1   **23.** 4   **25.** 3   **27.** 6   **29.** 7   **31.** 1   **33.** 0   **35.** 2   **37.** 4   **39.** 6   **41.** 6   **43.** 5   **45.** 4   **47.** 0   **49.** 7   **51.** 5   **53.** 7   **55.** 8   **57.** 5

Pages 50–51   Lesson 2.2
Exploratory   1. Begin at zero and count around the clock by fives three times.   **3.** For clock system n, find the "ordinary" product of the numbers. Then divide the number by n. The remainder is the answer in clock n.   **5.** Find two numbers with an "ordinary" product that is one greater than a multiple of 50.   **7.** 2   **9.** 1   **Written   1.** 4   **3.** 4   **5.** 2   **7.** 0   **9.** 0   **11.** 3   **13.** {1, 3, 5}   **15.** ∅   **17.** {0, 2, 4}   **19.** {2}   **21.** {0, 1, 3, 4}   **23.** 4   **25.** 6   **27.** 3   **29.** 2   **31.** Multiplication is commutative in clock systems because it is based on "ordinary" multiplication which is commutative.   **33.** 38   **35.** 49   **37.** Wednesday   **39.** Saturday   **41.** January   **43.** October   **Mixed Review   1.** {6}   **2.** {2}   **3.** {8}   **4.** {0}   **5.** {2}   **6.** {1, 3}   **7.** {6}   **8.** {10}   **9.** If $n < 3$ or $n \geq 5$, then $n \neq 3$ and $n \neq 4$.   **10.** If no integers are even, then some squares are not rectangles.   **11.** no   **12.** yes   **13.** $z = 5$ **14.** $a \geq 2$

Pages 54–55   Lesson 2.3
Exploratory   1. $10 = 7 + 3$ **3.** $3x = 2x + x$ **5.** 7   **7.** 7   **9.** $12 = 4 \times 3$ **11.** $8y = 4 \times 2y$ **13.** 2   **15.** meaningless   **Written   3.** 0   **5.** 2   **7.** 9   **9.** 6   **11.** 6   **13.** {4}   **15.** {2}   **17.** {0}   **19.** {10}   **21.** 4   **23.** 4   **25.** 3   **27.** meaningless   **29.** 2, 5   **31.** 0, 2, 4   **33.** 4

Pages 59–60   Lesson 2.4
Exploratory   1. the integers without 0   **3.** the rationals without 1, −1   **5.** the whole numbers greater than 5   **7.** {1, 3, 5}   **9.** yes   **Written   1.** yes, both   **3.** yes, neither   **5.** yes, both   **7.** yes, neither   **9.** yes, both   **11.** $4\frac{1}{2}$ **13.** $\frac{3}{8}$ **15.** 2.8   **17.** yes   **19.** yes   **21.** yes   **23.** 0   **25.** $\frac{7\pi}{24}$ **27.** π   **29.** 6.93   **31.** 8.73   **33.** 3   **35.** 59   **37.** 48   **39.** −1   **41.** $\frac{9}{16}$ **43.** 24   **45.** 2   **47.** 50   **49.** yes   **51.** no   **53.** yes   **55.** yes   **57.** no   **59.** no

Pages 63–64   Lesson 2.5
Exploratory   1. −7   **3.** $-\frac{3}{4}$ **5.** $-2 + \sqrt{11}$ **7.** $-\frac{2}{5}$ **9.** $-\frac{1}{101}$ **Written   1.** 0   **3.** 1   **5.** 3   **7.** 0   **9.** 4   **11.** 6   **13.** 5   **15.** 3   **17.** 5   **19.** 1   **21.** does not exist   **23.** 7   **25.** ⌒   **27.** ⌒   **29.** ☉   **31.** ∿   **33.** −3   **35.** $-\frac{1}{4}$ **37.** 0   **39.** −8   **41.** $\frac{1}{32}$ **43.** does not exist   **Mixed Review   3.** 1   **4.** 5   **5.** −11   **6.** −16   **7.** 24   **8.** −20   **9.** 4   **10.** 2   **11.** −1   **12.** 1   **13.** −3   **14.** −5   **15.** yes   **16.** yes   **17.** yes   **18.** yes   **19.** max, ?   **20.** {2}   **21.** {1}   **22.** {6}   **23.** {0, 1}   **24.** {0}   **25.** ∅

Page 67   Lesson 2.6
Exploratory   1. yes   **3.** No, zero has no inverse.   **Written   1.** No, zero has no inverse.   **3.** yes   **5.** No, zero has no inverse.   **7.** No, subtraction is not associative.   **9.** no, not an operational system   **11.** no, not an operational system   **13.** yes   **15.** No, 2, 3, and 4 have no inverse.   **17.** No, 2, 4, and 6 have no inverse.   **19.** No, 3 and 6 have no inverse. Only clock n systems in which n is a prime number are groups.   **21.** yes   **23.** yes   **25.** No, θ has no inverse.

Pages 69–70   Lesson 2.7
Exploratory   1. When two rational numbers are added, the sum is one and only one rational number.   **3.** yes, 0   **5.** $(Q, +)$ is a group because it is an operational system, addition is associative, there is an identity element, and every element has an inverse.   **Written   1.** no   **3.** no   **5.** no   **7.** no   **9.** no   **11.** yes   **13.** no   **15.** no   **17.** yes   **19.** $2m + 2n + 2(m + n)$ **21.** $(E, +)$ is an operational system, is associative, and has an identity. To show that an inverse for $2m$ exists, let i be the inverse, and by definition, $2m + i = 0$. Therefore, $i = -2m$. Because the inverse is an even number, $(E, +)$ is a group.   **23.** Let $3m$ and $3n$ be any two multiples of 3. The sum, $3m + 3n$, is also a multiple of 3 because $3m + 3n = 3(m + n)$.   **27.** multiples of 28, yes

Pages 73–75   Lesson 2.8
Exploratory   1. The group may not be commutative.   **3.** No, zero has no inverse.   **5.** no   **7.** Yes, if an element of the system appears more than once in any row or column, the Cancellation Law will not hold.   **9.** {6}

Written 1. $\{-0.091\}$ 3. $\{-1\frac{1}{10}\}$ 5. $\{90\}$
7. (G, *) is a commutative group.   9. Definition of identity   11. Associativity of * in a group   13. Right Operation   15. Right Operation   17. Associativity, Definition of inverse, Definition of identity   19. (G, *) is a commutative group.

Page 78   Lesson 2.9
Exploratory 1. $\begin{pmatrix} 1 & 2 & 3 \\ 2 & 1 & 3 \end{pmatrix}$, t
3. $\begin{pmatrix} 1 & 2 & 3 \\ 1 & 2 & 3 \end{pmatrix}$, e 5. $\begin{pmatrix} 1 & 2 & 3 \\ 1 & 3 & 2 \end{pmatrix} \circ \begin{pmatrix} 1 & 2 & 3 \\ 1 & 3 & 2 \end{pmatrix}$
Written 1. $\begin{pmatrix} 1 & 2 & 3 \\ 3 & 1 & 2 \end{pmatrix}$ 3. $\begin{pmatrix} 1 & 2 & 3 \\ 1 & 2 & 3 \end{pmatrix}$
5. $\begin{pmatrix} 1 & 2 & 3 \\ 1 & 2 & 3 \end{pmatrix}$ 7. $\begin{pmatrix} 1 & 2 & 3 \\ 1 & 2 & 3 \end{pmatrix}$ 9. $\begin{pmatrix} 1 & 2 & 3 \\ 3 & 1 & 2 \end{pmatrix}$
11. $\begin{pmatrix} 1 & 2 & 3 \\ 1 & 3 & 2 \end{pmatrix}$

Page 80   Problem Solving Application
1. $1063   3. 24 dimes and 14 quarters   5. 118   7. 12 glazed at 33¢, 3 plain at 23¢; 6 glazed at 27¢, 9 plain at 37¢   9. 853

Page 83   Chapter Review
1. 3   3. 0   5. $\{4\}$ 7. $\{5\}$ 9. 6   11. 6
13. $\{2\}$ 15. $\{4\}$ 17. 5   19. 3   21. $\{2\}$
23. $\{3\}$ 25. yes, both   27. yes, both   29. no
31. does not exist   33. 1   35. yes   37. no, not an operational system   39. no, no identity element   41. no, not an operational system
43. $\{-5.74\}$ 45. $\{\frac{8}{9}\}$ 47. Left Operation
49. Right Cancellation   51. $\begin{pmatrix} 1 & 2 & 3 \\ 1 & 2 & 3 \end{pmatrix}$

CHAPTER 3   A SYSTEM WITH TWO OPERATIONS

Page 87   Lesson 3.1
Exploratory 1. No, there are no inverses.
3. yes   5. no, zero has no inverse   7. 3
9. 127   11. $-\frac{1}{4}$ 13. 0.999   15. -27 17. $\frac{2}{3}$
19. 160   21. $\frac{5}{3}$ 23. $\frac{10}{7}$ **Written** 1. 2   3. 1
5. 3   7. 9   9. 4   11. 0, 6   13. 1   15. 9
17. 1   19. 9   21. 3   23. 10   25. 1   27. $-\frac{2}{3}$
29. $-\frac{1}{2}$ 31. 5   33. $\frac{5}{16}$ 35. 12   37. 6

Pages 90–91   Lesson 3.2
Exploratory 5. $3z^2 + 3b$ 7. $-5a - 5b$
9. $-5a^2 - 10b$ 11. $7x^2 + 7x$
Written 1. no, yes, no   7. $x = 2$
9. $x = -\frac{4}{3}$ 11. $\frac{5}{8}$ 13. 29   15. 50   17. -16
19. 4   21. 3.5

Pages 94–95   Lesson 3.3
Exploratory 1. no, only one operation   3. no, no multiplicative inverse   5. yes   7. yes
Written 1. $x = \frac{1}{2}$ 3. $x = -12$ 5. $\alpha = 3$
7. $y = \frac{9}{4}$ 9. $q = -0.2125$ 11. $x = \frac{1}{2}, -\frac{1}{2}$
13. ∅   15. ∅   17. $x = 0$ 19. $x = 1$ 21. $B = 21\sqrt{5}$ 23. $x = \sqrt{2}$ 25. $x = \pm\sqrt{13}$
27. $x = \pm\sqrt{2}$ 29. $x = \frac{2}{3}, -\frac{2}{3}$ **Mixed Review**
1. $x = -1$ 2. $x = 5\sqrt{3}$ 3. $x = -\frac{5}{3}$
4. $x = -\frac{3}{13}$ 5. $x = -\frac{1}{3}$ 6. $x = 3\frac{19}{33}$
7. $x = -\frac{2\sqrt{6}}{9}$ 8. $x = -10.125$ 9. $x \neq 5$ and $y \neq 7$ 10. n is composite, and n is odd or n is not divisible by 5.   11. If a is odd and $b = 0$, then a is divisible by 3.   12. D   13. A
14. C   15. F   16. D   17. F   18. E   19. F
20. D   21. D   22. A   23. D   24. no   25. yes

Page 99   Lesson 3.4
Exploratory 3. no   5. yes   7. no
Written 1. Multiply both sides by $\frac{1}{3}$. In Z_5, $\frac{1}{3} = 2$. Multiplying in Z_5, $2 \cdot 2 = 4$.   3. Add (-8) to both sides. Associative rule; Inverse; Identity; In Z_{13}, $3 + 5 = 8$.   5. $x = 3$
7. $x = 4$ 9. $x = 5$ 11. $x = 1$ 13. $x = 3$
15. Z_8 is not a field.

Page 103   Lesson 3.5
Exploratory 1. 0   3. 0   5. 1   7. 9   9. 2
Written 1. $a + b$ 3. $a + (b + c)$
5. $(b + b) + b$ 7. $a \cdot b$ 9. $a \cdot (b \cdot c)$

11. $c^{-1} \cdot c^{-1}$ for $c \neq 0$ **13.** Given; Right Operation; Associativity; Definition of additive inverse; Definition of additive identity

Page 106   Lesson 3.6
Exploratory   7. $x + 1 = 0$ or $x - 9 = 0$
9. $x - 4 = 0$ or $x + 1 = 0$
Written   1. $\{-2, 9\}$ **3.** $\{-5, -6\}$
5. $\{-9, -10\}$ **7.** $\{5, -5\}$ **9.** $\{3, -16\}$
11. $\left\{\frac{2}{9}, -\frac{2}{9}\right\}$ **13.** $\{4\}$ **15.** $\{1, 8\}$ **17.** $\{3, 4\}$
19. Z_8 is not a field.   **21.** Z_{15} is not a field.

Pages 109–110   Lesson 3.7
Exploratory   1. $7 + (-2)$ **3.** $9 + 3$
5. $\pi r^2 + (-1)$ **7.** $x^2 \cdot \frac{1}{y}$ **9.** $x^3 \cdot \frac{1}{x}$ **11.** $6^{-1} \cdot 7$
Written   1. 1   **3.** 2   **5.** 3   **7.** 6   **9.** 2   **11.** 5
13. $2\sqrt{3}$ **15.** $\sqrt{3}$ **17.** $2\sqrt{3}$ **19.** $\frac{\sqrt{13}}{2}$
21. $x = \frac{9}{5}$ **23.** $x = \frac{32}{225}$ **25.** $x = -3.25$
27. $x = \frac{3\sqrt{6}}{7}$ **29.** Definition of additive inverse; Field Combination Theorem 2; Definition of additive inverse; Distributive law in $(F, +, \cdot)$; Right Cancellation
Mixed Review   1. b   **2.** d   **3.** c   **4.** a   **5.** b
6. c   **7.** b   **8.** c

Pages 113–114   Lesson 3.8
Exploratory   1. $-3, -2, 0$ **3.** $-\frac{1}{2}, \frac{1}{2}, 0.\overline{5}$
5. $-\frac{31}{10}, -3, -(1.6)^2$ **7.** $0.002, \frac{5{,}002}{10{,}000}, \frac{502}{1{,}000}$
9. $\frac{2}{7}, \frac{13}{41}, \frac{1}{3}$ **11.** a field with an order relationship   **13.** If $a < b$, $a + c < b + c$.
15. If $a < b$ and $0 < c$, then $ac < bc$.
Written   1. Additive Property; Associative law; Definition of additive inverse; Definition of additive identity; Left Operation; Associative law; Definition of multiplicative inverse; Definition of multiplicative identity   **3.** $\sqrt{2}$, $1.\overline{4}$, 1.45 **5.** $\sqrt{2} + 1$, 3.1, π **7.** $\sqrt{5}$, 2.25, $2.\overline{5}$ **9.** $x < \frac{3}{2}$ **11.** $a > -\frac{1}{5}$ **13.** $c \geq d$ **15.** If all of the ordered field requirements are not satisfied, then F is not an ordered field.
17. $z < -8$ **19.** $k < -4$ **21.** $k < -1$

23. $x > \frac{4}{13}$
25. Given: $a < b$ Prove: $0 < b - a$

STATEMENTS	REASONS
1. $a < b$	1. Given
2. $a + (-a) < b + (-a)$	2. Additive Property
3. $0 < b + (-a)$	3. Definition of inverse
4. $0 < b - a$	4. Definition of subtraction

29. either $2 < 4$, $2 = 4$, or $2 > 4$ **31.** $2 < 1$
33. The field $(Z_7, +, \cdot)$ is not ordered.

Page 116   Problem Solving Application
Exploratory   1. $\{7\}$ **3.** $\left\{\frac{55}{2}\right\}$ **Written   1.** 5 cows   **3.** 13.5 days   **5.** about 20 seasons
7. 9   **9.** There are 99 such numbers.
11. 77 coins

Page 119   Chapter Review
1. yes   **3.** No, zero has no inverse.
5. $3x + 12$ **7.** $-3x + 12$ **9.** no   **11.** no
13. 6   **15.** 3   **17.** Left Operation; Associative law; Inverse; Identity   **19.** $a \cdot (b \cdot c)$
21. $(a \cdot b) \cdot c$ **23.** $\{-10, -6\}$ **25.** $\{7, -6\}$
27. $xy^4 + (-x^2y^2)$ **29.** $7 \cdot \frac{1}{4}$ **31.** $6 \cdot 2$
33. $x < -3$ **35.** $\frac{1}{8}$, 0.1251, $\frac{11}{79}$

CHAPTER 4   INTRODUCTION TO EUCLIDEAN GEOMETRY

Page 124   Lesson 4.1
Exploratory   1. It is almost impossible to define the word "point" without using the word "point."   **3.** A plane can be thought of as a flat surface that extends forever in all directions.
5. true; definition of collinear   **7.** true; definition of coplanar   **9.** true; B lies on $\overline{AD}$.
11. false; C is not between A and B.   **13.** false; $\overline{EC} \cap \overline{CF} = $ point C **15.** false; E is not between C and F.   **17.** true; definition of adjacent angles   **Written   1.** J   **9.** G   **11.** G, F, N   **13.** N, F, G, B, H   **15.** G, B, H, C, J, D, K   **17.** M, L, E, K, D, J   **19.** N, M, L, E, K

Page 127   Lesson 4.2
Exploratory   1. 1, Axiom 4−3   **3.** 1, Axiom 4−2   **5.** no, Axiom 4−5   **7.** T is the midpoint of $\overline{RS}$, definition of bisector   **Written**
1. always; Axiom 4−2   **3.** always; Axiom 4−2   **5.** always; Axiom 4−4   **7.** never; If the points are collinear, they are in more than one plane. If the points are noncollinear, they are in one and only one plane.   **9.** 4   **11.** 2
13. No, because if A is between B and C, AB can still be equal to $\frac{1}{2}$ AC.   **15.** No, because $\overline{AB}$ can intersect $\overline{CD}$ at any point.   **19.** none
21. E

Pages 130–131   Lesson 4.3
Exploratory   1. true; indicated in figure
3. false; $\angle ABE \not\equiv \angle EBD$ **5.** true; definition of congruent angles   **Written   9.** 46, 44   **11.** 42, 48   **13.** 77, 103   **15.** 32   **17.** 43   **19.** x = 12
21. $12 < x < 24\frac{6}{7}$ **23.** x = 5   **25.** x = 20
27. x = 7   **29.** 139°   **31.** 37°   **33.** x = 31°
35. x = 17°   **Mixed Review   1.** x = −1
2. x = −27   **3.** x = −1   **4.** x = $\frac{1}{39}$
5. x = 0.6   **6.** x = 0.25   **7.** x = 6 or x = −1
8. x = $-\frac{5}{2}$ or x = 2   **9.** S or T   **10.** T   **11.** P
12. $\overleftrightarrow{PQ}$ **13.** There are none.   **14.** Yes; Axiom 4−3   **15.** x = 17

Pages 133–134   Lesson 4.4
Exploratory   1. subtraction   **3.** transitive
5. reflexive   **7.** substitution   **9.** symmetric
11. multiplication   **13.** Division by 0 is undefined.   **Written   1.** Segment addition axiom   **3.** Addition axiom   **5.** Definition of midpoint   **7.** Definition of complementary
9. B is the midpoint of $\overline{AC}$.; Given, Definition of Midpoint, Segment Addition Axiom, Substitution, Multiplication Axiom   **11.** Given, Distributive Law, Subtraction Axiom, Division Axiom

Pages 136–138   Lesson 4.5
Exploratory   1. Theorem 4−5   **3.** Theorem 4−6   **5.** 90   **7.** 27   **9.** 180   **11.** If two angles are vertical, then they are congruent.   **13.** If two angles are adjacent, then they have a common side.   **Written   1.** 38   **3.** 91   **5.** 110

7. 70   **9.** 110   **11.** Given, Definition of supplementary, Substitution, Given, Definition of congruent angles, Subtraction axiom, Definition of congruent angles   **13.** If the antecedent is assumed to be true and the consequent is false, then the conditional statement is false.

17. a.

STATEMENTS	REASONS
1. $AB = AB$	1. Reflexive axiom
2. $\overline{AB} \cong \overline{AB}$	2. Definition of congruent segments

b.

STATEMENTS	REASONS
1. $\overline{AB} \cong \overline{CD}$	1. Given
2. $AB = CD$	2. Definition of congruent segments
3. $CD = AB$	3. Symmetric axiom
4. $\overline{CD} \cong \overline{AB}$	4. Definition of congruent segments

c.

STATEMENTS	REASONS
1. $\overline{AB} \cong \overline{CD}$, $\overline{CD} \cong \overline{EF}$	1. Given
2. $AB = CD$, $CD = EF$	2. Definition of congruent segments
3. $AB = EF$	3. Transitive axiom
4. $\overline{AB} \cong \overline{EF}$	4. Definition of congruent segments

Pages 140–142   Lesson 4.6
Exploratory   1. 136   **3.** 50   **5.** true   **7.** true
9. false   **11.** false   **13.** true
Written   9. Supplement axiom, Given, Theorem 4−8, Theorem 4−9

Pages 145–146   Lesson 4.7
Exploratory   5. $\angle 2, \angle 8; \angle 4, \angle 5$ **7.** $\angle 1, \angle 5; \angle 2, \angle 6; \angle 3, \angle 8; \angle 4, \angle 7$ **9.** Axiom 4−15   **11.** Theorem 4−11   **13.** Theorem 4−4, Axiom 4−15   **15.** Axiom 4−15
17. Theorem 4−4, Axiom 4−15   **Written**
1. Axiom 4−14   **3.** Theorem 4−10   **5.** n ∥ p; Axiom 4−15   **7.** n ∥ p; Axiom 4−15   **9.** none
11. none   **13.** Given, Definition of perpendicular lines, Theorem 4−7, Axiom 4−15

Pages 148–150   Lesson 4.8
Exploratory   1. $m\angle 2 = 130, m\angle 3 = 130, m\angle 4 = 50, m\angle 5 = 130, m\angle 6 = 50, m\angle 7 =$

50, m∠8 = 130   **3.** m∠1 = 60, m∠2 = 120, m∠4 = 60, m∠5 = 120, m∠6 = 60, m∠7 = 60, m∠8 = 120   **5.** m∠1 = 35, m∠2 = 145, m∠3 = 145, m∠5 = 145, m∠6 = 35, m∠7 = 35, m∠8 = 145   **7.** m∠1 = 156, m∠2 = 24, m∠3 = 24, m∠4 = 156, m∠5 = 24, m∠6 = 156, m∠7 = 156, m∠8 = 24   **Written 1.** n ∥ k   **3.** ℓ ∥ m   **5.** ℓ ∥ m   **7.** n ∥ k   **9.** ∠1 ≅ ∠2; Theorem 4–12   **11.** ∠1 ≇ ∠2; They are not alternate interior angles.   **13.** Theorem 4–13   **15.** Axiom 4–14   **17.** ∠4 + ∠3, ∠6   **19.** 60

25.

STATEMENTS	REASONS
1. ∠3 ≅ ∠4, n ∥ k	1. Given
2. ∠2 ≅ ∠3	2. Theorem 4–12
3. ∠2 ≅ ∠4	3. Theorem 4–6

Mixed Review   1. x < 6   **2.** a > –3   **3.** n < 6   **4.** t > 3   **5.** 71, 71   **6.** 81, 81   **7.** 51, 129   **8.** 16, 164   **9.** 53, 53   **10.** 61, 29

Page 153   Lesson 4.9
Exploratory   1. x ≠ y   **3.** $\overline{AB}$, ⊥ $\overline{CD}$ **5.** a = b   **7.** Jim is not taller than Joe.   **9.** Angle X is not an acute angle.   **Written 1.** Assume ∠1 ≅ ∠2. Therefore, by Theorem 4–13, ℓ ∥ m. However, this contradicts the given statement that ℓ ∦ m. Therefore, if ∠1 ≇ ∠2, then ℓ ∦ m.   **3.** Assume n ⊥ m. We are given that n ⊥ ℓ. Therefore, by Theorem 4–11, ℓ ∥ m. But, it is given that ℓ ∥ m. Therefore, if ℓ ∥ m and n ⊥ ℓ, then n is not perpendicular to m.   **5.** one   **9.** Theorem 4–16   **11.** Theorem 4–16 with V on same side of $\overline{PQ}$ as R

Pages 155–156   Lesson 4.10
Exploratory   1. never; An acute triangle always has 3 acute angles.   **3.** never; Theorem 4–19   **5.** sometimes; The two remote interior angles can have a sum less than 90.   **Written 1.** m∠B = 50   **3.** m∠C = 66, m∠A = 76, m∠B = 38   **5.** 64   **7.** m∠ABD = 25, m∠CBD = 52   **9.** 96   **11.** 50

Page 159   Lesson 4.11
Exploratory   1. yes; definition of convex polygon   **3.** yes; definition of convex polygon   **5.** no; definition of convex polygon   **7.** ∠DEA, ∠EAB   **9.** The measures are equal because the angles are vertical.   **Written 1.** 1080   **3.** 3240   **5.** 7740   **7.** 108   **9.** 144   **11.** 8   **13.** 10   **15.** 9   **17.** 90   **19.** 24   **21.** no; Solve $\frac{180(n-2)}{n}$ = 50. Since n is not a whole number, 50 cannot be the measure of an interior angle.   **23.** yes; 6 sides   **25.** no; Solve $\frac{180(n-2)}{n}$ = 181. n is a negative number

Page 161   Problem Solving Application
1. 8.5, 16.5   **3.** 43   **5.** 22.5   **7.** 5, 6, 7   **9.** 88%   **11.** 5 pounds of apples, 10 pounds of bananas   **13.** Theo: 20, Tina: 16   **15.** passenger: 63 mph, express: 78 mph

Page 165   Chapter Review
1. always; meaning of between   **3.** never; The rays overlap.   **5.** never; Axioms are accepted as true.   **7.** sometimes; if B is between A and C   **9.** always; definition of complementary   **11.** never; The bisector of ∠QRS contains R, not Q.   **13.** sometimes; if a ≠ 0   **15.** sometimes; if the angles are formed by intersecting lines   **17.** always; definition of perpendicular lines   **19.** sometimes; They can be skew.

21.

STATEMENTS	REASONS
1. ℓ ∥ m, n ∥ k	1. Given
2. ∠2 ≅ ∠3, ∠3 ≅ ∠5	2. Axiom 4–15
3. ∠2 ≅ ∠5	3. Theorem 4–6

23. Assume n ∥ k. Therefore, by Axiom 4–14, ∠2 ≅ ∠3. By Theorem 4–4, ∠1 ≅ ∠2. Then, by Theorem 4–6, ∠1 ≅ ∠3. However, this contradicts the given statement that ∠1 ≇ ∠3. Thus n ∦ k.   **25.** 32.5   **27.** 7

CHAPTER 5   CONGRUENT TRIANGLES

Pages 169–170   Lesson 5.1
Exploratory   1. Congruent figures have the same shape and the same size.   **3.** Corresponding parts of congruent triangles are congruent.   **5.** $\overline{BC}$ **7.** $\overline{AC}$ **9.** $\overline{AB}$ **11.** $\overline{XW} \cong \overline{NQ}, \overline{WY} \cong \overline{QM}, \overline{XY} \cong \overline{NM}$ **13.** ∠X   **15.** true   **17.** true   **19.** false   **21.** true

23. true **Written 1.** $\angle BAC \cong \angle DAC$, $\angle B \cong \angle D$, $\angle BCA \cong \angle DCA$, $\overline{AB} \cong \overline{AD}$, $\overline{BC} \cong \overline{DC}$, $\overline{AC} \cong \overline{AC}$ **3.** $\angle BAC \cong \angle DCA$, $\angle B \cong \angle D$, $\angle BCA \cong \angle DAC$, $\overline{AB} \cong \overline{CD}$, $\overline{BC} \cong \overline{DA}$, $\overline{AC} \cong \overline{CA}$ **5.** $\angle EDA \cong \angle FBC$, $\angle DEA \cong \angle BFC$, $\angle EAD \cong \angle FCB$, $\overline{DE} \cong \overline{BF}$, $\overline{EA} \cong \overline{FC}$, $\overline{DA} \cong \overline{BC}$ **7.** $\angle CAB \cong \angle DAE$, $\angle ACB \cong \angle ADE$, $\angle CBA \cong \angle DEA$, $\overline{AC} \cong \overline{AD}$, $\overline{CB} \cong \overline{DE}$, $\overline{AB} \cong \overline{AE}$ **9.** They seem to be congruent. **15.** $x = 3$ **17.** $x = 7$ **19.** $x = 8$ **21.** no **23.** yes

Pages 173–174   Lesson 5.2
Exploratory 5. SAS **7.** SAS **9.** yes; ASA **11.** No, sides can be different lengths.
Written 1. $\overline{AB} \cong \overline{DE}$, $\overline{AC} \cong \overline{DF}$ **3.** $\angle B \cong \angle E$, $\angle C \cong \angle F$ **5.** $\angle C \cong \angle F$ **7.** Given, Theorem 4–5, SAS

11.

STATEMENTS	REASONS
1. $\overline{AB} \parallel \overline{CD}$, $\overline{AB} \cong \overline{CD}$	1. Given
2. $\angle DCA \cong \angle BAC$	2. Theorem 4–12
3. $\overline{AC} \cong \overline{CA}$	3. Theorem 4–5
4. $\triangle ABC \cong \triangle CDA$	4. SAS

Pages 176–179   Lesson 5.3
Exploratory 1. $\angle M \cong \angle C$, $\angle L \cong \angle B$, $\angle K \cong \angle A$, $\overline{ML} \cong \overline{CB}$, $\overline{LK} \cong \overline{BA}$, $\overline{KM} \cong \overline{AC}$ **3.** $\angle BAC \cong \angle FEC$, $\angle ABC \cong \angle EFC$, $\angle BCA \cong \angle FCA$, $\overline{AB} \cong \overline{EF}$, $\overline{BC} \cong \overline{FC}$, $\overline{AC} \cong \overline{EC}$ **5.** $\angle BAC \cong \angle LMK$, $\angle ABC \cong \angle MLK$, $\angle BCA \cong \angle LKM$, $\overline{AB} \cong \overline{ML}$, $\overline{BC} \cong \overline{LK}$, $\overline{AC} \cong \overline{MK}$ **7.** $x = 4$, $DE = 12$, $EF = SM = 14$, $FD = MR = 16$ **9.** $z = 6$, $EF = SM = 12$, $FD = MR = 6$, $DE = RS = 11$ **11.** We are given that $\angle QMP \cong \angle RMP$ and $\angle MPQ \cong \angle MPR$. By Theorem 4–5, $\overline{MP} \cong \overline{MP}$. Therefore, by ASA, $\triangle QMP \cong \triangle RMP$. Then by CPCTC, $\overline{PQ} \cong \overline{PR}$.
Written 1. $AB = DE = 6$, $BC = EF = 15$, $CA = FD = 10$ **3.** $AB = DE = 16$, $BC = EF = 14$, $CA = FD = 19$ **5.** Given, Definition of perpendicular lines, Theorem 4–7, Given, Definition of angle bisector

7.

STATEMENTS	REASONS
1. $\overline{ST} \cong \overline{UT}$, $\angle 3 \cong \angle 4$	1. Given
2. $\overline{RT} \cong \overline{RT}$	2. Theorem 4–5
3. $\triangle TSR \cong \triangle TUR$	3. SAS
4. $\angle 1 \cong \angle 2$	4. CPCTC

19. $x = 2$, $y = 3$, $AB = DE = 6$, $BC = EF = 12$ **21.** $x = 7$, $y = 2$, $AB = DE = 32$, $BC = EF = 31$ **Mixed Review 1.** $x = 21$, $m\angle 4 = 58$ **2.** $x = 20$, $m\angle 4 = 77$ **3.** $x = 16$, $m\angle 4 = 85$ **4.** $x = 8$, $m\angle 4 = 69$ **5.** $x = 7$, $m\angle 4 = 72$ **6.** $x = 8$, $m\angle 4 = 71$ **7.** $x = 11$, $m\angle 4 = 72$ **8.** 36 **9.** 116 **10.** 67, 49, 64 **11.** $\overline{AB} \cong \overline{DE}$, $\overline{BC} \cong \overline{EF}$ **12.** $\angle A \cong \angle D$, $\angle C \cong \angle F$ **13.** $\overline{AB} \cong \overline{DE}$, $\overline{BC} \cong \overline{EF}$ **14.** $\overline{AB} \cong \overline{DE}$, $\angle A \cong \angle D$ or $\overline{BC} \cong \overline{EF}$, $\angle C \cong \angle F$

Pages 182–183   Lesson 5.4
Exploratory 1. ASA involves two angles and the included side; AAS involves two angles and a nonincluded side. **5.** No, congruent parts do not correspond. **9.** yes, ASA **11.** yes, AAS
Written 1. yes, AAS **3.** yes, ASA **5.** no, not enough information **7.** The angle is not included between the sides. **9.** yes, AAS **11.** no **13.** yes **15.** no **17.** yes

Pages 186–188   Lesson 5.5
Exploratory 1. $\triangle ABC$, $\triangle ABE$, $\triangle ADC$ **3.** $\triangle DBC$, $\triangle ABC$ **5.** $\overline{AB}$, $\overline{BE}$ **7.** $\overline{EC}$, $\overline{CB}$ **9.** $\triangle BDC \cong \triangle CEB$, $\triangle ABE \cong \triangle ACD$, $\triangle DFB \cong \triangle EFC$ **11.** $\overline{DC} \cong \overline{EB}$, $\overline{CB} \cong \overline{BC}$, $\overline{DB} \cong \overline{EC}$, $\angle CDB \cong \angle BEC$, $\angle DCB \cong \angle EBC$, $\angle CBD \cong \angle BCE$ **13.** We are given that $\triangle ABE \cong \triangle ACD$. Therefore, by CPCTC, $\angle ABE \cong \angle ACD$. We are also given that $\triangle DCB \cong \triangle EBC$. Therefore, by CPCTC, $\angle CDB \cong \angle BEC$ and $\overline{DB} \cong \overline{EC}$. Thus, by ASA, $\triangle DFB \cong \triangle EFC$. **Written 1.** $\triangle MPV$, $\triangle MRV$; $\triangle MPL$, $\triangle MRL$ **3.** $\triangle PVL$, $\triangle RVL$; $\triangle MPL$, $\triangle MRL$ **5.** yes; ASA **7.** yes; SAS

9.

STATEMENTS	REASONS
1. $\overline{BC} \cong \overline{ED}$ ($BC = ED$)	1. Given
2. $BC + CD = CD + ED$	2. Addition axiom
3. $BC + CD = BD$ $CD + ED = EC$	3. Segment addition axiom
4. $BD = EC$ ($\overline{BD} \cong \overline{EC}$)	4. Substitution
5. $\angle A \cong \angle F$, $\angle B \cong \angle E$	5. Given
6. $\triangle ADB \cong \triangle FCE$	6. AAS

Pages 190–191   Lesson 5.6
Exploratory   7. 3 of each   **Written   1.** 65, 65
3. 70, 40   **5.** 96, 96, 60   **7.** 28, 28, 45

9. STATEMENTS	REASONS
1. $\overline{AB} \cong \overline{AC}$, $\angle EDB \cong \angle FDC$, D is the midpoint of $\overline{BC}$.	1. Given
2. $\angle 3 \cong \angle 4$	2. Theorem 5–2
3. $\overline{BD} \cong \overline{CD}$	3. Definition of midpoint
4. $\triangle EBD \cong \triangle FCD$	4. ASA
5. $\overline{EB} \cong \overline{FC}$	5. CPCTC

Pages 193–194   Lesson 5.7
Exploratory   5. yes, HL   **7.** yes, SAS   **9.** no, not enough information   **Written   1.** No, HL is only true for right triangles.

5. STATEMENTS	REASONS
1. $\overline{AB} \cong \overline{AC}$, $\overline{MD} \perp \overline{AB}$, $\overline{ME} \perp \overline{AC}$, M is the midpoint of $\overline{BC}$.	1. Given
2. $\angle BDM$ and $\angle CEM$ are right angles.	2. Definition of perpendicular lines
3. $\angle BDM \cong \angle CEM$	3. Theorem 4–7
4. $\overline{BM} \cong \overline{CM}$	4. Definition of midpoint
5. $\angle B \cong \angle C$	5. Theorem 5–2
6. $\triangle BDM \cong \triangle CEM$	6. AAS
7. $\overline{DM} \cong \overline{EM}$	7. CPCTC

Pages 197–198   Lesson 5.8
Exploratory   1. always; addition property
3. always; addition property   **5.** always; transitive property   **7.** always; addition property   **9.** always; multiplication property
Written   1. $y \geq -10\frac{1}{2}$ **3.** $y < 6\frac{1}{2}$ **5.** $y \geq 1\frac{2}{3}$
9. $AB > CD$; $AB > EF$ **13.** no; $a = a$
17. $\angle T, \angle S$ **19.** $\overline{YZ}, \overline{XY}, \overline{XZ}$ **21.** $\overline{PR}, \overline{PQ}, \overline{QR}$

Page 200   Lesson 5.9
Exploratory   1. Yes, because M, N, and P form a triangle and Theorem 5–7 applies.   **3.** Yes, if M, N, and P are collinear and M is between N and P.   **Written   1.** yes   **3.** no   **5.** no
7. yes   **9.** 1, 7   **11.** $|x - y|$, $x + y$

Pages 202   Problem Solving Application
Exploratory   1. yes   **3.** no   **5.** no   **7.** yes
9. yes   **11.** yes   **13.** yes   **15.** no
Written   1. no   **3.** no   **5.** no   **7.** yes   **9.** B
11. D   **13.** C

Pages 204–205   Chapter Review
1. $\overline{AB} \cong \overline{US}$, $\overline{BC} \cong \overline{ST}$, $\overline{AC} \cong \overline{UT}$, $\angle A \cong \angle U$, $\angle B \cong \angle S$, $\angle C \cong \angle T$

3. STATEMENTS	REASONS
1. L is the midpoint of $\overline{KM}$, $\overline{JK} \perp \overline{KM}$, $\overline{NM} \perp \overline{KM}$	1. Given
2. $\overline{KL} \cong \overline{ML}$	2. Definition of midpoint
3. $\angle 2$ and $\angle 4$ are right angles.	3. Definition of perpendicular lines
4. $\angle 2 \cong \angle 4$	4. Theorem 4–7
5. $\angle KLJ \cong \angle MLN$	5. Theorem 4–4
6. $\triangle KLJ \cong \triangle MLN$	6. ASA

7. yes; by Theorem 4–5, $\overline{AD} \cong \overline{AD}$. Therefore, by SAS, $\triangle ACD \cong \triangle ABD$. Then, by CPCTC, $\overline{AC} \cong \overline{AB}$ (or $AC = AB$).   **13.** always; definition of greater than   **15.** no

CHAPTER 6   APPLICATIONS OF CONGRUENT TRIANGLES

Pages 210–211   Lesson 6.1
Exploratory   1. true; definition of a parallelogram   **3.** true; Theorem 6–1
5. false; not corresponding angles   **7.** true; Theorem 4–15   **9.** $\angle RUT$ **11.** $\overline{UT}$ **13.** $\overline{SW}$
Written   1. 133, 47, 133   **3.** 93, 87, 93
5. 117, 63, 117   **7.** 105, 75   **9.** $81\frac{1}{3}$, $98\frac{2}{3}$
11. 37, 96, 37, 96   **13.** 28, 16, 28, 16
15. 12, 6   **17.** 98, 266   **19.** CPCTC   **21.** false
23. false   **25.** false   **27.** If quadrilateral $ABCD$ is a square, then it is a rectangle. true   **29.** If quadrilateral $ABCD$ is a parallelogram, then it is a trapezoid. false   **31.** If quadrilateral $ABCD$ is a rectangle, then it is a rhombus. false

Pages 216–217   Lesson 6.2
Exploratory   1. $\overline{AB} \parallel \overline{DC}$, $\overline{AD} \parallel \overline{BC}$, $\overline{AB} \cong \overline{DC}$, $\overline{AD} \cong \overline{BC}$, $\angle ADC \cong \angle ABC$, $\angle DAB \cong \angle DCB$,

$\overline{AC}$ bisects $\overline{DB}$, $\overline{DB}$ bisects $\overline{AC}$ **3.** true; definition of a parallelogram   **5.** false; The sides must also be congruent or both pairs of opposite sides must be parallel.   **7.** true; definition of a parallelogram   **9.** true; CPCTC and Theorem 6–6   **11.** false; Both pairs of opposite angles must be congruent.
13. Theorem 6–5   **15.** Since $\angle 1 \cong \angle 2$, $\overline{AB} \parallel \overline{CD}$ by Theorem 4–13. Also, since $\angle 3 \cong \angle 4$, $\overline{AD} \parallel \overline{BC}$ by Theorem 4–13. Thus, by the definition of a parallelogram, $ABCD$ is a parallelogram.   **17.** By CPCTC, $\overline{AD} \cong \overline{BC}$ and $\angle 3 \cong \angle 2$. Since $\angle 3 \cong \angle 2$, $\overline{AD} \parallel \overline{BC}$ by Theorem 4–13. Thus, by Theorem 6–5, $ABCD$ is a parallelogram.   **Written**   **1.** $6\frac{1}{2}$
3. $x = 3$, $y = -4$ **5.** $x = 3$, $y = 17$
7. $x = 1$, $y = 3$ **9.** If a quadrilateral is a parallelogram, then the diagonals of the parallelogram separate it into two congruent triangles. If a diagonal of a quadrilateral separates it into two congruent triangles, then the quadrilateral is a parallelogram. true; CPCTC and Theorem 6–5   **11.** If a quadrilateral is a parallelogram, then the diagonals bisect each other. If the diagonals of a quadrilateral bisect each other, then the quadrilateral is a parallelogram. true; Theorem 6–7   **15.** No; An isosceles trapezoid meets these requirements and is not a parallelogram.

19.

STATEMENTS	REASONS
1. $\angle 2 \cong \angle 3$, $\overline{AD} \cong \overline{CE}$, $\overline{DC} \cong \overline{AB}$	1. Given
2. $\overline{BC} \cong \overline{EC}$	2. Theorem 5–3
3. $\overline{BC} \cong \overline{AD}$	3. Theorem 4–5
4. $ABCD$ is a parallelogram.	4. Theorem 6–6

Pages 220–221   Lesson 6.3
Exploratory   5. true; All 4 of the large triangles formed are congruent to each other. Thus, all 4 angles of the parallelogram are congruent. Therefore, the angles must all be right angles, and the parallelogram is a rectangle.   **7.** true; All 4 of the small triangles formed are congruent by AAS. Therefore, all of the sides are congruent and the parallelogram is a rhombus.   **9.** true; definition of a square

11. false; All rhombuses do not have 4 right angles.   **13.** false; All rhombuses do not have 4 right angles.   **15.** true; All of the angles are right angles, so any two are supplementary.
17. false; This is only true for rectangles.
19. true; Theorem 6–9   **21.** false; This is only true for squares.   **23.** true; Theorem 6–1
25. true; Theorem 6–11   **27.** opposite sides parallel and congruent, opposite angles congruent, diagonals bisect each other, each diagonal separates it into two congruent triangles   **29.** all properties of a parallelogram, four congruent sides, perpendicular diagonals, each diagonal bisects two opposite angles
31. four congruent sides, each diagonal bisects two opposite angles, perpendicular diagonals
Written   1. 21, 21, 21, 21   **3.** 20, 20, 20, 20
5. 32, 32, 32, 32

11.

STATEMENTS	REASONS
1. Rectangle $ABCD$	1. Given
2. $\overline{DC} \cong \overline{AB}$	2. Theorem 6–2
3. $\angle DCE$ and $\angle ABE$ are right angles.	3. Definition of a rectangle
4. $\overline{DE} \cong \overline{AE}$	4. Given
5. $\triangle DCE \cong \triangle ABE$	5. HL
6. $\angle EDC \cong \angle EAB$	6. CPCTC

Pages 224–225   Lesson 6.4
Exploratory   1. $\overline{AB}$, $\overline{DC}$ **3.** $\angle D$, $\angle C$; $\angle A$, $\angle B$ **5.** $\overline{AD}$, $\overline{BC}$ **7.** 120, 120, 60, 60   **9.** 56, 56, 124, 124   **11.** 41, 41, 139, 139   **Written**

1.

STATEMENTS	REASONS
1. $ABCD$, an isosceles trapezoid with $\overline{AB} \parallel \overline{DC}$	1. Given
2. $\overline{AD} \cong \overline{BC}$	2. Definition of an isosceles trapezoid
3. $\angle ADC \cong \angle BCD$	3. Theorem 6–12
4. $\overline{DC} \cong \overline{CD}$	4. Theorem 4–5
5. $\triangle ACD \cong \triangle BDC$	5. SAS

Mixed Review   1. $MN = MP = 29$, $NP = 21$
2. $MN = MP = 26$, $NP = 16$ **3.** $MN = MP = 19$, $NP = 18$ **4.** $MN = MP = 36$, $NP = 24$ **5.** $\angle B \cong \angle D$ **6.** $\overline{BD}$ bisects $\overline{AC}$.
7. $\overline{AB} \cong \overline{CD}$ or $\overline{AD} \parallel \overline{BC}$ **8.** $\overline{AD} \parallel \overline{BC}$ or $\overline{AB} \cong \overline{CD}$ **9.** $\overline{EF}$, $\overline{DE}$, $\overline{DF}$ **10.** $3 < x < 6$

Pages 229–230   Lesson 6.5
Exploratory   1. Rulers and protractors are used in drawings but may not be used in constructions.   **3.** A figure was constructed to fulfill a requirement.

Pages 232–234   Lesson 6.6
Exploratory   1. The bisector that is constructed is also perpendicular to the segment.   **5.** A segment drawn from any vertex to the midpoint of the opposite side. 3; yes   **Written   25.** By construction, $\overline{AF} \perp \overline{BC}$.   **27.** By construction, $\angle RPS \cong \angle 1$. Therefore, $\overleftrightarrow{PR} \parallel \ell$ by Theorem 4–13.

Page 236–237   Lesson 6.7
Exploratory   1. $\overline{DE} \parallel \overline{BC}$ **3.** $\angle ADE \cong \angle ABC$
5. 20   **7.** $6t + 8$ **Written   1.** 18, 36   **3.** 15, 30   **5.** 16, 32   **7.** 4, 8 or 9, 18   **9.** $13\frac{1}{2}$
11. 19   **13.** 29   **15.** 32, 32, 32   **17.** 288, 144
19. 36 cm

Pages 239   Problem Solving Application
1. d   **3.** c   **7.** Richard, Sondra, Tom

Page 241   Chapter Review
1. always   **3.** sometimes   **5.** 100, 80, 100, 80
7. 130, 50, 130, 50   **9.** 116, 32
13. $x = 8, y = 4, z = -14$
15. $m\angle ADC = 117.5, m\angle DAF = 62.5$
19. 8, 16

CHAPTER 7   SIMILARITY

Pages 246–247   Lesson 7.1
Exploratory   1. $\frac{2}{5}$ **3.** $\frac{3}{2}$ **5.** $\frac{3}{5}$ **7.** $\frac{2}{3}$ **9.** $\frac{5}{3}$
11. $\angle A, \angle E; \angle B, \angle F; \angle C, \angle G; \angle D, \angle H$
13. $\overline{AB}, \overline{EF}; \overline{BC}, \overline{FG}; \overline{CD}, \overline{GH}; \overline{DA}, \overline{HE}$ **15.** $2\frac{1}{2}$
17. 2   **Written   1.** $y = \frac{10}{3}; 5, 2; y, 3$
3. $t = \frac{2}{5}; 3, 2; 5, 3t$ **5.** $y = \frac{5}{2}; y + 1, 5; y, 3$
7. $t = -\frac{29}{4}; 3t - 7, 2; 3 + 2t, 5$ **9.** $t = 3$ or 5; 3, 5; $t, 8 - t$ **11.** Yes, because their corresponding angles have equal measures.
13. A, A; D, B; E, C   **15.** $\frac{AD}{AB} = \frac{DE}{BC}; \frac{AD}{AB} = \frac{AE}{AC};$
$\frac{AE}{AC} = \frac{DE}{BC}$ **17.** no   **19.** no   **21.** no   **23.** 40, 50   **25.** 42, 48   **27.** 75, 105   **29.** 126, 54
31. 40, 40, 100   **33.** 36, 54, 90   **35.** 50
37. 6%

Pages 250–251   Lesson 7.2
Exploratory   1. yes; interchanging property
3. no; using the means-extremes property, $xy = 48$ **5.** yes; addition property   **7.** $\frac{3}{5}$ **9.** $\frac{3}{5}$
11. $\frac{p + q + 1}{q + 1}$ **13.** $\frac{MT}{AB}$ **15.** $\frac{7}{2}$ **17.** $\frac{7}{3}$ **19.** $\frac{2}{3}$
Written   1. 6   **3.** 8   **5.** 10   **7.** $2\sqrt{5}$
9. $100\sqrt{2}$ **11.** 6   **13.** $\frac{1}{2}$ **15.** $\sqrt{rs}$ **17.** $5\frac{1}{7}, 6\frac{6}{7}$
19. $\frac{2\sqrt{3}}{5}$ **21.** 10   **23.** 3   **25.** 10   **27.** If $\frac{a}{b} = \frac{c}{d}$, $ad = bc$ by the means-extremes property. By the division axiom, $\frac{ad}{ac} = \frac{bc}{ac}$ and $\frac{d}{c} = \frac{b}{a}$ or $\frac{b}{a} = \frac{d}{c}$.   **29.** If $\frac{a}{b} = \frac{c}{d}$, by the addition axiom, $\frac{a}{b} + \frac{b}{b} = \frac{c}{d} + \frac{d}{d}$. Therefore, $\frac{a + b}{b} = \frac{c + d}{d}$.
31. If b is the mean proportional between a and c, then $\frac{a}{b} = \frac{b}{c}$. By the means-extremes property, $b^2 = ac$ and $b = \sqrt{ac}$.
Mixed Review   1. c   **2.** b   **3.** a   **4.** b   **5.** d
6. d   **7.** a

Pages 254–255   Lesson 7.3
Exploratory   1. yes; AA   **3.** No, there are not 2 pairs of congruent angles.
5. $\triangle ADE \sim \triangle ABC$ by AA since $\angle A \cong \angle A$ and $\angle ADE \cong \angle B$ (or $\angle AED \cong \angle C$).
7. 14   **9.** $16\frac{2}{3}$ **11.** 21   **13.** 2
Written   1. $\triangle BDE \sim \triangle BAC$; AA

Pages 258–260   Lesson 7.4
Exploratory   1. MK; Theorem 7–2   **3.** JK, JN; Theorem 7–2   **5.** 5   **7.** 10   **9.** $12\frac{1}{2}$ **11.** 3
Written   1. QR; Theorem 7–3   **3.** UT − TS; Theorem 7–3   **5.** 18   **7.** 20   **9.** 6   **11.** 7
13. 4.09 m   **15.** 25.2   **17.** 4, 20
29. $52\frac{1}{2}$ **31.** $12\frac{19}{27}$ **33.** $\frac{AB}{AC} = \frac{BD}{DC}$ **35.** $15\frac{1}{5}$
37. $7\frac{3}{5}$

Pages 263–264   Lesson 7.5

Exploratory   **1.** $\triangle DGE \sim \triangle FGD$, $\triangle DGE \sim \triangle FDE$, $\triangle DGF \sim \triangle EDF$ **3.** $\triangle DGE \sim \triangle FGD$: $\overline{DG}, \overline{FG}$; $\overline{GE}, \overline{GD}$; $\overline{DE}, \overline{FD}$; $\triangle DGE \sim \triangle FDE$: $\overline{DG}, \overline{FD}$; $\overline{GE}, \overline{DE}$; $\overline{DE}, \overline{FE}$; $\triangle DGF \sim \triangle EDF$: $\overline{DG}, \overline{ED}$; $\overline{GF}, \overline{DF}$; $\overline{DF}, \overline{EF}$ **5.** 6   **7.** 4   **9.** $3\sqrt{2}$ **11.** EF, EG; Theorem 7–7   **13.** DG; Theorem 7–5   **15.** DF; Theorem 7–7   **17.** $4\sqrt{2}$ **19.** $10\sqrt{2}$ **21.** $7\sqrt{2}$ **23.** $13 + 2\sqrt{2}$ **25.** $-2\sqrt{2}$ **27.** 5   **29.** $2\frac{1}{2}$ **31.** $\frac{1}{2}$

Written   **1.** 9   **3.** 6   **5.** $3\sqrt{15}$ **7.** $20\frac{1}{4}$ **9.** 12   **11.** $4\sqrt{5}$ **13.** 16   **15.** 3, 12   **17.** 7, 28

Pages 267–268   Lesson 7.6

Exploratory   **1.** 13   **3.** 8   **5.** 6   **7.** yes   **9.** yes   **11.** yes   **13.** yes   **Written**   **1.** 29   **3.** $\sqrt{67}$ **5.** $6\sqrt{5}$ **7.** $\sqrt{26.24}$ **9.** $2\sqrt{2}$ **11.** $\sqrt{58.32}$ **13.** 52   **15.** $8\sqrt{10}$ **17.** 9   **19.** $\sqrt{61}$ **21.** $6\sqrt{3}$ **23.** $\sqrt{s^2 - \left(\frac{s}{2}\right)^2}$ or $\frac{s\sqrt{3}}{2}$ **25.** 60   **27.** 26   **29.** 17

Pages 270–271   Lesson 7.7

Exploratory   **1.** $x = 5\sqrt{2}$, $y = 5\sqrt{2}$ **3.** $x = 6\sqrt{2}$, $y = 6$ **5.** $x = 8\sqrt{3}$, $y = 16$ **7.** $x = 2$, $y = 1$ **9.** $x = 12\sqrt{2}$, $y = 12$ **11.** $x = \frac{4\sqrt{3}}{3}$, $y = \frac{8\sqrt{3}}{3}$ **Written**   **1.** $6\sqrt{2}$ cm, 8.5 cm   **3.** $20\sqrt{2}$ ft, 28.3 ft   **5.** $6\sqrt{6}$ m, 14.7 m   **7.** $2\sqrt{3}$ in.   **9.** $7\sqrt{3}$ mm   **11.** $\frac{15\sqrt{3}}{2}$ cm   **13.** 16, $8\sqrt{3}$, 24, $16\sqrt{3}$, 32   **15.** 6, 3, 9, $3\sqrt{3}$, $6\sqrt{3}$ **17.** 5   **19.** $\frac{21}{2}$ cm   **21.** $40\sqrt{3}$ **23.** $39\sqrt{3}$ **25.** 90

Page 274   Lesson 7.8

Exploratory   **1.** Construct $\angle G \cong \angle D$ and $\angle H \cong \angle F$. The triangle formed is similar to $\triangle DEF$.

Pages 276–277   Lesson 7.9

Exploratory   **1.** true   **3.** false   **5.** false   **7.** false   **9.** {(2, 3), (2, 4), (2, 9), (2, 16), (3, 4), (3, 9), (3, 16), (4, 9), (4, 16), (9, 16)}   **11.** {(2, 2), (3, 3), (4, 4), (9, 9), (16, 16)}   **13.** {(4, 2)}   **15.** {(2, 4), (3, 9), (4, 16)}

Written   **1.** $T = \{\triangle ADC, \triangle CBA, \triangle AEB, \triangle BEC, \triangle CED, \triangle DEA, \triangle DAB, \triangle BCD\}$ **3.** $C = \{(\triangle ADC, \triangle ADC), (\triangle ADC, \triangle CBA), (\triangle CBA, \triangle ADC), (\triangle CBA, \triangle CBA), (\triangle AEB, \triangle AEB), (\triangle AEB, \triangle CED), (\triangle CED, \triangle AEB), (\triangle CED, \triangle CED), (\triangle BEC, \triangle BEC), (\triangle BEC, \triangle DEA), (\triangle DEA, \triangle BEC), (\triangle DEA, \triangle DEA), (\triangle DAB, \triangle DAB), (\triangle DAB, \triangle BCD), (\triangle BCD, \triangle DAB), (\triangle BCD, \triangle BCD)\}$ **5.** $P = \{(\overline{AB}, \overline{CD}), (\overline{CD}, \overline{AB}), (\overline{AD}, \overline{BC}), (\overline{BC}, \overline{AD})\}$ **7.** {(J, E), (Z, F), (Z, C)}   **9.** {(M, J), (M, Z), (M, N), (N, J), (N, Z), (F, C), (C, F), (L, B)}   **11.** {(M, E), (M, F), (M, C), (M, B), (M, L), (N, E), (N, F), (N, C)}   **13.** {(E, F), (E, C), (E, B), (E, L), (F, E), (F, B), (F, L), (C, E), (C, B), (C, L), (B, E), (B, F), (B, C), (L, E), (L, F), (L, C)}

Page 280   Lesson 7.10

Exploratory   **1.** true, definition of a relation   **3.** false, (4, 3) is not in R.   **5.** false; (3, 4) is in R, but (4, 3) is not in R.   **7.** false; R is not symmetric or transitive.   **9.** none

Written   **1.** transitive   **3.** reflexive, transitive   **5.** transitive   **7.** yes

Page 282   Problem Solving Application

1. 18 ft by 40 ft   **3.** 19 cm   **5.** 11 in. by 11 in.   **7.** 8 cm, 16 cm, 13 cm   **9.** 7 m   **11.** 60 yd by 80 yd

Page 285   Chapter Review

1. $\frac{32}{11}$ **3.** 9, -1 **5.** 10   **7.** FG, JG **9.** $\sqrt{42}$ **11.** $6\frac{2}{3}$ **13.** 2   **15.** 105.44 yd   **19.** $R = \{(0, 0), (0, 3), (0, 6), (3, 3), (3, 0), (3, 6), (6, 6), (6, 0), (6, 3), (1, 1), (1, 4), (1, 7), (4, 4), (4, 1), (4, 7), (7, 7), (7, 1), (7, 4), (2, 2), (2, 5), (2, 8), (5, 5), (5, 2), (5, 8), (8, 8), (8, 2), (8, 5)\}$ **21.** yes, because *is parallel to* is reflexive, symmetric, and transitive

CHAPTER 8   GEOMETRY WITH COORDINATES

Pages 289–290   Lesson 8.1

Exploratory   **1.** $(-3, 3)$ **3.** $(5, 4)$ **5.** $(0, 1)$

7. $(-3, -2)$ **9.** $(0, -3)$ **11.** 3 **13.** It is zero. **Written** **17.** I **19.** II **21.** 1 **23.** 5 **25.** 11 **27.** 5 **29.** 6 **31.** 16 **33.** 20 **35.** $(5, 5)$ **37.** $(-3, -6)$ **39.** $(5, 2)$

Page 293   Lesson 8.2
Exploratory **1.** yes; $2x + y = 5$ **3.** yes; $2x - y = 0$ **5.** no **7.** no **9.** no **11.** $(3, 2)$, $(-1, 6)$ **13.** $(1, -1)$, $(5, -5)$
Written **19.** $9; -3$ **21.** $\frac{1}{2}; 3$ **23.** $0; 0$
25. 6 **27.** -3 **29.** 3 **31.** 3 **33.** $\{(2, 3)\}$
35. $\{(1, 4)\}$ **37.** $\{(-1, -2)\}$

Pages 295–297   Lesson 8.3
Exploratory **1.** $y = x - 3$ **3.** $y = 2x + \frac{3}{4}$
5. $y = -4x - 2$ **7.** adding **9.** adding
11. subtracting **Written** **1.** $(-1, 8)$
3. $(3, -1)$ **5.** $(14, 4)$ **7.** $(2, 2)$ **9.** $\left(\frac{4}{3}, \frac{4}{3}\right)$
11. $(0, 0)$ **13.** $(8, -1)$ **15.** $(0, 3)$
17. $(6, -2)$ **19.** $(5, -1)$ **21.** $(2, 2)$
23. $\left(\frac{3}{4}, \frac{2}{3}\right)$ **25.** $\left(\frac{1}{2}, \frac{4}{3}\right)$ **27.** $(2.25, 3.6)$
29. $(1.25, 2.75)$ **31.** $(3, -1)$ **33.** $(-1, 5)$
35. $(13, 7)$ **37.** $\left(\frac{7}{3}, -\frac{2}{3}\right)$ **39.** $(1, 4)$
41. $\left(-5, \frac{2}{3}\right)$ **43.** $(32, -58)$ **45.** $(0.5, 4)$
47. 36, 12 **49.** Lindsey is 34 years old, Kyung is 17 years old. **51.** 28 m, 112 m
53. 5 quarters, 9 dimes **55.** 55 mm by 88 mm **57.** $m\angle P = m\angle R = 146$, $m\angle Q = m\angle S = 34$ **Mixed Review**
1. $x = 25, y = 5\sqrt{26}$ **2.** $x = 12, y = 6\sqrt{5}$
3. $x = 15, y = 9$ **4.** $ST = 30$ **5.** $RU = 4$, $RW = 8$ **6.** $RU = 3, RT = 12$ **7.** $64\sqrt{2}$ cm
8. $54\sqrt{3}$ m

Pages 300–301   Lesson 8.4
Exploratory **1.** Form a right triangle with $(2, 1)$ and $(5, 9)$ the endpoints of the hypotenuse and the vertex of the right angle at $(5, 1)$. The legs are 3 units and 8 units in length. Using the Pythagorean Theorem, you find the measure of the segment is $\sqrt{3^2 + 8^2}$ or $\sqrt{73}$. **3.** The vertex of the right angle is $(5, 1)$. The legs measure 4 and 7 and the segment measures $\sqrt{65}$. **5.** 5 **7.** 10 **9.** 10 **11.** 5
Written **1.** $2\sqrt{5}$ **3.** $3\sqrt{2}$ **5.** $\sqrt{34}$ **7.** $5\sqrt{2}$
9. $\sqrt{170}$ **11.** $\frac{1}{4}\sqrt{65}$ **15.** 5, 5, 8 **17.** 6, $\sqrt{34}, \sqrt{34}$ **19.** $2\sqrt{5}, 2\sqrt{10}, 2\sqrt{5}$
21. $2\sqrt{10}, 2\sqrt{5}, 2\sqrt{5}$ **23.** $AB = CD = \sqrt{34}, AD = BC = 7$ **25.** $AB = CD = \sqrt{61}$, $AD = BC = 5$ **27.** $AB = CD = 5, AD = BC = 5$ **29.** $AC = \sqrt{41}, BD = 5\sqrt{5}$
31. $AC = 2\sqrt{26}, BD = 2\sqrt{17}$ **33.** $AC = 6$, $BD = 8$ **35.** yes **37.** no **39.** yes

Pages 304–305   Lesson 8.5
Exploratory **1.** $(2, 3)$ **3.** $(4, 5)$ **5.** $(3, 7)$
7. $(-2, 7)$ **9.** $\left(7\frac{1}{2}, -8\right)$ **11.** $\left(-\frac{1}{2}, -1\right)$
Written **1.** $\left(-4, -3\frac{1}{2}\right)$ **3.** $\left(\frac{1}{2}, \frac{1}{6}\right)$
5. $(1.2, 0.9)$ **7.** $(-\sqrt{2}, 4\sqrt{5})$ **9.** $(0, 0)$
11. $(14, 1)$ **13.** $(4, -5)$ **15.** $(6, 1)$
17. $(-3, 23)$ **19.** $(1, -1)$ **21.** $(-3, 7)$
23. $(2, 3), (4, 1), (6, 4)$ **25.** $\left(\frac{1}{2}, 6\right), \left(1, -\frac{1}{2}\right)$, $\left(2\frac{1}{2}, \frac{1}{2}\right)$ **27.** $\left(2\frac{1}{2}, -3\right), \left(7, -6\frac{1}{2}\right)$, $\left(4\frac{1}{2}, -7\frac{1}{2}\right)$ **29.** yes **31.** no **33.** $(1, 3)$; $(3, 1)$; $DE = 2\sqrt{2}; BC = 4\sqrt{2}; DE = \frac{1}{2}BC$
35. $(-1, 3); (-1, -2); DE = 5; BC = 10$; $DE = \frac{1}{2}BC$ **39.** $(3, 1); \sqrt{10}; 2\sqrt{10}$; The median is half the hypotenuse. **41.** $(3, 0)$; 3; 6; The median is half the hypotenuse.

Pages 308–309   Lesson 8.6
Exploratory **1.** positive **3.** positive
5. negative **7.** zero **9.** $\frac{1}{3}$ **11.** -1 **13.** no slope **Written** **17.** $\frac{1}{5}$ **19.** 0 **21.** $\frac{d-b}{c-a}$
23. -1 **25.** -2 **27.** 1 **29.** no slope
31. no slope **33.** 6 **35.** 8 **37.** -3
39. $k = -4$ **Mixed Review** **1.** $x = 12$, $y = -2$ **2.** $x = \frac{38}{7}, y = \frac{13}{7}$ **3.** $x = 7$,

$y = -5$ **4.** both   **5.** (12, −19)   **6.** (−1, 4)
7. $AC = 8$, $BC = 15$ **8.** 12   **9.** $k = 7$
10. cheeseburger: $1.39, shake: $0.99

Pages 312–313   Lesson 8.7
Exploratory   1. They are the same.   **3.** It is zero.   **5.** neither   **7.** parallel   **9.** parallel
11. parallel   **Written   1.** parallel
3. perpendicular   **5.** neither   **7.** parallel
9. perpendicular   **11.** no   **13.** yes   **15.** no
17. rhombus   **19.** square   **21.** other (rectangle)
23. no   **25.** no   **27.** $\overline{AB}, \frac{1}{3}; \overline{AC}, -\frac{11}{4}; \overline{BC},$
−12   **29.** $\overline{AB}, -\frac{9}{8}; \overline{AC}, -\frac{1}{2}; \overline{BC}, -\frac{4}{3}$
31. $-3, \frac{4}{11}, \frac{1}{12}$ **33.** $\frac{8}{9}, 2, \frac{3}{4}$ **35.** 6   **37.** 1
39. −10   **41.** −5   **43.** no   **45.** yes
47. parallelogram   **49.** square

Pages 317–318   Lesson 8.8
Exploratory   1. $y = -x + 3$ **3.** $y = x + 1$
5. $y = 6$ **7.** $y = \frac{1}{6}x + \frac{1}{3}$ **9.** $y = -3x - \frac{4}{5}$
Written: 1. 1; 2   **3.** −3; 1   **5.** $\frac{3}{5}$; 2   **7.** 1;
−4   **9.** 2; −4   **19.** $y = -2x + 5$
21. $y = \frac{1}{2}x - 3$ **23.** $\frac{1}{2}$, (3, 1)   **25.** $-\frac{3}{5}$,
(−5, −3)   **27.** 1, (−6, −1)   **29.** $y = 2x + 3$
31. $y = -2x + 10$ **33.** $y = -\frac{2}{3}x - \frac{13}{3}$
35. $y = x + 1$ **37.** $y = -\frac{1}{2}x$
39. $y = \frac{1}{3}x + \frac{14}{3}$ **41.** $y = x$ **43.** $y = 1$
45. $y = 0$ **47.** parallel   **49.** perpendicular
51. parallel   **53.** $y = x + 1$
55. $y = -\frac{2}{3}x - \frac{22}{3}$ **57.** $y = -\frac{1}{2}x$
59. $y = -x + 5$ **61.** $y = \frac{3}{2}x + \frac{7}{2}$ **63.** (3, 3);
$y = -x + 6$; $m = -1$; $y = -x + 6$; $y = x$;
(3, 3)   **69.** (4, 2)

Pages 322–323   Lesson 8.9
Exploratory   1. Let A be the point with coordinates (−3, 1), B be the point with coordinates (−1, 5), C be the point with coordinates (3, −2). $AB = CD = 2\sqrt{5}$. $BC = AD = 3\sqrt{5}$. Slope of $\overline{AB}$ = slope of $\overline{CD}$ = 2; Therefore, $\overline{AB} \parallel \overline{CD}$. Slope of $\overline{BC}$ = slope of $\overline{AD} = -\frac{1}{2}$; Therefore, $\overline{BC} \parallel \overline{AD}$ and $\overline{AB} \perp \overline{BC}$, $\overline{BC} \perp \overline{CD}$, $\overline{CD} \perp \overline{AD}$, and $\overline{AD} \perp \overline{AB}$.   **5.** 35
7. 30   **Written   3.** Let A be the point with coordinates (3, −4), B be the point with coordinates (−2, −3), C be the point with coordinates (2, −10), and D be the point with coordinates (7, −11). Slope of $\overline{AB}$ = slope of $\overline{CD} = -\frac{1}{5}$. Slope of $\overline{BC}$ = slope of $\overline{AD} = -\frac{7}{4}$.
7. 24   **9.** 88   **11.** 16   **13.** 18   **15.** 45
17. Let A be the point with coordinates (0, 0), B be the point with coordinates (2, 3), C be the point with coordinates (10, 3), and D be the point with coordinates (11, 0). $BC = 8$; $AD = 11$; slope of $\overline{BC}$ = slope of $\overline{AD} = 0$.
23. 30   **25.** 99   **27.** 13   **29.** 32   **31.** $27\frac{1}{2}$
33. $41\frac{1}{2}$ **35.** $18\frac{1}{2}$ **39.** $\frac{1}{2}$ **41.** 8   **43.** 8
45. 10   **47.** 12   **49.** $37\frac{1}{2}$ **51.** $58\frac{1}{2}$ **53.** 45

Page 325   Problem Solving Application
1. 15°C   **3.** 32°C   **5.** 176°F   **7.** 50°F
9. $F = \frac{9}{5}C + 32$ or $C = \frac{5}{9}(F - 32)$
11. about 5 km   **13.** about 3 seconds
15. $d = \frac{3}{2}t$ or $t = \frac{2}{3}d$ **17.** 156 pounds

Page 327   Chapter Review
11. (4, 2)   **13.** (2, 1)   **15.** $\left(\frac{14}{5}, \frac{4}{5}\right)$ **17.** 5
19. 10   **21.** 17   **23.** (7, 2)   **25.** (1, 6)
27. $\left(-1, -\frac{3}{2}\right)$ **29.** $\frac{3}{2}$ **31.** no slope   **33.** $\frac{2}{3}$
35. neither   **37.** parallel   **39.** $y = 2x - 5$
41. $y = \frac{1}{5}x + \frac{7}{5}$ **43.** $y = -\frac{4}{5}x - \frac{1}{5}$ **45.** 5
47. $22\frac{1}{2}$ **49.** 16

CHAPTER 9   USING COORDINATES

Pages 333–334   Lesson 9.1
Exploratory   1. C(4, 4), D(0, 4)   **3.** B(8, 0), D(0, 5)   **5.** B(7, 0), D(4, 5)   **7.** C(k + j, h)
9. C(m + j, n)   **11.** neither   **13.** neither

SELECTED ANSWERS 551

15. neither **Written** **1.** $A(-b, 0)$, $C(b, 2b)$, $D(-b, 2b)$ **3.** $A(-f, 0)$, $B(f, 0)$, $D(-f, g)$ **5.** Let A be the point with coordinates $(0, 0)$, B be the point with coordinates $(4, 0)$, C be the point with coordinates $(4, 3)$, and D be the point with coordinates $(0, 3)$. The slope of $\overline{AB}$ = the slope of $\overline{CD}$ = 0. Therefore, $\overline{AB} \parallel \overline{CD}$. The slope of $\overline{AD}$ = the slope of $\overline{BC}$, which is undefined. Therefore, $\overline{AD} \parallel \overline{BC}$.
11. $Q(2a, 0)$, $S(0, 2b)$ **13.** yes **15.** $\dfrac{c}{a+b}$
17. Since $\dfrac{c}{a+b} \cdot \dfrac{c}{b-a} = \dfrac{c^2}{-(a^2-b^2)}$ and $c^2 = a^2 - b^2$, you can substitute c^2 for $a^2 - b^2$. So, $\dfrac{c^2}{-(a^2-b^2)} = \dfrac{c^2}{-c^2}$ or -1. Since the product of the slopes of $\overline{AC}$ and $\overline{BD}$ is -1, $\overline{AC} \perp \overline{BD}$.
19. Given: Rectangle $ABCD$ with vertices $A(0, 0)$, $B(0, a)$, $C(b, a)$, $D(b, 0)$ **Prove:** $\overline{AC} \cong \overline{BD}$ **Proof:** By the distance formula, $AC = \sqrt{(b-0)^2 + (a-0)^2}$ or $\sqrt{a^2+b^2}$; $BD = \sqrt{(b-0)^2 + (0-a)^2}$ or $\sqrt{a^2+b^2}$. By substitution, $AC = BD$ or $\overline{AC} \cong \overline{BD}$.

Pages 337–338 Lesson 9.2
Exploratory **1.** $B(0, a)$ **3.** $A(4a, 0)$
5. $C(2a + \frac{1}{2}c, w)$ **7.** both **9.** neither
Written **1.** $2\sqrt{b^2 + c^2}$
3. $2\sqrt{a^2 + b^2 + c^2 - 2ab}$ **5.** $(a, 0)$ **7.** The midpoint of $\overline{AB}$ is $E(-1, 6)$. The midpoint of $\overline{BC}$ is $F\left(\dfrac{5}{2}, 5\right)$. The slope of $\overline{EF}$ is $\dfrac{6-5}{-1-\frac{5}{2}}$ or $\dfrac{-2}{7}$. The slope of $\overline{AC}$ is $\dfrac{3-5}{4-(-3)}$ or $\dfrac{-2}{7}$. Since the slopes of $\overline{EF}$ and $\overline{AC}$ are equal, $\overline{EF} \parallel \overline{AC}$.
9. $\sqrt{b^2 + c^2}$ **11.** d
13. $\sqrt{b^2 + c^2 + \dfrac{d^2}{4} - db}$ **17.** $A(0, 8)$, $B(2b, 2c)$, $C(2a, 0)$. The midpoint M_1 of $\overline{AB}$ is (b, c), the midpoint M_2 of $\overline{BC}$ is $(a + b, c)$. The measure of $\overline{AC}$ is $2a$. The measure of $\overline{M_1M_2}$ is $\sqrt{(a+b-b)^2 + (c-c)^2}$ or $\sqrt{(a+0)^2 + (0)^2}$ or $\sqrt{a^2}$ or a. The measure of $\overline{M_1M_2}$ is $\frac{1}{2}(2a)$ or a. **19. Given:** Right triangle ABC with vertices $A(0, 2b)$, $B(2a, 0)$, $C(0, 0)$

Prove: $AM = BM = CM$ where M is the midpoint of $\overline{AB}$ **Proof:** By the midpoint formula, M has coordinates (a, b).
$AM = \sqrt{(a-0)^2 + (b-2b)^2}$ or $\sqrt{a^2+b^2}$.
$BM = \sqrt{(a-2a)^2 + (b-0)^2}$ or $\sqrt{a^2+b^2}$.
$CM = \sqrt{(a-0)^2 + (b-0)^2}$ or $\sqrt{a^2+b^2}$.
Therefore, $AM = BM = CM$.

Page 341 Lesson 9.3
Exploratory **1.** a circle with center at point P and a radius of 3 cm **3.** the perpendicular bisector of $\overline{PQ}$ **5.** a pair of lines each of which bisects pairs of vertical angles formed by lines r and s **Written** **1.** two parallel lines on either side of the given line and each is 4 cm from it **3.** the intersection point of the diagonals of the square **5.** the perpendicular bisector of the line segment having the two given points as endpoints **7.** the perpendicular bisector of the diagonal connecting the two vertices **9.** The circular portions are quarter circles with a vertex as center and a 2 in. radius. The line segments in the figure are 5 in. long and parallel to the sides of the square.
11. a sphere with radius of 5.5 cm and center at the given point **13.** two planes each containing the line of intersection of the two given planes and halfway between them **15.** a zigzag path of points equidistant from points on the random paths of the floats **17.** a spiral not including the center of a circle and points within the circle whose radius is the same as the radius of the label

Pages 345–346 Lesson 9.4
Exploratory **1.** $x = 4$ **3.** $y = 10$ **5.** $x = 15$
7. $x = 5$ and $x = -5$ **9.** $x = 12$ and $x = 4$
Written **1.** a circle with a radius of 1 unit and center of $(0, 0)$ **7.** $x^2 + y^2 = 9$
9. $x^2 + y^2 = 100$ **11.** $x^2 + y^2 = 20$ **13.** yes
15. no **17.** no **19.** yes **21.** yes **23.** $(6, 3)$, 8 units **25.** $(3, -8)$, $\sqrt{5}$ units **27.** $(0, -7)$, $2\sqrt{2}$ units **29.** $(x-4)^2 + (y-6)^2 = 81$
31. $(x-1)^2 + (y+2)^2 = 169$
33. $(x+4)^2 + (y+6)^2 = 11$
35. 81.64 units, 530.66 units **37.** 20.83 units, 34.54 units **39.** $x = 1$ **41.** $x = 3.5$
43. $y = 2x + 1$ **45.** $x = -1$ **47.** $y = 6$
49. $y = -1$ **51.** $(-1, 4)$; $\dfrac{1}{2}$; $y = -2x + 2$

53. (2, 4); $\frac{1}{3}$; $y = -3x + 10$ **55.** (3, 5); -2; $y = -2x + 11$

Pages 349–350   Lesson 9.5
Exploratory   1. the intersection of a circle with a radius of 6 cm and center at the given point on the line and the two lines on either side of the given line 4 cm from the given line **3.** the two intersection points of a circle with a radius of 3 cm and center at the given point and the two lines on either side of the given line 3 cm from the given line **5.** the 2 intersection points of a circle with a radius of 5 cm and center at P and the line 4 cm from ℓ on the same side of ℓ as P **7.** the 3 intersection points of the circle with a radius of 11 cm and center at P and the two lines 3 cm from ℓ on each side of ℓ
Written   1. the two intersection points of a circle with a radius of 5 cm and center at point A and the line parallel to and halfway between lines ℓ and m **3.** the two intersection points of a circle with a radius of 13 cm and center at point Q and a circle with a radius of 5 cm and center at point R **5.** the 2 intersection points of the circle with a radius of 2 cm and center at point A and a circle with a radius 2 cm and center at point B **7.** the intersection point of the perpendicular bisector of $\overline{AC}$ and the bisector of $\angle ABC$ **9.** The perpendicular bisectors are parallel. **11.** $y = x$, $y = -x$ **13.** (2, 2), (8, 2), (2, 8), (8, 8) **15.** (3, 4), (3, -4), (-3, 4), (-3, -4) **17.** (6, 8), (6, -8), (-6, 8), (-6, -8) **19.** (1, $\sqrt{3}$), (1, $-\sqrt{3}$) **21.** $x^2 + y^2 = 16$ **23.** 4 **25.** 1 **27.** 0 **29.** 2   **Mixed Review   1.** 12 **2.** $22\frac{1}{2}$
3. the interior of a circle having a radius of 3 cm **4.** a plane perpendicular to the line segment formed by the 2 points
5. $(x + 1)^2 + (y - 4)^2 = 121$ **6.** $y = -\frac{1}{7}x$
7. $k = 1$ **8.** $k = -1$ **9.** $3x + 4y = 23$
10. $k = -4$ or $k = 3$

Pages 352–353   Lesson 9.6
Written   1. (x, y) is $y + 1$ units from the x-axis. **5.** $y = \frac{1}{4}x^2$ **7.** $y = \frac{1}{4}x^2 - 2x + 4$

9. $y = \frac{1}{10}x^2 - \frac{1}{5}x - \frac{2}{5}$ **11.** (x, y) is $x + 2$ units from the y-axis. **15.** $x = \frac{1}{8}y^2 - \frac{1}{4}y + \frac{1}{8}$
17. $x = \frac{1}{2}y^2 - 4y + \frac{23}{2}$
19. $x = -\frac{1}{10}y^2 + \frac{2}{5}y + \frac{11}{10}$

Page 357   Lesson 9.7
Exploratory   1. Always true; this is illustrated by a folded piece of paper. The two parts represent two planes and the fold, a line.
3. Always true; there is only one plane through a line and a point not on the line. This plane also contains the one line through the given point parallel to the first line. **5.** Always true; the plane containing a line and a point outside the line also contains a line through the point intersecting the line. **7.** Sometimes true; an illustration of this is a wobbly four-legged chair. Because three points determine a plane, no plane containing a fourth point outside the first plane will contain all 4 points.
Written   1. (0, 12, 0), (0, 12, 3), (8, 12, 0), (8, 12, 3) **3.** $z = 3$ **5.** $x = 0$, $x = 8$ **9.** The diagonals intersect at (4, 6, 3) and are congruent because they are diagonals of a rectangle and diagonals of a rectangle are congruent. **11.** False; an illustration is the pages of a book. **13.** True; the coordinate planes have the equations $z = 0$ (xy-plane), $y = 0$ (xz-plane), $x = 0$ (yz-plane)

Page 359   Problem Solving Application
Exploratory   1. yes, 5 **3.** no **5.** yes, 50
7. yes, -1 **Written   1.** 32 **3.** 24 **5.** $-2\frac{2}{7}$
7. 4 hours 16 minutes **9.** $\frac{1}{2}$ **11.** $21\frac{1}{3}$ m³
13. 640 cycles/sec

Page 361   Chapter Review
1. The slope of $\overline{AB}$ = the slope of $\overline{CD}$ = 0. Therefore, $\overline{AB} \parallel \overline{CD}$. The slope of $\overline{AD}$ = the slope of $\overline{BC}$ = 2. Therefore, $\overline{AD} \parallel \overline{CD}$.
3. Given: isosceles trapezoid $ABCD$ with vertices $A(0, 0)$, $B(a, 0)$, $C(a - b, c)$, $D(b, c)$
Prove: $\overline{AC} \cong \overline{BD}$ **Proof:** $AC = \sqrt{(a - b)^2 + c^2} = \sqrt{a^2 + b^2 + c^2 - 2ab}$

and $BD = \sqrt{(b-a)^2 + c^2} = \sqrt{a^2 + b^2 + c^2 - 2ab}$.
Since $AC = BD$, $\overline{AC} \cong \overline{BD}$. **5.** $\sqrt{a^2 + b^2}$
7. $\left(\dfrac{a}{2}, \dfrac{b}{2}\right)$ **11.** the perpendicular bisector of the line segment whose endpoints are the two given points **13.** a sphere with a radius of 5 cm and center at the given point **15.** yes **17.** (1, 3), 4 units **19.** $3x - 4y = -29$ **23.** $y = \dfrac{1}{4}x^2 - 1$ **25.** $x = -\dfrac{1}{6}y^2 - \dfrac{2}{3}y - 2\dfrac{1}{6}$ **27.** false

CHAPTER 10 SOLVING QUADRATIC EQUATIONS

Pages 367–368 Lesson 10.1
Exploratory **1.** a polynomial with two unlike terms **3.** $(a + b)(a - b) = a^2 - b^2$
5. $a(b + c) = ab + ac$ **7.** $8x^3$ **9.** x^3y^4z
11. $-3x - 12$ **13.** $x^3 - x^2$ **15.** x^2 **17.** $3x$
19. $50xz$ **21.** $2yz$ **Written** **1.** $8c - 16$
3. $2y^2 - 8y^3$ **5.** $4x^2y^2 + 4x^3y$ **7.** $x^2 + 7x + 10$ **9.** $x^2 - 81$ **11.** $x^2 - 25$ **13.** $2w^2 + 19w + 35$ **15.** $8x^2 - 2x - 1$ **17.** $6y^2 - 7y - 20$ **19.** $8(c - 2)$ **21.** $5x(x - 1)$
23. $2xy(1 + y)$ **25.** $13xy(z - 2)$
27. $xy(y^2z + 3x)$ **29.** $100x^2(1 - x)$
31. $(x - 2)(x + 2)$ **33.** $(m + 1)^2$
35. $(x - 9)(x + 9)$ **37.** $(x - 5)(x + 3)$
39. $(2c - 1)(2c + 1)$ **41.** $(x - 8)(x + 3)$
43. $(2c + 1)(c + 5)$ **45.** $(5x - 1)(x + 4)$
47. $(2x - 3)(x - 4)$ **49.** $9(y - 1)(y + 1)$
51. $x(x - 2)$ **53.** $3(b - 4)(b + 4)$
55. $-2(x^2 - x - 15)$ **57.** $h^2(h - 1)$
59. $2m^4(m + 1)(m - 1)$ **61.** $4x(d - 3) \cdot (d + 2)$ **63.** $y^4(y - 1)(y + 1)$ **65.** The terms are the same, but they are "reversed."
67. $(x + r)(x + s) = x(x + s) + r(x + s) = x^2 + xs + rx + rs = x^2 + (r + s)x + rs$

Pages 371–372 Lesson 10.2
Exploratory **1.** $x = 0$ or $x - 6 = 0$; $\{0, 6\}$
3. $2y = 0$ or $y + 4 = 0$; $\{-4, 0\}$ **5.** $x + 1 = 0$ or $x + 5 = 0$; $\{-5, -1\}$ **7.** $c - 5 = 0$ or $c - 3 = 0$; $\{3, 5\}$ **9.** $2y + 1 = 0$ or $y - 2 = 0$; $\left\{-\dfrac{1}{2}, 2\right\}$ **11.** $3x - 4 = 0$ or $4x - 3 = 0$; $\left\{\dfrac{3}{4}, \dfrac{4}{3}\right\}$ **13.** $3x^2 - 3x - 1 = 0$
15. $x^2 + 5x + 1 = 0$ **17.** $x^2 + 5x - 13 = 0$
19. 3, -5, -1 **21.** 4, 0, -1 **23.** a **25.** c
Written **1.** $\{-4, -1\}$ **3.** $\{0, 7\}$
5. $\{-5, -2\}$ **7.** $\{-4, 4\}$ **9.** $\{-5, 10\}$
11. $\{-2\}$ **13.** $\left\{0, \dfrac{1}{2}\right\}$ **15.** $\{-1, 9\}$ **17.** $\{-3\}$
19. $\{-2, 10\}$ **21.** $\{-5, 4\}$ **23.** $\{-10, 3\}$
25. $\left\{-\dfrac{2}{3}, -1\right\}$ **27.** $\left\{\dfrac{3}{2}, 2\right\}$ **29.** $\left\{\dfrac{1}{4}, 4\right\}$
31. $\left\{-\dfrac{3}{2}, 7\right\}$ **33.** $\left\{-\dfrac{7}{3}, 2\right\}$ **35.** $\{-7, 7\}$
37. $\{-5, -1\}$ **39.** $\{3, 4\}$ **41.** $\{4, 8\}$
43. $\left\{-\dfrac{1}{3}, \dfrac{1}{3}\right\}$ **45.** $\left\{-\dfrac{1}{2}, \dfrac{1}{2}\right\}$ **47.** $x^2 + 4x + 3 = 0$ **49.** $x^2 - 3x + 2 = 0$ **51.** $x^2 - \dfrac{9}{2}x - \dfrac{5}{2} = 0$ **53.** $x^2 - \dfrac{5}{12}x + \dfrac{1}{24} = 0$ **55.** $x^2 + \dfrac{2}{3}x + \dfrac{1}{12} = 0$ **57.** $x^2 + 2x + 1 = 0$

Page 376 Lesson 10.3
Exploratory **1.** the line about which a parabola is symmetric **3.** the equation of the axis of symmetry of a parabola **5.** when a is positive; when a is negative. **7.** $x = 0$
9. $x = 1$ **11.** no **13.** yes **15.** no
Written **13.** $x = 0$, (0, 0) **15.** $x = -2$, $(-2, -13)$ **17.** $x = 2$, (2, 3) **19.** $x = \dfrac{3}{2}$, $\left(\dfrac{3}{2}, -\dfrac{9}{2}\right)$ **21.** $x = \dfrac{1}{6}$, $\left(\dfrac{1}{6}, \dfrac{11}{12}\right)$

Pages 379–380 Lesson 10.4
Exploratory **1.** Graph the associated parabola. Then the x-coordinates of the points where the parabola crosses the x-axis are the solutions.
3. if $ax + bx + c$ is a perfect square trinomial
5. $\{-1, 1\}$ **7.** $\emptyset$ **9.** $\{2\}$ **Written** **1.** $\{-2, 2\}$
3. $\{-1, 4\}$ **5.** $\{-4, 1\}$ **7.** $\{2, 3\}$ **9.** $\left\{-\dfrac{1}{2}, \dfrac{1}{2}\right\}$
11. $\{-5\}$ **13.** $\{1, 6\}$ **15.** $\{4\}$ **17.** $\left\{-2\dfrac{1}{5}, 2\dfrac{1}{5}\right\}$

$y - 2 = 0$; $\left\{-\dfrac{1}{2}, 2\right\}$
$4x - 3 = 0$; $\left\{\dfrac{3}{4}, \dfrac{4}{3}\right\}$
11. $3x - 4 = 0$ or

19. $\{-2\frac{1}{10}, 3\frac{1}{10}\}$ **21.** $\{-\frac{7}{10}, \frac{7}{10}\}$
23. $\{-\frac{7}{5}, 3\frac{2}{5}\}$ **25.** $\{-\frac{7}{4}, \frac{1}{4}\}$ **27.** $\{-\frac{7}{5}, -\frac{1}{5}\}$
29. ∅

Pages 383–384   Lesson 10.5
Exploratory 1. it does not always give exact solutions; to get exact solutions
3. $x = \pm\sqrt{121}$ **5.** $c = \left(\frac{m}{2}\right)^2$ **7. (a)** Write the equation. **(b)** Subtract 3 from both sides.
Written 1. $\{-1 \pm \sqrt{2}\}$ **3.** $\{5 \pm \sqrt{13}\}$
5. $\{-6 \pm \sqrt{6}\}$ **7.** $\left\{\frac{2 \pm \sqrt{2}}{2}\right\}$
9. $\left\{\frac{-9 \pm 4\sqrt{6}}{12}\right\}$ **11.** $\left\{\frac{-2 \pm 3\sqrt{6}}{6}\right\}$ **13.** 9
15. 169 **17.** $\frac{9}{4}$ **19.** ± 10 **21.** ± 22
23. $\{-12, 8\}$ **25.** $\{-2, 5\}$ **27.** $\{2, 4\}$
29. $\{-8, 6\}$ **31.** $\{3, 7\}$ **33.** $\{-15, 12\}$
35. $\{-6 \pm 4\sqrt{2}\}$ **37.** $\left\{\frac{-5 \pm \sqrt{57}}{2}\right\}$
39. $\left\{\frac{3 \pm \sqrt{5}}{2}\right\}$ **41.** $\{-\frac{3}{2}, \frac{1}{3}\}$ **43.** $\{-7, \frac{3}{2}\}$
45. $\{-3, \frac{5}{3}\}$ **47.** ∅ **49.** ∅ **51.** ∅

Pages 387–388   Lesson 10.6
Exploratory 1. If $ax^2 + bx + c = 0$, then $x = \frac{-b \pm \sqrt{b^2 - 4ac}}{2a}$. the coefficient of x^2, the coefficient of x, and the constant term, respectively **3.** The discriminant tells you if the equation has real roots. **5.** $2x^2 - 5x - 1 = 0$, $a = 2$, $b = -5$, $c = -1$ **7.** $x^2 + 12x - 2 = 0$, $a = 1$, $b = 12$, $c = -2$
9. $x^2 + 2x = 0$, $a = 1$, $b = 2$, $c = 0$
11. -7, no real roots **13.** 0, has real roots
15. 72, has real roots **17.** The roots are rational. **19.** sum $= \frac{13}{2}$, product $= 1$
21. sum $= \frac{4}{3}$, product $= \frac{5}{3}$ **23.** sum $= -4$, product $= -20$ **Written 1.** $\{-5, 6\}$
3. $\{-5, 3\}$ **5.** $\{4, 6\}$ **7.** $\{1, 4\}$ **9.** $\{-\frac{5}{3}, 4\}$
11. $\{-\frac{3}{2}, \frac{5}{3}\}$ **13.** $\{-\frac{5}{2}, \frac{1}{7}\}$ **15.** $\{-\frac{2}{5}, \frac{1}{4}\}$
17. $\{-\frac{3}{4}, \frac{5}{6}\}$ **19.** $\left\{\frac{1 \pm \sqrt{113}}{4}\right\}$ **21.** $\left\{\frac{8 \pm 2\sqrt{29}}{13}\right\}$
23. $\{0, 13\}$ **25.** $\left\{\frac{7 \pm 3\sqrt{21}}{14}\right\}$ **27.** $\{-4, \frac{8}{7}\}$
29. $\{\pm\frac{\sqrt{30}}{2}\}$ **31.** has real roots

Pages 392–393   Lesson 10.7
Exploratory 1. Let w represent the width. $w(w + 5) = 24$ **3.** Let w represent the width. $w(4w + 1) = 18$ **5.** Let x be the smaller number. $x(x + 2) = 35$ **7.** Let y be the smaller number. $y(y + 5) = 24$ **Written**
1. 3 in. by 8 in. **3.** 2 m by 9 m **5.** $-7, -5$; 5, 7 **7.** $-8, -3$; 3, 8 **9.** $-14, -7$; 7, 14
11. $-14, -13$; 33, 34 **13.** $ST = 12$, $RD = 8$ **15.** $TM = 6$, $RU = 15$ **17.** $1\frac{1}{2}$ in.
19. 4 ft **21.** 10 m **23.** 2.52 **25.** 1.38
27. 27 **Mixed Review 1.** $x = 0, (0, -2)$
2. $x = -4, (-4, -22)$ **3.** $x = \frac{2}{3}, \left(\frac{2}{3}, \frac{8}{3}\right)$
4. $\{4, 5\}$ **5.** $\{-\frac{1}{2}, 2\}$ **6.** $\{-\frac{3}{4}, \frac{1}{5}\}$
7. $\{-5 \pm 2\sqrt{6}\}$ **8.** $\{4 \pm 3\sqrt{2}\}$
9. $\left\{\frac{-9 \pm \sqrt{21}}{10}\right\}$ **10.** $x^2y^2(x - 8)(x + 8)$
11. ∅ **12.** (3, 2) **13.** $y = \frac{1}{4}x^2 - \frac{3}{2}x + \frac{9}{4}$
14. ± 28 **15.** $z = -3$

Page 396   Lesson 10.8
Exploratory 1. to find all values of x and y that make both equations true at the same time **3.** The solution set is ∅. **5.** You will get a negative discriminant. **Written 1.** $\{(0, 0), (2, 4)\}$ **3.** $\{(0, 0), (2, 2)\}$
5. $\{(-2, -5), (1, -8)\}$ **7.** $\{(1, -3), (6, 2)\}$
9. $\left\{\left(-1\frac{1}{2}, -1\frac{3}{4}\right), (0, -1)\right\}$ **11.** ∅ **13.** ∅
15. ∅ **17.** $\{(-3, 9), (0, 0)\}$ **19.** $\{(-1, 5), (0, 4)\}$ **21.** $\{(0, 2), (5, -3)\}$ **23.** ∅
25. $\{(2, 5), (4, 9)\}$ **27.** ∅

Page 399   Lesson 10.9
Exploratory   1. $2 \cdot 2 \cdot 7$ **3.** $z(z-1)$
5. $3x(x+5)$ **7.** $(t-5)(t+5)$
9. $(4n-3)(n+4)$ **11.** $16a$ **13.** $x-5$
15. $6(c-4)$ **17.** $x-4$ **19.** $7y$ **21.** $y+7$
Written   1. $\dfrac{1}{2a}$ **3.** $-5n$ **5.** $1-2q$ **7.** $x+1$
9. 4 **11.** $\dfrac{-3}{2b+1}$ **13.** $\dfrac{-1}{c+2}$ **15.** $-\dfrac{2(x-1)}{x+1}$
17. $\dfrac{p-q}{p+q}$ **19.** $\dfrac{-5x^4}{2x-5}$ **21.** $\dfrac{4}{x+3}$ **23.** $\dfrac{j+k}{j-k}$
25. $-\dfrac{a+3}{a+2}$ **27.** -1 **29.** $\dfrac{4y-1}{8y-1}$

Page 402   Lesson 10.10
Exploratory   1. $\dfrac{1}{2}$ **3.** $\dfrac{2b}{3a}$ **5.** $\dfrac{8}{9}$ **7.** $\dfrac{5x}{y}$
9. $\dfrac{x}{2y} \cdot \dfrac{4y}{-x}$ **11.** $5k \cdot \dfrac{4}{9}$ **Written   1.** $\dfrac{135x^3}{8}$ **3.** $\dfrac{1}{4}$
5. $\dfrac{45x^3}{128y^2}$ **7.** $\dfrac{x^4 - 2x^3 + 2x - 1}{xy}$ **9.** $\dfrac{x+1}{x-1}$
11. $-\dfrac{2(x-2)}{x}$ **13.** $\dfrac{4b(x+15)}{x(x+10)}$ **15.** $8(y-1)$
17. $\dfrac{2(x+3)}{(x-3)(2x-1)}$ **19.** $\dfrac{(x-6)(x+2)(x+3)}{x+12}$
21. $(a-b)$ **23.** $\dfrac{w-4}{w-10}$

Pages 405–406   Lesson 10.11
Exploratory   1. 24 **3.** $21ab$ **5.** $120m^2n^3$
7. $x-3$ **9.** $6a$ **11.** t^2 **13.** $x^2 - x$
15. $2y^2 - 3y + 1$ **17.** -4 **Written   1.** $\dfrac{19}{24}$
3. $\dfrac{3b-4a}{ab}$ **5.** $\dfrac{29x}{28}$ **7.** $\dfrac{14n-5}{6n^2}$ **9.** $\dfrac{k(5x+1)}{7x^2}$
11. $\dfrac{x^2+3}{x-5}$ **13.** $\dfrac{13x+3y}{18}$ **15.** $\dfrac{2(x^2+4)}{(x-2)(x+2)}$
17. $\dfrac{27-5x}{(x-4)(x-3)}$ **19.** $\dfrac{y^2+3y+6}{y(y-2)(y+2)}$
21. $\dfrac{2(5-3x)}{(2x+1)(2x-1)}$ **23.** $-\dfrac{x^2-5x+1}{(x-1)^2}$
25. $\dfrac{a^3+7a^2+8a+12}{(a+2)^2(a-2)}$ **27.** $\dfrac{10a^2+11a+1}{(3a-2)(2a+1)}$
Mixed Review   1. c **2.** b **3.** d **4.** a **5.** b **6.** d

Pages 408–409   Chapter Review
1. $(x-4)(x+3)$ **3.** $2(a+7)(a+1)$
5. $\{-2\}$ **7.** $\{-1, 1\}$ **9.** $\left\{-\dfrac{3}{4}, 1\right\}$
15. $\{-3, 3\}$ **17.** $\{0, 5\}$ **19.** $\{-5, 3\}$ **21.** $\emptyset$
23. $\{-10, 6\}$ **25.** $\left\{-\dfrac{1}{2}\right\}$ **27.** 3 cm by 15 cm

29. 24 or -13 **31.** $\{(0, 0), (2, 4)\}$ **33.** $\{(1, 1), (5, 1)\}$ **35.** $\dfrac{4ab^2c^4 - 12a^3bc + 13}{2c}$ **37.** $\dfrac{x+3}{x(x-6)}$
39. $\dfrac{3axy}{10}$ **41.** $\dfrac{(x+4)^2}{(x+2)^2}$ **43.** $\dfrac{7ab(x-9)}{3(x-5)}$ **45.** 2
47. $\dfrac{10axy-3}{6x^2y}$ **49.** $\dfrac{8x-9}{x^2-4}$

CHAPTER 11   PROBABILITY AND COMBINATORICS

Pages 414–415   Lesson 11.1
Exploratory   1. $\{A, B, C, D, E, F, G, H, I, J\}$
3. $\{$red, yellow, blue$\}$ **5.** $\{0, 1, 2, 3, \ldots, 14\}$ **7.** $\dfrac{1}{4}$ **9.** $\dfrac{1}{2}$ **Written   1.** $\dfrac{1}{18}$ **3.** $\dfrac{1}{9}$ **5.** $\dfrac{1}{12}$
7. $\dfrac{5}{12}$ **9.** $\dfrac{1}{6}$ **11.** $\dfrac{5}{18}$ **15.** $P(E)$ is greater than $\dfrac{1}{2}$.
17. $\dfrac{1}{4}$ **19.** $\dfrac{7}{8}$ **21.** $\dfrac{3}{50}$ **23.** $\dfrac{4}{25}$ **25.** 1 **27.** $\dfrac{1}{16}$
29. 0 **31.** $\dfrac{1}{2}$

Page 418   Lesson 11.2
Exploratory   1. choosing an A **3.** choosing Montana, Minnesota, Michigan, or Maine
5. tossing a (4, 4) **7.** choosing a club that is a face card **9.** choosing a heart **Written   1.** $\dfrac{1}{13}$
3. $\dfrac{4}{13}$ **5.** $\dfrac{7}{13}$ **7.** $\dfrac{8}{13}$ **9.** $\dfrac{1}{6}$ **11.** $\dfrac{2}{9}$ **13.** $\dfrac{5}{12}$
15. $\dfrac{1}{2}$ **17.** 1 **19.** $\dfrac{3}{5}$ **21.** $\dfrac{1}{2}$ **23.** $\dfrac{2}{5}$

Pages 421–423   Lesson 11.3
Exploratory   1. 20 **3.** 16 **5.** $\dfrac{1}{12}$ **7.** $\dfrac{1}{9}$ **9.** $\dfrac{9}{25}$
Written   1. $\dfrac{1}{36}$ **3.** $\dfrac{1}{144}$ **5.** $\dfrac{1}{216}$ **7.** $\dfrac{2}{5}$ **9.** $\dfrac{7}{15}$
11. $\dfrac{4}{9}$ **13.** $\dfrac{1}{9}$ **15.** $\dfrac{7}{12}$ **17.** $\dfrac{16}{81}$ **19.** $\dfrac{32}{243}$ **21.** $\dfrac{1}{8}$
23. $\dfrac{1}{4}$ **25.** $\dfrac{1}{2}$ **27.** $\dfrac{2}{7}$ **29.** $\dfrac{2}{7}$ **31.** $\dfrac{1}{7}$ **33.** 0
35. $\dfrac{1}{12}$ **37.** $\dfrac{15}{512}$ **39.** $\dfrac{7}{15}$ **Mixed Review   1.** 1
2. $\dfrac{m+4}{4(m+2)^2}$ **3.** 3 **4.** $5+g$ **5.** $\dfrac{1-5c}{c-4}$
6. $\dfrac{2m^2+7m-1}{2m^2+7m+5}$ **7.** $\dfrac{7}{12}$ **8.** $\dfrac{1}{3}$ **9.** $\dfrac{2}{3}$ **10.** $\dfrac{19}{36}$
11. 15 **12.** $\dfrac{1}{15}$

Page 426   Lesson 11.4
Exploratory 1. 120   3. 5040   5. 720   7. 1
Written 1. 24   3. 5040   5. 24   7. 72   9. 6
11. 4320   13. 120   15. 24   17. yes   19. 24

Pages 429–430   Lesson 11.5
Exploratory 1. 60   3. 4   5. 2520
7. 970,200   9. 6   11. 60   **Written** 1. 60
3. 120   5. 120   7. 40   9. 210   11. 420
13. 6720   15. 56   17. 5,760,000
19. 5,184,000   21. 17,576,000
23. 15,625,000   25. 60,840,000   27. 10^9
29. 205,320   31. 8   33. $\frac{1}{20}$

Pages 432–433   Lesson 11.6
Exploratory 1. 120   3. 3   5. 12   7. 60
Written 1. 40,320   3. 181,440   5. 1
7. 360   9. 10,080   11. 30   13. 45,360
15. 151,200   17. 27,720   19. 126
21. 151,000   23. 6   25. $\frac{1}{3}$

Page 436   Lesson 11.7
Exploratory 1. 1   3. 1   5. 1
Written 1. combination   3. combination
5. combination   7. combination
9. combination   11. permutation   13. The number of combinations of *n* things taken *n* at a time.   15. There is only one way to choose nothing from *n* things.

Pages 438–440   Lesson 11.8
Exploratory 1. R, S, T; R, S, W; S, T, W; R, T, W   3. S, T, W   5. A, B, C, D; A, B, C, E; A, B, C, F; A, B, D, E; A, B, D, F; A, B, E, F; A, C, D, E; A, C, D, F; A, C, E, F; A, D, E, F; B, C, D, E; B, C, D, F; B, C, E, F; B, D, E, F; C, D, E, F   7. 15   9. 56   11. 126   13. 6
Written 1. 20   3. 455   5. 36   7. 7   9. 21
11. 36   13. 8   15. 792   17. 126   19. 351
21. 56   23. 252   25. 2520   27. 28,561
29. 720   31. 35   33. $\frac{n}{2}(n-3)$ **Mixed Review**
1. $\frac{5}{39}$ 2. $\frac{14}{39}$ 3. $\frac{20}{39}$ 4. 5040   5. 35,280
6. 10,080   7. 1440   8. 14,400   9. 480
10. 907,200   11. 18   12. 8 blue, 12 gold

Pages 443–445   Lesson 11.9
Exploratory 1. $\frac{2}{3}$ 3. $\frac{8}{11}$ 5. $\frac{43}{64}$ 7. 1

Written 1. 462   3. 210   5. $\frac{5}{11}$ 7. 20   9. 18
11. 7   13. $\frac{1}{28}$ 15. 252   17. 6   19. 15
21. 66   23. 56   25. 220   27. $\frac{7}{8}$ 29. $\frac{63}{64}$
31. $\frac{1023}{1024}$ 33. $\frac{91}{216}$; $\frac{671}{1296}$; $1 - \left(\frac{5}{6}\right)^n$ 35. $\frac{127}{128}$
37. $\frac{19}{33}$ 39. $\frac{1}{1938}$ 41. 0   43. $\frac{120}{323}$

Page 448   Problem Solving Application
Written 1. 241   3. 314   5. 304   7. 304
9. 303   11. 10   13. 11   15. 11   17. 21
19. 32   21. 16   23. $\frac{2}{5}$ 25. $\frac{3}{5}$

Page 450   Chapter Review
1. {(1, H), (1, T), (2, H), (2, T), (3, H), (3, T), (4, H), (4, T)}   3. $\frac{1}{4}$ 5. $\frac{1}{2}$ 7. $\frac{7}{13}$ 9. 60   11. 6
13. 120   15. 120   17. 24   19. 120   21. 2450
23. 840   25. 360   27. 2520   29. 1260   31. 1
33. 1   35. combination   37. 4   39. 120
41. 792   43. 2002   45. 300   47. 21

CHAPTER 12   TRANSFORMATION GEOMETRY

Pages 456–457   Lesson 12.1
Exploratory 1. Yes; domain = {100, 35, 20, 0}, range = (212, 95, 68, 32}   3. No   5. Yes; domain = {5, 9, 17}, range = {8, 6}   7. yes
9. no   **Written** 1. {−3, −2, −1, 0, 1, 2}
3. −1   5. −5   7. −3   9. −1, 1   11. −5
13. 1   15. −11   23. 7   25. 11
29. {7, 9, 11, 13}   31. {7, 9, 11, 13}   33. 3
35. 4   37. 7   39. 9

Pages 460–461   Lesson 12.2
Written 1. point *R*   3. point *M*   5. $\overline{NQ}$
7. quadrilateral *QNLK*   19. *B* is between *A* and *C* as *B'* is between *A'* and *C'*. *C* is between *B* and *D* and *C'* is between *B'* and *D'*.   21. yes
23. orientation   25. b   27. a

Pages 463–464   Lesson 12.3
Written 3. no   5. B, C, D, E, H, I, K, O, X
Mixed Review 1. $\frac{5}{16}$ 2. $\frac{1}{4}$ 3. $\frac{3}{16}$ 4. a, b
5. b   6. a, b   7. (3, 1)   8. $y = 3x - 11$

9. 350   **10.** 5 or 1; no; −1 has more than one preimage.

Pages 467–468   Lesson 12.4
Exploratory   1. (−2, 3)   **3.** $(-2\frac{1}{2}, 7)$
5. (4, −4)   **7.** (−3, −5)   **9.** $(-7, -3\frac{1}{2})$
11. neither   **13.** x-axis   **15.** both
Written   1. (1, −3)   **3.** (−3, −4)
5. $(3\frac{1}{2}, -3\frac{1}{2})$ **7.** (x, −y)   **9.** (0, 4)
11. (7, 2)   **13.** A'(−4, 2), B'(−1, 5), AB = A'B' = $3\sqrt{2}$ **15.** A'(3, 4), B'(1, −5), AB = A'B' = $2\sqrt{17}$ **19.** A'(−3, −4), B'(5, −6), AB = $2\sqrt{17}$ = $2\sqrt{17}$ **21.** A'(2, 4), B'(5, 1), AB = $3\sqrt{2}$, A'B' = $3\sqrt{2}$ **23.** A'(4, −3), B'(6, 5), AB = A'B' = $2\sqrt{17}$ **25.** (7, 4)
27. (10, −4)   **29.** (8 − x, y)   **31.** A'(3, −2), B'(−2, −9)   **33.** A'(3, 22), B'(−2, 15)
35. (x, −4 − y)   **37.** (x, 10 − y)   **41.** 1, 1

Pages 471–472   Lesson 12.5
Exploratory   7. no   **9.** yes   **11.** no
Written   1. K   **3.** $\overline{FI}$ **5.** $\angle JMN$ **7.** H   **9.** F
11. $\overline{FI}$ **13.** $\angle GHI$ **15.** (0, 12)   **17.** (−6, 1)
19. (2, 9)   **21.** (9, −2)   **23.** $T_{1,3}$ **25.** $T_{-5, 12}$
27. AB = $2\sqrt{5}$, A'B' = $2\sqrt{5}$ **29.** AA' = $\sqrt{17}$, BB' = $\sqrt{17}$ **31.** The diagonals of a parallelogram bisect each other.

Page 475   Lesson 12.6
Exploratory   1. a rotation with center C through an angle of 30° counterclockwise   **3.** V
Written   1. (0, 4)   **3.** (−2, 2)   **5.** (−y, x)
7. (−x, −y)   **13.** no   **15.** yes   **17.** yes
19. yes   **21.** yes   **23.** yes   **25.** yes

Pages 478–479   Lesson 12.7
Exploratory   1. (−2, −3)   **3.** (5, 0)
5. (−1, −5)   **7.** (−5, −7)   **Written**
5. A'(1, −5), B'(3, 4); slope of $\overline{AB}$ = $\frac{9}{2}$, slope of $\overline{A'B'}$ = $\frac{9}{2}$ **7.** translation   **9.** rotation
11. rotation   **15.** point, line   **17.** line   **19.** line
21. point, line   **23.** P(3, 2)   **25.** P(3, −4)
27. (−2, 5)   **29.** (−10, 7)   **Mixed Review**
1. Δ, Φ, Λ, Π, Θ, Ω   **2.** Φ, Θ, Σ   **3.** Φ, Θ
4. Γ, ϑ   **5.** Δ, Φ, Θ   **6.** (−1, −1)

7. (−7, −7)   **8.** (3, −4)   **9.** (5, −4)
10. (4, −3)   **11.** (−4, 3)   **12.** (3, −7)
13. (−3, 3)   **14.** (2, 1)   **15.** (−1, 2)
16. (2, −7)   **17.** (−2, 1)   **18.** $T_{8, -2}$
19. x = −1   **20.** Urn X: 7 red, 14 blue; Urn Y: 28 red, 14 blue

Pages 482–483   Lesson 12.8
Exploratory   1. line reflection R_ℓ followed by line reflection R_m **3.** a half-turn around the origin followed by translation $T_{2, 3}$ **9.** No. If the lines are not perpendicular, the images of composite reflections are different depending upon the order of the reflections.
Written   13.–24. AB = A'B' = $\sqrt{13}$
31. counterclockwise   **33.** no   **35.** yes
39. No. The line reflection is not a direct isometry though a translation is. The composition will not be a direct isometry.
41. Yes. Each translation is a direct isometry. Any number taken consecutively will therefore preserve orientation.

Page 487   Lesson 12.9
Exploratory   1. Since P → P' so that ℓ is the perpendicular bisector of $\overline{PP'}$ and P' → P" so that ℓ is the perpendicular bisector of $\overline{P'P''}$, P and P" must be the same point.   **3.** In $T_{a, b}$, (x, y) → (x + a, y + b). In $T_{-a, -b}$, (x, y) → (x − a, y − b). In $T_{-a, -b} \circ T_{a, b}$, (x, y) → (x + a − a, y + b − b) or (x, y) → (x, y). Hence every point maps into itself.   **5.** Yes, $T_{0,0}$. In $T_{0,0}$, (x, y) → (x + 0, y + 0) or (x, y) → (x, y).   **7.** No. In a reflection, only points on the line map into themselves. Since the whole plane must be reflected, it is impossible for a single reflection to map points into themselves.
Written   15. R_{ℓ_2} **17.** I   **19.** Associative Property   **21.** I   **23.** I   **25.** Yes, because for every element a of the set, there is an element a^{-1} such that $a \circ a^{-1}$ = I.   **27.** Yes.

Page 490   Lesson 12.10
Written   1. 3 or −3   **3.** $\frac{1}{5}$ or $-\frac{1}{3}$
5. (32, −24)   **7.** (−32, 24)   **11.** yes   **13.** no
15. yes

Page 492   Problem Solving Application
Written   2. about 27.7 miles

Pages 494–495   Chapter Review
1. yes; domain = {1, 2, 3}; range = {a, b}
3. no **5.** 9 **7.** 2 **9.** 4 **13.** no **15.** (2, −3)
17. (1, 4) **19.** (2, 11) **21.** (4, 7) **23.** (−3, 1)
25. (−1, −4) **27.** (−3, −4) **29.** (1, 1)
31. (−2, 0) **33.** (−9, 2) **35.** R_{ℓ_1} **37.** (6, 14)

CHAPTER 13   INTRODUCTION TO TRIGONOMETRY

Pages 499–500   Lesson 13.1
Exploratory 1. $\overline{PR}$ **3.** $\overline{QR}$ **5.** $\overline{PQ}$ **7.** 15
9. 16 **11.** $\sqrt{7}$ **Written 1.** KL **3.** KL **5.** RS
7. RT **9.** RS **11.** 5(RS) **13.** $3\frac{1}{3}$ **15.** $\frac{1}{3}$
17. 10 **19.** △KLQ ~ △KMR ~ △KNS ~ △KPT because when parallel lines are cut by a transversal, corresponding angles are congruent. Also, ∠K ≅ ∠K.

Pages 503–504   Lesson 13.2
Exploratory 1. ∠EFQ, ∠GHQ, ∠KLQ, ∠MRQ, ∠STQ, ∠VWQ **3.** ST **5.** EF **7.** GH
9. VW; GH = 2, QH = 8, QW = 20, VW = 5. Since $\frac{2}{8} = \frac{5}{20}$, the answer checks.
Written 1. 1 **3.** 1.7321 **5.** 0.3057 **7.** $\overline{AD}$, $\overline{DG}$ **9.** $\overline{BE}$, $\overline{AB}$ **11.** $\overline{KG}$, $\overline{DK}$ **13.** $\overline{AK}$, $\overline{KD}$
21. 0.4663 **23.** 0.0524 **25.** 1.8040 **27.** 24
29. 54 **31.** 56

Pages 507–510   Lesson 13.3
Exploratory 7. 0.7002 **Written 1.** 4
3. 142.82 **5.** 93.76 **7.** 47.67 **9.** 0.94
11. 1360.93 **13.** 17 **15.** 34 **17.** 39 **19.** $\frac{7}{8}$
21. 2.7 **23.** 31 **25.** 15 **27.** 27 **29.** 52.16
31. 77 **33.** not a right triangle **35.** 4 **37.** 5
39. 39 **41.** 21 **43.** 119.29 cm² **45.** 16.55 ft
47. 369 ft **49.** 46.19 ft **51.** 14.56 ft
53. 133.27 ft **55.** 44 ft **57.** 10° **Mixed Review 1.** (5, 1) **2.** (9, 6) **3.** (3, −5)
4. (−8, −4) **5.** $\left(-\frac{8}{3}, -\frac{4}{3}\right)$ **6.** (2, 1)
7. $5\sqrt{5}$ **8.** 4 **9.** about 3.36 **10.** about 72°
11. $\frac{1286}{1287}$

Pages 513–515   Lesson 13.4
Exploratory 1. $\overline{RS}$ **3.** $\overline{RT}$ **5.** RT **7.** $\frac{ST}{RS}$
Written 1. 0.6428 **3.** 0.7660 **5.** 0.7071
7. 0.1736 **9.** 0.9848 **11.** 0.8480
13. 0.6428 **15.** 0.7660 **17.** 0.7071
19. 0.1736 **21.** 0.9848 **23.** 0.8480
25. 0.1219 **27.** 0.9659 **29.** 0.3827
31. 0.8339 **33.** $\frac{5}{12}$ **35.** $\frac{12}{13}$ **37.** $\frac{5}{13}$ **39.** 23
41. 10 **43.** $\frac{3}{5}$ **45.** $\frac{4}{5}$ **47.** $\frac{4}{5}$ **49.** $\frac{3}{5}$ **51.** 37
53. 53 **55.** 20.07 **57.** 12 **59.** 2.93 **61.** 37
63. 1697.13 m **65.** 489.05 m **67.** 4.14 ft
69. 1.30 m **71.** AB = BC = 10.19, BD = 4.79 **73.** 62 **75.** BD = 43.19, BC = 31.07; perimeter = 122.14, area = 932.1 **77.** 39

Pages 519–520   Lesson 13.5
Exploratory 1. a circle with a radius of 1 unit
Written 1. 0.5 **3.** 0.5 **5.** −0.3420
7. −0.9397 **9.** −0.8660 **11.** −0.9397
13. 1 **15.** 0 **17.** −1 **19.** 0 **21.** 140, 400
23. 240, 660 **25.** −330° **27.** −300° **29.** if the angle has measure a, (3 cos a, 3 sin a)

Page 522   Problem Solving Application
Written 1. 334 ft **3.** 470 m **5.** 16 ft
7. 123 cm **9.** 12 cm **11.** m∠QMT = 64, m∠QNT = 76, m∠QPT = 53

Pages 524–525   Chapter Review
1. AC **3.** EB **5.** $2\frac{1}{2}$ **7.** $4\frac{1}{2}$ **13.** $\overline{CD}$, $\overline{BD}$
15. $\overline{ED}$, $\overline{BD}$ **17.** $\overline{BE}$, $\overline{BA}$ **19.** 0.5774
21. 0.7400 **23.** 10 **25.** 39 **27.** 1.79
29. 959.60 ft **31.** 0.9925 **33.** 0.2588
35. $\frac{4\sqrt{41}}{41}$ **37.** $\frac{5\sqrt{41}}{41}$ **39.** $\frac{4}{5}$ **41.** 39 **43.** 17
45. 2.93 **47.** sin 90° = 1, sin 180° = 0, sin 270° = −1, sin 360° = 0, cos 90° = 0, cos 180° = −1, cos 270° = 0, cos 360° = 1
51. 0.5 **53.** 1 **55.** 0.7071 **57.** −0.5
59. −0.7071 **61.** −0.7071

Index

A

Abscissa, 288
Addition
    angles, 128
    Axiom, 132
    in clock systems, 44–47
    Property, 111, 248
    Property of Inequality, 196
Additive inverse, 107
Adjacent angles, 123
    exterior sides, 123
Adjacent side, 498
Alternate exterior angles, 143
Alternate interior angles, 143
Altitudes, 189, 231
Angle Addition Axiom, 128
Angle Measure Axiom, 128
Angle of depression, 509
Angle of elevation, 509
Angles, 123
    Addition Axiom, 128
    adjacent, 123
    alternate exterior, 143
    alternate interior, 143
    Angle Measure Axiom, 128
    axioms about, 128–130
    base, 179, 189–190
    bisectors, 129, 189, 227–228
    complementary, 129
    congruent, 128, 227
    corresponding, 143
    exterior, 143
    exterior of, 123
    included, 169
    initial side, 517
    interior, 143
    interior of, 123
    measures greater than 360°, 519
    negative measures, 519
    obtuse, 518
    of elevation, 509
    of depression, 509
    of polygons, 157–159
    of triangles, 154–156
    Protractor Axiom, 128
    reflex, 518
    remote exterior, 154–155
    right, 129, 139
    sides, 123
    standard position, 517
    supplementary, 129
    Supplementary Axiom, 130
    terminal side, 517
    theorems about 135–139, 147–148, 154–155
    vertex, 123
    vertical, 123, 136
Antecedents, 6
Area
    of rectangles, 389
    using coordinates, 319–323
Arrow diagrams, 454
Associative property, 58–59, 65–69
Axioms, 71, 125–131, 144, 171–174, 192
Axis of symmetry, 375

B

Between, 122
Bi-conditionals, 7–9
    equivalences, 8
Binomials, 364
    product of, 364–365
Bisectors
    of angles, 129, 189, 227–228
    of line segments, 126, 231
    perpendicular, 140, 231

C

Calculators, 171, 291, 300, 373, 377, 412, 437, 469, 519, A2-A9
Chapter Review, 40, 83, 118, 165, 204, 241, 285, 327, 361, 408, 450, 494, 524
Chapter Summary, 38, 81, 117, 162, 203, 240, 283, 326, 360, 407, 449, 493, 523
Chapter Test, 42, 84, 120, 166, 206, 242, 286, 328, 362, 410, 452, 496, 526
Circles, 226
    circumscribed, 273
    equations of, 342–344
    unit, 517–520
Circumscribed circles, 273
Clock systems, 43–55
    equations in, 46, 50, 53–54
    operations in, 44–54
Closure, 57
Collinear, 122
Combinations, 434–439
Combinatorics, 425–452
    combinations, 434–439
    Counting Principle, 419
    factorial, 425
    permutations, 425–432
Commutative groups, 66
Commutative property, 58
Compass, 226
Complementary angles, 129
Completing the square, 381–384
Composite mapping, 456
Compositions, 480–483
Compound loci, 347–350
Conclusions, 10
Conditionals, 6–9
    bi-conditionals, 7–9
    contrapositives, 17–18
    converses, 17
    inverses, 17
Congruent angles, 128, 227
Congruent segments, 126, 226
Congruent triangles, 167, 242
    axioms about, 171–174
    corresponding parts, 168, 175–179
    overlapping, 184–188
    right, 192–194
    theorem about, 180–181, 191
Conjunct, 2
Conjunctions, 2–3
    bi-conditionals, 7–9
    negation of, 24–25
Connectives, 2
Consequent, 6
Constructions, 226–234, 272–274
    altitudes, 231
    angle bisectors, 227–228
    circumscribed circles, 273
    congruent angles, 227
    congruent segments, 226
    dividing segments, 272
    images, 459
    medians, 232
    parallel lines, 232
    perpendicular bisectors, 231
    perpendicular lines, 228–231
    segment bisectors, 231
    similar triangles, 272
Contrapositives, 17
    law of, 18–19

Converses, 17
Convex polygons, 157–158
Coordinate geometry, 287–362
Coordinates, 275, 288
Coordinate system, 288
Coplanar, 122
Corollary, 155
Corresponding angles, 143
Corresponding parts, 168, 175–179
Cosine ratio, 511–515, 517–520
Counting Principle, 419–422
Cross multiplication, 245

D

De Morgan Laws, 25–27
Diagonals, 446
Difference of two squares, 365
Dilations, 474–476
    rule for, 475
    scale factor, 474
Diophantus, 313
Direct isometry, 483
Directrix, 351
Discriminate, 386
Disjunct, 2
Disjunctions, 2–3
    negation of, 24–27
Distance formula, 299–301
Distributive Law, 88–93
Division
    Axiom, 132
    by zero, 53
    definition of, 107
    in clock systems, 53–54
Dodecagon, 158
Domain, 4, 454–455
Double negation, 8

E

Elimination method, 294–296
Endpoints, 122
Equations
    graphing, 291–293
    in clock systems, 46, 50, 53–54
    linear, 291–293
    of a circle, 342–344
    point-slope, 316
    quadratic, 369–410
    slope-intercept, 314–316
    solving, 86–87, 94, 98, 105–106
    systems of, 292, 394–396
Equiangular triangles, 155
Equivalence relations, 278–280
    reflexive relation, 278
    symmetric relation, 278–279
    transitive relation, 279–280
Equivalences, 8
Euclidean geometry, 121–274
Existential quantifiers, 31–32
    negations of, 34
Exterior angles, 143
Exterior Angle Theorem, 155
Extremes, 245

F

Factorial, 425
Factoring, 365–372
Field Combination Theorems, 102, 104, 108, 109
Fields, 92–114
    Additive Property, 111
    finite, 96–99
    Multiplicative Property, 111
    ordered, 111–114
    theorems, 101–105, 108–109
    Transitive Property, 111
    Trichotomy Property, 111
First Octant, 355
Focus, 351

G

Geometric mean, 248, 262
Geometric Mean Property, 249
Geometry
    coordinate, 287–328
    Euclidean, 121–274
    transformation, 453–496
    trigonometry, 497–526
Glide reflections, 483
Golden ratio, 397
Golden rectangles, 397
Graphs, 288–293
    abscissa, 288
    area, 319–323
    circles, 342–344
    coordinates, 288
    dilations, 489
    distance formula, 299–301
    First Octant, 355
    half-turns, 477
    inequalities, 292
    linear equations, 291–293
    line reflections, 465–468
    midpoint formula, 315–317
    octants, 355
    ordered pairs, 288
    ordinate, 288
    parabolas, 292, 351–353, 373–380
    parallel lines, 310, 312–313
    perpendicular lines, 311–313
    points, 289–290
    point-slope equation, 316
    quadrants, 288
    rectangular coordinate system 289
    rotations, 477
    slope, 306–316
    slope-intercept equation, 304–316
    systems of equations, 292, 394–396
    three-dimensional space, 354–357
    translations, 470–471
    x-axis, 288
    x-coordinate, 288
    x-intercept, 291
    y-axis, 288
    y-coordinate, 288
    y-intercept, 291
    z-coordinate, 355
Groups, 65–78, 92
    commutative, 66
    finite, 65–67
    infinite, 68–70
    Left Cancellation, 73
    Left Operation, 72
    non-commutative, 76–78
    Reflexive Property of Equality, 71
    Right Cancellation, 72
    Right Operation, 71
    Substitution Property of Equality, 71
    theorems about, 71–75

H

Half-turns, 476–478
    point symmetry, 477
    rule for, 477
Heptagons, 157
Hexagons, 157
Hypotenuse, 192, 498

INDEX 561

I

Identity element, 61, 65–69
Identity transformation, 485
Images, 454
Included angle, 169
Included side, 169
Independent events, 419
Indirect proof, 28, 151–153, 197
Inequalities, 111–114, 195–200
    Addition Property, 196
    graphs of, 292
    Multiplication Property, 196
    Transitive Property, 196
    Trichotomy Property, 196
Infinite groups, 68–70
Initial side, 517
Inscribed polygons, 273
Integers, 4
Interchanging Property, 248
Interior angles, 143
Inverses
    in logic, 17
    in mathematical systems, 61–69
Isometry, 482
    direct, 483
Isosceles triangles, 189–191
    base angles, 189–190
    right, 269
    theorems about, 190
Isosceles trapezoid, 222–225

J

Journals, 33, 76, 112, 143, 195, 235, 273, 323, 354, 403, 433, 460, 501

L

Law of detachment, 13–16
Law of disjunctive inference, 10–12
Law of double negation, 8
Law of syllogism, 20–23
Law of the contrapositive, 18–19
Left Cancellation, 73
Left Operation, 72
Legs, 192, 498
Linear equations
    graphing, 291–293
    in clock systems, 46, 50, 53–54
    in two variables, 291
    point-slope, 316
    slope-intercept, 314–316
    solving, 86–87, 94, 98
    systems of, 292, 294–296
Line of reflection, 458
Line of symmetry, 462
Line reflections, 458–468
    construction of, 459
    in coordinate plane, 465–468
    over horizontal line, 466
    over y-axis, 465
Lines, 122
    axioms about, 125
    parallel, 143–153, 232, 257, 310
    perpendicular, 138–139, 228–231, 311
    skew, 143
    transversals, 143
Line segments, 122
    Addition Axiom, 126
    axioms about, 125–126
    bisectors, 126, 231
    congruent, 126, 226
    divided proportionally, 249
    dividing, 272
    measurement, 125
    midpoint, 126, 303–305
    perpendicular bisectors, 231
Line symmetry, 462–463
Loci, 339–353
    circles, 338, 342–344
    compound, 347–350
    in coordinate plane, 342–346
    parabolas, 351–353
Logic, 1–42
    antecedents, 6
    bi-conditionals, 7–9
    conclusions, 10
    conditionals, 6–9
    conjunct, 2
    conjunctions, 2–3
    connectives, 2
    consequent, 6
    contrapositives, 17–18
    converses, 17
    De Morgan Laws, 25–27
    disjunct, 2
    disjunctions, 2–3
    domain, 4
    equivalences, 8
    existential quantifiers, 31–32, 34
    indirect argument, 28
    inverses, 17
    law of detachment, 13–16
    law of disjunctive inference, 10–12
    law of double negation, 8
    law of syllogism, 20–23
    law of the contrapositive, 18–19
    negations, 2
    open sentences, 4
    premises, 10
    proofs, 10–23, 28
    quantifiers, 30–35
    Reductio ad Absurdum, 29
    solution sets, 4
    statements, 2
    truth tables, 2–3, 6–8
    truth values, 2–3, 6–8
    universal quantifiers, 30–31, 33, 35

M

Mappings, 454–457
    arrow diagrams, 454
    composite, 456
    domain, 454–455
    images, 454
    one-to-one, 455
    preimages, 454
    range, 454–455
Mathematical Excursions, 28, 55, 100, 277, 313, 346, 397, 433, 446, 516
Mathematical systems, 43–120
    Addition Property, 111
    additive inverse, 107
    associative property, 58–59, 65–69
    clock systems, 43–55
    closure property, 57
    commutative property, 58
    Distributive Law, 88–93
    division, 107
    Field Combination Theorems, 102, 104, 108–109
    fields, 92–113
    groups, 65–78, 92
    identity element, 61, 65–69
    inverse, 61–69
    Left Cancellation, 73
    Left Operation, 72
    multiplicative inverse, 107
    Multiplicative Property, 110

non-commutative groups, 76–78
operational systems, 57, 65–69
operation on a set, 56
operations, 56–63
order relation, 110
Reflexive Property of Equality, 71
Right Cancellation, 72
Right Operation, 71
Substitution Property of Equality, 71
subtraction, 107
transformations, 484–487
Transitive Property, 110
Trichotomy Property, 110
two-fold operational system, 92
with two operations, 85–120
Matrix logic, 36
Mean proportional, 248, 262
Means, 245
Means-Extremes Property, 245
Measurement, 125
of angles, 128
Medians, 189, 232
Midpoint, 126
formula, 303–305
Mixed Review, 23, 51, 64, 95, 110, 131, 150, 179, 225, 251, 297, 309, 350, 393, 406, 423, 440, 464, 479, 510
Monomials, 364
numerical coefficients, 364
variables, 364
Multiplication
Axiom, 132
in clock systems, 48–51
of binomials, 364
of polynonials, 364–365
Property of Inequality, 196
Property, 111
Multiplicative inverse, 107

N

Natural numbers, 4
Negation, 2
De Morgan Laws, 25–27
double, 8
of quantified statements, 33–34
Noncollinear, 122
Non-commutative groups, 78–80

Noncoplanar, 122
Numerical coefficients, 364

O

Octants, 355
One-to-one mapping, 455
Open sentences, 4
Operational systems, 57, 65–69
Operation on a set, 56
Operations, 56–63
associative property, 58–59, 65–69
closure property, 57
commutative property, 58
identity element, 61, 65–69
inverse, 61–69
Opposite side, 498
Order, 483
Ordered fields, 111–114
Additive Property, 111
Multiplicative Property, 111
theorem, 106–111
Transitive Property, 111
Trichotomy Property, 111
Ordered pairs, 275, 288
Order relation, 111
Ordinate, 288
Orientation, 483
Origin, 288
Outcome sets, 412

P

Parabolas, 292, 351–353, 373–380
axis of symmetry, 375
directrix, 351
focus, 351
maximum point, 375
minimum point, 375
vertex, 375
Parallel lines, 143–153, 257, 312–313
construction of, 232
slopes of, 310
Parallelograms, 209–221, 332
rectangles, 218–221
rhombuses, 218–221
squares, 219–220
Pentagons, 157, 225
Performance Assessment, 38, 81, 117, 162, 203, 240, 283, 326, 360, 407, 449, 493, 523
Perimeters, 257–258

Permutations, 425–432
with repetitions, 431–432
Perpendicular bisectors, 140, 231
Perpendicular lines, 139–140
construction of, 228–231
slopes of, 311
theorem about, 140
Planes, 122
axioms about, 125
Point reflection, 476
Points, 122
Point-slope equation, 316
Polygons
angles of, 157–159
congruent triangles, 208–242
convex, 157–158
diagonals, 446
dodecagons, 158
heptagons, 157
hexagons, 157
inscribed, 273
parallelograms, 209–221, 332
pentagons, 157, 225
quadrilaterals, 157, 209–225
rectangles, 218–221, 331
regular, 158
rhombuses, 218–221
right triangles, 155, 261–271, 498–515
similar, 244–246, 252–255
similar triangles, 252–255
squares, 219–220
triangles, 154–156, 235–237
Polynomials, 364–368
binomials, 364
factoring, 365–368
monomials, 364
multiplying, 364–365
trinomials, 364, 366
Portfolio Suggestions, 38, 81, 117, 162, 203, 240, 283, 326, 360, 407, 449, 493, 523
Postulates, 125
Preimages, 454
Premises, 10
Probability, 412–422, 441–452
Counting Principle, 419–422
independent events, 419
of certain events, 413
of impossible events, 413
"or" statements, 416–418
outcome sets, 412
tree diagrams, 420
Problem Solving,
eliminate possibilities, 36

graphs and estimation, 324
guess and check, 79
identify subgoals, 115
indirect reasoning, 238
inverse variation, 358
list possibilities, 447
perimeter and area, 281
quadratic equations, 389
right triangles, 521
transformations, 491
use equations, 160
use figures, 201
Projects, A10-A16
Proofs, 10–23, 135–199
    coordinate, 330–339
    indirect, 28, 151–153
    Reductio ad Absurdum, 28
Proportions, 244–264
    Addition Property, 248
    cross multiplication, 245
    extremes, 245
    geometric mean, 248, 262
    Geometric Mean Property, 249
    Interchanging Property, 248
    mean proportional, 248, 262
    Means-Extremes Property, 245
    Reciprocal Property, 248
    Subtraction Property, 248
Protractor Axiom, 128
Protractors, 226
Pythagoras, 266
Pythagorean Theorem, 265–271
Pythagorean Triples, 267

Q

Quadrants, 288
Quadratic equations, 369–410
    completing the square, 381–384
    constant terms, 369
    problem solving, 389–392
    quadratic formula, 385–388
    solving by factoring, 105–106, 369–372
    solving by graphing, 377–380
    sum and product of roots, 387
    systems of equations, 393–396
Quadratic formula, 385–388
    discriminant, 386
Quadratic-linear systems, 394–395
Quadrilaterals, 157
    coordinate proofs, 331–334

parallelograms, 209–221, 332
rectangles, 218–221, 331
rhombuses, 218–221
squares, 219–220
trapezoids, 222–225
Quantifiers, 30–35
    existential, 31–32, 34
    negations of, 33–35
    universal, 30–31, 33, 35

R

Radius, 226
Range, 454–455
rational expressions, 398
    adding and subtracting, 403
    multiplying and dividing, 400
    simplifying, 398
Rational numbers, 4
Ratios, 244–247
    golden, 397
Rays, 123
    opposite, 123
Real numbers, 4
    Addition Axiom, 132
    Division Axiom, 132
    Multiplication Axiom, 132
    Reflexive Axiom, 132
    Substitution Axiom, 132
    Subtraction Axiom, 132
    Symmetric Axiom, 132
    Transitive Axiom, 132
Reciprocal Property, 248
Rectangles, 218–221, 331
    area of, 389
    golden, 397
    squares, 219–220
Rectangular coordinate systems, 289
Reductio ad Absurdum, 29
Reflections, 458–468
    in coordinate plane, 465–468
    over horizontal line, 466
    over y-axis, 465
Reflex angles, 518
Reflexive Axiom, 132
Reflexive Property of Equality, 71
Reflexive relations, 278
Regular polygons, 158
Relations, 275–280
    equivalence, 278–280
Remote exterior angles, 154–155
Resistance, 423

Rhombuses, 218–221
    squares, 219–220
Right angles, 129
    theorems about, 139
Right Cancellation, 72
Right Operation, 71
Right triangles, 155, 261–271
    adjacent side, 498
    congruent, 192–194
    hypotenuse, 192, 498
    isosceles, 269
    legs, 192, 498
    opposite side, 498
    Pythagorean Theorem, 265–271
    30°-60°-90°, 269–270
    trigonometry, 498–515
Rotational symmetry, 453, 474
Rotations, 453, 473–478
    center of, 473
    half-turns, 476–478
    point symmetry, 477
Rulers, 226

S

Scale factors, 488
Segment Addition Axiom, 126
Similarity, 243–286
    Addition Property, 248
    cross multiplication, 245
    geometric mean, 248
    Geometric Mean Property, 249
    in constructions, 272
    Interchanging Property, 248
    mean proportional, 248
    Means-Extremes Property, 245
    perimeters, 257–258
    polygons, 244–246, 252–255
    proportions, 249–264
    Pythagorean Theorem, 265–271
    ratios, 244–247
    Reciprocal Property, 248
    right triangles, 261–271
    Subtraction Property, 248
    triangles, 252–271
Similar polygons, 244–246, 252–255
Similar triangles, 252–255
    construction, 272
Sine ratio, 511–515, 517–520
Skew lines, 143

Slope, 306–316
    negative, 307
    of parallel lines, 310
    of perpendicular lines, 311
    positive, 307
Slope-intercept equation, 314–316
Solution sets, 4
Squares, 219–220
    diagonals, 269
Standard position, 517
Statements, 2
    antecedent, 6
    bi-conditionals, 7–9
    conclusions, 10
    conditionals, 6–9
    conjunct, 2
    conjunctions, 2–3
    consequent, 6
    contrapositives, 17
    converses, 17
    disjunct, 2
    disjunctions, 2–3
    existential quantifiers, 31–32, 34
    inverses, 17
    negations of, 2, 24–25
    premises, 10
    quantifiers, 30–35
    universal quantifiers, 30–31, 33, 35
Straightedges, 226
Substitution Axiom, 132
Substitution method, 294–296
Substitution Property of Equality, 71
Subtraction
    Axiom, 132
    definition of, 107
    in clock systems, 52–53
    Property, 248
Supplementary angles, 129
Supplement Axiom, 130
Symmetric Axiom, 132
Symmetric relation, 278–279
Symmetry
    line, 462–463
    parabolas, 375
    point, 477
    rotational, 462–463
Systems of equations, 294–297, 394–396
    elimination method, 294–297
    graphing, 292, 394
    quadratic-linear system, 394–395

substitution method, 294–297

T

Tangent ratio, 501–511
Technology, *see calculators*
Terminal side, 517
Theorems, 71, 125, 135–199
    about angles, 135–139, 147–148, 154–155
    about congruent triangles, 180–181, 191
    about isosceles triangles, 190
    about parallel lines, 144–153
    about perpendicular lines, 140
    about right angles, 139
    about triangles, 154–155, 180–181, 190–191, 196–199
    Exterior Angle, 155
Three-dimensional space, 354–357
Transformation geometry, 453–496
    arrow diagrams, 454
    composite mappings, 456
    compositions, 480–483
    coordinate planes, 465–472, 477, 481, 489
    dilations, 488–490
    direct isometry, 483
    domain, 454–455
    glide reflections, 483
    half-turns, 476–478
    identity, 485
    images, 454
    isometry, 482
    line of reflection, 458
    line of symmetry, 462
    line reflections, 458–468
    line symmetry, 462–463
    mappings, 454–455
    mathematical system, 484–487
    one-to-one mapping, 455
    order, 483
    orientation, 483
    point reflection, 476
    point symmetry, 478
    preimages, 454
    range, 454–455
    rotational symmetry, 453–474
    rotations, 453, 473–478

scale factors, 488
    transformations, 453, 458–490
    translations, 469–472
Transformations, 453, 458–490
    See Transformational Geometry
Transitive Axiom, 132
Transitive Property, 111
Transitive Property of Inequality, 196
Transitive relation, 279–280
Translations, 469–472
    in coordinate plane, 470–471
    rule for, 470
Transversals, 143
Trapezoids, 222–225
    base angles, 222
    bases, 222
    isosceles, 222–225
    legs, 222
Tree diagrams, 420
Triangles, 235–237, 256–274
    altitudes, 189, 231
    angle bisectors, 189
    angles of, 154–156
    axioms about, 171–174
    congruent, 167–242
    coordinate proofs, 335–338
    corresponding parts, 168, 175–179
    hypotenuse, 192
    included angle, 169
    included side, 169
    isosceles, 179, 189–191, 269
    legs, 192, 498
    medians, 189, 232
    Pythagorean Theorem, 265–271
    remote exterior angles, 154–155
    right, 155, 261–271, 498–515
    similar, 252–255
    theorems about, 154–155, 180–181, 190–191, 196–199
    30°-60°-90°, 269–270
    trigonometry, 498–515
Trichotomy Property, 111
Trichotomy Property of Inequality, 196
Trigonometric ratios
    cosine, 511–515, 518–520
    sine, 511–515, 517–520
    tangent, 501–511

INDEX   565

Trigonometry, 497–524
    adjacent side, 498
    angle of depression, 509
    angle of elevation, 509
    cosine ratio, 511–515,
        518–520
    hypotenuse, 498
    initial side, 517
    legs, 498
    opposite side, 498
    reflex angles, 518
    right triangles, 498–515
    sine ratio, 511–515,
        517–520
    standard position, 517
    table, 528
    tangent ratio, 501–511
    terminal side, 517
    unit circle, 517–520
Trinomials, 364
    factoring, 366
Truth tables, 2–3
    for bi-conditionals, 7
    for conditionals, 6
    for conjunctions, 2
    for disjunctions, 2

Truth value, 2
    for bi-conditionals, 7–8
    for conditionals, 6–7
    for conjunctions, 2–3
    for contrapositives, 17
    for converses, 17
    for disjunctions, 2–3
    for inverses, 17
Two-fold operational system, 92

U

Undefined terms, 122
Unit circle, 517–520
Universal quantifiers, 30–31,
    33, 35
    negations of, 33, 35

V

Vertex, 123
    of a parabola, 375
Vertical angles, 123, 136

W

Whole numbers, 4

X

x-axis, 288
x-coordinate, 288
x-intercept, 291

Y

y-axis, 288
y-coordinate, 288
y-intercept, 291

Z

z-coordinate, 355

Photo Credits

Cover, © Richard Fukahara/Westlight;

1, © Bettmann Archive; **13**, Doug Martin; **25**, Culver Pictures; **43**, © T. Tracy/FPG International, Inc; **65**, © Harry Hartman/Bruce Coleman, Inc.; **85**, © Gail Shumway/FPG International, Inc.; **121**, Cessna Aircraft Company; **147**, Aaron Haupt Photography; **167**, © Duomo; **207**, © Superstock, Inc.; **243**, © Superstock, Inc; **248**, Steve Lissau; **267**, file photo; **287**, ©Heribert Schwarte/FPG International, Inc.; **329**, NASA: **363**, Duomo/ © Mitchell Layton; **397 (t)** © FPG International, Inc., **(b)** Shostal; **411**, © Tom Murphy/Superstock, Inc.; **428**, © Richard Davis/Photo Researchers, Inc.; **433**, Ron Rovtar; **434**, Doug Martin; **436**, Ron Rovtar; **453**, © Ken Lax/Photo Researchers, Inc.; **458**, © C.G. Maxwell/FPG International, Inc.; **462**, file photo; **478**, Ron Rovtar; **497**, Steve Lissau; **505**, Ted Rice; **516**, NASA; **A1**, James Blinn/JPL; **A10**, StudiOhio; **A11**, Matt Meadows; **A12**, **A13**, StudiOhio; **A14**, Kenji Kerins; **A15**, Doug Martin; **A16**, Courtesy JPL.

GLENCOE
Technology Activities and Extended Projects
for Integrated Mathematics, Course 2

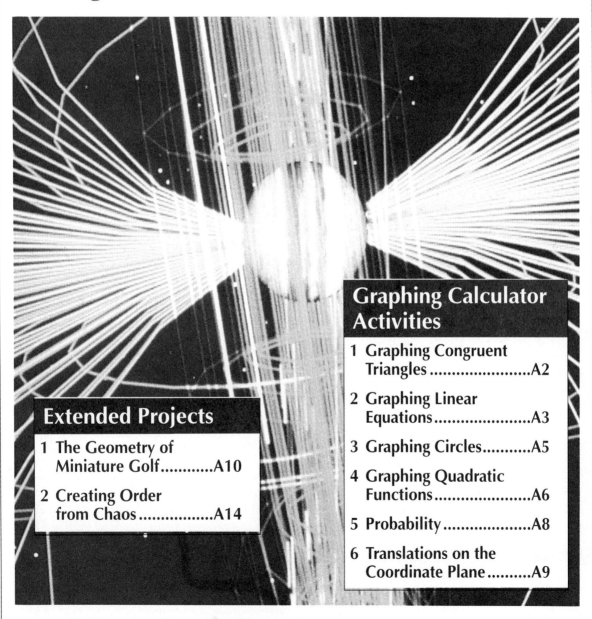

Graphing Calculator Activities

1. Graphing Congruent Triangles A2
2. Graphing Linear Equations A3
3. Graphing Circles A5
4. Graphing Quadratic Functions A6
5. Probability A8
6. Translations on the Coordinate Plane A9

Extended Projects

1. The Geometry of Miniature Golf A10
2. Creating Order from Chaos A14

ACTIVITY 1: Graphing Congruent Triangles

Use with: Lesson 5.2, pages 171–174

The activities on pages A2–A9 use a TI-82 graphing calculator. If you have a different type of programmable graphing calculator, consult the user's guide to adapt the keystrokes for use on your calculator.

Example: Graph $\triangle ABC$ with vertices $A(0, 2)$, $B(6, 2)$, and $C(5, 4)$ and $\triangle XYZ$ with vertices $X(7, 5)$, $Y(13, 5)$, and $Z(12, 7)$. Do the two triangles appear to be congruent?

- Set your viewing window for [0, 14] by [0, 14] with scale factors of 1. If you are not familiar with setting the viewing window, see Example 1 on page A3.
- Draw the first side of $\triangle ABC$ by using the LINE feature in the DRAW menu. This feature will draw a line between the two sets of coordinates indicated.

    To draw AB: Press [2nd] [DRAW] 2. The prompt **Line(** appears. Press 0 [,] 2 [,] 6 [,] 2 [)] to enter the coordinates of the two points. The line will appear on your coordinate screen.

- To complete the triangle, press [2nd] [DRAW] 2 and use the arrow keys to move the cursor to one endpoint of the segment. Press [ENTER]. Then move the cursor to the location of the third vertex. Watch the coordinates given at the bottom of the screen to approximate the position of the point. Press [ENTER] twice. The segment will appear. Finally, move the cursor to the other endpoint of the segment and press [ENTER] to complete the triangle.

- Repeat the steps to draw the second triangle. It appears that $\triangle ABC \cong \triangle XYZ$.

1. $\angle A \cong \angle X$, $\angle B \cong \angle Y$, $\angle C \cong \angle Z$, $\overline{AB} \cong \overline{XY}$, $\overline{BC} \cong \overline{YZ}$, $\overline{AC} \cong \overline{XZ}$

Write
1. List the corresponding parts of $\triangle ABC$ and $\triangle XYZ$ that are congruent.
2. How could you use the **Line(** command instead of the cursor to draw the other sides of the triangle? **Use the command and the coordinates to draw each segment.**

Draw Graph $\triangle ABC$ and $\triangle XYZ$ for the coordinates given. Do the two triangles appear to be congruent?

3. $A(-8, 1)$, $B(-8, -6)$, $C(-5, -6)$, $X(3, 5)$, $Y(3, -2)$, $Z(6, 5)$ **yes**
4. $A(-3, 8)$, $B(-3, 0)$, $C(-5, 4)$, $X(4, -3)$, $Y(4, -10)$, $Z(2, -6)$ **no**

ACTIVITY 2 Graphing Linear Equations

Use with: Lesson 8.2, pages 291–293

The graphing calculator is a powerful tool for studying graphs of a wide variety of functions. The basic steps of setting the **viewing window** and entering an equation to be graphed are the same for many types of graphs.

Example 1: Set the viewing window.

The viewing window for a graph is that portion of the coordinate plane displayed on the graphics screen. When the window is described, we use a notation that denotes the range for the x-coordinates and the range for the y-coordinates. A viewing window of [–8, 12] by [–10, 8] means that the values for x lie between –8 and 12 and the values for y lie between –10 and 8.

- To set the values for your viewing window, press [WINDOW] . A screen like the one at the right will appear. To change the values, press the down arrow key. The first line will be highlighted. Change the values to the appropriate number for the minimum value of x. Be sure to use the [(–)] key to enter a negative sign.

WINDOW FORMAT
XMIN = –10
XMAX = 10
XSCL = 1
YMIN = –10
YMAX = 10

    After entering the number, press [ENTER] to go to the next line. Xscl and Yscl describe scale factor increments you want to use on the axes. Unless the range is very large, you will probably use 1 most of the time.

The window [–10, 10] by [–10, 10] is called the **standard viewing window**. It can be automatically set by pressing [ZOOM] and 6.

If you press [ZOOM] and 5, the screen will show increments on the x- and y-axes that are the same size. When you use this feature, your graph will appear similar to how it appears on standard grid paper.

Now let's take a look at some linear functions and their graphs.

Example 2: Graph $y = 3x + 5$ using the standard viewing window.

- Make sure your calculator is set for the standard viewing window.
- Enter the appropriate key sequence.
  Press [Y=] 3 [X,T,θ] [+] 5 [GRAPH].

Example 3: Graph $y = -x + 20$.

- Use the standard viewing window.
- Press [Y=] [(-)] [X,T,θ] [+] 20 [GRAPH].
- Notice none of the graph is shown when you graph this equation in the standard viewing window, so we must change the range values to view a complete graph.

There are several viewing windows that will allow us to view the complete graph. Change your viewing window to [–10, 25] by [–5, 25] with scale factors of 1 for each scale. Now graph by pressing [GRAPH].

Graph Use your graphing calculator to graph each equation. State the range values that you used to view a complete graph. Sketch each graph on your own paper. **See student's work.**

1. $y = 6x + 7$
2. $y = -5x - 6$
3. $y = -3x + 12$
4. $y = -7x + 22$
5. $y = 5x - 35$
6. $y = 0.1x - 1$

Write 7. Observe your graphs. When do the graphs slant upward and when do they slant downward? **Upward for positive coefficient of x; downward for negative coefficient of x**

ACTIVITY 3 Graphing Circles

Use with: Lesson 9.4, pages 342–346

The TI-82 graphing calculator has a special feature that allows you to graph a circle without knowing the equation of the circle.

Example: Graph a circle whose center is located at (3, 4) with a radius of 5.

- First set your viewing window to the standard viewing screen. Clear the Y= list of all other equations.
- To select the **Circle(** function, press [2nd] [DRAW] and make sure DRAW is highlighted at the top of the screen. The Circle(function is number 9 on the menu. Either press 9 or use the arrow keys to scroll down to item 9 and press [ENTER].

  The Circle(prompt appears on the screen. Enter the coordinates of the center and the radius, and graph the circle.

  Circle(3 [,] 4 [,] 5 [,] [)] [ENTER].

  Notice that the circle looks like an ellipse, because the width of the increments on the x-axis are wider on the screen than those on the y-axis.

- Press [ZOOM] and 5 to change the viewing window.

  Press [2nd] [QUIT] to return to the home screen.

  Press [ENTER] to redraw the circle.

  Selecting the Circle(function and pressing [CLEAR] erases the circle graph.

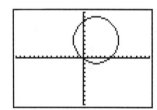

Graph Use what you learned in Lesson 9.4 to graph the circle given by each equation.

1. $(x - 2)^2 + (y - 4)^2 = 25$
2. $(x - 6)^2 + (y - 3)^2 = 64$
3. $(x + 5)^2 + (y - 3)^2 = 121$
4. $(x - 3)^2 + (y + 8)2 = 5$
5. $(x + 6)^2 + (y + 10)^2 = 18$
6. $x^2 + (y + 7)^2 = 8$

See students' work.

Write 7. Describe the differences and similarities in the graphs of $y^2 + x^2 = 25$ and $(y + 4)^2 + (x - 5)^2 = 25$. **The same shape and size but the second graph is shifted left 4 units and up 5 units.**

ACTIVITY 4 Quadratic Equations

Use with: *Lessons 10.3–10.4, pages 373–380*

In this activity, we will study quadratic equations. The graphs of quadratic equations are parabolas.

Example 1: Graph $y = x^2$ using the standard viewing window.

- First, clear the graphics screen by pressing [Y=], then use the arrow keys and the [CLEAR] key to select and clear any equations from the Y= list.
- Make sure the viewing window is properly set.
- Press [Y=] [X,T,θ] [x^2] [GRAPH].

Example 2: Graph $y = -0.15x^2 + 30$.

Let's try graphing in the standard viewing window.
- Clear the graphics window before graphing.
- Press [Y=] [(−)] 0.15 [X,T,θ] [x^2] [+] 30 [GRAPH].

Nothing appears on the graphics screen. We must change the range parameters to be able to view a complete graph.

- Set the viewing window to [−20, 20] by [−5, 35] with scale factors of 5 for both axes.

   Press [GRAPH] to graph the equation again.

There are three possible outcomes when solving a quadratic equation. The equation will have either two real solutions, one real solution, or no real solutions. A graph of each of these outcomes is shown below.

two real solutions

one real solution

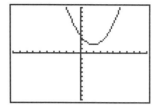
no real solutions

Example 3: Solve $3x^2 + 6x - 1 = 0$ using the graphing calculator.

- Use the standard viewing window and clear the graphics screen before graphing the equation, $y = 3x^2 + 6x - 1$.
- The TI-82 has a special function to find the roots. Press [2nd] [CALC] 2. Move the cursor along the graph until it is left of the first x-intercept. Press [ENTER] to set the lower bound. Move the cursor past the x-intercept and press [ENTER] to set the upper bound. Press [ENTER] to find the first root. $x \approx -2.154701$
- Repeat this procedure to find the second root, $x \approx 0.15470054$.

Graph Graph each equation. Tell how many real roots are possible. See students' graphs.

1. $y = 3.2x^2 + 9.2$ **0**
2. $y = x^2 - 5x + 6$ **2**
3. $y = 6x^2 + 108x + 480$ **2**
4. $y = 2x^2 + 9x - 18$ **2**
5. $y = 2x^2 - x - 15$ **2**
6. $y = 1.3x^2 - 3.8x + 5.1$ **0**

Calculate Find the solutions of each quadratic equation accurate to four decimal places by using your graphing calculator.   1.5000, 0.2460

7. $2x^2 - x - 15 = 0$ **3.000, –2.5000**
8. $0.2x^2 - 0.3492x - 0.0738 = 0$

no real solutions
9. $1.2x^2 - 3.6x + 5.8 = 0$
10. $3x^2 + 3.08x - 1.36 = 0$ **0.$\overline{3}$, –1.3600**
11. $x^2 + 31.54x + 229.068 = 0$
    –20.2000, –11.3400
12. $35x^2 + 66x - 65 = 0$
    –2.6000, 0.7142

ACTIVITY 5 Probability

Use with: Lesson 11.1, pages 412–415

The graphing calculator program at the right will generate random numbers. You can use the program to simulate real events like rolling a die or tossing a coin.

- To enter the program, press [PRGM] and highlight NEW.

- Enter the program name RANDNUM and press [ENTER]. A colon appears on the next line. Enter the commands by pressing [PRGM] and selecting the correct command from the CTL or I/O lists.

  You enter letters by using [ALPHA] and the desired key. The A-LOCK function helps when typing in several capital letters.

  The arrow is entered by pressing [STO▶]. The Rand command can be accessed by pressing [MATH], highlighting PRB, and selecting 1.

For more information on programming, consult your calculator's user's guide.

```
PRGM 1: RANDNUM
:CLRHOME
:DISP "LEAST INTEGER"
:INPUT S
:DISP "GREATEST
 INTEGER"
:INPUT L
:DISP "NUMBER OF
 VALUES TO GENERATE"
:INPUT A
:0→B
:LBL 1
:B + 1→B
:INT ((L − S + 1) RAND +
 S)→R
:DISP R
:PAUSE
:IF A ≠ B
:GOTO 1
```

Example: Use the RANDNUM program to simulate rolling a die 50 times. Make a table to show the results.

- Run the program by pressing [PRGM], selecting RANDNUM, and pressing [ENTER].
- You get a prompt which says "LEAST INTEGER?" Since the least number on your cube is 1, press 1 and then [ENTER].
- Another prompt asks you to enter the greatest integer, which in this case is 6. Enter 6.
- A third prompt asks you the number of values. You wish to roll the cube 50 times, so enter 50.
- When you press [ENTER], a number appears. This is the result of your first roll. Record this entry and continue pressing [ENTER] until you have generated all 50 values.

Write

1. Do you think that each number of a die has an equal chance of occurring when you throw it? Does your 50 rolls substantiate your theory? **Yes; See students' work.**

2. How could you use the program to simulate spinning a game spinner that had seven equal-sized regions 15 times? **The greatest integer would be 7, and the number of times is 15.**

ACTIVITY 6 Translations

Use with: Lesson 12.5, pages 469–472

You can use the DRAW command to graph a polygon on a coordinate grid. In the same manner as you created the polygon, you can use your knowledge of translations to create the translated image on the same coordinate grid.

Example: Graph a parallelogram with vertices $A(0, 0)$, $B(5, 0)$, $C(7, 4)$, and $D(2, 4)$. Then graph the image of $A'B'C'D'$ under $T_{-4,-5}$.

- First clear any previously-entered equations from the Y= list.
- Graph each segment of the parallelogram by entering the Line(command followed by the pairs of coordinates of the endpoints of each segment. You can return to the home screen by pressing [CLEAR].
  [2nd] [DRAW] 2 0 [,] 0 [,] 5 [,] 0 [ENTRY] draws segment AB.
- Instead of re-entering the Line(command for each additional segment, you can use [2nd] [ENTRY] to repeat the expression and then use the arrow keys to move to the parts you want to edit.
- To graph the translation, simply repeat the step you did to draw the parallelogram, but from each x-coordinate, subtract 4, and from each y-coordinate, subtract 5.
  For $\overline{A'B'}$, [2nd] [DRAW] 2 0 − 4 [,] 0 − 5 [,] 5 − 4 [,] 0 − 5 [ENTRY].

Continue this process for the other segments in $A'B'C'D''$.

Graph Graph each polygon and identify it. Then graph its translated image. Sketch both figures on your own graph paper. See students' work.

1. $A(3, 4)$, $B(7, 8)$, $C(6, -2)$, $T_{2, 3}$ **triangle**
2. $D(0, 0)$, $E(2, -6)$, $F(6, -6)$, $G(8, 0)$, $T_{-4, -3}$ **trapezoid**
3. $H(-3, 0)$, $I(0, -3)$, $J(3, 0)$, $K(0, 3)$, $T_{-2, -2}$ **square**
4. Create your own parallelogram, $T_{3, -5}$ See students' work.

Write 5. How could you use the techniques in this activity to graph an image and its reflection over a given line? Subtract or add the appropriate amount to the coordinate using the formulas in Lesson 12.4.

PROJECT 1: The Geometry of Miniature Golf

"Follow the bouncing ball!"

That old saying originally applied to a ball on a movie screen that guided viewers, word by word, through song lyrics printed on the screen. But it could apply equally well to sports.

Think of examples of bouncing balls in sports—a baseball rebounding off a bat or careening off an outfield wall, a tennis ball flying off a racquet, a basketball banking off a backboard into a basket. In each case, the ball hits an obstruction and flies off at an angle. Players who are successful in these sports must understand angles and must learn to control the angles at which balls hit and reflect off of the obstructions that lie in their paths.

Basketball

Billiards

Raquetball

Table Tennis

Miniature golf is designed around the idea of a ball bouncing off an obstruction. The fun of the game lies in analyzing angles and hitting the ball so it ends up exactly where the player wants—in the cup!

PROJECT 1

Analyzing Miniature Golf

In this project, your group will investigate how a ball bounces in miniature golf. You will work out strategies to determine the angle at which the ball should be hit so it will bounce off one or more obstructions and drop into the cup. Then you will construct a model of a miniature golf hole and prepare an instruction guide for players that describes how to play the hole.

PROJECT 1: Getting Started

Have your group follow these steps.

1. Discuss the game of miniature golf. Ask individuals who have played the game to describe their strategies for playing.
2. Obtain several golf balls and putters.
3. Arrange an area where you can bounce off walls, 2-by-4s, or other obstructions you set up for the purpose. Use a cardboard circle to represent the cup.
4. Work together to learn how and where a golf ball should be hit so it ends up in the cup. You might use protractors or other measuring devices. Design situations like those shown below to test your theories. For each situation, ask yourself: At what angle should I hit the ball to make a hole-in-one?

1 bounce (2 methods)

2 bounces

3 bounces

As you analyze angles, keep in mind the importance of the stroke of the golf club. How does the way that you hold the club and the speed at which you hit the ball affect the way the ball bounces off each obstruction?

5. Discuss your results with your group. Then list conclusions your group has reached as a result of your investigation.
6. Work with your group to design the final, most challenging hole for next year's "World Championship of Miniature Golf." Decide on the shape of the green, the obstructions you will place in the ball's path, the placement of the tee and the cup, and so on. Discuss interesting miniature golf course features you have seen. For example, some courses place a windmill in the middle of one of the greens. The blades of the windmill turn continually, nearly touching the ground. Players must time their stroke so that the ball reaches the windmill when a blade is not obstructing the path. When creating your design, don't overlook such artistic matters as the landscaping of the hole, the color scheme you will use, and the overall attractiveness of the hole for players.
7. Build a model of the hole. Try to make the model look as much like an actual miniature golf course hole as possible.
8. Finally, prepare a guide for people who will play your hole in the championship. Suggest strategies for playing the hole. Describe ways to analyze the angles so that a player can figure out how to put the ball in the cup from the tee or anywhere else.

Extensions

1. Billiards players use diamonds printed on the cushions of a billiards table to determine where to hit the cue ball so that it strikes the other balls.

   Research the "diamond system" and report on how the diamonds are used to analyze the angle of a shot.

2. Choose a sport that interests you. Describe the importance of angles in the sport and explain how a player who understands how to play the angles has an advantage over players who do not.

3. Research and report on the law of physics which states that the angle of incidence is congruent to the angle of reflection.

Culminating Activities

Show what you have learned in this project by completing one of these activities.

1. Participate in a class model miniature golf tournament. Each group gives an oral presentation, displays and describes the design of its golf hole, points out special or unique features of the hole, and outlines methods for playing it. Each group might also prepare and distribute copies of its guidelines for playing the hole it designed. Students can then move from hole to hole to inspect each design and, if design permits, to play the holes.

2. Prepare a handbook entitled "How to Win at Miniature Golf." In your handbook, explain the rules for playing the angles that you have discovered in your investigation.

PROJECT 2: Creating Order From Chaos

The inclination to sort and classify is probably as old as human nature itself. Archaeologists have discovered fuel logs and animal bones neatly stacked by prehistoric peoples according to size. Today we arrange compact discs alphabetically and classify mountains by height. Behind the tendency of prehistoric and contemporary people alike is the human need to make sense of and gain some control over an apparently disorderly world.

Classifications can be harmful, as when people categorize or stereotype each other by race or religion. Other classifications of individuals can be sources of great insight. Some psychologists use a 4-dimensional system to classify people according to whether they are extroverted or introverted (E or I), sensing or intuitive (S or N), feeling or thinking (F or T), and judging or perceiving (J or P). Sixteen classifications result. Research suggests that about 13% of the population are type ESTJ and 1% are type INFP. Knowledge of one's type can be a source of self-understanding and an incentive for personal growth.

Every branch of science has numerous classification systems. Astronomers classify stars by their surface temperatures. Geologists classify rocks by their crystal form, hardness, specific gravity, and several other categories. The mineral galena, for example, is cubical, 2.5 in hardness, and 7.5 in specific gravity. In a 3-dimensional [crystal form, hardness, specific gravity] classification system, galena could be classified as a [Cu, H2.5, SG7.5] mineral. The three dimensions in the classification distinguish galena from nearly every other mineral.

3-Dimensional Classification System

Create a Classification System

In this project, you will collect and analyze data in an area of interest to you. Then you will create a classification system that could be of use to people doing research in the area you have chosen. You will analyze your data to determine the frequency of each of the categories in your system and any other statistics you feel would be of value to people using your system.

Getting Started

Follow these steps to carry out your project.

- Brainstorm with your group to choose an area of mutual interest for which you will create your classification system. The area can be from science, social studies, language arts, art, sports, or another topic you prefer. Be sure that a classification system is not already in use in the area you choose.
- Choose a number of important and distinguishing categories that you can use to classify items in your area. Examples of categories are specific gravity (rocks) and surface temperature (stars). You should choose an ample number of categories, since after collecting data you may find that one or more categories are not as distinguishing as you had anticipated. These can always be dropped.
- Collect data on your categories.
- Define your classification system. With the perfect system, an item's classification should place it together with other items that are similar to it in important and useful ways. Your system should have at least three distinguishing dimensions or categories.
- Calculate the portion of the entire population that is represented by each classification in your system. For example, in the psychology classification system discussed above, 13% of the population is classified ESTJ. Compile any other statistics that might help you demonstrate the importance and usefulness of your system.

PROJECT 2

Extensions

1. Research and report on the taxonomic system for classifying plants and animals.
2. An N-dimensional classification system has a classification in the first dimension, b in the second, c in the third, and so on, through z in the Nth. How many distinct N-dimensional classifications can be delineated using the system?
3. The Richter scale is a 1-dimensional classification system that classifies earthquakes according to their energy. Choose a familiar 1-dimensional system and propose a second dimension that could be used to classify objects of study more precisely.

Culminating Activities

Show what you have learned in this project by completing one of the following activities.

1. Write a report describing your work on this project. Explain the importance of your classification system. Describe the system and tell how you collected your data and compiled your statistics. Give examples of how your system can be used.
2. Make an oral report to the class describing your classification system. Bring several examples of items in your area of study and demonstrate how each can be classified using your system.